CONDENSED MATERIALS GROUP INDEX TO
SIX-VOLUME RETRIEVAL GUIDE SUPPLEMENT I*

*Material classes such as woods, pharmaceuticals, foods, natural products, etc., or references concerning major compendia, are not reported in this six-volume **Supplement I.** Special bibliographic searches on these may be requested from TPRC directly.

CONDENSED INSTRUCTIONS ON USE OF VOLUME 6*

INQUIRY EXAMPLE 1: Our technical staff would like to find **all** references which give information/data on the **thermal conductivity** of **Foamed Concrete.**

SEARCH STRATEGY AND RESULTS: On page A1 of the Materials Directory, looking under **Concrete, Foamed** you will find the notation "see Concrete, Expanded . . . . . . 525–." Now go to **Concrete, Expanded** and note property codes **a, d, e** with substance number 526-0034. Continue the search in Part B, Chapter 1— Thermal Conductivity, page B1. Here, under substance number 526-0034, you will find fifteen entries, which may be examined to determine which TPRC numbers are to be selected. Finally, find the bibliographic citations in Part C. Using this same substance number 526-0034, a total of twenty-five more entries can be found in the **Basic Edition** (Plenum, 1967). Books 1, 2, and 3 of the **Basic Edition** correspond directly to Parts A, B, and C of this six-volume **Supplement I.**

INQUIRY EXAMPLE 2: My engineering staff would like **thermal diffusivity** data on **Plaster of Paris.**

SEARCH STRATEGY AND RESULTS: On page A2 of the Materials Directory, you will find **Plaster of Paris,** but, property **d** is not listed. However, note other cross references, such as: "see also Gypsum Board . . . . . 661–." Now in this same volume, on page A47, you will find **Gypsum Board** property **d** and substance number 661-0470. Continue the search in Part B, Chapter 4—Thermal Diffusivity, page B17. Here you will find substance 661-0470 and there is one reference, TPRC number 37058. Continue to Part C to find the bibliographic citation for 37058.

Here are the relationships of classes to volumes in this publication series that must be borne in mind when you notice a "see also" substance class cross reference: 100–127, Vol. 1; 200–227, 606, and 631, Vol. 2; 300–482, Vol. 3; 501, 504, 507, and 521, Vol. 4; 511 and 516, Vol. 5; 526–551, 621, 651, and 661, Vol. 6.

INQUIRY EXAMPLE 3: The materials engineer at our company would like to locate references on **thermal conductivity** for crumpled **Aluminum Foil.**

SEARCH STRATEGY AND RESULTS: On page A2 of the Materials Directory, note the substance name **Aluminum Foil, Corrugated** or **Crumpled,** property **a** and substance number 528-0164. Continue the search on page B1, and you will find four entries opposite substance number 528-0164 with TPRC numbers 39866, 40228, 54566, and 55057. The complete bibliographic citation can then be found in Part C in this volume.

INQUIRY EXAMPLE 4: References are needed for **reflectance** for **PV-100 White Paint.**

SEARCH STRATEGY AND RESULTS: On page A40 you will find **Paint, White PV-100** with properties **g, h, j** and substance number 551-0949. Go to page B31 for substance 551-0949 and you will find reference TPRC number 43527. Continue the search in Part C to locate the bibliographic citation for 43527.

INQUIRY EXAMPLE 5: Our aerospace engineering staff would like to find a certain technical paper. The only information known about this paper is that it contains data on **charring materials** and the author's last name is **Wilson.**

SEARCH STRATEGY AND RESULTS: On page D26 of the Author Index, you will find the following: **WILSON D R** 52260; **WILSON E G** 63421; **WILSON R G** 35530 36511 37899; **WILSON W G** 40082 48324 50306. Now, go to Part C to look up the bibliographic citations for these nine different TPRC numbers. Through a process of elimination you will find that only the paper with TPRC **35530** fits the description given and that the author's name is **R. G. Wilson.**

INQUIRY EXAMPLE 6: Our design group would like references on the properties of carbon-reinforced **Poly(adipic acid-1,6-hexanediamine) Resin.**

SEARCH STRATEGY AND RESULTS: You will find this substance's name, properly systematized as **Resin, Poly(adipic acid-1,6-hexanediamine),** starts at the bottom of page A15 and continues at the top of page A16 with property codes **a** and **n** and substance number 531-1507. Conclude the search in the manner described in the examples above.

---

*For additional details on the use of this volume, see the introductory remarks for Parts A, B, C, and D, scan the **Contents,** and also note the **Condensed Materials Group Index** on the opposite page.

thermophysical properties research literature retrieval guide

supplement I
1964-1970

volume 6
coatings, systems, and composites

Y. S. TOULOUKIAN
Editor

J. KOOLHAAS GERRITSEN
Technical Editor

W. H. SHAFER
Managing Editor

thermophysical properties research literature retrieval guide

supplement I
1964-1970

A Comprehensive Compilation of Scientific and Technical Literature by the Thermophysical Properties Research Center (TPRC), Purdue University

Y. S. Touloukian
Editor

J. Koolhaas Gerritsen
Technical Editor

W. H. Shafer
Managing Editor

See inside back page for CONDENSED MATERIALS INDEX
SIX-VOLUME RETRIEVAL GUIDE SUPPLEMENT

New literature on thermophysical properties is being constantly accumulated at TPRC. Contact TPRC and use its interim updating search services for the most current scientific information

thermophysical properties research literature retrieval guide

supplement I
1964-1970

volume 6
coatings, systems, and composites

Y. S. TOULOUKIAN
Editor

J. KOOLHAAS GERRITSEN
Technical Editor

W. H. SHAFER
Managing Editor

PART A. MATERIALS DIRECTORY
PART B. SEARCH PARAMETERS
PART C. BIBLIOGRAPHY
PART D. AUTHOR INDEX

SPRINGER SCIENCE+BUSINESS MEDIA, LLC 1973

Library of Congress Catalog Card Number 60-14226
ISBN 978-1-4757-6838-1   ISBN 978-1-4757-6836-7 (eBook)
DOI 10.1007/978-1-4757-6836-7

© 1973 Springer Science+Business Media New York
Originally published by Plenum Press, London in 1973

CONTENTS OF VOLUME 6

FOREWORD
(To Basic Edition 1967)

The "Thermophysical Properties Research Literature Retrieval Guide" makes it possible for an individual scientist or engineer, working in his office, to search quickly the contents of world journals, reports, and books for references to articles containing information on seven groups of specified thermophysical properties (16 properties) of all substances.

Why is the publication of such a work an important event? Science and technology in this eighth decade of the twentieth century face a present and a future filled with challenge and opportunity to serve mankind. Vistas of a better life in the future through scientific research and engineering enterprise can be projected in almost every area of man's material life. And yet science itself has created and is creating obstacles to its own progress — to the realization of its dreams and potential. So vast is the fund of existing knowledge from which the science and technology of the future must grow that man literally knows not what he knows. He must grope in a morass of paper for needed knowledge already obtained by others. Factual information gleaned from nature by tedious and exacting effort is hidden in that jungle known as the world literature. Science has been better at creating new knowledge than in organizing that already on hand. True, guideposts of various kinds exist to show the way to the seeker of scientific facts, but frequently the route is long and roundabout. Without the excellent abstracting services that cover wide domains of science the situation would be chaotic. But there are few literature searching aids that selectively and completely condense the reference material of narrow fields so that the scientific specialist or technical specialist or technical librarian may conduct a fast and thorough search for needed information without the necessity of first laboriously scanning the pertinent abstract publications or original sources. Fortunately, the problem of rapid, effective information services of specialized areas is now receiving attention in several quarters.

The Thermophysical Properties Research Center (TPRC) is a leader among the groups that have plunged into the task of improving the information highways of science and technology. In its "Thermophysical Properties Research Literature Retrieval Guide" it has presented to the scientific public a new approach to the search for scientific information. Its methodology is unique. It brings to the desk of the user a vast amount of information on thermophysical properties. The literature of a delineated field has been combed meticulously by an expert staff for all information on the properties of interest. The whereabouts in the literature of each piece of pertinent information has been stored permanently on magnetic tape — and in the "Retrieval Guide." In a sense it is a marriage of machine storage and retrieval with conventional publication in book form. Now, scientific workers need not repeat the laborious preliminary examination of the literature. They may use the "Retrieval Guide" for retrospective search and contact TPRC for mechanized retrieval for the more recent unpublished portion. The tremendous saving of time for the individual searcher is obvious. Perhaps equally important is the significant amount of literature reported which is not cited in abstracting journals. The editors and staff of TPRC are to be congratulated for devising, producing, and maintaining a valuable and unique tool for the engineer, research scientist, compiler, and technical librarian.

Guy Waddington

Director, Office of Critical Tables
National Academy of Sciences
National Academy of Engineering
National Research Council

PREFACE

The phenomenal growth of science and technology has brought about a universal appreciation of the fact that present limitations in many technical developments are often a direct result of the paucity of knowledge on material properties. Engineering developments in the years ahead will be closely linked to the research that is done today to contribute to a better understanding of the properties of matter, of which thermophysical properties constitute a major segment.

While research on the properties of materials continues, adequate steps are not being taken to ensure that this invaluable body of information be coordinated, synthesized, organized, and disseminated to the ultimate user, namely, the individual scientist and engineer.

It is generally agreed that the present level of research support on thermophysical properties of matter falls short of existing needs and anticipated future demands; but what is even more disturbing is the fact that engineering groups across the nation are using no more than a fraction of the information already available, either because it is in a form not directly useful to them or, often because its existence is not generally known. As a result, such information remains buried in the world's scientific literature. The repercussions of this latter condition are indeed serious since it leads to unintentional duplication of research effort with the resultant waste of time and scientific manpower.

In conjunction with its research activities, TPRC screens the world's literature and collects published information on a wide range of materials in the field of thermophysics. This information concerns data, theoretical estimation methods, and experimental measurement techniques. Technical papers come from journals, abstracting services, reports, doctoral dissertations, masters theses, and many other sources. The full evaluation and analysis of the collected raw data are needed before publications on recommended values can be prepared. Such effort is obviously time consuming and expensive and therefore this critical evaluation is currently performed at a rather modest funding level. The end result is that much of the available world literature is not being processed and distilled.

As a complementary effort to its Data Tables Series, TPRC published in 1967 a work entitled "Thermophysical Properties Research Literature Retrieval Guide." This three-book work reported 33,700 references on seven thermophysical property groups and about 45,000 materials. This Basic Edition systematically covered the world's unclassified literature published essentially between 1920 and mid-1964, in many instances going much earlier.

The present work, referred to as Supplement I to the Basic Edition, reports an additional 26,000 references on sixteen thermophysical properties of 20,000 materials, covering the years from mid-1964 to 1971. An additional 9,000 synonyms and trade names are cross-referenced to assist the user in identifying the material or substance of interest.

Supplement I follows the same format of presentation as the Basic Edition. However, it has been restructured for improved user convenience in that a series of six Retrieval Guides have been designed for various material classes. As a result, each user group can purchase, at reasonable cost, selected volumes of specific interest, as well as the complete six-volume set.

It is sincerely hoped that the Basic Edition (1967) and Supplement I (1973) of the "Thermophysical Properties Research Literature Retrieval Guide" will constitute a permanent and valuable contribution to science and technology as well as to scientific documentation. These volumes, and those to follow, should prove to be an invaluable source of information to every scientist and engineer, with a scope of knowledge humanly impossible to master for any one individual or any group of individuals. Perhaps even more important, it is hoped that a wealth of information, heretofore unknown, will have been made available to many, including the specialist. The TPRC staff is most anxious to receive comments, suggestions, and criticism from all users of these volumes. All communications will be gratefully appreciated. Specific information concerning TPRC's operations, services, publications, and research activities can be obtained by communicating with the Director.

The preparation of these volumes was made possible through the collective financial support received from a large number of governmental, industrial, and nonprofit research organizations. Their interest and support are gratefully acknowledged.

In closing I wish to acknowledge the individual and collective accomplishments of TPRC's Scientific Documentation Division: Mr. G. Kvakovszky and Mrs. V. Ramdas, Technical Coding; Mrs. M. R. Troyer and the late Mrs. N. Y. Moore, Documentalists; Mrs. B. M. Schick, Literature Searcher; Miss J. Baker and Mrs. N. Phillips, Clerical Operations; Mrs. S. J. Creamer and Mrs. J. A. Brittingham, Library.

Special thanks are extended to Mr. J. W. Phillips, TPRC Computer Programmer, and the staff at R. R. Donnelley and Sons, Chicago, Illinois, who were responsible for the computer-assisted phototypesetting of this new Retrieval Guide.

Y. S. Touloukian

Director, Thermophysical Properties
Research Center
Distinguished Atkins Professor
of Engineering

Part A
MATERIALS DIRECTORY

USE OF MATERIALS DIRECTORY

I. ORGANIZATION AND GENERAL CONSIDERATIONS

The organization of thermophysical properties information at TPRC is by material. A *Condensed Materials Index* to the six-volume Retrieval Guide Supplement is located on the inside cover in the back of this volume.

In order to index the world literature the classification system must be general and systematic, and yet flexible. Therefore, TPRC has adopted a highly structured classification scheme which arranges materials into logical groups that have closely related chemical composition. However, certain materials do not lend themselves to a purely chemical classification and a more logical method is to classify them instead into compatible groups either by the physical form and/or by their use and application. These materials are normally those of engineering interest and consist primarily of what is referred to as systems and composites.

Within this system of classification by material name, there exists a class of pertinent information for which no specific material name would be appropriate, e.g., "Theory of the Thermal Conductivity" or "New Technique for the Thermal Diffusivity Measurement of Solids" or "Emissivity of a Black-Body Cavity." Literature covering this class of publications is not reported in this volume but is available at TPRC and a special computer search and retrieval can be made upon request.

The index to materials is given alphabetically by name in the main body of Part A. Substances which are not specifically identifiable are listed alphabetically first, followed by specific substances, again in alphabetical order. It will be noted that more than 9000 synonyms, trade names, equivalents, and cross references have been incorporated in the *Materials Directories* of the six volumes of this set. A cross reference is preceded by the words "See Also" and a synonym, trade name, or equivalent is followed by the word "See." Examples on the use of each volume are given on the inside back cover.

It may be of interest to point out at this juncture that the user of this index should not be concerned with the structure of the seven-digit TPRC substance number associated with each material. These numbers serve to uniquely identify a material and their structure is only of internal significance to TPRC.

II. DEFINITIONS AND LIMITATIONS USED IN MATERIAL CLASSIFICATION

For the effective use of the material index of this volume certain definitions and limitations of terms as accepted by TPRC should be understood. These are briefly listed below:

1. *Impurities*—For the purposes of classification, TPRC defines the "impurity" limits as follows:

   Elements. Total of impurities must be ≤ 0.5 percent and individual impurity ≤ 0.2 percent by weight. A metallic element having impurities in excess of these limits is considered to be an alloy whereas a nonmetallic element is considered to be a mixture.

   Compounds. Total of impurities must be ≤ 5.0 percent and individual impurity ≤ 2.0 percent by weight. A compound with impurities in excess of these limits is considered to be a mixture.

2. *Doping*—Doped materials are entered separately in the directory without specification of the dopant, for example: "Silicon, Doped."

3. *Isotopes*—Isotopes are listed under the corresponding elements as a single entry. Exceptions are the isotopes of hydrogen and helium, which are designated separately as deuterium, tritium, and helium-3.

4. *Names of Chemical Compounds*—In naming compounds, TPRC follows the rule of the International Union of Pure and Applied Chemistry. In addition, TPRC uses the Annual Indices as issued by the Chemical Abstracts Service to update and add new chemical names.

5. *Systems and Composites*—A "system" is defined as a combination of materials with clearly defined boundaries existing between them or a distinct configuration of homogeneous materials. Systems are classified into three groups, consisting of metallic, nonmetallic, and metallic–nonmetallic components, respectively.

   There are certain considerations which require mention in systems classification, namely:

   a. *Moisture Content.* The term "moist" encompasses all degrees of water content up to and including the saturation point. Moistness is ignored in material classification.

   b. Because of the broad basis of the content of the systems groupings, a more narrowly defined group of systems is used, such as Processed Composites: (Insulations, Paper and Wood Products, and Wall Structures). The user should be aware of the possibility of finding a material listed in a special group even though it would also fulfill the criteria for the general systems grouping or other materials groups.

III. EXCLUSIONS TO MATERIAL AND PROPERTY COVERAGE

While TPRC attempts to cover the world research literature on all matter for the sixteen thermophysical properties it monitors, for reasons of scientific and technical rationale and practical expediency it has become necessary to put a number of constraints on certain classes of materials in the coverage of specific properties.

1. GENERAL CONSTRAINTS

a. Nonoriginal papers, promotional literature, and product catalogs are excluded. However, extensive review articles, major handbooks, and data compilations are included.
b. Data reported under unsteady-state and nonequilibrium conditions are excluded.
c. Data reported in arbitrary units or relative ratio without specification of reference used are excluded.
d. Data on all polymeric liquid mixtures or solutions of undefined character are excluded.

2. SPECIFIC CONSTRAINTS RELATIVE TO CERTAIN PROPERTIES

Property	Constraint
a. Emittance and reflectance	Wavelength range outside of 10 to 2 × 10⁵ cm⁻¹ or 0.5 to 1000 μm or 1.24 × 10⁻³ to 25 eV.
b. Absorptance, transmittance, and absorptance-to-emittance ratio	Same wavelength range as above in addition to exclusion of liquids, gases, organic compounds, and inorganic complexes.
c. Diffusion coefficient	All diffusion involving solids and diffusion of subatomic particles.
d. Thermal linear and volumetric expansion	All liquids, gases, and irreversible processes

IV. USEFUL REFERENCES FOR MATERIALS IDENTIFICATION

The user will find it necessary and useful at times to use handbooks and dictionaries to obtain a generally accepted name as formula for a given material. The latest editions of the following selected references were found useful by TPRC in the classification and identification of materials:

1. *Chemical Synonyms and Trade Names: A Dictionary and Commercial Handbook,* W. Gardner, Technical Press, Ltd., London, England.
2. *The Condensed Chemical Dictionary,* Reinhold Publishing Corporation, New York, N. Y.
3. *Engineering Alloys,* N. E. Woldman, Reinhold Publishing Corporation, New York, N. Y.
4. *Handbook of Material Trade Names,* O. T. Zimmerman and I. Lavine, Industrial Research Service, Inc., Dover, N. H.
5. *Synthetic Organic Chemical Manufacturers Association Handbook: Commercial Organic Chemical Names,* Chemical Abstract Service, American Chemical Society, Columbus, Ohio.
6. *Modern Plastics Encyclopedia* (annual), Modern Plastics, Inc., New York, N.Y.
7. *Handbook of Designations and Specifications for Soviet Nonmetallic Polymeric Materials,* Project White Stork, Wright-Patterson Air Force Base, Ohio.
8. *Handbook of Soviet Alloy Compositions,* ASTIA, Arlington, Virginia.
9. *The Merck Index of Chemicals and Drugs,* Merck and Co., Inc., Rahway, N. J.
10. *The Ring Index,* A. M. Patterson, L. T. Capell, and D. F. Walker, Chemical Abstracts Service, American Chemical Society, Columbus, Ohio.

VOLUME 6. COATINGS, SYSTEMS, AND COMPOSITES

Substance Name	Property	Number
Concrete, Perlite	a	526-0017
Concrete, Perlite, Moist	a	526-0213
Concrete, Pumice	a	526-0019
Concrete, Pumice, Moist	a	526-0221
Concrete, Quartzite	n	526-0168
Concrete, Quartzite, Moist	n	526-0239
Concrete, Sand, Expanded	a	526-0223
Concrete, Sand, Expanded, Moist	a	526-0224
Concrete, Sandstone	n	526-0237
Concrete, Sandstone, Moist	n	526-0241
Concrete, Scoria		
see Concrete, Slag		526-
Concrete, Silicate	a	526-0230
Concrete, Slag	a n	526-0020
Concrete, Slag, Expanded	a	526-0015
Concrete, Slag, Expanded, Moist	a	526-0018
Concrete, Slag, Moist	n	526-0243
Concrete, Slag, Volcanic	a	526-0231
Concrete, Slate, Expanded	a	526-0080
Concrete, Stone	a e n	526-0198
Ferron		
see Plaster, Insulating		526-
Foamed		
see Expanded		
Mortar	a e h n	526-0074
Mortar (Asbestos - Cement, Portland - Vermiculite)	a	526-0209
Mortar, Moist	a e n	526-0194
Mortar, Silicon Oxide	n	526-0200
Neat Cement		
see Cement		526-
Phomene		
see Concrete, Expanded		526-
Plaster Of Paris	a	526-0122
see also Calcium Sulfate		106-
see also Gypsum		521-
see also Gypsum Board		661-
Plaster, Insulating	a	526-0046
Plaster, White	g	526-0195
Siporex		
see Concrete, Expanded		526-
Terra Alba		
see Plaster Of Paris		526-
Zonolite Concrete		
see Concrete (Cement, Portland - Vermiculite, Expanded)		526-

METALLIC SYSTEMS – CONTACTS, HONEYCOMBS, MULTILAYERS, AND OTHERS CLASS 528

Substance Name	Property	Number
A G 3		
see Contact, French Aluminum Alloy A G 3 - Steel		528-
Aluminum Alloy X 7002, Beryllium Reinforced	n	528-0425
Aluminum Alloy 2024, Boron Reinforced	n	528-0426
Aluminum Alloy, Stainless Steel A M S 355 Reinforced	n	529-0427
Aluminum Foil, Corrugated Or Crumpled	a	528-0164
Aluminum, Boron Reinforced	n	528-0411
Aluminum, Copper Covered	a e	528-0421
Cable, Aluminum	a e	528-0378
Cable, Aluminum, Steel Reinforced, Zinc Covered	a e	528-0380
Cable, Steel, Aluminum Covered	a e	528-0379
Contact, Aluminum	c	528-0128
Contact, Aluminum - Brass	c	528-0385
Contact, Aluminum - Carbon Steel	c	528-0374
Contact, Aluminum - Copper	c	528-0308
Contact, Aluminum - Iron	c	528-0057
Contact, Aluminum - Molybdenum	c	528-0325
Contact, Aluminum - Stainless Steel	c	528-0007
Contact, Aluminum - Thorium	c	528-0334
Contact, Aluminum - Titanium	c	528-0335
Contact, Aluminum - Uranium	c	528-0058
Contact, Aluminum Alloy L 70	c	528-0329
Contact, Aluminum Alloy 1100 - Platinum	c	528-0341
Contact, Aluminum Alloy 1100 - Uranium, Nickel Covered	c	528-0049
Contact, Aluminum Alloy 1100 - Uranium, Tin Covered	c	528-0052

Substance Name	Property	Number
Contact, Aluminum Alloy 2024	c	528-0266
Contact, Aluminum Alloy 2024 - Iron	c	528-0398
Contact, Aluminum Alloy 2024 - Magnesium Alloy Z H 62 A	c	528-0039
Contact, Aluminum Alloy 2024 - Stainless Steel 303	c	528-0347
Contact, Aluminum Alloy 6061	c	528-0240
Contact, Aluminum Alloy 6061 - Stainless Steel 304	c	528-0318
Contact, Aluminum Alloy 7075	c	528-0020
Contact, Aluminum Alloy, Aluminum Covered	c	528-0381
Contact, Armco T - 303 C G Stainless Steel	c	528-0324
Contact, Bismuth Telluride - Iron	c	528-0343
Contact, Brass	c	528-0238
Contact, Brass - Copper	c	528-0207
Contact, Brass - Steel	c	528-0250
Contact, Brass Alloy 3 - Carbon Steel AISI 1020	c	528-0327
Contact, British Aluminum Alloy B S He20 - En 58 F Steel	c	528-0400
Contact, British Aluminum Alloy D T D 610	c	528-0363
Contact, British Steel En 58 F	c	528-0399
Contact, Cadmium Arsenide	c	528-0351
Contact, Cadmium Arsenide, With Filler	c	528-0361
Contact, Constantan	c	528-0060
Contact, Constantan - Copper	c	528-0079
Contact, Copper	a cd	528-0137
Contact, Copper - Copper Alloy (18 Ni)	c	528-0333
Contact, Copper - Indium	c	528-0382
Contact, Copper - Lead	c	528-0176
Contact, Copper - Lead Alloy (6 Bi)	c	528-0332
Contact, Copper - Russian Aluminum Alloy D 1	c	528-0270
Contact, Copper - Russian Brass L 80	c	528-0390
Contact, Copper - Russian Steel St 0.8 Kp, With Filler	c	528-0416
Contact, Copper - Russian Steel St 30	c	528-0208
Contact, Copper - Tin	c	528-0177
Contact, Copper - Tool Steel	c	528-0407
Contact, Copper, With Filler	c	528-0313
Contact, Duralumin	c	528-0353
Contact, Duralumin - Russian Brass L 80	c	528-0355
Contact, Duralumin - Russian Steel St 30	c	528-0354
Contact, French Aluminum Alloy A G 3 - Steel	c	528-0362
Contact, Gallium Arsenide	c	528-0096
Contact, Gold - Platinum Alloy (13 Rh)	c	528-0405
Contact, Indium - Silicon	c	528-0342
Contact, Iron	c	528-0184
Contact, Lead	c	528-0251
Contact, Magnesium Alloy (Zr) - Steel	c	528-0383
Contact, Magnesium Alloy (Zr) - Uranium	c	528-0386
Contact, Magnesium Alloy A Z 31 B	c	528-0267
Contact, Magnesium Alloy A Z 31 B - Stainless Steel 303	c	528-0412
Contact, Magnox A-12 - Uranium	c	528-0281
Contact, Mercury - Stainless Steel 304	c	528-0367
Contact, Mercury, Doped - Stainless Steel 304	c	528-0366
Contact, Metal, Normal - Metal, Superconducting	c	528-0113
Contact, Metal, Superconducting	e	528-0326
Contact, Metallic	a c	528-0143
Contact, Metallic With Filler	c	528-0346
Contact, Mild Steel	c	528-0123
Contact, Mild Steel, With Filler	c	528-0312
Contact, Molybdenum	c	528-0403
Contact, Molybdenum - Stainless Steel	c	528-0404
Contact, Molybdenum - Stainless Steel, With Filler	c	528-0360
Contact, Molybdenum, With Filler	c	528-0358
Contact, Potassium - Russian Steel 1 Kh18 N9 T	c	528-0369
Contact, Russian Aluminum Alloy D 1	c	528-0210
Contact, Russian Aluminum Alloy D 1 - Russian Brass 80	c	528-0212
Contact, Russian Aluminum Alloy D 16 - Steel St 45	c	528-0226
Contact, Russian Brass L 80	c	528-0356
Contact, Russian Brass L 80 - Russian Steel St 30	c	528-0357
Contact, Russian Copper Alloy M2	c	528-0387
Contact, Russian Steel E Zh 1	c	528-0225
Contact, Russian Steel E Zh 2	c	528-0215
Contact, Russian Steel E Zh 2, With Filler	c	528-0216
Contact, Russian Steel Kh18 N10 T	c	528-0331
Contact, Russian Steel Kh18 N10 T - Sodium	c	528-0406

Substance Name	Property	Number
Contact, Russian Steel St 3	c	528-0415
Contact, Russian Steel St 3, With Filler	c	528-0417
Contact, Russian Steel St 30	c	528-0213
Contact, Russian Steel St 45	c	528-0224
Contact, Russian Steel 1 Kh18 N9 T	c	528-0222
Contact, Russian Steel 1 Kh18 N9 T – Sodium	c	528-0370
Contact, Russian Steel 30 Kh G S A	c	528-0322
Contact, Stainless Steel	c	528-0282
Contact, Stainless Steel – Stainless Steel 316, With Filler	c	528-0359
Contact, Stainless Steel 302 – Tool Steel	c	528-0408
Contact, Stainless Steel 303	c	528-0188
Contact, Stainless Steel 304	c	528-0269
Contact, Stainless Steel 347	c	528-0047
Contact, Stainless Steel 416	c	528-0019
Contact, Steel	c	528-0036
Contact, Steel (0.22 C, 0.45 Mn)	c	528-0397
Contact, Steel AISI C1020	c	528-0423
Contact, Steel AISI 4140	a c	528-0010
Contact, Tin	c	528-0236
Contact, Titanium	c	528-0311
Contact, Uranium	c	528-0309
Contact, Zinc	c	528-0307
Composite		
see Honeycomb		
see Multilayer		528–
Copper Alloy (Be), Niobium Covered, Inconel Covered	a n	528-0348
Copper Alloy (Zr), Stainless Steel Covered	n	528-0349
Copper, Nickel Covered	n	528-0015
Cylinder, Hollow, Stainless Steel 302	j	528-0095
Dumet Wire		
see Iron Alloy (42 Ni), Copper Covered		528–
Fiber Optic, Arsenic Gallium Tellurium Alloy	j	528-0396
Fiber Optic, Arsenic Germanium Selenium Alloy	j	528-0394
Fiber Optic, Arsenic Selenium Tellurium Alloy	j	528-0395
Filter, Interference	hij	528-0256
Filter, Zinc Selenide	j	528-0345
Fluidized Bed, Aluminum	a e	528-0301
Fluidized Bed, Copper	a e	528-0410
Fluidized Packed Bed, Steel Spheres – Copper – Nickel	d	528-0103
Honeycomb	a	528-0294
Honeycomb, Aluminum Alloy	a e	528-0024
Honeycomb, Aluminum Alloy, No Faces	a	528-0297
Honeycomb, Haynes Alloy 25, Brazed With Coast Metals 52	a g	528-0183
Honeycomb, Stainless Steel	a	528-0134
Honeycomb, Stainless Steel 301,		
Stainless Steel Faces	d	528-0109
Indium Alloy (7 Pb), Manganese Covered	a	528-0422
Interface, Aluminum Alloy – Beryllium	p	528-0336
Interface, Bismuth – Copper	p	528-0373
Interface, Chromium – Silver	p	528-0181
Interface, Copper – Lead	p	528-0255
Interface, Gallium – Mercury	p	528-0372
Interface, Metal – Metal	p	528-0099
Interface, Metal – Metal, Alkali	p	528-0424
Interface, Molybdenum – Tin	p	528-0187
Interface, Tin – Tungsten	p	528-0257
Iron Alloy (Ni), Copper Alloy Covered	n	528-0344
Iron Alloy (1.2 Ni, 0.54 C), Iron Alloy		
(1.2 Ni, 0.29 C) Covered	n	528-0391
Iron Alloy (42 Ni), Copper Covered	n	528-0330
Iron, Nickel Covered	a	528-0012
Joint		
see Contact		528–
Laminate (Except Reinforced Resins)		
see Multilayer		528–
L 70		
see Contact, Aluminum Alloy L 70		528–
Laminate		
see Multilayer		528–
Molybdenum Alloy, Chromalloy G Covered	n	528-0321
Molybdenum Alloy, Durak Mg Covered	n	528-0320
Molybdenum Alloy, Molybdenum Silicide Covered	n	528-0319
Molybdenum Alloy, Siliconized	n	528-0323
Molybdenum, Chromium Covered	n	528-0338

Substance Name	Property	Number
Multilayer, Aluminum	a g	528-0303
Multilayer, Aluminum – (Aluminum – Tin – Uranium)	d	528-0392
Multilayer, Beryllium – Lead	a d	528-0409
Multilayer, Brass – Stainless Steel	a d	528-0315
Multilayer, Copper – Lead		528-0139
Multilayer, Copper – Stainless Steel	a d	528-0314
Multilayer, Gold – Magnesium	m	528-0145
Multilayer, Gold – Tin	a	528-0418
Multilayer, Indium	a	528-0401
Multilayer, Indium Alloy (Bi) – Manganese	a	528-0420
Multilayer, Manganese – Tin	a	528-0419
Multilayer, Metallic	a j	528-0106
Multilayer, Russian Stainless Steel Kh18 N10 T	a	528-0337
Multilayer, Russian Steel 1 Kh18 N9 T	a	528-0166
Multilayer, Steel, Spaced	a	528-0328
Multilayer, Tantalum	a	528-0368
Niobium Alloy, Chromium Titanium Disilicide Covered	a e n	528-0339
Packed Bed, Aluminum Cylinders	a	528-0114
Packed Bed, Copper	a	528-0371
Packed Bed, Iron Spheres	a	528-0132
Packed Bed, Lead Spheres	a	528-0116
Packed Bed, Nickel Catalyst Pellets	a	528-0105
Packed Bed, Stainless Steel Spheres	a	528-0306
Packed Bed, Steel Cylinders	a	528-0185
Packed Bed, Steel Spheres	a	528-0147
Rectifier	a c	528-0465
Russian Steel Kh18 N10 T Foil, Crumpled	a	528-0428
Shot		
see Spheres		528–
Silver, Inconel – Silver (35) Alloy Covered	n	528-0429
Silver, Inconel Covered	a	528-0352
Silver, Stainless Steel Covered	n	528-0350
Solar Cell (Also Irradiated)	ghi	528-0186
Solar Cell, Silicon	ghij	528-0253
Solar Cell, Silicon, Aluminum Covered	h j	528-0254
Solar Cell, Silicon, Copper Covered	h	528-0364
Solar Cell, Silicon, Nickel Covered	h	528-0161
Solar Cell, Silicon, Silver Covered	h	528-0365
Space Metal		
see Honeycomb, Stainless Steel 301,		
Stainless Steel Faces		528–
Stainless Steel 304, Aluminum Covered	a	528-0317
Stainless Steel 304, Magnesium Covered	a	528-0316
Stainless Steel 316, Vanadium Alloy (20 Ti)		
Covered	n	528-0402
Stainless Steel, Brass Covered	a d	528-0414

NONMETALLIC SYSTEMS – CONTACTS, FILTERS, HONEYCOMBS, INTERFACES, MULTILAYERS, AND OTHERS   CLASS 531

Adduct, Urea – Hexadec-1-ene	e	531-1458
A C X– G-79		
see Resin, Phenolic, Graphite Reinforced		
(Series Of Additives)		531–
A G-4 S		
see Resin, Phenolic, Glass Reinforced		531–
A P F G		
see Resin, Phenolic, Glass Reinforced		531–
Acacia Gum		
(Syn. For Gum, Arabic)		
see Interface, (Gum, Arabic + Kerosene) – Water		531–
Aceto-caustin		
see Interface, (Dichloroethane + Trichloro-		
-acetic Acid) – (Trichloroacetic Acid +		
Water)		531–
Adduct, Urea – Decane	e	531-0380
Adduct, Urea – Dodecane	e	531-0381
Adduct, Urea – Dec-1-ene	e	531-1384
Adduct, Urea – Eicos-1-one	e	531-1467
Adduct, Urea – Eicosane	e	531-0412
Adduct, Urea – Hexadecane	e	531-0383
Adduct, Urea – Undecane	e	531-1278

Substance Name	Property	Number
Aerated		
see Expanded		531-
Air Gap	a	531-0697
Air On Oil	b	531-0972
Air On Sodium Chloride	b	531-0973
Air On Solid	b	531-1152
Aircomb		
see Honeycomb, Paper, Phenolic Resin Impregnated		531-
Alathon G-0530		
see Resin, Polyethylene, Glass Reinforced		531-
Aluminum Barium Silicate		
see Powder, Barium Aluminosilicate, Ethylene Adsorbate		531-
Aluminum Cadmium Silicate		
see Powder, Cadmium Aluminosilicate, Ethylene Adsorbate		531-
(Aluminum Oxide - Silicon Oxide), 3,6-bis(Dimethylamino)-acridine Adsorbate	h	531-1602
Aluminum Oxide - Vacuum	a	531-1444
Aluminum Oxide, Carbon Oxide (C O) Adsorbate	j	531-1676
Aluminum Oxide, Expanded - Vacuum	a	531-1445
Aluminum Oxide, Expanded, Chromium Oxide And Nickel Oxide Impregnated	g	531-1604
Aluminum Oxide, Fibrous - Silicon Oxide, Fibrous - Vacuum	a	531-1440
Aluminum Oxide, Fibrous - Vacuum	a	531-1439
Aluminum Oxide, Glass Covered	n	531-1704
Aluminum Oxide, Silicon Oxide Covered	n	531-1705
Aluminum Oxide, 2- Propanol Adsorbate	j	531-1414
Aluminum Oxide, 3,6-bis(Dimethylamino)acridine Adsorbate	h	531-1583
Aluminum Phosphate, Silicon Oxide Filled	n	531-1603
Aluminum Silver Silicate		
see Powder, Silver Aluminosilicate, Ethylene Adsorbate		531-
Amfab T V 20-60		
see Resin, Polytetrafluoroethylene, Glass Reinforced		531-
Ammonium Perchlorate, Poly-2-methylpropene Bonded	a	531-1792
Ammonium Perchlorate, Polyurethane Bonded	a	531-1793
Androstan-3-(α)-ol-17-one		
(Syn. For- Androsterone)		
see Interface, (Androsterone + Petrolatum-liquid) - Water		531-
5 α- Androstane		
(Syn. For- Androstane)		
see Interface, (Androstane + Petrolatum-liquid) - Water)		531-
3 β- Androstanol-17-one		
(Syn. For- Epiandrosterone)		
see Interface, (Epiandrosterone + Petrolatum-liquid Water		531-
Delta Superscript 4- Androsten-17 β-ol-3-one		
(Syn. For- Testosterone)		
see Interface, (Petrolatum-liquid + Testosterone) - Water		531-
Apple - Gas	a	531-0370
Apple (Fresh, Dried, Etc.) - Vacuum	a	531-1368
Arabic Gum		
(Syn. For- Gum, Arabic)		
see Interface, (Gum, Arabic + Kerosene) - Water		531-
Argon On Carbon	b	531-1359
Argon On Glass	b	531-1262
Argon On Mica	b	531-1212
Argon On Polytetrafluoroethylene	b	531-1616
Argon On Solid	b	531-1153
Argon On Uranium Carbide, Oxidized	b	531-0905
Asbestos, Fibrous - Vacuum	a	531-1682
Astrolite		
see Resin, Phenolic, Glass Reinforced		531-
Avco Phenolic Fiberglass		
see Resin, Phenolic, Glass Reinforced		531-
Avcoat 5026-39		
see Honeycomb, Nylon - Phenolic Resin, Glass Reinforced, Epoxy Resin Filled		531-
Basalt - Vacuum	a de	531-1337
Basalt, Fibrous - Vacuum	a	531-1752
Benzene On Glass	b	531-1450
Benzoresorcinol	b	531-9900
see Interface, 2,4- Dihydroxybenzophenone - Gas		531-
4- Benzoylresorcinol		
see Interface, 2,4- Dihydroxybenzophenone - Gas		531-
Bioquin		
(Syn. For- 8- Quinolinol)		
see Interface, Hydrocarbon, Aliphatic, Alkane - 8- Quinolinol		531-
Boron Iodide On Pyrex	b	531-1117
Bromine On Glass	b	531-1448
C N P		
see Resin, Phenolic, Nylon Reinforced		531-
C T L 39-9		
(Syn. For- Resin, Phenylsilane)		
see Resin, Phenylsilane, Glass Reinforced		531-
Cab-o-sil		
see Powder, Silicon Oxide - Gas		531-
(Calcium Oxide, Fibrous + Zirconium Oxide, Fibrous) - Vacuum	a	531-1443
Caproic Ether		
see Interface, Ethyl Caproate - Water		531-
Capronic Ether		
see Interface, Ethyl Caproate - Water		531-
Carb- l- Tex		
see Carbon Cloth, Carbon Bonded		531-
Carbanil		
see Interface, Phenylisocyanate - Water		531-
Carbon Cloth, Carbon Bonded	n	531-1665
Carbon Oxide (C O_2) On Glass	b	531-1469
Carbon Oxide (Co_2) On Solid	b	531-0377
Carbon Resistor		
see Carbon, Resin Bonded		531-
Carbon, Resin Bonded	a e n	531-0974
β- Carotene, Acetone Adsorbate	j	531-1030
Catalyst, (Aluminum Oxide - Chromium Oxide) ← Butane	m	531-0798
Catalyst, (Aluminum Oxide - Chromium Oxide) ← Nitrogen	m	531-0803
Catalyst, (Aluminum Oxide - Chromium Oxide) ← 1- Butene	m	531-0795
Celkate T-21		
see Powder, (Magnesium Oxide + Silicon Oxide) - Vacuum		531-
Cellular		
see Expanded		531-
Ceramic Insulation, Dynaflex - Gas	a	531-0923
Ceramic Insulation, Dynaflex - Vacuum	a	531-1214
Ceramic, ($Al_2 O_3 + B_2 O_3 + Na_2 O + Si O_2$), Fibrous - Vacuum	a	531-1772
Ceramic, ($Al_2 O_3 + Ca O + Fe_2 O_3 + Si O_2 + Ti O_2$) - Gas	d	531-1740
Ceramic, ($Al_2 O_3 + Fe_2 O_3 + Si O_2 + Ti O_2$) - Gas	d	531-1742
Ceramic, ($Al_2 O_3 + Fe_2 O_3 + Si O_2 + Ti O_2 +$ Alkalies) - Gas	d	531-1741
Ceramic ($Al_2 O_3 + Si O_2$), Fibrous - Vacuum	a	531-1771
Ceramic, Brick - Gas	a	531-1775
Ceramic, Carbon - Gas	a	531-1499
Ceramic, Ceramic And Silicon Oxide Reinforced, Bonded	d	531-0079
Ceramic, Dinas - Vacuum	a	531-1079
Ceramic, Fireclay, Porous - Gas	a d	531-0799
Ceramic, Forsterite, Light Weight - Vacuum	a	531-1080
Ceramic, Magnesite - Vacuum	a	531-1078
Cercor (Corning)		
see Honeycomb, Glass		531-
Char From Plastic Material - Gas	a	531-1309
Char From Plastic Material - Vacuum	a	531-1294
Chopped Nylon Phenolic (C N P)		
see Resin, Phenolic, Nylon Reinforced		531-
Clathrate, Ethylene Oxide - Water	n	531-1800
Clathrate, Hydroquinone - Hydrocyanic Acid	e	531-1053
Clathrate, Hydroquinone - Hydrogen Chloride	e	531-0838
Clathrate, Hydroquinone - Krypton	e	531-0533
Clathrate, Hydroquinone - Methanol (3 $C_6 H_4$ (O H)$_2$ (C H_3 O H)$_{0.974}$)	e	531-0309
Clathrate, Hydroquinone - Nitrogen	e	531-0840

Substance Name	Property	Number
Interface	h j p	531-0714
Interface, (Acetic Acid + Benzene) – (Acetic Acid + Water)	p	531-1386
Interface, (Acetic Acid + Benzene) – Ammonia	p	531-0632
Interface, (Acetic Acid + Benzene) – Water	p	531-1388
Interface, (Acetic Acid + Carbon Tetrachloride) – (Acetic Acid + Water)	p	531-0946
Interface, (Acetic Acid + Carbontetrachloride) – Water	p	531-1403
Interface, (Acetic Acid + Chlorobenzene) – (Acetic Acid + Water)	p	531-0938
Interface, (Acetic Acid + Chloroform) – (Acetic Acid + Water)	p	531-0934
Interface, (Acetic Acid + Cyclohexane) – (Acetic Acid + Water)	p	531-0943
Interface, (Acetic Acid + Dichloroethane) – (Acetic Acid + Water)	p	531-0939
Interface, (Acetic Acid + Pyridine) – Ammonia	p	531-0645
Interface, (Acetic Acid + Toluene) – (Acetic Acid + Water)	p	531-0935
Interface, (Acetic Acid + Toluene) – (Acetic Acid + Sucrose + Water)	p	531-0936
Interface, (Acetic Acid + Toluene) – Water	p	531-1397
Interface, (Acetic Acid + Water) – Benzene	p	531-1387
Interface, (Acetic Acid + Water) – Carbon Tetrachloride	p	531-1398
Interface, (Acetic Acid + Water) – Toluene	p	531-1393
Interface, (Acetone + Benzene + Water), Aqueous Phase – Organic Phase	p	531-0624
Interface, (Acetone + Hexane + Water), Aqueous Phase – Organic Phase	p	531-0450
Interface, (Acetone + 1,1,2- Trichloroethane + Water), Aqueous Phase – Organic Phase	p	531-0621
Interface, (Acetone + 4- Methyl-2-pentanone + Water), Aqueous Phase – Organic Phase	p	531-0977
Interface, (Agar + Kerosene) – Water	p	531-1531
Interface, (Agar + Oil, Olive) – Water	p	531-1536
Interface, (Aluminum Oxide + Cryolite) – Gas	p	531-1064
Interface, (Androstane + Petrolatum–liquid) – Water)	p	531-1013
Interface, (Androsterone + Petrolatum–liquid) – Water	p	531-1014
Interface, (Benzene + (Didecylamine Sulfate – Uranyl Sulfate) Complex) – (Sulfuric Acid + Water)	p	531-1557
Interface, (Benzene + Didecylamine) – Water	p	531-1553
Interface, (Benzene + Didecylamine Nitrate) – (Nitric Acid + Water)	p	531-1554
Interface, (Benzene + Didecylamine Sulfate) – (Sulfuric Acid + Water)	p	531-1555
Interface, (Benzene + Didecylamine Sulfate) – Sodium Sulfate + Sulfuric Acid + Water)	p	531-1556
Interface, (Benzene + Didecylamine Sulfate + (Didecylamine Sulfate – Uranyl Sulfate) Complex) – (Sulfuric Acid + Water)	p	531-1558
Interface, (Benzene + Hexane + Tetradecafluorohexane), Benzene Phase – Tetradecafluorohexane Phase	p	531-0627
Interface, (Benzene + Methanol + Water), Aqueous Phase – Organic Phase	p	531-0625
Interface, (Benzene + Propanol + Water), Aqueous Phase – Organic Phase	p	531-0930
Interface, (Benzene + Propionic Acid) – (Propionic Acid + Water)	p	531-1404
Interface, (Benzene + Propionic Acid) – Water	p	531-1411
Interface, (Benzene + Tetraheptylammonium Chloride) – (Lithium Chloride + Water)	p	531-1549
Interface, (Benzene + Water) – Butyric Acid	p	531-1121
Interface, (Benzene + Water) – Caproic Acid	p	531-1123
Interface, (Benzene + Water) – Caprylic Acid	p	531-1124
Interface, (Benzene + Water) – Valeric Acid	p	531-1122
Interface, (Benzene + Water), Aqueous Phase – Organic Phase	p	531-0623
Interface, (Benzoyl Peroxide + Polyvinyl Alcohol + Water) – Styrene	p	531-1224
Interface, (Benzoyl Peroxide + Polyvinyl Alcohol + Water) – Toluene	p	531-1226
Interface, (Benzoyl Peroxide + Water) – Styrene	p	531-1222
Interface, (Benzoyl Peroxide + Water) – Toluene	p	531-1223
Interface, (Butane + Decane) – Methane	p	531-1538
Interface, (Butylacetate + Formic Acid) – (Formic Acid + Water)	p	531-0937
Interface, (Butylacetate + Nitric Acid) – (Nitric Acid + Water)	p	531-0954
Interface, (Calcium Chloride + Water) – Water Adsorbed On Glass	p	531-1141
Interface, (Calcium Chloride + Water) – Stearic Acid	f	531-0918
Interface, (Calcium-p-undecylaminobenzenesulfonate + Water) – Heptane	p	531-1355
Interface, (Carbon Sulfide + Hexane + Tetradecafluoro-hexane), Carbon Sulfide Phase – Tetradecafluoro-hexane Phase	p	531-0628
Interface, (Carbon Tetrachloride + Cyclohexylamine) – (Cyclohexylamine + Water)	p	531-0951
Interface, (Carbon Tetrachloride + Diethylamine) – (Diethylamine + Water)	p	531-0953
Interface, (Carbon Tetrachloride + Heptane) – Water	p	531-0997
Interface, (Carbon Tetrachloride + Oil, Mineral) – Water	p	531-1606
Interface, (Carbon Tetrachloride + Propanol + Water), Aqueous Phase – Organic Phase	p	531-0626
Interface, (Carbon Tetrachloride + Propionic Acid) – (Propionic Acid + Water)	p	531-0945
Interface, (Chloroethane + Isovaleric Acid) – (Isovaleric Acid + Water)	p	531-0948
Interface, (Chlorophyll + 1– Dodecanol + Petrolatum–liquid) – Water	p	531-0980
Interface, (Chlorophyll – Oil, Mineral) – Water	p	531-0881
Interface, (Cholesterol + Dioctadecyl Phosphite) – Water	p	531-1495
Interface, (Cholesterol + Dioctadecyl Phosphite + Dodecane) –(Sodium Chloride + Water)	p	531-0821
Interface, (Cholesterol + Dodecane) – (Hexadecyltri-methylammonium Bromide + Sodium Chloride + Water)	p	531-0488
Interface, (Cholesterol + Dodecane + Dodecyl Phosphate) – (Sodium Chloride + Water)	p	531-0804
Interface, (Cholesterol + Hexadecyltrimethylammonium Bromide) – (Sodium Chloride + Water)	p	531-1496
Interface, (Cholesterol + Petrolatum–liquid) – Water	p	531-1019
Interface, (Cholesterol, Oxidized + Octane) – (Sodium Chloide + Water)	p	531-0914
Interface, (Cortisol + Petrolatum–liquid) – Water	p	531-1018
Interface, (Cortisone + Petrolatum–liquid) – Water	p	531-1012
Interface, (Cyclohexane + Propionic Acid) – (Propionic Acid + Water)	p	531-0944
Interface, (Cyclohexylamine + Toluene) – (Cyclohexylamine + Water)	p	531-0952
Interface, (Decane + Dodecanol) – Bis-(2-ethylhexyl) Sodium Sulfosuccinate	p	531-0614
Interface, (Decane + Dodecanol + Oil, Mineral) - Bis-(2-ethylhexyl)sodium Sulfosuccinate	p	531-0603
Interface, (Decane + Isosorbide Mono-trans-13-docosenoate) – (Sodium Chloride + Water)	p	531-1339
Interface, (Decane + Isosorbide Monolaurate) – (Sodium Chloride + Water)	p	531-1333
Interface, (Decane + Isosorbide Monooleate) – (Sodium Chloride + Water)	p	531-1338
Interface, (Decyl Sodium Sulfate + Sodium Chloride + Water) – Hexadecane	p	531-1045
Interface, (Decyl Sodium Sulfate + Sodium Chloride + Water) – Dodecane	p	531-1046
Interface, (Decyl Sodium Sulfate + Sodium Chloride + Water) – Octane	p	531-1047
Interface, (Dichloroethane + Formic Acid) – (Formic Acid + Water)	p	531-0940
Interface, (Dichloroethane + Propionic Acid) – (Propionic Acid + Water)	p	531-0941
Interface, (Dichloroethane + Trichloro-acetic Acid) – (Trichloroacetic Acid + Water)	p	531-0942
Interface, (Didodecylammonium Nitrate + Water) – Xylene	p	531-0959
Interface, (Diethylamine + Toluene) –	p	531-0950

Substance Name	Property	Number
(Diethylamine + Water)	p	531-0950
Interface, (Diolein + Dodecane) - (Sodium Chloride + Water)	p	531-1494
Interface, (Distearin + Octane) - (Sodium Chloride + Water)	p	531-0552
Interface, (Dodecane + Lecithin) - (Sodium Chloride + Water)	p	531-0489
Interface, (Dodecanol + Oil, Mineral - -bis-(2-ethylhexyl) Sodium Sulfosuccinate	p	531-0602
Interface, (Dodecanol + Oil, Mineral) - (Potassium + Water)	p	531-0594
Interface, (Dodecanol + Oil, Mineral) - Water	p	531-0601
Interface, (Dodecanol + Oil, Mineral) - Water, Sea	p	531-0616
Interface, (Dodecanol + Water) - Xylene	p	531-0689
Interface, (Dodecyl Sodium Sulfate + Sodium Chloride + Water) - Tetradecane	p	531-1048
Interface, (Dodecyl Sodium Sulfate + Sodium Chloride + Water) - Octane	p	531-1050
Interface, (Dodecyl Sodium Sulfate + Sodium Chloride + Water) - Hexane	p	531-1051
Interface, (Dodecyl Sodium Sulfate + Sodium Chloride + Water) - Petrolatum-liquid	p	531-1371
Interface, (Dodecyl Sodium Sulfate + Water) - Petrolatum-liquid	p	531-1370
Interface, (Dodecyl Sodium Sulfate + Water) - Heptadecane	p	531-0224
Interface, (Dodecyl Sodium Sulfate + Water) - Hexane	p	531-0246
Interface, (Dodecyl Sodium Sulfate + Water) - Nonane	p	531-0270
Interface, (Dodecyl Sodium Sulfate + Water) - Octane	p	531-0240
Interface, (Dodecyl Sodium Sulfate + Water) -1- Hexene	p	531-0265
Interface, (Dodecyl Sodium Sulfate + Water) -1- Octane	p	531-0241
Interface, (Dodecylammonium Nitrate + Water) - Xylene	p	531-0993
Interface, (Dye + Surfactant + Water) - Oil	p	531-1638
Interface, (Epiandrosterone + Petrolatum-liquid) - Water	p	531-1015
Interface, (Estradiol + Petrolatum-liquid) - Water	p	531-1010
Interface, (Estriol + Petrolatum-liquid) - Water	p	531-0990
Interface, (Estrone + Petrolatum-liquid) - Water	p	531-1011
Interface, (Ethyl Ether + Formic Acid) - (Formic Acid + Water)	p	531-0932
Interface, (Ethylene Oxide + p-fluorobutylphenol) - Oil, Mineral	p	531-1537
Interface, (Formic Acid + Water) - Sodium Formate	p	531-1008
Interface, (Gelatin + Kerosene) - Water	p	531-1530
Interface, (Gelatin + Oil, Olive) - Water	p	531-1535
Interface, (Gelatin + Water) - Gas	p	531-0460
Interface, (Glycerol + Water) - Heptane	p	531-0995
Interface, (Gum, Arabic + Kerosene) - Water	p	531-1528
Interface, (Gum, Arabic + Oil, Olive) - Water	p	531-1533
Interface, (Gum, Tragacanth + Kerosene) - Water	p	531-1527
Interface, (Gum, Tragacanth + Oil, Olive) - Water	p	531-1532
Interface, (Heptane + Hexadecane) - Water	p	531-0921
Interface, (Heptane + Petrolatum-liquid) - Water	p	531-0999
Interface, (Heptane + Propionic Acid) - (Propionic Acid + Water	p	531-0933
Interface, (Heptane + Valeric Acid) - (Valeric Acid + Water)	p	531-0947
Interface, (Hexadecane + Octane) - Water	p	531-0922
Interface, (Hexane + Propanol + Water), Aqueous Phase - Organic Phase	a p	531-0910
Interface, (Hexane + Tributyl Phosphate) - (Lithium Nitrate + Water)	p	531-1500
Interface, (Hexane + Tributyl Phosphate) - Water	p	531-1514
Interface, (Inorganic Electrolyte + Water) - Oil, Lecithin	p	531-1446
Interface, (Kerosene + Magnetite + Oleic Acid) - Water	p	531-1156
Interface, (Kerosene + Oleic Acid) - Water	p	531-1155
Interface, (Kerosene + Pectin) - Water	p	531-1529
Interface, (Lauric Acid + Water) - Xylene	p	531-0924
Interface, (Nitrobenzene + Octadecyltrimethyl Ammonium Picrate) - Water	p	531-0362
Interface, (Oil (Series Of Additives)) - (Sulfuric Acid + Water)	p	531-1591
Interface, (Oil, Mineral + Tetrachloroethylene) - Water	p	531-1607
Interface, (Oil, Olive + Pectin) - Water	p	531-1534

Substance Name	Property	Number
Interface, (Petrolatum - Liquid) - (Sulfuric Acid + Water)	p	531-1589
Interface, (Petrolatum-liquid + Progesterone) - Water	p	531-1017
Interface, (Petrolatum-liquid + Steroid D, H, A.) - Water	p	531-1025
Interface, (Petrolatum-liquid + Testosterone) - Water	p	531-1016
Interface, (Polyvinyl Alcohol + Water) - Styrene	p	531-1220
Interface, (Polyvinyl Alcohol + Water) - Toluene	p	531-1225
Interface, (Potassium Carbonate + Water) - Water, Adsorbed On Glass	p	531-1140
Interface, (Propionic Acid + Tetrabromoethane) - (Propionic Acid + Water)	p	531-0949
Interface, (Propionic Acid + Toluene) - (Propionic Acid + Water)	p	531-1412
Interface, (Propionic Acid + Toluene) - Water	p	531-1432
Interface, (Propionic Acid + Water) - Toluene	p	531-1413
Interface, (Sodium Cholate + Water) - Tetradecane	p	531-0760
Interface, (Sodium Cholate + Water) -p- Xylene	p	531-0765
Interface, (Sodium Cholate + Water) -p- Xylene, (Distearin + Monostearin + Stearic Acid) Modified	p	531-0771
Interface, (Sodium Fluoride + Zirconium Fluoride) - Gas	p	531-1061
Interface, (Sodium Fluoride + Zirconium Fluoride + Zirconium Oxide) - Gas	p	531-1062
Interface, (Tridodecylammonium Nitrate + Water) - Xylene	p	531-0956
Interface, (1- Propanol + Water) - Toluene	p	531-1374
Interface, (1,1,2- Trichloroethane + Water), Aqueous Phase - Organic Phase	p	531-0619
Interface, (1,4- Butanediamine + Water) - Chloroform	p	531-1415
Interface, (11- Deoxycorticosterone + Petrolatum-liquid) - Water	p	531-1024
Interface, (17- Hydroxypregn-4-ene-3,20-dione + Petrolatum-liquid) - Water	p	531-1023
Interface, (2- Butanone + Hexane + Water), Aqueous Phase - Organic Phase	p	531-0926
Interface, (2- Butanone + 1,1,2- Trichloroethane + Water), Aqueous Phase - Organic Phase	p	531-0622
Interface, (2- Propanol + Water) - Toluene	p	531-1375
Interface, (3-α- Hydroxy-5-β- -androstan-17-one + Petrolatum-liquid) - Water	p	531-1021
Interface, (3-β- Hydroxy-5-β- -androstan-17-one + Petrolatum-liquid) - Water	p	531-1022
Interface, (4,4'-(1,2- Diethylethylene) Diphenol + Petrolatum-liquid) - Water	p	531-1020
Interface, Acetic Acid - Ammonia	p	531-1648
Interface, Acetic Acid - Cyclohexane	p	531-1728
Interface, Acetic Acid - Ethanol	p	531-1731
Interface, Acetic Acid - Water	p	531-1725
Interface, Acetone - Carbon Tetrachloride - Water	p	531-0917
Interface, Aluminum Oxide - Vacuum	p	531-1710
Interface, Ammonia - Benzene	p	531-0630
Interface, Ammonia - Pyridine	p	531-0634
Interface, Aniline - Benzyl Acetate	p	531-0982
Interface, Aniline - Water	p	531-1206
Interface, Anisole - (Surfactant + Water)	p	531-1523
Interface, Anisole - Water	p	531-1525
Interface, Anthranilic Acid - (Anthranilic Acid + Water)	p	531-1073
Interface, Benzaldehyde - Water	p	531-1184
Interface, Benzene - (Dodecyl Sodium Sulfate + Sodium Chloride + Water)	p	531-1718
Interface, Benzene - (Dodecyl Sodium Sulfate + Water)	p	531-0202
Interface, Benzene - (Lithium Caprylate + Lithium Chloride + Water)	p	531-1236
Interface, Benzene - (Lithium Caprylate + Water)	p	531-1235
Interface, Benzene - (Propionic Acid + Water)	p	531-1406
Interface, Benzene - (Sodium Caprylate + Sodium Chloride + Water)	p	531-1234
Interface, Benzene - (Sodium Caprylate + Water)	p	531-1233
Interface, Benzene - (Sodium Chloride + Water)	p	531-1734
Interface, Benzene - (Sodium Cholata + Water)	p	531-0732
Interface, Benzene - (Sulfuric Acid + Water)	p	531-1552
Interface, Benzene - (2- Methyl-2-propanol + Water)	p	531-1373
Interface, Benzene - (2- Propanol + Water)	p	531-1372

Substance Name	Property	Number
Interface, 2- Hexanone - Water	p	531-1119
Interface, 2- Hydroxy-4-(Octyloxy)benzophenone - Gas	p	531-1131
Interface, 2- Hydroxy-4-methoxybenzophenone - Gas	p	531-1130
Interface, 2- Methoxyethanol - Water	p	531-1068
Interface, 2- Methyl-1-pentene - Water	p	531-1303
Interface, 2- Methyl-1-propanol - Water	p	531-1196
Interface, 2- Methylbutane - Water	p	531-1167
Interface, 2- Methylheptane - Water	p	531-1207
Interface, 2- Methylheptane-octyl Phenoxyethanol - Water	p	531-0916
Interface, 2- Methylpentane - Water	p	531-1239
Interface, 2- Octanone - Water	p	531-1210
Interface, 2- Pentanone - Water	p	531-1098
Interface, 2,2- Dimethylbutane - Water	p	531-1237
Interface, 2,2- Dimethylpentane - Water	p	531-1296
Interface, 2,2'- Dihydroxy-4-methoxybenzophenone - Gas	p	531-1129
Interface, 2,2,3- Trimethylbutane - Water	p	531-1295
Interface, 2,2,3- Trimethylpentane - Water	p	531-1298
Interface, 2,2,4- Trimethyl-1-pentene - Water	p	531-1306
Interface, 2,2,4- Trimethyl-2-pentene - Water	p	531-1308
Interface, 2,2,4- Trimethylpentane - Water	p	531-1293
Interface, 2,3- Dimethylbutane - Water	p	531-1238
Interface, 2,3- Dimethylpentane - Water	p	531-1297
Interface, 2,3,4- Trimethylpentane - Water	p	531-1299
Interface, 2,4- Dihydroxybenzophenone - Gas	p	531-1128
Interface, 2,4- Dimethylpentane - Water	p	531-1243
Interface, 2,4- Pentanedione - Water	p	531-1151
Interface, 2,6- Dimethyl-4-heptanone - Water	p	531-1148
Interface, 2,7- Dimethyloctane - Water	p	531-1208
Interface, 3- Ethyl-2-methylpentane - Water	p	531-1292
Interface, 3- Ethylhexane - Water	p	531-1273
Interface, 3- Ethylpentane - Water	p	531-1242
Interface, 3- Methyl-1-butanol - Water	p	531-1198
Interface, 3- Methyl-1-pentene - Water	p	531-1302
Interface, 3- Methylheptane - Water	p	531-1291
Interface, 3- Methylhexane - Water	p	531-1255
Interface, 3- Methylpentane - Water	p	531-1241
Interface, 3- Methylpentene - Nitroethane	p	531-1437
Interface, 4- Methyl-1-pentene - Water	p	531-1301
Interface, 4- Methyl-2-hexyne - Water	p	531-1307
Interface, 4-(2- Ethylhexyloxy)-2-hydroxybenzophenone - Gas	p	531-1132
Iodine On Glass	b	531-1449
Irish Refrasil		
see Resin, High Char, Silicon Oxide Reinforced		531-
Isoandrosterone		
see Interface, (Epiandrosterone +	b	531-
(Syn. For- Epiandrosterone)		
Petrolatum-liquid) - Water		531-
Isobutylethylene		
see Interface, 4- Methyl-1-pentene - Water		531-
Isohexylene		
see Interface, 4- Methyl-1-pentene - Water		531-
Isosorbide Monobrassidate		
(Syn. For- Isosorbide Mono-trans-13-docosenoate)		
see Interface, (Decane + Isosorbide Mono-trans-13--docosenoate) - (Sodium Chloride + Water)		531-
K A S T		
see Resin, Phenolic, Glass Reinforced		531-
Kaolinite, Fibrous - Vacuum	a	531-1686
Kaowool		
see Ceramic (Al₂ O₃ + Si O₂), Fibrous - Vacuum		531-
Kelecin		
(Syn. For- Lecithin)		
see Interface, (Dodecane + Lecithin) - (Sodium Chloride + Water)		531-
Kendalls Desoxy Compound B		
(Syn. For- 11- Deoxycorticosterone)		
see Interface, (11- Deoxycorticosterone + Petrolatum-liquid) - Water		531-

Substance Name	Property	Number
Ketohydroxyestrin		
(Syn. For- Estrone)		
see Interface, (Estrone + Petrolatum-liquid) - Water		531-
Kodak Wratten No. 47		
see Filter, Optical		531-
Kodak Wratten No. 61		
see Filter, Optical		531-
Krypton On Carbon	b	531-1340
Krypton On Glass	b	531-1263
Krypton On Uranium Carbide, Oxidized	b	531-0906
see Composite		
see Multilayer		
see Resin, Reinforced		531-
Lava - Vacuum	a	531-1770
Lecithol		
(Syn. For- Lecithin)		
see Interface, (Dodecane + Lecithin) - (Sodium Chloride + Water)		531-
Lockheat 1 Or 2 (Lockheed)		
see Ceramic, Ceramic And Silicon Oxide Reinforced, Bonded		531-
Lunar Material - Vacuum	a	531-1611
M C C-50 P Thermoelectric Material (Monsanto)		
see Graphite, Boron Carbide Covered		531-
M C C-60 N Thermoelectric Material (Monsanto)		
see Graphite, Silicon Carbide Covered		531-
M M M 3270 Screen		
see Resin, Clear, Glass Bead Embedded		531-
M M M 3280 Screen		
see Resin, Clear, Glass Bead Embedded, White Adhesive Reinforced		531-
M X 2600		
see Resin, Phenolic, Silcon Oxide Reinforced (Series Of Additives)		531-
M X 2646		
see Resin, Phenolic, Polyamide Modified, Silicon Oxide Reinforced		531-
M X 4926		
see Resin, Phenolic, Carbon Reinforced (Series Of Additives)		531-
Magnesium Oxide, 3,6-bis(Dimethylamino)acridine Adsorbate	h	531-1610
Malachite Green		
see Diatomite, C. I. Basic Green 4 Adsorbate		531-
Martin S L A-561		
see Resin, Silicone, Cork, Phenolic Resin And Silicon Oxide Reinforced		531-
Mc Donnel S-6		
see Honeycomb, Phenolic Resin, Glass Reinforced, Silicone Resin Filled		531-
Menformone		
(Syn. For- Estrone)		
see Interface, (Estrone + Petrolatum-liquid) - Water		531-
Metal Oxide, Water Adsorbate	p	531-1721
Methane On Carbon	b	531-1383
4- Methoxy-2,2'-dihydroxybenzophenone		
see Interface, 2,2'- Dihydroxy-4-methoxybenzophenone - Gas		531-
Micarta		
see Resin, Phenolic, Reinforced		531-
Microcline		
see Powder, Potassium Aluminosilicate, Ethylene Adsorbate		531-
Microflect Screen (Radiant)		
see Resin, White, Glass Bead Embedded		531-
Mineral Wool - Vacuum	a	531-1230
Mitten, Plastic Interlined (Tri-lok)	a	531-1667
Mitten, Plastic Interlined (Tri-lok) - Gas	a	531-1666
Multilayer (General)	a de gh j n	531-0633
Multilayer, (Glycerol + Water) - Silicon Oxide	a	531-1378
see also Composite		531-
see also Resin, Reinforced		531-
Multilayer, Aluminum Phosphate - Glass	a e gh j n	531-1381
Multilayer, Asbestos - Asbotermite	a	531-1146
Multilayer, Asbestos - Asbotermite - Diatomite	a	531-1145

Substance Name	Property	Number
Multilayer, Asbestos – Asbotermite – Nyuvela	a	531-1144
Multilayer, Asbestos – Diatomite – Paper	a	531-1143
Multilayer, Asbestos – Diatomite – Peat	a	531-1142
Multilayer, Asbestos – Nyuvela	a	531-1147
Multilayer, Asphalt – Felt	n	531-1433
Multilayer, Asphalt – Glass Felt	n	531-1435
Multilayer, Barium Fluoride – Resin, Epoxy	j	531-1380
Multilayer, Carbon Fabric – Vacuum	a	531-1118
Multilayer, Carbon Fabric, Resin Bonded, Carbon Filled	a	531-1784
Multilayer, Cellulose	a h	531-0976
Multilayer, Ceramic	n	531-0476
Multilayer, Coal Tar Pitch – Organic Felt	n	531-1434
Multilayer, Cork – Glass – Hyperlon – Polytetrafluoroethylene – Gas	a	531-1135
Multilayer, Cotton Leaf	h j	531-1027
Multilayer, Cotton-wool	a	531-0639
Multilayer, Cotton-wool – Gas	a	531-0658
Multilayer, Dacron – Nylon – Polyamide – Polychloroprene – Polyurethane – Vacuum	a	531-1622
Multilayer, Dacron – Nylon – Polyamide – Polychloroprene – Polyurethane – Gas	a	531-1623
Multilayer, Dacron – Vacuum	a	531-1635
Multilayer, Dimplar	a	531-1660
Multilayer, Fabric	h j	531-1797
Multilayer, Fireclay Brick	a d	531-1653
Multilayer, Fireclay Brick, Salts Impregnated	a d	531-1654
Multilayer, Glass	hij	531-0909
Multilayer, Glass – Lacquer, Linen Reinforced	a	531-1782
Multilayer, Glass – Phenolic Resin, Cotton Reinforced	a	531-1781
Multilayer, Glass – Poly(Ethylene Glycol – Terephthalic Acid) – Polyurethane, Expanded	a	531-1657
Multilayer, Glass – Poly(Ethylene Glycol – Terephthalic Acid)	a	531-1658
Multilayer, Glass – Polyvinyl Alcohol, Iodine Treated	j	531-0883
Multilayer, Glass – Water	a	531-0806
Multilayer, Glass Fabric – Gas	a	531-1165
Multilayer, Glass Fabric – Vacuum	a	531-1163
Multilayer, Glass Paper	a	531-0975
see also Fibrous Glass		507-
Multilayer, Glass Paper – Gas	a	531-1164
Multilayer, Glass Paper – Vacuum	a	531-1634
Multilayer, Glass, Fibrous – Vacuum	a	531-1057
Multilayer, (Glycerol + Water) – Silicon Oxide		531-1378
Multilayer, Glass, Spaced	a d	531-0807
Multilayer, Graphite – Gas	a	531-1424
Multilayer, Graphite – Tantalum Carbide	d	531-1162
Multilayer, Graphite – Vacuum	a	531-1425
Multilayer, Graphite Fabric – Vacuum	a	531-1396
Multilayer, Magnesium Oxide – Polymethyl Methacrylate	h	531-0513
Multilayer, Mica – Water	a e	531-0589
Multilayer, Micarta	a d n	531-0434
Multilayer, Nylon – Poly(Ethylene Glycol – Terephthalic Acid)	a	531-1659
Multilayer, Nylon – Polyamide – Polychloroprene – Gas	a	531-1702
Multilayer, Nylon – Polyamide – Polychloroprene – Vacuum	a	531-1703
Multilayer, Nylon – Polyamide – Polyethylene – Gas	a	531-1621
Multilayer, Nylon – Polyamide – Polyethylene – Vacuum	a	531-1620
Multilayer, Paper	a j	531-0135
Multilayer, Paper, Corrugated	a	531-1455
Multilayer, Perlon Stocking	j	531-1314
Multilayer, Phenolic Resin	g	531-0093
Multilayer, Plastic	h	531-0692
Multilayer, Plastic, Corrugated	a	531-1456
Multilayer, Poly(Ethylene Glycol – Terephthalic Acid) – Silk	a	531-1661
Multilayer, Poly(Ethylene Glycol – Terephthalic Acid) – Vacuum	a	531-1431
Multilayer, Poly(Ethylene Glycol – Terephthalic Acid) – Polyurethane, Expanded	a	531-1656
Multilayer, Polyethylene, Spaced	h	531-0587
Multilayer, Polytetrafluoroethylene	d	531-1280
Multilayer, Polytetrafluoroethylene – Methane	m	531-0514

Substance Name	Property	Number
Multilayer, Polyurethane, Expanded – Vacuum	a d	531-0979
Multilayer, Resin, Epoxy – Zinc Sulfide	j	531-1379
Multilayer, Rubber	d	531-1281
Multilayer, Rubber, Perforated	a	531-0891
Multilayer, Rubber, Perforated, Water Filled	a	531-0646
Multilayer, Silicon Oxide	j	531-0505
Multilayer, Silicon Oxide – Vanadium Oxide	j	531-0928
Multilayer, Silicon Oxide – Xylose	j	531-0512
Multilayer, Silicon Oxide Paper	a	531-0986
Multilayer, Silicon Oxide, Fibrous – Vacuum	a	531-1785
Multilayer, Sodium Tetraborate	a d	531-1395
Multilayer, Zirconium Oxide – Vacuum	a	531-1276
Naphthalene On Glass	b	531-1451
Narmco 4028 see Resin, Phenolic, Carbon Reinforced		531-
Narmco 534 (Syn. For– Resin, Phenylsilane) see Resin, Phenylsilane, Glass Reinforced		531-
Neon On Carbon	b	531-1376
Neon On Glass	b	531-1085
Neon On Solid	b	531-0376
Nerofil see Powder, Carbon – Gas		531-
Niobium Carbide, Graphite Reinforced	n	531-1392
Nitrogen On Glass	b	531-1084
Nitrogen Oxide (N_2 O) On Glass	b	531-1647
O C L No. 207, Blue Interference Filter see Filter, Metallic Oxides – Silicon Oxide, Multilayer		531-
Octabenzone see Interface, 2- Hydroxy-4-(Octyloxy)benzophenone – Gas		531-
Oestrone (Syn. For– Estrone) see Interface, (Estrone + Petrolatum-liquid) – Water		531-
Ovalecithin (Syn. For– Lecithin) see Interface, (Dodecane + Lecithin) – (Sodium Chloride + Water)		531-
Oxine (Syn. For– 8- Quinolinol) see Interface, Hydrocarbon, Aliphatic, Alkane – 8- Quinolinol		531-
Oxybenzone see Interface, 2- Hydroxy-4-methoxybenzophenone – Gas		531-
Oxychinolin (Syn. For– 8- Quinolinol) see Interface, Hydrocarbon, Aliphatic, Alkane – 8- Quinolinol		531-
Oxygen On Glass	b	531-1264
Oxyquinoline (Syn. For– 8- Quinolinol) see Interface, Hydrocarbon, Aliphatic, Alkane – 8- Quinolinol		531-
P T K see Resin, Phenolic, Cotton Reinforced		531-
Packed Bed (General)	a d	531-0384
Packed Bed, Aloxite see Packed Bed, Aluminum Oxide Spheres – Gas		531-
Packed Bed, Aluminum Oxide Spheres – Gas	a	531-0421
Packed Bed, Aluminum Oxide Spheres – Vacuum	a	531-1042
Packed Bed, Catalytic	a	531-1268
Packed Bed, Cement Clinkers	a	531-0003
Packed Bed, Cement Clinkers – Gas	a	531-0347
Packed Bed, Cement Clinkers – Water	a	531-0449
Packed Bed, Ceramic – Acetone	d	531-1697
Packed Bed, Ceramic – Benzene	a de	531-1694
Packed Bed, Ceramic – Ethanol	a de	531-1695
Packed Bed, Ceramic – Water	a de	531-1696
Packed Bed, Ceramic Rings	a	531-1627
Packed Bed, Charcoal	a	531-0227
Packed Bed, Diatomite Cylinders	a	531-1039
Packed Bed, Glass Fibers – Dioctyl Phthalate	a	531-1249
Packed Bed, Glass Helixes – Water	a	531-0065

Substance Name	Property	Number
Packed Bed, Glass Raschig Rings - Water	a	531-0067
Packed Bed, Glass Spheres	a	531-0242
Packed Bed, Glass Spheres - Gas	a	531-0010
Packed Bed, Glass Spheres - Oil	a	531-1798
Packed Bed, Glass Spheres - Vacuum	a cd	531-0683
Packed Bed, Glass Spheres - Water	a	531-0394
Packed Bed, Intalox Saddles	a	531-1628
Packed Bed, Magnesium Oxide	a	531-0656
Packed Bed, Phenolic Spheres - Gas	a	531-0448
Packed Bed, Phenolic Spheres - Vacuum	a	531-0576
Packed Bed, Poly(Ethylene Glycol - Terephthalic Acid)resin Fibers - Dioctyl Phthalate	a	531-1250
Packed Bed, Poly(Methyl Methacrylate) Resin Spheres	a	531-0984
Packed Bed, Poly(Methyl Methacrylate) Resin Spheres - Gas	a	531-0981
Packed Bed, Polyamide Resin Fibers - Dioctyl Phthalate	a	531-1251
Packed Bed, Porcelain Spheres	e	531-0002
Packed Bed, Porcelain Spheres - Gas	a	531-1493
Packed Bed, Raschig Rings	a	531-0005
Packed Bed, Raschig Rings - Gas	a	531-0049
Packed Bed, Raschig Rings - Water	a	531-0457
Packed Bed, Silicon Oxide Spheres	a	531-0737
Packed Bed, Slag Spheres	a	531-1438
Packed Bed, Slag Spheres - Gas	a	531-1809
Packed Bed, Zirconium Oxide Spheres - Gas	a	531-1040
Panelite		
see Resin, Phenolic, Fabric Reinforced		531-
Paper, Crumpled Or Corrugated	a	531-1133
Pear (Fresh, Dried, Etc.) - Gas	a	531-0367
Pear (Fresh, Dried, Etc.) - Vacuum	a	531-1367
Pentadecylene		
see Interface, 1- Pentadecene - Water		531-
Perfluorodibutyl Ether		
(Syn. For- Octadecafluorobutyl Ether)		
see Interface, Heptane - Octadecafluorobutyl Ether		531-
Perfluorotoluene		
(Syn. For- Octafluorotoluene)		
see Interface, Octafluorotoluene - Toluene		531-
Perfluorotributylamine		
(Syn. For- Heptacosafluorotributylamine)		
see Interface, Heptacosafluorotributylamine - Heptane		531-
Petrolatum Jelly, Glass Embedded	a	531-1692
Phenolite G-7		
see Resin, Silicone, Glass Reinforced		531-
Phenylcarbonimide		
see Interface, Phenylisocyanate - Water		531-
Phosphatidyl Choline		
(Syn. For- Lecithin)		
see Interface, (Dodecane + Lecithin) - (Sodium Chloride + Water)		531-
Phospholutein		
(Syn. For- Lecithin)		
see Interface, (Dodecane + Lecithin) - (Sodium Chloride + Water)		531-
Powder - Gas	a	531-0652
Powder - Vacuum	a c e g	531-0511
Powder, (Aluminum Oxide + Perlite) - Vacuum	a	531-0586
Powder, (Aluminum Oxide + Silicon Oxide) - Vacuum	a	531-0583
Powder, (Aluminum Oxide + Silicon Oxide), Hydrogen Adsorbate	h	531-0669
Powder, (Aluminum Oxide + Silicon Oxide), Oxygen Adsorbate	h	531-0663
Powder, (Barium Oxide + Calcium Oxide + Strontium Oxide) - Vacuum	a	531-1539
Powder, (Carbon + Diatomite) - Vacuum	a	531-1680
Powder, (Carbon + Perlite) - Vacuum	a	531-0826
Powder, (Carbon + Silicon Oxide) - Vacuum	a	531-0827
Powder, (Magnesium Oxide + Silicon Oxide) - Vacuum	d	531-1751
Powder, Aluminum Oxide - Gas	a d	531-0373
Powder, Aluminum Oxide - Liquid	a	531-1783
Powder, Aluminum Oxide - Vacuum	a	531-1043
Powder, Aluminum Silicate - Gas	a	531-1477
Powder, Barium Aluminosilicate, Ethylene Adsorbate	j	531-0344
Powder, Basalt - Vacuum	a	531-0342
Powder, Beryllium Oxide - Gas	a	531-1617
Powder, Boron Nitride - Gas	a	531-1802

Substance Name	Property	Number
Powder, Boron Nitride - Vacuum	a	531-1801
Powder, Cadmium Aluminosilicate, Ethylene Adsorbate	j	531-0315
Powder, Calcium Aluminosilicate, Ethylene Adsorbate	j	531-0346
Powder, Carbon - Gas	a	531-0447
Powder, Carbon - Vacuum	a d	531-0574
Powder, Char From Plastic Material - Gas	a	531-1520
Powder, Charcoal - Vacuum	a	531-0571
Powder, Chondrite - Vacuum	a	531-1360
Powder, Diatomite - Gas	a	531-0061
Powder, Diatomite - Vacuum	a	531-0573
Powder, Glass - Gas	a	531-0119
Powder, Glass - Liquid	a	531-1769
Powder, Glass - Vacuum	a	531-1107
Powder, Glass - Water	a	531-0487
Powder, Granite - Vacuum	a c	531-1385
Powder, Granodiorite - Vacuum	a	531-1361
Powder, Graphite - Gas	a	531-0651
Powder, Graphite - Vacuum	a	531-1498
Powder, Lithium Aluminosilicate, Ethylene Adsorbate	j	531-0320
Powder, Lithium Montmorillonite, 2-methyl-2-propanol Adsorbate	i	531-0790
Powder, Magnesium Carbonate - Vacuum	a	531-1228
Powder, Magnesium Oxide - Gas	a d	531-0112
Powder, Magnesium Oxide - Vacuum	a	531-1229
Powder, Magnesium Silicate, Copper - (2,9- Dimethyl-1,10-phenanthroline) Complex Adsorbate	hi	531-1377
Powder, Moon Rock - Vacuum	a	531-1462
Powder, Olivine - Vacuum	a	531-1358
Powder, Perlite - Gas	a	531-0349
Powder, Perlite - Vacuum	a d	531-0348
Powder, Perlite, Expanded - Gas	a	531-1106
Powder, Perlite, Expanded - Vacuum	a	531-1109
Powder, Plutonium Oxide - Uranium Oxide - Gas	a	531-1149
Powder, Potassium Aluminosilicate, Ethylene Adsorbate	j	531-0329
Powder, Pumice - Gas	a	531-1105
Powder, Pumice - Vacuum	a	531-1108
Powder, Sand - Gas	a	531-0648
Powder, Sand - Vacuum	a d	531-0331
Powder, Sand - Water	a h	531-0456
Powder, Sand, Fluid Saturated	a	531-1768
Powder, Silicon Oxide - Gas	a	531-0172
Powder, Silicon Oxide - Vacuum	a cd g	531-0171
Powder, Silicon Oxide, Hexadecyltrimethylammonium Bromide Adsorbate	j	531-0438
Powder, Silver Aluminosilicate, Ethylene Adsorbate	j	531-0312
Powder, Sodium Aluminosilicate, Ethylene Adsorbate	j	531-0324
Powder, Sodium Montmorillonite, 2-methyl-2-propanol Adsorbate	i	531-0791
Powder, Steatite - Vacuum	a	531-0577
Powder, Tektite - Vacuum	a	531-1357
Powder, Tellurine - Vacuum	a	531-0985
Powder, Tellurine - Mineral Oil	a	531-0987
Powder, Terpoly(Acrylonitrile - Methyl Methacrylate - Styrene) - Gas	a	531-0471
Powder, Terpoly(Acrylonitrile - Methyl Methacrylate - Styrene) - Vacuum	a	531-0472
Powder, Thorium Carbide - Uranium Carbide, Carbon Coated - Gas	a	531-1312
Powder, Thorium Carbide, Carbon Coated - Gas	a	531-1253
Powder, Thorium Oxide - Gas	a	531-1463
Powder, Thorium Oxide - Vacuum	a	531-1464
Powder, Uranium Carbide, Carbon Coated - Gas	a	531-1252
Powder, Uranium Oxide - Gas	a	531-0081
Powder, Uranium Oxide, Graphite Coated - Gas	a	531-1075
Powder, Vermiculite - Gas	a	531-1480
see also Mica, Vermiculite		521-
Powder, Zirconium Oxide - Gas	d	531-1240
Pregn-4-ene-3,20-dione		
(Syn. For- Progesterone)		
see Interface, (Petrolatum-liquid + Progesterone) - Water		531-

Substance Name	Property	Number
4- Pregnen-21-ol-3,20-dione		
(Syn. For- 11- Deoxycorticosterone)		
see Interface, (11-deoxycorticosterone +		
Petrolatum–liquid) – Water		531-
Delta Superscript 4- Pregnene-3,20-dione		
(Syn. For- Progesterone)		
see Interface, (Petrolatum–liquid + Progesterone) –		
Water		531-
Progestarol		
(Syn. For- Progesterone)		
see Interface, (Petrolatum–liquid + Progesterone) –		
Water		531-
17 β- Progesterone		
(Syn. For- Progesterone)		
see Interface, (Petrolatum–liquid + Progesterone) –		
Water		531-
Progestin		
(Syn. For- Progesterone)		
see Interface, (Petrolatum–liquid + Progesterone) –		
Water		531-
Progestone		
(Syn. For- Progesterone)		
see Interface, (Petrolatum–liquid + Progesterone) –		
Water		531-
Pumice – Vacuum	a	531-1774
Purple Blend (N A S A-langley)		
see Resin, Silicone, Phenolic Resin And		
Silicon Oxide Reinforced		531-
Pyrocarb		
see Multilayer, Carbon Fabric, Resin Bonded,		
Carbon Filled		531-
Pyropreg A C X- G-79		
see Resin, Phenolic, Graphite Reinforced		
(Series Of Additives)		531-
Quinophenol		
(Syn. For- 8- Quinolinol)		
see Interface, Hydrocarbon, Aliphatic, Alkane –		
8- Quinolinol		531-
Radiant Microflect Screen		
see Resin, White, Glass Bead Embedded		531-
Raven Beaded Screen		
see Resin, White, Glass Bead Embedded,		
Cloth Reinforced		531-
Raven Silverlite Screen		
see Resin, Gray, Glass Bead Embedded, Cloth Reinforced		531-
Raypan		
see Insulation Raypan – Vacuum		531-
Reflector	h	531-1663
Reichsteins Substance Q		
(Syn. For- 11- Deoxycorticosterone)		
see Interface, (11- Deoxycorticosterone +		
Petrolatum–liquid) – Water		531-
Resbenzophenone		
see Interface, 2,4- Dihydroxybenzophenone – Gas		531-
Resin, Acrylic Epoxy Polyester, Glass Reinforced	a e n	531-1244
Resin, Aromatic, Glass Reinforced	e n	531-0957
Resin, Aromatic, Glass Reinforced – Gas	a	531-0931
Resin, Aromatic, Glass Reinforced – Vacuum	a	531-0967
Resin, Carbon Reinforced	a n	531-1755
Resin, Clear, Glass Bead Embedded	h	531-0547
Resin, Clear, Glass Bead Embedded,		
White Adhesive Reinforced	h	531-0543
Resin, Copoly(Acrylonitrile – Styrene), Glass Reinforced		
(Series Of Additives)	n	531-1700
Resin, Corrugated	a	531-0782
Resin, Elastomer Modified,		
Silicon Oxide Reinforced (Series Of Additives)	a de	531-0014
Resin, Epoxy – Furan, Boron Nitride Filled	a	531-1803
Resin, Epoxy – Phenolic – Gas	a	531-1760
Resin, Epoxy – Phenolic, Glass Reinforced	a de n	531-0066
Resin, Epoxy – Phenolic, Glass Reinforced – Gas	a	531-1761
Resin, Epoxy – Phenolic, Silicon Oxide Reinforced	a e	531-1330
Resin, Epoxy – Silicone, Expanded,		
Fiber Reinforced	a e ghi	531-1642

Substance Name	Property	Number
Resin, Epoxy – Silicone, Expanded, Fiber Reinforced – Gas	e	531-1644
Resin, Epoxy – Silicone, Expanded, Fiber Reinforced – Vacuum	a	531-1646
Resin, Epoxy, Beryllium Oxide And Glass Reinforced	a e n	531-1363
Resin, Epoxy, Boron Nitride Reinforced	a	531-1519
Resin, Epoxy, Expanded – Vacuum	a	531-1809
Resin, Epoxy, Glass Reinforced	a de no	531-0568
Resin, Epoxy, Glass Reinforced – Gas	a	531-0857
Resin, Epoxy, Glass Reinforced – Vacuum	a	531-0971
Resin, Epoxy, Glass Reinforced (Series Of Additives)	a e	531-1336
Resin, Epoxy, Graphite Reinforced	a de n	531-1518
Resin, Epoxy, Nylon Reinforced	a de	531-1473
Resin, Epoxy, Perlite Reinforced	a e n	531-0960
Resin, Epoxy, Perlite Reinforced – Gas	a	531-0958
Resin, Epoxy, Perlite Reinforced – Vacuum	a	531-0968
Resin, Epoxy, Phenolic Resin Reinforced	a e n	531-0966
Resin, Epoxy, Phenolic Resin Reinforced – Gas	a	531-0964
Resin, Epoxy, Phenolic Resin Reinforced – Vacuum	a	531-0970
Resin, Epoxy, Potassium Titanate Reinforced	a e n	531-0963
Resin, Epoxy, Potassium Titanate Reinforced – Gas	a	531-0961
Resin, Epoxy, Potassium Titanate Reinforced – Vacuum	a	531-0969
Resin, Epoxy, Silicon Oxide Reinforced – Vacuum	a	531-1632
Resin, Epoxy, Silicon Oxide Reinforced (Series Of Additives)	a	531-1421
Resin, Flan, Cotton Reinforced	a de	531-1474
Resin, Glass Reinforced	a ghij	531-0122
Resin, Gray, Glass Bead Embedded, Cloth Reinforced	h	531-0542
Resin, High Char, Magnesium Hydroxide Reinforced	a de	531-0036
Resin, High Char, Silicon Oxide Reinforced	a de	531-0033
Resin, Melamine, Cotton Reinforced	a de	531-1472
Resin, Melamine, Glass Reinforced	a de	531-0198
Resin, Phenolic – Nylon, Phenolic Resin Reinforced	a de n	531-1765
Resin, Phenolic, Aluminum Oxide Reinforced	a g	531-1315
Resin, Phenolic, Asbestos Reinforced	a de g i k n	531-0301
Resin, Phenolic, Asbestos Reinforced – Gas	a	531-0871
Resin, Phenolic, Asbestos Reinforced, Moist		n	531-1501
Resin, Phenolic, Asbestos, Glass And Graphite Reinforced	a e g i	531-1356
Resin, Phenolic, Carbon Reinforced	a de n	531-0060
Resin, Phenolic, Carbon Reinforced (Series Of Additives)	a e m	531-0717
Resin, Phenolic, Cotton Reinforced	a de	531-0195
Resin, Phenolic, Fabric Reinforced		n	531-0023
Resin, Phenolic, Fabric Reinforced – Vacuum	a	531-1581
Resin, Phenolic, Fiber Reinforced (Series Of Additives)	a e g	531-1515
Resin, Phenolic, Glass Reinforced	a de g n	531-0303
Resin, Phenolic, Glass Reinforced – Gas	a	531-0851
Resin, Phenolic, Graphite Reinforced	a de g no	531-0620
Resin, Phenolic, Graphite Reinforced (Series Of Additives)	a de g n	531-1390
Resin, Phenolic, Graphite Reinforced – Gas	a	531-0855
Resin, Phenolic, Modified, Glass Reinforced	a	531-0550
Resin, Phenolic, Nylon Reinforced	a de g n	531-0679
Resin, Phenolic, Nylon Reinforced – Gas	a	531-0853
Resin, Phenolic, Paper Reinforced	a de	531-0192
Resin, Phenolic, Polyamide Modified, Silicon Oxide Reinforced	a de gh	531-0017
Resin, Phenolic, Refractory Reinforced	a	531-1056
Resin, Phenolic, Reinforced		531-0093
Resin, Phenolic, Silcon Oxide Reinforced (Series Of Additives)	a de gh n	531-1391
Resin, Phenolic, Silicon Oxide Reinforced	a de gh n	531-0569
Resin, Phenolic, Silicone Resin Impregnated	a	531-1517
Resin, Phenolic, Zirconium Oxide Reinforced	d g	531-0040
Resin, Phenolic, Zirconium Oxide Reinforced (Series Of Additives)	g	531-0047
Resin, Phenylsilane, Glass Reinforced	a e n	531-0304
Resin, Phenylsilane, Glass Reinforced – Gas	a	531-0861
Resin, Phenylsilane, Silicon Oxide Reinforced	a	531-0296
Resin, Poly–(1,6-hexanediamine – Sebacic Acid), Glass Reinforced (Series Of Additives)		n	531-1585
Resin, Poly-(4, 4 '-isopropylidenediphenol Carbonate), Glass Reinforced (Series Of Additives)	a e n	531-1287
Resin, Poly-(4,4 '- Isopropylidenediphenol Carbonate), Carbon Reinforced	a	531-1509
Resin, Poly(Adipic Acid – 1,6- Hexanediamine),	a n	531-1507

Substance Name	Property	Number
Technikota Beaded Screen		
see Resin, White, Glass Bead Embedded		531-
Tekstolite P T K		
see Resin, Phenolic, Cotton Reinforced		531-
Tekstolot		
see Resin, Phenolic, Cotton Reinforced		531-
tertryl (Syn. For- Testosterone)		
see Interface, (Petrolatum–liquid + Testosterone) – Water		531-
Testiculosterone		
see Interface, (Petrolatum–liquid + Testosterone) – Water		531-
trans- Testosterone		
see Interface, (Petrolatum–liquid + Testosterone) – Water		531-
Testrone		
see Interface, (Petrolatum–liquid + Testosterone) – Water		531-
α- Tetradecylene		
see Interface, 1– Tetradecene – Water		531-
Theelin		
(Syn. For- Estrone)		
see Interface, (Estrone + Petrolatum–liquid) – Water		531-
Thelykinin		
(Syn. For- Estrone)		
see Interface, (Estrone + Petrolatum–liquid) – Water		531-
Thermo–lag T-500-111 – H. C. (Emerson)		
see Honeycomb, Nylon – Phenolic Resin, Glass Reinforced, Thermolag T-500 Covered		531-
Thermoflex		
see Ceramic (Al₂ O₃ + Si O₂), Fibrous – Vacuum		531-
Tissuglas		
see Multilayer, Glass Paper		531-
Tokokin		
(Syn. For- Estrone)		
see Interface, (Estrone + Petrolatum–liquid) – Water		531-
Tri–lok		
see Mitten, Plastic Interlined (Tri–lok)		531-
Undecylenic Acid		
see Interface, 10– Undecenoic Acid – Water		531-
Uranium Carbide, Carbon Covered (Also Irradiated) ← Silver	m	531-1110
Uranium Carbide, Carbon Covered (Also Irradiated) ← Strontium	m	531-1111
Uranium Carbide, Carbon Covered (Aslo Irradiated) ← Barium	m	531-1112
Uranium Carbide, Carbon Covered (Also Irradiated) ← Neodymium	m	531-1113
Uranium Carbide, Carbon Covered (Also Irradiated) ← Carium	m	531-1114
Uvinul 400		
see Interface, 2,4– Dihydroxybenzophenone – Gas		531-
V F T– S		
see Resin, Phenolic, Modified, Glass Reinforced		531-
V P S–4		
see Resin, Acrylic Epoxy Polyester, Glass Reinforced		531-
Venetian Blind	hij	531-1254
Versamid		
(Syn. For- Polyamide)		
see Multilayer, Nylon – Polyamide – Polychloroprene – Gas		531-
Vitellin		
(Syn. For- Lecithin)		
see Interface, (Dodecane + Lecithin) – (Sodium Chloride + Water)		531-
W B C 2230		
W B C 2234		
see Resin, High Char, Silicon Oxide Reinforced		531-
W B C 5217		
see Resin, High Char, Magnesium Hydroxide Reinforced		
see Resin, Phenolic, Silicon Oxide Reinforced		531-
Waste, Radioactive, Glass Covered	a	531-1650
Water On Ice	b	531-1754

Substance Name	Property	Number
White Magic I I Screen (Da– Lite)		
see Resin, White, Glass Bead Embedded		531-
X R 2015 (U. S. Polymeric Co.)		
see Resin, Elastomer Modified, Silicon Oxide Reinforced (Series Of Additives)		531-
Xenon On Carbon	b	531-1331
Xenon On Glass	b	531-1265
Xenon On Uranium Carbide, Oxidized	b	531-0907
Y N 25		
see Resin, Phenolic, Nylon Reinforced		531-
Zeolite, Deuterium Faujasite, 1–hexene Adsorbate	j	531-0822
Zeolite, Hydrogen Faujasite, 1–hexene Adsorbate	j	531-0823
Zinc Oxide, (Oxygen + Hydrogen) Adsorbate	h	531-0835
Zinc Oxide, Oxygen Adsorbate	h	531-0834
Zirconium Carbide, Fibrous – Gas	a	531-1612
Zirconium Hydride, Porous - Gas	a d	531-1796
Zirconium Hydride, Porous – Vacuum	a d	531-1795
Zirconium Oxide, Expanded, Carbon Impregnated	a	531-0588
Zirconium Oxide, Expanded, Graphite Impregnated	a	531-1389
Zirconium Oxide, Fibrous – Vacuum	a	531-1773
Zirconium Oxide, Water Adsorbate	j	531-1674
Zirconium Oxide, Water, Dideuterated Adsorbate	j	531-1675

METALLIC – NONMETALLIC SYSTEMS –CONTACTS, FILTERS, HONEYCOMBS, INTERFACES, MULTILAYERS, AND OTHERS   CLASS 541

Substance Name	Property		Number
Aerated			
see Expanded			541-
Air On Copper	b		541-1166
Air On Mercury	b		541-0652
Air On Nickel	b		541-1227
Air On Platinum	b		541-0227
Air On Tungsten	b		541-0689
Aircraft	a	g	541-0981
Aluminized Poly(Ethylene Glycol – Terephthalic Acid), Cones On Thermistor, Gold Coated, Aluminum Base		j	541-1011
Aluminum Alloy (1 Cd, 5 Zn), Glass Fiber Reinforced		n	541-1032
Aluminum Alloy 1100, (Boron Fiber, Silicon Carbide Covered) Reinforced		n	541-1346
Aluminum Alloy 1100, Glass Fiber Reinforced		o	541-1043
Aluminum Alloy 2014, Glass Fiber Reinforced		no	541-1030
Aluminum Alloy 2014, Glass Fiber Reinforced, Aluminum Alloy 1100 (Cadmium And Zinc Added) Covered		n	541-1027
Aluminum Alloy 2024, (Boron Fiber, Silicon Carbide Covered) Reinforced		n	541-1345
Aluminum Oxide, Molybdenum Fiber Reinforced	a		541-0082
Aluminum Oxide, Platinum Covered – Vacuum	a e		541-0914
Aluminum, (Tungsten, Silicon Carbide Covered) Reinforced		n	541-1162
Aluminum, Aluminum Oxide Covered	a		541-1152
Aluminum, Asbestos Covered, Corrugated	a		541-0619
Aluminum, Clay Filled		n	541-0616
Aluminum, Glass Reinforced	a		541-1327
Aluminum, Silicon Oxynitride Covered	a		541-1153
Argon On Alkali Metal	b		541-0984
Argon On Aluminum	b		541-0391
Argon On Aluminum, Oxidized	b		541-0610
Argon On Carbon Oxide (C O₂) Covered Copper	b		541-1339
Argon On Copper	b		541-0697
Argon On Germanium	b		541-1103
Argon On Gold	b		541-0588
Argon On Hydrogen Covered Tungsten	b		541-1034
Argon On Lithium	b		541-0378
Argon On Molybdenum	b		541-0381
Argon On Nickel	b		541-0446
Argon On Nitrogen Covered Tungsten	b		541-1033
Argon On Oxygen Covered Tungsten	b		541-1022
Argon On Platinum	b		541-0325
Argon On Platinum Covered Tungsten	b		541-0932
Argon On Potassium	b		541-0377
Argon On Potassium Covered Tungsten	b		541-0803
Argon On Silicon	b		541-0653
Argon On Silver	b		541-0584

Substance Name	Property	Number
Argon On Sodium	b	541-0383
Argon On Tantalum	b	541-1104
Argon On Tungsten	b	541-0340
Auresin		
see Glass, Gold Covered		541-
Base, Coating Covered	a e n	541-1295
Base, Fluid Adsorbate	j	541-1216
Battery Cell	a e	541-1094
Battery Cell, Cadmium - Nickel	a	541-0866
Black Body, Razor Blade Edges, Blackened	g i k	541-0651
Black Body, V-groove, Magnesium Alloy H Z-318, Blackened	g	541-1083
Bladder Cloth		
(Syn. For- Fabric, Rubberized)		
see Multilayer, Aluminum - Fabric, Rubberized -		
Glass - Nylon - Polyimide - Vacuum		541-
Boron Iodide On Tungsten	b	541-0778
Boron, Water Adsorbate	j	541-1266
Brass, Aluminum Oxide Covered	n	541-1023
Brass, Titanium Oxide Covered	n	541-1024
Brass, Zirconium Silicate Covered	n	541-1025
Bromodifluoromethane On Platinum	b	541-0490
Bromotrifluoromethane On Platinum	b	541-0485
Cadmium On Nitrogen Covered Tungsten	b	541-1170
(Calcium Oxide + Uranium Oxide + Zirconium Oxide),		
Stainless Steel 304 Covered - Vacuum	a	541-1223
(Calcium Oxide + Zirconium Oxide), Stainless		
Steel 304 Covered - Gas	a	541-1157
(Carbon Oxide (C O) + Oxygen) On Gallium Arsenide	b	541-1313
Carbon Oxide (C O) On Gallium Arsenide	b	541-1312
Carbon Oxide (C O) On Platinum	b	541-0737
Carbon Oxide (C O) On Zinc Selenide	b	541-1315
Carbon Oxide (C O₂) On Metal	b	541-0760
Carbon Oxide (C O₂) On Platinum	b	541-0928
Carbon Oxide (C O₂) On Platinum Covered Tungsten	b	541-0931
Carbon Oxide (C O₂) On Tungsten	b	541-0770
Carbon Oxide On Gold	b	541-0575
Carbon Tetrachloride On Platinum	b	541-0484
Carbon Tetrafluoride On Platinum	b	541-0508
Carbon, Tungsten Covered	a	541-1364
Catalyst, Aluminum Oxide - Chromium	h	541-0278
Catalyst, Aluminum Oxide - Chromium, Ammonia Adsorbate	c h	541-0282
Catalyst, Aluminum Oxide - Chromium, Methanol Adsorbate	h	541-0314
Catalyst, Aluminum Oxide - Chromium, Water Adsorbate	h	541-0281
Catalyst, Nickel On Silicon Oxide, Formic Acid Adsorbate	j	541-0664
Catalyst, Silver - Gas	a	541-0759
Catalyst, Silver - Vacuum	a	541-0758
Cavity, Cone, Soot Lined Iron	g	541-0454
Cavity, Cylinder, Glass, Aluminum Covered	j	541-0209
Cavity, Cylinder, Glass, Silver Covered	j	541-0211
Cavity, Cylinder, Paper Covered Brass, Black	gh	541-0394
Cavity, Cylinder, Paper Covered Brass, Green	gh	541-0393
Cavity, Cylinder, Paper Covered Brass, White	gh	541-0392
Cavity, Cylinder, Paper Covered Brass, Yellow	gh	541-0395
Cavity, Cylinder, Soot Covered Iron	g	541-0455
Cavity, Groove, Aluminum Sides, Black Painted		
Aluminum Bottom	g	541-0754
Cellular		
see Expanded		
Ceramic (Al₂ O₃ + Al P O₄),		
Metal Reinforced	a e g n	541-1358
Ceramic (Zr O₂ + H₂ F P O₃),		
Metal Reinforced	a e g n	541-1359
Ceramic (Zr O₂ + H₂ Si F₆),		
Metal Reinforced	n	541-1362
Ceramic, Metal Fiber Reinforced	a	541-0833
Cesium Chloride On Carbon Covered Iridium	b	541-1343
Chloranilic Acid		
(Syn. For- 2,5- Dichloro-5,6-dihydro-p-		
-benzoquinone)		
see Interface, 2,5- Dichloro-5,6-dihydroxy-p-benzoquin		
Mercury		541-
Chlorodifluoromethane On Platinum	b	541-0491
Chlorofluoromethane On Platinum	b	541-0509
Chlorotrifluoromethane On Gold	b	541-0583

Substance Name	Property	Number
Chlorotrifluoromethane On Platinum	b	541-0507
Chromium On Aluminum Oxide	b	541-0685
Composite, Aluminum Alloy 2024, Covered -		
(Copoly-1,1-difluoroethylene -		
Hexafluoropropene) (Insulation)	a	541-0915
Composite, Aluminum Alloy 2024, Covered - Glass		
Fibrous - Polytetrafluoroethylene	a	541-1089
Composite, Metal - Organic	n	541-1347
Composite, Reinforced	a n	541-0108
see also Honeycomb		541-
see also Multilayer		541-
see also Resin, Reinforced		541-
Composite, Steel, Stainless 17-7 P H - U I Ceramic		
Adhesive (Numbered Series)	n	541-1060
Condenser, Mica	a	541-0789
Contact, (Aluminum - Aluminum Oxide Cermet) - Uranium		
Carbide - Gas	c	541-0990
Contact, (Aluminum + Carbon), Powder - Vacuum	c	541-0843
Contact, (Aluminum - Aluminum Oxide Cermet) -		
Uranium Carbide	c	541-0771
Contact, (Aluminum - Aluminum Oxide Cermet) - Iron	c	541-0997
Contact, (Beryllium Oxide + Uranium Oxide) - Nickel A		
(Also Irradiated)	c	541-0904
Contact, (Carbon + Copper), Powder - Vacuum	c	541-0844
Contact, (Carbon + Graphite), Powder - Vacuum	c	541-0845
Contact, Almag 35 - Aluminum Alloy 6061 - Vacuum	c	541-0838
Contact, Aluminum - Aluminum, Anodized, With Filler	c	541-0462
Contact, Aluminum - Copper - Vacuum	c	541-1013
Contact, Aluminum - Helium 4, Liquid	c	541-1291
Contact, Aluminum - Iron - Vacuum	c	541-1120
Contact, Aluminum - Iron, With Filler	c	541-0834
Contact, Aluminum - Molybdenum - Vacuum	c	541-0891
Contact, Aluminum - Stainless Steel - Vacuum	c	541-0326
Contact, Aluminum - Uranium - Gas	c	541-0262
Contact, Aluminum - Vacuum	c	541-0197
Contact, Aluminum Alloy (99 Al) - Stainless Steel 304 -		
Vacuum	c	541-1121
Contact, Aluminum Alloy L 70, With Filler	c	541-0998
Contact, Aluminum Alloy 1100 - Platinum - Vacuum	c	541-1450
Contact, Aluminum Alloy 1100 - Platinum, With		
Filler - Vacuum	c	541-1057
Contact, Aluminum Alloy 1100 - Vacuum	c	541-0695
Contact, Aluminum Alloy 2024 - Iron - Vacuum	c	541-1118
Contact, Aluminum Alloy 2024 - Magnesium Alloy 2 H 62 A,		
With Filler	c	541-0541
Contact, Aluminum Alloy 2024 - Stainless Steel 303 - Vacuum	c	541-0860
Contact, Aluminum Alloy 2024 - Stainless Steel 304 - Vacuum	c	541-0204
Contact, Aluminum Alloy 2024 - Vacuum	c	541-0199
Contact, Aluminum Alloy 2024, With Filler - Vacuum	c	541-0302
Contact, Aluminum Alloy 6061 - Iron - Gas	c	541-0895
Contact, Aluminum Alloy 6061 - Magnesium Alloy A Z 91 C -		
Vacuum	c	541-0839
Contact, Aluminum Alloy 6061 - Mica	c	541-1252
Contact, Aluminum Alloy 6061 - Vacuum	c	541-0192
Contact, Aluminum Alloy 6061, With Filler	c	541-1350
Contact, Aluminum Alloy 6061, With Filler - Vacuum	c	541-0283
Contact, Aluminum Alloy 7075 - Vacuum	c	541-0286
Contact, Aluminum Alloy 7075, With Filler	c	541-0025
Contact, Aluminum Alloy 7075, With Filler - Vacuum	c	541-0287
Contact, Aluminum Oxide -		
Russian Steel Kh18 N10 T - Vacuum	c	541-1408
Contact, Aluminum Oxide - Copper	c	541-1018
Contact, Aluminum Oxide - Copper, With Filler	c	541-1309
Contact, Aluminum Oxide - Gold, With Filler	c	541-0464
Contact, Aluminum Oxide - Iridium	c	541-1000
Contact, Aluminum Oxide - Iron - Gas	c	541-0903
Contact, Aluminum Oxide - Magnesium Alloy (Zr)	c	541-1386
Contact, Aluminum Oxide - Molybdenum - Gas	c	541-1412
Contact, Aluminum Oxide - Molybdenum - Vacuum	c	541-1409
Contact, Aluminum Oxide - Nickel - Vacuum	c	541-1384
Contact, Aluminum Oxide - Russian Steel Kh18 N10 T - Gas	c	541-1411
Contact, Aluminum Oxide - Stainless Steel 304	c	541-1255
Contact, Aluminum Oxide - Stainless Steel 304 - Gas	c	541-0901
Contact, Aluminum Oxide - Tin, With Filler	c	541-0461

A19

Substance Name	Property	Number
Contact, Aluminum, Anodized – Tin, With Filler	c	541-0460
Contact, Aluminum, With Filler	c	541-0403
Contact, Aluminum, With Filler – Vacuum	c	541-0847
Contact, Beryllium – Vacuum	c	541-0200
Contact, Beryllium Oxide – Russian Steel Kh18 N10 T – Vacuum	c	541-1407
Contact, Beryllium Oxide – Russian Steel Kh18 N10 T – Gas	c	541-1410
Contact, Brass – Carbon	c	541-1113
Contact, Brass – Copper – Vacuum	c	541-0840
Contact, Brass – Steel – Gas	c	541-0397
Contact, Brass – Steel – Vacuum	c	541-0396
Contact, Brass – Vacuum	c	541-0692
Contact, Brass (62 Cu, 35 Zn, 3 Pb) – Vacuum	c	541-0400
Contact, Brass (62 Cu, 35 Zn, 3 Pb), With Filler – Vacuum	c	541-0402
Contact, Brass (70 Cu, 30 Zn) – Stainless Steel 304 – Vacuum	c	541-1127
Contact, Brass, Leaded – Vacuum	c	541-1012
Contact, Brass, With Filler	c	541-0089
Contact, Brass, With Filler – Vacuum	c	541-0693
Contact, British Steel En 8 – Steel (C, Cr) – Gas	c	541-1059
Contact, British Steel En 8 – Steel (C, Cr) – Vacuum	c	541-1058
Contact, Chromium Potassium Sulfate – Copper, With Filler	c	541-0361
Contact, Copper – Gas	c	541-1020
Contact, Copper – Helium 3, Liquid, Solid Or Gas	c	541-0830
Contact, Copper – Helium 4, Liquid Or Gas	c	541-0835
Contact, Copper – Photographic Film	c	541-0785
Contact, Copper – Platinum – Vacuum	c	541-1119
Contact, Copper – Poly(Ethylene Glycol – Terephthalic Acid), With Filler	c	541-0779
Contact, Copper – Poly(Ethylene Glycol – Terephthalic Acid)	c	541-0781
Contact, Copper – Silicon Oxide	c	541-1017
Contact, Copper – Sodium Chloride	c	541-1016
Contact, Copper – Stainless Steel – Vacuum	c	541-1324
Contact, Copper – Uranium Nitride – Vacuum	c	541-1422
Contact, Copper – Vacuum	c	541-0157
Contact, Copper Beryllium 25 – Stainless Steel 304 – Vacuum	c	541-1123
Contact, Copper Oxide – Duralumin	c	541-0957
Contact, Copper Oxide – Russian Brass L 80	c	541-0925
Contact, Copper Oxide – Russian Steel St 30	c	541-0946
Contact, Copper, With Filler	a cd	541-0773
Contact, Duralumin – Gas	c	541-0958
Contact, Gold – Gas	c	541-0413
Contact, Gold – Helium 4, Liquid	c	541-1403
Contact, Gold – Helium 4, Liquid, With Filler	c	541-1404
Contact, Gold – Silicon Oxide, With Filler	c	541-0463
Contact, Graphite – Iron	c	541-1062
Contact, Graphite – Magnesium Alloy (Zr)	c	541-1387
Contact, Graphite – Tungsten	c	541-0182
Contact, Helium 3, Liquid – Metal	c	541-0920
Contact, Helium 4, Liquid – Indium	c	541-1039
Contact, Helium 4, Liquid – Lead	c	541-1040
Contact, Helium 4, Liquid – Metal	c	541-0919
Contact, Helium 4, Liquid – Tin	c	541-1038
Contact, Incoloy 825 – Stainless Steel 304 – Vacuum	c	541-1125
Contact, Inconel X 750 – Stainless Steel 304 – Vacuum	c	541-1126
Contact, Inconel 600 – Stainless Steel 304 – Vacuum	c	541-1124
Contact, Indium – Sapphire	c	541-1277
Contact, Indium – Silicon, With Filler	c	541-1061
Contact, Indium – Vacuum	c	541-0841
Contact, Iron – Gas	c	541-0894
Contact, Iron – Mica – Gas	c	541-0902
Contact, Iron – Stainless Steel 304 – Gas	c	541-0900
Contact, Iron – Vacuum	c	541-0893
Contact, Iron, With Filler	c	541-0772
Contact, Lead – Gas	c	541-0415
Contact, Lead – Sapphire	c	541-1265
Contact, Magnesium Alloy A Z 31 B – Stainless Steel 303 – Vacuum	c	541-0861
Contact, Magnesium Alloy A Z 31 B – Vacuum	c	541-0155
Contact, Magnesium Alloy A Z 31 B, With Filler – Vacuum	c	541-0836
Contact, Magnox – Uranium Carbide – Vacuum	c	541-0995
Contact, Magnox A–12 – Uranium – Gas	c	541-0249
Contact, Martin Hardcoat On Aluminum Alloy 6061 – Gas	c	541-0896
Contact, Martin Hardcoat On Aluminum Alloy 6061 With Filler – Gas	c	541-0897
Contact, Metal – Ball Bearing – Metal – Gas	c	541-1053
Contact, Metal – Gas	c	541-1299
Contact, Metal – Nonmetal	c	541-0084
Contact, Metal – Nonmetal – Vacuum	c	541-1117
Contact, Metal – Vacuum	c	541-0153
Contact, Mica – Stainless Steel 304 – Gas	c	541-0899
Contact, Mica – Titanium	c	541-1351
Contact, Molybdenum – Stainless Steel 304 – Vacuum	c	541-1391
Contact, Molybdenum – Uranium Nitride – Vacuum	c	541-1420
Contact, Molybdenum – Vacuum	c	541-1390
Contact, Monel K 500 – Stainless Steel 304 – Vacuum	c	541-1122
Contact, Nickel – Stainless Steel – Vacuum	c	541-1323
Contact, Nickel A – Uranium Oxide (Also Irradiated)	c	541-0870
Contact, Niobium – Uranium Carbide	c	541-0862
Contact, Niobium – Vacuum	c	541-0398
Contact, Nylon Mesh, Aluminized – Vacuum	c	541-0846
Contact, Oxide Coated Stainless Steel – Gas	c	541-0898
Contact, Oxidized Steel 1020	c	541-1353
Contact, Potassium – Russian Steel 1 Kh18 N9 T – Gas	c	541-1102
Contact, Potassium Alloy (44 Na) – Stainless Steel – Gas	c	541-0774
Contact, Russian Aluminum Alloy D 1 – Gas	c	541-0152
Contact, Russian Steel Kh21 G7 A N5 – Vacuum	c	541-1021
Contact, Russian Steel St 20 – Vacuum	c	541-1283
Contact, Russian Steel St 30 – Vacuum	c	541-1078
Contact, Russian Steel 1 Kh18 N9 T – Kh21 G7 A N5 – Vacuum	c	541-1282
Contact, Russian Steel 1 Kh18 N9 T – Vacuum	c	541-0988
Contact, Russian Steel 30 Kh G S A – Vacuum	c	541-0987
Contact, Russian Titanium Alloy V T1–1 – Vacuum	c	541-1077
Contact, Silicon Oxide – Tin, With Filler	c	541-0459
Contact, Silicon Oxide, With Filler	c	541-0458
Contact, Silicone Rubber, With Beryllium Copper Fins – Vacuum	c	541-0848
Contact, Silicone Rubber, With Brass Wire Brush – Vacuum	c	541-0849
Contact, Silver, With Gold Flashed Beryllium Copper Fins – Vacuum	c	541-0850
Contact, Stainless Steel – Uranium Carbide – Gas	c	541-0993
Contact, Stainless Steel – Uranium Oxide – Gas	c	541-0176
Contact, Stainless Steel – Uranium Zirconium Hydride	c	541-1272
Contact, Stainless Steel – Uranium Zirconium Hydride – Gas	c	541-1273
Contact, Stainless Steel – Vacuum	c	541-1320
Contact, Stainless Steel A M S 5613 – Vacuum	c	541-0083
Contact, Stainless Steel 17–4 Ph – Vacuum	c	541-0289
Contact, Stainless Steel 302 – Uranium Nitride – Vacuum	c	541-1316
Contact, Stainless Steel 303 – Vacuum	c	541-0401
Contact, Stainless Steel 304 – Gas	c	541-0892
Contact, Stainless Steel 304 – Uranium Oxide – Vacuum (Also Irradiated)	c	541-0905
Contact, Stainless Steel 304 – Uranium Oxide (Also Irradiated)	c	541-0863
Contact, Stainless Steel 304 – Vacuum	c	541-0154
Contact, Stainless Steel 304, Aluminum Plated – Vacuum	c	541-0202
Contact, Stainless Steel 304, Magnesium Plated – Stainless Steel 304 – Vacuum	c	541-0201
Contact, Stainless Steel 304, With Filler	c	541-0542
Contact, Stainless Steel 304, With Filler – Vacuum	c	541-0913
Contact, Stainless Steel 416 – Vacuum	c	541-0193
Contact, Stainless Steel 416, With Filler	c	541-1300
Contact, Stainless Steel 416, With Filler – Vacuum	c	541-0679
Contact, Steel – Vacuum	c	541-0166
Contact, Steel (0.22 C, 0.45 Mn) – Gas	c	541-1271
Contact, Steel AISI C 1020 – Vacuum	c	541-0691
Contact, Steel AISI C 1020, With Filler – Vacuum	c	541-1301
Contact, Steel Wire Mesh – Vacuum	c	541-0842
Contact, Steel 4140, With Filler	c	541-1044
Contact, Titanium – Vacuum	c	541-0889
Contact, Titanium Alloy 6 Al – 4 V – Vacuum	c	541-0890
Contact, Titanium, With Filler	c	541-1352
Contact, Tungsten – Vacuum	c	541-1140
Contact, Uranium Carbide – Zircaloy 2 – Gas	c	541-0992
Contact, Uranium Nitride – Vanadium Alloy (15 Cr, 5 Ti) – Vacuum	c	541-1421
Contact, Uranium Oxide – Zircaloy-2 – Gas	c	541-0236

Substance Name	Property	Number
Contact, Uranium Oxide – Zircaloy-2 (Also Irradiated)	c	541-0263
Contact, Uranium Oxide – Zirconium	c	541-0629
Contact, Uranium Oxide – Zirconium Alloy (Cu)	c	541-1385
Contact, Uranium Oxide – Zirconium, With Filler	c	541-0628
Copper On Aluminum Oxide, Carbon Oxide (C O) Adsorbate	j	541-1335
Copper On Silicon Oxide, Carbon Oxide (C O) Adsorbate	j	541-1336
Copper, Carbon Oxide (C O) Adsorbate	ij	541-1238
Copper, Frost Covered	a	541-1139
Copper, Indium Covered, Phenolic Resin, Vinyl Modified Covered	a	541-0111
Copper, Oxidized	a	541-1131
Deuterium On Nickel	b	541-0440
Deuterium On Platinum	b	541-0735
Deuterium On Tungsten	b	541-0947
Dexter Paper		
(Syn. For- Glass)		
see Multilayer, Aluminum – Glass		541-
Dibromodichloromethane On Platinum	b	541-0488
Dibromodifluoromethane On Platinum	b	541-0486
Dibromomethane On Platinum	b	541-0489
Dichlorodifluoromethane On Platinum	b	541-0506
Dichlorofluoromethane On Platinum	b	541-0505
Dichloromethane On Platinum	b	541-0483
E I 736		
see Interface, Russian Alloy E I 736 – Slag		
(Al₂ O₃, Ca F₂)		541-
Ethylene On Platinum	b	541-0526
F M D M (Russian)		
see Resin, Epoxy – Furan, Iron Filled		541-
Fiber Optic, Magnesium Fluoride – Silicon Oxide, Stainless Steel Covered	j	541-1019
Film, Molybdenum – Glass Cermet, Metal Brazed	a	541-1363
Filter, Copper Screen – Poly(Ethylene Glycol – Terephthalic Acid), Multilayer	j	541-1154
Filter, Germanium – Silicon Oxide, Multilayer	j	541-1169
Filter, Glass – Silver, Multilayer	j	541-1072
Filter, Infrared	j	541-1128
Filter, Metallic Screen – Poly(Ethylene Glycol – Terephthalic Acid)	j	541-1318
Filter, Metallic Screen – Polyethylene, Black	j	541-1317
Filter, Metallic Screen – Polyethylene, Black – Thallium Bromide Iodide	j	541-1319
Filter, Multilayer	j	541-1164
Filter, Polyethylene – Silicon, Multilayer	j	541-1165
Firerod		
see Heater, Cartridge, Nickel Alloy Sheath, Magnesium Oxide Packed		541-
Fluidized Bed, Steel Spheres – Aluminum Oxide	d	541-0409
Fluidized Bed, Steel Spheres – Copper – Nickel – Gas	d	541-0399
Fluidized Bed, Steel Spheres – Glass	d	541-0410
Foamed		
see Expanded		541-
Foundry Mold, Aluminum Alloy Filled	a	541-0762
Foundry Mold, Cast Iron Filled	a	541-0761
Fuel, Nuclear	a de n	541-0019
Gadolinium Molybdenum Oxide, Silver Covered	e	541-1097
Gas On Metal	b	541-0627
Glass, Gold Covered	a	541-1280
Gold, Oxidized	a	541-0853
Graphite T S 574		
see Graphite, Molybdenum Sulfide And Silver Impregnated		541-
Graphite, Aluminum And Silicon Impregnated (Also Irradiated)	e	541-1414
Graphite, Chromium – Zirconium Boride Cermet Covered	n	541-1042
Graphite, Germanium – Silicon Covered	a de n	541-1130
Graphite, Glass – Molybdenum Silicide Cermet Covered	n	541-0411
Graphite, Magnesium Impregnated (Also Irradiated)	e	541-1396
Graphite, Metal Impregnated	n	541-0910
Graphite, Molybdenum Sulfide And Silver Impregnated	a d n	541-0962
Graphite, Silicon Covered	n	541-0417
Hastelloy X, Aluminum – Nickel – Zirconium Oxide Cermet Covered	a	541-1297

Substance Name	Property	Number
Hastelloy X, Zirconium Oxide Covered	a	541-1298
Heater, Cartridge, Nickel Alloy Sheath, Magnesium Oxide Packed	g	541-0412
(Helium + Nitrogen) On Tungsten	b	541-1158
Helium On Alkali Metal	b	541-0982
Helium On Aluminum	b	541-0355
Helium On Aluminum, Oxidized	b	541-0609
Helium On Argon Covered Tungsten	b	541-0425
Helium On Beryllium	b	541-0356
Helium On Carbon Oxide (C O₂) Covered Tungsten	b	541-0950
Helium On Copper	b	541-1167
Helium On Deuterium Covered Tungsten	b	541-0944
Helium On Ethane Covered Tungsten	b	541-0952
Helium On Ethylene Covered Tungsten	b	541-0959
Helium On Germanium	b	541-1106
Helium On Hydrogen Covered Tungsten	b	541-0942
Helium On Hydrogen On Potassium Covered Tungsten	b	541-0818
Helium On Lithium	b	541-0358
Helium On Methane Covered Tungsten	b	541-0951
Helium On Molybdenum	b	541-0359
Helium On Nickel	b	541-0344
Helium On Nitrogen Covered Tungsten	b	541-0949
Helium On Oxygen Covered Tungsten	b	541-0948
Helium On Oxygen On Hydrogen Covered Tungsten	b	541-0953
Helium On Oxygen On Potassium Covered Tungsten	b	541-0800
Helium On Platinum	b	541-0330
Helium On Platinum Covered Tungsten	b	541-0929
Helium On Potassium	b	541-0357
Helium On Potassium Covered Tungsten	b	541-0819
Helium On Potassium On Hydrogen Covered Tungsten	b	541-0804
Helium On Potassium On Oxygen Covered Tungsten	b	541-0820
Helium On Sodium	b	541-0360
Helium On Steel	b	541-0418
Helium On Tantalum	b	541-1105
Helium On Tungsten	b	541-0315
Helium On Tungsten, Impure	b	541-0768
Helium 3 On Potassium	b	541-1342
Helium 3 On Tungsten	b	541-0389
Helium 4 On Potassium	b	541-1341
Helium 4 On Tungsten	b	541-0390
Honeycomb	a	541-0644
Honeycomb, Aluminum Alloy, Aluminum Alloy Faces, Epoxy Resin Bonded	a	541-0880
Honeycomb, Aluminum Alloy, Aluminum Alloy Faces, Epoxy Resin Bonded	a	541-0881
Honeycomb, Fiberglass, Aluminum And Phenolic Resin Glass Reinforced Faces	a	541-0013
Honeycomb, Flexcore, Aluminum Faces	a	541-0757
Honeycomb, Flexcore, Aluminum Faces – Vacuum	a	541-1086
Honeycomb, Haynes Alloy 25, Brazed – Vacuum	a	541-0700
Honeycomb, Inconel X, Coated, Aluminum Oxide And Aluminum Silicate Filled, Stainless Steel 17-7 Ph Faces	a	541-1233
Honeycomb, Inconel, Coated, Aluminum Oxide And Aluminum Silicate Filled, Beryllium And A-286 Honeycomb Faces	a	541-1234
Honeycomb, Inconel, Coated, Aluminum Oxide And Min-k Filled, Beryllium And A-286 Honeycomb Faces	a	541-1237
Honeycomb, Inconel, Coated, Aluminum Silicate And Silicon Oxide Filled, Beryllium And A-286 Honeycomb Faces	a	541-1236
Honeycomb, Inconel, Coated, Aluminum Silicate And Zirconium Oxide Filled, Beryllium And A-286 Honeycomb Faces	a	541-1235
Honeycomb, Narmco Resin, Glass Reinforced, Aluminum Alloy Face	a	541-1088
Honeycomb, Nickel, Epoxy Resin Bonded	a n	541-0696
Honeycomb, Phenolic Resin, Aluminum Alloy Faces	e	541-0023
Honeycomb, 60-10 M 252 Alloy, Zirconium Oxide – Zirconium Hydride, Expanded, Filled	a	541-1035
(Hydrogen + Neon) On Gold Alloy (Pd)	b	541-1307
(Hydrogen + Neon) On Gold	b	541-1303
(Hydrogen + Neon) On Tungsten	b	541-0767
Hydrogen On Aluminum	b	541-0926
Hydrogen On Gold	b	541-0824
Hydrogen On Hydrogen Covered Iron	b	541-0936
Hydrogen On Monatomic Hydrogen Covered Tungsten	b	541-0456
Hydrogen On Nickel	b	541-0439

Substance Name	Property	Number
Interface, Graphite – Tin Alloy (1.9 Ti)	p	541-1041
Interface, Hafnium – Gas	p	541-0688
Interface, Heptane – Mercury	p	541-0876
Interface, Hexane – Mercury	p	541-0555
Interface, Indium – Gas	p	541-0662
Interface, Indium – Magnesium Oxide	p	541-1205
Interface, Indium – Silicon Oxide (Si O₂)	p	541-1211
Interface, Indium – Vacuum	p	541-0659
Interface, Indium – Water, Adsorbed On Glass	p	541-0884
Interface, Iron – Gas	p	541-0420
Interface, Iron – Silicon Oxide (Si O₂)	p	541-0973
Interface, Iron – Slag (Al₂ O₃, B₂ O₃, Ca O, Mg O, Mn O, Si O₂)	p	541-1074
Interface, Iron – Slag (Al₂ O₃, Ca O, Ca F₂, Mg O, Si O₂)	p	541-0407
Interface, Iron – Slag (Al₂ O₃, Ca O, Ca F₂, Mg O, Na₃ Al F₆, Si O₂)	p	541-0408
Interface, Iron – Slag (Al₂ O₃, Ca O, Ca F₂, Mg O, Mn O, Si O₂)	p	541-1073
Interface, Iron – Slag (Al₂ O₃, Ca O, Ca F₂, Mn O, Si O₂)	p	541-1075
Interface, Iron – Slag (Al₂ O₃, Ca O, Si O₂)	p	541-1076
Interface, Iron – Slag (Al₂ O₃, Ca O, Si O₂, Ti O₂)	p	541-1424
Interface, Iron – Slag (Al₂ O₃, Mg O, Si O₂)	p	541-1431
Interface, Iron Alloy (0 – 100 Pd) – Gas	p	541-1049
Interface, Iron Alloy (0 – 1.2 C, 0 – 1.2 P) – Gas	p	541-1183
Interface, Iron Alloy (0 – 3.32 C) – Silicon Oxide (Si O₂)	p	541-0974
Interface, Iron Alloy (0 – 4.6 C) – Zirconium Oxide – Gas	p	541-0908
Interface, Iron Alloy (0 – 40 W) – Slag (Al₂ O₃, Ca O, Ca F₂, Mg O, Si O₂)	p	541-0589
Interface, Iron Alloy (0 – 50 Cr) – Slag (Al₂ O₃, Ca O, Ca F₂, Mg O, Si O₂)	p	541-0591
Interface, Iron Alloy (0 – 50 Mo) – Slag (Al₂ O₃, Ca O, Ca F₂, Mg O, Si O₂)	p	541-0597
Interface, Iron Alloy (0.035 – 1 P) – Melt (Mn O, Si O₂)	p	541-0654
Interface, Iron Alloy (0.035 – 1 P) – Melt (Mn O, P₂ O₅, Si O₂)	p	541-0655
Interface, Iron Alloy (0.2 – 2.2 C, 0.65 – 2.2 Si) – Slag (Al₂ O₃, Fe O, Mg O, Si O₂)	p	541-1430
Interface, Iron Alloy (0.1 – 3.2 Si) – Gas	p	541-0422
Interface, Iron Alloy (0.2 – 3.91 Ni) – Gas	p	541-0423
Interface, Iron Alloy (0.21 Zr) – Gas	p	541-0698
Interface, Iron Alloy (0.23 – 2.78 C) – Gas	p	541-0421
Interface, Iron Alloy (0.42–1.1 S, Carbon Saturated) – Slag (Al₂ O₃, Ca O, Si O₂)	p	541-1426
Interface, Iron Alloy (0.5 – 8.3 S) – Slag (Al₂ O₃, Ca O, Si O₂)	p	541-1425
Interface, Iron Alloy (0.61 C, 0.67 Mn, 0.27 Si, 0.25 S) – Magnesium Oxide – Gas	p	541-0916
Interface, Iron Alloy (0.61 C, 0.67 Mn, 0.27 Si, 0.25 S) – Slag (Ca O, Mg O, Si O₂) – Gas	p	541-0924
Interface, Iron Alloy (0.61 C, 0.67 Mn, 0.27 Si, 0.25 S) – Slag (Al₂ O₃, Ca O, Mg O, Si O₂) – Gas	p	541-0991
Interface, Iron Alloy (0.61 C, 0.67 Mn, 0.27 Si, 0.25 S) – Gas	p	541-0996
Interface, Iron Alloy (0.61 C, 0.67 Mn, 0.27 Si, 0.25 S) – Refractory, Fireclay – Gas	p	541-0917
Interface, Iron Alloy (10 – 20 Ti) – Slag (Al₂ O₃, Ca O, Si O₂)	p	541-1423
Interface, Iron Alloy (2.5 C) – Gas	p	541-0824
Interface, Iron Alloy (3 C, 1 Mn) – Slag (Ca O, Fe O, Mg O, Mn O, Si O₂)	p	541-1338
Interface, Iron Alloy (3 C, 1 Mn) – Slag (Ca O, Mg O, Mn O, Si O₂)	p	541-1337
Interface, Iron Alloy (3.5 C, 1 Si) – Gas	p	541-1305
Interface, Iron Alloy (3.5 C, 3.3 Si) – Gas	p	541-1306
Interface, Iron Alloy (3.9 C) – Gas	p	541-0822
Interface, Iron Alloy (4 C, 1.2 Si) – Vacuum	p	541-0969
Interface, Iron Alloy (4 C, 1.2 Si, 0 – 1 Y) – Vacuum	p	541-0970
Interface, Lead – Gas	p	541-0605

Substance Name	Property	Number
Interface, Lead – Lead Chloride	p	541-0618
Interface, Lead – Sodium Chloride, Sodium Fluoride Added	p	541-0641
Interface, Lead – Sodium Chloride, Sodium Sulfate Added	p	541-0640
Interface, Lead – Sodium Hydroxide	p	541-0172
Interface, Lead – Vacuum	p	541-0857
Interface, Lead Alloy (0 – 100 Sn) – Vacuum	p	541-0711
Interface, Lead Alloy (0 – 5 Na) – Sodium Chloride	p	541-0639
Interface, Lead Alloy (39 Sn) – Gas	p	541-0719
Interface, Lithium – Lithium Chloride	p	541-1014
Interface, Lithium Fluoride – Mercury	h	541-1135
Interface, Magnesium – Gas	p	541-0826
Interface, Magnesium – Vacuum	p	541-1181
Interface, Magnesium Oxide – Tin	p	541-1204
Interface, Mercury – (N-methylacetamide + Potassium Hexafluorophosphate)	p	541-1256
Interface, Mercury – (Octane + Nonylphenol)	p	541-1382
Interface, Mercury – (Octane + Oleic Acid)	p	541-1381
Interface, Mercury – (Perchloric Acid + Water)	p	541-1247
Interface, Mercury – (Potassium Chloride + Water)	p	541-0827
Interface, Mercury – (Potassium Permanganate + Water)	p	541-0276
Interface, Mercury – (Sodium Chloride + Water)	p	541-0791
Interface, Mercury – (Sodium Perchlorate + Water)	p	541-1284
Interface, Mercury – (Sodium Thiocyanate + Water)	p	541-0665
Interface, Mercury – (Valeric Acid + Water)	p	541-0872
Interface, Mercury – (1- Propanol + Sodium Sulfate + Water)	p	541-1099
Interface, Mercury – (1,2- Propanediol + Sodium Sulfate + Water)	p	541-1101
Interface, Mercury – (3- Pentanol + Perchloric Acid + Water)	p	541-1248
Interface, Mercury – Benzene – Water	p	541-1190
Interface, Mercury – Gas	p	541-0529
Interface, Mercury – Nitrobenzene	p	541-0568
Interface, Mercury – Octadecane	p	541-1202
Interface, Mercury – Octane	p	541-0556
Interface, Mercury – Paraffin Wax	p	541-0548
Interface, Mercury – Pentane	p	541-0674
Interface, Mercury – Polyethylene	p	541-0549
Interface, Mercury – Polytetrafluoroethylene	p	541-0550
Interface, Mercury – Propane	p	541-1180
Interface, Mercury – Propanol	p	541-0560
Interface, Mercury – Sodium Chloride	h	541-1133
Interface, Mercury – Solution	p	541-1138
Interface, Mercury – Stearic Acid	p	541-0673
Interface, Mercury – Thallium Bromide Iodide	h	541-1132
Interface, Mercury – Toluene	p	541-0553
Interface, Mercury – Vacuum	h	541-0709
Interface, Mercury – Water	p	541-0569
Interface, Mercury – Water, Adsorbed On Glass	p	541-0885
Interface, Mercury – Xylene	p	541-0554
Interface, Mercury Alloy (0 – 51 In) – (Perchloric Acid + Water)	p	541-0923
Interface, Mercury Alloy (29 Tl) – Gas	p	541-0786
Interface, Mercury Alloy (5.4 Na) – Gas	p	541-0787
Interface, Mercury, Doped – Gas	p	541-0864
Interface, Metal – Slag	p	541-1198
Interface, Metal – Vacuum	p	541-1182
Interface, Molybdenum – Vacuum	p	541-1379
Interface, Nickel – Gas	p	541-0049
Interface, Nickel – Oil	h	541-1281
Interface, Nickel – Slag (Al₂ O₃, Ca O, Mg O)	p	541-1186
Interface, Nickel – Vacuum	p	541-0085
Interface, Nickel – Zirconium Oxide	p	541-0763
Interface, Nickel Alloy (0 – Unsp. C) – Vacuum	p	541-0480
Interface, Nickel Alloy (0 – 2 In) – Gas	p	541-0522
Interface, Nickel Alloy (0 – 2 Sn) – Gas	p	541-0523
Interface, Nickel Alloy (0 – 32 Si) – Gas	p	541-0613
Interface, Nickel Alloy (0.004 – 0.87 Ti) – Vacuum	p	541-0525
Interface, Nickel Alloy (0.027 – 8.72 Cr) – Vacuum	p	541-0524
Interface, Nickel Alloy (0.05 Ti) – Gas	p	541-0093
Interface, Nickel Alloy (0.09 Ti) – Gas	p	541-0101
Interface, Nickel Alloy (0.49 Ti) – Gas	p	541-0102
Interface, Nickel Alloy (0.53 Cr) – Gas	p	541-0132
Interface, Nickel Alloy (0.97 Ti) – Gas	p	541-0103

Substance Name	Property	Number
Vacuum	a g	541-0406
Multilayer, Aluminum – Dimplar – Polyimide – Vacuum	a g	541-0477
Multilayer, Aluminum – Dimplar – Vacuum	g	541-0452
Multilayer, Aluminum – Epoxy Resin – Nickel – Phenolic Resin, Expanded – Poly(Ethylene Glycol – Terephthalic Acid)	a e n	541-0694
Multilayer, Aluminum – Epoxy Resin, Glass Reinforced – Polyurethane	a	541-0668
Multilayer, Aluminum – Fabric, Rubberized – Glass – Nylon – Polyimide – Vacuum	a	541-1006
Multilayer, Aluminum – Fabric, Rubberized – Nylon – Poly(Ethylene Glycol - Terephthalic Acid) – Polyester – Vacuum	a	541-1005
Multilayer, Aluminum – Fibrous Material – Vacuum	a	541-1275
Multilayer, Aluminum – Glass	a g	541-0223
Multilayer, Aluminum – Glass – Gas	a	541-0977
Multilayer, Aluminum – Glass – Poly(Ethylene Glycol – Terephthalic Acid) – Polyurethane	a	541-0661
Multilayer, Aluminum – Glass – Poly(Ethylene Glycol – Terephthalic Acid) – Polyurethane – Gas	a ghi	541-0707
Multilayer, Aluminum – Glass – Poly(Ethylene Glycol – Terephthalic Acid)	a	541-0865
Multilayer, Aluminum – Glass – Poly(Ethylene Glycol – Terephthalic Acid) – Polyurethane – Vacuum	a	541-0906
Multilayer, Aluminum – Glass – Poly(Ethylene Glycol – Terephthalic Acid) – Vacuum	a	541-0964
Multilayer, Aluminum – Glass – Poly(Ethylene Glycol – Terephthalic Acid) – Polytetrafluoroethylene	g i k	541-1268
Multilayer, Aluminum – Glass – Polyester – Vacuum	a	541-1091
Multilayer, Aluminum – Glass – Polyester, Corrugated Or Crumpled	a	541-0715
Multilayer, Aluminum – Glass – Polyimide – Silicone	a	541-1392
Multilayer, Aluminum – Glass – Polyimide – Vacuum	a	541-0705
Multilayer, Aluminum – Glass – Silver	h	541-1264
Multilayer, Aluminum – Glass – Vacuum	a	541-0121
Multilayer, Aluminum – Glass, Metal Opacified – Vacuum	a	541-0404
Multilayer, Aluminum – Graphite – Polyimide – Vacuum	a	541-0704
Multilayer, Aluminum – Matt	a e	541-0883
Multilayer, Aluminum – Nylon	a	541-0225
Multilayer, Aluminum – Nylon – Gas	a	541-1270
Multilayer, Aluminum – Nylon – Poly(Ethylene Glycol – Terephthalic Acid)	a g	541-0062
Multilayer, Aluminum – Nylon – Poly(Ethylene Glycol – Terephthalic Acid) – Vacuum	a	541-1004
Multilayer, Aluminum – Nylon – Poly(Ethylene Glycol – Terephthalic Acid) – Gas	a	541-1267
Multilayer, Aluminum – Nylon – Polyester	a	541-0712
Multilayer, Aluminum – Nylon – Vacuum	a	541-0124
Multilayer, Aluminum – Phenolic Resin	a	541-0718
Multilayer, Aluminum – Poly(Ethylene Glycol – Terephthalic Acid) – Gas	a gh	541-1007
Multilayer, Aluminum – Poly(Ethylene Glycol – Terephthalic Acid) – Silver – Vacuum	g	541-1246
Multilayer, Aluminum – Poly(Ethylene Glycol – Terephthalic Acid) – Polyurethane	a	541-1399
Multilayer, Aluminum – Poly(Ethylene Glycol – Terephthalic Acid)	a	541-0451
Multilayer, Aluminum – Poly(Ethylene Glycol – Terephthalic Acid) – Polyurethane – Vacuum	a	541-0672
Multilayer, Aluminum – Poly(Ethylene Glycol – Terephthalic Acid), Crumpled Or Corrugated Or Perforated	a e	541-0882
Multilayer, Aluminum – Poly(Ethylene Glycol – Terephthalic Acid) – Vacuum	a e g i	541-0976
Multilayer, Aluminum – Poly(Ethylene Glycol – Terephthalic Acid), Crumpled – Gas	a	541-0980
Multilayer, Aluminum – Poly(Ethylene Glycol – Terephthalic Acid) – Silk	a	541-1143
Multilayer, Aluminum – Poly(Ethylene Glycol – Terephthalic Acid) – Polytetrafluoroethylene	a	541-1173
Multilayer, Aluminum – Polyester – Gas	a	541-1221
Multilayer, Aluminum – Polyester – Polyurethane	a	541-0716
Multilayer, Aluminum – Polyester – Vacuum	a	541-1090
Multilayer, Aluminum – Polyester, Corrugated Or Crumpled	a	541-0710
Multilayer, Aluminum – Polyimide	g	541-0476
Multilayer, Aluminum – Polyimide – Gas	a	541-1009
Multilayer, Aluminum – Polyimide – Silicon Oxide – Vacuum	a	541-0706
Multilayer, Aluminum – Polyimide – Vacuum	a h	541-1084
Multilayer, Aluminum – Polytetrafluoroethylene – Vacuum	ghi k	541-0135
Multilayer, Aluminum – Polyurethane, Glass Reinforced	a	541-0670
Multilayer, Aluminum – Resin, Fabric Reinforced	a de	541-0044
Multilayer, Aluminum – Silicon Oxide	a	541-1087
Multilayer, Aluminum – Silicon Oxide, Metal Opacified – Vacuum	a	541-1201
Multilayer, Aluminum – Silk – Gas	a	541-1008
Multilayer, Aluminum – Silk – Vacuum	a	541-1393
Multilayer, Aluminum – Vacuum	a	541-0966
Multilayer, Aluminum – Vinyl Resin	a	541-0713
Multilayer, Aluminum – Zirconium Oxide	a	541-1176
Multilayer, Aluminum Alloy 1145 – Glass – Vinyl Resin – Vacuum	a	541-1219
Multilayer, Aluminum Alloy 1145 – Glass – Vinyl Resin – Substance, Expanded – Vacuum	a	541-1222
Multilayer, Aluminum Alloy 1145 – Glass – Vinyl Resin – Gas	a	541-1220
Multilayer, Aluminum Alloy 1145 – Nylon – Vacuum	a	541-1218
Multilayer, Aluminum Alloy 1145 – Phenolic Resin – Vacuum	a	541-1215
Multilayer, Aluminum Alloy 3003 – Cybond 4000 Adhesive (Also With Additives)	a	541-1179
Multilayer, Aluminum Oxide – Inconel	n	541-1036
Multilayer, Aluminum Oxide – Kovar – Steel	n	541-1037
Multilayer, Aluminum Oxide – Niobium	c	541-1331
Multilayer, Aluminum Oxide – Stainless Steel 302 – Vacuum	a	541-0690
Multilayer, Carbon – Tantalum – Vacuum	a e	541-0869
Multilayer, Copper – Copper Iodide – Copper Sulfide – Platinum	a	541-1397
Multilayer, Copper – Glass – Vacuum	a	541-1296
Multilayer, Copper – Silicon Oxide	a g	541-0780
Multilayer, Copper – Silicon Oxide – Vacuum	a g	541-0832
Multilayer, Copper – Silicon Oxide, Metal Opacified – Vacuum	a	541-0427
Multilayer, Copper – Zirconium Oxide	a	541-1177
Multilayer, Dynaflex – Platinum Alloy (10 Rh)	a	541-0623
Multilayer, Electroluminescent	g i	541-0518
Multilayer, Electroluminescent Material – Glass – Iron – transparent Conductor	g i k	541-0516
Multilayer, Electroluminescent Material – Metal – Poly (Methyl Methacrylate) – transparent Conductor	g i k	541-0517
Multilayer, Glass – Indium Alloy (Bi) – Manganese	a	541-1405
Multilayer, Glass – Platinum	hij	541-1413
Multilayer, Glass – Silicone	j	541-1159
Multilayer, Glass – Stainless Steel	a	541-0823
Multilayer, Glass – Stainless Steel 316 – Vacuum	a	541-0703
Multilayer, Glass, Metal Opacified	a g	541-0825
Multilayer, Glass, Metal Opacified – Vacuum	a	541-0405
Multilayer, Gold – Poly(Ethylene Glycol – Terephthalic Acid) – Vacuum	a g i	541-1085
Multilayer, Gold – Polyimide – Gas	a	541-1010
Multilayer, Gold – Polyimide – Silicon Oxide	a	541-1357
Multilayer, Gold – Polyimide – Vacuum	a ghi	541-0450
Multilayer, Graphite – Phenolic Resin, With Beryllium – Titanium – Zirconium Brazed Niobium – Zirconium Alloy Joint	a	541-0571
Multilayer, Graphite – Phenolic Resin, With Copper – Silver – Titanium Brazed 316 Stainless Steel Joint	a	541-0570
Multilayer, Graphite – Tantalum (Insulation)	a	541-0975
Multilayer, Hafnium Oxide – Molybdenum	g	541-0677
Multilayer, Hastelloy C – Vacuum	a	541-1141
Multilayer, Lead – Poly(Ethylene Glycol – Terephthalic Acid) – Vacuum	a	541-0994
Multilayer, Magnesium Oxide – Steel	a	541-1095
Multilayer, Metal – Non-metal – Vacuum	a g	541-1245
Multilayer, Molybdenum – Silicon Oxide – Vacuum	a e	541-0867
Multilayer, Molybdenum – Zirconium Oxide – Vacuum	a	541-1045
Multilayer, Nickel – Polyimide – Vacuum	g	541-0453
Multilayer, Nickel – Silicon Oxide	g	541-0821
Multilayer, Nickel – Silicon Oxide – Vacuum	a g	541-0831
Multilayer, Nickel – Silicon Oxide, Metal Opacified – Vacuum	a g	541-0431
Multilayer, Nickel – Zirconium Oxide	a	541-1178

Substance Name	Property	Number
Multilayer, Platinum – Silicon Oxide	a	541–0540
Multilayer, Platinum – Silver – Silver Iodide – Silver Sulfide	a	541–1398
Multilayer, Platinum Alloy (10 Rh) – Silicon Oxide	a	541–0989
Multilayer, Poly(Ethylene Glycol – Terephthalic Acid) – Tin – Vacuum	a g	541–0474
Multilayer, Russian Stainless Steel Kh18 N10 T – Gas	a	541–0656
Multilayer, Silicon Oxide – Stainless Steel	a	541–0829
Multilayer, Silicon Oxide – Tantalum – Vacuum	a g	541–0912
Multilayer, Silicon Oxide, Metal Opacified	a g	541–0828
Multilayer, Silicon Oxide, Metal Opacified – Vacuum	a	541–0012
Multilayer, Silicon Steel – Filler	a	541–0343
Multilayer, Stainless Steel – Zirconium Oxide	a	541–1136
Multilayer, Stainless Steel 302 – Vacuum	a	541–0437
Multilayer, Tantalum – Thorium Oxide	a	541–1174
Multilayer, Thorium Oxide – Tungsten	a	541–1175
Multilayer, Titanium – Zirconium Oxide	a	541–0434
N R C–2		
see Multilayer, Aluminum – Poly(Ethylene Glycol – Terephthalic Acid)		541–
Neon On Alkali Metal	b	541–0983
Neon On Aluminum	b	541–0362
Neon On Aluminum, Oxidized	b	541–0723
Neon On Beryllium	b	541–0363
Neon On Deuterium Covered Tungsten	h	541–0946
Neon On Ethylene Covered Tungsten	b	541–0960
Neon On Gold	b	541–1274
Neon On Gold Alloy (Pd)	b	541–1290
Neon On Hydrogen Covered Iron	b	541–0933
Neon On Hydrogen Covered Palladium	b	541–1302
Neon On Hydrogen Covered Palladium Alloy (Au)	b	541–1304
Neon On Hydrogen Covered Tungsten	b	541–0943
Neon On Iron	b	541–0364
Neon On Lithium	b	541–0374
Neon On Molybdenum	b	541–0375
Neon On Nickel	b	541–0441
Neon On Nitrogen Covered Iron	b	541–0935
Neon On Nitrogen Covered Tungsten	b	541–0954
Neon On Oxygen Covered Iron	b	541–0934
Neon On Oxygen Covered Tungsten	b	541–0955
Neon On Palladium	b	541–1269
Neon On Palladium Alloy (Au)	b	541–1279
Neon On Platinum	b	541–0724
Neon On Potassium	b	541–0365
Neon On Potassium Covered Tungsten	b	541–0801
Neon On Sodium	b	541–0376
Neon On Tungsten	b	541–0339
Nichrome V, Enamel Covered	n	541–1214
Nickel Alloy (18 Cr, Al, Y), Thorium Oxide Reinforced	g	541–1401
Nickel Alloy (20 Cr), Thorium Oxide Reinforced	g	541–1400
Nickel, (Tungsten, Silicon Carbide Covered) Reinforced	n	541–1081
Nickel, Carbon Black Covered	d	541–0466
Nickel, Carbon Oxide (C O) Adsorbate	j	541–1287
Nickel, Porous – Gas	a	541–0941
Niobium, Aluminum Oxide Covered	a	541–1217
Nite-lite (Sylvania)		
see Multilayer, Electroluminescent Material – Glass – Iron – transparent Conductor		541–
Nitrogen On Aluminum	b	541–0927
Nitrogen On Copper	b	541–1168
Nitrogen On Germanium	b	541–1108
Nitrogen On Gold	b	541–0537
Nitrogen On Nickel	b	541–0443
Nitrogen On Nitrogen Covered Tungsten	b	541–0963
Nitrogen On Oxygen Covered Iron	b	541–0938
Nitrogen On Platinum	b	541–0736
Nitrogen On Silver	b	541–0625
Nitrogen On Tantalum	b	541–1107
Nitrogen On Tungsten	b	541–0769
(Nitrogen Oxide (N O) + Nitrogen Oxide (N O_2) + Oxygen) On Gold	b	541–1213
Nitrogen Oxide (N_2 O) On Gold	b	541–0576
Nitrogen Oxide (N_2 O) On Platinum	b	541–0478
Nitrogen, Monatomic On Gold	b ghi	541–0699
Nitrogen, Monatomic On Silver	b	541–0636
Nitrogen, Triatomic On Gold	b	541–0708
Nitrogen, Triatomic On Silver	b	541–0637
Oil Drilling Pipe	a	541–1402
Oxygen On Aluminum	b	541–0722
Oxygen On Gallium Arsenide	b	541–1310
Oxygen On Germanium	b	541–1110
Oxygen On Gold	b	541–0545
Oxygen On Nickel	b	541–0445
Oxygen On Oxygen Covered Iron	b	541–0939
Oxygen On Platinum	b	541–0738
Oxygen On Tantalum	b	541–1109
Oxygen On Tungsten	b	541–0444
Oxygen On Zinc Selenide	b	541–1314
Packed Bed, Copper Spheres – Water	a	541–1239
Packed Bed, Glass Spheres – Steel Spheres	a	541–1114
Packed Bed, Glass Spheres – Steel Spheres – Gas	a	541–1115
Packed Bed, Glass Spheres – Steel Spheres – Vacuum	a	541–1116
Packed Bed, Iron Spheres – Gas	a	541–0565
Packed Bed, Iron Spheres – Vacuum	a	541–0118
Packed Bed, Lead Spheres – Gas	a	541–0519
Packed Bed, Nickel Catalyst – Gas	a	541–0414
Packed Bed, Steel Spheres – Gas	a	541–0113
Packed Bed, Steel Spheres – Oil	a	541–1418
Packed Bed, Steel Spheres – Vacuum	a c	541–0338
Packed Bed, Steel Spheres – Water	a	541–0587
Platinum, Ethylene Adsorbate	j	541–1355
Plutonium Uranium Carbide, Niobium Covered, Helium Filled	a	541–1224
Plutonium Uranium Carbide, Stainless Steel 316 Covered, Helium Filled	a	541–1225
Potassium Phosphorus Hexafluoride (Syn. For– Potassium Hexafluorophosphate)		
see Interface, (Dimethylacetamide + Potassium Hexafluorophosphate) – Mercury		541–
Powder – Gas	a	541–0546
Powder, (Aluminum + Silicon Oxide) – Vacuum	a d	541–1244
Powder, (Bronze + Silicon Oxide) – Vacuum	a	541–1344
Powder, (Tungsten + Zirconium Oxide, Hollow) – Vacuum	a	541–0868
Powder, Aluminum – Gas	a n	541–0433
Powder, Aluminum – Vacuum	n	541–1286
Powder, Copper – Gas	a	541–0436
Powder, Iron – Gas	a	541–0578
Powder, Iron – Vacuum	a	541–0577
Powder, Magnesium – Gas	a	541–0442
Powder, Nickel – Gas	a	541–0448
Powder, Uranium – Gas	a	541–0424
Powder, Uranium, Oxidized – Gas	a	541–0428
Powder, Zirconium – Gas	a	541–0430
Powder, Zirconium, Oxidized – Gas	a	541–0429
Propane On Stainless Steel	b	541–0457
Resin, Aluminum Reinforced	g	541–0547
Resin, Epoxy – Furan, Iron Filled	a n	541–1429
Resin, Epoxy, Aluminum Reinforced	a e	541–0720
Resin, Epoxy, Aluminum Reinforced, Aluminum Filled	a	541–0721
Resin, Epoxy, Aluminum Reinforced, Aluminum Oxide Filled	a	541–0756
Resin, Epoxy, Aluminum Reinforced, Titanium Filled	a	541–1226
Resin, Epoxy, Boron Reinforced	a n	541–1308
Resin, Epoxy, Copper Reinforced	a	541–1228
Resin, Epoxy, Copper Reinforced, Titanium Filled	a	541–1230
Resin, Epoxy, Glass Reinforced, Aluminum Filled	a e	541–0753
Resin, Epoxy, Lead Balls Embedded	a	541–0837
Resin, Epoxy, Titanium Reinforced	a	541–1385
Resin, Epoxy, Tungsten Reinforced	a n	541–1349
Resin, Phenolic, Aluminum Reinforced	a	541–1328
Resin, Phenolic, Graphite Reinforced, Silicon – Silicon Carbide Coated	a	541–0802
Resin, Phenolic, Nickel Reinforced	a	541–1415
Resin, Polyester, Aluminum Covered	a	541–1142
Resin, Polyethylene, Copper Filled	a n	541–1330
Resin, Polyimide, Aluminum Covered	a	541–1145
Resin, Polyimide, Gold Covered	a	541–1146
Resin, Polyimide, Silver Covered	a	541–1147
Resin, Polyimide, Silver Covered, Silicon Oxide Overcoat	a	541–1356
Resin, Polyimide, Silver Filled	a	541–1329

APPLIED COATINGS   CLASS 551

Substance Name	Property	Number
A F–66		
see Paint (Ti O₂ + Silicone) (Also Irradiated)		551–
A F–67		
see Paint (Ti O₂ + Silicone) (Also Irradiated)		551–
A F–68–5		
see Paint (Ti O₂ + Silicone) (Also Irradiated)		551–
A G 0.6		
see Anodized Aluminum Alloy A G 0.6 (French)		551–
A I 93 Coating		
see Ceramic (Sn O₂), Aluminum Phosphate Bonded		
(Also Irradiated)		551–
A M F Kote		
see Molybdenum Silicon Intermetallic		551–
A M F–34		
see Aluminum – Boron		551–
A N– T T– V–116		
see Paint (Alkyd Resin), Clear		551–
A R F–2		
see Paint (Zn O + Potassium Silicate)		
(Also Irradiated)		551–
A 5		
see Anodized Aluminum Alloy A 5 (French)	a d	551–
171– A–152 (Fuller)		
see Paint (Al + Silicone) (Also Irradiated)		551–
A–418		
see Ceramic N B S A–418 (Al₂ O₃ + B₂ O₃ +		
Ba O + Ca O + Clay + Cr₂ O₃ +		
Si O₂ + Zn O + Zr O₂)		551–
Ablative Material	gh	551–2957
Acetate	i	551–2493
Acetylene Black		
see Carbon Black		551–
Acetylene Soot		
see Carbon Black		551–
Acid Acriflavine		
see Acriflavine Hydrochloride		551–
Acid Trypaflavine		
see Acriflavine Hydrochloride		551–
Acme Quality Aluminum 803 Spray Enamel		
see Paint (Al)		551–
Acme Quality Black 801		
see Paint, Black 801 (Acme Quality)		551–
Acme Quality White 800		
see Paint, White 800 (Acme Quality)		551–
Acriflavine Hydrochloride	h	551–2314
Adhesive, S R 585 (G E)	i	551–2610
Adhesive, Silastic 280 (Dow Corning)	i	551–2612
Aerated		
see Expanded		551–
Algae	h	551–2886
Algae, Moist	h	551–2956
Alkyd Oil		
see Paint (Alkyd Resin), Clear		551–
Alkydal L52		
see Paint (Alkyd Resin), Clear		551–
Allyltrichlorosilane	j	551–2302
Alodyne Finished Aluminum (Also Irradiated)	hi	551–1384
Alodyne Finished Aluminum Alloy	ghi k	551–1258
Aluminized Epoxy		
see Paint (Al + Epoxy Resin)		551–
Aluminized Fabric	g	551–2065
Aluminized Resin, Poly(Ethylene Glycol –		
Terephthalic Acid) (Also Irradiated)	ghi	551–3296
Aluminized Resin, Polyester	g	551–3046
Aluminized Resin, Polytetrafluorothylene		
(Also Irradiated)	ghi k	551–2772
Aluminum – Aluminum Oxide, Multilayer	h	551–2363
Aluminum – Beryllium – Niobium	ghi	551–1373
Aluminum – Beryllium – Tantalum	ghi	551–1471
Aluminum – Beryllium – Titanium	ghi	551–1661
Aluminum – Boron	gh	551–2573
Aluminum – Ceramic Powder	g	551–3071
Aluminum – Cerium Oxide	h	551–3043
Aluminum – Chromium – Molybdenum Disilicide	g	551–2303

Substance Name	Property	Number
Aluminum – Disilicide (Boeing)	g	551–1755
Aluminum – Germanium – Silicon Oxide, Multilayer	h	551–2873
Aluminum – Lithium Fluoride	h	551–3042
Aluminum – Magnesium Fluoride, Multilayer	h	551–2990
Aluminum – Molybdenum – Tin	g	551–2603
Aluminum – Nickel – Silicon	gh	551–2326
Aluminum – Niobium	ghi	551–0258
Aluminum – Silicon	gh	551–0283
Aluminum – Tantalum	ghi	551–0260
Aluminum – Tin	g	551–2009
Aluminum – Titanium	h	551–0264
Aluminum (Also Irradiated)	ghijk	551–1858
Aluminum Alloy (Fe)	g	551–2799
Aluminum Alloy (Si)	g	551–3494
Aluminum Alloy 6061 (Also Irradiated)	g i k	551–2492
Aluminum Alloy 7075 (Also Irradiated)	hi	551–2491
Aluminum Arsenic Gallium Intermetallic	h	551–3083
Aluminum Lithium Silicate (Also Irradiated)	ghi	551–1326
Aluminum Magnesium Oxide	g i	551–2316
Aluminum Nickel Intermetallic	ghi k	551–2298
Aluminum Niobium Intermetallic	gh	551–2478
Aluminum Nitride	h	551–3299
Aluminum Oxide – Aluminum Titanium Oxide	g	551–0762
Aluminum Oxide – Iron Titanium Oxide	g	551–2669
Aluminum Oxide – Magnesium Fluoride – Zirconium Oxide,		
Multilayer	h	551–3410
Aluminum Oxide – Titanium Oxide	gh	551–2784
Aluminum Oxide (Also Irradiated)	a ghijk	551–0487
Aluminum Radiation Shields, Super Insulating (Linde)	g	551–3098
Aluminum Sodium Fluoride		
(Syn. For– Chiolite)		
see Chiolite – Antimonite		551–
Aluminum Tantalum Intermetallic	gh	551–2474
Aluminum Titanium Intermetallic	gh	551–2301
Alzak Process		
see Anodized Aluminum (Also Irradiated)		551–
Ammonia	h	551–2541
Ammonium Bromide	j	551–3293
Ammonium Chloride	j	551–2308
Ammonium Manganese Pyrophosphate – Barium Sulfate	h	551–2828
Ammonium Manganese Pyrophosphate – Barium Sulfate – Iron		
Phosphate	h	551–2818
Amphibole, Hornblende	j	551–2400
Anodized Aluminum (Also Irradiated)	ghijk	551–0278
Anodized Aluminum Alloy A G 0.6 (French)	h	551–2913
Anodized Aluminum Alloy A 5 (French)	h	551–2870
Anodized Aluminum Alloy 1100	g i k	551–2240
Anodized Aluminum Alloy 1170	g	551–3347
Anodized Aluminum Alloy 2014	g i	551–1601
Anodized Aluminum Alloy 2024	ghi	551–0605
Anodized Aluminum Alloy 5052	g i	551–3324
Anodized Aluminum Alloy 5557	ghi	551–3323
Anodized Aluminum Alloy 6061 (Also Irradiated)	ghi k	551–0924
Anodized Aluminum Alloy 7075	g i	551–1417
Anodized Aluminum, Blackened	ghi	551–1606
Anodized Anticorodal	hi	551–3256
Anodized Beryllium	gh	551–3326
Anodized Beryllium Alloy (Cu)	ghi	551–1425
Anodized Beryllium Alloy Q M V	h	551–1261
Anodized Brass	h	551–3496
Anodized Duralumin	h	551–3279
Anodized Magnesium	g	551–3010
Anodized Magnesium, Black	h	551–2586
Anodized Metal, Black	i	551–2786
Anodized Metal, Clear	g	551–2787
Anodized Reflectal	h	551–2535
Anodized Remiral	h	551–2555
Anodized Silicon	j	551–2807
Anodized Silicon – Phosphorus	m	551–2809
Anodized Silicon, Doped	j	551–2808
Anodized Tantalum	h jk	551–2052
Anodized Tantalum Oxide	g i	551–2863
Anodized Titanium	gh k	551–0926
Anodized Zirconium	h	551–2625

Substance Name	Property	Number
Cesium Azide – Lead Bromide	j	551-2932
Cesium Azide – Mercury Iodide	j	551-3308
Cesium Bromide	j	551-2906
Cesium Bromide – Lead Nitride	j	551-2931
Cesium Bromide – Lead Tellurium Intermetallic – Zinc Sulfide, Multilayer	j	551-2969
Cesium Bromide – Lead Tellurium Intermetallic, Multilayer	jk	551-2968
Cesium Chloride	j	551-2907
Cesium Chloride – Lead Nitride	j	551-2930
Cesium Fluoride	j	551-2901
Cesium Iodide	h j	551-2905
Cesium Iodide – Copper Azide	j	551-2928
Cesium Iodide – Germanium, Multilayer	j	551-2981
Cesium Iodide – Lead Tellurium Intermetallic – Zinc Sulfide, Multilayer	j	551-2984
Cesium Iodide – Lead Tellurium Intermetallic, Multilayer	j	551-2982
Chalk	g	551-2616
Char From Ablative Material	gh	551-2962
Chiolite – Antimonite	h	551-2462
Chromalloy W– X see Silicon Tungsten Intermetallic		551-
Chromalloy W–3 see Molybdenum Silicon Intermetallic		551-
Chromate	g k	551-0680
Chromate Conversion see Chromate		551-
Chromium	ghijk	551-0637
Chromium – Disilicide (Boeing)	g	551-1751
Chromium – Silver	h	551-3002
Chromium Black	ghi	551-1226
Chromium Carbide (Also Irradiated)	ghij	551-1328
Chromium Chloride	j	551-3335
Chromium Copper Oxide	g	551-2771
Chromium Nickel Intermetallic (Also Irradiated)	g i	551-1329
Chromium Oxide – Nickel Oxide	g	551-1694
Chromium Oxide – Nickel Oxide – Silicon Oxide, Aluminum Phosphate Bonded	g	551-1472
Chromium Oxide (Also Irradiated)	ghij	551-2623
Chromium Silicon Intermetallic	gh	551-2306
Chromium Silicon Titanium Intermetallic	gh	551-2615
Chromium Titanium Intermetallic	gh	551-2853
Coating (General) (Also Irradiated)	a ghijk	551-1571
Coating A I 93 see Ceramic (Sn O₂), Aluminum Phosphate Bonded (Also Irradiated)		551-
Coating, A M F (For Niobium)	g	551-3104
Coating, A M F (For Tantalum)	g	551-3106
Coating, Absorbing	h j	551-1734
Coating, Antireflection	h j	551-2287
Coating, Barrier Layer Anodize (Boeing)	i	551-3181
Coating, Clear	h	551-3078
Coating, Conversion, Ematal Type (L M S C)	i	551-2288
Coating, Corning E. C. 1004	j	551-2427
Coating, Dielectric	gh j	551-2286
Coating, Flocculated	j	551-3317
Coating, Heat Resistant	gh j	551-3412
Coating, High Emittance (Also Irradiated)	gh	551-3207
Coating, High Reflection	h	551-2577
Coating, High Reflection, Blue	h	551-2583
Coating, High Reflection, Gray	h	551-2576
Coating, High Reflection, Green	h	551-2581
Coating, High Reflection, White	h	551-2578
Coating, High Refractive Index Over Low Refractive Index	h j	551-2459
Coating, Hy-cal Black (Also Irradiated)	i	551-3097
Coating, Inhomogeneous	h	551-2404
Coating, L E V 31	g j	551-3259
Coating, L E V 32	ghij	551-3258
Coating, L M S C Solar Reflector (Also Irradiated)	g i	551-3079
Coating, Low Emittance (Also Irradiated)	hi	551-3092
Coating, Low Refractive Index Over High Refractive Index	h	551-2413
Coating, M M S-450	g	551-2488
Coating, Multilayer	h j	551-2130
Coating, Nonabsorbing Base	j	551-2484
Coating, Nonscattering	h j	551-2275

Substance Name	Property	Number
Coating, Optical Solar Reflector (Also Irradiated)	ghi k	551-2147
Coating, Pigmented	h	551-2344
Coating, Selective Reflection	h j	551-2401
Coating, Solar Absorber, Infrared Reflector	i	551-2952
Coating, Thermal Control (Also irradiated)	ghijk	551-2566
Coating, Thermophototropic	hij	551-3424
Coating, W D-40	ghi k	551-2988
Cobalt	gh	551-2848
Cobalt – Disilicide (Boeing)	g	551-1747
Cobalt Fluoride	j	551-2927
Cobalt Oxide	ghij	551-0620
Cobalt Sulfide Dyed Anodized Aluminum	g i k	551-3250
(Cobalt-(2-quinolyl Pyridine – 2-aldehyde- –hydrazone)) Complex see 2– Quinolyl Pyridine–2-aldehydehydrazone– Cobalt Complex		551-
Connecticut No. 3 see Tape, Hard Rubber, Optical Pressure Sensitive (Connecticut No. 3) (Also Irradiated)		551-
Coomassie Brilliant Blue R see C. I. Acid Blue 83		551-
Copoly- see Resin		551-
Copoly(Formaldehyde Melamine – Vinyl Butyral) Resin see Resin, Melamine Poly(Vinyl Butyral) (Series Of Additives)		551-
Copper (Also Irradiated)	ghij	551-1953
Copper Alloy (Ni)	j	551-2925
Copper Azide – Rubidium Iodide	j	551-2923
Copper Bromide	h j	551-2351
Copper Chloride	h j	551-2353
Copper Chromate see Chromium Black		551-
Copper Iodide	h j	551-2354
Copper Iodide Phosphorus Selenide	j	551-3394
Copper Oxalate	h	551-3381
Copper Oxide	ghij	551-0582
Copper Phosphorus Selenide	h j	551-3392
Copper Phosphorus Selenide Sulfide	j	551-3393
Copper Phthalocyanine Blue (Syn. For- Phthalocyaninatocopper) see Paint (2- Phthalocyaninatocopper + Acrylic Resin)		551-
Copper Selenium Intermetallic	j	551-2926
Copper Sulfate	j	551-3215
Copper Sulfide	gh j	551-0548
Cordoscreen (Technikote) see Resin, White		551-
Corning E. C. 1004 see Coating, Corning E. C. 1004		551-
Cotton	h	551-2753
Cryodeposit	hi	551-2589
Cryolite	h	551-2018
Cryolite – Lead Oxide, Multilayer	hij	551-3382
Cryolite – Zinc Sulfide, Multilayer	h j	551-2887
Cyanamid S7094-3 see Paint (Ti O₂ + Melamine Poly(Vinyl Butyral)) (Also Irradiated)		551-
Cyanamid S7094-4 see Paint (Ti O₂ + Terpoly(Ethyl Acrylate + Methacrylic Acid + Methyl Methacrylate))		551-
Cyclopentane	j	551-2938
Cyclopentane, Decadeuterated	j	551-2939
Cymel + Poly(Vinylbutyral) see Resin, Melamine Poly(Vinyl Butyral) (Series Of Additives)		551-
Cymel 300 see Resin, Hexamethoxymethylmelamine		551-
D C Q 92-009 see Resin, Silicone, Clear		551-
D C 705 see Resin, Silicone (Also Irradiated)		551-

Substance Name	Property	Number
D C 90–090		
see Ablative Material		551–
D P		
(Syn. For– Dioctadecyl Phosphite)		
see Paint (Dioctadecyl Phosphite)		551–
Da– Lite Mat White Screen		
see Resin, White		551–
Da– Lite Wonderlite Screen		
see Paint (Al)		551–
Dag E C 1652		
see Carbon Black		551–
Dag E C 1789		
see Calcium Fluoride		551–
Dayview Type 1 Screen (Kodak)		
see Coating, High Reflection		551–
Detecto Temp		
(Syn. For– Paint (Detectotemp.)		
see Paint (Detecto Temp + Acrylic Resin +		
Potassium Silicate)		551–
Diatomite	h	551–3144
Dibenzoyl		
see Benzil		551–
Dicalite		
see Diatomite		551–
Didymium	h	551–3493
Didymium Fluoride	j	551–3368
o–(1,8– Dihydroxy–3,6–disulfo–2–naphthyl)		
–azobenzenearsenic Acid	h	551–2746
o–(1,8– Dihydroxy–3,6–disulfo–2–naphthyl)		
–azobenzenearsenic Acid – Lanthanum	h	551–2725
1,2– Dihydroxyanthraquinone – Lanthanum	h	551–2570
1,2– Dihydroxyanthraquinone	h	551–2744
Diphenyl Diketone		
see Benzil		551–
Diphenylglyoxal		
see Benzil		551–
Disil		
see Disilicide (Boeing)		551–
Disilicide (Boeing)	g	551–1539
Disilicide (Boeing) – Iron	g	551–1748
Disilicide (Boeing) – Manganese	g	551–1752
Disilicide (Boeing) – Niobium	g	551–2641
Ditzler D M A 311		
see Paint (Ti O₃ + Acrylic Resin + Nitrocellulose)		551–
Dow Corning X P – 310		
see Paint (Al + Silicone) (Also Irradiated)		551–
Dow Plasticizer 2		
(Syn For– Bis–(p–tert-butylphenyl) Phenyl		
Phosphate)		
see Paint (Mg C O₃ + 2 Pb C O₃ . Pb(O H)₂ +		
Bis–(p–tert-butylphenyl) Phenyl Phosphate +		
Isobutyl Methacrylate)		551–
Dow Plasticizer 7		
(Syn. For– Tris–(p–tert Butylphenyl) Phosphate)		
see Paint (Mg C O₃ + Ethyl Cellulose +		
Tris–(p–tert-butylphenyl) Phosphate)		551–
Dow 15 Finished Magnesium Alloy H M 21 A	g l k	551–1266
Dow 17 Finished Magnesium	h	551–0780
Dow 17 Finished Magnesium Alloy A Z 31	hi	551–1421
Dow 7 Finished Magnesium	g	551–1738
Dow 7 Finished Metal	g	551–2785
Du Pont 29–915		
see Paint, White Dupont 29–915		551–
Du Pont 65–3010		
see Paint, White		551–
Duco Wrought Iron 71		
see Paint, Black Duco Wrought Iron 71 (Dupont)		551–
Dulite li Black Oxide	g	551–3052
Dulite, Oxide Conversion, Steel Base		
see Oxidized Steel		551–
Durak B (Chromizing Co.)	g	551–1162
Durak Ka	gh	551–2324
Duralumin D–16	gh	551–2755
Dysprosium	h	551–3051

Substance Name	Property	Number
Dysprosium Oxide	h	551–2531
Eastman White Reflectance Paint (Distill. Prod. Ind.)		
see Paint (Ba S O₄ + Polyvinyl Alcohol)		551–
Ebonol C		
see Oxidized Copper		551–
Ebonol S		
see Oxidized Steel		551–
Echo A–12 Laminate		
see Aluminized Resin, Poly(Ethylene Glycol – Terephthalic Acid) (Also Irradiated)		551–
Eloxized		
see Anodized		551–
Elvacite 6011 Coating		
see Resin, Poly(Methyl Methacrylate)		551–
Ematal		
see Anodized Aluminum (Also Irradiated)		551–
Ematal Type Coating (L M S C)		
see Coating, Conversion, Ematal Type (L M S C)		551–
Enameloid (Sherwin– Williams)		
see Paint, Black Enameloid (Sherwin – Williams)		551–
Eosin B	h	551–3014
Eosin Y	h	551–3012
Erbium	l	551–2525
Erbium – Gold	h	551–2449
Eriochrome Cyanine		
see C. I. Mordant Blue 3		551–
Ethanol	h	551–3398
Ethyl–(4–(p–(Ethyl–(m–sulfobenzyl)amino)– –α–(p–phenetidino)phenyl)benzylidene) – 2,5–cyclohexadien–1–ylidene)–(m– –sulfobenzyl)–ammonium Hydroxide Inner Salt		
see C. I. Acid Blue 83		551–
Europium	h j	551–2395
Europium Oxide	h	551–2167
Exon 461		
(Syn. For– Copoly(Chloroethylene – Chlorotrifluoroethylene)		
see Paint (Zn O + Copoly(Chloroethylene – Chlorotrifluoroethylene))		551–
F C E–11 Continental Coatings Corporation		
see Barium Titanium Oxide		551–
F C T–11 Continental Coatings Corporation		
see Iron Titanium Oxide		551–
F C Z–11 Continental Coatings Corporation		
see Zirconium Silicate (Also Irradiated)		551–
Fayalite	h j	551–3076
see also Olivine, Fayalite		521–
see also Iron Silicate		110–
Fiber Optic	h	551–3074
Film, Photographic		
see Gelatin – Silver Bromide		551–
Flavine		
see Acriflavine Hydrochloride		551–
Fluorine – Tin Oxide	gh j	551–3275
Foamed		
see Expanded		551–
Forsterite	h	551–3194
see also Olivine, Forsterite		521–
Frit A–418		
see Ceramic N B S A–418 (Al₂ O₃ + B₂ O₃ + Ba O + Ca O + Clay + Cr₂ O₃ + Si O₂ + Zn O + Zr O₂)		551–
Fuller Black Velvet Heavy Duty Plastic Enamel 1518		
see Paint (Resin), Black Velvet 1518 (W. P. Fuller		551–
Fuller Dielectric 168– W–20		
see Paint, White Fuller Dielectric 168– W–20		551–
Fuller Gloss White 517– W–1		
see Paint (Ti O₂ + Alkyd Silicone) (Also Irradiated)		551–
Fuller 171 W560		
see Paint (Acrylic Resin), White (Also Irradiated)		551–
Fuller 171– A–152 Aluminum Silicone Paint		
see Paint (Al + Silicone) (Also Irradiated)		551–

Substance Name	Property	Number
Fuller 172- A-1	h	551-9900
see Paint (Al + Silicone) (Also Irradiated)		551-
Fuller 517-w-1 White Silicone Paint		
see Paint (Ti O₂ + Alkyd Silicone)		
(Also Irradiated)		551-
Furnace Lining	i	551-3348
G. E. S S 4090		
see Resin, Silicone, Clear		551-
G. E. L T V 602 Silicone Potting Compound		
see Resin, Silicone, Clear		551-
G. E. S-4044 'r		
see Paint, S-4044 'r (G. E.)		551-
G. E. 7031 Varnish		
see Paint (Phenolic Vinyl Resin), Black		551-
Gadolinium	h j	551-2824
Gadolinium Iron Oxide	j	551-2924
Gadolinium Oxide	j	551-3372
Gallium	hij	551-1394
Gallium Nitride	h	551-2550
Gelatin	j	551-2207
Gelatin - Silver Bromide	h j	551-1511
Germanium	ghijk	551-1768
Germanium – Magnesium Fluoride	j	551-2408
Germanium – Zinc Sulfide, Multilayer	j	551-2976
Germanium Iodide	j	551-2985
Germanium Oxide	h	551-2604
Germanium Tellurium Intermetallic	h j	551-1818
Glass	gh k	551-2759
Glass (Pb O + Si O₂)	j	551-2373
Glass Resin 100 (Owens Illinois)		
see Resin, Silicone (Also Irradiated)		551-
Glass Resin 901 (Owens Illinois)		
see Resin, Silicone (Also Irradiated)		551-
Glass, Black	h	551-2383
Glass, Blue	h	551-2382
Glass, Brown	h	551-2377
Glass, Corning 0120, Reduced	h	551-2364
Glass, Gray	h	551-2379
Glass, Green	h	551-2380
Glass, Non-copper	h	551-2297
Glass, Pink	h	551-2378
Glass, Pittsburg 3235 Borosilicate	h	551-3011
Glass, Red	h	551-2381
Glass, Vitrolite	h	551-2489
see also Vitrolite Glass		507-
Glass, White	h	551-2376
Glasurit, Black		
see Paint (Epoxy Resin), Black (Also Irradiated)		551-
Glasurit, White		
see Paint (Epoxy Resin), White (Also Irradiated)		551-
Glidair Zapon 131- B-190 B		
see Paint, Black Glidair Zapon 131- B-190 B		
(Glidden)		551-
Glidden R G I-22576		
see Paint, White R G I-22576 (Glidden)		551-
Glidden R G I-22818		
see Paint, Black R G I-22818 (Glidden)		551-
Glidden 131- B-216		
see Paint, Black 131- B-216 (Glidden)		551-
Glidden 2295		
see Paint, White 2295 (Glidden)		551-
Glidden 5064		
see Paint, White 5064 (Glidden)		551-
Glidden 9099		
see Paint, Black 9099 (Glidden)		551-
Glycinethymol Blue	h	551-2743
Glycinethymol Blue - Lanthanum	h	551-2569
Glyptal Black 10021 A Paint (G. E.)		
see Paint (Alkyd Resin), Black		551-
Glyptal Red Enamel 1201		
see Paint (Alkyd Resin), Red		551-
Glyptal, Brown Paint (G. E.)		
see Paint (Alkyd Resin), Brown		551-
Goethite	h	551-3000

Substance Name	Property	Number
Gold – Platinum - Titanium Oxide	h j	551-2546
Gold – Titanium Oxide	h j	551-2534
Gold (Also Irradiated)	ghijk	551-0223
Gold ← Chromium	m	551-2703
Gold ← Gold	m	551-2702
Gold ← Iron	m	551-2704
Gold ← Nickel	m	551-2705
Gold Alloy (Ag)	h	551-0269
Gold Alloy (Al)	h	551-2834
Gold Alloy (Fe)	h	551-3314
Gold Black	ghij	551-0799
Gold Black Alloy (Cu)	h	551-3333
Gold Black Alloy (Ni)	h	551-3336
Gold Black Alloy (Pd)	h	551-3337
Gold Resinate	g	551-3073
Graphite	gh jk	551-2572
Graphite – Titanium Oxide	h	551-3318
H T 424		
see Tape, Adhesive, Epoxy Phenolic, H T 424		551-
H-10 (Hughes Aircraft)		
see Paint (Clay + Silicone)		551-
Hafnium Carbide	g	551-1378
Hafnium Oxide	gh j	551-2170
Hafnium Oxide Yttrium Oxide Stabilized	g	551-1618
Hafnium Oxide, Calcium Oxide Stabilized	g	551-2148
Hanovia Liquid Bright Palladium		
see Paint, Hanovia Liquid Bright Palladium		551-
Hanovia 6518		
see Paint (Au)		551-
Heatrem		
see Paint (Al)		551-
Hematite	h	551-3001
Hexacosane, Copoly(Ethylene - Vinyl Acetate) Bonded	h	551-2674
Hexacosane, Polystyrene Bonded	h	551-2673
Hexadecanol	h	551-2368
Hi-heat Silicone Aluminum		
see Paint (Al + Silicone) (Also Irradiated)		551-
High- Heat Aluminum Paint 324 (Sprayon)		
see Paint (Al)		551-
Hy- Cal Black		
see Coating, Hy-cal Black (Also Irradiated)		551-
Hycar		
(Syn. For- Polyacrylic Hycar)		
see Paint (Ti O₂ + Polyacrylic Hycar)		551-
Hydrobromic Acid	j	551-2452
Hydrobromic Acid, Monodeuterated	j	551-2453
Hydrochloric Acid	j	551-2450
Hydrochloric Acid, Monodeuterated	j	551-2451
Hydrogen Sulfide - Nitrogen	i	551-3266
3- Hydroxy-4-(4-sulfo-1-naphthyl)azo)-		
2,7-naphthalenedisulfonic Acid		
(Syn. For- C. I. Acid Red 27)		
see Paint (C. I. Acid Red 27 + Acrylic Resin)		551-
Imron		
see Resin, Polyurethane		551-
Inconel	gh	551-2823
Indium	hij	551-1460
Indium – Lithium Fluoride	hij	551-3003
Indium Iodide	j	551-2972
Indium Oxide	h j	551-2429
Indium Phosphide	j	551-2385
Indium Sulfide	j	551-2975
Ink, Fluorescent	h	551-3416
Insulube		
see Paint (Molydisulfide + Polymer)		551-
Intermetallic	g	551-2730
Iridite		
see Chromate		551-
Iridium	gh j	551-2842
Iron	h j	551-0618
Iron Idodide	i	551-3084
Iron Nickel Oxide	h j	551-3377
Iron Oxide (Also Irradiated)	ghij	551-0643

Substance Name	Property	Number
Iron Titanate		551-
see Iron Titanium Oxide		551-0678
Iron Titanium Oxide	g	
see also Titanium Beach Sand		521-
Iron Yttrium Oxide	h j	551-3068
Japalac 1208 (Glidden)		
see Paint, Black Japalac 1208 (Glidden)		551-
Jet Dry 78 (Globe)		
see Paint, Black Jet Dry 78 (Globe)		551-
Kanigen Coating		
see Cermet, Nickel – Nickel Phosphide		551-
Kel- F Latex K F 8213		
see Resin, Copoly(Chlorotrifluoroethylene –		
1,1– Difluoroethylene)		551-
Kel- F Resin 800		
see Resin, Copoly(Chlorotrifluoroethylene –		
1,1– Difluoroethylene)		551-
Kemacryl Aluminum		
see Paint (Al + Acrylic Resin) (Also Irradiated)		551-
Kemacryl Lacquer, White		
see Paint (Ti O₂ + Acrylic Resin) (Also Irradiate		551-
Kemacryl M49 B C12		
see Paint (Acrylic Resin), Black		551-
Kemacryl M49 W C17		
see Paint (Ti O₂ + Acrylic Resin) (Also Irradiate		551-
Kerpo 25 Aluminum		
see Paint (Al)		551-
Karpogrey Q. D. Spray Metallic Enamel		
W B– S– N52 E-4 (Kerr)		
see Paint (Metallic), Gray		551-
Kodak Black Dull Brushing Lacquer No. 4		
see Paint, Black Kodak 4		551-
Kodak Dayview Type 1 Screen		
see Coating, High Reflection		551-
Kodox 515		
see Zinc Oxide (Also Irradiated)		551-
Koropon Super 765		
see Paint, Koropon Super 765		
(Pacific Paint And Varnish)		551-
Krylon Bright Silver Aluminum Paint 1401		
see Paint (Al)		551-
Krylon 1301		
see Paint, Clear 1301 (Krylon)		551-
Krylon 1501		
see Paint (Acrylic Resin), White (Also Irradiated)		551-
Krylon 1502 Series		
see Paint (Acrylic Resin), White (Also Irradiated)		551-
Krylon 1602, Black		
see Paint (C + Silicates + Lacquer), Black		551-
Krylon, Crystal Clear		
see Paint (Acrylic Resin), Clear		551-
Krylon, White		
see Paint (Acrylic Resin), White (Also Irradiated)		551-
L E V 31		
see Coating, L E V 31		551-
L E V 32		
see Coating, L E V 32		551-
I-7 (Mc Donnell Co.)		
see Silicide		551-
Lacquer A N– T T– V-116 (Pratt And Lambert)		
see Paint (Alkyd Resin), Clear		551-
Lacquer A-10		
see Paint (Al + Acrylic Resin) (Also Irradiated)		551-
Lacquer Ditzler D M A 311		
see Paint (Ti O₂ + Acrylic Resin + Nitrocellulose)		551-
Lacquer 1234		
see Paint (Al)		551-
Lacquer, Kemacryl, Black		
see Paint (C + Acrylic Resin), Black		551-
Lacquer, Kemacryl, White		
see Paint (Ti O₂ + Acrylic Resin) (Also Irradiate		551-
Laminar X-500		
see Paint (Ti O₂ + Polyurethane)		551-

Substance Name	Property	Number
Lampblack		551-
see Carbon Black		
Lanthanum – Methylthymol Blue	h	551-2567
Lanthanum – Pyridylazoresorcinol	h	551-2727
Lanthanum – Pyrocatechol Violet	h	551-2568
Lanthanum – Xylenol Orange	h	551-2726
Lanthanum Fluoride (Also Irradiated)	h j	551-1767
Lanthanum Oxide	h	551-2995
Lapis Lazuli		
see Paint (Ultramarine + Nitrocellulose)		551-
L B–2 Aluminide Coating		
see Aluminum Niobium Intermetallic		551-
Lead	h j	551-1621
Lead Azide	h j	551-2810
Lead Bromide	j	551-2933
Lead Chloride	hij	551-2339
Lead Chromate + Lead Molybdate + Lead Sulfate		
(Syn. For- Molybdate Orange)		
see Paint (Molybdate Orange + Alkyd Styrene Resin)		551-
Lead Fluoride	h j	551-2348
Lead Fluoride – Sodium Hexafluoroaluminate, Multilayer	h	551-3301
Lead Fluoride – Sodium Hexafluroaluminate – Zinc		
Selenide – Zinc Sulfide, Multilayer	j	551-3300
Lead Iodide	hij	551-2445
Lead Molybdenum Oxide	j	551-2565
Lead Oxide	hi	551-2169
Lead Selenium Intermetallic	h j	551-1810
Lead Selenium Tellurium Intermetallic	h	551-2557
Lead Stearate	h	551-2943
Lead Sulfide	ghijk	551-2627
Lead Sulfide Dyed Anodized Aluminum	g i	551-2332
Lead Tellurium Intermetallic	h j	551-0988
Lead Tellurium Intermetallic – Zinc Sulfide, Multilayer	j	551-2978
Lead Tellurium Tin Intermetallic	h j	551-2594
Leonita 201- S		
(Syn. For- Acrylic Epoxy Silicone)		
see Paint (Ti O₂ + Acrylic Epoxy Silicone)		551-
Lichen	h	551-2590
Liquid Bright Platinum Ga (Johnson, Matthey Co.)		
see Platinum		551-
Lithium Bromide	j	551-2896
Lithium Chloride	j	551-2900
Lithium Fluoride	ghij	551-2369
Lithium Fluoride – Potassium Fluoride	j	551-2371
Lithium Iodide	j	551-2802
Lockheed Sodium Silicate D – Ultrox Paint		
see Paint (Zr Si O₄ + Sodium Silicate) (Also Irra		551-
Lutetium	h j	551-3049
Luxuria Screen (trans Lux)		
see Paint (Al)		551-
M M S–450		
see Coating, M M S–450		551-
M S D–105		
see Paint (Zn O + Potassium Silicate)		
(Also Irradiated)		551-
M 49– B C9 (Sherwin Williams)		
Magnesium	h j	551-1464
Magnesium Carbonate	h	551-2544
Magnesium Fluoride – Silver – Zinc Sulfide, Multilayer	j	551-2588
Magnesium Fluoride – Silver, Multilayer	j	551-2585
Magnesium Fluoride – Zinc Sulfide, Multilayer	h j	551-2888
Magnesium Fluoride (Also Irradiated)	ghij	551-1838
Magnesium Oxide	ghij	551-1508
Magnesium Oxide – Thallium Chloride	j	551-2899
Magnesium Titanium Zinc Oxide	h	551-3195
Magnesium Zirconium Oxide	g	551-2868
Magnolia Plastics Magnobond Coating		
see Resin, Epoxy – Polysulfide		551-
Manganese Iodide	i	551-3087
Manganese Oxide	hij	551-0347
Manganese Phosphate	g	551-2966
Marine Corps T T– p-664		
see Paint, Green, Marine Corps T T– p-664		551-

Substance Name	Property	Number
Martin Emissivity Coating		
see Resin, Acrylic		551-
Mat-1		
see Paint (Ti O₂ + Acrylic Resin)		
(Also Irradiated)		551-
Mc Cormick Blue Dye	h	551-2388
Mc Donnell L B-2		
see Aluminum Niobium Intermetallic		551-
Mc Donnell I-7		
see Silicide		551-
Mc Donnell T S-137		
see Niobium Silicon Intermetallic		551-
Mercury	h	551-1392
Mercury Bromide	j	551-3345
Mercury Chloride	j	551-3344
Mercury Electrical Coating 168- W-20		
see Paint, White Mercury Electrical Coating 168- W-20		
(W. P. Fuller)		551-
Mercury Fluoride	j	551-3343
Mercury Iodide	j	551-3307
Mercury Sulfide	j	551-2362
Mercury Tellurium Intermetallic	j	551-2859
Metal	ghij	551-0325
Metal Oxide	g i k	551-3353
Metaltone Hammer Finish Silver Paint		
(Ideal Chem.)		
see Paint (Ag)		551-
Metco X P - 1121		
see Aluminum Oxide - Titanium Oxide		551-
Metco X P 1106		
see Cermet, Cobalt - Chromium Carbide		551-
Metco X P 1110		
see Cermet, Cobalt - Tungsten Carbide		551-
Metco X p-1114		
see Titanium Oxide		551-
Metco 101		
see Cermet, Titanium - Aluminum Oxide		551-
Metco 105		
see Aluminum Oxide (Also Irradiated)		551-
Metco 201		
see Zirconium Oxide (Also Irradiated)		551-
Metco 404		
see Aluminum Nickel Intermetallic		551-
Metco 43 C		
see Nickel Alloy (Cr)		551-
Methanol	h	551-3397
Methylthymol Blue	h	551-2728
Micobond		
see Paint (Phenolic Vinyl Resin), Black		551-
Molybdenum	ghij	551-2991
Molybdenum Oxide	ij	551-2247
Molybdenum Silicon - Boron Zironium Intermetallic	g	551-3327
Molybdenum Silicon Intermetallic	gh	551-0287
Molybdenum Sulfide	hi	551-2624
Molybdenum Sulfide, Ceramic Bonded	g	551-2825
Monarch 71 (Cabot)		
see Carbon Black		551-
Moss	h	551-2885
Mystik 7420		
see Tape, Copper, Mystik (Numbered Series)		551-
Mystik 7452		
see Tape, Aluminum, Mystik (Numbered Series)		551-
Mystik 7455		
see Tape, Aluminum - Glass, Mystik (Numbered Series)		
		551-
N A A-85 Coating (North American Aviation, Inc.)		
see Aluminum - Ceramic Powder		551-
N B S		
see Ceramic N B S A-418 (Al₂ O₃ + B₂ O₃ +		
Ba O + Ca O + Clay + Cr₂ O₃ +		
Si O₂ + Zn O + Zr O₂)		551-
N B S		
see Ceramic N B S N-143 (Ba O + Be O + Ca O +		
Ce O₂ + Clay + P₂ O₅ + Si O₂ + Zn O)		551-

Substance Name	Property	Number
N R C-2		
see Aluminized Resin, Poly(Ethylene Glycol -		
Terephthalic Acid) (Also Irradiated)		551-
Naphthacene	j	. 551-2595
Naphthol Red		
(Syn. For- C. I. Acid Red 27)		
see Paint (C. I. Acid Red 27 + Acrylic Resin)		551-
Neo- Spectra Mark I I		
see Carbon		551-
Neodymium	j	551-2007
Neodymium Oxide	h	551-2171
Neon	h	551-3325
Nernst Glower		
see Rare Earth Oxides		551-
Nesa Coating		
see Tin Oxide		551-
Nickel	gh j m	551-0233
Nickel Alloy (Cr)	g	551-1498
Nickel Black	ghi k	551-3253
Nickel Chrome Spinel		
see Chromium Oxide - Nickel Oxide		551-
Nickel Oxide	gh j	551-0346
Nickel Sulfide Dyed Anodized Aluminum	g i k	551-3249
(Nickel-(2-quinolyl Pyridine -		
2-aldehydehydrazone)) Complex		
see 2- Quinolyl Pyridine-2-aldehydehydrazone-		
Nickel Complex		551-
Nickel, Dark Over Nickel, Bright (Tabor)	gh	551-0231
Nickelized Silicon Oxide	j	551-2312
Niobium Carbide	g	551-2989
Niobium Oxide - Silicon Oxide Multilayer	h	551-3332
Niobium Silicon Intermetallic	g	551-2811
Niobium Zinc Intermetallic	gh	551-2855
Nitrogen Oxide	j	551-2448
Nylon	ghi	551-3463
O S R		
see Silver (Also Irradiated)		551-
Oil	gh	551-2618
Oil, Corn	g	551-2639
Oil, Mineral	g	551-2637
Optiglow Screen (Radiant)		
see Paint (Al)		551-
Oxide	ghij	551-0690
Oxidized (Chromium - Nickel - Vanadium)	g i	551-3474
Oxidized Alloy	g	551-0498
Oxidized Aluminum	ghij	551-0468
Oxidized Aluminum - Beryllium - Niobium	gh	551-1863
Oxidized Aluminum - Beryllium - Tantalum	gh	551-1933
Oxidized Aluminum - Beryllium - Titanium	gh	551-1934
Oxidized Aluminum - Molybdenum - Tin	g	551-2605
Oxidized Aluminum - Niobium	gh	551-1785
Oxidized Aluminum - Tantalum	gh	551-1766
Oxidized Aluminum Alloy 1100	g	551-3156
Oxidized Aluminum Alloy 2024	g	551-0479
Oxidized Aluminum Nickel Intermetallic	g	551-2821
Oxidized Aluminum Niobium Intermetallic	g	551-2814
Oxidized Aluminum Tantalum Intermetallic	gh	551-2830
Oxidized Arsenic Gallium Intermetallic	h	551-3287
Oxidized Basalt	h	551-3045
Oxidized Beryllium	g	551-0536
Oxidized Boron Titanium Intermetallic	g	551-2851
Oxidized Boron Zirconium Intermetallic	g	551-2813
Oxidized Brass	ghi	551-0539
Oxidized Brass L S-62	g	551-2279
Oxidized Carbon	gh	551-2997
Oxidized Carbon Steel	g	551-3276
Oxidized Cermet, Aluminum - Aluminum Oxide	g	551-3278
Oxidized Cermet, Nickel - Thorium Oxide	g	551-2598
Oxidized Chromium	gh	551-2543
Oxidized Chromium Silicon Intermetallic	g	551-2852
Oxidized Chromium Silicon Titanium Intermetallic	g	551-2606
Oxidized Chromium Titanium Intermetallic	g	551-2854
Oxidized Coal	h	551-2349
Oxidized Cobalt Alloy N-155	g	551-2486

Substance Name	Property	Number
Paint (B₄ Si + Cr B + Aluminum Phosphate)	g	551-2456
Paint (B₄ Si + Cr₂ O₃ + Aluminum Phosphate)	ghi	551-2444
Paint (B₄ Si + Fe₂ O₃ + Aluminum Phosphate)	g	551-2420
Paint (B₄ Si + Ni O + Aluminum Phosphate)	ghi	551-2441
Paint (B₄ Si + Ti Cr₂ + Aluminum Phosphate)	ghi	551-2454
Paint (Ba P₂ O₆ + Zn O + Silicate)	h	551-3126
Paint (Ba S O₄)	h j	551-2706
Paint (Ba S O₄ + Carboxymethylcellulose)	h	551-3224
Paint (Ba S O₄ + Gelatin)	h	551-3226
Paint (Ba S O₄ + Oil)	h	551-2875
Paint (Ba S O₄ + Polyvinyl Alcohol)	h	551-3222
Paint (Ba Ti O₃ + Silicone)	h	551-0775
Paint (Ba Ti O₃ + Ti O₂ + Epoxy Resin)	h	551-2513
Paint (Bi₂ O₂ C O₃ + Acrylic Epoxy Silicone)	h	551-2360
Paint (Black Oxide + Si O₂)	g	551-2091
Paint (Butadiene - Vinyl Toluene), White	h	551-2659
Paint (C), Black	h	551-1609
Paint (C + Acrylic Resin), Black	g i k	551-0767
Paint (C + Alkyd Resin), White To Gray	h	551-2683
Paint (C + Alkyd Silicone)	hi	551-3093
Paint (C + Clay + Ti O₂ + Alkyd Phthalic Resin + Alkyd Silicone)	hi	551-1463
Paint (C + Epoxy Phenolic Resin)	ghi	551-3090
Paint (C + Epoxy Resin)	hi	551-3091
Paint (C + Epoxy Silicone)	hi	551-3095
Paint (C + Molybdate Orange + Pb Cr O₄ + Nitrocellulose)	h	551-2504
Paint (C + Nitrocellulose)	gh	551-0440
Paint (C + Pb Cr O₄ + Ti O₂ + Ultramarine + Nitrocellulose)	h	551-2506
Paint (C + Polyester) (Also Irradiated)	hi	551-3380
Paint (C + Polyfluoroethylene)	ghi	551-3086
Paint (C + Polyisoprene)	g	551-2719
Paint (C + Si O₂)	g i	551-3206
Paint (C + Silicates + Lacquer), Black	ghi	551-0906
Paint (C + Silicone)	ghi k	551-1944
Paint (C + Sodium Silicate)	g	551-0779
Paint (C + Ti O₂ + Alkyd Resin), White To Gray	h	551-2698
Paint (C + Ti O₂ + Nitrocellulose)	h	551-2494
Paint (C + Ti O₂ + Silicone) (Also Irradiated)	ghi k	551-3037
Paint (C + Ti O₂ + Ultramarine + Nitrocellulose)	h	551-2503
Paint (C. I. Acid Red 27 + Acrylic Resin)	h j	551-2628
Paint (Ca + Ti + Alkyd Resin), White	h	551-2685
Paint (Ca C O₃ + Alkyd Resin), White	h	551-2687
Paint (Ca C O₃ + Clay + Si O₂ + Ti O₂ + Alkyd Resin)	h	551-3448
Paint (Ca C O₃ + Mg Si O₃ + Ti O₂ + Alkyd Resin)	h	551-3450
Paint (Ca C O₃ + Si O₂ + Ti O₂ + Alkyd Resin)	h	551-3446
Paint (Ca C O₃ + Ti O₂ + Alkyd Resin)	h	551-3449
Paint (Ca F₂ + Potassium Silicate) (Also Irradiated)	i	551-3018
Paint (Ca F₂ + Sodium Silicate) (Also Irradiated)	i	551-3017
Paint (Ca O + Si O₂ + Zr O₂ + Potassium Silicate)	i	551-3169
Paint (Ca S O₄ + Oil)	h	551-2634
Paint (Ca S O₄ + Ti O₂ + Alkyd Resin), White	h	551-2697
Paint (Ca Si O₃)	g	551-3187
Paint (Ca Si O₃ + Polytetrafluoroethylene)	h	551-2352
Paint (Ca Si O₃ + Potassium Silicate)	g i	551-3172
Paint (Ca Si O₃ + Sn O₂ + Zr O₂ + Aluminum Phosphate)	g i	551-3184
Paint (Ca Si O₃ + Sodium Silicate)	i	551-0534
Paint (Ca Si O₃ + Zn O + Silicate)	h	551-3125
Paint (Ca Si O₃ + Zr Si O₄ + Potassium Silicate) (Also Irradiated)	i	551-3027
Paint (Ca W O₄)	g	551-3186
Paint (Ca₃(P O₄)₂ + Potassium Silicate)	g i	551-3174
Paint (Cd Zn S)	g	551-3185
Paint (Ce O₂ + Aluminum Phosphate)	gh	551-0824
Paint (Ce O₂ + Silicone)	h	551-2530
Paint (Chalk)	g	551-2617
Paint (Chromium Cobalt Nickel Spinel + Aluminum Phosphate)	g	551-3146
Paint (Chromium Cobalt Nickel Spinel + Si O₂ + Aluminum Phosphate)	g i	551-3204
Paint (Chromium Iron Nickel Spinel + Si O₂ + Potassium Silicate)	g i	551-3177
Paint (Chromium Iron Nickel Spinel + Si O₂ + Aluminum Phosphate)	g i	551-3202
Paint (Chromium Nickel Spinel + Si O₂ + Aluminum Phosphate)	g i	551-3192
Paint (cis-13- Docosanoic Acid + Poly(Vinyl Butyral))	i	551-2911
Paint (Clay + Sb₂ O₃)	i	551-3232
Paint (Clay + Silicone)	hi k	551-3201
Paint (Clay + Ti O₂ + Alkyd Silicone), White	hi	551-1462
Paint (Co O + Aluminum Phosphate)	h	551-2155
Paint (Co O + Cr₂ O₃ + Tempilaq + Silicone)	g	551-2475
Paint (Co O + Potassium Silicate)	h	551-2157
Paint (Co₂ O₃ + Cr₂ O₃ + Ni O + Aluminum Phosphate)	g i	551-3147
Paint (Co₂ O₃ + Cu O + Ni O + Fe₂ O₃)	g i	551-3161
Paint (Co₃ O₄ + Aluminum Phosphate)	gh	551-2158
Paint (Co₃ O₄ + Ni O + Aluminum Phosphate)	g	551-2418
Paint (Co₃ O₄ + Potassium Silicate)	h	551-2164
Paint (Cr B + Aluminum Phosphate)	g	551-2423
Paint (Cr B + Cr₂ O₃ + Aluminum Phosphate)	g	551-2443
Paint (Cr B + Fe₂ O₃ + Aluminum Phosphate)	g	551-2422
Paint (Cr B + Ni O + Aluminum Phosphate)	g	551-2440
Paint (Cr B + Ti Cr₂ + Aluminum Phosphate)	ghi	551-2447
Paint (Cr B₂ + Aluminium Phosphate)	g	551-2424
Paint (Cr B₂ + Fe₂ O₃ + Aluminum Phosphate)	g	551-2425
Paint (Cr₂ O₃ + Aluminum Phosphate)	gh	551-0823
Paint (Cr₂ O₃ + Fe₂ O₃ + Aluminum Phosphate)	g	551-2446
Paint (Cr₂ O₃ + Fe₃ O₄ + Ni O + Si O₂)	g	551-2805
Paint (Cr₂ O₃ + H₃ B O₃ + Si O₂)	g	551-2483
Paint (Cr₂ O₃ + Ni O + Aluminum Phosphate)	g	551-2419
Paint (Cr₂ O₃ + Ni O + Si O₂ + Aluminum Phosphate)	g	551-1473
Paint (Cr₂ O₃ + Ni O + Tempil + Si O₂)	g	551-2482
Paint (Cr₂ O₃ + Ni O + Tempil + Silicone)	g	551-2883
Paint (Cr₂ O₃ + Ni O + Tempilaq + Silicone)	g	551-2477
Paint (Cr₂ O₃ + Potassium Silicate)	h	551-2172
Paint (Cr₂ O₃ + Tempil + Si O₂)	g	551-2481
Paint (Cr₂ O₃ + Tempilaq)	g	551-2471
Paint (Cr₂ O₃ + Tempilaq + Silicone)	g	551-2472
Paint (Cr₂ O₃ + Ti Cr₂ + Aluminum Phosphate)	g	551-2442
Paint (Cr₂ O₃ + Ti O₂ + Silicone), Green	hi	551-1125
Paint (Cu₂ (O H)₂ C O₃ + Ti O₂ + Alkyd Resin), White To Gray	h	551-2699
Paint (Detecto Temp + Acrylic Resin + Potassium Silicate)	g	551-2466
Paint (Detecto Temp + Si O₂)	g	551-2465
Paint (Detecto Temp + Zr O₂ + Acrylic Resin + Potassium Silicate)	g	551-2468
Paint (Diatomite + Aluminum Phosphate)	hi	551-1192
Paint (Diatomite + Potassium Silicate)	hi	551-1191
Paint (Diatomite + Zn O + Silicate)	h	551-3121
Paint (Diatomite + Zn O + Silicone)	h	551-3140
Paint (Dioctadecyl Phosphite)	i	551-3247
Paint (Dioctadecyl Phosphite + Alkyd Resin + Poly (Vinyl Butyral))	i	551-2917
Paint (Dioctadecyl Phosphite + Poly(Vinyl Butyral))	hij	551-2909
Paint (Dioctadecyl Phosphite + Silicone)	i	551-2912
Paint (Dy₂ O₃ + Silicone)	h	551-2520
Paint (Dy₂ O₃ + Zr O₂ + Silicone)	h	551-2522
Paint (Enstatite + Potassium Silicate)	h	551-2350
Paint (Epoxy Resin)	g i k	551-2347
Paint (Epoxy Resin), Black (Also Irradiated)	ghi k	551-3054
Paint (Epoxy Resin), Clear	j	551-1142
Paint (Epoxy Resin), White (Also Irradiated)	ghi k	551-0466
Paint (Fe P O₄ + Mn N H₄ P₂ O₇ + Glue)	h	551-2800
Paint (Fe P O₄ + Mn N H₄ P₂ O₇ + Zn O + Oil)	h	551-2817
Paint (Fe₂ O₃)	h	551-2542
Paint (Fe₂ O₃ + Alkyd Resin)	h	551-0588
Paint (Fe₂ O₃ + Alkyd Styrene Resin)	h	551-3512
Paint (Fe₂ O₃ + Aluminum Phosphate)	gh	551-2230
Paint (Fe₂ O₃ + Lithium Aluminum Silicate + Ni O + Sodium Silicate) (Also Irradiated)	i	551-0602
Paint (Fe₂ O₃ + Molybdate Orange + Alkyd Styrene Resin)	h	551-3514
Paint (Fe₂ O₃ + Ni O + Aluminum Phosphate)	g	551-2417
Paint (Fe₂ O₃ + Ni O + Zr O₂ + Potassium Silicate) (Also Irradiated)	i	551-3038
Paint (Fe₂ O₃ + Ni₃ Al + Aluminum Phosphate)	g	551-2431

Substance Name	Property	Number
Paint (Fe₂ O₃ + Potassium Silicate)	h	551-2246
Paint (Fe₂ O₃ + Ta Al₃ + Aluminum Phosphate)	g	551-2433
Paint (Fe₂ O₃ + Ti Cr₂ + Aluminum Phosphate)	g	551-2435
Paint (Fe₂ O₃ + Ti O₂ + Silicone), Red	ghi	551-1130
Paint (Fe₃ O₄ + Other Oxides | Silicone), Black	g i k	551-1135
Paint (Fe₄(Fe(C N)₆)₃ White To Gray To Blue	h	551-2679
Paint (Gd₂ O₃ + Silicone)	h	551-2521
Paint (Gd₂ O₃ + Zr O₂ + Silicone)	h	551-2529
Paint (Ge O₂ + Aluminum Phosphate)	g i	551-3188
Paint (Hexacosane + Copoly(Ethylene - Vinyl Acetate))	h	551-3154
Paint (Hexacosane + Polystyrene)	hi	551-3153
Paint (K P O₃ + Potassium Silicate)	i	551-3148
Paint (K₂ Si O₄ + Ti O₂ + Zn O + Poly(Methyl Methacrylate))	h	551-3199
Paint (Kaolin + Potassium Silicate)	g i	551-2993
Paint (La F₃ + Zn O + Silicate)	h	551-3137
Paint (La₂ O₃ + Potassium Silicate) (Also Irradiated)	ghi	551-1181
Paint (La₂ O₃ + Sodium Silicate) (Also Irradiated)	i	551-3021
Paint (Li B O₂ + Zn O + Silicate)	h	551-3136
Paint (Li F + Potassium Silicate)	hi	551-3167
Paint (Li F + Sodium Silicate) (Also Irradiated)	i	551-3019
Paint (Linseed Oil)	j	551-3319
Paint (Lithium Aluminum Silicate + Potassium Silicate) (Also Irradiated)	ghi k	551-2611
Paint (Lithium Aluminum Silicate + Sodium Silicate) (Also Irradiated)	ghi	551-0533
Paint (Lithium Aluminum Silicate + Zn O + Silicate)	h	551-3123
Paint (Lithium Aluminum Silicate + Zr Si O₄ + Sodium Silicate) (Also Irradiated)	i	551-3025
Paint (Magnesium Silicate + Molybdate Orange + Pb Cr O₄ + Phthalocyanine Blue + Quinacridone Red + Ti O₂ + Copoly (Alkyd Phthalic Styrene)), Olive Drab	h	551-3110
Paint (Magnesium Silicate + Molybdate Orange + Copoly(Alkyd Phthalic Styrene))	h	551-3354
Paint (Magnesium Silicate + Pb Cr O₄ + Copoly(Alkyd Phthalic Styrene))	h	551-3356
Paint (Magnesium Silicate + Phthalocyanine Blue + Copoly(Alkyd Phthalic Styrene))	h	551-3364
Paint (Magnesium Silicate + Quinacridone Red + Copoly(Alkyd Phthalic Styrene))	h	551-3351
Paint (Magnesium Silicate + Ti O₂ + Copoly(Alkyd Phthalic Styrene))	h	551-3365
Paint (Metallic), Gray	h	551-0432
Paint (Mg Al₂ O₄ + Potassium Silicate)	hi	551-1184
Paint (Mg C O₃ + Ethyl Cellulose + Tris-(p-tert-butylphenyl) Phosphate)	h	551-2781
Paint (Mg C O₃ + Ti O₂ + Epoxy Resin) (Also Irradiated)	hi	551-2507
Paint (Mg C O₃ + Ti O₂ + Silicone)	h	551-2518
Paint (Mg C O₃ + 2 Pb C O₃ . Pb(O H)₂ + Bis-(p-tert-butylphenyl) Phenyl Phosphate + Isobutyl Methacrylate)	h	551-2777
Paint (Mg C O₃ + 2 Pb C O₃ . Pb(O H)₂ + Ethyl Cellulose + Tris-(p-tert-butylphenyl) Phosphate)	h	551-2775
Paint (Mg C O₃ + 2 Pb C O₃ . Pb(O H)₂ + Ethyl Cellulose + Vinyl Resin)	h	551-2776
Paint (Mg C O₃ + 2 Pb C O₃ . Pb(O H)₂ + Ethyl Cellulose + Tris-(p-tert-butylphenyl) Phosphate + Vinyl Resin)	h	551-2779
Paint (Mg C O₃ + 2 Pb C O₃ . Pb₀.₅H)₂ + Butyl Stearate + Isobutyl Methacrylate)	h	551-2778
Paint (Mg F₂ + Zn O + Silicate)	h	551-3135
Paint (Mg H P O₄ + Zn O + Silicate)	h	551-3130
Paint (Mg O)	h	551-3468
Paint (Mg O + Acrylic Epoxy Silicone)	h	551-1223
Paint (Mg O + Acrylic Resin) (Also Irradiated)	h	551-3157
Paint (Mg O + Ethyl Cellulose + Tris-(p-tert-butylphenyl) Phosphate)	h	551-2780
Paint (Mg O + Oil)	h	551-2876
Paint (Mg O + Polyisoprene)	g	551-2718
Paint (Mg O + Potassium Silicate)	g i	551-3182
Paint (Mg O + Si O₂ + Zr O₂ + Potassium Silicate)	i	551-3170
Paint (Mg O + Silicone)	hi	551-2285
Paint (Mg Si O₃ + Alkyd Resin), White	h	551-2688
Paint (Mg Si O₃ + Potassium Silicate)	hi	551-1185
Paint (Mica + Alkyd Resin), Tan	h	551-2692
Paint (Mica + Ti O₂ + Epoxy Resin)	h	551-2511
Paint (Mica + Ti O₂ + Silicone) (Also Irradiated)	i	551-3033
Paint (Mn N H₄ P₂ O₇ + Glue)	h	551-2819
Paint (Mn N H₄ P₂ O₇ + Zn O + Oil)	h	551-2820
Paint (Molochite + Aluminum Phosphate)	hi	551-1194
Paint (Molochite + Potassium Silicate)	hi	551-1193
Paint (Molochite + Zn O + Potasssium Silicate)	h	551-1195
Paint (Molybdate Orange + Alkyd Styrene Resin)	h	551-3513
Paint (Molybdate Orange + Nitrocellulose)	h	551-2498
Paint (Molybdate Orange + Ti O₂ + Nitrocellulose)	h	551-2496
Paint (Molydisulfide + Polymer)	g i	551-2345
Paint (Na F + Potassium Silicate) (Also Irradiated)	i	551-3020
Paint (Na₃ B₄ O₇ + Zn O + Silicate)	h	551-3127
Paint (Na₂ W O₄ + Zn O + Silicate)	h	551-3134
Paint (Nb₂ O₅ + Zn O + Silicate)	h	551-3132
Paint (Ni O + Aluminum Phosphate)	gh	551-2254
Paint (Ni O + Potassium Silicate)	h	551-2248
Paint (Ni O + Ti Cr₂ + Aluminum Phosphate)	g	551-2437
Paint (Ni₃ Al + Aluminum Phosphate)	g	551-2428
Paint (Nitroalkydepoxy Resin), White	h	551-3375
Paint (Nitrocellulose)	j	551-0485
Paint (Nitrocellulose), Black	h	551-0489
Paint (Nitrocellulose), White	h	551-0490
Paint (Nitrocellulose + Acrylic Resin), Black	h	551-3159
Paint (Nitrocellulose + Acrylic Resin), Gray	h	551-3160
Paint (Oil)	j	551-1071
Paint (Oil), White	h	551-0654
Paint (Oleoresin), White	h	551-2663
Paint (Oleoresin - Rubber, Chlorinated), White	h	551-2676
Paint (Paraffin + Copoly(Ethylene - Vinyl Acetate))	i	551-2914
Paint (Paraffin + Copoly(Styrene - Vinyl Toluene))	i	551-2915
Paint (Pb C O₃ + Gel)	h	551-2797
Paint (Pb C O₃ + Melamine Poly(Vinyl Butyral))	hi	551-2838
Paint (Pb C O₃ + Solvent 827)	h	551-2796
Paint (Pb Cr O₄ + Nitrocellulose)	h	551-2497
Paint (Pb Cr O₄ + Ti O₂ + Nitrocellulose)	h	551-2495
Paint (Pb O + Alkyd Resin), Yellow	h	551-2694
Paint (Pb S O₄ + Oil)	h	551-2882
Paint (Pb Ti O₃ + Melamine Poly(Vinyl Butyral))	hi	551-2844
Paint (Pb Ti O₃ + Ti O₂ + Epoxy Resin)	h	551-2514
Paint (Pentaphthalic Resin), Black	h	551-3374
Paint (Permanent Vat Violet + Alkyd Resin), White To Pink	h	551-2682
Paint (Petalite + Potassium Silicate)	h	551-1182
Paint (Phenolic Vinyl Resin), Black	ghi k	551-2739
Paint (Poly(Vinyl Butyral), White	h	551-3376
Paint (Polychloroprene)	h	551-3286
Paint (Polytetrafluoroethylene), Clear	h	551-1208
Paint (Polytetrafluoroethylene + Silicone)	h	551-3141
Paint (Polytetrafluoroethylene + Zn O + Silicone)	h	551-3142
Paint (Polyurethane), Black	g i k	551-2390
Paint (Polyurethane), Clear	h j	551-3070
Paint (Polyurethane), Green	h	551-3067
Paint (Polyurethane), White	ghi k	551-2678
Paint (Potassium Silicate) (Also Irradiated)	hij	551-1206
Paint (Resin), Black Velvet 1518 (W. P. Fuller)	h	551-0429
Paint (Sb₂ O₃)	i	551-3231
Paint (Sb₂ O₃ + Acrylic Epoxy Silicone)	h	551-1225
Paint (Sb₂ O₃ + Alkyd Resin), White	h	551-2686
Paint (Sb₂ O₃ + Melamine Poly(Vinyl Butyral))	hi	551-2837
Paint (Sb₂ O₃ + Potassium Silicate)	ghi k	551-1917
Paint (Sb₂ S₃ + Acrylic Resin), Black	h k	551-3158
Paint (Se + Polyisoprene)	g i	551-2721
Paint (Si C + Si O₂)	g	551-2104
Paint (Si C + Silicone)		551-1112
Paint (Si C + Steatite + Si O₂)	g	551-2804
Paint (Si C + U O₂ + Si O₂)	g	551-2803

Substance Name	Property	Number
Paint (Si O) (Also Irradiated)	h	551-3271
Paint (Si O₂ + Alkyd Resin), White	h	551-2689
Paint (Si O₂ + Potassium Silicate)	ghi	551-1186
Paint (Si O₂ + Ti O₂ + Alkyd Resin)	h	551-3447
Paint (Si O₂ + Ti O₂ + Aluminum Phosphate)	ghi	551-3233
Paint (Si O₂ + Ti O₂ + Silicone) (Also Irradiated)	ghi	551-3260
Paint (Si O₂ + Ti O₃ + Zr O₂ + Aluminum Phosphate)	g i	551-3242
Paint (Si O₂ + Zn O + Zr O₂ + Potassium Silicate)	i	551-3168
Paint (Si O₂ + Zr O₂ + Aluminum Phosphate)	ghi	551-3241
Paint (Si O₂ + Zr O₂ + Potassium Silicate)	g i k	551-3075
Paint (Silicate), White	ghi	551-2031
Paint (Silicone)	gh j	551-2758
Paint (Silicone), Black	ghi k	551-1116
Paint (Silicone), Clear	ghi	551-1207
Paint (Silicone), White	ghi k	551-2736
Paint (Sn O₂ + Aluminum Phosphate) (Also Irradiated)	ghi k	551-1188
Paint (Sn O₂ + Potassium Silicate)	ghi	551-1189
Paint (Sn O₂ + Silicate)	g i	551-3094
Paint (Sn O₂ + Silicone) (Also Irradiated)	i	551-3034
Paint (Sn O₂ + Sodium Silicate) (Also Irradiated)	i	551-3028
Paint (Sn O₂ + Zn O + Silicate)	h	551-3120
Paint (Sr Mo O₄ + Acrylic Resin)	ghi	551-2894
Paint (Sr Mo O₄ + Potassium Silicate)	ghi	551-2890
Paint (Sr Ti O₃ + Ti O₂ + Epoxy Resin)	h	551-2515
Paint (Stearic Acid + Poly(Vinyl Butyral))	i	551-2916
Paint (Ta₂ O₅ + Aluminum Phosphate)	h	551-2251
Paint (Ta₂ O₅ + Potassium Silicate)	hi	551-3166
Paint (Ta₂ O₅ + Zn O + Silicate)	h	551-3129
Paint (Tempil + Si O₂)	g	551-2479
Paint (Thalo Blue + Alkyd Resin), White To Blue	h	551-2680
Paint (Thalo Green + Alkyd Resin), White To Green	h	551-2681
Paint (Ti Al₂ + Aluminum Phosphate)	g	551-2432
Paint (Ti Cr₂ + Aluminum Phosphate)	g	551-2434
Paint (Ti O₂)	ghi	551-1908
Paint (Ti O₂ + (C₄ H₉ O)₄ Ti + Silicone)	i	551-1230
Paint (Ti O₂ + Acrylic Epoxy Resin)	g i k	551-3107
Paint (Ti O₂ + Acrylic Epoxy Silicone)	h	551-1221
Paint (Ti O₂ + Acrylic Resin) (Also Irradiated)	ghi k	551-1175
Paint (Ti O₂ + Acrylic Resin + Nitrocellulose)	h	551-0559
Paint (Ti O₂ + Acrylic Resin + Silicone), White	h	551-1634
Paint (Ti O₂ + Alkyd Formaldehyde Melamine Resin)	h	551-3228
Paint (Ti O₂ + Alkyd Formaldehyde Urea Resin)	h	551-3229
Paint (Ti O₂ + Alkyd Resin) (Also Irradiated)	hi	551-2684
Paint (Ti O₂ + Alkyd Resin + Rubber, Chlorinated)	h	551-2701
Paint (Ti O₂ + Alkyd Silicone) (Also Irradiated)	ghi k	551-3040
Paint (Ti O₂ + Aluminum Phosphate + Potassium Silicate) (Also Irradiated)	h	551-3100
Paint (Ti O₂ + Castor Oil Resin)	hi	551-2391
Paint (Ti O₂ + Cellulose Acetate + Epoxy Resin)	h	551-2510
Paint (Ti O₂ + Cellulose, Cyanoethylated)	i	551-2845
Paint (Ti O₂ + Copoly(Acryionitrile - 1, 1- Dichloroethylene))	h	551-2599
Paint (Ti O₂ + Copoly(Ethyl Acrylate + Methyl Methacrylate))	i	551-2754
Paint (Ti O₂ + Epoxy Polyamide Resin)	h	551-3230
Paint (Ti O₂ + Epoxy Resin) (Also Irradiated)	ghi k	551-1141
Paint (Ti O₂ + Epoxy Resin + Melamine Poly (Vinyl Butyral) + Terpoly(Ethyl Acrylate + Methacrylic Acid + Methyl Methacrylate))	i	551-2769
Paint (Ti O₂ + Epoxy Resin + Terpoly(Ethyl Acrylate - Methacrylic Acid - Methyl Methacrylate))	i	551-2580
Paint (Ti O₂ + Epoxyamine Resin)	h	551-2516
Paint (Ti O₂ + Formaldehyde Melamine Resin)	h	551-1023
Paint (Ti O₂ + Formaldehyde Urea Resin)	h	551-0905
Paint (Ti O₂ + Gel)	h	551-2795
Paint (Ti O₂ + Lead Silicate + Polychloroprene)	h	551-3284
Paint (Ti O₂ + Melamine Poly(Vinyl Butyral)) (Also Irradiated)	ghijk	551-1805
Paint (Ti O₂ + Melamine Poly(Vinyl Butyral) + Terpoly(Ethyl Acrylate + Methacrylic	i	551-2767

Substance Name	Property	Number
Acid + Methyl Methacrylate))	i	551-2767
Paint (Ti O₂ + Nitrocellulose)	hi	551-2283
Paint (Ti O₂ + Oil)	h	551-2880
Paint (Ti O₂ + Poly-1,1-difluoroethylene)	hi	551-3149
Paint (Ti O₂ + Poly(Methyl Methacrylate))	h	551-3214
Paint (Ti O₂ + Polyacrylic Hycar)	h	551-3285
Paint (Ti O₂ + Polychloroethylene Silicone)	h	551-3395
Paint (Ti O₂ + Polychlorotrifluoroethylene)	h	551-2335
Paint (Ti O₂ + Polyimide)	h	551-2336
Paint (Ti O₂ + Polyurethane)	ghi	551-0908
Paint (Ti O₂ + Potassium Silicate) (Also Irradiated)	ghi k	551-1811
Paint (Ti O₂ + Potassium Silicate), Boron Treated	g	551-2765
Paint (Ti O₂ + Si O₂)	g i	551-3208
Paint (Ti O₂ + Silicone) (Also Irradiated)	ghi k	551-2751
Paint (Ti O₂ + Sodium Silicate) (Also Irradiated)	i	551-3023
Paint (Ti O₂ + Solvent 827)	h	551-2794
Paint (Ti O₂ + Terpoly(Ethyl Acrylate + Methacrylic Acid + Methyl Methacrylate))	i	551-2337
Paint (Ti O₂ + Ultramarine + Alkyd Resin), Blue	h	551-2700
Paint (Ti O₂ + Ultramarine + Nitrocellulose)	h	551-2501
Paint (Ti O₂ + Vinyl Resin)	h	551-1118
Paint (Ti O₂ + Zn O + Alkyd Resin), White	h	551-2696
Paint (Ti O₂ + Zn O + Polychloroprene)	h	551-3283
Paint (Ti O₂ + Zn O + Silicate + Silicone) (Also Irradiated)	h	551-3313
Paint (Ti O₂ + Zn O + Silicone)	hi	551-3196
Paint (Ti O₂ + Zr O₂)	g	551-3220
Paint (Ultramarine + Nitrocellulose)	h	551-2499
Paint (Vinyl Resin), White	h	551-2675
Paint (Wollastonite + Potassium Silicate)	h	551-1180
Paint (Y₂ O₃ + Potassium Silicate) (Also Irradiated)	hi	551-3164
Paint (Y₂ O₃ + Sodium Silicate) (Also Irradiated)	i	551-3029
Paint (Zinc Chromate)	ghi	551-3461
Paint (Zinc Chromate + Aldehyde Resin + Alkyl Resin + Phenol, Oil Modified)	h	551-3065
Paint (Zinc Titanate + Potassium Silicate)	i	551-2768
Paint (Zirconium Spinel + Potassium Silicate)	g i	551-3212
Paint (Zirconium Spinel + Potassium Silicate)	i	551-3163
Paint (Zirconium Spinel + Zr O₂)	ghi	551-3211
Paint (Zirconium Spinel + Zr Si O₄)	g i	551-3217
Paint (Zn O)	ghi	551-1521
Paint (Zn O + (C₄ H₉ O)₄ Ti + Silicone)	g i k	551-1234
Paint (Zn O + Acrylic Epoxy Silicone)	h	551-1222
Paint (Zn O + Alkyd Resin)	h	551-2690
Paint (Zn O + Aluminum Phosphate + Potassium Silicate) (Also Irradiated)	h	551-3099
Paint (Zn O + Catalyst + Silicone)	i	551-2609
Paint (Zn O + Copoly-(1,1- Difluoroethylene - Hexafluoropropene))	h	551-1215
Paint (Zn O + Copoly(Chloroethylene - Chlorotrifluoroethylene))	h	551-1220
Paint (Zn O + Copoly(Chlorotrifluoroethylene - 1,1- Difluoroethylene))	h	551-1216
Paint (Zn O + Copoly(Chlorotrifluoroethylene - 1,1- Difluoroethylene) + Polytetrafluoroethylene)	h	551-2358
Paint (Zn O + Copoly(Ethyl Acrylate + Methyl Methacrylate))	i	551-2756
Paint (Zn O + Copoly(Hexafluoropropene - Tetrafluoroethylene))	h	551-1214
Paint (Zn O + Melamine Poly(Vinyl Butyral))	hi	551-1799
Paint (Zn O + Oil)	h i	551-2877
Paint (Zn O + Poly-(1,1- Difluoroethylene))	hi	551-1218
Paint (Zn O + Poly(Methyl Methacrylate))	i	551-3171
Paint (Zn O + Poly(Vinyl Butyral))	i	551-2582
Paint (Zn O + Polytetrafluoroethylene)	h	551-1213
Paint (Zn O + Polyurethane)	h	551-2517
Paint (Zn O + Potassium Silicate) (Also Irradiated)	ghi k	551-0601
Paint (Zn O + Potassium Silicate + Silicone)	gh	551-2822
Paint (Zn O + Silicate) (Also Irradiated)	h	551-3117
Paint (Zn O + Silicone) (Also Irradiated)	ghi k	551-1545
Paint (Zn O + Sodium Silicate) (Also Irradiated)	hi	551-1205
Paint (Zn O + Terpoly(Ethyl Acrylate + Methacrylic Acid + Methyl Methacrylate))	i	551-2760

Substance Name	Property	Number
Paint (Zn O + Zn S + Silicate)	h	551-3118
Paint (Zn O + Zr O₂ + Silicate)	h	551-3119
Paint (Zn O + Zr Si O₄ + Silicate)	h	551-3124
Paint (Zn O + Zr(O H)₄ + Silicate)	h	551-3128
Paint (Zn O, Doped + Silicone)	hi	551-2322
Paint (Zn S)	g i	551-2764
Paint (Zn S + Acrylic Epoxy Silicone)	h	551-2357
Paint (Zn S + Alkyd Silicone)	hi	551-2384
Paint (Zn S + Bakelite Epoxy Resin) (Also Irradiated)	h	551-3081
Paint (Zn S + Cellulose, Cyanoethylated)	i	551-2846
Paint (Zn S + Copoly(Chloroethylene – Chlorotrifluoroethylene))	h	551-1219
Paint (Zn S + Copoly(Chlorotrifluoroethylene – 1,1– Difluoroethylene))	h	551-1217
Paint (Zn S + Copoly(Difluoroethylene – Hexafluoropropene))	h	551-2359
Paint (Zn S + Melamine Poly(Vinyl Butyral) (Also Series Of Additives)	hij	551-1802
Paint (Zn S + Oil)	h	551-2878
Paint (Zn S + Potassium Silicate) (Also Irradiated)	hi	551-1190
Paint (Zn S + Silicone) (Also Irradiated)	hi	551-1227
Paint (Zn S + Sodium Silicate) (Also Irradiated)	g i	551-3022
Paint (Zn Ti O₃ + Copoly(Ethyl Acrylate + Methyl Methacrylate))	i	551-2757
Paint (Zn Ti O₃ + Melamine Poly(Vinyl Butyral))	i	551-2574
Paint (Zn Ti O₃ + Poly(Vinyl Butyral))	i	551-2770
Paint (Zn₂ Si O₄)	g	551-3183
Paint (Zn₂ Ti O₂ + Silicone)	hi	551-3509
Paint (Zn₂ Ti O₄ + Potassium Silicate)	hi	551-3203
Paint (Zn₂ Ti O₄ + Silicone)	hi	551-3175
Paint (Zr O₂)	ghi	551-3219
Paint (Zr O₂ + Aluminum Phosphate)	ghi	551-1198
Paint (Zr O₂ + Polytetrafluoroethylene)	h	551-1211
Paint (Zr O₂ + Potassium Silicate)	ghi	551-1197
Paint (Zr O₂ + Si O₂)	h	551-1199
Paint (Zr O₂ + Silicate)	hi	551-3138
Paint (Zr O₂ + Silicone) (Also Irradiated)	hi	551-3035
Paint (Zr O₂ + Sodium Silicate)	h	551-1201
Paint (Zr Si O₄) (Also Irradiated)	g i	551-3216
Paint (Zr Si O₄ + Potassium Silicate (Also Irradiated)	ghi k	551-1202
Paint (Zr Si O₄ + Silicate)	i	551-2766
Paint (Zr Si O₄ + Sodium Silicate) (Also Irradiated)	g i k	551-3041
Paint (1– Octadecanol + Alkyd Resin + Melamine Poly(Vinyl Butyral))	i	551-2919
Paint (1– Octadecanol + Alkyd Resin + Poly (Vinyl Butyral))	i	551-2920
Paint (2 Pb C O₃ . Pb(O H)₂ + Ethyl Cellulose + Tris-(p-tert-butylphenyl) Phosphate	h	551-2782
Paint (2 Pb C O₃, Pb(O H)₂ + Alkyd Resin), White	h	551-2693
Paint (2– Phthalocyaninatocopper + Acrylic Resin)	h j	551-2632
Paint, A – 31 – 45 see Paint (Ti O₂)		551-
Paint, A R F–2 see Paint (Zn O + Potassium Silicate) (Also Irradiated)		551-
Paint B–1060 A (Boeing) see Paint (Zn O)		551-
Paint, Black	ghijk	551-0501
Paint, Black + Zinc Chromate	hi	551-3462
Paint, Black (M M M)	ghi	551-2600
Paint, Black Da Cote (Murphy)	h	551-2664
Paint, Black Duco Wrought Iron 71 (Dupont)	h	551-0441
Paint, Black Enameloid (Sherwin – Williams)	h	551-2667
Paint, Black Fuller	g	551-2835
Paint, Black Glidair Zapon 131– B–190 B (Glidden)	h	551-2653
Paint, Black Japalac 1208 (Glidden)	h	551-2652
Paint, Black Jet Dry 78 (Globe)	h	551-2656
Paint, Black Kodak 4	h	551-2657
Paint, Black Optical	gh	551-1974
Paint, Black P T 404 (Product Techniques)	g	551-3006
Paint, Black Parsons Optical	ghijk	551-0435

Substance Name	Property	Number
Paint, Black R G I–22818 (Glidden)	hi	551-3451
Paint, Black Ripolin 1505	g	551-3103
Paint, Black Sergeant	hi	551-2394
Paint, Black Sicon (Midland)	ghi	551-1730
Paint, Black Thermoflat (Catalac)	hi	551-3270
Paint, Black Thermolag 500	a e gh n	551-2608
Paint, Black Velvet (M M M)	e ghi	551-0436
Paint, Black Zapon (Glidden) see Paint (C + Nitrocellulose)		551-
Paint, Black 131– B–216 (Glidden)	h	551-2649
Paint, Black 801 (Acme Quality)	h	551-2643
Paint, Black 9099 (Glidden)	hi	551-3455
Paint, Blue	h j	551-2839
Paint, Brown	h	551-0506
Paint, C A 10163 (Sherwin – Williams)	gh	551-3238
Paint, C B – lt – 15 – 10 see Paint (C), Black		551-
Paint, C C L Solar Green (Contains Antimony Sulfide)	h	551-3139
Paint, Clear Rust– Oleum 200	hi	551-3454
Paint, Clear 1301 (Krylon)	h	551-2658
Paint, Cream	h	551-0965
Paint, Dreem 13 N 27 E S4 (Brooklyn)	h	551-2646
Paint, Flat Black Acrylic (Sherwin Williams M49– B C9) see Paint (Acrylic Resin), Black		551-
Paint, Gold	ghi	551-3218
Paint, Gray	ghi	551-0976
Paint, Gray (M M M)	e	551-2602
Paint, Gray Spraint 63 (Minit Spray)	h	551-2670
Paint, Gray 325 (Sprayon)	h	551-2672
Paint, Gray 326 (Sprayon)	h	551-2671
Paint, Green	ghi	551-0505
Paint, Green Rust– Oleum 205	hi	551-3459
Paint Green M I I– E-46061	h	551-3112
Paint Green Marine Corps T T– p–664	h	551-2937
Paint, Hanovia Liquid Bright Palladium	g i k	551-2921
Paint, H–10 (Hughes Aircraft) see Paint (Clay + Silicone)		551-
Paint, J 4881 – J 4877 (Rinshed – Mason)	gh	551-3239
Paint, J 4881 – J 4878 (Rinshed – Mason)	gh	551-3237
Paint, J 15934 (Rinshed – Mason) see Paint (Al + Silicone) (Also Irradiated)		551-
Paint, Koropon Super 765 (Pacific Paint And Varnish)	g i k	551-3382
Paint, Laminar X–4–83 Gray Poly	g	551-3061
Paint, M49 W C17 (Sherwin Williams) see Paint (Ti O₂ + Acrylic Resin) see		551-
Paint, Laminar X–500 4– B–2 Black Poly	g	551-3060
Paint, Nuclear Enterprises 561	h	551-2538
Paint, Olive Drab	h	551-2554
Paint, Olive Drab (M M M)	h	551-3101
Paint, Olive Drab, T T– E-516	h	551-3143
Paint, Orange	h	551-1042
Paint, P T 404 (Product Techniques) see Paint, Black P T 404 (Product Techniques)		551-
Paint, Polyurethane, Clear		551-3070
Paint, Porcelain Enamel	h	551-2720
Paint, Products Techniques 201 see Paint (Epoxy Resin)		551-
Paint, Q 36 K 802 (Rinshed – Mason) see Paint (C + Silicone)		551-
Paint, Red	ghij	551-2841
Paint, Red Dulux	h	551-2278
Paint, Red Rust– Oleum 1210	h	551-2666
Paint, Red Rust– Oleum 215	hi	551-3458
Paint, S–13 see Paint (Zn O + Silicone) (Also Irradiated)		551-
Paint, S–4044 Y (G. E.)	j	551-2761
Paint, Silver Gray Rust– Oleum 208	hi	551-3457
Paint, Silver Herbol	g	551-3262
Paint, Silverbrite No. 55 (Sherwin Williams) see Paint (Al)		551-

Substance Name	Property	Number
R-900		
see Titanium Oxide		551-
R-905 (Du Pont)		
see Paint (Ti O_2)		551-
Radiant Optiglow Screen		
see Paint (Al)		551-
Rare Earth Oxides	g	551-1172
Raven Mat White Fiberglass Screen		
see Resin, White		551-
Raven Mat White Screen		
see Resin, White		551-
Raven Silver Lenticular Screen		
see Paint (Al)		551-
Raven 40		
see Carbon		551-
Regal (Numbered Series) (Cabot)		
see Carbon Black		551-
Resin, Acrylic	g	551-1790
Resin, Blue	h	551-2587
Resin, Copoly-(1,1-difluoroethylene – Hexafluoropropene)	hij	551-2998
Resin, Copoly(Acrylonitrile – Butadiene), Black	h	551-2738
Resin, Copoly(Acrylonitrile – Butadiene), Blue	h	551-2735
Resin, Copoly(Butadiene – Styrene)	h	551-2741
Resin, Copoly(Chlorotrifluoroethylene – 1,1- Difluoroethylene)	h	551-3311
Resin, Copoly(Chlorotrifluoroethylene – 1,1- Difluoroethylene), White	h .	551-3312
Resin, Copoly(Fluoroethylene – Propene)	j	551-2601
Resin, Copoly(Isoprene – 2-methylpropene)	h	551-2742
Resin, Epoxy	g i	551-2635
Resin, Epoxy – Polysulfide	g	551-3072
Resin, Epoxy – Silicone	g	551-3066
Resin, Epoxy, Starch Filled	h	551-3115
Resin, Hexamethoxymethlmelamine	i	551-2918
Resin, J-1190 (Armstrong Cork)	j	551-3289
Resin, Melamine Poly(Vinyl Butyral) (Series Of Additives)	i	551-1789
Resin, Poly-(1,1- Dichloroethylene)	j	551-2827
Resin, Poly-(1,6- Hexanediamine – Sebacic Acid)	j	551-3039
Resin, Poly-(2- Chloro-p- Xylene) (Series Of Additives)	j	551-3007
Resin, Poly-(3,4- Dichlorostyrene)	j	551-3504
Resin, Poly-(4,4'- Isopropylidenediphenol Carbonate) (Also Irradiated)	ghij	551-3261
Resin, Poly(Adipic Acid – 1,6-hexanediamine)	j	551-3026
Resin, Poly(Ethylene Glycol)	j	551-3507
Resin, Poly(Ethylene Glycol – Terephthalic Acid) (Also Irradiated)	ghij	551-2020
Resin, Poly(Methyl Methacrylate)	h j	551-3506
Resin, Poly(o-chlorostyrene)	j	551-3503
Resin, Poly(p- Chlorostyrene)	j	551-1237
Resin, Poly(p-methyl Styrene)	j	551-3501
Resin, Poly(Vinyl Butyral) (Also Irradiated)	hi	551-1796
Resin, Poly(Vinyl Carbazole)	j	551-3505
Resin, Poly(Vinyl Cinnamate)	j	551-2398
Resin, Polyacrylonitrile	j	551-2986
Resin, Polyadenylic Acid Potassium Salt	j	551-3357
Resin, Polyamide	g ijk	551-3056
Resin, Polybutadiene	j	551-2523
Resin, Polychloroethylene	g ij	551-1890
Resin, Polychloroprene	gh	551-2713
Resin, Polychloroprene, Black	h	551-2734
Resin, Polychloroprene, Blue	h	551-2733
Resin, Polycytidylic Acid Potassium Salt	j	551-3359
Resin, Polyethylene	g j	551-3355
Resin, Polyethylene, Clear	g	551-3252
Resin, Polyethylene, Red	g	551-3255
Resin, Polyethylene, White	g	551-3257
Resin, Polyfluoroethylene	hij	551-2508
Resin, Polyguanylic Acid Sodium Salt	j	551-3358
Resin, Polyhexahydro-2 H- Azepin-2-one	j	551-3016
Resin, Polyimide	gh j	551-3108
Resin, Polystyrene	h j	551-3057
Resin, Polystyrene, Water Adsorbate	h	551-2430
Resin, Polytetrafluoroethylene	ghij	551-2266
Resin, Polytetrafluoroethylene, Glass Reinforced	g	551-3009
Resin, Polyurethane	ghij	551-2631
Resin, Polyurethane, White	g i k	551-3281
Resin, Polyuridylic Acid Ammonium Salt	j	551-3380
Resin, Silicone (Also Irradiated)	ghijk	551-1689
Resin, Silicone, Clear	g j	551-2762
Resin, Silicone, Expanded	h	551-3113
Resin, Silicone, Gray	h	551-2750
Resin, Silicone, Sodium Chloride Filled	h	551-3116
Resin, Terpoly-(1,1- Difluoroethylene – Hexafluoropropene – Unknown Monomer)	g	551-2722
Resin, Vinyl, Black	h	551-2752
Resin, Vinyl, Blue	h	551-2732
Resin, Vinyl, Clear	h	551-2343
Resin, Vinyl, Green	h	551-2340
Resin, Vinyl, Red	h	551-2341
Resin, Vinyl, White	h	551-2342
Resin, White	h	551-2584
Resoflex R296		
see Paint (Alkyd Resin), Clear		551-
Rhenium	g	551-3274
Rhodium	ghi	551-2717
D- Ribose-5-phosphate Barium Salt	j	551-3361
Rinshed- Mason J 4881- J 4877		
see Paint J 4881- J 4877 (Rinshed – Mason)		551-
Rinshed- Mason J 4881- J 4878		
see Paint J 4881- J 4878 (Rinshed – Mason)		551-
Ripolin 1505		
see Paint, Black Ripolin 1505		551-
Rokide (General)	g i k	551-3352
Rokide A ($Al_2 O_3$)	a e ghi k n	551-0154
see also Aluminum Oxide		102-
Rokide C ($Al_2 O_3 + Ca O + Cr_2 O_3 + Fe_2 O_3 + Mg O + Na_2 O + Si O_2 + Ti O_2$)	a e g n	551-2763
Rokide Coating		
see Rokide (General)		551-
Rokide M A ($Al_2 O_3 + Cr_2 O_3 + Fe_2 O_3 + Mg O + Na_2 + Si O_2 + Ti O_2$)	a e g i n	551-1290
Rokide Z ($Zr O_2$)	a e ghi k n	551-0153
see also Zirconium Oxide		122-
Rokide Z S ($Al_2 O_3 + Ca O + Fe_2 O_3 + Na O_2 + Si O_2 + Ti O_2 + Zr O_2$)	a e ghi n	551-1289
Rubber	h	551-3309
Rubber, Natural	gh	551-2707
Rubber, Polyisoprene	g	551-2708
Rubber, White	h	551-3310
Rubidium	h	551-1673
Rubidium Azide	j	551-2935
Rubidium Bromide	h j	551-2893
Rubidium Chloride	j	551-2897
Rubidium Fluoride	j	551-2902
Rubidium Iodide	h j	551-2526
Rust- Oleum 200		
see Paint, Clear Rust- Oleum 200		551-
Rust- Oleum 205		
see Paint, Green Rust- Oleum 205		551-
Rust- Oleum 208		
see Paint, Silver Gray Rust- Oleum 208		551-
Rust- Oleum 215		
see Paint, Red Rust- Oleum 215		551-
Rust- Oleum 225		
see Paint, White Rust- Oleum 225		551-
Rust-oleum Red 1210		
see Paint, Red Rust- Oleum 1210		551-
Ruthenium	h	551-3342
S A 9184 (Andrew Brown Co.)		
see Paint, White		551-
S E 551 – N (G E)		
see Resin, Silicone (Also Irradiated)		551-
S p-500		
see Zinc Oxide (Also Irradiated)		551-
S R 585 (G E)		
see Adhesive, S R 585 (G E)		551-

Substance Name	Property	Number
S–13		
see Paint (Zn O + Silicone) (Also Irradiated)		551–
S–13 M		
see Paint (Zn O + Silicone) (Also Irradiated)		551–
S–4044(G. E.)		
see Paint, S–4044 'r (G. E.)		551–
S–6100		
see Ceramic S–6100 (Solar Aircraft)		551–
Samarium	j	551–3280
Samarium Oxide	h	551–2476
Sandoz – B K Dyed Anodized Aluminum	g i	551–2333
Sandoz– O A Dyed Anodized Aluminum	g i k	551–3251
Saran White Paint		
see Paint (Ti O₂ + Copoly(Acrylonitrile – 1,		
1- Dichloroethylene))		551–
Sauereisen No. 1		
see Ceramic (Si O₂ + Sodium Silicate)		551–
Sauereisen No. 19		
see Ceramic (Clay + Ponolith + Si O₂ +		
Sodium Silicate)		551–
Sauereisen No. 6		
see Ceramic (Mullite + Sodium Silicate)		551–
Sauereisen No. 74		
see Ceramic (Al₂ O₃ + Sodium Silicate)		551–
Scotch Tape 425		
see Tape, Aluminum, Scotch (Numbered Series)		551–
Scotch 425		
see Tape, Aluminum, Scotch (Numbered Series)		551–
Selenium	hij	551–0850
Selenium Zinc Intermetallic	h j	551–1684
Semon Bache Opal Screen		
see Silicon Oxide (Also Irradiated)		551–
Sherwin- Williams C A 10163		
see Paint C A 10163 (Sherwin – Williams)		551–
Sicon White		
see Paint (Ti O₂ + Silicone) (Also Irradiated)		551–
Silastic 280		
see Adhesive, Silastic 280 (Dow Corning)		551–
Silicide	g	551–0689
Siliclad		
see Resin, Silicone (Also Irradiated)		551–
Silicon	gh j	551–2313
Silicon – Tantalum	ghi	551–0253
Silicon Alloy (Fe)	g	551–3063
Silicon Carbide	ghij	551–2614
Silicon Carbide – Silicon Oxide, Aluminum Phosphate Bonded	g	551–1386
Silicon Carbide, Aluminun Phosphate Bonded	g	551–0814
Silicon Monox		
see Silicon Oxide (Also Irradiated)		551–
Silicon Nitride	hij m	551–3213
Silicon Oxide – Tantalum Oxide, Multilayer	h	551–3331
Silicon Oxide – Thorium Oxide, Multilayer	h	551–3409
Silicon Oxide – Titanium Oxide	h	551–2527
Silicon Oxide – Titanium Oxide, Multilayer	h	551–3330
Silicon Oxide (Also Irradiated)	e ghijk	551–0577
Silicon Oxide ← Phosphorus	m	551–2464
Silicon Tantalum Intermetallic	gh	551–2304
Silicon Titanium Intermetallic	g	551–2310
Silicon Tungsten Intermetallic	gh	551–2309
Silicon Zirconium Intermetallic	h	551–3264
Silicon, Doped	gh	551–0226
Silicone Rubber	gh	551–2138
Siliconized Graphite	g	551–0351
Silver (Also Irradiated)	ghijk	551–2416
Silver Alloy (Au)	h	551–3055
Silver Alloy (In)	h j	551–2361
Silver Black	h j	551–3338
Silver Bromide	h j	551–2638
Silver Chloride	hij	551–2630
Silver Iodide	j	551–2647
Silver Nitrate	j	551–3291
Silver Sulfide	ghi	551–0702
Silvergain Screen (Stewart Film.)		
see Paint (Al)		551–
Skyspar Enamel, White		
see Paint (Ti O₂ + Epoxy Resin)		
(Also Irradiated)		551–
Skyspar, White (A–423)		
see Paint (Ti O₂ + Epoxy Resin)		
(Also Irradiated)		551–
Socofal		
see Aluminum (Also Irradiated)		551–
Sodium	h j	551–1722
Sodium Bromide	j	551–2895
Sodium Carbonate	h	551–2961
Sodium Chloride	ghijk	551–0648
Sodium Dichromate	g i	551–2334
Sodium Fluoride		551–2904
Sodium Hexafluoroaluminate – Zinc Sulfide, Multilayer	h	551–2193
Sodium Hexafluoroaluminate – Zinc Sulfide	h	551–2407
Sodium Iodide	j	551–2892
Sodium Salicylate	h j	551–3481
Sodium Selenite	j	551–3077
Sodium Silicate, Expanded	h	551–3114
Soiled Surface (Also Irradiated)	hij	551–2415
Solar Aircraft S–6100		
see Ceramic S–6100 (Solar Aircraft)		551–
Solar Cell, Silicon, Nickel Electrode Base		
(International Rectifier Co.)		
see Silicon, Doped		551–
Solar S10 – 33 A		
see Aluminum (Also Irradiated)		551–
Solar S11 – 86 B		
see Paint (Ag)		551–
Spheron 6 (Cabot)		
see Carbon Black		551–
Spraint 63 (Minit Spray)		
see Paint, Gray Spraint 63 (Minit Spray)		551–
Sprayon 325		
see Paint, Gray 325 (Sprayon)		551–
Sprayon 326		
see Paint, Gray 326 (Sprayon)		551–
Stearic Acid	h	551–2987
Steel	h j	551–2281
Sterling (Numbered Series) (Cabot)		
see Carbon Black		551–
Stewart Film. Matte Blue Screen		
see Resin, Blue		551–
Stewart Film. Silvergain Screen		
see Paint (Al)		551–
Strontium	h	551–2396
Strontium Fluoride	h	551–2234
Strontium Nitrate	h	551–2959
Strontium Oxide	i	551–1572
Strontium Titanium Oxide	g i	551–0813
Sulfur	h j	551–2533
Sulfur Over Zinc Sulfide	j	551–2532
Super Kemtone 793 (Sherwin– Williams)		
see Paint, White Super Kemtone 793		
(Sherwin – Williams)		551–
Supercarbovar (Cabot)		
see Carbon Black		551–
Superlith X X X N		
see Zinc Sulfide		551–
Sylcor R–505 C		
see Aluminum – Tin		551–
Sylcor R–512 A		
see Niobium Silicon Intermetallic		551–
T R W Coating (Thompson– Ramo– Wooldridge, Tapco		
Division)		
see Chromium Silicon Titanium Intermetallic		551–
T S–137 (Mc Donnell Co.)		
see Niobium Silicon Intermetallic		551–
T T – P28		
see Paint (Al + Silicone) (Also Irradiated)		551–
T T– T–425 A		
see Titanium Oxide		551–

Substance Name	Property	Number
–benzothiazolinium Iodide – 1,1' Diethyl–2,		
2' Quinocyanine Iodide Dyed	j	551–2319
Zinc Sulfide	h j	551–2367
Zinc Sulfide, Doped	k	551–3069
Zinc Zirconium Silicate	h	551–3191
Zirconium	h	551–1865
Zirconium Carbide (Also Irradiated)	ghi	551–1327
Zirconium Molybdenum Oxide	j	551–2564
Zirconium Oxide (Also Irradiated)	a ghij	551–1321
Zirconium Oxide, Yttrium Oxide Stabilized	g	551–3477
Zirconium Silicate (Also Irradiated)	ghi	551–1331

BULK AND PROCESSED FABRICS   CLASS 621

Substance Name	Property	Number
Agave Sisalana		
(Syn. For– Sisal)		
see Sisal Fabric		621–
(Aluminum – Silk – Tetoron) Fabric, Silver	h	621–0081
(Aluminum – Tetoron) Fabric, Silver	h	621–0080
Aluminum Screen	j	621–0096
Balloon Fabric	ghij	621–0275
Benax 2 A 1		
(Syn. For– Additive To Wool)		
see Wool Fabric (With Additive), Moist		621–
β Cloth		
see Glass Fabric		621–
Burlap	h	621–0004
Burlap, Chlorophyll Stained	h	621–0108
Canvas, Green	h	621–0041
Canvas, Olive	h	621–0098
Carbon Fabric	a g	621–0073
Coat Fabric	a	621–0090
Copper Screen	j	621–0029
(Cotton – Rayon) Fabric, Black	h	621–0076
(Cotton – Wool) Fabric	a	621–0091
Cotton Fabric	a h	621–0062
Cotton Fabric, Black	h	621–0352
Cotton Fabric, Blue	h	621–0292
Cotton Fabric, Duck	h	621–0282
Cotton Fabric, Gray	h	621–0393
Cotton Fabric, Herringbone	h	621–0092
Cotton Fabric, Khaki	h	621–0295
Cotton Fabric, Olive Drab	h	621–0014
Cotton Fabric, White	hij	621–0040
Dowfax 2 A 1		
(Syn. For– Additive To Wool)		
see Wool Fabric (With Additive), Moist		621–
Fabric (General)	a ghij	621–0217
Fabric, Black	h	621–0088
Fabric, Gray	hij	621–0083
Fabric, Green	h	621–0106
Fabric, Movie Screen	h	621–0099
Fabric, Olive	h	621–0103
Fabric, Red	h	621–0102
Fabric, Tan	h	621–0107
Fabric, White	h	621–0101
Fabric, Yellow	h	621–0105
Fiber	a	621–0077
Fortisan–acetate		
see Rayon Fabric		621–
Glass Fabric	a ghi n	621–0028
Gold Screen	h	621–0042
Graphite Fabric	a g j	621–0308
Gunny		
see Burlap		621–
Hessian		
see Burlap		621–
Jute		
see Burlap		621–
Karma Fabric	g i	621–0003
Linen Fabric	a	621–0018
Metallic Screen	hij	621–0188
Nickel Screen	h j	621–0214
(Nylon – Polypropene) Fabric	h	621–0104

Substance Name	Property	Number
Nylon Fabric	a	621–0227
Oil Cloth, Silk	j	621–0085
Perlon Stocking	j	621–0043
Petryanov Fabric	a	621–0093
Plastic Screen	j	621–0097
Pluton Cloth	a	621–0306
(Poly(Ethylene Glycol – Terephthalic Acid) –		
Serge – Wool) Fabric	a de	621–0125
(Poly(Ethylene Glycol – Terephthalic Acid) –		
Serge – Wool) Fabric, Moist	a de	621–0126
Poly(Ethylene Glycol – Terephthalic Acid) Fabric	a de	621–0122
Poly(Ethylene Glycol – Terephthalic Acid) Fabric, Moist	a de	621–0123
Polyamide Fabric	a d h j	621–0023
Polyamide Fabric, Green	h j	621–0115
Polyamide Fabric, Orange	h j	621–0114
Polyamide Fabric, White	h j	621–0112
(Rayon – Silk) Fabric, Black – Gold	h	621–0078
(Rayon – Silk) Fabric, Black – Silver	h	621–0082
(Rayon – Silk) Fabric, Pink – Silver	h	621–0079
Rayon Fabric	i	621–0002
Rayon, Cupra		
see Rayon Fabric		621–
Rayon, Lusterless		
see Rayon Fabric		621–
Rayon, Viscose		
see Rayon Fabric		621–
Serge		
see Wool Fabric, Serge		621–
Silicon Oxide (Si O₂) Fabric, Green	h	621–0110
Silk Fabric	a	621–0001
Sisal Fabric	h	621–0111
Stainless Steel Screen	a g	621–0316
Titanium Screen	a	621–0086
Tungsten Screen	a	621–0087
Velvet, Black	h	621–0075
Viscose Fabric		
see Rayon Fabric		621–
Viscose Rayon		
see Rayon Fabric		621–
Wool Fabric	a de	621–0024
Wool Fabric (With Additive), Moist	e	621–0189
Wool Fabric, Blue	h	621–0283
Wool Fabric, Covert	h	621–0376
Wool Fabric, Elastique	h	621–0377
Wool Fabric, Flannel	h	621–0012
Wool Fabric, Melton	h	621–0383
Wool Fabric, Moist	a de	621–0124
Wool Fabric, Olive Drab	h	621–0017
Wool Fabric, Serge	h	621–0378
Wool Fabric, Tweed	a	621–0089
Wool Felt	a	621–0100
Worsted Fabric	a	621–0129
Worsted Fabric, Blue	h	621–0094
Worsted Fabric, Khaki	h	621–0218

BULK SURFACE FINISHES–ADHESIVES, COATINGS, INKS, LACQUERS, PAINTS, VARNISHES, AND OTHERS   CLASS 651

Substance Name	Property	Number
A A–1063– D Wax		
see Wax, Synthetic		651–
Acrawax C		
see Wax, Synthetic		651–
Adhesive (General)	a de n p	651–0544
Adhesive, Apco 1219	j	651–0318
Adhesive, Armstrong A–4	n	651–0319
Adhesive, Bostik m–356	n	651–0289
Adhesive, Cybond 4000	a	651–0275
Adhesive, Cybond 4000 (Series Of Additives)	a	651–0276
Adhesive, Delta Bond 152	a	651–0285
Adhesive, Fortafix	a	651–0284
Adhesive, Pyroxylin	j	651–0070
Advawax 280		
see Wax, Synthetic	.	651–

Substance Name	Property	Number
Paraffin Scale		
see Wax, Paraffin		651–
Paraffin Scale		
see Wax, Ceresin		651–
Paraffin Scale		
see Wax, Ozokerite		651–
Parawax	p	651–0074
Pigment (General)	i	651–0467
Pigment, Basic Silicate White Lead	h	651–0053
Pigment, Iron Oxide	h	651–0011
Pigment, Ultramarine Blue	h	651–0341
Shofu Parrafin Wax		
see Wax, Paraffin		651–
Ultraflex		
see Wax, Synthetic	.	651–
Varnish	+	
see Paint		651–
Wax, Ceresin	n	651–0478
see also Wax, Ozokerite		651–
see also Wax, Paraffin		651–
Wax, Ozokerite	n	651–0479
see also Wax, Ceresin		651–
see also Wax, Paraffin		651–
Wax, Paraffin	a j nop	651–0476
see also Wax, Ceresin		651–
see also Wax, Ozokerite		651–
Wax, Paraffin (Series Of Additives)	j	651–0068
Wax, Paraffin, Chlorinated	p	651–0564
Wax, Synthetic	n	651–0025
Wood Oil		
(Syn. For– Tung Oil)		
see Paint, Alkyd Phenolic Resin – Tung Oil 7301 (G. E.)		651–
see Paint 7301 (G. E.)		651–

PROCESSED COMPOSITES – INSULATIONS, PAPER AND WOOD PRODUCTS, AND WALL STRUCTURES CLASS 661

Substance Name	Property	Number
A D I–17		
see (Aluminum – Aluminum Oxide – Asbestos – Carbon – Silicon Nitride, Fibrous), A D I–17 (Insulation Powder)		661–
Aerated		
see Expanded		
Aerocor		
see Glass Fiberboard		661–
Air Space, Bounded By Aluminum	a	661–0389
Air Space, Bounded By Aluminum, Asbestos Spaced	a	661–0451
Air Space, Bounded By Aluminum, Paper Spaced	a	661–0450
Air Space, Bounded By Glass	a g	661–0183
Air Space, Bounded By Wood	c	661–0186
(Aluminum – Aluminum Oxide – Asbestos – Carbon – Silicon Nitride, Fibrous), A D I–17 (Insulation Powder)	a	661–0152
(Aluminum – Glass Fiber Blanket – Plastic Laminate) (Insulation)	a	661–0448
Arbolit		
see Woodboard		661–
(Asbestos – Calcium Silicate) (Insulation)	a	661–0430
(Asbestos – Diatomite) (Insulation)	a	661–0164
(Asbestos – Magnesium Oxide) (Insulation)	a	661–0122
Asbestos – Vermiculite	a	661–0442
see also Mica, Vermiculite		
(Syn. For– Vermiculite)		521–
Asbestos Board	a de	661–0291
Asbestos Cement Board	a d	661–0119
Asbestos Felt	a	661–0010
Asbestos Millboard		
see Asbestos Board		661–
Asbestos Paper	a g	661–0285
Asbestos Paper, Corrugated	a	661–0297
Asbestos, Cement Covered	a	661–0124
Asbestos, Phenolic Resin Bonded (Insulation)	a	661–0249

Substance Name	Property	Number
B. P. Board		
see Fiberboard		661–
Basalt Board	a	661–0434
see also Basalt		521–
Basalt Paper	a	661–0483
Boron Nitride Felt	a	661–0461
Bridge, Wood	h	661–0187
Building Board	a	661–0287
Building Material (General)	a de ghi	661–0042
Bulrush, Cloth Bounded (Insulation)	a	661–0472
Cabots Quilt		
see Eelgrass, Burlap Bounded (Cabots Quilt)		661–
Calsilite (Insulation)	h	661–0465
Cane Fiberboard	a de h j	661–0060
Carbon Fiber Felt	g	661–0320
see also Carbon, (Fibrous)		101–
Carbon Paper	a	661–0419
Cardboard	a de	661–0051
Cardboard, White	h	661–0380
Cellufoam (Insulation)	a	661–0413
Cellufoam, Cellular		
see Wood Fiberboard		661–
Cellular		
see Expanded		
Celotex		
see Cane Fiberboard		661–
Ceramic Paper	a	661–0420
Cereal Fiberboard	a	661–0108
(Clay – Perlite), Expanded, Phosphate Bonded (Insulation)	a n	661–0433
Cork, Baked Slab	a	661–0041
Cork, Baked, Granulated	a	661–0375
Cork, Bonded		
see Corkboard		661–
Cork, Cement Covered	a	661–0188
Cork, Pitch Bonded	a	661–0318
Cork, Rubber Bonded	a h	661–0143
Corkboard	a de	661–0001
Corkboard, Expanded	a	661–0155
Corkboard, Moist	a	661–0350
(Cotton – Polyamide – Polyamide, Rubberized – Polyamide – Poly(Ethylene Glycol – Terephthalic Acid), Aluminized – Felt – Polyamide) (Insulation)		661–0349
(Cotton – Polyamide – Polyamide, Rubberized – Polyamide – Poly(Ethylene Glycol – Terephthalic Acid), Aluminized, Alternated With Cotton – Felt – Polyamide) (Insulation)	a	661–0392
(Cotton – Polyamide – Polyamide, Rubberized – Polyamide – Felt – Polyamide) (Insulation)	a	661–0393
(Cotton – Polyamide – Polyamide, Rubberized – Poly(Ethylene Glycol – Terephthalic Acid), Aluminized – Polyamide) (Insulation)	a	661–0396
Dry Zero (Insulation)	a	661–0412
Durestos		
see Asbestos, Phenolic Resin Bonded (Insulation)		661–
Eelgrass Blanket, Paper Bounded	a	661–0086
Eelgrass, Burlap Bounded (Cabots Quilt)	a	661–0347
Egyptian Fiber		
see Fiberboard		661–
Evansfoam (Insulation)	a	661–0456
Felt, Black	h	661–0334
Felt, Refractory, W R p– X– A Q		661–0422
Felt, White	h	661–0443
Fiberboard	a n	661–0008
Fiberfrax		
see Fiberfrax Felt		661–
Fiberfrax Felt	a h	661–0367
see also Ceramic, Fibrous, Fiberfrax ($Al_2 O_3$, $B_2 O_3$, $Na_2 O$, $Si O_2$)		504–
Fiberoid		
see Fiberboard		661–
Fibrofelt, Felted Vegetable Fibers	a	661–0355
Flaxlinum, Felted Vegetable Fibers	a	661–0354

Substance Name	Property	Number
Floor, Concrete, Filled Poly(Vinyl Acetate) Covered	d	661-0479
Floor, Concrete, Filled Rubber Covered	d	661-0478
Flooring, (Cement – Rubber) (Series Of Fillers)	a	661-0321
Flooring, Asphalt (Series Of Fillers)	a	661-0246
Flooring, Bituminous (Series Of Fillers)	a	661-0054
Flooring, Magnesite, Wood Dust Filled	a	661-0066
Flooring, Poly(Vinyl Acetate) (Series Of Fillers)	a de	661-0445
Flooring, Rubber (Series Of Fillers)	a de	661-0452
Foamed		
see Expanded		
Glass Fiber Blanket	a d g	661-0090
see also Fibrous Glass		507-
Glass Fiber, Resin Bonded	a	661-0222
Glass Fiberboard	a	661-0016
see also Fibrous Glass		507-
Graphite Fiber Felt	a g	661-0366
see also Graphite, (Fibrous)		101-
Gray Fiber		
see Fiberboard		661-
Gypsum Board	a de gh	661-0470
see also Gypsum		521-
see also Plaster Of Paris		526-
H-1502 Insulation, Expanded	a	661-0455
Hairfelt	a	661-0358
Hairfelt And Other Fibers, Paper Bounded		
(Keystone Hair)	a	661-0345
Hard Fiber		
see Fiberboard		661-
Hardboard	a	661-0039
Hastelloy X Wire Wool, Inconel Bonded (Insulation)	a	661-0307
Horn Fiber		
see Fiberboard		661-
Insul Board	a	661-0034
Insulation (General)	a de jk	661-0040
Insulation, Expanded	a	661-0468
Insulation, Fibrous	a	661-0295
Insulation, Granular	a	661-0327
Insulation, Moist	a	661-0047
Insulation, Multilayer	a	661-0397
Insulite	a	661-0058
J– M 1999, 2002 Insulation		
see Min– K (Asbestos – Binder – Opacifier –		
Silicon Oxide) (Numbered Series)		
(Also Irradiated)		661-
Kaolin Board	a	661-0444
see also Clay, Kaolin		521-
Kaolin Paper	a	661-0484
see also Clay, Kaolin		521-
Keystone Hair		
see Hairfelt And Other Fibers, Paper Bounded		
(Keystone Hair)		661-
Kimoloboard	a	661-0182
Kimsul, Wood Pulp Blanket	a	661-0360
I– K Insulation (Asbestos – Binder –		
Opacifier – Silicon Oxide)	a	661-0411
Leatheroid		
see Fiberboard		661-
Linofelt, Paper Bounded Vegetable Fibers	a	661-0346
Linoleum	i	661-0075
Lithboard (Binder – Mineral Wool – Vegetable Fibers)	a	661-0116
m-31 Insulation (Asbestos – Potassium Titanate –		
Silicon Oxide)	gh	001-0455
Magnesium Oxide Base (Insulation)	a	661-0243
see also Magnesium Oxide		119-
Marinite (Asbestos – Binder – Silicon Oxide)	a e	661-0417
Masonite	a h	661-0004
Masonite, Moist	a	661-0111
Mastic, Bitumen – Rubber	a de	661-0425
Microtherm 20 Cr (Asbestos – Binder –		
Chrome Oxide – Opacifier – Silicon Oxide)	a	661-0416
Min– K (Asbestos – Binder – Opacifier –		
Silicon Oxide) (Numbered Series)		
(Also Irradiated)	a	661-0017
Min– K (Asbestos – Binder – Opacifier – Silicon Oxide)		

Substance Name	Property	Number
(Numbered Series), Expanded	a	661-0474
Mineral Cotton Board		
see Mineral Wool Board (Also Irradiated)		661-
Mineral Wool Board (Also Irradiated)	a d	001-0003
see also Mineral Wool		501-
Mineral Wool Board, Asphalt Bonded	a d	661-0014
Mineral Wool Felt	a	661-0330
Mineral Wool, Bonded		
see Mineral Wool Board (Also Irradiated)		661-
P F Insulation (Numbered Series)		
see Glass Fiber, Resin Bonded		661-
Palarite		
see Silicon Oxide Fiber, Carbon Bonded		661-
Paper	a e gh j n	661-0050
Paper Board, White	h	661-0480
Paper, Bakelized	a	661-0038
Paper, Black	h j	661-0067
Paper, Bond, Blue	j	661-0439
Paper, Bond, Fluorescent	j	661-0441
Paper, Bond, Pink	j	661-0426
Paper, Bond, White	h j	661-0440
Paper, Bond, Yellow	j	661-0432
Paper, Building	a	661-0447
Paper, Condenser	e	661-0482
Paper, Electrical Conducting	g	661-0457
Paper, Felt, Creosote Coated	a	661-0449
Paper, Gray	h	661-0467
Paper, Green	h	661-0435
Paper, Kraft	h j	661-0415
Paper, Newsprint	g j	661-0438
Paper, Oil Impregnated	a d	661-0052
Paper, Pink	h	661-0436
Paper, Print	h	661-0170
Paper, Print, White	h	661-0446
Paper, Rayon Fiber, Pink	h	661-0009
Paper, Tissue, Pink	h	661-0030
Paper, Wax	a d	661-0410
Paper, White	ghij	661-0048
Paper, Writing	j	661-0171
Pegulan Floor Covering	g	661-0477
Pergaloid (German Wrapping)	a d	661-0429
Plaster Board	a e g	661-0019
Plywood	a d gh	661-0049
Plywood, Fir	h	661-0037
Plywood, Phenol Formaldehyde Resin Bonded	a de	661-0462
Polychoroethylene, Cement Bonded	a	661-0485
Pregwood	n	661-0437
Pressboard	a d	661-0137
Pressboard, Oil Impregnated	a	661-0431
Pulpboard	a	661-0104
Pyroid (Insulation)	a	661-0424
Pyrotex (Insulation)	a	661-0423
Q Felt		
see Quartz Fiber Felt		661-
Quartz Fiber Felt	a	661-0331
see also Quartz		521-
see also Silicon Oxide, Fibrous		122-
Red Fiber		
see Fiberboard		661-
Relin		
see Linoleum		661-
Remika (Insulation) (Czech)	a	661-0428
Rock Cork		
see Mineral Wool Board (Also Irradiated)		661-
Rock Wool Board		
see Mineral Wool Board (Also Irradiated)		661-
Roofing, Asphalt (Felt Saturated With Asphalt)	a	661-0113
Roofing, Iron, Red	h	661-0232
Roofing, Shingle	h	661-0315
Roofing, Shingle, Black	h	661-0453
Roofing, Shingle, Green	h	661-0454
Roofing, Tar	a	661-0418
Roofing, Tile, Red	h	661-0223
Satincote (Insulation)	a	661-0476

Substance Name	Property	Number
Satincote (Insulation), Moist	a	661–0351
Sawdust, Cement Bonded	a	661–0043
Sawdust, Magnesium Oxychloride Cement Bonded	a	661–0156
Shoe Sole Material	a	661–0398
Silicate Cotton Board		
see Mineral Wool Board (Also Irradiated)		661–
Silicon Oxide Paper	a	661–0421
Silicon Oxide Fiber, Carbon Bonded	a e	661–0471
Slag Wool Board		
see Mineral Wool Board (Also Irradiated)		661–
Stone Felt (Insulation)	a	661–0414
Stramit, Moist (Russian)	a	661–0395
Superex (Insulation)	a	661–0409
Teledeltos Paper		
see Paper, Electrical Conducting		661–
Textan, Rubber Composition (Insulation)	a	661–0114
Thermocole	a	661–0107
Tipersul		
(Syn. For– Potassium Titanate (Fibrous)		
see m–31 Insulation (Asbestos – Potassium Titanate – Silicon Oxide)		661–
transite		
see Asbestos Cement Board		661–
T30 I– R Laminate (Carborundum)		
see Insulation, Multilayer		661–
U Foam (Insulation)	a	661–0466
Vegetable Fiber Board		
see Fiberboard		661–
Velox		
see Paper, Print		661–
Vetresit (Insulation) (Swiss)	a	661–0427
Vulcanized Fiber		
see Fiberboard		661–
W R p– X– A Felt (Refractory Prod.)		
see Felt, Refractory, W R p– X– A Q		661–
Wall	a	661–0193
Wall, (Brick – Expanded Concrete)	a	661–0406
Wall, (Brick – Expanded Concrete), Moist	a	661–0407
Wall, (Ebonite, Expanded – Fiberboard)	a	661–0408
Wall, (Fiberglass Insulation)	a	661–0459
Wall, (Lath – Plaster – Mineral Wool) (Insulation)	a	661–0211
Wall, (Mineral Wool Insulation)	a	661–0460
Wall, Brick	a d g	661–0292
Wall, Brick, Moist	a	661–0405
Wall, Concrete (Cement – Clay – Clinker – Fuel Ash), Moist	a	661–0403
Wall, Concrete (Cement – Crushed Brick)	d	661–0260
Wall, Concrete (Cement – Gravel)	d	661–0056
Wall, Concrete (Cement – Gravel), Moist	a	661–0404
Wall, Concrete (Cement – Slag)	a	661–0399
Wall, Concrete (Cement – Slag), Moist	a	661–0402
Wall, Concrete, Expanded	a	661–0400
Wall, Concrete, Expanded, Moist	a	661–0401
Wall, Concrete, Expanded, Plastered	d	661–0078
Wallboard	a de	661–0117
Weatherwood (Insulation)		
see Wood Felt		661–
Whalebone Fiber		
see Fiberboard		661–
Wood Block (Pavement)	h	661–0245
Wood Board	a	661–0359
Wood Board, Moist	a	661–0394
Wood Board, Oil Impregnated	a	661–0473
Wood Felt	a	661–0356
Wood Felt, Moist	a	661–0475
Wood Fiber Blanket	a	661–0299
Wood Fiber Blanket, Paper Bounded	a	661–0109
Wood Fiberboard	a n	661–0002
Wood Fiberboard, Moist	a	661–0352
Wood Shaving Board	a	661–0106
Wood, Processed	a no	661–0481
Wool Blanket, Animal, Paper Bounded	a	661–0094
Wool Felt, Paper Bounded	a	661–0336
Zenitherm		
see (Asbestos – Diatomite) (Insulation)		661–

Substance Name	Property	Number
Zytel		
see (Cotton – Polyamide – Polyamide, Rubberized – Polyamide – Poly(Ethylene Glycol – Terephthalic Acid), Aluminized, Alternated With Cotton – Felt – Polyamide) (Insulation)		
Cotton – Felt – Polyamide) (Insulation)		661–

USE OF SEARCH PARAMETERS

This part of the Retrieval Guide in essence represents the computer memory search element of a retrieval system. Part B is arranged in 16 chapters, each chapter representing a specific property. Within each chapter substances are listed by their increasing seven-digit substance numbers, followed by physical state within each substance, subject area of the work, the language, the temperature range, and the year. The last column gives the TPRC serial number pertaining to the document retrieved. The sample code designations for the various search parameters need not be memorized since they are interpreted at the bottom of each page of this part of the volume.

It is readily evident that substance numbers link this portion of the volume with the *Materials Directory* (Part A), and the TPRC numbers provide the link to the *Bibliography* (Part C). The general procedure for retrieval of references in making literature searches is as follows:

1. Locate and record the seven-digit substance number from the *Materials Directory* for the material desired in a given search.
2. Locate and record selected TPRC numbers within the appropriate property chapter in this portion of the retrieval guide.
3. Use as many of the five search parameters desired for the search and record the corresponding TPRC numbers.
4. Use the selected TPRC numbers to locate the complete bibliographic citations in the *Bibliography* (Part C).

Chapter 1 Thermal Conductivity

Substance Number	Phys. State	Sub-ject	Lan-guage	Temper-ature	Year	TPRC Number
526-0004	S	D	E	N	1966	39355
0004	P	C	R	N	1968	51661
0004	P	C	E	N	1968	51002
0004	M	D	E	N	1966	42695
0005	P	C	F	N	1965	42373
0006	S	C	E	N	1968	53178
0006	S	C	E	N	1969	56053
0006	P	C	F	N	1965	42373
0006	M	C	E	U	1968	58400
0008	S	D	E	N	1966	34197
0010	P	D	E	N	1949	61706
0015	M	D	R	N	1969	57018
0017	S	D	E	N	1969	54375
0017	S	D	R	N	1966	59500
0017	S	D	E	N	1966	59501
0018	S	D	R	N	1965	39987
0018	S	D	E	N	1965	39988
0019	S	G	G	N	1964	37646
0019	S	D	E	N	1943	40228
0020	S	D	E	N	1943	40228
0023	S	D	E	N	1966	34197
0034	S	D	R	N	1963	35944
0034	S	D	F	N	1963	35945
0034	S	D	E	N	1943	40228
0034	S	D	E	N	1969	54423
0034	S	D	R	N	1969	58439
0034	P	C	F	N	1965	42373
0034	M	D	E	N	1964	37058
0034	M	D	E	N	1967	43052
0034	M	D	R	N	1967	45418
0034	M	D	R	N	1961	48286
0034	M	D	O	N	1966	49791
0034	M	D	R	N	1968	49830
0034	M	D	F	N	1968	53144
0034	M	D	E	N	1967	55009
0034	M	D	R	N	1969	57019
0040	S	D	E	N	1966	34197
0042	S	D	E	N	1966	34197
0046	S	D	E	N	1927	39852
0054	S	D	R	N	1965	39987
0054	S	D	E	N	1965	39988
0054	M	D	E	N	1967	43052
0054	M	D	R	N	1967	45418
0054	M	D	R	N	1961	48286
0054	M	D	R	N	1968	49830
0054	M	D	F	N	1968	53144
0057	S	D	E	N	1966	34197
0058	S	D	E	N	1943	40228
0059	S	D	J	N	1963	37596
0059	S	D	E	N	1943	40228
0067	S	D	E	N	1966	34401
0067	S	D	E	N	1953	39860
0067	S	D	E	N	1967	42907
0067	S	D	E	N	1966	43774
0067	S	D	E	N	1947	44094
0067	S	S	F	H	1965	47334
0067	S	D	E	N	1968	55303
0067	S	D	E	N	1953	55981
0067	S	D	E	N	1969	58439
0067	P	C	F	N	1965	42373
0069	S	D	E	N	1964	37058
0069	S	D	E	N	1966	39083
0074	P	D	G	N	1958	51235
0080	S	D	E	N	1943	40228
0082	S	D	R	N	1965	34669
0095	S	D	E	H	1967	48662
0095	S	D	E	H	1967	48663
0110	S	G	G	N	1964	37646
0122	S	D	R	N	1965	39987
0122	S	D	E	N	1965	39988
0122	S	D	E	N	1921	40025
0127	S	D	E	N	1965	34669
0128	S	C	E	N	1968	53178
0128	S	D	E	N	1969	56053
0128	P	D	E	N	1945	61037
0129	P	D	E	N	1945	61037
0132	S	D	E	H	1911	48358
0133	M	D	E	N	1967	43052
0133	M	D	R	N	1967	45418
0133	M	D	E	N	1961	48286
0135	S	D	F	N	1968	57389
0144	M	G	C	N	1964	37646
U155	S	D	R	N	1965	39987
0155	S	D	E	N	1965	39988
0155	S	D	E	N	1966	41593
0182	S	D	E	N	1965	34669
0182	S	D	E	N	1966	41593
0182	S	D	E	N	1969	58340
0183	S	D	E	N	1965	34669
0183	S	D	E	N	1965	34669
0184	S	D	E	N	1965	34669
0185	S	D	E	N	1965	34669
0186	S	D	E	N	1965	34669
0187	S	D	E	N	1965	34669
0188	S	D	E	N	1965	34838
0188	S	D	E	N	1966	41593

Substance Number	Phys. State	Sub-ject	Lan-guage	Temper-ature	Year	TPRC Number
526-0189	S	D	E	N	1965	34669
0190	M	D	E	N	1967	43052
0190	M	D	R	N	1967	45418
0190	M	D	R	N	1961	48286
0190	M	D	R	N	1967	48930
0191	S	D	R	H	1967	48930
0191	S	D	E	H	1967	48931
0192	S	D	R	N	1967	48930
0192	S	D	E	H	1967	48931
0194	M	D	G	N	1958	51235
0198	S	D	E	H	1911	48358
0203	S	D	E	N	1966	41593
0204	S	D	E	N	1966	41593
0207	S	D	R	N	1967	45279
0207	S	D	E	H	1967	45494
0207	S	D	R	H	1966	46646
0207	S	D	E	N	1967	50880
0207	S	D	F	N	1968	57389
0209	M	D	J	N	1966	49874
0210	S	D	R	N	1966	41240
0210	S	D	E	N	1966	63164
0211	S	D	R	N	1966	41240
0211	S	D	E	N	1966	63164
0213	M	D	E	N	1969	54375
0214	S	D	E	N	1964	39737
0215	S	D	E	N	1964	39737
0215	S	D	R	N	1969	55263
0216	S	D	E	N	1964	39737
0217	M	D	G	N	1963	36809
0217	M	D	R	N	1963	36809
0218	M	D	G	N	1963	36809
0218	M	D	R	N	1963	36809
0220	S	D	E	N	1966	41593
0220	S	D	E	N	1969	58340
0221	M	G	G	N	1964	37646
0222	M	G	G	N	1964	37646
0223	M	G	G	N	1964	37646
0223	M	D	R	N	1969	57018
0224	M	G	G	N	1964	37646
0225	M	G	G	N	1964	37646
0226	M	D	E	N	1969	57513
0228	S	D	F	N	1968	57389
0229	S	D	F	N	1968	57389
0230	S	D	E	N	1966	59500
0230	S	D	E	N	1966	59501
0231	S	D	R	N	1966	59500
0231	S	D	E	N	1966	59501
0251	S	D	E	N	1966	41593
0265	S	D	R	N	1967	63152
0265	S	D	E	N	1967	63153
0268	P	D	E	N	1949	61706
528-0010	S	D	G	N	1968	55220
0012	S	D	E	N	1938	40268
0024	S	S	E	N	1965	34753
0101	S	D	E	N	1964	36269
0101	S	E	F	N	1965	36824
0105	M	D	E	N	1965	36377
0106	S	T	G	U	1965	51575
0114	M	D	O	N	1965	34076
0116	P	C	E	N	1968	47499
0116	M	C	E	N	1969	41302
0116	M	D	R	N	1965	45325
0116	M	D	R	N	1969	52546
0116	M	D	R	N	1969	54346
0116	M	D	E	N	1969	62951
0132	M	G	F	N	1966	45257
0132	M	D	E	N	1969	54062
0132	M	D	R	N	1969	54346
0132	M	D	E	N	1969	62951
0134	S	D	E	N	1966	48843
0134	M	D	E	N	1966	48811
0137	S	D	E	N	1966	50663
0139	S	D	E	L	1966	34309
0143	S	D	E	U	1957	57439
0147	S	D	E	N	1968	51905
0147	S	D	E	N	1965	62180
0147	P	D	E	N	1967	52200
0147	M	D	E	N	1965	36377
0147	M	C	E	N	1009	41302
0104	S	D	E	N	1943	40228
0164	S	D	R	N	1969	54566
0164	S	D	E	N	1969	55057
0164	M	D	E	N	1933	39866
0166	S	D	R	N	1964	38609
0166	S	D	E	N	1964	38610
0175	S	D	R	N	1943	40228
0183	S	G	E	N	1963	33858
0183	M	D	E	H	1964	48507
0185	M	D	O	N	1965	34076
0294	S	S	E	N	1968	42148
0294	S	D	E	N	1958	56356
0297	M	D	E	N	1966	48811
0301	M	D	E	N	1970	57767
0303	S	D	E	H	1969	55048

Substance Number	Phys. State	Sub-ject	Lan-guage	Temper-ature	Year	TPRC Number
528-0000	P	C	E	N	1968	47499
0314	S	D	E	N	1968	50380
0315	S	D	E	N	1968	50380
0316	S	D	E	N	1966	48381
0317	S	D	E	N	1966	48381
0328	S	D	R	U	1969	56024
0328	S	D	E	U	1969	56025
0328	M	D	E	N	1935	47953
0337	M	D	R	N	1967	48670
0337	M	D	E	N	1967	48671
0339	S	D	E	H	1965	34059
0340	S	D	E	H	1965	34059
0348	S	D	E	N	1965	37995
0352	S	D	E	N	1965	37995
0368	M	D	E	H	1961	49007
0371	S	D	E	N	1968	52091
0375	S	D	E	H	1967	52873
0376	S	C	E	N	1965	35620
0377	S	C	E	N	1965	35620
0378	S	C	E	N	1965	35620
0379	S	C	E	N	1965	35620
0380	S	C	E	N	1965	35620
0393	S	D	E	N	1965	43295
0401	S	C	G	L	1964	38417
0409	S	C	E	N	1970	58436
0410	M	C	E	N	1970	67767
0414	S	D	E	N	1965	43295
0418	S	G	E	L	1965	44854
0419	S	G	E	L	1965	44854
0420	S	G	E	L	1969	54503
0420	S	D	E	L	1970	57838
0421	S	D	E	N	1967	40398
0422	S	T	E	L	1967	40056
0428	S	D	R	N	1969	54566
0428	S	D	E	N	1969	55057
0465	S	D	G	N	1968	55220
0466	S	D	G	N	1968	55220
531-0003	M	D	E	N	1965	36377
0005	M	D	E	N	1965	36377
0010	M	D	E	N	1965	36377
0010	M	C	E	N	1968	47499
0010	M	D	E	N	1968	51905
0010	M	D	E	N	1967	52200
0014	S	D	E	N	1967	48812
0017	S	D	E	N	1967	48812
0033	S	D	E	N	1967	48812
0036	S	D	E	N	1967	48812
0042	S	D	E	N	1955	40197
0048	S	D	E	N	1955	40197
0049	M	D	E	N	1965	36377
0056	S	D	E	N	1965	34048
0056	S	D	E	N	1966	40611
0060	S	D	E	N	1965	35530
0060	S	D	E	N	1965	36145
0060	S	D	E	N	1970	37053
0060	S	D	E	N	1964	37499
0060	S	D	E	N	1965	38028
0060	S	D	E	N	1962	40299
0060	S	D	E	N	1964	42359
0060	S	D	E	N	1967	44487
0060	S	D	E	N	1968	44806
0060	S	D	R	N	1968	49033
0060	S	D	E	N	1968	52893
0060	S	C	E	N	1965	63107
0060	S	D	E	N	1969	57946
0081	M	G	E	N	1957	45699
0081	M	D	E	N	1961	56385
0064	M	D	G	N	1966	40167
0064	M	D	R	H	1966	46589
0064	M	D	E	H	1966	46590
0064	M	D	E	H	1966	49759
0065	M	G	E	N	1964	40404
0066	S	D	E	N	1966	40611
0066	S	D	E	N	1968	57225
0067	M	G	E	N	1964	46604
0067	M	D	E	N	1965	62180
0068	S	D	E	N	1965	36876
0081	P	G	E	N	1958	36876
0081	P	D	E	H	1964	37274
0081	P	D	E	H	1967	45579
0081	M	G	E	H	1967	49986
0092	S	D	E	N	1965	35530
0092	S	D	E	N	1966	48811
0092	M	D	E	H	1966	58327
0112	M	D	E	H	1965	39242
0119	M	D	E	N	1966	39869
0119	M	D	E	N	1964	46653
0119	M	D	E	N	1967	49193
0119	M	D	U	N	1950	51604
0121	M	D	E	N	1961	56385
0122	S	D	E	N	1964	36269
0122	S	C	R	N	1965	43079
0122	S	D	E	N	1966	51533
0122	S	D	R	N	1970	62857

Phys. State: G. Gas; L. Liquid; M. Multiphase; P. Powder; S. Solid

Substance Number	Phys. State	Subject	Language	Temperature	Year	TPRC Number
531-0123	S	D	E	N	1963	34979
0135	M	D	J	N	1969	56793
0138	S	D	E	N	1959	50803
0157	L	D	E	N	1967	40915
0169	L	D	E	N	1967	40915
0171	P	D	E	N	1965	35535
0171	P	D	E	N	1964	38646
0171	P	C	E	N	1970	38913
0171	P	D	E	N	1960	42824
0171	P	D	E	N	1957	45699
0171	P	G	E	N	1963	49018
0171	P	D	E	N	1966	49738
0171	P	C	R	N	1970	62437
0171	P	D	E	N	1970	63135
0171	M	D	R	N	1965	34599
0171	M	D	R	N	1959	42504
0171	M	D	E	N	1964	45653
0171	M	G	E	N	1966	46250
0171	M	D	E	N	1965	47289
0171	M	D	F	N	1962	47926
0172	P	D	E	N	1966	47174
0172	P	G	E	N	1963	49018
0172	E	C	R	N	1967	49687
0172	M	D	E	N	1964	45653
0172	M	D	E	N	1957	45699
0172	M	C	E	N	1968	47499
0172	M	D	E	N	1961	56385
0174	S	D	E	N	1969	37570
0174	S	D	E	N	1969	55028
0174	S	D	E	N	1970	57639
0174	P	D	E	N	1967	49152
0174	P	D	E	N	1967	49250
0174	M	D	E	N	1968	48434
0174	M	D	E	N	1968	57191
0181	L	D	E	N	1967	40915
0182	L	D	E	N	1967	40915
0186	L	D	E	N	1967	40915
0188	S	S	E	N	1966	42284
0188	M	D	E	N	1961	39858
0188	M	D	E	N	1965	44339
0188	M	D	E	N	1968	51693
0192	S	D	E	N	1966	40611
0192	S	D	R	N	1967	46369
0192	S	D	J	N	1968	51800
0192	M	D	R	N	1967	46170
0193	L	D	E	N	1967	40915
0194	L	D	E	N	1967	40915
0195	S	D	E	N	1966	40611
0195	S	D	R	N	1967	46369
0195	S	D	R	N	1968	49033
0195	S	D	J	N	1968	51800
0195	S	D	R	N	1970	62857
0195	M	D	R	N	1967	46170
0198	S	D	E	N	1966	40611
0198	S	D	J	N	1968	51800
0201	S	D	E	N	1966	40611
0201	S	D	E	L	1967	44158
0201	S	D	J	N	1968	51800
0201	S	D	E	N	1970	61187
0201	S	C	E	N	1970	63135
0207	L	D	E	N	1967	40915
0227	S	D	E	N	1965	45991
0242	S	D	E	N	1968	51905
0242	S	G	E	N	1970	57661
0242	S	D	E	N	1970	61781
0242	S	D	E	N	1965	62180
0242	P	C	E	N	1968	47499
0242	P	C	R	U	1966	48084
0242	P	D	E	N	1967	52090
0242	P	D	E	N	1967	52200
0242	M	C	E	N	1961	34829
0242	M	D	E	N	1965	36377
0242	M	G	E	N	1964	40082
0242	M	G	F	N	1966	45257
0248	M	E	E	N	1965	36367
0254	M	D	R	N	1931	43022
0254	M	D	R	N	1931	43023
0269	S	D	E	N	1969	37570
0269	S	D	E	N	1968	50307
0269	S	D	E	N	1968	54818
0269	S	D	E	N	1969	55028
0269	S	D	E	N	1968	57191
0269	S	D	E	N	1970	57639
0269	P	D	E	N	1967	49152
0269	P	D	E	N	1967	49250
0269	M	D	E	N	1968	48434
0273	M	D	E	L	1961	43368
0288	S	D	G	N	1967	46921
0288	S	D	R	N	1968	54578
0291	S	D	E	N	1959	50803
0294	S	D	E	N	1964	35796
0295	S	D	E	N	1959	50803
0296	S	C	E	N	1965	63107
0301	S	D	E	N	1965	34048
0301	S	D	E	N	1962	40299
0301	S	D	E	N	1966	40611
0301	S	D	R	N	1968	49033
0301	S	D	J	N	1968	51800
0303	S	D	R	N	1965	34666
0303	S	S	E	L	1965	34753
0303	S	D	E	N	1964	36202

Substance Number	Phys. State	Subject	Language	Temperature	Year	TPRC Number
531-0303	S	D	E	N	1965	38402
0303	S	D	E	N	1966	40611
0303	S	G	R	N	1965	43097
0303	S	G	J	N	1967	46251
0303	S	G	E	N	1968	46252
0303	S	D	E	N	1966	48811
0303	S	D	R	N	1968	49033
0303	S	T	R	U	1968	49591
0303	S	D	J	N	1968	51800
0303	S	D	R	N	1970	62857
0303	S	C	E	N	1970	63135
0303	M	C	R	N	1969	57009
0304	S	D	E	N	1965	34048
0304	S	D	E	N	1964	37499
0321	S	S	E	N	1968	42148
0321	M	D	R	U	1964	53103
0321	M	D	E	U	1966	53104
0331	P	D	E	N	1962	39516
0331	M	D	R	N	1967	46172
0342	S	C	E	N	1969	56491
0342	P	D	E	N	1964	38646
0342	P	D	E	N	1964	45753
0342	P	G	E	N	1963	49023
0342	P	D	E	N	1966	49738
0342	P	D	E	N	1969	54820
0342	M	G	E	N	1966	46250
0347	M	D	E	N	1965	36377
0348	P	D	E	N	1965	42374
0348	P	D	E	N	1957	45699
0348	P	D	E	N	1964	45753
0348	P	G	E	N	1969	59176
0348	M	D	R	N	1959	42504
0348	M	D	E	N	1967	49193
0349	P	D	E	N	1964	38646
0349	E	C	R	N	1967	49687
0349	M	D	E	N	1957	45699
0349	M	C	E	N	1968	47499
0349	M	D	E	N	1967	49193
0349	M	D	E	N	1961	56385
0365	S	T	E	U	1969	50040
0367	M	D	E	N	1964	37447
0370	M	D	E	N	1964	37447
0373	M	D	E	L	1965	33811
0373	M	D	E	L	1965	38834
0384	S	E	E	N	1961	38320
0384	S	T	E	U	1966	44938
0384	P	E	E	U	1969	53594
0384	P	T	E	U	1970	60049
0384	M	C	E	U	1963	37726
0384	M	C	E	N	1970	38175
0384	M	C	E	N	1970	39257
0384	M	C	E	N	1969	41302
0384	M	T	E	U	1966	41466
0384	M	C	E	U	1961	55356
0384	M	T	E	U	1970	58726
0394	M	C	E	N	1961	34829
0394	M	D	E	N	1965	36360
0394	M	D	E	N	1965	36377
0394	M	G	E	N	1964	40082
0394	M	G	E	N	1964	40404
0394	M	D	E	N	1968	51693
0394	M	G	E	N	1968	56530
0394	M	D	E	N	1970	59139
0394	M	D	E	N	1965	62180
0408	M	D	E	N	1961	39858
0421	M	G	E	H	1965	38586
0421	M	G	R	H	1965	42531
0434	S	D	E	H	1960	52260
0447	M	D	E	N	1957	45699
0447	M	D	E	N	1961	56385
0448	M	D	E	N	1957	45699
0449	M	D	E	N	1965	36377
0456	M	D	E	N	1965	36377
0456	M	D	E	N	1970	57632
0457	M	D	E	N	1965	36377
0463	S	D	E	N	1966	48811
0463	S	D	E	N	1968	50306
0463	M	D	E	N	1966	58327
0471	M	D	R	N	1966	41839
0472	M	D	R	N	1966	41839
0475	S	T	E	U	1968	47661
0475	M	E	E	H	1964	40396
0487	M	D	E	N	1966	39869
0487	M	D	E	N	1970	57632
0502	S	S	E	N	1965	34753
0511	P	S	E	N	1967	44944
0511	P	T	E	U	1966	45543
0511	P	T	E	N	1968	52865
0511	M	T	R	U	1966	49166
0511	M	T	E	U	1966	54400
0515	S	D	E	L	1966	35110
0515	S	D	E	N	1968	37491
0515	S	D	E	N	1966	40611
0515	S	D	R	N	1967	45790
0515	S	D	J	N	1968	51800
0524	M	C	E	N	1970	57767
0525	S	D	E	N	1968	37491
0530	S	D	E	N	1964	37394
0530	S	D	E	H	1965	39264
0535	S	S	E	H	1965	34428
0536	S	S	E	H	1965	34428

Substance Number	Phys. State	Subject	Language	Temperature	Year	TPRC Number
531-0536	S	D	E	H	1964	36797
0536	M	D	G	N	1967	46106
0537	S	G	E	L	1965	34659
0537	S	D	E	L	1964	50019
0550	S	T	R	U	1968	49591
0553	M	C	E	H	1965	34752
0553	M	D	E	H	1965	35826
0553	M	D	E	H	1966	39358
0553	M	D	E	H	1967	46648
0553	M	D	E	H	1966	47098
0553	M	D	E	H	1968	48855
0553	M	D	E	N	1968	56351
0554	S	S	E	N	1965	34753
0554	S	D	E	N	1963	49933
0555	S	D	E	H	1965	39264
0568	S	D	E	N	1965	34048
0568	S	G	E	L	1965	34659
0568	S	D	E	L	1966	35110
0568	S	D	E	L	1965	35899
0568	S	D	E	N	1964	36202
0568	S	D	E	N	1964	37499
0568	S	D	E	N	1966	39083
0568	S	D	E	N	1965	39754
0568	S	D	E	L	1965	40148
0568	S	D	E	N	1966	40611
0568	S	D	E	N	1966	44554
0568	S	G	E	N	1966	44681
0568	S	G	J	N	1967	46251
0568	S	G	E	N	1968	46252
0568	S	D	E	N	1966	48811
0568	S	D	E	L	1964	50019
0568	S	D	J	N	1968	51800
0568	S	D	E	L	1969	57598
0568	S	D	R	N	1970	62857
0569	S	D	E	N	1964	36202
0569	S	D	E	N	1965	36419
0569	S	D	E	H	1968	36538
0569	S	D	E	N	1964	37499
0569	S	D	E	N	1962	40299
0569	S	D	E	N	1966	43475
0569	S	D	E	N	1968	52893
0569	S	C	E	N	1965	63107
0571	P	D	E	N	1957	45699
0573	P	D	E	N	1957	45699
0574	P	G	F	N	1966	45257
0574	P	D	E	N	1957	45699
0576	P	D	E	N	1957	45699
0577	P	D	E	N	1957	45699
0583	P	D	E	N	1957	45699
0586	P	D	E	N	1957	45699
0588	M	D	E	H	1964	35151
0589	L	D	R	N	1967	53012
0589	M	D	R	N	1964	38189
0592	M	D	E	N	1964	37447
0592	M	D	E	N	1968	50638
0592	M	D	E	N	1961	56385
0618	S	D	R	N	1967	46606
0618	S	D	E	N	1967	46607
0620	S	D	E	N	1965	36145
0620	S	D	E	N	1964	37255
0620	S	D	E	N	1964	37499
0620	S	D	E	N	1962	40299
0620	S	D	E	F	1963	40301
0620	S	D	E	H	1965	40305
0620	S	D	E	N	1964	41910
0620	S	D	E	N	1964	42359
0620	S	D	E	N	1966	43475
0620	S	D	E	N	1968	48806
0620	S	D	E	N	1968	52893
0620	S	C	E	N	1965	63107
0620	M	G	E	H	1969	57946
0631	M	D	E	N	1965	35536
0631	M	D	E	N	1964	37447
0631	M	D	G	N	1967	45209
0631	M	D	E	N	1968	50638
0631	M	D	G	N	1969	55608
0631	M	D	E	N	1958	56356
0631	M	G	E	N	1970	58744
0633	S	E	E	N	1965	37915
0633	S	E	E	N	1966	38190
0633	S	E	R	N	1966	41482
0633	S	T	R	U	1967	45155
0633	S	S	E	L	1966	46217
0633	S	E	E	N	1967	46639
0633	S	E	R	L	1967	46863
0633	S	T	R	U	1966	48092
0633	S	T	R	U	1966	48093
0633	S	E	E	N	1967	53612
0633	S	E	E	N	1968	54179
0633	S	S	E	L	1967	56366
0633	S	D	E	N	1969	56570
0633	S	G	E	N	1970	59608
0633	S	D	E	N	1970	63133
0639	S	D	E	N	1964	37242
0639	S	D	E	N	1938	41684
0646	M	C	E	N	1961	34829
0648	M	D	E	N	1965	36377
0651	M	D	E	H	1969	54298
0652	P	S	E	N	1966	42284
0652	M	T	E	N	1966	39869
0652	M	D	E	N	1966	39869

Phys. State: **G.** Gas; **L.** Liquid; **M.** Multiphase; **P.** Powder; **S.** Solid

Substance Number	Phys. State	Subject	Language	Temperature	Year	TPRC Number
531-0652	M	C	O	N	1966	41319
0652	M	C	E	N	1967	44183
0656	P	C	R	U	1966	48084
0658	G	D	E	N	1964	37242
0671	M	C	E	N	1967	43690
0671	M	G	E	N	1966	49383
0677	S	D	E	N	1966	40611
0679	S	D	R	N	1968	49033
0679	S	D	J	N	1968	51800
0683	S	D	E	N	1968	51905
0683	S	G	E	N	1970	57661
0683	S	D	E	N	1970	61781
0683	P	G	E	N	1963	49023
0683	P	D	E	N	1967	52090
0683	P	D	E	N	1967	52200
0683	M	G	F	N	1966	45257
0683	M	D	R	N	1967	46172
0683	M	G	E	N	1966	46250
0697	G	C	E	N	1965	42424
0717	S	D	E	N	1965	36419
0717	S	D	E	N	1964	37255
0717	S	G	E	N	1968	51884
0737	M	D	O	N	1965	34076
0782	S	D	E	N	1943	40228
0799	E	C	R	N	1967	49687
0799	M	D	E	N	1965	38402
0799	M	G	R	H	1965	42535
0799	M	G	E	H	1966	43468
0799	M	C	E	N	1968	47499
0806	M	C	E	N	1961	34829
0807	G	D	E	N	1947	48357
0807	M	C	E	N	1961	34829
0820	S	D	E	F	1967	33971
0826	P	D	E	N	1964	38646
0827	P	D	E	N	1964	38646
0831	S	D	E	N	1965	34485
0851	S	D	E	L	1966	42657
0853	S	D	E	L	1966	42657
0855	S	D	E	L	1966	42657
0857	S	D	E	L	1966	42657
0859	S	D	E	L	1966	42657
0861	S	D	E	L	1966	42657
0863	S	D	E	L	1966	42657
0865	S	D	E	L	1966	42657
0867	S	D	E	L	1966	42657
0869	S	D	E	L	1966	42657
0871	S	D	E	L	1966	42657
0891	M	C	E	N	1961	34829
0910	M	C	E	N	1968	47499
0923	S	G	E	H	1967	43744
0923	M	D	E	N	1967	46648
0927	M	D	E	H	1966	43100
0931	S	D	E	L	1966	42657
0955	S	C	R	N	1964	37200
0955	S	C	E	N	1964	37450
0955	S	D	E	N	1960	42824
0955	S	D	E	N	1967	44986
0955	S	G	E	N	1968	52206
0955	S	D	E	N	1969	53597
0955	M	D	R	N	1965	34599
0955	M	D	R	N	1959	42504
0955	M	D	E	N	1965	47289
0955	M	G	E	N	1967	52232
0955	M	G	S	N	1967	52232
0958	S	D	E	L	1966	42657
0960	S	D	E	L	1964	50019
0961	S	D	E	L	1966	42657
0963	S	D	E	L	1964	50019
0964	S	D	E	L	1966	42657
0966	S	D	E	L	1964	50019
0967	S	D	E	L	1966	42657
0968	S	D	E	L	1966	42657
0969	S	D	E	L	1966	42657
0970	S	D	E	L	1966	42657
0971	S	D	E	L	1966	42657
0974	S	G	E	L	1964	37035
0974	S	D	E	L	1962	42071
0974	S	C	E	L	1966	42558
0975	S	D	E	H	1968	54179
0976	S	D	E	N	1966	41168
0979	M	D	E	N	1968	50638
0979	M	D	E	N	1968	50953
0979	M	D	E	N	1967	50612
0981	M	C	E	N	1968	47499
0984	M	C	E	N	1968	47499
0985	M	C	E	N	1968	47499
0986	S	D	E	H	1968	54179
0987	M	C	E	N	1968	47499
1038	S	D	E	H	1965	35826
1038	S	D	E	H	1965	36420
1038	S	D	E	H	1966	39358
1038	S	D	E	N	1966	41169
1038	S	D	R	N	1966	43894
1038	S	D	E	N	1966	43895
1038	S	D	E	N	1967	44986
1038	S	D	E	H	1967	46648
1038	S	D	E	H	1966	47098
1038	S	D	E	H	1966	49761
1038	S	D	E	N	1968	51906
1038	S	D	E	N	1968	53086
1038	M	D	E	H	1964	45639
531-1038	M	D	E	H	1968	48855
1039	M	D	O	N	1965	34076
1040	M	G	E	N	1965	38586
1040	M	G	R	N	1965	42531
1042	G	G	E	N	1965	38586
1042	S	G	R	N	1965	42531
1043	P	D	E	H	1963	49027
1043	P	D	E	N	1968	51864
1043	P	D	E	N	1970	63135
1043	M	D	E	L	1965	38834
1054	S	C	E	N	1970	59568
1056	S	D	E	H	1966	48379
1057	S	D	E	N	1966	48383
1057	S	D	E	H	1969	55048
1057	S	C	E	N	1970	61186
1057	S	D	E	U	1970	61192
1057	M	D	R	N	1967	45570
1057	M	D	E	N	1969	58321
1057	M	D	E	N	1970	61704
1075	M	D	G	H	1968	50490
1077	M	G	E	N	1956	48292
1078	M	D	R	H	1960	51562
1079	M	D	R	H	1960	51562
1080	M	D	R	H	1960	51562
1093	S	D	E	N	1968	50306
1094	M	D	E	N	1968	50306
1095	S	D	E	H	1961	47215
1095	S	D	E	N	1968	49546
1095	S	D	E	N	1970	60164
1105	M	D	E	N	1967	49193
1106	M	D	E	N	1967	49193
1107	P	D	J	N	1963	36246
1107	P	D	E	N	1964	38646
1107	P	D	E	N	1966	41217
1107	P	D	E	N	1960	42824
1107	P	D	E	N	1964	45753
1107	P	D	E	N	1966	49738
1107	P	S	E	N	1968	51855
1107	P	D	E	N	1968	51904
1107	P	G	E	N	1969	52300
1107	P	G	E	N	1968	56537
1107	M	D	E	N	1964	45653
1107	M	D	E	N	1967	49193
1108	P	D	E	N	1964	38646
1108	P	D	E	N	1966	41217
1108	P	D	E	N	1964	45753
1108	P	D	E	N	1967	46040
1108	P	D	E	N	1967	46042
1108	P	D	E	N	1966	49738
1108	M	G	E	N	1966	46250
1108	M	D	E	N	1967	49193
1109	P	D	E	N	1965	35535
1109	M	D	E	N	1967	49193
1118	M	D	E	H	1968	48862
1120	S	D	R	H	1967	48415
1133	M	D	E	N	1899	48324
1133	M	D	R	N	1969	56318
1134	M	D	E	N	1966	48811
1135	M	D	E	N	1966	48811
1136	M	D	E	N	1966	48811
1137	M	D	E	N	1966	48811
1142	S	D	R	N	1932	47979
1143	S	D	R	N	1932	47979
1144	S	D	R	N	1932	47979
1145	S	D	R	N	1932	47979
1146	S	D	R	N	1932	47979
1147	S	D	R	N	1932	47979
1149	M	G	E	H	1967	49986
1157	S	D	E	N	1968	50308
1159	L	D	R	N	1967	47327
1163	M	D	R	N	1967	45570
1163	M	D	E	N	1970	61704
1164	M	D	R	N	1967	45570
1164	M	D	E	N	1970	61704
1165	M	D	R	N	1967	45570
1165	M	D	E	N	1970	61704
1214	S	D	E	H	1967	46648
1217	S	D	E	N	1965	34640
1218	S	D	E	N	1965	34640
1219	M	D	R	N	1967	46172
1227	M	D	R	N	1959	42504
1228	M	D	R	N	1959	42504
1229	M	D	R	N	1959	42504
1230	M	D	R	N	1959	42504
1244	S	D	R	N	1966	43080
1245	M	D	R	N	1966	45455
1245	M	C	E	N	1970	57767
1246	S	S	J	U	1967	45441
1246	S	E	E	H	1960	52260
1247	S	D	R	N	1966	45275
1249	S	G	E	N	1961	38320
1250	S	G	E	N	1961	38320
1251	S	G	E	N	1961	38320
1252	M	D	E	H	1967	45825
1252	M	D	E	H	1966	47236
1253	M	D	E	H	1967	45825
1253	M	D	E	H	1966	47236
1256	M	D	F	N	1967	45087
1257	M	D	F	N	1967	45087
1258	M	C	O	U	1967	45080
1266	M	D	R	N	1931	43023
531-1276	S	D	E	H	1968	48826
1287	S	D	G	N	1966	41631
1288	S	D	E	N	1962	47240
1289	S	D	E	N	1968	42639
1289	S	D	E	H	1966	49759
1204	S	D	F	N	1967	45925
1294	S	D	E	N	1968	57223
1309	S	D	E	N	1967	45925
1309	S	D	E	N	1968	57223
1309	S	D	E	H	1970	57644
1312	M	D	E	H	1967	45825
1315	M	D	E	H	1962	44560
1318	S	S	E	N	1966	42284
1321	M	D	E	L	1966	41182
1322	M	D	E	L	1966	41182
1330	S	D	E	N	1966	40476
1336	S	D	E	N	1965	39754
1337	P	D	E	N	1962	39516
1337	P	D	E	N	1970	59561
1356	S	D	E	N	1966	38611
1357	P	D	E	N	1964	38646
1357	P	D	E	N	1965	42374
1357	P	D	E	N	1964	45753
1357	P	G	E	N	1963	49023
1358	P	D	E	N	1964	38646
1358	P	D	E	N	1965	42374
1358	P	D	E	N	1964	45753
1358	P	G	E	N	1963	49023
1358	P	D	E	N	1969	54820
1360	P	D	E	N	1964	38646
1360	P	D	E	N	1964	45753
1300	P	G	E	N	1963	49023
1361	P	D	E	N	1964	38646
1361	P	D	E	N	1965	42374
1361	P	G	E	N	1963	49023
1363	S	D	E	N	1964	37499
1364	M	D	E	N	1964	37447
1367	M	D	E	N	1964	37447
1368	M	D	E	N	1964	37447
1369	S	D	E	N	1966	46059
1369	M	D	E	N	1964	37447
1378	S	G	E	N	1964	44867
1381	S	D	E	N	1964	36084
1385	P	D	E	N	1964	45753
1389	M	D	E	H	1963	37338
1390	S	D	E	N	1965	36419
1390	S	D	E	N	1964	37255
1390	S	D	E	N	1963	40301
1391	S	D	E	N	1965	36419
1391	S	D	E	N	1967	48812
1394	S	D	E	N	1965	36471
1395	S	D	E	N	1966	36480
1396	S	D	E	H	1965	36420
1399	S	D	E	H	1967	55203
1400	S	D	E	H	1967	55203
1401	S	D	E	H	1967	55203
1402	S	C	E	N	1964	46334
1402	S	C	E	N	1969	53853
1402	S	C	E	N	1968	58397
1405	S	D	E	H	1964	36797
1405	M	D	G	H	1969	57392
1420	S	D	E	N	1964	36285
1421	S	D	E	N	1964	36269
1424	M	D	E	H	1966	49759
1424	M	D	E	H	1970	57643
1425	S	D	E	H	1966	49759
1425	M	D	E	H	1970	57643
1426	S	D	E	N	1965	47397
1429	S	D	E	N	1965	47397
1430	S	D	E	N	1965	47397
1431	S	D	E	U	1968	47410
1438	M	D	E	N	1969	54062
1439	S	D	E	H	1965	36420
1439	M	D	E	H	1964	45639
1440	S	D	E	H	1965	36420
1441	S	D	E	H	1965	36420
1441	M	D	E	H	1964	45639
1441	M	D	G	H	1969	57392
1442	S	D	E	H	1965	36420
1443	S	D	E	H	1965	36420
1444	S	D	E	H	1965	36420
1444	S	D	E	N	1960	42824
1445	S	D	E	H	1965	36420
1447	S	D	E	N	1966	39873
1455	S	D	E	N	1950	40447
1456	S	D	E	N	1950	40447
1457	S	S	E	N	1965	35862
1457	S	S	E	H	1965	37590
1457	S	E	E	U	1964	43583
1457	S	T	E	U	1970	59599
1462	P	D	E	N	1970	56105
1463	M	D	E	H	1968	51903
1464	P	D	E	H	1968	51903
1471	S	G	E	N	1968	52206
1472	S	D	J	N	1968	51800
1473	S	D	J	N	1968	51800
1474	S	D	J	N	1968	51800
1475	S	D	J	N	1968	51800
1477	P	D	E	N	1966	47174
1480	P	D	E	N	1966	47174
1487	S	D	E	N	1959	56352

Phys. State: **G.** Gas; **L.** Liquid; **M.** Multiphase; **P.** Powder; **S.** Solid

Subject: C. Calculated Data + Theoretical; D. Data + Experimental; E. Experimental + Theoretical; G. General (Data + Theoretical + Experimental)

Substance Number	Phys. State	Subject	Language	Temperature	Year	TPRC Number
531-1492	S	S	E	L	1967	56366
1493	M	D	J	N	1969	54483
1498	P	G	F	N	1966	45257
1499	S	D	F	H	1969	57267
1505	S	D	E	N	1968	53519
1506	S	D	E	N	1968	53519
1507	S	D	E	N	1968	53519
1508	S	D	E	N	1968	53519
1509	S	D	E	N	1968	53519
1513	S	T	E	U	1966	54228
1515	S	D	E	N	1968	53094
1516	S	D	E	N	1968	53094
1518	S	D	R	N	1965	36612
1518	S	D	E	N	1965	41719
1518	S	D	E	N	1966	43475
1518	S	D	E	N	1966	53098
1518	S	D	E	L	1969	57598
1519	S	D	E	N	1966	44554
1520	M	D	E	H	1969	54298
1539	P	D	E	N	1969	56578
1540	M	G	E	N	1970	58744
1541	S	C	E	N	1969	56491
1542	M	G	E	N	1970	58744
1559	S	D	E	N	1969	56570
1580	S	D	E	L	1964	35796
1581	S	D	E	L	1964	35796
1582	S	D	E	L	1964	35796
1592	S	D	R	N	1966	44594
1605	S	D	E	N	1962	36109
1608	S	D	E	N	1966	38190
1608	S	D	R	N	1966	41482
1608	M	D	E	N	1961	56385
1609	M	D	E	N	1961	56385
1611	P	D	E	N	1970	59561
1612	S	D	R	H	1970	62228
1612	S	D	E	H	1970	62229
1617	M	D	E	H	1965	39242
1618	S	D	E	H	1965	39264
1620	S	D	E	N	1965	39745
1621	M	D	E	N	1965	39745
1622	S	D	E	N	1965	39745
1623	M	D	E	N	1965	39745
1625	S	D	E	H	1965	35739
1625	S	D	E	N	1966	40193
1626	S	D	E	N	1966	40193
1627	S	D	E	N	1964	40684
1628	S	D	E	N	1964	40684
1629	M	C	E	N	1955	40671
1630	M	C	E	N	1955	40671
1631	M	C	E	N	1955	40671
1632	S	D	E	N	1966	41165
1634	S	D	E	L	1966	41218
1634	S	D	E	N	1968	53086
1635	S	D	E	L	1966	41218
1639	M	D	E	N	1966	58327
1640	M	D	E	N	1966	58327
1641	M	D	E	N	1968	55829
1642	M	D	E	N	1968	55829
1645	M	D	E	N	1968	55829
1646	M	D	E	N	1968	55829
1650	S	D	F	N	1968	37370
1653	S	D	R	N	1969	55503
1654	S	D	R	N	1969	55503
1656	S	D	E	N	1970	57581
1657	S	D	E	N	1970	57581
1658	S	D	E	N	1970	57581
1659	S	D	E	N	1970	57581
1660	S	D	E	N	1970	57581
1661	S	D	E	N	1970	57581
1664	M	D	E	L	1970	35591
1666	M	D	E	N	1965	42759
1667	M	D	E	N	1965	42759
1668	S	C	E	N	1970	59568
1668	S	G	E	N	1969	61109
1671	S	D	O	N	1969	58147
1672	S	C	E	N	1965	63107
1680	S	D	E	N	1960	42824
1681	S	D	E	N	1960	42824
1682	S	D	E	N	1960	42824
1683	S	D	E	N	1960	42824
1684	S	D	E	N	1960	42824
1685	S	D	E	N	1960	42824
1686	S	D	R	H	1966	42004
1686	S	D	E	H	1966	43896
1692	S	C	E	N	1968	58397
1694	M	D	E	N	1967	42955
1694	M	D	R	N	1967	45153
1695	M	D	E	N	1967	42955
1695	M	D	R	N	1967	45153
1696	M	D	E	N	1967	42955
1696	M	D	R	N	1967	45153
1698	S	D	E	N	1966	41169
1699	S	D	E	N	1966	41169
1702	M	D	E	N	1965	39745
1703	S	D	E	N	1965	39745
1746	M	D	E	N	1968	50638
1747	M	D	E	N	1968	50638
1752	S	C	E	N	1970	38913
1752	S	C	R	N	1970	62437
1755	S	D	E	N	1967	60099
1757	S	D	E	N	1970	59299

Substance Number	Phys. State	Subject	Language	Temperature	Year	TPRC Number
531-1758	P	D	E	N	1970	59299
1759	P	D	E	N	1970	59299
1760	S	D	E	N	1968	57222
1761	S	D	E	N	1968	57222
1762	S	D	E	N	1968	57222
1763	S	D	E	N	1968	57222
1764	S	D	E	N	1968	57222
1765	S	D	E	N	1968	57222
1765	S	D	E	N	1968	57225
1766	S	D	E	N	1968	57225
1767	S	D	E	N	1968	57225
1768	M	D	E	N	1964	45653
1769	M	D	E	N	1964	45653
1769	M	D	E	N	1970	57632
1770	P	D	E	N	1964	45753
1771	M	D	E	H	1964	45639
1772	M	D	E	H	1964	45639
1773	M	D	E	H	1964	45639
1774	P	D	E	N	1964	45753
1775	M	G	E	N	1970	60608
1780	S	D	E	N	1965	43295
1781	S	D	E	N	1969	56570
1782	S	D	E	N	1969	56570
1783	M	D	E	N	1970	57632
1784	S	D	E	H	1970	57642
1785	S	C	E	N	1970	61186
1787	S	D	E	N	1968	55303
1792	S	D	G	N	1966	46022
1793	S	D	G	N	1966	46022
1795	M	D	E	N	1968	53113
1796	M	D	E	N	1968	53,113
1798	M	D	E	N	1965	62180
1801	P	D	R	N	1969	56847
1801	P	D	E	N	1969	63348
1802	P	D	R	N	1969	56847
1802	P	D	E	N	1969	63348
1803	S	D	R	N	1970	61567
1809	M	D	E	N	1969	54062
541-0002	S	D	E	N	1964	37631
0002	S	D	E	H	1966	40368
0008	S	D	E	H	1966	40368
0008	P	D	E	H	1966	40368
0008	P	D	E	N	1960	51576
0010	M	G	E	H	1966	34751
0010	M	G	E	H	1964	40396
0012	S	D	E	N	1968	54315
0012	M	D	E	N	1969	58320
0012	M	D	E	N	1969	58321
0013	S	D	E	N	1959	50803
0017	S	G	E	H	1966	34751
0017	M	G	E	H	1964	40396
0019	S	D	E	N	1964	35797
0019	S	D	G	U	1970	60314
0019	S	D	G	U	1970	61785
0019	M	T	R	U	1966	43999
0019	M	T	E	U	1966	44000
0044	S	D	E	N	1968	47768
0044	S	C	R	N	1966	53875
0044	S	C	E	N	1966	53876
0044	S	D	R	N	1969	62953
0044	S	D	E	N	1969	62954
0062	S	D	E	N	1970	57581
0062	M	D	E	N	1968	59204
0082	S	D	E	H	1960	52128
0104	M	D	E	N	1969	58320
0104	M	D	E	N	1969	58321
0108	S	T	E	U	1967	45491
0108	S	G	E	N	1969	55248
0111	S	D	E	L	1969	56168
0113	M	D	E	N	1965	36377
0113	M	C	E	N	1968	47499
0113	M	D	E	N	1968	51905
0113	M	D	E	N	1967	52200
0118	M	G	F	N	1966	45257
0121	S	D	E	L	1964	35796
0121	S	D	E	N	1966	38190
0121	S	D	E	L	1966	41218
0121	S	D	R	N	1966	41482
0121	S	D	E	N	1966	48383
0121	S	D	E	H	1968	51490
0121	S	D	E	N	1967	52118
0121	S	D	E	N	1968	54277
0121	S	D	E	H	1969	55048
0121	S	D	E	N	1970	57658
0121	S	C	E	N	1970	61186
0121	S	D	E	U	1970	61192
0121	M	D	R	N	1965	34599
0121	M	D	R	N	1966	42017
0121	M	D	R	N	1967	45570
0121	M	D	E	N	1965	47289
0121	M	D	E	N	1969	57851
0121	M	D	E	N	1969	58321
0121	M	D	E	N	1970	61704
0124	S	D	E	N	1965	39745
0223	S	D	R	N	1965	34599
0223	S	D	E	L	1966	41336
0223	S	D	E	N	1967	45482
0223	S	D	E	L	1967	47286
0223	S	D	E	N	1965	47289
0223	S	D	E	N	1968	49527

Substance Number	Phys. State	Subject	Language	Temperature	Year	TPRC Number
541-0223	S	D	E	H	1968	51490
0223	S	D	E	N	1969	53607
0223	S	D	E	H	1968	54179
0223	M	D	E	U	1961	47935
0223	M	D	E	N	1961	56385
0225	S	D	E	L	1967	47286
0338	S	D	E	N	1968	51905
0338	P	D	E	N	1967	52200
0338	M	D	R	N	1967	46172
0343	S	D	G	N	1966	44283
0404	M	D	E	H	1969	58320
0404	M	D	E	N	1969	58321
0405	S	D	E	N	1968	54315
0405	M	D	E	N	1969	58321
0406	S	D	E	N	1969	53611
0414	M	D	E	N	1965	36377
0424	M	D	E	N	1966	39869
0427	M	D	E	N	1969	58321
0428	M	G	E	N	1966	39869
0429	M	D	E	N	1966	39869
0430	M	D	E	N	1966	39869
0431	S	D	E	N	1968	54315
0431	M	D	E	N	1969	58321
0432	M	D	E	N	1969	58321
0433	M	D	E	N	1966	39869
0434	M	D	E	N	1969	57853
0436	M	D	E	N	1966	39869
0437	S	D	E	L	1967	44158
0442	M	D	E	N	1966	39869
0448	M	D	E	N	1966	39869
0450	S	D	E	U	1970	61192
0451	S	D	E	N	1970	59427
0451	M	D	G	L	1969	57859
0451	M	D	E	N	1968	59204
0474	S	D	E	N	1967	45669
0477	S	D	E	N	1967	45669
0519	M	C	E	N	1968	47499
0538	M	C	E	H	1965	34752
0540	S	C	E	H	1965	34752
0543	S	D	E	N	1963	45477
0544	S	D	E	N	1963	45477
0546	M	C	O	N	1966	41319
0546	M	C	E	N	1967	44183
0565	M	D	E	N	1969	54062
0570	S	D	E	N	1966	34740
0570	S	D	E	N	1968	42628
0571	S	D	E	N	1966	34740
0571	S	D	E	N	1968	42628
0572	S	G	E	H	1966	34751
0572	S	C	E	N	1969	55805
0574	M	G	E	H	1966	34751
0574	M	C	E	N	1969	55805
0577	P	C	E	N	1965	38586
0577	P	C	R	N	1965	42531
0578	M	G	E	N	1965	38586
0578	M	G	R	N	1965	42531
0587	M	D	E	N	1965	36377
0619	S	D	E	N	1943	40228
0620	S	D	E	H	1967	41136
0623	S	G	E	H	1967	43744
0644	S	S	E	N	1968	42148
0656	M	D	R	N	1967	48670
0656	M	D	E	N	1967	48671
0661	S	D	E	N	1966	42440
0668	S	D	E	N	1966	42440
0670	S	D	E	N	1966	42440
0672	S	D	E	L	1964	35796
0672	S	D	E	N	1970	57581
0672	S	D	E	N	1970	59427
0672	M	D	E	N	1965	37915
0672	M	D	E	N	1968	50953
0672	M	D	E	N	1969	53611
0672	M	D	E	N	1966	58327
0678	M	G	E	H	1967	34726
0690	S	D	E	L	1967	44158
0694	S	D	E	N	1966	40343
0696	S	D	E	N	1966	40343
0700	S	D	E	H	1966	33858
0700	S	D	E	N	1965	35894
0700	M	D	E	H	1964	48507
0703	S	D	E	N	1967	44986
0703	S	D	E	N	1966	48383
0704	S	D	E	N	1967	44986
0704	S	D	E	N	1966	48383
0705	S	D	E	N	1967	44986
0705	S	D	E	N	1966	48383
0705	S	D	E	N	1968	53086
0706	S	D	E	N	1967	44986
0706	S	D	E	N	1966	48383
0707	M	D	E	N	1967	47759
0710	S	D	E	L	1967	47286
0712	S	D	E	L	1967	47286
0713	S	D	E	L	1967	47286
0715	S	D	E	L	1967	47286
0716	S	D	E	L	1967	47286
0718	S	D	E	L	1967	47286
0720	S	G	E	N	1969	35338
0720	S	D	E	N	1965	39754
0720	S	G	E	N	1969	55248
0721	S	D	E	N	1965	39754
0753	S	D	E	N	1965	39754

Phys. State: **G.** Gas; **L.** Liquid; **M.** Multiphase; **P.** Powder; **S.** Solid

Substance Number	Phys. State	Subject	Language	Temperature	Year	TPRC Number
541-0756	S	D	E	N	1965	39754
0757	S	D	E	N	1966	39873
0758	M	D	E	N	1963	51271
0759	M	D	E	N	1963	51271
0761	S	C	R	H	1958	51560
0762	S	C	R	H	1958	51560
0766	P	D	E	N	1960	51576
0773	S	D	E	N	1966	36480
0773	M	S	E	L	1970	61245
0780	S	D	E	N	1967	49153
0780	S	D	E	H	1968	49527
0780	S	D	E	H	1968	51490
0780	S	D	E	N	1967	52118
0780	S	D	E	H	1968	54179
0780	S	D	E	H	1969	55048
0789	S	D	E	L	1937	51520
0790	S	D	E	L	1937	51520
0802	S	D	E	N	1965	36145
0821	S	D	E	H	1967	45482
0821	S	D	E	N	1967	49153
0821	S	D	E	H	1968	49527
0821	S	D	E	H	1968	51490
0821	S	D	E	N	1969	53607
0821	S	D	E	H	1968	54179
0821	S	D	E	H	1969	55048
0823	S	D	E	N	1968	51490
0825	S	D	E	N	1967	49153
0825	S	D	E	N	1968	49527
0825	S	D	E	N	1968	51490
0825	S	D	E	H	1968	54179
0828	S	D	E	N	1966	40527
0828	S	D	E	N	1968	51490
0828	S	D	E	H	1968	54179
0829	S	D	E	N	1968	51490
0831	S	D	E	H	1968	49527
0831	S	D	E	H	1967	52118
0831	S	D	E	N	1968	54315
0831	S	D	E	H	1969	55048
0831	S	D	E	N	1970	57658
0831	S	C	E	N	1970	61186
0831	M	D	E	N	1969	58321
0832	S	D	E	N	1967	46397
0832	S	D	E	H	1968	49527
0832	S	D	E	N	1967	52118
0832	S	D	E	H	1969	55048
0832	S	C	E	N	1970	61186
0832	M	D	E	N	1969	58321
0833	S	S	G	U	1966	51698
0833	S	S	E	U	1966	51699
0837	S	D	I	N	1967	49515
0853	S	D	E	L	1964	37409
0865	S	D	E	L	1967	47285
0865	S	D	E	L	1967	48844
0865	S	C	E	L	1969	53543
0865	S	D	E	N	1970	59607
0865	S	D	E	N	1970	61191
0865	M	D	E	N	1961	56385
0866	S	C	E	N	1964	48814
0866	M	D	E	N	1969	54242
0867	S	D	E	H	1968	48826
0867	M	D	E	H	1968	48862
0867	M	D	E	N	1969	58321
0868	M	D	E	H	1968	48862
0869	S	D	E	H	1968	48826
0869	M	D	E	H	1968	48862
0880	S	D	E	N	1966	40343
0880	M	D	E	N	1966	48811
0881	M	D	E	N	1966	48811
0882	S	D	E	N	1970	59607
0882	M	D	R	N	1967	45570
0882	M	D	E	N	1966	48811
0882	M	D	G	L	1969	57859
0883	M	D	E	N	1966	48811
0906	M	D	E	N	1967	47759
0912	S	D	E	H	1965	36420
0912	S	D	E	N	1967	46397
0912	S	D	E	N	1967	52118
0914	S	D	E	H	1965	36420
0915	S	D	E	N	1959	47714
0918	S	C	E	N	1964	46334
0918	S	C	E	N	1969	53853
0918	S	C	E	N	1968	58397
0940	M	D	E	N	1948	47531
0941	M	D	E	N	1948	47531
0962	S	D	E	N	1968	50308
0964	S	D	E	N	1966	41169
0964	S	D	R	L	1967	46864
0964	S	D	E	L	1968	50417
0964	S	D	E	N	1967	53612
0964	M	D	R	N	1966	42017
0965	M	C	E	N	1969	55805
0966	S	D	R	N	1967	46864
0966	S	D	E	N	1968	50417
0967	S	D	F	H	1966	46504
0971	S	C	E	N	1969	55805
0975	M	D	E	H	1961	49007
0976	S	D	E	N	1966	38190
0976	S	D	R	N	1966	41482
0976	S	D	E	N	1967	45669
0976	S	D	E	U	1968	47410
0976	S	D	E	N	1969	53611
541-0976	S	D	E	N	1967	53612
0976	S	D	E	U	1970	61192
0976	M	D	R	N	1967	45570
0976	M	D	E	N	1969	57851
0976	M	C	E	L	1969	59276
0977	S	D	E	N	1965	35537
0977	S	D	E	N	1967	46639
0977	M	D	R	N	1967	45570
0977	M	D	E	N	1970	61704
0978	M	D	E	H	1961	49007
0979	M	D	E	H	1961	49007
0980	M	D	R.	N	1967	45570
0981	S	T	E	H	1966	54318
0989	S	D	E	H	1966	49761
0994	S	D	E	U	1968	47410
1003	S	D	E	H	1961	46647
1004	S	D	E	L	1964	35796
1004	S	D	E	N	1965	39745
1004	S	D	E	N	1966	41169
1004	S	D	E	N	1967	46639
1004	S	D	E	N	1969	53611
1004	S	G	E	N	1970	59608
1005	S	D	E	N	1967	46639
1006	S	D	E	N	1967	46639
1007	S	D	E	N	1967	46639
1007	M	C	E	L	1969	59276
1007	M	E	E	L	1969	60025
1008	S	D	E	N	1967	46639
1009	S	D	E	N	1967	46639
1010	S	D	E	N	1967	46639
1035	M	D	E	H	1962	52417
1045	S	D	E	H	1968	48826
1045	M	D	E	H	1969	58320
1050	S	T	E	U	1964	37011
1050	S	T	E	U	1960	52125
1080	S	D	E	N	1966	35741
1082	S	E	E	U	1969	47673
1084	S	D	E	N	1967	44986
1084	S	D	E	N	1967	45669
1084	S	D	E	U	1970	61192
1084	M	D	E	N	1969	57851
1085	S	D	E	N	1967	45669
1086	S	D	E	N	1966	39873
1087	S	D	E	N	1966	39873
1087	S	D	E	H	1969	55048
1088	S	D	E	N	1959	47714
1089	S	D	E	N	1959	47714
1090	S	D	E	N	1965	35537
1091	S	D	E	N	1965	35537
1092	S	D	E	N	1965	35539
1094	M	S	E	N	1968	50531
1095	S	D	E	N	1936	50660
1096	S	D	E	N	1959	50688
1114	S	D	E	N	1968	51905
1114	P	D	E	N	1967	52200
1115	M	D	E	N	1968	51905
1115	M	D	E	N	1967	52200
1116	S	D	E	N	1968	51905
1116	P	D	E	N	1967	52200
1129	S	D	E	N	1967	52890
1130	S	D	E	N	1966	52891
1131	S	D	E	L	1968	51868
1136	S	D	E	N	1969	37375
1136	S	D	E	N	1968	54256
1136	M	D	E	N	1969	57853
1139	S	D	E	N	1969	55945
1139	S	T	E	N	1970	59140
1141	S	D	E	N	1966	36550
1142	S	D	E	N	1964	35796
1142	S	C	E	N	1968	56374
1143	S	D	E	N	1968	37491
1143	S	D	E	N	1970	59607
1143	S	D	E	N	1970	63133
1145	S	D	E	N	1967	44986
1145	S	D	E	N	1966	48383
1146	S	D	E	N	1967	44986
1146	S	D	E	N	1970	63135
1147	S	D	E	N	1967	44986
1148	S	D	E	N	1970	60715
1149	M	G	E	N	1970	60715
1150	S	G	E	N	1970	60715
1151	M	G	E	N	1970	60715
1152	S	C	E	U	1970	60107
1153	S	C	E	U	1970	60107
1155	S	C	E	U	1970	60107
1156	S	D	E	N	1966	34851
1157	S	D	E	N	1966	34851
1171	S	T	E	U	1966	54228
1173	S	D	E	N	1968	53087
1174	S	D	E	H	1968	54256
1175	S	D	E	H	1968	54256
1175	M	G	E	H	1967	34726
1176	S	D	E	N	1968	54256
1177	S	D	E	N	1968	54256
1178	S	D	E	N	1968	54256
1179	S	D	E	N	1969	54293
1201	S	D	E	N	1968	54315
1215	S	D	E	L	1964	35796
1217	S	D	R	H	1970	62054
1217	S	D	E	H	1970	62055
1218	S	D	E	L	1964	35796
541-1219	S	D	E	L	1964	35796
1220	S	D	E	L	1964	35796
1221	S	D	E	L	1964	35796
1222	S	D	E	L	1964	35796
1223	S	D	E	H	1964	35815
1224	S	D	E	H	1964	35812
1225	S	D	E	H	1964	35812
1226	S	G	E	N	1969	35338
1226	S	G	E	N	1969	55248
1228	S	G	E	N	1969	35338
1228	S	G	E	N	1969	55248
1229	S	D	E	N	1964	36290
1230	S	G	E	N	1969	35338
1230	S	G	E	N	1969	55248
1231	S	D	E	H	1964	36123
1232	S	D	E	H	1964	36123
1232	S	D	E	H	1964	37393
1233	S	D	E	N	1962	36109
1234	S	D	E	N	1962	36109
1235	S	D	E	N	1962	36109
1236	S	D	E	N	1962	36109
1237	S	D	E	N	1962	36109
1239	M	G	E	N	1968	56530
1239	M	D	E	N	1970	59139
1240	M	D	E	H	1964	37393
1240	M	D	E	H	1965	43489
1241	S	D	E	H	1964	37393
1244	P	G	E	N	1969	59176
1245	M	C	E	L	1969	60028
1267	M	D	E	N	1965	39745
1270	M	D	E	N	1965	39745
1275	S	D	E	N	1966	41210
1276	S	D	E	L	1966	41218
1280	S	D	G	N	1969	55722
1295	S	S	E	N	1969	57666
1296	M	D	R	N	1966	42017
1297	S	D	E	N	1970	40761
1298	S	D	E	N	1970	40761
1308	S	D	E	L	1969	57598
1322	S	D	E	N	1970	58453
1325	S	D	E	N	1969	58202
1327	M	C	R	N	1969	57009
1328	M	C	R	N	1969	57009
1329	S	D	R	N	1969	55826
1344	M	C	E	N	1970	38913
1344	M	C	R	N	1970	62437
1348	S	D	E	N	1960	42824
1356	S	D	E	N	1967	44986
1357	S	D	E	N	1967	44986
1358	S	S	E	F	1966	42294
1358	S	D	E	H	1962	46907
1358	S	D	E	H	1959	47025
1358	S	D	E	H	1962	48004
1359	S	D	E	H	1962	46907
1359	S	D	E	H	1959	47025
1359	S	D	E	H	1962	48004
1360	S	D	R	N	1965	42392
1360	S	D	E	N	1965	44182
1363	S	D	E	N	1968	50176
1364	S	D	E	N	1966	54194
1365	M	D	E	N	1966	48811
1392	S	D	E	N	1970	61187
1393	S	D	E	U	1970	61192
1395	S	G	E	N	1969	55248
1397	S	D	F	N	1970	59191
1398	S	D	F	N	1970	60706
1399	M	D	E	N	1968	59204
1402	M	T	R	N	1968	61051
1405	S	D	E	L	1968	56708
1406	S	E	E	N	1968	57455
1406	S	E	E	N	1970	57658
1415	S	C	E	N	1965	63107
1416	M	D	E	N	1968	53113
1417	M	D	E	N	1968	53113
1418	M	D	E	N	1965	62180
1429	S	D	R	N	1970	61567
551-0153	S	D	E	H	1959	47025
0153	S	S	E	H	1967	50724
0153	S	D	E	F	1955	60523
0154	S	D	E	H	1959	47025
0154	S	S	E	H	1967	50724
0154	S	D	E	F	1955	60523
0487	S	C	R	H	1966	42012
1289	S	D	E	H	1967	50724
1289	S	D	E	F	1955	60523
1290	S	S	E	U	1967	50724
1321	S	D	E	U	1968	51703
1571	S	G	E	N	1969	53583
1571	S	E	F	N	1967	53680
2438	S	C	E	L	1968	45517
2608	S	D	E	N	1965	39065
2763	S	S	E	H	1967	50724
621-0001	S	D	E	N	1968	37491
0018	S	D	E	N	1968	37491
0023	S	D	E	U	1964	38249
0024	S	C	E	N	1969	49616
0024	S	D	E	N	1969	56600
0028	S	D	E	N	1960	42824
0062	S	D	E	N	1968	37491

Phys. State: **G.** Gas; **L.** Liquid; **M.** Multiphase; **P.** Powder; **S.** Solid

Thermal Conductivity B6

Substance Number	Phys. State	Sub-ject	Lan-guage	Temper-ature	Year	TPRC Number
621-0073	S	D	E	N	1965	46221
0077	S	S	E	N	1944	48295
0086	S	D	E	N	1968	47768
0087	S	D	E	N	1968	47768
0089	S	D	E	N	1938	41684
0090	S	D	E	N	1938	41684
0091	S	D	E	N	1955	54296
0093	S	D	E	N	1969	56570
0100	S	D	E	N	1968	37491
0122	S	D	E	N	1969	56600
0123	S	D	E	N	1969	56600
0124	S	D	E	N	1969	56600
0125	S	D	E	N	1969	56600
0126	S	D	E	N	1969	56600
0129	S	D	E	N	1938	41684
0217	S	D	E	L	1968	37491
0217	S	D	O	N	1957	39734
0217	S	D	E	N	1937	48236
0227	S	C	E	N	1969	49616
0306	S	D	E	N	1968	47768
0308	S	D	E	N	1965	46221
0308	M	G	E	H	1969	57946
0316	S	D	E	N	1968	47768
0316	S	D	E	N	1970	59508
651-0071	S	D	R	N	1966	44457
0071	S	D	R	N	1968	55626
0071	S	D	E	N	1969	57533
0075	S	D	R	N	1963	49400
0075	S	D	E	N	1967	49401
0076	S	D	R	N	1966	44457
0079	S	D	R	N	1966	44457
0096	S	D	R	N	1966	44457
0096	S	D	R	N	1968	55626
0096	S	D	E	N	1969	57533
0097	S	D	R	N	1966	44457
0097	S	D	R	N	1968	55626
0097	S	D	E	N	1969	57533
0097	L	D	R	N	1968	55627
0097	L	D	E	N	1969	57537
0104	S	D	R	N	1966	44457
0106	S	D	R	N	1966	44457
0107	S	D	R	N	1966	44457
0129	S	D	R	N	1966	44457
0135	S	D	R	N	1966	44457
0135	S	D	R	N	1968	50095
0148	S	D	R	N	1966	44457
0148	S	D	R	N	1968	50095
0230	S	D	E	N	1968	53413
0233	S	D	R	N	1966	44457
0236	S	D	R	N	1966	44457
0275	S	D	E	N	1969	54293
0276	S	D	E	N	1969	54293
0277	S	C	E	N	1969	56569
0282	S	D	E	N	1969	56585
0283	S	D	R	N	1968	56209
0283	S	D	E	N	1969	56601
0284	S	D	E	L	1969	55497
0285	S	D	E	L	1969	55497
0290	S	D	R	N	1968	55626
0290	S	D	E	N	1969	57533
0291	S	D	R	N	1968	55626
0291	S	D	E	N	1969	57533
0292	S	D	R	N	1968	50095
0292	S	D	R	N	1968	55626
0292	S	D	E	N	1969	57533
0292	L	D	R	N	1968	55627
0292	L	D	E	N	1969	57537
0293	S	D	R	U	1968	53488
0293	S	D	E	U	1968	53903
0293	S	D	R	N	1968	55626
0293	S	D	E	N	1969	57533
0298	L	D	R	N	1968	55627
0298	L	D	E	N	1969	57537
0302	L	D	R	N	1968	55627
0302	L	D	E	N	1969	57537
0309	L	D	R	N	1968	55627
0309	L	D	E	N	1969	57537
0310	L	D	R	N	1968	55627
0310	L	D	E	N	1969	57537
0311	L	D	R	N	1968	55627
0311	L	D	E	N	1969	57537
0312	L	D	R	N	1968	55627
0312	L	D	E	N	1969	57537
0313	L	D	R	N	1968	55627
0313	L	D	E	N	1969	57537
0314	L	D	R	N	1968	55627
0314	L	D	E	N	1969	57537
0320	S	D	E	N	1970	58454
0321	S	D	E	N	1970	58454
0347	S	D	E	N	1970	58454
0364	S	D	E	N	1970	58454
0466	S	D	R	N	1968	50095
0476	S	D	R	N	1965	34196
0476	S	D	E	N	1965	34196
0476	S	D	E	N	1921	40025
0476	L	D	E	N	1968	51851
0476	M	D	E	N	1954	43591
0544	S	S	E	N	1968	42148
661-0001	S	D	E	N	1964	36199

Substance Number	Phys. State	Sub-ject	Lan-guage	Temper-ature	Year	TPRC Number
661-0001	S	D	E	N	1964	37058
0001	S	D	E	N	1968	37491
0001	S	D	E	N	1966	39442
0001	S	D	E	N	1953	39860
0001	S	D	E	N	1921	40025
0001	S	D	E	N	1967	42907
0001	S	D	E	N	1916	48034
0001	S	D	E	N	1938	48291
0001	S	D	E	N	1967	55009
0001	S	D	E	N	1953	55981
0001	M	D	F	N	1963	36058
0001	M	D	E	N	1933	39866
0002	S	C	E	N	1964	34115
0002	S	G	G	N	1964	37646
0002	S	D	E	N	1927	39852
0002	S	D	E	N	1943	40228
0003	S	D	E	N	1921	40025
0003	S	D	E	N	1916	48034
0003	M	D	F	N	1963	36058
0003	M	D	R	N	1968	51025
0004	S	D	E	N	1950	40447
0004	S	D	E	N	1947	40712
0004	S	D	E	N	1953	48387
0004	M	D	E	N	1933	39866
0008	S	D	E	N	1967	42907
0008	S	D	E	N	1967	43783
0008	S	D	E	N	1967	55009
0010	S	D	E	N	1966	44554
0014	M	D	R	N	1969	56318
0016	S	D	E	N	1968	37491
0017	S	D	E	N	1969	37570
0017	S	D	E	N	1968	48434
0017	S	D	E	N	1967	49250
0017	S	D	E	N	1968	50307
0017	S	D	E	N	1969	55028
0017	S	D	E	N	1968	57191
0017	S	D	E	N	1970	57639
0017	P	D	E	N	1967	49152
0017	P	D	E	N	1967	49250
0017	P	D	F	N	1968	54189
0019	S	D	E	N	1943	40228
0034	S	D	E	N	1938	48291
0038	S	D	G	N	1951	43837
0039	S	D	E	N	1943	40228
0039	S	D	E	N	1961	53006
0040	S	S	E	L	1967	33988
0040	S	D	E	N	1964	37058
0040	S	E	E	N	1966	38190
0040	S	S	E	N	1963	38525
0040	S	S	E	N	1932	39844
0040	S	S	E	U	1963	40300
0040	S	S	E	U	1965	40724
0040	S	E	R	N	1966	41482
0040	S	D	E	N	1960	42824
0040	S	E	I	U	1965	43636
0040	S	D	E	N	1919	43637
0040	S	E	F	N	1967	46184
0040	S	D	E	N	1958	47730
0040	S	E	E	N	1962	47928
0040	S	E	E	L	1961	47930
0040	S	E	E	N	1962	48330
0040	S	E	E	N	1968	51890
0040	S	S	E	N	1959	54286
0040	S	E	E	N	1969	54423
0040	S	T	E	U	1968	55228
0040	S	S	E	L	1967	56366
0040	S	S	E	N	1967	56379
0040	S	T	E	U	1970	57660
0040	S	D	E	U	1970	61193
0040	S	E	E	U	1965	62179
0040	G	D	E	N	1960	42824
0040	P	D	E	N	1960	42824
0040	M	D	E	N	1966	42708
0040	M	S	E	N	1966	42916
0041	S	D	E	N	1943	40228
0041	S	D	E	N	1950	40447
0042	S	E	R	N	1963	35944
0042	S	E	F	N	1963	35945
0042	S	E	E	N	1964	37058
0042	S	D	E	N	1954	40226
0042	S	E	E	N	1950	40447
0042	S	E	R	U	1967	41506
0042	S	E	R	N	1965	43097
0042	S	E	R	N	1965	43098
0042	S	D	E	N	1916	48034
0042	S	S	E	N	1966	52909
0042	S	E	R	N	1966	59500
0042	S	E	E	N	1966	59501
0042	P	E	E	N	1950	40447
0043	S	D	E	N	1943	40228
0047	M	D	E	N	1966	42708
0049	S	D	E	N	1943	40228
0050	S	D	E	N	1968	37491
0050	S	D	E	N	1964	41516
0050	M	D	E	N	1938	43605
0051	S	D	G	N	1949	57056
0052	S	C	E	N	1970	57663
0052	M	C	E	N	1968	47907
0052	M	D	G	N	1962	61705
0054	S	D	E	N	1943	40228
0058	S	D	E	N	1921	40025

Substance Number	Phys. State	Sub-ject	Lan-guage	Temper-ature	Year	TPRC Number
661-0058	S	D	E	N	1916	48034
0060	S	D	E	N	1943	40228
0060	S	D	E	N	1950	40447
0060	S	G	E	N	1956	40670
0060	S	D	E	N	1967	42907
0060	S	D	E	N	1967	55009
0060	M	D	E	N	1933	39866
0066	S	D	E	N	1943	40228
0086	S	D	E	N	1922	39850
0090	S	D	E	N	1964	36327
0094	S	D	E	N	1927	39852
0104	S	D	E	N	1916	48034
0106	S	D	E	N	1943	40228
0106	S	D	E	N	1967	43783
0107	S	D	E	N	1969	54423
0108	S	D	E	N	1927	39852
0109	S	D	E	N	1927	39852
0111	S	D	E	N	1947	40712
0113	S	D	E	N	1921	40025
0114	S	D	E	N	1921	40025
0116	S	D	E	N	1921	40025
0116	S	D	E	N	1916	48034
0117	S	D	E	N	1964	37058
0117	S	D	E	N	1921	40025
0119	S	D	E	N	1943	40228
0119	S	G	E	N	1956	40670
0119	S	D	E	N	1953	48387
0122	S	D	F	N	1967	46184
0124	S	D	E	N	1921	40025
0124	S	D	E	N	1943	40228
0124	S	D	E	N	1916	48034
0137	S	C	E	N	1970	57663
0143	S	D	E	N	1943	40228
0152	M	D	E	H	1961	49007
0155	S	D	F	N	1969	62533
0156	S	D	E	N	1943	40228
0164	S	D	E	N	1921	40025
0164	S	D	E	N	1969	53585
0164	M	D	I	N	1966	41292
0164	M	D	I	N	1966	44990
0182	S	D	E	N	1967	42907
0182	S	D	E	N	1967	55009
0183	S	D	E	N	1969	53608
0183	S	G	E	U	1969	53609
0183	M	C	O	N	1958	51572
0188	S	D	E	N	1943	40228
0193	S	E	I	U	1961	48349
0193	S	E	I	N	1967	51616
0193	S	E	E	U	1969	53608
0193	M	E	O	N	1963	37815
0193	M	D	I	N	1967	44995
0193	M	E	E	U	1924	47970
0211	S	D	E	N	1943	39833
0222	S	D	E	N	1962	48330
0222	S	D	E	N	1968	50174
0222	S	D	E	N	1969	59206
0243	S	D	E	N	1967	42907
0243	S	D	E	N	1967	55009
0246	S	D	E	N	1943	40228
0249	S	D	E	N	1960	41450
0285	S	D	E	N	1921	40025
0285	S	D	E	N	1958	47730
0285	S	D	E	N	1968	47768
0285	S	D	E	N	1916	48034
0285	M	D	E	N	1933	39866
0287	S	D	E	N	1927	39852
0291	S	D	E	N	1968	37491
0291	S	D	E	N	1921	40025
0291	S	G	E	N	1956	40670
0291	S	D	E	N	1968	47768
0291	S	D	E	N	1916	48034
0291	M	D	R	N	1967	46170
0292	S	D	E	N	1937	41679
0292	S	C	O	N	1967	56145
0295	S	S	E	N	1961	50659
0295	S	D	E	N	1970	61878
0295	M	S	F	U	1965	50884
0297	S	D	E	N	1943	40228
0297	S	D	E	N	1916	48034
0297	M	D	E	N	1921	40025
0299	M	D	E	N	1933	39866
0307	S	D	E	N	1965	34264
0318	S	D	E	N	1943	40228
0320	S	S	E	N	1965	34428
0320	S	D	E	H	1964	36797
0320	S	D	E	N	1964	41910
0320	S	D	E	N	1965	46221
0320	M	D	G	N	1967	46106
0321	S	D	E	N	1943	40228
0327	S	D	R	N	1964	38609
0327	S	D	E	N	1964	38610
0327	S	E	E	N	1953	40669
0330	S	D	E	N	1943	40228
0331	S	D	E	H	1965	36144
0336	S	D	E	N	1921	40025
0345	S	D	E	N	1921	40025
0345	S	D	E	N	1916	48034
0346	S	D	E	N	1921	40025
0346	S	D	E	N	1916	48034
0347	S	D	E	N	1921	40025
0347	S	D	E	N	1916	48034

Phys. State: **G.** Gas; **L.** Liquid; **M.** Multiphase; **P.** Powder; **S.** Solid

Substance Number	Phys. State	Sub-ject	Lan-guage	Temper-ature	Year	TPRC Number	Substance Number	Phys. State	Sub-ject	Lan-guage	Temper-ature	Year	TPRC Number	Substance Number	Phys. State	Sub-ject	Lan-guage	Temper-ature	Year	TPRC Number
661-0349	S	G	E	N	1966	44681	661-0468	M	D	E	N	1958	56356							
0350	S	G	E	N	1953	40669	0468	M	D	E	N	1970	57595							
0350	S	D	E	N	1949	42306	0468	M	E	E	L	1970	57647							
0350	S	D	E	N	1938	48291	0468	M	E	E	N	1969	57940							
0351	S	D	E	N	1948	40713	0468	M	D	E	N	1970	61878							
0351	S	D	E	N	1949	42306	0468	M	E	E	U	1965	62179							
0352	S	C	E	N	1964	34115	0470	S	D	E	N	1964	37058							
0352	S	D	E	N	1949	42306	0471	S	D	E	N	1970	60111							
0352	M	G	G	N	1964	37646	0472	S	D	E	N	1921	40025							
0354	S	D	E	N	1921	40025	0473	M	D	G	N	1962	61705							
0354	S	D	E	N	1916	48034	0474	M	D	E	N	1968	54818							
0355	S	D	E	N	1921	40025	0475	S	D	E	N	1947	40712							
0355	S	D	E	N	1916	48034	0475	S	D	E	N	1948	40713							
0356	S	D	E	N	1947	40712	0476	S	D	E	N	1948	40713							
0356	S	D	E	N	1948	40713	0481	S	S	E	N	1970	36021							
0358	S	D	E	N	1921	40025	0483	S	D	R	N	1970	59149							
0358	S	D	E	N	1916	48034	0483	S	D	R	N	1970	59150							
0358	M	D	E	N	1899	48324	0484	S	D	R	N	1970	59149							
0359	S	D	R	N	1965	39987	0484	S	D	E	N	1970	59150							
0359	S	D	E	N	1965	39988	0485	M	D	R	N	1969	56318							
0359	S	D	E	N	1967	43783														
0360	S	D	E	N	1938	48291														
0366	S	S	E	N	1965	34428														
0366	S	D	E	N	1965	46221														
0367	S	D	E	N	1966	44554														
0375	S	D	E	N	1943	40228														
0389	M	D	J	N	1964	36571														
0389	M	D	E	N	1933	39866														
0392	S	G	E	N	1966	44681														
0392	S	G	F	N	1966	44681														
0394	S	D	R	N	1965	39987														
0394	S	D	E	N	1965	39988														
0395	S	D	R	N	1965	39987														
0395	S	D	E	N	1965	39988														
0396	S	G	E	N	1966	44681														
0397	S	E	E	N	1966	44681														
0397	S	S	E	U	1967	45042														
0397	S	D	E	N	1968	47768														
0397	S	D	E	N	1968	53086														
0397	S	E	E	U	1968	54277														
0397	S	T	E	N	1968	56374														
0397	M	T	E	L	1968	58324														
0398	S	G	E	N	1966	44681														
0399	S	D	E	N	1966	42908														
0400	S	D	E	N	1966	42908														
0401	S	D	E	N	1966	42908														
0402	S	D	E	N	1966	42908														
0403	S	D	E	N	1966	42908														
0404	S	D	E	N	1966	42908														
0405	S	D	E	N	1966	42908														
0406	M	D	E	N	1966	45500														
0407	M	D	E	N	1966	45500														
0408	M	D	E	N	1966	42332														
0409	S	D	E	H	1931	39843														
0409	S	D	E	N	1949	48283														
0410	S	D	G	N	1949	57056														
0411	S	D	E	N	1952	49663														
0412	S	D	E	N	1938	48291														
0413	S	D	E	N	1938	48291														
0414	S	D	E	N	1938	48291														
0416	S	D	E	N	1967	49152														
0416	S	D	E	N	1967	50031														
0417	M	D	E	N	1966	48811														
0418	S	D	E	N	1916	48034														
0419	S	D	E	N	1968	47768														
0420	S	D	E	N	1968	47768														
0421	S	D	E	N	1968	47768														
0422	S	D	E	N	1968	47768														
0423	S	D	E	N	1968	47768														
0424	S	D	E	N	1968	47768														
0425	M	D	R	N	1967	45800														
0427	S	D	O	N	1960	44873														
0427	S	D	E	N	1961	44874														
0428	S	D	O	N	1960	44873														
0428	S	D	E	N	1961	44874														
0429	S	D	G	N	1949	57056														
0430	S	D	E	N	1969	53585														
0430	M	D	I	N	1966	41292														
0430	M	D	I	N	1966	44990														
0431	M	C	E	N	1968	47907														
0433	M	D	R	N	1968	50878														
0434	S	D	R	N	1970	59149														
0434	S	D	E	N	1970	59150														
0442	M	D	J	N	1966	49874														
0444	S	D	R	N	1970	59149														
0444	S	D	E	N	1970	59150														
0445	S	D	R	N	1969	58439														
0447	S	D	E	N	1927	39852														
0448	S	D	E	N	1964	36327														
0449	S	D	E	N	1927	39852														
0450	M	D	E	N	1933	39866														
0451	M	D	E	N	1933	39866														
0452	S	D	R	N	1969	58439														
0455	M	D	E	N	1967	45044														
0456	S	D	E	N	1962	47435														
0459	S	D	E	N	1968	51888														
0460	S	D	E	N	1968	51888														
0461	S	D	E	N	1966	44554														
0462	S	D	E	N	1964	37058														
0466	S	D	E	N	1969	54423														
0468	M	D	E	N	1960	42824														

Phys. State: G. Gas; L. Liquid; M. Multiphase; P. Powder; S. Solid

Chapter 2 Accommodation Coefficient

Substance Number	Phys. State	Sub-ject	Lan-guage	Temper-ature	Year	TPRC Number
531-0375	G	C	E	N	1967	47504
0375	M	C	E	N	1965	36440
0376	M	C	E	N	1965	36440
0377	M	C	E	N	1965	36440
0379	G	T	E	U	1968	50913
0379	M	C	E	U	1965	35129
0379	M	T	E	U	1965	36460
0379	M	T	E	U	1968	39483
0379	M	T	E	N	1966	40323
0379	M	T	E	U	1967	43452
0379	M	T	G	U	1935	43777
0379	M	T	E	U	1966	43778
0379	M	S	E	N	1967	45309
0379	M	T	E	U	1967	45492
0379	M	S	E	L	1967	46119
0379	M	E	E	U	1967	46400
0379	M	S	E	U	1967	46400
0379	M	T	E	U	1967	47079
0379	M	S	E	N	1967	47216
0379	M	T	E	U	1968	47743
0379	M	T	E	U	1968	49264
0379	M	E	E	U	1958	50301
0379	M	E	E	U	1958	51331
0379	M	E	E	U	1943	52655
0379	M	T	E	R	1966	54985
0379	M	T	E	U	1966	54986
0379	M	T	E	U	1970	59117
0379	M	E	E	N	1969	59169
0433	M	T	E	U	1966	34054
0904	M	C	F	N	1964	35268
0905	M	C	F	N	1964	35268
0906	M	C	F	N	1964	35268
0907	M	C	F	N	1964	35268
0972	M	C	E	N	1967	43692
0973	M	C	E	N	1967	43692
0973	M	D	E	N	1969	53052
1082	M	D	G	N	1955	43349
1082	M	G	G	N	1956	43357
1082	M	D	E	N	1958	50301
1082	M	D	E	N	1936	51602
1082	M	D	E	L	1937	51603
1082	M	D	E	L	1936	54137
1083	M	D	G	N	1955	43349
1083	M	G	G	N	1956	43357
1083	M	D	E	N	1936	51602
1083	M	D	E	L	1937	51603
1083	M	D	E	L	1936	54137
1084	M	D	G	N	1955	43349
1084	M	G	G	N	1956	43357
1084	M	D	E	N	1936	51602
1084	M	D	E	L	1936	54137
1085	M	D	G	N	1955	43349
1085	M	G	G	N	1956	43357
1085	M	D	E	N	1936	51602
1085	M	D	E	L	1937	51603
1085	M	D	E	L	1936	54137
1092	M	T	E	U	1932	51681
1117	M	D	E	N	1968	48697
1152	G	C	E	N	1967	47504
1153	G	C	E	N	1967	47504
1154	G	C	E	N	1967	47504
1212	M	G	E	N	1967	45253
1212	M	D	F	N	1967	49910
1213	M	G	E	N	1967	45253
1262	M	D	G	N	1955	43349
1262	M	G	G	N	1956	43357
1262	M	D	F	N	1967	49910
1263	M	D	G	N	1955	43349
1263	M	G	G	N	1956	43357
1264	M	D	G	N	1955	43349
1264	M	G	G	N	1956	43357
1265	M	G	G	N	1956	43357
1284	M	D	F	H	1970	58359
1331	M	D	F	H	1970	58359
1340	M	D	F	H	1970	58359
1340	M	G	F	N	1970	60695
1359	M	D	F	H	1970	58359
1376	M	D	F	H	1970	58359
1382	M	D	F	H	1970	58359
1382	M	G	F	N	1970	60695
1383	M	D	F	H	1970	58359
1448	M	D	E	N	1965	35659
1449	M	D	E	N	1965	35659
1450	M	D	E	N	1965	35659
1451	M	D	E	N	1965	35659
1452	M	D	E	N	1965	35659
1453	M	D	F	N	1970	60037
1468	M	D	F	N	1970	60037
1469	G	D	F	U	1970	60035
1618	M	D	E	N	1965	39094
1647	M	D	F	N	1970	58474
1647	M	D	F	N	1970	58891
1754	M	D	E	N	1969	38272
1791	P	D	E	N	1969	54820
541-0032	M	D	F	N	1966	40173
0032	M	D	F	H	1967	42840
0032	M	D	E	N	1967	51342
0032	M	D	F	U	1968	55357
0033	M	D	F	N	1966	40173
0033	M	D	F	H	1967	42840
0033	M	D	E	N	1967	51342
0033	M	D	F	U	1968	55357
0040	M	D	F	N	1966	40173
0040	M	D	F	H	1967	42840
0040	M	D	E	N	1967	51342
0040	M	D	F	U	1968	55357
0227	M	C	E	N	1964	38030
0227	M	G	E	N	1964	40587
0227	M	C	E	N	1952	40677
0315	S	S	E	L	1967	41113
0315	G	C	E	L	1937	51517
0315	M	D	E	N	1965	35851
0315	M	C	E	L	1963	37326
0315	M	D	E	N	1965	38716
0315	M	D	E	N	1966	39530
0315	M	D	E	L	1966	39881
0315	M	C	E	R	1939	40135
0315	M	C	E	H	1967	42096
0315	M	C	E	N	1966	43207
0315	M	C	E	N	1967	43452
0315	M	T	E	N	1967	45253
0315	M	G	G	N	1967	45745
0315	M	C	E	U	1967	46496
0315	M	C	E	N	1967	46879
0315	M	D	E	N	1957	47586
0315	M	G	E	N	1954	48863
0315	M	C	E	N	1967	49402
0315	M	D	E	N	1962	49580
0315	M	D	E	N	1954	50299
0315	M	C	E	N	1958	50301
0315	M	G	E	N	1957	50302
0315	M	D	E	N	1955	51276
0315	M	C	E	L	1962	51287
0315	M	C	J	F	1968	51373
0315	M	C	E	L	1932	51682
0315	M	C	E	L	1932	51683
0315	M	C	E	L	1933	51684
0315	M	D	E	L	1933	51685
0315	M	G	E	H	1962	51706
0315	M	G	E	U	1930	51708
0315	M	C	E	N	1967	51726
0315	M	C	E	N	1957	51727
0315	M	C	E	N	1967	52740
0315	M	C	E	N	1969	53075
0315	M	C	G	N	1936	54128
0315	M	C	E	U	1970	58897
0315	M	C	E	N	1969	59894
0325	M	D	E	N	1966	33923
0325	M	D	F	N	1964	36706
0325	M	D	F	N	1952	40677
0325	M	D	F	H	1967	42840
0325	M	D	E	N	1967	45307
0325	M	D	E	N	1967	45308
0325	M	D	F	N	1962	49580
0325	M	D	F	N	1967	49910
0325	M	D	E	N	1954	50299
0325	M	D	E	N	1949	50926
0325	M	D	E	N	1944	51274
0325	M	D	E	N	1950	51275
0325	M	D	E	N	1934	51516
0325	M	D	E	N	1943	52655
0325	M	D	E	N	1957	52740
0325	M	D	G	N	1954	54149
0330	M	C	E	N	1966	33923
0330	M	C	E	N	1966	40323
0330	M	C	E	N	1952	40677
0330	M	C	E	H	1967	42096
0330	M	D	E	N	1966	42599
0330	M	D	E	N	1962	49580
0330	M	D	E	N	1954	50299
0330	M	G	E	N	1949	50926
0330	M	D	E	N	1944	51274
0330	M	D	E	N	1950	51275
0330	M	D	E	L	1934	51516
0330	M	D	E	N	1937	51518
0330	M	D	E	N	1943	52655
0330	M	C	E	N	1957	52740
0330	M	C	E	N	1969	53075
0330	M	C	E	U	1970	58897
0339	M	D	E	N	1965	35851
0339	M	C	E	L	1963	37326
0339	M	D	F	F	1964	38309
0339	M	D	E	N	1965	38716
0339	M	D	E	L	1966	39881
0339	M	C	E	R	1939	40135
0339	M	C	E	H	1967	42096
541-0339	M	D	G	N	1961	43152
0339	M	C	E	N	1966	43207
0339	M	T	E	N	1967	45252
0339	M	G	G	N	1967	45745
0339	M	C	E	U	1967	46496
0339	M	D	E	N	1967	46879
0339	M	D	E	H	1967	46949
0339	M	D	E	N	1957	47586
0339	M	D	E	N	1967	47665
0339	M	G	E	N	1954	48863
0339	M	C	E	N	1967	49402
0339	M	D	E	N	1962	49580
0339	M	D	E	N	1954	50299
0339	M	C	E	L	1962	51287
0339	M	D	E	N	1933	51685
0339	M	G	E	H	1962	51706
0339	M	D	E	N	1967	51726
0339	M	D	E	N	1957	52740
0339	M	C	E	N	1969	53075
0339	M	C	G	U	1936	54128
0339	M	C	E	U	1970	58897
0339	M	C	E	N	1969	59894
0340	M	D	E	N	1965	35851
0340	M	C	E	L	1963	37326
0340	M	C	E	L	1964	38251
0340	M	D	E	L	1965	38716
0340	M	D	E	L	1966	39881
0340	M	C	E	R	1939	40135
0340	M	C	E	H	1967	42096
0340	M	C	E	N	1966	43207
0340	M	T	E	N	1967	45252
0340	M	G	G	N	1967	45745
0340	M	C	E	U	1967	46496
0340	M	D	E	N	1967	46879
0340	M	D	E	H	1967	46949
0340	M	D	E	N	1967	47665
0340	M	C	E	N	1954	48863
0340	M	C	E	N	1967	49402
0340	M	D	E	N	1962	49580
0340	M	D	E	N	1954	50299
0341	M	C	E	L	1966	39881
0341	M	C	E	H	1967	42096
0341	M	C	E	N	1966	43207
0341	M	T	E	N	1967	45252
0341	M	G	E	N	1967	45745
0341	M	C	E	U	1967	46496
0341	M	D	E	N	1967	46879
0341	M	D	E	H	1967	46949
0341	M	C	E	N	1967	49402
0341	M	D	E	L	1962	51287
0341	M	C	E	N	1967	51726
0341	M	D	E	N	1967	52740
0341	M	D	E	N	1969	53075
0341	M	C	E	U	1970	58897
0341	M	C	E	N	1969	59894
0344	M	D	F	N	1965	35886
0344	M	D	E	N	1965	35893
0344	M	C	E	N	1966	39881
0344	M	C	E	H	1967	42096
0344	M	C	E	U	1967	46496
0344	M	D	F	N	1967	46879
0344	M	D	E	L	1954	50299
0344	M	C	E	L	1932	51682
0344	M	C	E	N	1957	52740
0344	M	C	E	N	1969	53075
0344	M	C	E	L	1963	37326
0344	M	C	E	L	1969	55847
0346	M	C	E	N	1966	39881
0346	M	C	E	H	1967	42096
0346	M	C	E	N	1966	43207
0346	M	G	G	U	1967	45745
0346	M	C	E	U	1967	46496
0346	M	D	E	N	1967	46879
0346	M	D	E	N	1967	49402
0346	M	D	E	N	1967	51726
0346	M	D	E	N	1957	52740
0346	M	C	E	U	1969	58897
0346	M	C	E	U	1970	59894
0346	M	C	E	H	1967	42096
0355	M	C	E	H	1967	42529
0355	M	G	E	N	1967	45253
0355	M	D	E	N	1962	49580
0355	M	D	E	N	1954	50299
0355	M	C	E	N	1949	50926

Accommodation Coefficient

B10

Substance Number	Phys. State	Subject	Language	Temperature	Year	TPRC Number
541-0356	M	C	E	H	1967	42096
0356	M	G	E	N	1957	50302
0357	M	C	E	H	1967	42096
0357	M	D	E	N	1967	42529
0357	M	D	E	L	1958	50301
0358	M	C	E	H	1967	42096
0358	M	D	E	N	1967	42529
0359	M	C	E	H	1967	42096
0359	M	G	G	N	1967	45745
0360	M	C	E	H	1967	42096
0360	M	D	E	N	1958	50301
0362	M	C	E	H	1967	42096
0362	M	D	E	N	1967	42529
0362	M	D	E	N	1962	49580
0362	M	D	E	N	1954	50299
0362	M	G	E	N	1949	50926
0363	M	C	E	H	1967	42096
0363	M	G	E	N	1957	50302
0364	M	C	E	H	1967	42096
0365	M	C	E	H	1967	42096
0365	M	D	E	N	1967	42529
0365	M	D	E	N	1958	50301
0374	M	C	E	H	1967	42096
0374	M	D	E	N	1967	42529
0375	M	C	E	H	1967	42096
0375	M	G	G	N	1967	45745
0376	M	C	E	H	1967	42096
0376	M	D	E	N	1958	50301
0376	M	C	E	L	1962	51287
0377	M	C	E	H	1967	42096
0377	M	D	E	N	1967	42529
0377	M	D	E	N	1958	50301
0378	M	C	E	H	1967	42096
0378	M	D	E	N	1967	42529
0381	M	C	E	H	1967	42096
0381	M	G	G	N	1967	45745
0383	M	C	E	H	1967	42096
0383	M	D	E	N	1958	50301
0384	M	C	E	H	1967	42096
0384	M	D	E	N	1967	42529
0385	M	C	E	H	1967	42096
0385	M	G	G	N	1967	45745
0388	M	C	E	H	1967	42096
0389	M	C	E	L	1966	39881
0389	M	C	E	H	1967	42096
0389	M	D	E	N	1969	53354
0389	M	C	E	L	1969	54049
0389	M	D	E	L	1969	55636
0389	M	G	E	N	1969	59892
0389	M	G	E	N	1970	61753
0390	M	C	E	L	1966	39881
0390	M	C	E	H	1967	42096
0390	M	D	E	N	1969	53354
0390	M	C	E	L	1969	54049
0390	M	D	E	L	1969	55636
0390	M	G	E	N	1969	59892
0390	M	G	E	N	1970	61753
0391	M	D	F	N	1964	36706
0391	M	D	E	N	1970	38104
0391	M	D	F	H	1967	42840
0391	M	D	E	N	1967	42858
0391	M	G	E	N	1967	45253
0391	M	D	E	N	1962	49580
0391	M	D	E	N	1954	50299
0391	M	G	E	N	1949	50926
0418	M	G	F	N	1966	39285
0419	M	G	F	N	1966	39285
0425	M	C	E	U	1967	46496
0439	M	D	E	N	1957	52740
0440	M	D	E	N	1957	52740
0441	M	D	E	N	1957	52740
0443	M	D	E	N	1957	52740
0444	M	D	E	H	1967	46949
0444	M	D	E	N	1967	51727
0444	M	D	E	N	1957	52740
0445	M	D	E	N	1957	52740
0446	M	D	F	N	1965	35886
0446	M	D	E	N	1957	52740
0447	M	D	E	N	1957	52740
0456	M	D	E	N	1966	39530
0457	M	D	G	N	1966	34936
0478	M	D	G	N	1954	54149
0479	M	D	G	N	1954	54149
0483	M	D	S	N	1965	34064
0484	M	D	S	N	1965	34064
0485	M	D	S	N	1965	34064
0486	M	D	S	N	1965	34064
0487	M	D	S	N	1965	34064
0488	M	D	S	N	1965	34064
0489	M	D	S	N	1965	34064
0489	M	D	S	N	1964	36607
0490	M	D	S	N	1965	34064
0490	M	D	S	N	1964	36607
0491	M	D	S	N	1965	34064
0491	M	D	S	N	1964	36607
0505	M	D	S	N	1965	34064
0505	M	C	S	N	1964	36607
0506	M	D	S	N	1965	34064
0506	M	C	S	N	1964	36607
0507	M	D	S	N	1965	34064
0507	M	C	S	N	1964	36607
541-0507	M	D	G	N	1954	54149
0508	M	D	S	N	1965	34064
0508	M	C	S	N	1964	36607
0509	M	D	S	N	1965	34064
0509	M	C	S	N	1964	36607
0526	M	D	G	N	1954	54149
0530	M	D	G	N	1955	43349
0530	M	G	G	N	1967	45745
0530	M	D	G	N	1953	46011
0530	M	D	E	N	1967	46012
0537	M	D	G	N	1953	46011
0537	M	D	E	N	1967	46012
0537	M	D	F	U	1968	55357
0537	M	D	E	H	1969	59893
0545	M	D	G	N	1953	46011
0545	M	D	E	N	1967	46012
0575	M	D	G	N	1953	46011
0575	M	D	E	N	1967	46012
0576	M	D	G	N	1953	46011
0576	M	D	E	N	1967	46012
0583	M	D	G	N	1953	46011
0583	M	D	E	N	1967	46012
0584	M	D	F	N	1964	36706
0584	M	D	F	H	1967	42840
0584	M	D	F	N	1967	49910
0584	M	D	E	N	1969	57980
0588	M	D	F	N	1964	36706
0588	M	D	F	H	1967	42840
0588	M	D	F	N	1967	49910
0588	M	D	E	H	1969	59893
0609	M	C	F	N	1964	35268
0609	M	G	E	N	1949	50926
0610	M	C	F	N	1964	35268
0610	M	G	E	N	1949	50926
0611	M	C	F	N	1964	35268
0612	M	C	F	N	1964	35268
0624	M	D	F	H	1967	42840
0625	M	D	F	H	1967	42840
0625	M	D	F	U	1968	55357
0626	M	D	F	H	1967	42840
0627	M	T	F	U	1967	43464
0627	M	C	J	F	1968	51373
0627	M	T	E	N	1969	53075
0627	M	E	E	U	1969	59998
0627	M	T	R	U	1970	62230
0627	M	T	E	U	1970	62231
0627	M	T	R	U	1967	62244
0627	M	T	E	U	1967	62245
0636	M	D	F	H	1967	42840
0636	M	D	F	U	1968	55357
0637	M	D	F	H	1967	42840
0637	M	D	F	U	1968	55357
0638	M	D	F	H	1967	42840
0652	M	C	E	N	1967	43692
0653	M	C	E	N	1967	43692
0663	M	D	E	N	1967	42529
0666	M	D	E	N	1967	42529
0669	M	D	E	N	1967	42529
0682	M	D	E	U	1964	41953
0682	M	D	E	N	1969	54481
0682	M	D	E	N	1969	59169
0683	M	C	E	U	1966	41988
0684	M	C	E	U	1966	41988
0685	M	C	E	U	1966	41988
0689	M	G	R	N	1962	37792
0689	M	G	E	N	1964	37793
0689	M	D	E	N	1966	40526
0697	M	D	E	N	1968	49265
0697	M	D	F	N	1967	49910
0697	M	D	E	N	1969	59896
0697	M	G	E	N	1970	61044
0699	M	D	F	U	1968	55357
0708	M	D	F	U	1968	55357
0722	M	G	E	N	1949	50926
0722	M	G	E	N	1944	51274
0723	M	G	E	N	1949	50926
0724	M	C	E	N	1952	40677
0724	M	D	E	N	1954	50299
0724	M	G	E	N	1949	50926
0724	M	G	E	N	1944	51274
0724	M	D	E	N	1950	51275
0724	M	D	E	N	1943	52655
0724	M	D	E	N	1957	52740
0725	M	C	E	N	1952	40677
0725	M	G	E	N	1944	51274
0725	M	D	E	N	1950	51275
0725	M	D	E	N	1957	52740
0726	M	G	E	N	1944	51274
0726	M	D	E	N	1957	52740
0726	M	D	G	N	1954	54149
0734	M	C	E	N	1966	40323
0734	M	C	E	N	1952	40677
0734	M	D	E	N	1962	49580
0734	M	G	E	N	1944	51274
0734	M	D	E	N	1950	51275
0734	M	D	E	N	1934	51516
0734	M	D	E	N	1937	51518
0734	M	D	E	N	1943	52655
0734	M	D	E	N	1957	52740
0735	M	G	E	N	1944	51274
0735	M	D	E	N	1937	51518
541-0735	M	D	E	N	1943	52655
0735	M	D	E	N	1957	52740
0736	M	C	E	N	1966	40323
0736	M	C	E	N	1952	40677
0736	M	D	E	N	1962	49580
0736	M	G	E	N	1944	51274
0736	M	D	E	N	1950	51275
0736	M	D	E	N	1957	52740
0736	M	D	G	N	1954	54149
0737	M	G	E	N	1944	51274
0737	M	D	E	N	1950	51275
0738	M	C	E	N	1952	40677
0738	M	D	E	N	1950	51275
0738	M	D	E	N	1934	51516
0738	M	D	E	N	1943	52655
0738	M	D	E	N	1957	52740
0739	M	D	E	N	1950	51275
0739	M	D	E	N	1934	51516
0739	M	D	E	N	1943	52655
0760	G	C	G	N	1942	51525
0767	M	D	E	N	1935	51686
0768	M	G	E	U	1930	51708
0769	M	D	E	H	1967	46949
0769	M	D	E	N	1957	47586
0769	M	D	E	N	1962	49580
0769	M	C	J	F	1968	51373
0769	M	G	E	H	1962	51706
0769	M	D	E	N	1967	51727
0769	M	D	E	N	1957	52740
0770	M	D	E	N	1957	47586
0770	M	G	E	H	1962	51706
0776	M	D	E	N	1954	50299
0776	M	D	E	N	1957	52740
0777	M	D	E	N	1954	50299
0778	M	D	E	N	1968	48697
0778	M	D	E	N	1967	50288
0784	M	D	E	N	1957	47586
0784	M	D	E	N	1962	49580
0784	M	D	E	H	1932	50802
0784	M	D	E	N	1957	52740
0800	M	D	E	N	1958	50301
0801	M	D	E	N	1958	50301
0803	M	D	E	N	1958	50301
0804	M	D	E	N	1958	50301
0818	M	D	E	N	1958	50301
0819	M	D	E	N	1958	50301
0820	M	D	E	N	1958	50301
0926	M	D	E	N	1962	49580
0927	M	D	E	N	1962	49580
0928	M	D	E	N	1962	49580
0928	M	D	E	N	1943	52655
0928	M	D	G	N	1954	54149
0929	M	D	E	N	1962	49580
0930	M	D	E	N	1962	49580
0931	M	D	E	N	1962	49580
0932	M	D	E	N	1962	49580
0933	M	D	E	H	1952	51522
0934	M	D	E	H	1952	51522
0935	M	D	E	H	1952	51522
0936	M	D	E	H	1952	51522
0937	M	D	E	H	1952	51522
0938	M	D	E	H	1952	51522
0939	M	D	E	H	1952	51522
0942	M	C	R	N	1939	40135
0942	M	D	E	N	1957	47586
0942	M	D	E	N	1958	50301
0942	M	C	G	N	1936	54128
0943	M	D	E	N	1957	47586
0944	M	D	E	N	1957	47586
0945	M	D	E	N	1957	47586
0947	M	D	E	N	1957	47586
0947	M	D	E	N	1957	52740
0948	M	C	R	N	1939	40135
0948	M	D	E	N	1957	47586
0948	M	D	E	N	1958	50301
0948	M	C	G	N	1936	54128
0949	M	C	R	N	1939	40135
0949	M	D	E	N	1967	44514
0949	M	D	E	N	1957	47586
0949	M	C	G	N	1936	54128
0950	M	D	E	N	1957	47586
0951	M	D	E	N	1957	47586
0952	M	D	E	N	1957	47586
0953	M	D	E	N	1957	47586
0954	M	C	R	N	1939	40135
0954	M	D	E	N	1957	47586
0954	M	C	G	N	1936	54128
0955	M	D	E	N	1957	47586
0956	M	D	E	N	1957	47586
0959	M	D	E	N	1957	47586
0960	M	D	E	N	1957	47586
0963	M	D	E	N	1967	44514
0982	M	D	E	N	1967	46879
0983	M	D	E	N	1967	46879
0984	M	D	E	N	1967	46879
0985	M	D	E	N	1967	46879
0986	M	D	E	N	1967	46879
1022	M	C	R	N	1939	40135
1022	M	C	G	N	1936	54128
1033	M	C	R	N	1939	40135
1033	M	C	G	N	1936	54128

Phys. State: **G.** Gas; **L.** Liquid; **M.** Multiphase; **P.** Powder; **S.** Solid

Accommodation Coefficient

Substance Number	Phys. State	Sub-ject	Lan-guage	Temper-ature	Year	TPRC Number	Substance Number	Phys. State	Sub-ject	Lan-guage	Temper-ature	Year	TPRC Number	Substance Number	Phys. State	Sub-ject	Lan-guage	Temper-ature	Year	TPRC Number
541-1034	M	C	R	N	1939	40135														
1034	M	C	G	N	1936	54128														
1103	M	D	E	N	1967	51727														
1104	M	D	E	N	1967	51727														
1105	M	D	F	N	1967	51727														
1106	M	D	E	N	1967	51727														
1107	M	D	E	N	1967	51727														
1108	M	D	E	N	1967	51727														
1109	M	D	E	N	1967	51727														
1110	M	D	E	N	1967	51727														
1137	M	D	E	N	1969	54481														
1158	M	C	J	F	1968	51373														
1166	M	G	E	N	1968	53357														
1167	M	G	E	N	1968	53357														
1167	M	G	E	N	1970	61044														
1168	M	G	E	N	1968	53357														
1168	M	G	E	N	1970	61044														
1170	M	T	E	H	1969	54556														
1213	M	D	E	N	1969	56350														
1227	M	D	E	N	1965	35893														
1242	M	D	E	N	1969	59169														
1243	M	D	E	N	1969	59169														
1269	M	D	F	N	1964	38310														
1274	M	D	F	N	1964	38310														
1279	M	D	F	N	1964	38310														
1290	M	D	F	N	1964	38310														
1302	M	D	F	N	1964	38310														
1303	M	D	F	N	1964	38310														
1304	M	D	F	N	1964	38310														
1307	M	D	F	N	1964	38310														
1310	M	D	R	N	1970	59083														
1310	M	D	E	N	1970	59084														
1312	M	D	R	N	1970	59083														
1312	M	D	E	N	1970	59084														
1313	M	D	R	N	1970	59083														
1313	M	D	E	N	1970	59084														
1314	M	D	R	N	1970	59083														
1314	M	D	E	N	1970	59084														
1315	M	D	R	N	1970	59083														
1315	M	D	E	N	1970	59084														
1339	M	D	E	N	1969	59896														
1340	M	D	E	H	1969	59893														
1341	M	G	E	N	1969	59892														
1342	M	G	E	N	1969	59892														
1343	M	D	R	N	1970	60244														
1343	M	D	E	N	1970	60245														

Phys. State: G. Gas; L. Liquid; M. Multiphase; P. Powder; S. Solid

Chapter 3 Thermal Contact Resistance

Substance Number	Phys. State	Sub-ject	Lan-guage	Temper-ature	Year	TPRC Number
528-0007	S	D	E	N	1964	42596
0007	S	D	E	U	1966	46396
0010	S	D	E	N	1968	48824
0010	S	D	G	N	1968	55220
0019	S	G	E	N	1967	51444
0019	S	C	E	U	1969	53569
0019	S	C	E	N	1968	56535
0019	S	E	E	N	1970	61182
0020	S	D	E	N	1965	38405
0020	S	D	E	N	1964	40681
0020	S	D	E	N	1963	48494
0020	S	D	E	N	1968	48824
0020	S	C	E	U	1969	53572
0020	S	D	E	N	1955	57237
0020	M	D	E	N	1961	40083
0036	S	D	E	N	1969	55288
0039	S	D	E	N	1966	35023
0047	S	D	E	N	1966	35023
0049	S	D	E	N	1955	40018
0052	S	D	E	N	1955	40018
0057	S	D	E	U	1966	46396
0057	S	C	E	U	1963	48497
0057	S	G	J	N	1967	48688
0057	S	D	E	N	1963	52292
0057	S	C	E	N	1966	54664
0057	M	C	E	N	1957	49662
0058	S	G	J	N	1967	48688
0060	S	D	I	N	1965	40194
0079	S	D	I	N	1965	40194
0096	S	D	E	N	1967	41250
0113	S	T	E	L	1966	39919
0123	S	D	E	N	1964	40683
0123	S	D	E	N	1965	48505
0128	S	D	E	N	1965	36946
0128	S	D	E	N	1964	43790
0128	S	D	E	U	1966	46396
0128	S	G	J	N	1967	48687
0128	S	G	J	N	1967	48689
0128	S	G	J	N	1967	48690
0137	S	C	E	N	1969	41375
0137	S	D	R	N	1954	43552
0137	S	D	E	N	1964	43788
0137	S	D	R	N	1954	43997
0137	S	D	F	N	1967	46480
0137	S	G	J	N	1967	48687
0137	S	G	J	N	1967	48689
0137	S	G	J	N	1967	48690
0137	S	D	E	N	1968	48824
0137	S	D	E	N	1967	50418
0137	S	D	E	N	1967	53177
0137	M	D	R	H	1967	43124
0137	M	D	E	H	1967	43140
0137	M	D	E	L	1968	48580
0143	S	S	E	N	1965	33800
0143	S	T	E	N	1965	34514
0143	S	T	F	U	1967	35200
0143	S	T	F	U	1969	37583
0143	S	T	E	H	1964	37689
0143	S	E	E	L	1964	39075
0143	S	E	E	N	1966	39357
0143	S	E	E	N	1964	40318
0143	S	E	E	U	1966	40474
0143	S	E	E	N	1964	45768
0143	S	T	E	U	1967	45796
0143	S	E	E	U	1956	47035
0143	S	E	E	U	1967	47095
0143	S	E	E	U	1963	48227
0143	S	E	E	U	1963	48494
0143	S	E	E	U	1968	50024
0143	S	E	E	U	1970	50168
0143	S	E	E	N	1963	50689
0143	S	E	G	N	1963	50689
0143	S	E	E	N	1912	50704
0143	S	E	E	N	1968	50897
0143	S	S	E	N	1968	50898
0143	S	T	E	U	1968	51910
0143	S	T	E	N	1963	52292
0143	S	T	E	N	1968	53568
0143	S	S	E	L	1966	58874
0143	S	D	E	U	1964	40020
0143	M	D	R	U	1964	43548
0143	M	T	F	U	1967	46479
0143	M	D	E	N	1963	51300
0176	S	D	E	L	1966	34309
0176	S	D	E	L	1963	43616
0176	S	D	E	L	1967	47427
0177	S	D	E	L	1966	34309
0177	S	D	E	N	1963	43616
0184	S	G	J	N	1967	48687
0184	S	G	J	N	1967	48689
0184	S	G	J	N	1967	48690
0188	S	C	E	N	1968	36270
0188	S	C	E	N	1967	51444
0188	S	C	E	U	1969	53569
0188	S	C	E	U	1968	56535
0188	S	C	E	U	1970	59138
528-0207	S	D	E	N	1963	48506
0208	S	D	R	N	1954	43552
0210	S	D	R	N	1954	43552
0212	S	D	R	N	1954	43552
0213	S	D	R	N	1954	43552
0213	S	D	R	N	1954	43997
0215	S	D	R	N	1954	43552
0215	S	D	R	N	1954	43997
0216	S	D	R	N	1954	43552
0222	S	D	E	N	1964	37283
0222	S	C	E	N	1961	43634
0222	S	C	R	U	1966	47984
0222	M	D	E	N	1968	48990
0222	M	D	E	N	1968	63154
0223	S	C	E	N	1961	43634
0224	S	D	E	N	1969	37283
0224	S	C	E	N	1970	60045
0225	S	D	E	N	1964	37283
0226	S	D	E	N	1964	37283
0236	S	G	J	N	1967	48687
0238	S	G	J	N	1964	43790
0238	S	G	J	N	1967	48687
0238	S	G	J	N	1967	48689
0238	S	D	E	N	1967	50418
0238	S	D	E	N	1965	58822
0240	S	D	E	N	1965	38615
0240	S	D	E	N	1964	43788
0240	S	D	E	N	1955	57237
0250	M	G	F	N	1965	33905
0250	M	G	F	N	1966	39285
0251	S	G	J	L	1967	48687
0251	S	G	J	L	1967	48690
0251	S	D	E	N	1968	56695
0266	S	D	E	U	1969	37066
0266	S	G	E	N	1965	39352
0266	S	G	E	N	1969	41427
0266	S	D	E	N	1954	44435
0266	S	C	R	U	1967	47095
0266	S	C	R	U	1966	47984
0266	S	D	E	N	1967	50418
0266	S	G	E	N	1967	51444
0266	S	D	E	N	1968	60177
0266	M	D	E	N	1957	49660
0267	S	D	E	N	1964	43788
0269	S	D	E	N	1966	35023
0269	S	D	E	N	1964	43788
0270	S	D	R	N	1954	43552
0281	S	D	E	N	1961	43546
0282	S	C	E	N	1967	53177
0282	S	C	E	N	1966	54664
0282	S	D	E	N	1970	59259
0307	S	G	J	N	1967	48687
0308	S	G	J	N	1967	48688
0309	S	G	J	N	1967	48690
0311	S	D	E	N	1965	38615
0311	S	C	E	N	1969	41375
0311	S	C	E	U	1963	48497
0311	S	D	E	N	1963	52292
0311	M	C	E	N	1957	49662
0312	S	C	E	N	1964	40683
0312	S	D	E	N	1965	48505
0313	S	D	E	L	1961	40769
0313	M	D	E	N	1968	48580
0318	S	D	E	N	1966	39357
0318	S	D	E	N	1968	48225
0318	S	D	E	U	1968	50332
0318	S	G	E	N	1968	51101
0318	S	C	R	N	1965	49167
0322	S	D	E	N	1963	46516
0324	S	D	E	U	1963	48498
0325	S	D	E	N	1965	47754
0327	S	G	E	N	1964	37381
0329	S	D	E	N	1967	45766
0331	S	G	E	N	1967	47427
0332	S	D	E	L	1961	40769
0333	S	D	E	N	1964	42596
0334	S	D	E	N	1964	42596
0335	S	D	E	U	1969	53571
0341	S	D	E	U	1968	53358
0342	S	G	E	U	1968	53982
0343	S	D	E	N	1968	53358
0346	S	G	E	N	1966	58874
0347	S	G	E	N	1969	41427
0347	S	C	E	F	1965	43491
0351	S	D	E	N	1969	48448
0353	S	D	E	N	1964	37283
0353	S	D	R	N	1954	43997
0353	S	D	R	N	1968	55879
0353	S	C	E	N	1970	60045
0353	S	C	E	N	1954	43997
0354	S	D	R	N	1954	43997
0355	S	D	R	N	1954	43552
0356	S	D	R	N	1954	43997
0356	S	D	R	N	1954	43552
0357	S	D	R	N	1954	43997
0357	S	D	R	N	1954	43997
528-0358	S	D	E	N	1967	49310
0359	S	D	E	N	1967	49310
0360	S	D	E	N	1967	49310
0361	S	D	F	N	1969	48448
0362	S	G	F	N	1966	41389
0363	S	D	E	N	1955	48508
0366	M	D	E	N	1965	47259
0367	M	D	E	N	1965	47259
0369	M	D	R	N	1966	51633
0369	M	D	E	N	1968	51634
0370	M	D	R	N	1966	51633
0370	M	D	E	N	1968	51634
0374	S	D	E	N	1960	36088
0381	S	D	E	N	1966	41932
0382	S	D	E	L	1969	56027
0383	S	G	F	N	1966	41389
0385	S	D	E	N	1968	53717
0386	S	G	F	N	1966	41389
0387	S	D	E	N	1964	37283
0390	S	D	R	N	1954	43552
0397	S	D	E	N	1948	40678
0398	S	D	E	U	1969	37066
0398	S	G	E	N	1969	41427
0399	S	E	E	N	1970	58721
0400	S	E	E	N	1970	58721
0403	S	D	E	N	1970	59259
0404	S	D	E	N	1970	59259
0405	S	D	E	N	1970	41397
0406	M	D	R	N	1963	42690
0406	M	D	E	N	1966	42691
0407	S	D	E	N	1968	44453
0408	S	D	E	N	1968	44453
0412	S	C	E	N	1965	43491
0415	S	D	R	N	1964	37283
0415	S	D	R	N	1969	56230
0416	M	D	R	N	1969	56230
0417	M	D	R	N	1969	56230
0423	S	D	E	N	1965	58822
0465	S	D	G	N	1968	55220
0466	S	D	G	N	1968	55220
531-0171	M	C	R	N	1966	41480
0171	M	C	E	N	1966	59491
0444	M	C	E	N	1966	54664
0459	S	S	E	L	1966	58874
0459	M	S	E	L	1965	33800
0459	M	T	F	U	1966	39296
0511	M	T	R	U	1966	49166
0511	M	T	R	U	1966	54400
0683	M	C	E	N	1966	41480
0683	M	C	E	N	1966	59491
0770	S	T	F	U	1967	35200
0770	S	S	E	N	1966	36413
0770	S	E	E	N	1970	37173
0770	S	T	E	H	1964	37689
0770	S	T	E	N	1966	39357
0770	S	E	E	N	1963	40017
0770	S	T	E	U	1967	47570
0770	S	T	E	U	1967	48556
0770	S	T	E	U	1966	49725
0770	S	E	E	N	1912	50704
0770	S	T	E	U	1968	51702
0770	S	T	E	U	1910	51910
0770	S	E	E	U	1969	53571
0770	S	T	E	U	1969	54071
0770	S	T	E	U	1965	58876
0770	M	T	E	U	1967	52849
0770	M	T	F	U	1967	46479
0770	M	T	F	N	1912	50704
0912	S	D	E	L	1958	43547
0913	S	D	E	L	1958	43547
1099	S	D	E	N	1963	48506
1100	S	D	E	N	1963	48506
1101	S	D	E	N	1963	48506
1102	S	D	E	N	1963	48506
1103	P	D	E	N	1963	48506
1104	P	D	E	N	1963	48506
1115	S	G	E	N	1963	48509
1116	S	G	E	N	1963	48509
1161	S	T	J	N	1957	51798
1161	S	T	E	N	1968	51799
1270	M	D	E	L	1962	41664
1274	S	D	E	N	1968	48824
1274	S	D	E	N	1967	53177
1275	S	D	E	N	1968	48824
1385	M	C	R	N	1954	41480
1385	M	C	E	N	1966	59491
1407	S	D	E	N	1968	50868
1407	S	D	E	N	1968	57450
1408	S	D	E	N	1968	50868
1409	S	D	E	N	1968	50868
1410	M	D	E	N	1966	44718
1410	M	C	E	L	1970	61087
1465	S	T	E	U	1970	42688
1465	S	E	E	U	1968	51911

Phys. State: G. Gas; L. Liquid; M. Multiphase; P. Powder; S. Solid

Thermal Contact Resistance

Substance Number	Phys. State	Subject	Language	Temperature	Year	TPRC Number
531-1465	S	T	E	U	1969	53854
1466	S	D	E	L	1968	51912
1503	S	C	E	N	1969	49590
1512	S	D	E	N	1966	54227
1691	M	G	E	L	1968	59163
1790	M	D	E	L	1970	59274
1808	M	S	E	L	1969	58585
1808	M	S	E	L	1970	58986
541-0025	S	D	E	N	1965	38405
0025	S	D	E	N	1964	40681
0025	S	D	E	N	1964	45714
0025	S	D	E	N	1968	48824
0025	M	D	E	N	1961	40083
0083	S	G	E	N	1966	54664
0084	S	T	F	U	1967	35200
0084	S	T	E	H	1964	37689
0084	S	E	E	N	1966	39357
0084	S	E	E	L	1964	58931
0084	M	S	E	N	1964	40318
0084	M	T	F	U	1967	46479
0084	M	S	E	N	1912	50704
0084	M	E	E	L	1970	58314
0084	M	E	E	L	1964	58931
0089	S	G	J	N	1967	48687
0089	S	D	E	N	1965	58822
0152	S	D	R	N	1954	43552
0153	S	E	E	N	1964	43789
0153	S	C	E	N	1963	48500
0153	S	S	E	U	1967	48897
0153	S	E	E	U	1970	50168
0153	S	E	E	N	1968	51694
0153	S	T	E	N	1968	51909
0153	S	E	E	N	1968	51911
0153	S	T	E	N	1963	52292
0154	S	C	E	U	1969	37022
0154	S	D	E	N	1964	43791
0154	S	D	E	N	1968	47768
0154	S	D	E	N	1964	48116
0154	S	D	E	N	1966	48499
0154	S	G	E	N	1969	58560
0154	S	D	E	N	1964	58821
0154	S	C	E	N	1970	59610
0154	S	D	E	N	1970	61181
0154	S	D	E	L	1970	63132
0155	S	G	E	N	1963	40017
0155	S	G	E	N	1966	40664
0155	S	G	E	N	1967	47570
0155	S	D	E	N	1964	48116
0155	S	G	E	N	1967	48170
0155	S	G	E	U	1966	48496
0155	S	D	E	N	1964	48501
0155	S	G	E	N	1969	58560
0155	S	D	E	N	1964	58821
0155	S	C	E	N	1970	59610
0155	M	G	E	N	1965	36943
0157	S	D	E	N	1965	47488
0157	S	D	E	N	1964	48116
0157	S	C	E	N	1968	48831
0157	S	D	E	N	1967	50418
0157	S	D	E	N	1964	58821
0166	M	D	R	N	1966	44342
0166	M	D	E	N	1966	59492
0176	S	D	E	H	1962	45475
0176	M	D	E	N	1966	46882
0179	S	G	E	N	1963	40017
0182	S	C	E	N	1962	48495
0192	S	D	E	N	1964	40019
0192	S	D	E	N	1965	47488
0192	S	D	E	N	1964	48116
0192	S	D	E	N	1964	48501
0192	S	E	E	N	1965	48503
0192	S	D	E	N	1964	48504
0192	S	D	E	N	1964	58821
0192	S	D	E	L	1970	63132
0193	S	G	E	N	1965	34329
0193	S	C	E	U	1969	37022
0193	S	G	E	N	1967	47570
0193	S	G	E	N	1967	48170
0193	S	D	E	N	1966	48382
0193	S	C	E	U	1970	57652
0193	S	D	E	N	1965	58822
0193	S	G	E	N	1970	61182
0197	M	D	R	N	1966	44342
0197	M	D	R	N	1967	54143
0197	M	D	E	N	1966	59492
0199	S	G	E	N	1966	40664
0199	S	G	E	N	1969	41427
0199	S	G	E	N	1967	47570
0199	S	D	E	N	1968	47768
0199	S	G	E	N	1967	48170
0199	S	D	E	N	1966	48391
0199	S	G	E	N	1962	48502
0199	S	D	E	N	1967	50418
0199	S	D	E	N	1968	51913
0199	S	G	E	N	1966	54664
0199	S	D	E	N	1969	55640
0199	S	G	E	N	1969	58560
0199	S	C	E	N	1970	59610
0199	S	D	E	N	1970	61181
0199	M	G	E	N	1965	36943

Substance Number	Phys. State	Subject	Language	Temperature	Year	TPRC Number
541-0200	S	D	E	N	1966	48381
0201	S	D	E	N	1966	48381
0202	S	D	E	N	1966	48381
0204	S	D	E	N	1966	48381
0204	S	G	E	N	1969	58560
0236	S	D	E	N	1964	51247
0236	M	G	E	N	1963	40689
0236	M	D	E	N	1966	46882
0249	S	D	E	N	1961	43546
0249	M	D	R	N	1968	48990
0249	M	D	E	N	1968	63154
0262	M	D	R	N	1968	48990
0262	M	D	E	N	1968	63154
0263	S	D	E	N	1964	37438
0282	S	D	E	N	1964	48501
0283	S	D	E	L	1970	63132
0283	M	D	E	N	1964	48501
0286	S	G	E	N	1964	48493
0286	S	G	E	U	1966	48496
0287	S	G	E	N	1964	48493
0289	S	G	E	N	1964	48493
0302	S	D	E	N	1968	47768
0302	S	D	E	N	1966	48381
0302	S	G	E	N	1962	48502
0302	S	D	E	N	1969	55640
0302	S	D	E	N	1970	61181
0326	S	D	E	N	1968	61079
0338	M	C	R	N	1966	41480
0338	M	C	E	N	1966	59491
0361	S	D	E	L	1965	34706
0361	M	D	E	L	1968	48580
0396	M	G	F	N	1965	33905
0396	M	G	F	N	1966	39285
0397	M	G	F	N	1965	33905
0397	M	G	F	N	1966	39285
0398	S	C	E	N	1965	33813
0398	S	C	R	N	1965	33814
0400	S	G	E	N	1966	54664
0400	S	G	E	N	1969	58560
0400	M	G	E	N	1965	36943
0401	S	G	E	N	1969	37583
0401	S	G	E	N	1963	40017
0401	S	G	E	N	1966	40664
0401	S	G	E	N	1967	47570
0401	S	G	E	N	1967	48170
0401	S	G	E	U	1966	48496
0401	S	G	E	N	1966	54664
0401	S	C	E	U	1970	57652
0401	M	G	E	N	1965	36943
0402	M	G	E	N	1965	36943
0403	S	D	E	N	1965	36946
0403	S	G	J	N	1967	48687
0403	S	G	J	N	1967	48690
0403	S	G	E	N	1969	61793
0413	M	C	E	L	1968	50271
0415	M	C	E	L	1968	50271
0458	M	D	E	L	1966	34756
0459	M	D	E	L	1966	34756
0460	M	D	E	L	1966	34756
0461	M	D	E	L	1966	34756
0462	M	D	E	L	1966	34756
0463	M	D	E	L	1966	34756
0464	M	D	E	L	1966	34756
0541	M	D	E	N	1966	35023
0542	M	D	E	N	1966	35023
0628	S	C	F	U	1967	46479
0629	M	C	F	U	1967	46479
0679	S	D	E	N	1966	48382
0679	S	D	E	N	1965	58822
0691	S	D	E	N	1966	48382
0691	S	D	E	N	1965	58822
0692	S	G	E	N	1963	40017
0692	S	D	E	N	1966	48382
0692	S	D	E	N	1967	50418
0692	S	D	E	N	1965	58822
0692	S	C	E	N	1970	59610
0693	S	G	E	N	1963	40017
0693	S	D	E	N	1966	48382
0693	S	D	E	N	1965	58822
0695	S	D	E	N	1966	48382
0695	S	D	E	N	1965	58822
0771	S	C	E	N	1965	39353
0772	S	G	J	N	1967	48687
0772	S	G	J	N	1967	48690
0773	S	D	E	L	1970	37014
0773	S	G	J	N	1967	48687
0773	S	D	E	N	1968	48824
0773	S	D	E	L	1970	58563
0773	M	S	E	L	1970	61245
0774	M	D	F	N	1966	41927
0779	S	D	E	N	1965	38727
0781	S	D	E	N	1965	38727
0785	S	D	E	N	1965	38727
0830	S	D	E	L	1964	58931
0830	M	S	E	L	1964	38460
0830	M	D	E	L	1968	56697
0830	M	D	E	L	1964	58931
0834	M	C	E	N	1957	49662
0835	M	D	E	L	1968	56697
0835	M	E	E	L	1970	58314
0835	M	D	E	L	1964	58931

Substance Number	Phys. State	Subject	Language	Temperature	Year	TPRC Number
541-0836	S	D	E	N	1964	48501
0838	S	E	E	N	1964	43789
0838	S	D	E	N	1964	45498
0838	S	D	E	N	1964	48504
0839	S	E	E	N	1964	43789
0839	S	D	E	N	1964	45498
0839	S	D	E	N	1964	48504
0840	S	D	E	N	1963	48506
0841	S	D	E	N	1963	48506
0842	S	D	E	N	1963	48506
0843	P	D	E	N	1963	48506
0844	P	D	E	N	1963	48506
0845	P	D	E	N	1963	48506
0846	P	D	E	N	1963	48506
0847	P	D	E	N	1963	48506
0848	P	D	E	N	1963	48506
0849	P	D	E	N	1963	48506
0850	P	D	E	N	1963	48506
0860	S	G	E	N	1969	41427
0860	S	C	E	N	1965	48510
0860	S	D	E	N	1968	51913
0860	S	G	E	N	1966	54664
0861	S	C	E	N	1965	48510
0861	S	G	E	N	1966	54664
0862	S	D	E	U	1967	49160
0863	S	D	E	N	1964	37438
0870	S	D	E	N	1964	37438
0889	S	G	E	U	1966	48496
0889	S	C	E	N	1968	48831
0889	S	C	E	N	1963	52292
0890	S	G	E	U	1966	48496
0891	S	D	E	U	1963	48498
0892	M	D	E	N	1966	48499
0893	S	D	E	N	1966	48499
0893	S	D	E	N	1968	51911
0893	S	G	E	N	1970	59611
0894	M	D	E	N	1966	48499
0895	M	D	E	N	1966	48499
0896	M	D	E	N	1966	48499
0897	M	D	E	N	1966	48499
0898	M	D	E	N	1966	48499
0899	M	D	E	N	1966	48499
0900	M	D	E	N	1966	48499
0901	M	D	E	N	1966	48499
0902	M	D	E	N	1966	48499
0903	M	D	E	N	1966	48499
0904	S	D	E	N	1964	37438
0905	S	D	E	N	1964	37438
0913	S	D	E	N	1968	47768
0913	S	D	E	N	1970	61181
0913	S	D	E	L	1970	63132
0919	M	S	E	L	1969	55723
0919	M	S	E	L	1969	58585
0919	M	S	E	L	1970	58986
0920	M	S	E	L	1969	55723
0925	S	D	R	N	1954	43997
0946	S	D	R	N	1954	43997
0957	S	D	R	N	1954	43997
0958	M	D	R	N	1954	43997
0987	S	D	R	N	1965	49551
0987	S	D	E	N	1967	49552
0988	S	D	R	N	1965	49551
0988	S	D	E	N	1967	49552
0988	M	D	R	N	1966	44342
0988	M	D	R	N	1967	54143
0988	M	D	E	N	1966	59492
0990	S	D	E	N	1966	49762
0990	M	C	E	N	1965	39353
0990	M	D	E	N	1966	46882
0992	M	D	E	N	1966	46882
0993	M	C	E	N	1965	39353
0993	M	D	E	N	1966	46882
0995	M	D	E	N	1966	46882
0997	M	D	E	N	1966	46882
0998	S	D	E	N	1964	37381
1000	S	D	E	L	1964	37580
1012	S	G	E	N	1966	40664
1013	S	D	E	N	1968	51694
1016	S	D	E	L	1961	40769
1017	S	D	E	L	1961	40769
1017	S	D	E	N	1963	48294
1018	S	D	E	L	1961	40769
1020	M	D	E	L	1969	49542
1020	M	C	E	L	1968	50271
1021	M	D	R	N	1966	44342
1021	M	D	R	N	1967	54143
1021	M	D	E	N	1966	59492
1038	M	D	E	L	1962	41664
1038	M	E	E	L	1970	58314
1039	M	D	E	L	1962	41664
1039	M	E	E	L	1970	58314
1040	M	D	E	L	1962	41664
1040	M	E	E	L	1970	58314
1044	S	D	E	N	1968	48824
1053	M	T	E	U	1967	44428
1057	S	D	E	N	1970	60713
1058	M	G	E	U	1969	53570
1059	M	G	E	U	1969	53570
1061	S	G	E	L	1968	53355
1062	S	G	E	U	1968	53982
1077	M	D	R	N	1966	44342

Phys. State: **G.** Gas; **L.** Liquid; **M.** Multiphase; **P.** Powder; **S.** Solid

Substance Number	Phys. State	Sub-ject	Lan-guage	Temper-ature	Year	TPRC Number	Substance Number	Phys. State	Sub-ject	Lan-guage	Temper-ature	Year	TPRC Number	Substance Number	Phys. State	Sub-ject	Lan-guage	Temper-ature	Year	TPRC Number
541-1077	M	D	R	N	1967	54143														
1077	M	D	E	N	1966	59492														
1078	M	D	R	N	1966	44342														
1078	M	D	R	N	1967	54143														
1078	M	D	E	N	1966	59492														
1082	S	D	E	N	1966	43479														
1102	M	D	R	N	1966	51633														
1102	M	D	E	N	1966	51034														
1113	S	U	E	N	1968	51902														
1117	S	E	E	N	1968	51911														
1118	S	G	E	N	1969	41427														
1118	S	D	E	N	1968	51913														
1119	S	D	E	N	1970	60713														
1120	S	C	E	N	1963	52292														
1121	S	D	E	N	1968	52324														
1122	S	D	E	N	1968	52324														
1123	S	D	E	N	1968	52324														
1124	S	D	E	N	1968	52324														
1125	S	D	E	N	1968	52324														
1126	S	D	E	N	1968	52324														
1127	S	D	E	N	1968	52324														
1140	S	D	E	N	1970	37173														
1140	S	G	E	N	1970	59611														
1252	S	D	E	N	1965	38615														
1255	S	D	E	N	1966	39357														
1265	S	D	E	L	1970	57323														
1271	M	D	E	N	1948	40678														
1272	S	C	E	N	1963	40709														
1273	M	C	E	N	1963	40709														
1277	S	D	E	L	1970	57323														
1282	M	D	R	N	1967	54143														
1283	M	D	R	N	1967	54143														
1291	M	E	E	L	1970	58314														
1299	S	S	E	I	1966	58874														
1000	S	D	E	N	1965	58822														
1301	S	D	E	N	1965	58822														
1309	S	D	E	L	1970	61246														
1316	S	D	E	N	1968	44453														
1316	S	D	E	N	1969	55561														
1320	S	C	E	U	1966	58882														
1323	S	D	E	N	1968	61079														
1324	S	D	E	N	1968	61079														
1331	S	D	E	N	1968	57569														
1350	S	D	E	N	1965	38615														
1351	S	D	E	N	1965	38615														
1352	S	D	E	N	1965	55288														
1353	S	D	E	N	1969	55288														
1384	S	D	E	H	1970	59676														
1385	S	G	F	N	1966	41389														
1386	S	G	F	N	1966	41389														
1387	S	G	F	N	1966	41389														
1390	S	D	E	N	1964	43791														
1391	S	D	E	N	1964	43791														
1403	M	D	E	L	1968	56696														
1404	M	D	E	L	1968	56696														
1407	S	C	E	N	1968	61078														
1408	S	C	E	N	1968	61078														
1409	S	C	E	N	1968	61078														
1410	S	C	E	N	1968	61078														
1411	S	C	E	N	1968	61078														
1412	S	C	E	N	1968	61078														
1420	S	D	E	N	1969	55561														
1421	S	D	E	N	1969	55561														
1422	S	D	E	N	1969	55561														
1450	S	D	E	N	1970	60713														
661-0186	M	D	E	N	1962	39721														

Chapter 4 Thermal Diffusivity

Substance Number	Phys. State	Subject	Language	Temperature	Year	TPRC Number
526-0006	S	D	E	N	1969	56053
0034	S	D	E	N	1966	33864
0034	S	D	R	N	1969	58439
0034	M	D	E	N	1964	37058
0034	M	D	O	N .	1966	49791
0064	S	D	E	N	1966	33864
0067	S	D	E	N	1966	34401
0067	S	D	E	N	1969	56579
0067	S	D	R	N	1969	58439
0069	S	D	E	N	1964	37058
0069	S	D	R	N	1966	59494
0069	S	D	E	N	1966	59495
0082	S	D	E	N	1965	34669
0127	S	D	E	N	1965	34669
0128	S	D	E	N	1965	34669
0182	S	D	E	N	1969	58340
0182	S	D	E	N	1965	34669
0183	S	D	E	N	1965	34669
0184	S	D	E	N	1965	34669
0185	S	D	E	N	1965	34669
0186	S	D	E	N	1965	34669
0187	S	D	E	N	1965	34838
0188	S	D	E	N	1965	34669
0189	S	D	E	N	1965	34669
0226	M	D	R	N	1969	57513
528-0103	M	D	E	N	1965	36367
0109	S	D	E	N	1959	35374
0137	S	D	E	N	1966	50663
0314	S	D	E	N	1968	50380
0315	S	D	E	N	1968	50380
0392	S	D	E	N	1965	38707
0393	S	D	E	N	1965	43295
0409	S	C	E	N	1970	58436
0414	S	D	E	N	1965	43295
531-0014	S	D	E	N	1967	48812
0017	S	D	E	N	1967	48812
0033	S	D	E	N	1967	48812
0036	S	D	E	U	1963	40301
0040	S	D	R	N	1968	49033
0060	S	D	R	N	1968	52893
0060	S	D	E	H	1968	54195
0064	S	D	E	N	1968	57225
0066	S	D	E	N	1966	40418
0068	S	D	E	N	1966	40418
0079	S	D	E	N	1966	40418
0092	S	D	E	N	1966	40418
0112	M	C	R	H	1966	43092
0123	S	D	R	N	1966	40418
0171	M	D	R	N	1965	34599
0171	M	D	E	N	1965	47289
0192	S	D	J	N	1968	51800
0195	S	D	R	N	1968	49033
0195	S	D	J	N	1968	51800
0198	S	D	J	N	1968	51800
0201	S	D	J	N	1968	51800
0288	S	D	R	N	1968	54578
0301	S	D	J	N	1968	49033
0301	S	D	J	N	1968	51800
0301	S	D	E	H	1968	54307
0303	S	D	E	N	1965	34666
0303	S	D	E	N	1965	38402
0303	S	G	R	N	1965	43097
0303	S	G	J	N	1967	46251
0303	S	G	E	N	1968	46252
0303	S	D	R	N	1968	49033
0303	S	D	J	N	1968	51800
0331	P	D	E	N	1962	39516
0348	P	G	E	N	1969	59176
0373	M	C	R	H	1966	43092
0384	P	E	E	U	1969	53594
0434	S	D	R	N	1966	59494
0434	S	D	E	N	1966	59495
0462	S	D	E	N	1966	40418
0463	S	D	E	N	1966	40418
0515	S	D	R	N	1967	45790
0515	S	D	J	N	1968	51800
0554	S	D	E	N	1963	49033
0568	S	G	E	N	1963	35832
0568	S	G	J	N	1967	46251
0568	S	G	E	N	1968	46252
0568	S	D	J	N	1968	51800
0568	S	D	E	H	1968	54307
0569	S	D	E	H	1962	40298
0569	S	D	E	N	1968	52893
0574	P	G	E	N	1969	59176
0620	S	D	E	H	1962	40298
0620	S	D	E	U	1963	40301
0620	S	D	E	N	1968	52893
0633	S	T	R	U	1966	48092
0633	S	T	R	U	1966	48093
0679	S	D	R	N	1968	49033
0679	S	D	J	N	1968	51800
0679	S	D	E	H	1968	54307
531-0683	S	G	E	N	1970	57661
0799	M	D	E	N	1965	38402
0807	G	D	E	N	1947	48357
0955	S	D	E	N	1960	42824
0962	S	D	E	N	1966	52891
0979	M	C	E	U	1969	53606
1162	S	D	E	H	1967	46860
1240	M	C	R	H	1966	43092
1246	S	E	E	H	1960	52260
1247	S	D	E	R	1966	45275
1277	M	D	R	N	1964	35922
1277	M	D	E	N	1965	35923
1280	S	D	E	N	1969	53605
1281	S	C	E	U	1969	53606
1289	S	D	E	H	1968	54195
1337	P	D	E	N	1962	39516
1390	S	D	E	H	1962	40298
1391	S	D	E	N	1962	40298
1391	S	D	E	N	1967	48812
1395	S	D	E	N	1966	36480
1472	S	D	E	N	1968	51800
1473	S	D	J	N	1968	51800
1474	S	D	J	N	1968	51800
1475	S	D	E	N	1966	52891
1518	S	D	E	N	1966	53098
1592	S	D	R	N	1966	44594
1653	S	D	R	N	1969	55503
1654	S	D	R	N	1969	55503
1694	M	D	E	N	1967	42955
1694	M	D	R	N	1967	45153
1695	M	D	E	N	1967	42955
1695	M	D	R	N	1967	45153
1696	M	D	E	N	1967	42955
1696	M	D	R	N	1967	45153
1697	M	D	E	N	1967	42955
1697	M	D	R	N	1967	45153
1740	S	D	E	H	1970	59046
1740	S	D	E	H	1970	60836
1741	S	D	R	H	1970	59046
1741	S	D	R	H	1970	60836
1742	S	D	R	H	1970	59046
1742	S	D	R	H	1970	60836
1751	P	G	E	N	1969	59176
1756	S	E	E	N	1970	60313
1765	S	D	E	N	1968	57225
1766	S	D	E	N	1968	57225
1767	S	D	E	N	1968	57225
1780	S	D	E	N	1965	43295
1795	M	D	E	N	1968	53113
1796	M	D	E	N	1968	53113
541-0019	S	D	E	N	1964	35797
0019	S	D	G	U	1970	61785
0044	S	D	R	N	1966	59494
0044	S	D	E	N	1966	59495
0399	M	D	E	N	1965	36367
0409	M	D	E	N	1965	36367
0410	M	D	E	N	1965	36367
0465	S	D	E	N	1966	39675
0466	S	D	E	N	1966	39675
0773	S	D	E	N	1966	36480
0962	S	D	E	N	1968	53098
1015	S	D	E	H	1964	52271
1130	S	D	E	N	1966	52891
1244	P	G	E	N	1969	59176
1257	S	D	E	N	1965	38707
1258	S	D	E	N	1965	38707
1259	S	D	E	N	1965	38707
1361	S	D	E	H	1969	54072
1416	M	D	R	H	1968	53113
1417	M	D	E	N	1968	53113
621-0023	S	D	E	U	1964	38249
0024	S	D	E	N	1969	56600
0122	S	D	E	N	1969	56600
0123	S	D	E	N	1969	56600
0124	S	D	E	N	1969	56600
0125	S	D	E	N	1969	56600
0126	S	D	E	N	1969	56600
651-0071	S	D	R	N	1966	44457
0071	S	D	R	N	1968	55626
0071	S	D	R	N	1969	57533
0076	S	D	R	N	1966	44457
0096	S	D	R	N	1966	44457
0096	S	D	R	N	1968	55626
0096	S	D	E	N	1969	57533
0097	S	D	R	N	1966	44457
0097	S	D	R	N	1968	55626
0097	S	D	R	N	1969	57533
0097	L	D	R	N	1968	55627
0097	L	D	E	N	1969	57537
0282	S	D	E	N	1969	56585
0290	S	D	R	N	1968	55626
651-0290	S	D	E	N	1969	57533
0291	S	D	R	N	1968	55626
0291	S	D	E	N	1969	57533
0292	S	D	R	N	1968	55626
0292	S	D	E	N	1969	57533
0292	L	D	R	N	1968	55627
0292	L	D	E	N	1969	57537
0293	S	D	R	U	1968	53488
0293	S	D	E	U	1968	53903
0293	S	D	R	N	1968	55626
0293	S	D	E	N	1969	57533
0298	L	D	R	N	1968	55627
0298	L	D	E	N	1969	57537
0302	L	D	R	N	1968	55627
0302	L	D	E	N	1969	57537
0309	L	D	R	N	1968	55627
0309	L	D	E	N	1969	57537
0310	L	D	R	N	1968	55627
0310	L	D	E	N	1969	57537
0311	L	D	R	N	1968	55627
0311	L	D	E	N	1969	57537
0312	L	D	R	N	1968	55627
0312	L	D	E	N	1969	57537
0313	L	D	R	N	1968	55627
0313	L	D	E	N	1969	57537
0314	L	D	R	N	1968	55627
0314	L	D	E	N	1969	57537
0544	S	S	E	N	1968	42148
661-0001	S	D	E	N	1964	37058
0001	S	D	E	N	1966	39442
0003	M	D	R	N	1968	51025
0014	M	D	R	N	1969	56318
0040	S	D	E	N	1964	37058
0040	S	E	E	N	1966	59494
0040	S	E	E	N	1966	59495
0042	S	E	E	N	1964	37058
0042	S	E	R	N	1965	43097
0042	S	E	R	N	1965	43098
0049	S	D	E	N	1970	63039
0836	O	O	S	1	SDGN 1949	5
0052	S	C	E	N	1970	57663
0056	S	D	E	N	1966	33864
0060	S	G	E	N	1956	40670
0078	S	D	E	N	1966	33864
0090	S	D	E	N	1970	63039
0117	S	D	E	N	1964	37058
0119	S	G	E	N	1956	40670
0137	S	C	E	N	1970	57663
0260	S	D	E	N	1966	33864
0291	S	G	E	N	1956	40670
0292	S	D	E	N	1966	33864
0292	S	D	E	N	1970	63039
0410	S	D	G	N	1949	57056
0425	M	D	R	N	1967	45800
0429	S	D	G	N	1949	57056
0445	S	D	R	N	1969	58439
0452	S	D	E	N	1964	37058
0462	S	D	E	N	1969	58439
0470	S	D	E	N	1964	37058
0478	S	D	R	N	1969	58439
0479	S	D	R	N	1969	58439

Phys. State: G. Gas; L. Liquid; M. Multiphase; P. Powder; S. Solid

Chapter 5 Specific Heat

Substance Number	Phys. State	Sub-ject	Lan-guage	Temper-ature	Year	TPRC Number
526-0006	M	C	E	U	1968	58400
0034	S	D	R	N	1969	52430
0034	M	D	E	N	1964	37058
0034	M	D	O	N	1966	49791
0067	S	D	R	N	1969	58439
0069	S	D	E	N	1964	37058
0074	P	D	G	N	1958	51235
0082	S	D	E	N	1965	34669
0127	S	D	E	N	1965	34669
0132	S	D	E	N	1911	48358
0182	S	D	E	N	1965	34669
0182	S	D	E	N	1969	58340
0183	S	D	E	N	1965	34669
0184	S	D	E	N	1965	34669
0185	S	D	E	N	1965	34669
0186	S	D	E	N	1965	34669
0187	S	D	E	N	1965	34838
0188	S	D	E	N	1965	34669
0189	S	D	E	N	1965	34669
0194	M	D	G	N	1958	51235
0198	S	D	E	N	1911	48358
528-0024	S	S	E	N	1965	34753
0301	M	C	E	N	1970	57767
0326	S	T	E	L	1967	48008
0339	S	D	E	N	1965	34059
0340	S	D	E	H	1965	34059
0376	S	C	E	N	1965	35620
0377	S	C	E	N	1965	35620
0378	S	C	E	N	1965	35620
0379	S	C	E	N	1965	35620
0380	S	C	E	N	1965	35620
0410	M	C	E	N	1970	57767
0413	S	D	E	N	1969	58455
0421	S	D	E	N	1967	40398
531-0002	S	G	R	N	1969	58753
0014	S	D	E	N	1966	40399
0017	S	D	E	N	1966	40399
0033	S	D	E	N	1966	40399
0036	S	D	E	N	1968	40399
0060	S	D	E	N	1965	35530
0060	S	D	E	N	1970	37053
0060	S	D	E	N	1964	37499
0060	S	D	E	N	1967	44987
0060	S	D	E	N	1968	48806
0060	S	D	R	N	1968	49033
0060	S	D	E	N	1968	52893
0060	S	D	E	N	1968	52896
0066	S	D	E	N	1968	57224
0066	S	D	E	N	1968	57225
0068	S	D	E	N	1965	35530
0092	S	D	E	N	1965	35530
0092	M	D	E	N	1966	48811
0092	M	D	E	N	1966	58327
0123	S	D	E	N	1963	34979
0192	S	D	J	N	1968	51800
0195	S	D	R	N	1968	49033
0195	S	D	J	N	1968	51800
0198	S	D	J	N	1968	51800
0201	S	D	E	L	1966	36416
0201	S	D	E	L	1966	42657
0201	S	D	J	N	1968	51800
0248	M	T	E	U	1967	43733
0288	S	D	R	N	1968	54578
0301	S	D	E	L	1965	36416
0301	S	D	R	L	1966	42657
0301	S	D	R	N	1968	49033
0301	S	D	J	N	1968	51800
0303	S	D	S	L	1965	34753
0303	S	D	E	N	1964	36202
0303	S	D	E	L	1965	36416
0303	S	D	E	N	1965	38402
0303	S	D	E	N	1966	42657
0303	S	G	R	N	1965	43097
0303	S	G	J	N	1967	46251
0303	S	G	J	N	1968	46252
0303	S	D	J	N	1968	51800
0303	S	D	E	N	1968	52896
0304	S	D	F	L	1965	36416
0304	S	D	E	N	1964	37499
0304	S	D	E	L	1966	42657
0306	S	D	E	H	1967	41776
0380	S	D	E	L	1965	38743
0381	S	D	E	L	1965	38743
0383	S	D	E	L	1965	38743
0412	S	D	E	L	1965	38743
0463	M	D	E	N	1966	58327
0511	P	S	E	N	1967	44944
0515	S	D	E	L	1965	36416
0515	S	D	E	L	1966	42657
0515	S	D	J	N	1968	51800
0524	M	C	E	N	1970	57767
0525	S	D	E	N	1966	36416
531-0525	S	D	E	L	1966	42657
0533	M	C	E	L	1965	34005
0537	S	D	E	L	1965	34659
0568	S	D	E	L	1965	34659
0568	S	D	E	N	1964	36202
0568	S	D	E	L	1965	36416
0568	S	D	E	N	1964	37499
0568	S	D	E	N	1965	39754
0568	S	D	E	L	1966	42657
0568	S	G	J	N	1967	46251
0568	S	G	E	N	1968	46252
0568	S	D	J	N	1968	51800
0569	S	D	E	N	1964	36202
0569	S	D	E	N	1965	36419
0569	S	D	E	N	1964	37499
0569	S	D	E	N	1966	43475
0569	S	D	E	N	1968	52893
0585	S	D	E	L	1965	36416
0585	S	D	E	L	1966	42657
0589	L	D	R	N	1967	53012
0589	M	D	R	N	1964	38189
0617	S	D	E	N	1968	50083
0620	S	D	E	L	1965	36416
0620	S	D	E	N	1964	37255
0620	S	D	E	N	1964	37499
0620	S	D	E	H	1963	40301
0620	S	D	C	H	1965	40305
0620	S	D	E	L	1966	42657
0620	S	D	E	N	1966	43475
0620	S	D	E	N	1968	48806
0620	S	D	E	N	1968	52893
0633	S	S	E	L	1967	56366
0633	S	D	E	N	1969	56570
0677	S	D	E	N	1968	50083
0679	S	D	E	L	1965	36416
0679	S	D	E	L	1966	42657
0679	S	D	R	N	1968	49033
0679	S	D	J	N	1968	51800
0679	S	D	E	N	1968	52896
0717	S	D	E	N	1965	36419
0717	S	D	E	N	1964	37255
0838	M	G	E	L	1966	45460
0840	M	G	E	L	1966	45460
0908	M	T	E	U	1967	45972
0929	S	D	E	L	1965	36416
0929	S	D	E	L	1966	42657
0955	S	D	E	L	1960	42824
0957	S	D	E	L	1966	42657
0960	S	D	E	L	1966	42657
0962	S	D	E	H	1966	52891
0963	S	D	E	L	1966	42657
0968	S	D	E	L	1966	42657
0974	S	D	J	L	1968	56300
1038	S	D	R	N	1968	43894
1038	S	D	E	N	1966	43895
1053	M	D	E	L	1968	50110
1134	M	D	E	N	1966	48811
1159	L	D	R	N	1967	47327
1244	S	D	R	N	1966	43080
1245	M	C	E	N	1970	57767
1246	S	S	J	U	1967	45441
1246	S	E	E	H	1960	52260
1246	S	S	E	L	1967	56366
1278	S	D	E	N	1969	55120
1287	S	D	G	N	1966	41631
1288	S	D	E	N	1962	47240
1330	S	D	E	N	1966	40476
1336	S	D	E	N	1965	39754
1337	P	D	E	N	1962	39516
1356	S	D	E	N	1966	38611
1363	S	D	E	N	1964	37499
1381	S	D	E	N	1964	37509
1384	S	D	E	N	1969	55120
1390	S	D	E	N	1965	36419
1390	S	D	E	N	1963	40301
1391	S	D	E	N	1965	36419
1391	S	D	E	N	1966	40399
1420	S	D	E	N	1964	36285
1457	S	T	E	U	1070	59107
1457	S	I	E	U	1970	59599
1458	S	D	E	N	1969	55120
1467	S	D	E	N	1969	55120
1472	S	D	J	N	1968	51800
1473	S	D	J	N	1968	51800
1474	S	D	J	N	1968	51800
1475	S	D	J	N	1968	51800
1476	S	D	E	H	1966	52891
1515	S	D	E	N	1968	53094
1516	S	D	E	N	1968	53094
1518	S	D	E	N	1966	43475
1639	M	D	E	N	1968	58327
1641	M	D	E	N	1968	55829
1642	M	D	E	N	1968	55829
1643	M	D	E	N	1968	55829
1644	M	D	E	N	1968	55829
531-1671	S	D	O	N	1969	58147
1694	M	D	E	N	1967	42955
1694	M	D	R	N	1967	45153
1695	M	D	E	N	1967	42955
1695	M	D	R	N	1967	45153
1696	M	D	E	N	1967	42955
1696	M	D	R	N	1967	45153
1765	S	D	E	N	1968	57224
1765	S	D	E	N	1968	57225
1766	S	D	E	N	1968	57224
1766	S	D	E	N	1968	57225
1767	S	D	E	N	1968	57224
1767	S	D	E	N	1968	57225
1780	S	D	E	N	1965	43295
541-0019	S	D	E	N	1964	35797
0019	S	D	E	N	1967	47566
0019	S	E	E	U	1967	47991
0023	S	S	E	N	1965	34753
0044	S	C	R	N	1966	53875
0044	S	C	E	N	1966	53876
0620	S	D	E	H	1967	41136
0694	S	D	E	N	1966	40343
0720	S	D	E	N	1965	39754
0753	S	D	E	N	1965	39754
0867	M	D	E	H	1968	48862
0869	M	D	E	H	1968	48862
0882	M	D	E	H	1966	48811
0883	M	D	E	N	1966	48811
0914	S	G	E	H	1965	47486
0976	M	D	G	L	1969	57859
1050	S	T	E	N	1968	43687
1094	M	S	E	N	1968	50531
1097	S	D	E	N	1969	55899
1130	S	D	E	N	1966	52891
1295	S	S	E	H	1969	57666
1358	S	D	E	H	1962	46907
1358	S	D	E	H	1962	48004
1359	S	D	E	H	1962	46907
1359	S	D	E	H	1962	48004
1385	M	D	E	N	1966	48811
1396	S	D	I	N	1967	47994
1414	S	D	I	N	1967	47994
551-0153	S	S	E	H	1967	50724
0154	S	S	E	H	1967	50724
0436	S	G	E	N	1967	47270
0577	S	D	E	N	1964	36080
0577	S	D	E	N	1964	41386
1289	S	S	E	U	1967	50724
1290	S	S	E	N	1967	50724
2802	S	D	E	N	1965	34051
2608	S	D	E	N	1965	39065
2763	S	S	E	H	1967	50724
621-0024	S	D	E	N	1969	56600
0122	S	D	E	N	1969	56600
0123	S	D	E	N	1969	56600
0123	M	D	E	N	1969	56337
0124	S	D	E	N	1969	56600
0125	S	D	E	N	1969	56600
0126	S	D	E	N	1969	56600
0189	M	D	E	N	1969	56180
651-0071	S	D	R	N	1966	44457
0097	L	D	R	N	1969	57537
0282	S	D	E	N	1969	56585
0292	L	D	R	N	1968	57537
0293	S	D	R	U	1968	53488
0293	S	D	R	U	1968	53903
0298	L	D	E	N	1969	57537
0302	L	D	E	N	1969	57537
0309	L	D	E	N	1969	57537
0310	L	D	E	N	1969	57537
0311	L	D	E	N	1969	57537
0312	L	D	E	N	1969	57537
0313	L	D	E	N	1969	57537
0544	S	S	E	N	1968	42148
661-0001	S	D	E	N	1964	37058
0001	S	D	E	N	1966	39442
0019	S	D	E	N	1956	40670
0040	S	D	E	N	1964	37058
0040	S	S	E	U	1963	40300
0042	S	E	R	N	1965	43097
0050	M	D	R	N	1969	56636
0051	M	D	R	N	1969	56636
0060	S	D	E	N	1956	40670
0117	S	D	E	N	1956	37058
0291	S	D	E	N	1956	40670
0417	M	D	E	N	1966	48811
0425	M	D	R	N	1967	45800
0445	S	D	R	N	1969	58439
0452	S	D	R	N	1969	58439
0462	S	D	E	N	1964	37058

Phys. State: G. Gas; L. Liquid; M. Multiphase; P. Powder; S. Solid

Substance Number	Phys. State	Sub-ject	Lan-guage	Temper-ature	Year	TPRC Number	Substance Number	Phys. State	Sub-ject	Lan-guage	Temper-ature	Year	TPRC Number	Substance Number	Phys. State	Sub-ject	Lan-guage	Temper-ature	Year	TPRC Number
661-0470	S	D	E	N	1964	37058														
0471	S	D	E	N	1970	60111														
0482	M	D	R	N	1969	56636														

Phys. State: **G.** Gas; **L.** Liquid; **M.** Multiphase; **P.** Powder; **S.** Solid

Chapter 6 Viscosity

Substance Number	Phys. State	Sub-ject	Lan-guage	Temper-ature	Year	TPRC Number	Substance Number	Phys. State	Sub-ject	Lan-guage	Temper-ature	Year	TPRC Number	Substance Number	Phys. State	Sub-ject	Lan-guage	Temper-ature	Year	TPRC Number
526-0006	L	D	R	H	1968	50764														
0006	L	D	E	H	1968	52215														
531-0248	M	G	E	N	1967	44813														
0299	L	D	R	N	1966	40996														
0299	L	D	E	N	1966	40997														
0300	L	D	R	N	1966	40996														
0300	L	D	E	N	1966	40997														
0918	L	C	E	N	1967	43670														
0919	L	C	E	N	1967	43670														
1055	L	D	R	N	1967	46368														
1422	M	D	G	N	1969	58622														
1522	L	D	R	N	1969	56611														
1522	L	D	E	N	1969	56612														

Phys. State: G. Gas; L. Liquid; M. Multiphase; P. Powder; S. Solid

Chapter 7 Emittance

Substance Number	Phys. State	Subject	Language	Temperature	Year	TPRC Number
526-0009	S	D	R	N	1967	61948
0009	S	D	E	N	1967	61949
0067	S	D	E	N	1966	35414
0067	S	G	E	N	1964	35856
0067	S	D	E	N	1965	42764
0112	S	D	R	N	1967	61948
0112	S	D	E	N	1967	61949
0195	S	D	R	N	1964	50235
0264	S	D	R	N	1967	61948
0264	S	D	E	N	1967	61949
528-0183	S	D	E	H	1966	39356
0186	S	D	F	N	1966	42921
0186	S	D	E	U	1965	43794
0186	S	C	E	N	1958	44273
0186	S	G	E	U	1968	51637
0253	S	D	E	N	1959	36167
0303	S	D	E	N	1967	45669
531-0017	S	D	E	N	1967	49228
0040	S	D	E	H	1962	40298
0047	S	D	E	H	1962	40298
0056	S	D	E	N	1967	45553
0093	S	G	E	N	1964	35856
0093	S	D	E	N	1965	42764
0122	S	C	E	N	1958	44273
0122	S	S	E	H	1960	49983
0171	P	G	E	N	1967	34541
0230	L	D	E	N	1969	34391
0230	L	D	R	N	1969	46008
0301	S	D	E	N	1967	45553
0303	S	S	E	N	1965	34753
0303	S	D	E	H	1962	48104
0321	S	D	E	N	1967	51719
0511	P	E	E	N	1967	34541
0515	S	G	I	N	1954	47205
0569	S	D	E	H	1962	37332
0569	S	D	E	H	1962	40298
0569	S	D	E	H	1962	48104
0569	S	D	E	N	1967	49228
0620	S	D	E	H	1962	37332
0620	S	D	E	H	1962	40298
0620	S	D	E	U	1963	40301
0620	S	D	E	N	1965	40305
0620	S	D	E	N	1968	50141
0633	S	S	E	N	1965	34753
0633	S	S	E	L	1967	56366
0679	S	D	E	H	1962	48104
0831	S	D	E	N	1965	34485
0831	M	D	E	N	1965	43470
1315	M	D	E	H	1962	44560
1356	S	D	E	N	1966	38611
1381	S	D	E	N	1964	36084
1381	S	D	E	N	1964	37509
1390	S	D	E	H	1962	37332
1390	S	D	E	H	1962	40298
1390	S	D	E	N	1963	40301
1391	S	D	E	H	1962	37332
1391	S	D	E	H	1962	40298
1391	S	D	E	N	1967	49228
1454	S	D	E	N	1965	42781
1515	S	D	E	N	1968	53094
1516	S	D	E	N	1968	53094
1517	S	D	E	N	1968	53094
1559	S	D	E	N	1967	45553
1604	S	D	E	H	1962	36109
1641	M	D	E	N	1968	55829
1642	M	D	E	N	1968	55829
541-0062	S	D	E	U	1970	59973
0135	S	D	E	N	1967	45669
0223	S	D	E	N	1968	49527
0392	S	G	E	N	1965	36493
0393	S	G	E	N	1965	36493
0394	S	G	E	N	1965	36493
0395	S	G	E	N	1965	36493
0406	S	D	E	N	1967	45669
0412	S	D	E	N	1966	39303
0431	S	D	E	N	1968	54315
0449	S	C	E	N	1962	36118
0450	S	D	E	N	1967	45008
0450	M	D	E	N	1969	60027
0452	S	D	E	N	1967	45669
0453	S	D	E	N	1967	45669
0454	S	D	E	U	1966	34453
0455	S	D	E	U	1966	34453
0474	S	D	E	N	1967	45669
0476	S	D	E	N	1968	42593
0477	S	D	E	N	1967	45669
0482	S	D	F	N	1966	42921
0482	S	D	E	N	1970	43918
0482	S	C	E	N	1958	44273
0516	S	G	E	N	1964	34339
0517	S	G	E	N	1964	34339
0518	S	T	E	U	1964	34339
541-0539	S	D	E	H	1965	34752
0647	S	C	E	N	1958	44273
0620	S	D	E	H	1967	41136
0620	S	D	E	H	1963	48776
0651	S	D	E	N	1964	37040
0667	S	G	E	N	1968	48153
0671	S	D	E	N	1968	42593
0671	S	D	E	N	1967	47268
0677	S	D	E	H	1966	48361
0678	S	D	E	H	1966	48361
	S	D	E	N	1967	47275
0699	S	D	E	N		
0707	S	D	E	N	1967	47275
0754	S	G	E	U	1968	51127
0780	S	D	E	N	1968	49527
0821	S	D	E	N	1968	49527
0825	S	D	E	N	1968	49527
0828	S	D	E	N	1968	49527
0831	S	D	E	N	1967	46397
0832	S	D	E	N	1967	46397
0912	S	D	E	N	1967	46397
0976	S	D	E	N	1967	45669
0981	S	C	E	H	1965	36212
0999	S	D	E	N	1962	49984
1007	M	E	E	L	1969	60028
1050	S	T	E	U	1965	36209
1050	S	D	E	N	1968	50626
1050	S	T	E	U	1960	52125
1083	S	D	E	N	1969	55741
1085	S	D	E	N	1967	45669
1172	S	D	E	U	1968	53085
1245	M	D	E	N	1969	60027
1246	M	D	E	N	1969	60027
1268	S	D	E	N	1964	40358
1321	S	D	E	N	1970	59609
1321	M	D	E	N	1965	43470
1358	S	D	E	H	1962	46907
1358	S	D	E	H	1959	47025
1358	S	D	E	H	1962	48004
1359	S	D	E	H	1962	46907
1359	S	D	E	H	1959	47025
1359	S	D	E	H	1962	48004
1400	S	D	E	H	1971	62784
1401	S	D	E	H	1971	62784
551-0129	S	D	E	N	1961	52095
0153	S	D	E	N	1959	47025
0153	S	D	E	N	1968	50539
0153	S	D	E	N	1967	50724
0153	S	D	E	N	1961	52095
0153	S	D	E	F	1955	60523
0154	S	D	E	N	1965	34736
0154	S	C	E	N	1958	44273
0154	S	D	E	N	1959	47025
0154	S	D	E	N	1961	48258
0154	S	D	E	N	1962	49984
0154	S	D	E	N	1966	50298
0154	S	D	E	N	1968	50539
0154	S	S	E	N	1967	50724
0154	S	D	E	N	1961	52095
0154	S	D	E	F	1955	60523
0223	S	D	E	N	1965	34251
0223	S	D	E	N	1965	34297
0223	S	D	E	N	1965	34736
0223	S	D	E	N	1964	35796
0223	S	D	E	N	1964	36090
0223	S	D	E	N	1965	36154
0223	S	D	E	N	1965	36504
0223	S	D	E	N	1965	36519
0223	S	D	E	N	1965	38392
0223	S	D	E	N	1965	38990
0223	S	S	E	N	1930	40619
0223	S	D	F	N	1966	42921
0223	S	D	E	N	1966	42977
0223	S	D	E	N	1965	43484
0223	S	D	E	N	1965	43527
0223	S	D	E	N	1967	44754
0223	S	D	E	N	1966	44771
0223	S	D	E	N	1967	44986
0223	S	D	E	N	1967	45389
0223	S	D	E	L	1966	48363
0223	S	D	E	N	1966	48383
0223	S	D	E	N	1962	49984
0223	S	D	E	N	1967	50023
0223	S	D	E	N	1960	50317
0223	S	D	E	N	1961	52095
0223	S	D	E	N	1968	53087
0223	S	D	E	N	1969	55722
0223	S	C	E	N	1966	56378
0223	S	D	E	N	1969	57227
0223	S	D	E	N	1969	57597
0223	S	C	E	L	1968	58324
0223	S	D	E	N	1969	58330
0223	S	D	E	N	1970	59613
0223	S	D	E	L	1970	61547
551-0223	M	D	E	N	1969	57851
0226	S	D	E	N	1964	43112
0231	S	D	E	N	1964	43112
0231	S	D	E	N	1962	49984
0233	S	D	E	N	1963	37479
0233	S	D	E	N	1960	38317
0233	S	D	E	H	1966	40448
0233	S	D	E	H	1963	43735
0233	S	D	R	H	1967	44552
0233	S	D	E	N	1967	44754
0233	S	D	E	H	1967	45327
0253	S	D	E	N	1965	35840
0258	S	D	E	N	1965	35840
0260	S	D	E	N	1965	35840
0278	S	D	E	N	1968	35552
0278	S	D	E	N	1966	36740
0278	S	G	E	N	1964	37795
0278	S	D	E	N	1965	38109
0278	S	D	E	N	1965	38672
0278	S	D	E	N	1967	46397
0278	S	D	E	U	1969	41919
0278	S	D	E	N	1964	42291
0278	S	D	E	N	1966	42368
0278	S	D	E	N	1962	43493
0278	S	C	F	N	1967	46227
0278	S	D	E	N	1965	46810
0278	S	D	E	N	1967	47268
0278	S	D	E	N	1967	47276
0278	S	D	E	L	1966	48363
0278	S	D	E	N	1968	49665
0278	S	D	E	N	1968	52905
0278	S	D	E	N	1968	54153
0278	S	D	E	N	1963	54703
0278	S	D	E	N	1969	55741
0278	S	D	R	N	1969	58681
0278	S	D	E	N	1969	58682
0278	S	D	E	N	1970	59649
0283	S	D	E	N	1964	57399
0287	S	G	E	N	1966	33961
0287	S	D	E	N	1964	33974
0287	S	D	E	H	1965	36539
0287	S	D	E	H	1963	37477
0287	S	D	E	N	1965	39338
0287	S	D	E	N	1966	40304
0287	S	D	R	N	1966	41744
0287	S	D	E	N	1966	41745
0287	S	D	E	N	1963	43735
0287	S	D	E	N	1963	45545
0287	S	D	E	H	1967	45552
0287	S	D	E	N	1965	46366
0287	S	D	E	N	1959	47024
0287	S	D	E	H	1962	47737
0287	S	D	E	H	1962	47938
0287	S	D	E	H	1962	47952
0287	S	D	R	H	1968	53277
0287	S	D	E	N	1968	53923
0287	S	D	E	N	1970	63131
0325	S	D	E	N	1966	34724
0346	S	D	E	H	1964	37597
0351	S	D	E	H	1965	36144
0352	S	D	E	H	1965	36511
0352	S	D	E	N	1965	37899
0428	S	D	E	H	1969	49984
0428	S	D	E	N	1961	52095
0428	S	D	E	N	1961	52784
0435	S	D	E	N	1965	34832
0435	S	D	E	N	1965	35414
0435	S	G	E	N	1964	35856
0435	S	D	E	N	1964	36126
0435	S	D	E	N	1964	37795
0435	S	D	E	U	1969	41919
0435	S	D	E	N	1965	42764
0435	S	D	F	N	1966	42921
0435	S	D	E	N	1966	44771
0435	S	D	E	N	1967	53988
0435	S	D	E	N	1963	54703
0435	S	D	E	N	1969	55741
0435	S	D	E	N	1964	36126
0436	S	D	E	N	1967	44226
0436	S	G	E	U	1969	51637
0436	S	G	E	L	1969	54701
0436	S	D	G	N	1969	55722
0436	S	D	E	N	1970	56418
0438	S	D	E	N	1969	57227
0440	S	D	E	N	1968	39303
0447	S	D	R	N	1965	38903
0447	P	D	E	N	1963	54703
0460	S	D	F	N	1966	42921
0460	S	D	E	N	1962	49984
0460	S	D	E	N	1961	53498
0466	S	D	E	N	1964	35796
0466	S	D	E	N	1970	43918
0466	S	D	E	N	1966	44771
0466	S	D	E	N	1965	46810

Phys. State: G. Gas; L. Liquid; M. Multiphase; P. Powder; S. Solid

Substance Number	Phys. State	Subject	Language	Temperature	Year	TPRC Number
551-0466	S	G	E	U	1968	51637
0466	S	D	E	N	1961	52095
0466	S	D	E	N	1961	52784
0466	S	D	E	N	1963	54703
0466	S	D	E	N	1966	56502
0468	S	D	R	N	1966	39236
0468	S	D	E	N	1966	41003
0468	S	D	E	N	1966	44771
0468	S	D	R	N	1967	46981
0468	S	D	E	N	1967	46982
0468	S	D	E	N	1962	50354
0468	S	D	I	N	1968	50497
0479	S	D	E	N	1964	40327
0482	S	D	E	N	1964	37398
0482	S	D	E	H	1965	40305
0482	S	D	E	N	1967	41136
0482	S	D	E	N	1962	41768
0482	S	D	E	N	1966	41849
0482	S	D	E	N	1964	43388
0482	S	D	E	H	1967	45552
0482	S	D	E	N	1966	50298
0482	S	D	E	N	1967	53492
0482	S	D	E	N	1963	54703
0482	S	D	E	H	1970	63131
0487	S	D	E	N	1969	35545
0487	S	D	E	H	1965	35902
0487	S	D	E	N	1964	36090
0487	S	D	E	N	1964	36126
0487	S	D	E	N	1965	36683
0487	S	D	R	N	1965	36903
0487	S	D	E	N	1963	37479
0487	S	D	E	N	1963	37493
0487	S	D	E	N	1965	38627
0487	S	D	E	N	1965	38992
0487	S	D	E	H	1966	40448
0487	S	D	E	N	1969	41573
0487	S	C	R	H	1966	42012
0487	S	D	G	H	1966	42939
0487	S	D	E	N	1966	43484
0487	S	C	E	H	1966	43996
0487	S	D	R	H	1967	44552
0487	S	D	E	H	1967	45327
0487	S	D	E	N	1963	45545
0487	S	D	E	H	1968	46986
0487	S	D	E	N	1967	47262
0487	S	D	E	N	1967	47275
0487	S	G	E	H	1961	48293
0487	S	D	E	N	1966	48362
0487	S	D	E	N	1968	49666
0487	S	D	E	N	1968	52905
0487	S	D	E	N	1963	54703
0487	S	D	E	H	1969	55086
0487	S	D	E	N	1965	56504
0487	S	D	E	N	1969	57597
0487	S	D	E	N	1970	59649
0487	S	D	R	H	1970	62054
0487	S	D	E	H	1970	62055
0487	P	D	E	N	1961	38630
0487	P	D	E	H	1965	38992
0498	S	S	E	H	1966	33865
0500	S	D	E	N	1965	34832
0500	S	D	E	N	1965	36470
0500	S	D	F	N	1966	42921
0500	S	D	E	N	1966	44771
0500	S	D	E	N	1962	49984
0500	S	D	E	N	1960	50317
0500	S	D	E	N	1968	51867
0500	S	D	E	N	1961	52095
0501	S	D	E	N	1965	34817
0501	S	G	E	N	1964	35856
0501	S	G	E	N	1964	37795
0501	S	D	E	N	1965	38990
0501	S	G	E	N	1966	39952
0501	S	D	E	N	1964	40756
0501	S	D	E	N	1966	42368
0501	S	D	E	N	1965	42764
0501	S	D	E	N	1959	43297
0501	S	D	E	N	1966	44771
0501	S	D	E	N	1962	49984
0501	S	D	E	N	1967	50023
0501	S	D	E	N	1964	50235
0501	S	D	I	N	1968	50497
0501	S	D	I	N	1968	50498
0501	S	D	E	N	1961	52784
0501	S	D	E	U	1968	53085
0501	S	D	G	N	1969	54365
0501	S	D	G	N	1912	55091
0501	S	D	E	N	1968	55175
0501	S	D	E	N	1969	57943
0501	S	D	E	N	1969	59615
0505	S	D	E	N	1964	40327
0505	S	D	E	N	1966	44771
0507	S	D	E	N	1965	36470
0507	S	D	E	N	1966	44771
0507	S	D	E	N	1962	49984
0507	S	D	E	N	1961	52784
0507	S	D	E	N	1968	53087
0507	S	D	E	N	1968	55175
0507	S	D	E	N	1969	59615
0528	S	D	G	N	1969	54365
0528	S	D	G	N	1969	55722
551-0533	S	D	E	N	1966	48377
0536	S	D	R	N	1967	46981
0536	S	D	E	N	1967	46982
0539	S	D	R	N	1970	62042
0539	S	D	E	N	1970	62043
0548	S	D	E	N	1964	35796
0567	S	D	E	N	1964	35821
0567	S	G	E	N	1967	47998
0570	S	D	E	N	1965	36523
0570	S	D	E	N	1961	52095
0570	S	D	E	N	1968	53087
0577	S	D	E	N	1964	35796
0577	S	D	E	N	1964	36126
0577	S	D	E	N	1959	36167
0577	S	D	E	N	1959	36386
0577	S	D	E	N	1965	36466
0577	S	G	E	N	1965	36499
0577	S	D	E	N	1964	37040
0577	S	D	E	N	1963	37194
0577	S	D	E	N	1965	38392
0577	S	C	E	H	1966	39622
0577	S	D	E	N	1966	40343
0577	S	C	E	N	1966	42737
0577	S	D	E	N	1966	42914
0577	S	D	E	N	1964	43112
0577	S	D	E	N	1966	43484
0577	S	D	E	N	1967	44986
0577	S	D	E	N	1967	45389
0577	S	D	E	N	1965	46810
0577	S	D	E	N	1967	47275
0577	S	D	E	N	1967	48032
0577	S	D	E	N	1969	52920
0577	S	D	E	N	1965	56504
0577	S	D	E	N	1970	57797
0577	S	D	E	N	1970	59649
0582	S	D	E	N	1965	34297
0582	S	D	E	N	1964	35796
0582	G	D	E	N	1959	51573
0601	S	D	E	N	1965	34832
0601	S	D	E	U	1967	34978
0601	S	D	E	N	1965	36154
0601	S	D	E	N	1965	38990
0601	S	D	E	N	1965	40136
0601	S	D	E	N	1964	40358
0601	S	D	E	N	1962	41123
0601	S	D	E	U	1969	41919
0601	S	D	E	N	1966	42505
0601	S	G	E	N	1966	43481
0601	S	D	E	N	1966	43482
0601	S	D	E	N	1965	43527
0601	S	D	E	N	1966	43530
0601	S	D	E	N	1966	44272
0601	S	D	E	N	1967	47272
0601	S	D	E	N	1966	48377
0601	S	D	E	N	1966	48378
0601	S	G	E	U	1968	51637
0601	S	D	E	N	1969	55070
0601	S	D	E	N	1969	55741
0601	S	D	E	N	1970	56418
0601	S	D	E	N	1969	57227
0601	S	D	G	N	1969	59668
0605	S	D	E	N	1962	43493
0620	S	D	E	N	1964	37597
0620	S	D	E	N	1960	38317
0620	S	D	E	U	1969	41919
0620	S	D	E	N	1969	55741
0620	G	D	E	N	1959	51573
0623	S	D	E	N	1960	50317
0637	S	D	E	N	1964	40413
0637	S	D	E	N	1965	42351
0643	S	D	E	N	1964	35796
0643	S	D	E	N	1964	36090
0643	S	D	E	H	1961	47554
0643	S	G	R	N	1970	62938
0643	S	G	E	N	1970	62939
0644	S	D	E	N	1964	35821
0644	S	D	E	H	1967	48487
0646	S	D	E	N	1965	34059
0648	S	D	E	N	1962	53496
0660	S	D	E	N	1959	36807
0660	S	D	E	N	1957	40798
0660	S	C	E	N	1969	45847
0661	S	D	E	N	1959	36807
0661	S	C	E	N	1969	45847
0664	S	D	E	N	1964	50235
0678	S	D	E	N	1965	35889
0678	S	D	E	H	1964	37275
0678	S	D	E	N	1964	38618
0678	S	D	E	N	1966	40400
0678	S	D	E	N	1965	42780
0678	S	D	E	N	1966	43483
0678	S	D	E	N	1967	46336
0678	S	D	E	N	1968	47423
0678	S	D	E	N	1969	52160
0678	S	D	E	H	1969	54983
0680	S	D	E	N	1966	50298
0684	S	S	E	N	1966	33865
0686	S	S	E	N	1966	34724
0689	S	D	E	N	1963	37480
0689	S	D	E	H	1967	45552
0690	S	S	E	H	1966	34724
551-0693	S	D	E	N	1966	40400
0693	S	D	E	N	1968	47423
0693	S	D	E	N	1969	52160
0702	S	D	E	N	1962	54959
0704	S	D	E	N	1963	37479
0704	S	D	E	H	1966	40448
0704	S	D	E	N	1962	54959
0705	S	D	E	N	1963	37479
0705	S	D	E	H	1966	40448
0705	S	D	E	N	1962	54959
0716	S	G	E	H	1965	34683
0716	S	D	E	N	1964	43388
0716	P	D	E	H	1962	36109
0723	S	D	R	N	1966	39236
0723	S	D	E	N	1966	41003
0723	S	D	E	N	1924	43414
0723	S	D	R	N	1967	46981
0723	S	D	E	N	1967	46982
0723	S	D	E	H	1961	47554
0723	S	D	E	N	1960	50317
0723	S	D	G	H	1966	50684
0723	S	D	E	N	1962	54959
0723	S	D	R	N	1969	58681
0723	S	D	E	N	1969	58682
0734	S	D	E	H	1965	40305
0734	S	D	E	N	1966	41849
0736	S	C	E	N	1969	45847
0747	S	D	E	H	1966	40324
0749	S	D	E	N	1964	43112
0752	S	D	R	N	1967	46981
0752	S	D	E	N	1967	46982
0752	S	D	R	H	1968	50943
0752	S	D	E	H	1968	50944
0752	S	D	R	N	1969	58681
0752	S	D	E	N	1969	58682
0753	S	D	E	N	1964	37886
0753	S	D	R	N	1967	46981
0753	S	D	E	N	1967	46982
0753	S	D	R	H	1968	50943
0753	S	D	E	H	1968	50944
0753	S	D	R	N	1969	58681
0753	S	D	E	N	1969	58682
0754	S	D	R	N	1967	46981
0754	S	D	E	N	1967	46982
0754	S	D	R	N	1969	58681
0754	S	D	E	N	1969	58682
0762	S	D	E	N	1965	35889
0762	S	D	E	N	1966	40400
0762	S	D	E	N	1965	42780
0762	S	D	E	N	1968	47423
0762	S	D	E	N	1969	52160
0767	S	D	E	N	1965	34736
0779	S	D	E	N	1962	51506
0799	S	D	E	N	1964	36126
0812	S	D	E	N	1965	35889
0812	S	D	E	H	1964	37275
0812	S	D	E	N	1965	38109
0812	S	D	E	N	1964	38618
0812	S	D	E	N	1966	40400
0812	S	D	E	N	1965	42780
0812	S	D	E	N	1966	43483
0812	S	D	E	N	1967	46336
0812	S	D	E	N	1965	46810
0812	S	D	E	N	1968	47423
0812	S	D	E	N	1969	52160
0813	S	D	E	H	1964	37275
0813	S	D	E	N	1965	38109
0813	S	D	E	N	1965	46810
0814	S	D	E	H	1964	37275
0823	S	D	E	N	1966	40351
0824	S	D	E	N	1966	40351
0826	S	D	E	N	1965	38109
0826	S	D	E	N	1965	46810
0826	S	D	E	N	1969	52160
0853	S	D	R	N	1967	46981
0853	S	D	E	N	1967	46982
0888	S	D	E	N	1964	43112
0888	S	D	E	U	1967	45665
0888	S	D	E	H	1961	47554
0888	S	D	E	N	1966	50298
0904	S	D	I	N	1967	47994
0906	M	D	E	N	1969	57851
0908	S	D	E	N	1965	38990
0911	S	D	E	H	1965	34341
0911	S	G	E	H	1965	34762
0911	S	D	R	N	1967	46981
0911	S	D	E	N	1967	46982
0911	S	D	E	N	1964	50235
0911	S	D	R	N	1969	58681
0911	S	D	E	N	1969	58682
0924	S	D	E	N	1964	42291
0924	S	D	E	N	1962	43493
0924	S	D	E	N	1963	54703
0924	S	D	E	N	1966	56502
0926	S	D	E	N	1963	54703
0929	S	D	R	N	1967	46981
0929	S	D	E	N	1967	46982
0929	S	D	R	N	1969	58681
0929	S	D	E	N	1969	58682
0935	S	D	R	N	1966	39236
0935	S	D	E	N	1966	41003

Phys. State: **G.** Gas; **L.** Liquid; **M.** Multiphase; **P.** Powder; **S.** Solid

Substance Number	Phys. State	Subject	Language	Temperature	Year	TPRC Number
551-0935	S	D	E	N	1964	43112
0935	S	D	E	N	1960	46167
0935	S	D	R	N	1967	46981
0935	S	D	E	N	1967	46982
0935	S	D	R	N	1969	58681
0935	S	D	E	N	1969	58682
0949	S	D	E	N	1965	36154
0949	S	D	E	N	1965	43527
0959	S	D	R	N	1966	39236
0959	S	D	E	H	1966	40324
0959	S	D	E	N	1966	41003
0959	S	D	E	N	1964	43112
0959	S	D	E	N	1924	43414
0959	S	D	R	N	1967	46981
0959	S	D	E	N	1967	46982
0959	S	D	E	N	1964	50235
0959	S	D	E	N	1960	50317
0959	S	D	E	N	1962	50354
0959	S	D	I	N	1968	50498
0959	S	D	E	N	1967	52118
0959	S	D	E	N	1968	54153
0959	S	D	E	N	1970	57887
0959	S	D	R	N	1969	58681
0959	S	D	E	N	1969	58682
0959	S	G	E	N	1970	62590
0960	S	D	R	N	1966	39236
0960	S	G	E	H	1966	39602
0960	S	D	E	H	1966	40324
0960	S	D	E	N	1966	41003
0960	S	G	E	H	1966	45459
0960	S	D	R	N	1967	46981
0960	S	D	E	N	1967	46982
0960	S	D	E	N	1966	48367
0960	S	D	E	N	1967	49153
0960	S	D	E	N	1960	50317
0960	S	D	R	N	1969	58681
0960	S	D	E	N	1969	58682
0976	S	D	E	N	1966	44771
1064	S	D	E	N	1965	34341
1064	S	D	E	N	1965	38109
1064	S	D	E	N	1966	42635
1064	S	D	E	N	1966	47527
1064	S	D	E	N	1968	48184
1064	S	D	E	N	1962	54959
1064	S	D	G	N	1912	55091
1064	S	D	E	N	1970	57578
1076	S	S	E	N	1965	39238
1076	S	S	E	N	1967	46337
1076	S	D	E	N	1964	50235
1076	S	S	E	N	1968	53122
1076	S	S	E	N	1969	57666
1112	S	D	E	N	1963	37194
1116	S	D	E	N	1963	54703
1130	S	D	E	N	1960	35546
1131	P	D	E	N	1961	38630
1135	S	D	E	N	1963	37194
1135	S	D	E	H	1962	41768
1135	S	D	E	U	1967	45665
1135	S	D	E	N	1966	50298
1140	S	D	E	H	1964	37275
1141	S	D	E	N	1964	34333
1141	S	D	E	N	1965	34736
1141	S	D	E	N	1965	36404
1141	S	D	E	N	1965	36523
1141	S	D	E	N	1966	36740
1141	S	D	E	N	1964	37040
1141	S	D	E	N	1964	37327
1141	S	D	E	N	1965	38109
1141	S	D	E	N	1965	39745
1141	S	D	E	N	1966	48377
1141	S	D	E	N	1962	49984
1162	S	D	E	H	1963	37477
1172	S	D	E	N	1963	37479
1172	S	D	E	H	1966	39303
1172	S	D	E	H	1966	40448
1172	S	G	G	H	1934	46996
1175	S	D	E	N	1965	34736
1175	S	D	E	N	1966	48372
1175	S	D	E	N	1962	49984
1179	S	D	E	N	1965	34832
1179	S	D	E	N	1962	41123
1181	S	D	E	N	1965	36523
1186	S	D	E	N	1962	41123
1188	S	D	E	N	1965	40136
1188	S	D	E	N	1962	41123
1188	S	D	E	N	1966	43530
1189	S	D	E	N	1965	40136
1189	S	D	E	N	1962	41123
1189	S	D	E	N	1966	48378
1197	S	D	E	N	1962	41123
1197	S	D	E	N	1966	48377
1198	S	D	E	N	1966	40351
1198	S	D	E	N	1962	41123
1202	S	D	E	N	1966	43482
1202	S	D	E	N	1966	48377
1202	S	D	E	N	1965	52152
1202	S	D	E	N	1969	55070
1207	S	D	E	N	1966	43530
1226	S	D	E	N	1964	35796
1226	S	D	E	H	1964	37275
1226	S	D	E	U	1969	41919
551-1226	S	D	R	N	1968	50642
1226	S	D	E	N	1970	59672
1234	S	D	E	N	1965	36525
1258	S	D	E	N	1963	37193
1258	S	D	E	N	1964	37443
1258	S	D	E	N	1965	40136
1258	S	G	E	N	1964	40414
1258	S	D	E	N	1963	42295
1258	S	D	E	N	1967	47268
1266	S	D	E	N	1965	34736
1266	S	D	E	N	1962	49984
1289	S	D	E	N	1965	38109
1289	S	D	E	N	1965	46810
1289	S	D	E	N	1968	50539
1289	S	S	E	N	1967	50724
1289	S	D	E	F	1955	60523
1290	S	D	E	N	1965	38109
1290	S	D	E	N	1965	46810
1290	S	S	E	H	1967	50724
1321	S	D	E	H	1965	35902
1321	S	D	E	N	1964	36090
1321	S	D	R	N	1965	36903
1321	S	D	E	N	1963	37479
1321	S	D	E	N	1963	37493
1321	S	D	E	N	1965	38627
1321	S	D	E	H	1966	40448
1321	S	D	G	H	1966	42939
1321	S	D	E	N	1963	45545
1321	S	D	E	H	1969	55086
1323	S	D	E	H	1964	35556
1323	S	D	E	N	1964	36090
1323	S	D	E	N	1963	37479
1323	S	D	E	H	1966	40448
1323	S	D	E	N	1963	45545
1324	S	D	E	N	1963	45545
1325	S	D	E	N	1964	36090
1325	S	D	E	H	1959	52421
1326	S	D	E	N	1963	45545
1327	S	D	E	N	1963	45545
1327	S	D	R	H	1969	54747
1327	S	D	E	H	1969	55575
1328	S	D	E	N	1963	37479
1328	S	D	E	H	1966	40448
1328	S	D	E	N	1963	45545
1329	S	D	E	N	1963	45545
1330	S	D	E	N	1964	33974
1330	S	D	E	N	1963	45545
1330	S	D	E	N	1965	46366
1331	S	D	R	N	1965	36903
1331	S	D	E	N	1965	38109
1331	S	D	E	N	1965	38627
1331	S	D	E	N	1963	45545
1331	S	D	E	N	1965	46810
1333	S	D	E	N	1963	45545
1362	S	D	E	N	1967	47263
1373	S	D	E	N	1965	35840
1377	S	D	E	N	1964	36090
1378	P	D	E	H	1964	37378
1386	S	D	E	H	1964	37275
1417	S	D	E	N	1962	43493
1425	S	D	E	N	1964	36090
1430	S	D	E	N	1965	36144
1430	S	D	R	N	1966	41413
1440	S	D	E	N	1965	40136
1440	S	D	E	N	1961	52095
1471	S	D	E	N	1965	35840
1472	S	D	E	H	1964	37275
1473	P	D	E	N	1961	38630
1498	S	D	E	N	1963	37493
1508	S	D	E	N	1964	36126
1520	S	G	E	U	1968	51637
1521	S	D	E	N	1961	41149
1539	S	D	E	N	1965	36144
1539	S	D	E	H	1962	37342
1545	S	D	E	N	1965	36404
1545	S	D	E	N	1965	36525
1545	S	D	E	N	1965	39745
1545	S	D	E	U	1969	41919
1545	S	D	E	U	1969	41945
1545	S	D	E	N	1966	42914
1545	S	G	E	N	1966	43481
1545	S	D	E	N	1966	43482
1545	S	D	E	N	1965	43527
1545	S	D	E	N	1966	43530
1545	S	D	E	N	1966	44270
1545	S	D	E	N	1966	48377
1545	S	D	E	N	1968	49665
1545	S	D	E	N	1968	53089
1545	S	D	E	N	1969	55070
1545	S	D	E	N	1969	55741
1545	S	D	E	N	1969	57227
1571	S	D	E	N	1965	34312
1571	S	S	E	N	1965	34736
1571	S	E	E	U	1964	36423
1571	S	S	E	H	1965	37590
1571	S	S	E	H	1966	39081
1571	S	C	E	H	1966	39622
1571	S	C	E	U	1965	48433
1571	S	S	E	H	1960	49983
1571	S	S	E	N	1959	52098
1571	S	D	E	N	1963	52788
551-1571	S	C	E	N	1967	52885
1571	S	S	E	U	1968	54883
1571	S	S	E	L	1967	56366
1571	S	S	E	N	1969	57666
1571	S	S	E	N	1967	58165
1571	S	S	E	U	1070	50100
1571	P	S	E	N	1968	53122
1574	S	D	E	N	1965	36504
1574	S	D	E	N	1965	36516
1574	S	D	E	N	1967	44754
1574	S	D	E	N	1960	50317
1575	S	D	E	N	1965	36504
1575	S	D	R	N	1969	58681
1575	S	D	E	N	1969	58682
1591	S	D	E	N	1964	36126
1591	S	D	E	H	1962	49984
1591	S	D	G	N	1912	55091
1601	S	D	E	N	1962	43493
1606	S	D	E	N	1965	46810
1618	S	D	E	H	1964	37378
1618	S	D	E	H	1969	54983
1659	S	D	E	N	1966	48378
1661	S	D	E	N	1965	35840
1663	S	D	E	N	1964	43388
1674	S	G	E	N	1970	61042
1689	S	D	E	U	1969	41919
1689	S	D	E	N	1970	43918
1689	S	D	E	N	1969	57227
1689	S	D	E	N	1969	57597
1689	S	D	E	N	1970	57854
1694	S	D	E	H	1964	37275
1730	S	D	E	N	1965	36154
1730	S	G	E	N	1966	39952
1732	S	G	E	N	1964	37795
1732	S	D	E	N	1967	53986
1738	S	D	E	N	1965	36519
1739	S	D	E	N	1966	44771
1747	S	D	E	H	1962	37342
1748	S	D	E	H	1962	37342
1751	S	D	E	H	1962	37342
1752	S	D	E	H	1962	37342
1755	S	D	E	H	1962	37342
1765	S	D	E	N	1965	35840
1766	S	D	E	N	1965	35840
1768	S	D	E	N	1964	37040
1768	S	D	E	N	1962	54959
1790	S	D	E	N	1966	43528
1805	S	D	E	N	1965	39745
1811	S	D	E	N	1965	34832
1811	S	D	E	N	1962	41123
1811	S	D	E	N	1969	41573
1858	S	D	E	N	1966	34449
1858	S	D	E	N	1964	35796
1858	S	D	E	N	1965	36504
1858	S	D	E	N	1965	36516
1858	S	D	E	N	1966	40343
1858	S	D	E	N	1964	40663
1858	S	D	E	U	1969	41919
1858	S	D	F	N	1966	42921
1858	S	D	E	N	1966	43484
1858	S	D	E	N	1965	43527
1858	S	D	E	N	1967	44754
1858	S	D	E	N	1966	44771
1858	S	D	E	N	1967	44986
1858	S	D	E	N	1967	45389
1858	S	D	E	N	1969	46420
1858	S	D	E	N	1965	46810
1858	S	D	E	N	1966	47527
1858	S	D	E	N	1967	48032
1858	S	D	E	N	1966	48383
1858	S	D	R	N	1968	50642
1858	S	D	E	N	1968	50896
1858	S	D	E	N	1968	51867
1858	S	D	E	U	1968	53085
1858	S	D	E	N	1968	53086
1858	S	D	E	N	1968	53087
1858	S	D	E	N	1968	53616
1858	S	D	E	N	1969	55741
1858	S	C	E	N	1969	56347
1858	S	D	E	N	1969	57227
1858	S	D	E	N	1970	59607
1858	S	D	G	N	1969	59668
1858	S	D	F	H	1970	60242
1858	S	D	F	N	1970	63133
1858	M	D	G	N	1969	57859
1863	S	D	E	N	1965	35840
1884	S	D	E	N	1961	52095
1890	S	D	E	U	1969	41919
1900	S	D	E	N	1964	43112
1902	S	D	E	N	1964	43112
1908	S	D	E	N	1962	41123
1908	S	D	E	N	1961	41149
1917	S	D	E	N	1964	34333
1917	S	D	E	N	1966	36740
1917	S	D	E	N	1966	48377
1933	S	D	E	N	1965	35840
1934	S	D	E	N	1965	35840
1935	S	D	E	N	1965	35840
1938	S	D	E	N	1969	35545
1938	S	D	E	N	1965	35818
1938	S	D	E	N	1965	36683

Phys. State: **G.** Gas; **L.** Liquid; **M.** Multiphase; **P.** Powder; **S.** Solid

Substance Number	Phys. State	Subject	Language	Temperature	Year	TPRC Number
551-1938	S	D	E	N	1965	38392
1938	S	D	E	N	1965	40303
1938	S	D	E	N	1965	42781
1938	S	D	E	N	1964	43423
1942	S	D	E	N	1965	35840
1943	S	D	E	N	1965	34736
1943	S	D	E	U	1967	34978
1943	S	D	E	N	1965	36154
1943	S	D	E	N	1965	36523
1943	S	D	E	N	1964	37040
1943	S	D	E	N	1965	40136
1943	S	D	E	N	1966	44771
1943	S	D	E	U	1967	45665
1943	S	D	E	N	1962	49984
1943	S	D	E	N	1966	56502
1943	S	D	E	N	1969	57597
1943	S	D	E	N	1970	59613
1944	S	D	E	N	1965	34832
1944	S	D	E	N	1965	39263
1944	S	D	E	N	1959	46025
1953	S	D	E	N	1965	36504
1953	S	D	E	N	1967	45389
1953	S	D	E	H	1962	49984
1974	S	D	E	N	1964	35796
2008	S	D	E	N	1965	35840
2009	S	D	E	N	1966	40304
2020	S	D	E	N	1964	37327
2020	S	D	E	N	1966	44771
2020	M	D	G	N	1969	57859
2031	S	D	E	N	1968	53491
2045	S	D	E	N	1965	35840
2071	S	D	E	N	1961	52095
2071	S	D	E	N	1961	52784
2071	S	D	E	N	1969	57597
2091	S	D	E	N	1963	37194
2104	S	D	E	N	1965	36144
2104	S	D	E	N	1963	37194
2104	S	D	E	N	1963	37480
2116	S	D	E	N	1965	35840
2132	S	D	E	N	1966	44272
2132	S	D	E	N	1965	52152
2132	S	D	E	N	1969	55070
2138	S	D	E	N	1964	37327
2147	S	D	E	N	1969	55070
2147	S	D	G	N	1969	59668
2148	S	D	E	H	1969	54983
2158	S	D	E	N	1966	40351
2166	S	D	E	H	1965	35902
2168	S	D	E	H	1969	54983
2170	S	D	E	H	1969	54983
2193	S	D	E	N	1966	44771
2230	S	D	E	N	1965	36469
2230	S	D	E	N	1966	40351
2240	S	D	E	N	1964	37840
2240	S	D	E	N	1964	42291
2249	S	D	R	N	1966	39236
2249	S	D	E	N	1966	41003
2253	S	D	R	N	1966	39236
2253	S	D	E	N	1966	41003
2254	S	D	E	N	1966	40351
2266	S	D	E	N	1964	37327
2266	S	D	E	N	1967	49229
2266	S	D	E	N	1968	53087
2267	S	D	R	N	1966	39236
2267	S	D	E	N	1966	41003
2277	S	D	R	N	1966	39236
2277	S	D	E	N	1966	41003
2277	S	D	R	N	1968	50002
2277	S	D	E	N	1968	50921
2277	S	D	R	N	1970	62042
2277	S	D	E	N	1970	62043
2279	S	D	R	N	1966	39236
2279	S	D	E	N	1966	41003
2280	S	D	E	N	1967	53492
2282	S	D	R	N	1966	39236
2282	S	D	E	N	1966	41003
2286	S	C	E	L	1970	61190
2289	S	D	E	N	1964	33974
2289	S	D	E	N	1965	46366
2290	S	D	E	N	1964	33974
2290	S	D	E	N	1965	46366
2292	S	D	E	N	1964	33974
2292	S	D	E	N	1965	46366
2296	S	D	E	N	1964	33974
2296	S	D	E	N	1965	46366
2298	S	D	E	N	1964	33974
2298	S	D	E	N	1965	46366
2298	S	D	E	N	1969	57597
2301	S	D	E	N	1964	33974
2301	S	D	E	N	1965	46366
2303	S	D	R	N	1966	41744
2303	S	D	E	N	1966	41745
2304	S	D	E	N	1964	33974
2305	S	D	E	N	1961	52095
2306	S	D	E	N	1964	33974
2306	S	D	E	N	1965	46366
2309	S	D	E	H	1967	45552
2310	S	D	E	N	1964	33974
2310	S	D	E	N	1965	46366
2313	S	D	E	N	1964	43112
2313	S	D	E	N	1969	56874

Substance Number	Phys. State	Subject	Language	Temperature	Year	TPRC Number
551-2315	S	D	E	N	1965	38992
2316	S	D	R	N	1965	36903
2316	S	D	E	N	1965	38627
2316	S	D	E	N	1963	45545
2321	S	D	E	N	1964	33974
2321	S	D	R	H	1968	53277
2321	S	D	E	H	1968	53923
2323	S	D	E	N	1950	46912
2324	S	D	E	H	1963	43421
2326	S	D	E	H	1963	43421
2328	S	D	E	H	1963	43421
2331	S	D	E	U	1967	45665
2332	S	D	E	U	1967	45665
2333	S	D	E	U	1967	45665
2334	S	D	E	U	1967	45665
2345	S	D	E	N	1966	44771
2346	S	D	E	N	1966	44771
2347	S	D	E	N	1966	44771
2347	S	D	E	N	1961	52095
2347	S	D	E	N	1961	52784
2366	S	D	E	N	1964	43112
2369	S	D	E	N	1970	38036
2369	S	D	E	N	1967	46135
2387	S	G	E	N	1964	37795
2387	S	D	E	N	1964	40756
2389	S	D	E	N	1964	40756
2390	S	D	E	N	1964	40756
2416	S	D	E	N	1964	35796
2416	S	D	E	N	1965	36504
2416	S	D	E	N	1969	37446
2416	S	D	E	N	1965	38392
2416	S	D	E	N	1966	40343
2416	S	S	E	N	1930	40619
2416	S	D	E	U	1969	41919
2416	S	D	E	N	1966	43484
2416	S	D	E	N	1966	44771
2416	S	D	E	N	1967	44986
2416	S	D	R	N	1967	45382
2416	S	D	E	N	1967	45389
2416	S	D	E	N	1967	47275
2416	S	D	E	N	1967	48032
2416	S	D	E	L	1966	48363
2416	S	D	E	N	1966	48383
2416	S	D	E	N	1968	49666
2416	S	D	E	N	1968	53086
2416	S	D	G	N	1969	54365
2416	S	D	E	L	1968	58326
2416	S	D	G	N	1969	59668
2416	M	D	E	N	1969	57851
2417	S	D	E	N	1966	40351
2418	S	D	E	N	1966	40351
2419	S	D	E	N	1966	40351
2420	S	D	E	N	1966	40351
2421	S	D	E	N	1966	40351
2422	S	D	E	N	1966	40351
2423	S	D	E	N	1966	40351
2424	S	D	E	N	1966	40351
2425	S	D	E	N	1966	40351
2428	S	D	E	N	1966	40351
2431	S	D	E	N	1966	40351
2432	S	D	E	N	1966	40351
2433	S	D	E	N	1966	40351
2434	S	D	E	N	1966	40351
2435	S	D	E	N	1966	40351
2436	S	D	E	U	1968	53085
2437	S	D	E	N	1966	40351
2438	S	D	E	N	1959	36807
2438	S	D	E	N	1964	40663
2438	S	D	E	H	1961	47554
2438	S	D	E	N	1968	53616
2438	S	D	E	N	1968	54153
2439	S	D	I	N	1967	46508
2440	S	D	E	N	1966	40351
2441	S	D	E	N	1966	40351
2442	S	D	E	N	1966	40351
2443	S	D	E	N	1966	40351
2444	S	D	E	N	1966	40351
2446	S	D	E	N	1966	40351
2447	S	D	E	N	1966	40351
2454	S	D	E	N	1966	40351
2456	S	D	E	N	1966	40351
2457	S	D	E	N	1966	40351
2465	S	D	E	H	1965	40314
2466	S	D	E	H	1965	40314
2467	S	D	E	H	1965	40314
2468	S	D	E	H	1965	40314
2469	S	D	E	H	1965	40314
2471	S	D	E	H	1965	40314
2472	S	D	E	H	1965	40314
2474	S	G	E	N	1966	33961
2474	S	D	E	N	1964	33974
2474	S	D	E	N	1965	46366
2475	S	D	E	H	1965	40314
2477	S	D	E	H	1965	40314
2478	S	D	E	N	1964	33974
2478	S	D	E	N	1964	37399
2478	S	D	E	N	1964	40327
2478	S	D	E	H	1967	45552
2478	S	D	E	N	1965	46366
2479	S	D	E	H	1965	40314
2480	S	G	E	H	1966	33961

Substance Number	Phys. State	Subject	Language	Temperature	Year	TPRC Number
551-2480	S	D	E	N	1964	33974
2481	S	D	E	H	1965	40314
2482	S	D	E	H	1965	40314
2483	S	D	E	H	1965	40314
2486	S	D	E	H	1966	40324
2488	S	D	E	N	1964	40327
2492	S	D	E	N	1969	57597
2524	S	G	E	N	1964	34339
2543	S	D	E	N	1964	43112
2543	S	D	R	N	1967	46981
2543	S	D	E	N	1967	46982
2543	S	D	R	N	1969	58681
2543	S	D	E	N	1969	58682
2545	S	D	F	N	1966	42921
2547	S	D	F	N	1966	42921
2547	S	D	E	N	1966	44771
2548	S	D	F	N	1966	42921
2549	P	D	E	N	1961	38630
2556	S	D	E	N	1964	37886
2558	S	D	E	N	1964	37886
2559	S	D	E	N	1964	37886
2560	S	D	E	N	1964	37886
2562	G	G	E	N	1965	33749
2563	S	D	E	H	1964	37488
2566	S	S	E	N	1964	36423
2566	S	S	E	U	1965	36532
2566	S	S	E	U	1965	36536
2566	S	D	E	N	1963	41266
2566	S	S	E	N	1963	45545
2566	S	S	E	N	1960	51531
2566	S	S	E	N	1967	54175
2566	S	D	E	N	1962	55198
2566	S	S	E	U	1970	59975
2572	S	D	E	N	1963	54703
2572	S	D	E	N	1969	59615
2573	S	D	E	N	1964	37399
2591	S	D	E	N	1963	37493
2592	S	D	E	N	1963	37493
2597	S	D	E	N	1965	34059
2598	S	D	E	N	1965	34059
2600	S	D	E	N	1965	34832
2600	S	D	E	N	1965	36154
2600	S	D	F	N	1966	42921
2600	S	D	E	N	1968	53087
2603	S	D	E	N	1965	34059
2603	S	D	E	H	1965	43458
2605	S	D	E	N	1965	34059
2606	S	D	E	N	1965	34059
2608	S	D	E	N	1965	39065
2608	S	D	E	N	1962	50354
2611	S	D	E	N	1965	36523
2611	S	D	E	N	1966	43530
2613	S	D	E	N	1964	36090
2614	S	D	E	N	1964	33974
2614	S	D	E	N	1965	46366
2614	S	D	R	H	1969	54568
2614	S	D	E	H	1969	55069
2614	S	D	E	H	1963	58962
2615	S	D	E	H	1965	34059
2615	S	D	E	H	1965	34121
2615	S	D	E	N	1964	37399
2615	S	D	E	N	1966	40304
2615	S	D	E	H	1965	43458
2615	S	D	E	N	1965	46366
2616	S	D	E	N	1965	34137
2617	S	D	E	N	1965	34137
2618	S	D	E	N	1965	34138
2621	S	D	E	N	1961	41149
2623	S	G	E	H	1965	34762
2623	S	D	E	H	1965	35902
2623	S	D	E	N	1963	37479
2623	S	D	E	N	1965	40136
2623	S	D	E	H	1966	40448
2623	S	D	E	H	1964	43422
2623	S	D	E	N	1966	48378
2627	S	G	E	N	1964	34339
2627	S	S	E	N	1966	52913
2629	S	D	E	N	1965	34258
2629	S	D	E	N	1968	48184
2631	S	D	E	N	1966	43528
2635	S	D	E	N	1966	43528
2637	S	G	E	N	1964	35856
2637	S	D	E	N	1965	42764
2637	L	D	E	N	1965	35414
2639	S	G	E	N	1964	35856
2639	S	D	E	N	1965	42764
2639	L	D	E	N	1965	35414
2640	S	D	E	N	1961	52095
2640	M	D	E	N	1965	35414
2641	S	D	E	H	1962	37342
2644	S	D	G	N	1969	54365
2654	S	D	E	N	1968	51867
2669	S	D	E	H	1964	37275
2677	S	D	E	N	1965	40136
2677	S	D	E	N	1961	52095
2677	S	D	E	N	1961	52153
2677	S	D	E	N	1961	52784
2678	S	D	E	N	1961	52095
2707	S	D	E	N	1964	37327
2708	S	D	E	N	1964	37327
2713	S	D	E	N	1964	37327

Phys. State: **G.** Gas; **L.** Liquid; **M.** Multiphase; **P.** Powder; **S.** Solid

Substance Number	Phys. State	Sub-ject	Lan-guage	Temper-ature	Year	TPRC Number
551-2714	S	D	E	N	1963	37479
2714	S	D	E	H	1966	40448
2715	S	D	E	H	1947	34226
2715	S	D	E	H	1964	37378
2716	S	D	E	N	1964	37327
2717	S	D	E	N	1965	36504
2717	S	D	E	N	1965	38392
2717	S	D	F	N	1966	42921
2718	S	D	E	N	1964	37327
2719	S	D	E	N	1964	37327
2721	S	D	E	N	1964	37327
2722	S	D	E	N	1964	37327
2730	S	S	E	H	1966	34724
2736	S	D	E	N	1965	34736
2736	S	D	E	N	1962	49984
2739	S	D	E	N	1965	34736
2739	S	D	E	N	1965	39263
2739	S	D	E	N	1962	49984
2751	S	D	E	N	1965	34736
2751	S	D	E	N	1960	35546
2751	S	D	E	N	1964	35796
2751	S	D	E	N	1964	37040
2751	S	D	E	N	1963	37194
2751	S	D	E	N	1965	38990
2751	S	D	E	N	1964	40358
2751	S	D	E	N	1966	43482
2751	S	D	E	N	1966	43530
2751	S	D	E	N	1966	48372
2751	S	D	E	N	1968	49665
2751	S	D	E	N	1962	49984
2751	S	D	E	N	1965	52152
2751	S	D	E	N	1969	55070
2751	S	D	E	N	1969	57227
2755	S	C	R	N	1964	34769
2755	S	C	E	N	1965	34770
2758	S	D	E	N	1965	34817
2758	S	D	E	N	1964	40327
2758	S	D	E	N	1965	42781
2759	S	D	E	N	1965	34817
2759	S	D	E	N	1965	43471
2759	S	D	R	N	1968	50642
2759	S	D	E	N	1963	54703
2762	S	D	E	N	1966	43528
2763	S	D	E	N	1965	34820
2763	S	D	E	N	1965	36469
2763	S	D	E	N	1966	43495
2763	S	D	E	N	1966	50298
2763	S	S	E	H	1967	50724
2764	S	D	E	N	1968	50731
2765	S	D	E	N	1965	34832
2771	S	D	E	N	1965	36469
2772	S	D	E	N	1965	36470
2772	S	D	E	U	1968	53085
2772	S	D	E	N	1968	53087
2772	S	D	E	N	1969	57597
2773	S	D	E	N	1961	35120
2783	S	D	E	N	1965	36470
2784	S	C	E	N	1965	34835
2784	S	D	E	N	1963	37479
2784	S	D	E	H	1966	40448
2785	S	D	E	N	1965	36470
2787	S	D	E	N	1965	36470
2799	S	D	E	N	1959	36807
2803	S	D	E	N	1963	37194
2804	S	D	E	N	1963	37194
2805	S	D	E	N	1963	37194
2811	S	D	E	N	1963	37194
2811	S	D	E	H	1963	43735
2811	S	D	E	H	1967	45552
2813	S	D	E	N	1964	33974
2814	S	D	E	N	1964	33974
2821	S	D	E	N	1964	33974
2822	S	D	E	N	1969	55741
2823	S	D	E	N	1969	35545
2823	S	D	E	N	1967	53492
2823	S	D	E	N	1969	55741
2825	S	D	E	N	1967	49310
2830	S	D	E	N	1964	33974
2831	S	D	E	N	1965	36154
2835	S	D	E	N	1964	36126
2836	S	D	E	N	1964	36126
2840	S	D	E	N	1966	44771
2841	S	D	E	N	1966	44771
2841	S	D	E	N	1969	59615
2842	S	D	F	H	1968	54810
2848	S	D	E	N	1963	37479
2848	S	D	E	H	1966	40448
2850	S	D	E	N	1964	33974
2851	S	D	E	N	1964	33974
2852	S	D	E	N	1964	33974
2853	S	D	E	N	1964	33974
2854	S	D	E	N	1964	33974
2855	S	D	E	N	1964	33974
2856	S	D	E	N	1964	33974
2860	S	D	E	N	1964	35821
2863	S	D	E	N	1964	36090
2865	S	D	E	N	1960	50317
2866	S	D	E	N	1962	50354
2867	S	D	E	H	1960	50391
2868	S	D	E	N	1963	37479
2868	S	D	E	H	1966	40448

Substance Number	Phys. State	Sub-ject	Lan-guage	Temper-ature	Year	TPRC Number
551-2872	S	D	E	N	1963	37479
2872	S	D	E	H	1966	40448
2874	S	D	E	N	1964	33974
2874	S	D	E	N	1963	37479
2874	S	D	E	H	1966	40448
2879	S	D	E	N	1963	37479
2879	S	D	E	H	1966	40448
2879	L	D	E	U	1964	36128
2881	S	D	E	N	1963	37479
2881	S	D	E	H	1966	40448
2881	S	D	E	H	1966	49719
2883	S	D	E	H	1965	40314
2890	S	D	E	N	1969	35545
2894	S	D	E	N	1969	41573
2910	S	C	E	N	1969	52592
2921	S	D	E	N	1961	52095
2922	S	D	E	N	1961	52095
2940	S	D	E	H	1968	51207
2940	S	D	E	H	1968	54282
2957	S	D	E	N	1968	56375
2962	S	D	E	H	1968	56375
2965	S	D	E	N	1958	56416
2966	S	D	E	N	1969	49589
2966	S	D	E	N	1970	59672
2971	S	D	E	N	1965	36154
2971	S	D	E	N	1965	38990
2971	S	D	E	N	1968	53087
2988	S	D	E	N	1961	52784
2988	S	D	E	N	1962	53496
2988	S	D	E	N	1961	53497
2989	S	D	R	H	1968	54120
2989	S	D	E	H	1968	54231
2991	S	D	E	N	1964	36090
2991	S	D	E	H	1963	37477
2991	S	D	E	N	1963	37479
2991	S	D	E	H	1966	40448
2993	S	D	E	N	1968	53087
2997	S	D	E	H	1965	36511
2997	S	D	E	H	1965	37899
2999	S	C	E	U	1965	36491
3006	S	D	E	N	1965	36516
3006	S	D	E	N	1965	40136
3009	S	D	E	N	1965	36517
3010	S	G	E	N	1968	48153
3022	S	D	E	N	1965	36523
3037	S	D	E	N	1959	46025
3040	S	D	E	N	1969	35545
3040	S	G	E	N	1964	37795
3040	S	D	E	N	1964	40358
3040	S	D	E	N	1969	41573
3040	S	D	E	N	1966	48372
3041	S	D	E	N	1965	34736
3041	S	D	E	N	1965	36523
3041	S	D	E	N	1962	49984
3044	S	D	R	N	1969	58681
3044	S	D	E	N	1969	58682
3046	S	D	E	N	1964	35796
3047	S	D	E	H	1970	62781
3052	S	D	E	N	1964	35796
3054	S	D	E	N	1965	34832
3054	S	D	E	U	1967	34978
3054	S	D	E	N	1964	35796
3054	S	D	E	N	1965	36154
3054	S	G	E	N	1966	39952
3054	S	D	F	N	1966	42921
3054	S	D	E	N	1965	43527
3054	S	D	E	N	1970	43918
3054	S	D	E	N	1967	47268
3054	S	D	E	L	1966	48363
3054	S	D	E	N	1963	54703
3054	S	D	E	N	1966	56502
3054	S	D	E	N	1969	57227
3056	S	D	E	N	1965	39745
3056	S	D	E	N	1967	47268
3060	S	D	E	N	1964	35796
3061	S	D	E	N	1964	35796
3063	S	D	E	H	1970	63131
3064	S	D	E	H	1970	63131
3066	S	D	E	N	1966	43528
3071	S	D	E	H	1963	37889
3072	S	D	E	N	1966	43528
3073	S	D	E	N	1965	38392
3075	S	D	E	N	1966	43530
3079	S	D	E	N	1966	48372
3080	S	D	E	N	1966	43530
3080	S	D	E	N	1966	48377
3086	S	D	E	N	1965	39263
3090	S	D	E	N	1965	39263
3094	S	D	E	N	1966	48377
3096	S	D	E	N	1965	40136
3096	S	D	E	N	1962	41123
3096	S	D	E	N	1966	48378
3098	S	D	E	L	1966	48363
3102	S	D	E	N	1966	39362
3103	S	D	E	N	1966	39362
3104	S	D	E	N	1965	39338
3105	S	D	E	N	1966	48378
3106	S	D	E	N	1965	39338
3107	S	D	E	N	1965	39745
3108	S	D	E	N	1966	43528
3146	S	D	E	N	1965	40136

Substance Number	Phys. State	Sub-ject	Lan-guage	Temper-ature	Year	TPRC Number
551-3147	S	D	E	N	1966	43530
3152	S	D	E	N	1965	40136
3155	S	D	E	N	1964	40358
3156	S	D	E	N	1963	40699
3156	S	D	E	N	1970	57887
3161	S	D	E	N	1962	41123
3165	S	D	E	N	1962	41123
3172	S	D	E	N	1962	41123
3173	S	D	E	N	1962	41123
3174	S	D	E	N	1962	41123
3177	S	D	E	N	1962	41123
3180	S	D	E	N	1962	41123
3182	S	D	E	N	1962	41123
3182	S	D	E	N	1961	41149
3183	S	D	E	N	1968	50731
3184	S	D	E	N	1962	41123
3185	S	D	E	N	1968	50731
3186	S	D	E	N	1968	50731
3187	S	D	E	N	1968	50731
3188	S	D	E	N	1962	41123
3192	S	D	E	N	1962	41123
3202	S	D	E	N	1962	41123
3204	S	D	E	N	1962	41123
3206	S	D	E	N	1962	41123
3207	S	D	E	N	1968	50896
3208	S	D	E	N	1962	41123
3209	S	D	E	N	1962	41123
3209	S	D	E	N	1966	43530
3211	S	D	E	N	1961	41149
3212	S	D	E	N	1961	41149
3216	S	D	E	N	1961	41140
3217	S	D	E	N	1961	41149
3218	S	D	I	N	1968	50497
3218	S	D	E	N	1968	55175
3219	S	D	E	N	1961	41149
3220	S	D	E	N	1961	41149
3233	S	D	E	N	1961	41149
3234	S	D	E	N	1968	50539
3235	S	D	E	N	1968	50539
3236	S	D	E	N	1968	50539
3237	S	D	E	N	1968	50539
3238	S	D	E	N	1968	50539
3239	S	D	E	N	1968	50539
3240	S	D	E	N	1968	50539
3241	S	D	E	N	1961	41149
3242	S	D	E	N	1961	41149
3243	S	D	E	N	1961	41149
3244	S	D	E	N	1961	41149
3245	S	D	E	N	1961	41149
3249	S	D	E	U	1967	45665
3249	S	D	E	N	1966	50298
3250	S	D	E	U	1967	45665
3250	S	D	E	N	1966	50298
3251	S	D	E	U	1967	45665
3251	S	D	E	N	1966	50298
3252	S	D	G	N	1969	54365
3253	S	D	E	N	1964	35796
3253	S	D	E	U	1967	45665
3253	S	D	E	N	1966	50298
3254	S	D	E	N	1966	50298
3255	S	D	G	N	1969	54365
3257	S	D	G	N	1969	54365
3258	S	D	E	N	1967	49229
3258	S	D	E	N	1969	55080
3259	S	D	E	N	1967	49229
3260	S	D	E	N	1968	49665
3261	S	D	E	N	1966	43528
3262	S	D	G	N	1969	54365
3272	S	D	E	N	1967	47268
3274	S	D	E	N	1968	48139
3275	S	D	R	N	1966	41413
3276	S	D	G	H	1966	50684
3277	S	D	E	N	1969	57597
3278	S	D	I	N	1967	47994
3281	S	D	E	N	1969	57597
3290	S	D	E	N	1969	57597
3294	S	D	I	N	1967	47994
3295	S	D	I	N	1967	47994
3296	S	D	E	N	1964	35796
3296	S	D	E	N	1964	36113
3296	S	D	E	N	1967	48032
3296	S	D	E	N	1966	56502
3296	S	C	E	L	1970	61190
3310	S	D	E	N	1969	41573
3320	S	D	E	H	1965	42974
3323	S	D	E	N	1962	43493
3324	S	D	E	N	1962	43493
3326	S	D	E	N	1969	52160
3326	S	D	R	N	1969	58681
3326	S	D	E	N	1969	58682
3327	S	D	E	N	1969	52160
3328	S	D	E	N	1965	35889
3328	S	D	E	N	1969	52160
3334	S	D	E	N	1961	52784
3339	S	D	E	N	1924	43414
3340	S	D	E	N	1924	43414
3347	S	D	E	N	1966	56502
3349	S	D	E	N	1962	54959
3350	S	D	E	N	1962	54959
3352	S	D	E	N	1962	47962
3353	S	D	E	N	1962	47962

Phys. State: G. Gas; L. Liquid; M. Multiphase; P. Powder; S. Solid

Substance Number	Phys. State	Sub- ject	Lan- guage	Temper- ature	Year	TPRC Number	Substance Number	Phys. State	Sub- ject	Lan- guage	Temper- ature	Year	TPRC Number	Substance Number	Phys. State	Sub- ject	Lan- guage	Temper- ature	Year	TPRC Number
551-3355	S	G	E	N	1964	35856														
3355	S	D	E	N	1964	37327														
3355	S	D	E	N	1965	42764														
3355	S	D	E	N	1967	47038														
3362	S	D	E	N	1961	52784														
3363	S	D	E	N	1961	52784														
3366	S	D	E	N	1963	54703														
3367	S	D	E	H	1968	54816														
3390	S	D	E	H	1962	46797														
3391	S	D	E	H	1962	46797														
3404	S	D	E	H	1961	47554														
3412	S	D	E	N	1947	44094														
3412	S	S	E	N	1961	51595														
3460	S	D	G	N	1969	54365														
3461	S	D	E	N	1964	40327														
3461	S	D	E	N	1966	44771														
3463	S	D	E	N	1964	37327														
3474	S	D	E	N	1960	38317														
3477	S	D	E	H	1964	37378														
3494	S	D	I	N	1967	46508														
3508	S	D	E	N	1964	39347														
621-0003	M	D	E	N	1970	57854														
0028	S	D	E	N	1967	49229														
0073	S	S	E	U	1965	34428														
0217	M	E	E	N	1967	48888														
0275	S	D	E	N	1964	35796														
0308	S	S	E	U	1965	34428														
0316	S	D	E	N	1959	43297														
661-0019	S	D	R	N	1967	61948														
0019	S	D	E	N	1967	61949														
0042	S	S	E	N	1959	52098														
0048	S	D	E	N	1969	59615														
0049	S	D	E	N	1965	35414														
0049	S	G	E	N	1964	35856														
0049	S	D	E	N	1965	42764														
0049	S	D	E	N	1970	63039														
0050	S	D	E	N	1962	50354														
0090	S	D	E	N	1970	63039														
0183	S	D	E	N	1970	63039														
0285	S	D	E	N	1961	35120														
0285	S	D	E	N	1962	51506														
0292	S	D	E	N	1970	63039														
0320	S	S	E	U	1965	34428														
0366	S	S	E	U	1965	34428														
0438	S	D	E	N	1962	50354														
0457	S	D	E	N	1962	50354														
0458	S	D	E	N	1967	46041														
0470	S	D	R	N	1967	61948														
0470	S	D	E	N	1967	61949														
0477	S	D	G	N	1969	55722														

Phys. State: G. Gas; L. Liquid; M. Multiphase; P. Powder; S. Solid

Chapter 8 Reflectance

Substance Number	Phys. State	Sub-ject	Lan-guage	Temper-ature	Year	TPRC Number
526-0004	S	D	E	N	1965	39092
0004	S	G	E	N	1920	46922
0004	S	D	E	N	1966	52897
0006	S	D	R	N	1968	59347
0006	S	D	E	N	1968	59348
0006	P	D	E	N	1966	38731
0067	S	D	E	N	1965	34262
0067	S	G	E	N	1965	35027
0067	S	D	E	N	1965	39092
0067	S	D	E	N	1969	56057
0067	S	D	E	N	1970	60103
0074	S	D	E	N	1964	36100
0132	S	D	E	N	1969	57187
0206	S	D	E	N	1966	52897
528-0161	S	C	R	N	1967	49124
0161	S	C	E	N	1967	49418
0186	S	D	F	N	1965	35934
0186	S	D	E	U	1965	43794
0186	S	D	E	N	1968	49665
0186	S	D	E	N	1969	58335
0253	S	D	R	N	1969	38116
0253	S	D	E	N	1969	38117
0253	S	D	R	N	1960	40746
0253	S	D	R	N	1966	45829
0253	S	D	E	N	1967	45830
0253	S	C	R	N	1967	49124
0253	S	C	E	N	1967	49418
0253	S	D	R	N	1970	60817
0253	S	D	E	N	1971	60818
0254	S	D	R	N	1969	38116
0254	S	D	E	N	1969	38117
0254	S	C	R	N	1967	49124
0254	S	C	E	N	1967	49418
0256	S	T	G	U	1965	43024
0364	S	D	R	N	1969	38116
0364	S	D	R	N	1969	38117
0365	S	D	R	N	1969	38116
0365	S	D	E	N	1969	38117
531-0017	S	D	E	N	1967	49228
0122	S	D	E	N	1967	43820
0122	S	S	E	N	1960	49983
0125	S	D	E	N	1966	40523
0328	S	D	E	N	1966	52897
0328	S	D	E	N	1969	54712
0328	S	C	R	N	1969	57276
0328	S	C	E	N	1969	57277
0328	S	D	E	N	1970	59835
0440	S	D	E	N	1965	35523
0440	M	D	E	N	1965	35523
0456	S	D	E	N	1968	52046
0513	P	D	E	N	1966	34814
0536	C	D	E	N	1966	33854
0539	S	D	E	N	1966	33854
0542	S	D	E	N	1966	33854
0543	S	D	E	N	1966	33854
0547	S	D	E	N	1966	33854
0569	S	D	E	N	1967	49228
0587	S	D	E	N	1965	34834
0633	S	T	E	N	1969	52925
0633	S	T	R	U	1970	60823
0633	S	T	E	U	1970	60824
0663	S	D	E	N	1966	33924
0669	S	D	E	N	1966	33924
0692	S	D	E	N	1965	39754
0714	S	T	E	U	1968	47049
0831	S	D	E	N	1965	34485
0831	M	D	E	N	1965	43470
0834	S	G	E	N	1967	34545
0835	S	G	G	N	1967	34545
0880	S	D	E	N	1966	33880
0880	S	D	E	N	1966	39008
0909	S	D	G	N	1952	43597
0976	S	D	E	N	1966	42169
1027	S	C	E	N	1968	42185
1027	M	C	E	N	1968	50190
1055	S	G	E	N	1966	48365
1091	L	D	R	N	1959	51598
1091	L	D	E	N	1959	55080
1091	M	D	E	N	1968	59160
1254	S	D	E	N	1966	40908
1254	S	D	E	N	1966	40909
1319	S	D	E	N	1965	42966
1323	M	D	E	N	1966	41373
1377	S	D	E	N	1966	45984
1381	S	D	F	N	1964	36084
1391	S	D	E	N	1967	49228
1422	M	D	E	N	1964	36268
1423	M	D	E	N	1964	36268
1454	S	D	E	N	1965	42781
1454	S	D	E	N	1962	55198
1454	S	D	E	N	1969	57221
1459	M	D	E	N	1970	49757
1460	S	D	E	N	1970	61942
531-1460	M	D	E	N	1970	49757
1461	S	D	E	N	1970	56100
1461	M	D	E	N	1970	49757
1579	M	D	E	N	1968	60204
1583	M	D	E	N	1969	60204
1587	M	D	E	N	1969	55500
1588	M	D	E	N	1969	55500
1593	M	D	E	N	1969	60204
1602	M	D	E	N	1969	60204
1610	M	D	E	N	1969	60204
1641	M	D	E	N	1968	55829
1642	M	D	E	N	1968	55829
1663	S	C	E	U	1970	58591
1750	M	D	E	N	1968	59160
1797	S	D	R	N	1967	53123
1799	S	D	R	N	1970	58767
1804	S	D	E	N	1970	61942
1805	S	D	E	N	1970	61942
1806	S	D	E	N	1970	61942
541-0135	S	D	E	N	1967	45669
0278	S	D	R	N	1966	40863
0278	S	D	E	N	1966	40864
0281	M	D	R	N	1966	40863
0281	M	D	E	N	1966	40864
0282	M	D	R	N	1966	40863
0282	M	D	E	N	1966	40864
0314	M	D	R	N	1966	40863
0314	M	D	E	N	1966	40864
0392	S	G	E	N	1965	36493
0393	S	G	E	N	1965	36493
0394	S	G	E	N	1965	36493
0395	S	G	E	N	1965	36493
0450	S	D	E	N	1967	45669
0482	S	D	E	N	1965	34045
0482	S	D	F	N	1965	35934
0650	M	D	E	N	1966	42612
0699	S	D	E	N	1967	47275
0707	S	D	E	N	1967	47275
0709	M	G	E	N	1969	51470
0999	S	E	F	U	1968	44629
0999	S	T	R	N	1967	63247
0999	S	T	E	N	1967	63248
1007	M	E	E	L	1969	60026
1050	S	D	E	N	1968	53086
1084	S	D	E	N	1967	45669
1093	S	D	E	N	1969	56056
1132	M	G	E	N	1969	51470
1133	M	G	E	N	1969	51470
1134	M	G	E	N	1969	51470
1135	M	G	E	N	1969	51470
1264	S	C	R	N	1969	27935
1264	S	C	E	N	1969	42684
1281	S	D	G	N	1969	55348
1321	M	D	E	N	1965	43470
1366	S	D	E	N	1965	39092
1366	S	D	E	N	1970	60103
1413	S	D	R	N	1970	61438
1413	S	D	E	N	1970	61439
551-0129	S	D	E	N	1966	34814
0129	S	D	E	N	1965	44137
0129	S	D	E	N	1955	47062
0153	S	D	E	N	1966	50539
0154	S	D	E	N	1960	40746
0154	S	D	E	N	1961	48258
0154	S	D	E	N	1968	50539
0223	S	D	E	N	1961	33844
0223	S	D	E	N	1965	34045
0223	S	D	E	N	1964	34325
0223	S	G	E	N	1965	34683
0223	S	D	E	N	1965	35806
0223	S	D	E	N	1965	35818
0223	S	D	E	N	1964	35856
0223	S	D	F	N	1965	35934
0223	S	D	F	N	1965	35988
0223	S	D	E	N	1964	36090
0223	S	D	E	N	1963	36146
0223	S	C	E	N	1963	36197
0223	S	G	E	N	1965	36425
0223	S	G	F	N	1964	36678
0223	S	D	E	N	1969	37122
0223	S	D	E	N	1964	37150
0223	S	D	E	N	1968	37491
0223	S	D	E	N	1964	37790
0223	S	D	E	N	1964	38143
0223	S	C	E	N	1965	38505
0223	S	G	E	N	1966	39120
0223	S	D	E	N	1967	39212
0223	S	D	E	N	1965	39754
0223	S	C	E	N	1957	39840
0223	S	D	E	N	1963	40467
0223	S	D	E	N	1960	40746
0223	S	D	E	N	1967	41128
0223	S	D	E	N	1966	42740
551-0223	S	D	E	N	1966	42977
0223	S	D	E	N	1965	43527
0223	S	D	E	N	1948	43655
0223	S	D	E	N	1962	43736
0223	S	D	F	N	1953	44281
0223	S	C	E	N	1966	44304
0223	S	C	E	N	1966	44314
0223	S	G	F	N	1962	46817
0223	S	D	F	N	1953	46914
0223	S	C	G	N	1951	46924
0223	S	D	G	N	1938	46997
0223	S	D	E	N	1955	47062
0223	S	D	E	N	1965	47434
0223	S	D	E	N	1968	49206
0223	S	D	E	N	1968	49327
0223	S	G	E	N	1966	50297
0223	S	D	E	N	1969	51937
0223	S	C	F	U	1968	51949
0223	S	D	E	N	1968	52182
0223	S	G	E	N	1968	53974
0223	S	D	E	N	1969	56054
0223	S	D	E	N	1969	56125
0223	S	C	E	N	1966	56378
0223	S	D	E	N	1969	57227
0223	S	D	E	N	1969	57408
0223	S	D	E	N	1970	57794
0223	S	D	E	N	1970	57800
0223	S	D	E	N	1969	58328
0223	S	D	E	N	1969	58335
0223	S	D	E	N	1968	58406
0223	S	D	E	N	1968	59160
0223	S	D	E	N	1970	59319
0223	S	G	E	N	1970	60616
0223	S	D	E	N	1970	61184
0223	S	D	E	N	1970	61719
0223	S	G	E	N	1970	61720
0223	L	G	E	N	1968	54201
0226	S	D	E	N	1964	43112
0233	S	D	E	N	1961	33844
0233	S	D	E	N	1964	37790
0233	S	T	R	N	1966	39580
0233	S	T	E	N	1966	39581
0233	S	D	E	N	1964	43112
0233	S	D	E	N	1965	47434
0233	S	D	R	N	1968	49788
0233	S	D	E	N	1968	51769
0233	S	D	E	N	1969	53763
0233	S	D	E	N	1968	56525
0253	S	D	E	N	1965	35840
0258	S	D	E	N	1965	35840
0260	S	D	E	N	1965	35840
0264	S	D	E	N	1965	35840
0269	S	C	E	N	1966	44314
0278	S	D	E	N	1964	34325
0278	S	D	E	N	1969	35545
0278	S	D	E	N	1968	35552
0278	S	D	E	N	1968	36046
0278	S	D	E	U	1965	36700
0278	S	D	E	U	1967	36884
0278	S	D	E	N	1965	38672
0278	S	D	E	N	1966	40294
0278	S	D	E	N	1969	41589
0278	S	D	E	N	1969	41590
0278	S	D	E	N	1964	42291
0278	S	D	E	N	1962	43493
0278	S	D	E	N	1965	44096
0278	S	C	E	N	1966	44155
0278	S	D	E	H	1965	44224
0278	S	D	E	N	1967	47276
0278	S	D	E	N	1967	47756
0278	S	D	E	N	1968	48912
0278	S	D	E	N	1968	49665
0278	S	D	E	N	1968	52905
0278	S	D	E	N	1968	54153
0278	S	D	E	N	1968	54303
0278	S	D	E	N	1963	54703
0278	S	D	E	N	1969	55741
0283	S	D	E	N	1964	37399
0287	S	D	E	N	1964	33974
0287	S	D	E	N	1965	46366
0313	S	D	F	N	1964	37832
0313	S	D	F	N	1964	38140
0313	S	D	F	N	1967	41348
0325	S	C	E	U	1965	36337
0325	S	T	G	U	1885	48344
0325	S	T	E	U	1969	52830
0325	S	T	E	N	1970	59527
0325	S	T	R	N	1958	61716
0325	S	T	E	N	1970	61717
0337	S	D	E	N	1965	35527
0337	S	D	E	N	1961	40793
0337	S	D	E	N	1964	43112
0346	S	D	E	N	1964	36090
0346	S	D	E	N	1964	37390

Phys. State: G. Gas; L. Liquid; M. Multiphase; P. Powder; S. Solid

Substance Number	Phys. State	Sub-ject	Lan-guage	Temper-ature	Year	TPRC Number
551-0346	S	D	E	N	1964	37398
0346	S	D	E	N	1966	40528
0346	S	D	E	N	1969	45437
0347	S	D	E	N	1964	37398
0352	S	D	E	H	1965	36511
0352	S	D	E	H	1965	37899
0427	S	D	E	N	1964	34680
0428	S	D	E	N	1964	34680
0429	S	D	E	N	1964	34680
0432	S	D	E	N	1964	34680
0434	S	D	E	N	1964	34680
0435	S	D	E	N	1964	34680
0435	S	D	E	N	1965	35723
0435	S	D	F	N	1965	35934
0435	S	D	E	N	1965	45840
0435	S	D	E	N	1967	52871
0435	S	D	E	N	1969	55741
0436	S	D	E	N	1964	34680
0436	S	D	E	N	1965	39263
0440	S	D	E	N	1964	34680
0440	S	C	E	N	1966	39479
0441	S	D	E	N	1964	34680
0447	S	D	E	N	1961	35580
0447	S	C	E	N	1965	36751
0447	S	D	E	N	1964	37398
0447	S	D	E	N	1966	37504
0447	S	D	E	N	1963	39496
0447	S	G	G	N	1965	40164
0447	S	D	E	N	1966	40528
0447	S	D	E	N	1966	44271
0447	S	D	R	N	1966	44373
0447	S	D	J	N	1968	54519
0447	S	D	E	N	1967	55869
0447	P	G	E	N	1966	42783
0447	P	D	E	N	1966	48232
0447	P	G	E	N	1967	48233
0460	S	D	E	N	1961	53498
0466	S	G	E	N	1964	34682
0466	S	D	E	N	1966	34814
0466	S	D	E	N	1964	37355
0466	S	D	E	N	1967	45529
0468	S	D	E	N	1966	34624
0468	S	S	E	N	1960	44278
0468	S	D	E	N	1963	44319
0468	S	D	E	N	1967	45738
0468	S	D	E	N	1967	49153
0468	S	D	E	N	1968	49647
0468	S	D	E	N	1967	52118
0468	S	G	E	N	1968	53974
0468	S	D	E	N	1969	57413
0468	S	D	E	N	1970	60699
0482	S	D	E	N	1970	58661
0487	S	D	E	N	1964	34325
0487	S	G	E	N	1965	34683
0487	S	D	E	N	1965	35028
0487	S	D	E	N	1969	35545
0487	S	D	E	N	1964	35819
0487	S	D	E	N	1968	36046
0487	S	D	E	N	1964	36090
0487	S	D	E	N	1965	36320
0487	S	C	R	N	1965	36356
0487	S	C	E	N	1965	36357
0487	S	D	E	N	1959	36386
0487	S	G	E	N	1965	36425
0487	S	D	E	N	1965	36504
0487	S	D	E	N	1965	36533
0487	S	D	E	N	1965	36683
0487	S	D	E	N	1964	37398
0487	S	D	E	N	1965	40303
0487	S	D	E	N	1966	40528
0487	S	D	E	N	1969	41573
0487	S	D	E	N	1966	44271
0487	S	S	E	N	1960	44278
0487	S	D	E	N	1967	47275
0487	S	D	E	N	1966	48362
0487	S	G	E	N	1967	49254
0487	S	D	E	N	1968	49666
0487	S	D	E	N	1968	52905
0487	S	D	E	N	1963	54703
0487	S	D	E	N	1969	56122
0487	S	D	E	N	1965	56504
0487	S	D	E	N	1970	59141
0487	S	G	E	N	1968	59164
0487	S	D	E	N	1970	59649
0487	S	D	E	N	1970	60580
0487	P	G	E	N	1966	34908
0487	P	D	E	N	1969	59686
0489	S	D	R	N	1969	56239
0490	S	D	R	N	1969	56239
0500	S	D	E	N	1966	33854
0500	S	D	E	N	1966	33879
0500	S	D	E	N	1955	34038
0500	S	D	E	N	1964	34680
0500	S	D	E	N	1963	40467
0500	S	D	E	N	1960	40746
0500	S	D	E	N	1955	47062
0500	S	D	E	N	1969	57187
0501	S	D	E	N	1966	33879
0501	S	D	E	N	1966	33880
0501	S	D	E	N	1955	34038
0501	S	D	E	N	1965	34041

Substance Number	Phys. State	Sub-ject	Lan-guage	Temper-ature	Year	TPRC Number
551-0501	S	D	E	N	1966	39008
0501	S	D	E	N	1967	39460
0501	S	D	E	N	1963	40467
0501	S	D	E	N	1966	41146
0501	S	D	R	N	1967	45226
0501	S	D	E	N	1955	47062
0501	S	D	E	N	1968	52199
0501	S	D	E	N	1966	52897
0501	S	D	E	N	1967	52899
0501	S	D	R	N	1969	56239
0501	S	D	E	N	1969	57187
0501	M	D	E	N	1966	33880
0501	M	D	E	N	1966	39008
0505	S	D	E	N	1969	52594
0505	S	D	E	N	1966	52897
0505	S	D	E	N	1967	52899
0505	S	D	E	N	1969	56057
0506	S	D	E	N	1966	52897
0507	S	D	E	N	1966	33879
0507	S	D	E	N	1955	34038
0507	S	D	E	N	1964	36634
0507	S	D	E	N	1964	37351
0507	S	D	E	N	1966	37595
0507	S	D	E	N	1965	39754
0507	S	D	E	N	1957	39842
0507	S	D	E	N	1966	41146
0507	S	D	E	N	1955	47062
0507	S	D	E	U	1968	48810
0507	S	D	E	N	1967	52855
0507	S	D	E	N	1969	57187
0507	S	D	E	N	1964	57860
0507	S	D	E	N	1970	58535
0507	S	D	E	N	1970	60103
0507	S	D	E	N	1970	61184
0528	S	G	E	N	1967	45871
0533	S	D	E	N	1963	45545
0539	S	D	E	N	1968	49251
0548	S	D	F	N	1969	38833
0548	S	D	R	N	1968	48803
0548	S	D	E	N	1968	49363
0548	S	D	F	N	1969	53707
0559	S	D	E	N	1965	39754
0577	S	D	E	N	1966	33854
0577	S	D	E	N	1965	34385
0577	S	D	E	N	1968	36046
0577	S	D	E	N	1964	36126
0577	S	C	R	N	1965	36356
0577	S	C	E	N	1965	36357
0577	S	D	E	N	1959	36386
0577	S	G	E	N	1965	36425
0577	S	D	E	N	1965	36466
0577	S	D	E	N	1965	36504
0577	S	D	E	N	1964	36634
0577	S	D	E	N	1965	36688
0577	S	D	F	N	1965	36963
0577	S	D	R	N	1969	38116
0577	S	D	E	N	1969	38117
0577	S	C	E	N	1964	38253
0577	S	D	E	N	1965	39262
0577	S	D	F	N	1966	39286
0577	S	D	F	N	1966	39288
0577	S	D	F	N	1966	39717
0577	S	D	E	N	1965	39754
0577	S	D	E	N	1966	40343
0577	S	D	E	N	1966	40528
0577	S	D	R	N	1966	40954
0577	S	D	E	N	1966	40955
0577	S	D	E	N	1966	42914
0577	S	D	E	N	1964	43112
0577	S	D	E	N	1965	43456
0577	S	D	E	N	1962	43736
0577	S	D	R	N	1966	44625
0577	S	D	E	N	1969	45808
0577	S	D	R	N	1966	45829
0577	S	D	E	N	1967	45830
0577	S	D	E	N	1955	47062
0577	S	D	E	N	1967	47275
0577	S	D	E	N	1967	48032
0577	S	D	E	N	1968	48228
0577	S	D	E	N	1966	48366
0577	S	D	E	N	1968	48573
0577	S	D	R	N	1968	51944
0577	S	D	E	N	1968	51945
0577	S	D	E	N	1969	52920
0577	S	D	E	N	1968	53513
0577	S	D	E	N	1968	54292
0577	S	D	E	N	1965	55084
0577	S	D	E	N	1969	57597
0577	S	D	E	N	1970	57797
0577	S	D	E	N	1970	58678
0577	S	D	E	N	1970	59141
0577	S	D	E	N	1970	59649
0577	S	C	E	N	1970	59985
0577	S	D	E	N	1970	63134
0577	P	D	E	N	1969	59686
0582	S	D	E	N	1955	34038
0582	S	D	E	N	1966	34487
0582	S	D	E	N	1964	37398
0582	S	D	F	L	1966	39181
0582	S	D	E	N	1966	39391
0588	S	D	E	N	1968	53459

Substance Number	Phys. State	Sub-ject	Lan-guage	Temper-ature	Year	TPRC Number
551-0601	S	D	E	N	1962	33902
0601	S	D	E	N	1966	34449
0601	S	G	E	N	1966	34455
0601	S	D	E	N	1965	36525
0601	S	D	E	U	1967	36884
0601	S	D	E	N	1968	40103
0601	S	D	E	N	1966	40294
0601	S	D	E	N	1964	40415
0601	S	D	E	N	1969	41610
0601	S	D	E	N	1962	42292
0601	S	D	E	N	1966	43482
0601	S	D	E	N	1965	43527
0601	S	D	E	N	1965	43840
0601	S	D	E	N	1965	44096
0601	S	D	E	H	1965	44224
0601	S	D	E	N	1966	44272
0601	S	D	E	N	1967	45664
0601	S	D	E	N	1967	47272
0601	S	D	E	N	1967	47278
0601	S	D	E	N	1966	48231
0601	S	D	E	N	1966	48232
0601	S	D	E	N	1966	48375
0601	S	D	E	N	1966	48492
0601	S	D	E	N	1964	50030
0601	S	D	E	N	1968	53088
0601	S	D	E	N	1968	53089
0601	S	D	E	N	1969	55070
0601	S	D	E	N	1969	55085
0601	S	D	E	N	1969	55741
0601	S	D	E	N	1969	57227
0601	S	D	E	N	1970	59637
0601	S	C	E	N	1970	59646
0601	S	D	G	N	1969	59668
0605	S	D	E	N	1962	43493
0618	S	D	F	N	1961	38059
0620	S	D	E	N	1964	37398
0620	S	D	R	N	1966	41412
0620	S	D	E	N	1967	49397
0620	S	D	R	N	1968	55318
0620	S	D	E	N	1969	55741
0623	S	D	E	N	1964	36090
0623	S	D	F	N	1965	36947
0623	S	D	E	N	1964	37390
0623	S	C	F	N	1966	44306
0623	S	D	F	N	1966	44924
0623	S	D	E	N	1968	46790
0623	S	D	E	N	1967	49397
0623	S	G	E	N	1966	50297
0623	S	G	F	N	1968	55320
0623	S	D	E	N	1968	56525
0631	S	D	E	N	1967	41161
0631	S	D	E	N	1955	43603
0637	S	D	E	N	1964	35458
0637	S	D	E	N	1965	36688
0637	S	D	R	N	1964	38299
0637	S	C	E	N	1957	39840
0637	S	G	E	N	1966	39976
0637	S	D	E	N	1964	40413
0637	S	D	E	N	1965	40514
0637	S	D	E	N	1965	42351
0637	S	D	E	N	1966	42978
0637	S	D	G	N	1965	43102
0637	S	D	E	N	1964	43112
0637	S	D	R	N	1967	45014
0637	S	D	E	N	1967	45015
0637	S	D	F	N	1954	46856
0637	S	D	F	N	1953	46914
0637	S	D	E	N	1955	47062
0637	S	D	F	N	1968	48131
0637	S	D	E	N	1968	52031
0637	S	D	E	N	1969	55549
0637	S	D	E	N	1968	56849
0643	S	D	E	N	1964	36090
0643	S	D	E	N	1964	36100
0643	S	D	E	U	1967	36884
0643	S	D	E	N	1964	37398
0643	S	D	E	N	1966	40294
0643	S	D	E	N	1964	43423
0643	S	D	E	N	1967	43801
0648	S	D	E	N	1968	50111
0648	S	D	E	N	1962	53496
0648	S	D	E	N	1969	57408
0654	S	D	R	N	1969	56239
0657	S	D	E	N	1966	38731
0657	S	D	E	N	1967	47034
0657	S	D	E	N	1968	50239
0657	S	D	E	N	1968	51914
0657	S	D	E	N	1967	52855
0657	S	D	E	N	1969	54056
0657	S	D	E	N	1970	56770
0657	S	D	E	N	1970	61721
0657	S	D	E	N	1970	63384
0657	L	G	E	N	1968	54201
0657	L	D	E	N	1969	54887
0659	S	D	E	N	1971	36066
0659	S	D	R	N	1970	45752
0659	S	D	E	N	1969	55822
0659	S	D	E	N	1969	59832
0684	S	T	E	U	1965	36496
0684	S	G	E	N	1967	56505
0690	S	T	E	U	1969	52929

Phys. State: G. Gas; L. Liquid; M. Multiphase; P. Powder; S. Solid

Substance Number	Phys. State	Subject	Language	Temperature	Year	TPRC Number
551-0702	S	D	E	N	1969	53485
0702	S	D	E	N	1970	57953
0723	S	D	F	N	1961	38059
0723	S	D	E	N	1965	39092
0734	S	D	E	N	1965	35020
0734	S	D	E	N	1967	39212
0749	S	D	E	N	1964	43112
0751	S	D	E	N	1965	39754
0775	S	D	E	N	1968	55175
0780	S	D	E	N	1969	58335
0799	S	D	E	N	1965	35723
0799	S	C	E	N	1953	41767
0799	S	D	E	N	1967	52871
0799	P	D	R	N	1965	36352
0799	P	D	E	N	1965	36353
0823	S	D	E	N	1965	35907
0824	S	D	E	N	1965	35907
0850	S	D	E	N	1968	47949
0850	S	D	E	N	1968	50974
0850	S	D	E	N	1968	51925
0850	S	G	E	N	1968	56550
0850	M	D	F	N	1967	45678
0853	S	D	E	N	1961	33844
0854	S	D	E	N	1966	39494
0854	S	C	E	N	1967	40269
0854	S	D	R	N	1968	51944
0854	S	D	E	N	1968	51945
0854	S	D	E	N	1969	59832
0888	S	D	E	N	1964	43112
0888	S	D	R	N	1967	45226
0905	S	D	E	N	1960	35546
0906	S	D	E	N	1964	34680
0906	S	D	E	N	1965	46109
0908	S	D	E	N	1960	35546
0908	S	D	E	N	1965	39754
0908	S	D	E	N	1968	52906
0911	S	D	E	N	1970	60699
0924	S	D	E	N	1960	40746
0924	S	D	E	N	1962	43493
0924	S	D	E	N	1963	54703
0926	S	D	E	N	1962	36477
0929	S	D	E	N	1965	34394
0935	S	C	E	N	1965	38391
0935	S	D	E	N	1965	45840
0949	S	D	E	N	1965	43527
0959	S	D	E	N	1961	33844
0959	S	D	E	N	1964	43112
0959	S	D	E	N	1967	49153
0959	S	D	E	N	1967	52118
0959	S	D	E	N	1970	58662
0959	S	D	E	N	1970	62417
0960	S	D	E	N	1967	49153
0960	S	D	E	N	1967	52118
0962	S	G	E	N	1967	56505
0965	S	D	E	N	1969	52594
0976	S	D	E	N	1966	33879
0976	S	D	E	N	1966	33880
0976	S	G	E	N	1928	40591
0976	S	D	E	N	1955	47062
0976	S	D	E	N	1966	52897
0976	S	D	E	N	1969	57187
0988	S	S	E	N	1970	57511
0989	S	D	E	N	1968	50246
0989	S	C	G	N	1966	57510
0989	S	D	E	N	1968	59010
0989	S	C	E	N	1970	61485
1023	S	D	E	N	1960	35546
1042	S	D	E	N	1955	34038
1042	S	D	E	N	1966	52897
1042	S	D	E	N	1970	60103
1064	S	D	E	N	1961	34072
1064	S	D	R	N	1961	43599
1064	S	D	G	N	1910	43611
1064	S	D	E	N	1967	47314
1064	S	D	E	N	1967	52871
1064	S	D	E	N	1911	54642
1076	S	T	E	U	1965	34327
1076	S	D	E	N	1953	36071
1076	S	D	E	N	1965	39092
1076	S	D	E	N	1960	40746
1076	S	T	E	U	1967	43811
1076	S	T	G	U	1931	47003
1076	S	D	E	N	1955	47062
1076	S	E	F	N	1954	51292
1076	S	E	E	N	1957	51307
1076	S	D	E	N	1953	51599
1076	S	C	E	N	1967	52903
1076	S	S	E	N	1969	57666
1076	S	C	E	N	1970	58452
1116	S	D	F	N	1965	35934
1118	S	D	E	N	1960	35546
1125	S	D	E	N	1960	35546
1130	S	D	E	N	1960	35546
1140	S	S	E	N	1967	43519
1141	S	D	E	N	1964	34333
1141	S	D	E	N	1965	39754
1144	S	D	E	N	1965	46109
1175	S	D	E	N	1964	36634
1175	S	D	E	U	1967	36884
1175	S	D	E	N	1965	39754
1175	S	D	E	N	1966	40294

Substance Number	Phys. State	Subject	Language	Temperature	Year	TPRC Number
551-1175	S	D	E	N	1966	42999
1179	S	D	E	N	1968	36046
1179	S	D	E	N	1965	36525
1179	S	D	E	N	1969	41589
1179	S	D	E	N	1969	41590
1179	S	D	E	N	1962	42292
1179	S	D	E	N	1968	54303
1179	S	D	E	N	1969	55221
1179	S	D	E	N	1970	59648
1180	S	D	E	N	1965	36525
1180	S	D	E	N	1962	42292
1181	S	D	E	N	1965	36525
1181	S	D	E	N	1962	42292
1182	S	D	E	N	1962	33902
1184	S	D	E	N	1965	36525
1184	S	D	E	N	1962	42292
1185	S	D	E	N	1965	36525
1186	S	D	E	N	1962	33902
1186	S	D	E	N	1965	36525
1186	S	D	E	N	1962	42292
1188	S	D	E	N	1965	35907
1188	S	D	E	N	1962	41123
1189	S	D	E	N	1962	33902
1189	S	D	E	N	1965	35907
1189	S	D	E	N	1965	36525
1190	S	D	E	N	1962	33902
1190	S	D	E	N	1965	36525
1190	S	D	E	N	1962	42292
1191	S	D	E	N	1962	33902
1191	S	D	E	N	1965	36525
1191	S	D	E	N	1962	42292
1191	S	D	E	N	1964	50030
1192	S	D	E	N	1962	33902
1193	S	D	E	N	1962	33902
1193	S	D	E	N	1965	36525
1193	S	D	E	N	1962	42292
1194	S	D	E	N	1962	33902
1195	S	D	E	N	1962	33902
1197	S	D	E	N	1962	33902
1197	S	D	E	N	1965	36525
1197	S	D	E	N	1962	42292
1197	S	D	E	N	1965	43840
1197	S	D	E	N	1964	49031
1197	S	D	E	N	1964	50030
1198	S	D	E	N	1962	33902
1198	S	D	E	N	1965	35907
1198	S	D	E	N	1965	36525
1198	S	D	E	N	1962	42292
1199	S	D	E	N	1965	36525
1199	S	D	E	N	1962	42292
1201	S	D	E	N	1962	42292
1202	S	D	E	N	1962	33902
1202	S	D	E	N	1965	36525
1202	S	D	E	N	1962	42292
1202	S	D	E	N	1966	43482
1202	S	D	E	N	1965	52152
1202	S	D	E	N	1967	52864
1202	S	D	E	N	1969	55070
1205	S	D	E	N	1962	33902
1206	S	D	E	N	1962	33902
1206	S	D	E	N	1969	59196
1206	S	D	E	N	1970	59647
1207	S	D	E	N	1965	36525
1207	S	D	E	N	1962	42292
1208	S	D	E	N	1965	36525
1208	S	D	E	N	1962	42292
1211	S	D	E	N	1965	36525
1211	S	D	E	N	1962	42292
1212	S	D	E	N	1962	42292
1213	S	D	E	N	1965	36525
1213	S	D	E	N	1962	42292
1214	S	D	E	N	1965	36525
1214	S	D	E	N	1962	42292
1215	S	D	E	N	1965	36525
1215	S	D	E	N	1962	42292
1216	S	D	E	N	1965	36525
1216	S	D	E	N	1962	42292
1217	S	D	E	N	1965	36525
1217	S	D	E	N	1962	42292
1218	S	D	E	N	1962	42292
1219	S	D	E	N	1962	42292
1220	S	D	E	N	1962	42292
1221	S	D	E	N	1962	42292
1222	S	D	E	N	1965	36525
1222	S	D	E	N	1962	42292
1223	S	D	E	N	1965	36525
1223	S	D	E	N	1962	42292
1224	S	D	E	N	1962	42292
1225	S	D	E	N	1962	42292
1226	S	D	R	N	1967	45226
1227	S	D	E	N	1962	33902
1227	S	D	E	N	1965	36525
1227	S	D	E	N	1962	42292
1232	S	D	E	N	1964	33974
1258	S	D	E	N	1963	37193
1258	S	D	E	N	1969	41334
1258	S	D	E	N	1963	42295
1261	S	D	E	N	1962	36477
1289	S	D	E	N	1968	50539
1321	S	G	E	N	1965	34683
1321	S	D	E	N	1964	35819

Substance Number	Phys. State	Subject	Language	Temperature	Year	TPRC Number
551-1321	S	D	E	N	1964	36090
1321	S	D	E	N	1964	37398
1321	S	D	E	N	1965	40303
1321	S	D	E	N	1966	40528
1321	S	D	E	N	1967	47278
1321	P	D	E	N	1967	45664
1321	P	G	E	N	1967	48233
1321	P	D	E	N	1969	59686
1323	S	D	E	N	1964	36090
1323	S	D	E	N	1964	37390
1325	S	D	E	N	1964	36090
1326	S	D	E	N	1963	45545
1327	S	D	E	N	1964	33974
1328	S	D	R	N	1970	61693
1328	S	D	E	N	1970	61694
1330	S	D	E	N	1964	33974
1330	S	D	E	N	1965	46366
1331	S	D	E	N	1964	49031
1370	S	D	E	N	1964	36414
1373	S	D	E	N	1965	35840
1377	S	D	E	N	1965	35818
1377	S	D	E	N	1964	36090
1377	S	D	E	N	1964	37390
1384	S	D	E	N	1970	63134
1392	M	D	E	N	1957	46842
1394	S	D	E	N	1968	37034
1394	S	D	E	N	1967	42092
1394	S	D	E	N	1970	60718
1394	M	D	E	N	1957	46842
1421	S	D	E	N	1960	40746
1425	S	D	E	N	1964	36090
1425	S	D	E	N	1962	36477
1425	S	D	E	N	1965	36533
1430	S	D	E	N	1964	37398
1430	S	D	E	N	1964	37790
1430	S	D	E	N	1966	40528
1430	S	D	R	N	1966	41413
1430	S	D	R	N	1968	52080
1430	S	D	E	N	1968	52081
1430	S	D	R	N	1968	55318
1460	S	D	E	N	1964	37501
1460	S	G	E	N	1965	38649
1460	S	D	F	N	1966	40124
1460	S	G	E	N	1964	40182
1460	S	D	E	N	1967	42092
1460	S	D	E	N	1968	56848
1460	S	D	E	N	1970	58623
1462	S	D	E	N	1960	48099
1463	S	D	E	N	1960	48099
1464	S	C	E	N	1964	37546
1464	S	D	F	N	1967	42325
1464	S	D	F	N	1967	44041
1471	S	D	E	N	1965	35840
1508	S	S	E	N	1966	33840
1508	S	D	E	N	1964	33974
1508	S	D	E	N	1965	34045
1508	S	D	E	N	1966	34550
1508	S	D	E	N	1953	36071
1508	S	D	E	N	1964	36090
1508	S	D	E	N	1965	36320
1508	S	D	E	H	1964	36414
1508	S	D	E	N	1969	37877
1508	S	D	E	N	1960	38776
1508	S	D	E	N	1966	39951
1508	S	D	E	N	1966	40528
1508	S	D	E	N	1966	41146
1508	S	D	E	N	1966	42740
1508	S	D	E	H	1966	43494
1508	S	D	G	N	1952	43597
1508	S	D	E	N	1929	43646
1508	S	G	E	N	1966	44933
1508	S	D	E	N	1962	45667
1508	S	D	E	N	1965	46109
1508	S	D	E	N	1931	47008
1508	S	D	E	N	1955	47062
1508	S	G	E	N	1968	47917
1508	S	D	E	N	1964	49006
1508	S	D	E	N	1968	51403
1508	S	D	E	N	1968	52199
1508	S	D	E	N	1969	52594
1508	S	D	E	N	1969	54291
1508	S	D	E	N	1969	55534
1508	S	D	E	N	1930	55924
1508	S	D	E	N	1957	56499
1508	S	D	E	N	1970	62782
1508	P	G	E	N	1966	34908
1508	P	D	E	N	1966	39220
1508	M	C	E	N	1966	39040
1511	S	C	E	N	1968	50367
1511	M	D	G	N	1970	59331
1521	S	D	E	N	1967	47278
1521	S	D	E	N	1967	48234
1535	S	D	E	N	1966	37595
1545	S	D	E	N	1962	33902
1545	S	D	E	N	1966	34449
1545	S	D	E	N	1968	36046
1545	S	D	E	N	1965	36525
1545	S	D	E	N	1965	39754
1545	S	D	E	N	1964	40416
1545	S	D	E	N	1969	41589
1545	S	D	E	N	1969	41610

Phys. State: **G.** Gas; **L.** Liquid; **M.** Multiphase; **P.** Powder; **S.** Solid

Substance Number	Phys. State	Subject	Language	Temperature	Year	TPRC Number
551-1545	S	D	E	U	1969	41945
1545	S	D	E	N	1962	42292
1545	S	D	E	N	1966	43482
1545	S	D	E	N	1965	43527
1545	S	D	E	N	1965	44096
1545	S	D	E	H	1965	44224
1545	S	D	E	N	1966	44270
1545	S	D	E	U	1964	44949
1545	S	D	E	N	1967	45664
1545	S	D	E	N	1967	47278
1545	S	D	E	N	1967	47279
1545	S	D	E	N	1967	47756
1545	S	D	E	N	1968	48228
1545	S	D	E	N	1966	48231
1545	S	G	E	N	1967	48233
1545	S	D	E	N	1967	48234
1545	S	D	E	N	1966	48366
1545	S	D	E	N	1966	48374
1545	S	D	E	N	1966	48375
1545	S	D	E	N	1966	48492
1545	S	D	E	N	1964	49031
1545	S	D	E	N	1968	49665
1545	S	D	E	N	1964	50030
1545	S	D	E	N	1968	53089
1545	S	D	E	N	1968	54303
1545	S	D	E	N	1969	55070
1545	S	D	E	N	1965	55084
1545	S	D	E	N	1969	55085
1545	S	D	E	N	1969	55221
1545	S	D	E	N	1969	55741
1545	S	D	E	N	1969	58335
1545	S	D	E	N	1970	59637
1545	S	D	E	N	1970	59648
1545	S	D	E	N	1968	60177
1545	M	D	E	N	1964	49031
1571	S	D	E	N	1967	33824
1571	S	T	E	U	1966	34406
1571	S	E	E	N	1965	34683
1571	S	T	G	U	1965	35455
1571	S	E	E	N	1965	35527
1571	S	E	F	N	1964	36178
1571	S	T	R	U	1965	36354
1571	S	T	E	U	1965	36355
1571	S	E	E	U	1964	36423
1571	S	E	E	U	1967	36884
1571	S	E	F	N	1964	37146
1571	S	T	F	N	1963	37633
1571	S	D	R	N	1969	38116
1571	S	D	E	N	1969	38117
1571	S	T	R	U	1965	38248
1571	S	T	E	U	1965	38924
1571	S	E	E	N	1966	39014
1571	S	E	E	N	1966	39185
1571	S	T	R	U	1966	39653
1571	S	T	E	U	1966	39654
1571	S	T	E	U	1966	40620
1571	S	S	E	N	1955	40651
1571	S	T	E	U	1967	40922
1571	S	S	R	U	1966	41739
1571	S	T	F	U	1966	41989
1571	S	E	E	N	1960	44278
1571	S	E	E	N	1961	44317
1571	S	E	E	N	1963	44319
1571	S	T	E	N	1968	48547
1571	S	S	E	N	1960	49983
1571	S	D	E	N	1968	51492
1571	S	T	E	U	1969	53985
1571	S	C	E	N	1965	54294
1571	S	S	E	U	1968	54883
1571	S	S	E	N	1967	58165
1571	S	T	E	N	1970	61718
1571	S	S	E	N	1970	61779
1571	L	T	E	N	1965	54313
1574	S	D	E	N	1966	34814
1574	S	D	E	N	1964	36090
1574	S	S	E	N	1960	44278
1574	S	D	E	N	1961	44317
1574	S	D	F	N	1953	46914
1574	S	D	E	N	1965	47434
1574	S	D	E	N	1967	49397
1574	S	G	E	N	1966	50297
1574	S	D	E	N	1969	56008
1574	S	D	E	N	1969	58328
1574	S	D	E	N	1970	58590
1574	S	G	E	N	1970	60616
1575	S	D	E	N	1965	36466
1575	S	D	E	N	1965	36504
1591	S	D	G	N	1910	43611
1591	S	D	E	N	1970	60590
1606	S	D	E	N	1970	59649
1609	S	D	E	N	1965	39754
1621	S	D	F	N	1966	39717
1621	S	D	F	N	1968	48131
1631	S	D	E	N	1960	48099
1634	S	D	E	N	1968	52906
1661	S	D	E	N	1965	35840
1672	S	D	E	N	1968	42451
1673	S	D	E	N	1970	60498
1674	S	G	E	N	1970	44376
1674	S	D	E	N	1970	60498
1684	S	D	E	N	1968	50379

Substance Number	Phys. State	Subject	Language	Temperature	Year	TPRC Number
551-1689	S	D	E	N	1964	34819
1689	S	D	E	N	1969	41421
1689	S	D	E	U	1969	41945
1689	S	D	E	N	1965	42290
1689	S	D	E	N	1966	48492
1689	S	D	E	N	1967	52151
1689	S	D	E	N	1970	59141
1716	S	D	E	N	1967	33799
1716	S	D	E	N	1950	35074
1716	S	D	E	N	1965	36342
1716	S	D	E	N	1964	37501
1716	S	D	F	N	1966	39286
1716	S	D	F	N	1967	39773
1716	S	G	E	N	1964	40182
1716	S	D	E	N	1955	47062
1722	S	D	F	N	1965	36963
1722	S	D	E	N	1967	43185
1722	S	D	E	N	1967	46004
1722	S	D	E	N	1967	47578
1728	S	D	F	N	1965	40280
1730	S	D	E	N	1964	34680
1730	S	D	E	N	1965	39263
1730	S	D	E	N	1964	44415
1734	S	T	F	U	1965	36823
1734	S	T	E	U	1964	38296
1734	S	T	G	U	1888	48345
1734	S	C	R	N	1969	57363
1734	S	C	E	N	1969	57364
1739	S	D	E	N	1966	34548
1739	S	D	E	N	1964	37501
1739	S	D	E	N	1965	39086
1739	S	G	E	N	1964	40182
1765	S	D	E	N	1965	35840
1766	S	D	E	N	1965	35840
1767	S	D	F	N	1965	43021
1768	S	D	E	N	1964	35819
1768	S	D	E	N	1965	35880
1768	S	D	E	N	1964	36036
1768	S	G	F	N	1964	36678
1768	S	G	E	N	1965	36684
1768	S	D	E	N	1965	36688
1768	S	C	E	N	1964	37546
1768	S	D	E	N	1970	37677
1768	S	D	F	N	1963	38629
1768	S	G	E	N	1966	39439
1768	S	D	E	N	1965	40303
1768	S	D	E	N	1965	43043
1768	S	D	E	N	1965	43457
1768	S	D	E	N	1967	44080
1768	S	D	F	N	1967	46313
1768	S	D	E	N	1949	47204
1796	S	D	E	N	1969	41421
1796	S	D	E	N	1967	52151
1799	S	D	E	N	1963	36112
1802	S	D	E	N	1963	36112
1805	S	D	E	N	1963	36112
1805	S	D	E	U	1967	36884
1805	S	D	E	N	1965	39036
1805	S	D	E	N	1966	40294
1810	S	D	E	N	1960	39894
1811	S	D	E	N	1965	35907
1811	S	D	E	N	1969	41573
1811	S	D	E	N	1967	45529
1818	S	D	E	N	1968	49962
1818	S	D	E	N	1969	56033
1857	S	D	E	N	1965	44137
1857	S	D	E	N	1955	47062
1857	S	G	G	N	1960	51581
1858	S	D	E	N	1961	33844
1858	S	G	E	N	1966	33896
1858	S	D	E	N	1965	34040
1858	S	D	E	N	1965	34041
1858	S	D	E	N	1965	34345
1858	S	D	E	N	1965	34385
1858	S	D	E	N	1966	34454
1858	S	D	E	N	1966	34456
1858	S	D	E	N	1966	34624
1858	S	G	E	N	1965	34683
1858	S	D	E	N	1965	34817
1858	S	D	E	N	1966	34925
1858	S	D	E	N	1965	35561
1858	S	D	E	N	1965	35806
1858	S	D	E	N	1965	35818
1858	S	D	F	N	1965	35934
1858	S	D	E	N	1968	36046
1858	S	D	E	N	1964	36090
1858	S	D	E	N	1964	36113
1858	S	D	E	N	1963	36146
1858	S	D	E	N	1965	36229
1858	S	D	E	N	1965	36471
1858	S	D	E	N	1965	36504
1858	S	D	E	N	1965	36530
1858	S	D	R	N	1964	37016
1858	S	D	E	N	1969	37122
1858	S	D	E	N	1964	37150
1858	S	D	E	N	1964	37355
1858	S	D	E	N	1968	37491
1858	S	C	E	N	1964	37546
1858	S	D	E	N	1966	39013
1858	S	D	E	N	1966	39014
1858	S	D	E	N	1967	39212

Substance Number	Phys. State	Subject	Language	Temperature	Year	TPRC Number
551-1858	S	D	E	N	1965	39262
1858	S	D	E	N	1965	39754
1858	S	D	E	N	1957	39842
1858	S	D	F	N	1966	39944
1858	S	D	E	N	1966	40343
1858	S	D	E	N	1969	40347
1858	S	D	E	N	1966	40529
1858	S	D	E	N	1956	40608
1858	S	D	E	N	1965	41122
1858	S	D	E	N	1969	41421
1858	S	D	E	N	1968	42270
1858	S	D	E	N	1965	42781
1858	S	D	E	N	1967	42890
1858	S	D	E	N	1967	42894
1858	S	D	E	N	1964	43112
1858	S	D	E	N	1965	43527
1858	S	D	G	N	1941	43657
1858	S	D	E	H	1965	44224
1858	S	S	E	N	1960	44278
1858	S	C	E	N	1966	44304
1858	S	D	E	N	1961	44317
1858	S	D	E	N	1963	44319
1858	S	D	F	N	1966	44924
1858	S	D	E	N	1967	45738
1858	S	D	E	N	1969	45808
1858	S	D	E	N	1955	47062
1858	S	D	E	N	1965	47434
1858	S	D	E	N	1967	48032
1858	S	D	E	N	1966	48362
1858	S	D	E	N	1966	48375
1858	S	D	E	N	1968	48567
1858	S	D	E	N	1968	48573
1858	S	G	E	U	1967	49042
1858	S	D	E	N	1968	49177
1858	S	D	E	N	1968	49206
1858	S	G	E	N	1967	49254
1858	S	D	E	N	1968	49291
1858	S	D	E	N	1969	49627
1858	S	D	E	N	1968	49647
1858	S	D	R	N	1968	50016
1858	S	G	E	N	1966	50297
1858	S	D	E	N	1968	52057
1858	S	D	E	N	1967	52151
1858	S	D	E	N	1968	52199
1858	S	D	E	N	1967	52855
1858	S	D	E	N	1967	52917
1858	S	D	E	N	1969	52920
1858	S	D	E	N	1968	53088
1858	S	G	E	N	1968	53974
1858	S	D	E	N	1968	54292
1858	S	D	E	N	1969	56006
1858	S	D	E	N	1969	56122
1858	S	D	E	N	1969	56833
1858	S	D	E	N	1968	56848
1858	S	D	R	N	1969	57005
1858	S	D	E	N	1969	57233
1858	S	D	E	N	1969	57413
1858	S	D	E	N	1970	57437
1858	S	C	E	N	1970	57793
1858	S	D	E	N	1970	57794
1858	S	E	E	N	1970	57801
1858	S	D	E	N	1970	57802
1858	S	D	E	N	1970	59244
1858	S	D	E	N	1970	59649
1858	S	D	G	N	1969	59668
1858	S	G	E	N	1970	60616
1858	S	C	R	N	1967	62794
1858	S	C	E	N	1967	62795
1858	S	G	E	N	1968	63393
1858	S	G	E	N	1971	63421
1858	L	C	F	N	1969	38996
1861	S	D	E	N	1943	36754
1861	S	C	E	N	1964	37546
1861	S	D	E	N	1968	49959
1861	S	D	E	N	1951	55224
1863	S	D	E	N	1965	35840
1865	S	C	E	N	1970	59712
1884	S	D	E	N	1970	56051
1900	S	D	E	N	1964	43112
1902	S	D	E	N	1964	43112
1908	S	D	E	N	1960	38776
1908	S	D	E	N	1961	41149
1908	S	D	E	N	1967	45212
1908	S	D	E	N	1964	49006
1908	S	D	E	N	1968	51403
1917	S	D	E	N	1964	34333
1926	S	D	E	N	1931	47008
1933	S	D	E	N	1965	35840
1934	S	D	E	N	1965	35840
1935	S	D	E	N	1965	35840
1938	S	D	E	N	1966	34456
1938	S	C	E	N	1966	34621
1938	S	D	E	N	1966	34624
1938	S	D	E	N	1969	35545
1938	S	D	E	N	1965	35818
1938	S	D	E	N	1964	35819
1938	S	D	E	N	1965	36683
1938	S	D	E	N	1966	39185
1938	S	D	E	N	1965	39754
1938	S	D	E	N	1965	40303
1938	S	D	E	N	1966	40528

Phys. State: **G.** Gas; **L.** Liquid; **M.** Multiphase; **P.** Powder; **S.** Solid

B33

Reflectance

Substance Number	Phys. State	Subject	Language	Temperature	Year	TPRC Number
551-1938	S	C	E	N	1947	40783
1938	S	D	E	N	1965	42781
1938	S	D	E	N	1967	42890
1938	S	E	F	N	1964	43128
1938	S	D	E	N	1964	43423
1938	S	D	E	N	1967	43684
1938	S	D	F	N	1967	44048
1938	S	S	E	N	1960	44278
1938	S	D	E	N	1966	44875
1938	S	D	E	N	1969	49627
1938	S	D	E	N	1969	52645
1938	S	D	E	N	1966	52900
1938	S	D	E	N	1969	56122
1938	S	D	E	N	1969	57031
1938	S	D	E	N	1969	57413
1938	S	C	E	N	1970	57793
1938	S	D	E	N	1970	59141
1938	S	D	E	N	1970	59649
1938	S	D	E	N	1970	61782
1939	S	D	E	N	1969	59686
1942	S	D	E	N	1965	35840
1943	S	D	F	N	1965	35934
1943	S	D	E	N	1968	36046
1943	S	D	E	N	1963	40420
1944	S	D	E	N	1965	39263
1944	S	D	E	N	1963	40420
1944	S	D	E	N	1959	46025
1953	S	D	E	N	1961	33844
1953	S	D	E	N	1950	35074
1953	S	G	F	N	1964	36678
1953	S	D	F	N	1964	36710
1953	S	D	E	N	1968	37491
1953	S	D	E	N	1964	37790
1953	S	D	F	N	1964	37831
1953	S	G	E	N	1966	39976
1953	S	D	E	N	1963	40420
1953	S	D	E	N	1966	42740
1953	S	D	E	N	1965	43413
1953	S	D	G	N	1948	43655
1953	S	D	E	N	1962	43736
1953	S	G	E	N	1966	44305
1953	S	C	E	U	1968	47918
1953	S	D	F	L	1965	49118
1953	S	C	E	N	1969	53071
1974	S	D	E	N	1931	47008
1974	S	D	E	N	1969	52594
1992	S	D	E	N	1965	35723
2008	S	D	E	N	1965	35840
2018	S	C	R	N	1969	57363
2018	S	C	E	N	1969	57364
2020	S	D	E	N	1965	39754
2020	S	D	E	N	1968	49665
2022	S	D	E	N	1969	57187
2031	S	D	E	N	1968	53491
2045	S	D	E	N	1965	35840
2052	S	D	E	N	1964	36090
2052	S	D	E	N	1970	62501
2095	S	D	E	N	1963	40420
2116	S	D	E	N	1965	35840
2130	S	S	E	N	1969	62978
2132	S	D	E	N	1966	44272
2132	S	D	E	N	1965	52152
2132	S	D	E	N	1969	55070
2138	S	D	E	N	1963	40420
2147	S	D	E	N	1969	55070
2147	S	D	G	N	1969	59668
2155	S	D	E	N	1965	35907
2156	S	D	E	N	1966	33966
2156	S	D	F	N	1964	37153
2156	S	D	R	N	1969	57102
2156	S	D	E	N	1969	57103
2156	S	D	E	N	1969	59832
2157	S	D	E	N	1965	35907
2158	S	D	E	N	1965	35907
2164	S	D	E	N	1965	35907
2165	S	C	F	N	1963	37633
2165	S	D	E	N	1966	40528
2165	S	D	E	N	1931	47008
2166	S	D	E	H	1966	34724
2166	S	D	E	N	1966	40528
2167	S	D	E	N	1966	40528
2167	S	C	E	N	1969	55037
2168	S	D	E	N	1965	35119
2168	S	D	E	N	1965	35818
2168	S	D	E	N	1964	35819
2168	S	D	E	N	1964	36090
2168	S	D	E	N	1965	36683
2168	S	D	E	N	1964	37390
2168	S	D	E	N	1964	37398
2168	S	D	E	N	1965	40303
2168	S	D	E	N	1966	40528
2168	S	D	E	N	1968	56848
2169	S	S	E	N	1967	35434
2169	S	C	E	N	1966	36893
2169	S	D	E	N	1966	40528
2169	S	C	R	N	1966	42262
2170	S	D	E	N	1966	40528
2171	S	D	E	N	1966	40528
2172	S	D	E	N	1965	35907
2193	S	D	R	N	1970	59389
2193	S	D	E	N	1970	59390

Substance Number	Phys. State	Subject	Language	Temperature	Year	TPRC Number
551-2217	S	D	E	N	1965	34040
2217	S	D	E	N	1965	34041
2217	S	D	E	N	1965	34345
2217	S	D	E	N	1967	43801
2217	S	D	E	N	1967	47034
2217	S	D	E	N	1967	47207
2217	S	D	E	N	1968	50239
2217	S	D	E	N	1968	51914
2217	S	D	E	N	1968	56357
2217	S	D	E	N	1970	56770
2217	S	D	J	N	1970	59290
2217	S	D	E	N	1970	63384
2217	G	D	E	N	1965	58823
2217	M	D	E	N	1966	33962
2217	M	D	E	N	1965	34345
2230	S	D	E	N	1965	35907
2234	S	D	E	N	1966	40528
2246	S	D	E	N	1965	35907
2248	S	D	E	N	1965	35907
2251	S	D	E	N	1965	35907
2253	S	D	R	N	1969	55812
2253	S	D	E	N	1969	55813
2254	S	D	E	N	1965	35907
2266	S	D	E	N	1965	39263
2266	S	D	E	N	1966	45987
2266	S	D	E	N	1964	49031
2275	S	C	G	N	1970	59192
2276	S	D	F	N	1970	60240
2278	S	D	E	N	1969	52594
2281	S	C	F	N	1969	38143
2283	S	C	E	N	1966	39479
2285	S	D	E	N	1962	42292
2286	S	T	E	N	1966	34628
2286	S	D	E	N	1965	35527
2286	S	E	E	N	1969	37122
2286	S	T	R	U	1966	40952
2286	S	T	E	U	1966	40953
2286	S	T	E	U	1968	47502
2286	S	T	E	N	1968	50975
2286	S	G	E	N	1966	52103
2286	S	D	E	N	1970	57794
2286	S	D	E	N	1970	58712
2287	S	D	R	N	1969	34367
2287	S	D	E	N	1969	34368
2287	S	T	R	N	1966	40099
2287	S	T	E	N	1966	40100
2287	S	T	R	U	1966	40600
2287	S	T	E	U	1966	40601
2287	S	C	E	N	1947	40783
2287	S	D	E	N	1967	44702
2287	S	D	E	N	1966	44875
2287	S	D	E	N	1970	57796
2287	S	D	E	N	1970	58663
2289	S	D	E	N	1964	33974
2289	S	D	E	N	1965	46366
2290	S	D	E	N	1964	33974
2290	S	D	E	N	1965	46366
2292	S	D	E	N	1964	33974
2292	S	D	E	N	1965	46366
2296	S	D	E	N	1964	33974
2296	S	D	E	N	1965	46366
2297	S	D	R	N	1966	41087
2297	S	D	E	N	1966	41088
2298	S	D	E	N	1964	33974
2298	S	D	E	N	1965	46366
2301	S	D	E	N	1964	33974
2301	S	D	E	N	1965	46366
2304	S	D	E	N	1964	33974
2304	S	D	E	N	1965	46366
2305	S	D	E	N	1965	44137
2305	S	D	E	N	1955	47062
2305	S	D	E	N	1970	56051
2306	S	D	E	N	1964	33974
2306	S	D	E	N	1965	46366
2309	S	D	E	N	1964	33974
2309	S	D	E	N	1965	46366
2313	S	D	E	N	1966	34623
2313	S	D	E	N	1964	35819
2313	S	C	E	N	1964	37546
2313	S	D	E	N	1966	39494
2313	S	D	E	N	1965	40303
2313	S	D	E	N	1967	42851
2313	S	D	E	N	1964	43112
2314	S	D	R	N	1966	40565
2314	S	D	E	N	1966	40566
2315	S	D	R	N	1967	33884
2315	S	D	E	N	1967	33885
2315	S	D	E	N	1965	35069
2315	S	D	E	N	1964	37398
2315	S	D	E	N	1969	41610
2315	S	D	E	N	1966	44271
2315	S	D	E	N	1931	47008
2315	S	D	R	N	1968	50282
2315	S	D	E	N	1969	55085
2315	S	D	E	N	1970	59637
2315	S	D	E	N	1970	59647
2315	P	G	E	N	1966	34908
2315	P	D	E	N	1966	43525
2315	P	D	E	N	1966	48232
2315	P	G	E	N	1967	48233
2315	P	D	E	N	1969	59686

Substance Number	Phys. State	Subject	Language	Temperature	Year	TPRC Number
551-2322	S	D	E	N	1968	48228
2324	S	D	E	H	1963	43421
2326	S	D	E	H	1963	43421
2327	S	D	E	N	1968	50365
2328	S	D	E	H	1963	43421
2330	S	D	E	N	1967	45873
2330	S	D	E	N	1969	55551
2335	S	D	E	N	1967	45212
2336	S	D	E	N	1967	45212
2338	S	D	E	N	1965	44137
2338	S	D	E	N	1970	56051
2339	S	G	E	N	1965	36684
2340	S	D	E	N	1967	44524
2340	S	D	E	N	1955	47062
2341	S	D	E	N	1967	44524
2342	S	D	E	N	1967	44524
2342	S	D	E	N	1955	47062
2343	S	D	E	N	1967	44524
2343	S	D	E	N	1955	47062
2344	S	S	E	U	1959	44767
2348	S	G	E	N	1965	36686
2349	S	D	G	N	1968	49510
2350	S	D	E	N	1962	42292
2351	S	D	E	N	1963	44971
2352	S	D	E	N	1962	42292
2353	S	D	E	N	1963	44971
2354	S	D	E	N	1963	44971
2354	S	G	F	N	1959	59536
2355	S	D	E	N	1970	58663
2356	S	D	E	N	1965	44137
2356	S	D	E	N	1970	56051
2357	S	D	E	N	1962	42292
2358	S	D	E	N	1962	42292
2359	S	D	E	N	1962	42292
2360	S	D	E	N	1962	42292
2361	S	C	E	N	1963	36197
2361	S	G	E	L	1967	42889
2363	S	D	E	N	1968	49647
2364	S	E	E	N	1965	38737
2365	S	C	J	N	1968	53166
2366	S	D	E	N	1964	43112
2367	S	D	E	N	1964	34325
2367	S	D	R	N	1965	35077
2367	S	D	E	N	1965	35078
2367	S	D	R	N	1969	38116
2367	S	D	E	N	1969	38117
2367	S	D	R	N	1966	41764
2367	S	D	E	N	1966	41765
2367	S	S	E	N	1960	44278
2367	S	D	E	N	1967	49233
2367	S	C	R	N	1968	49361
2367	S	C	E	N	1968	49362
2367	S	G	E	N	1967	52883
2367	S	D	E	N	1968	53509
2367	S	D	G	N	1969	55348
2367	S	D	E	N	1969	56007
2367	S	C	R	N	1969	57363
2367	S	C	E	N	1969	57364
2367	P	D	E	N	1969	59686
2368	L	D	E	N	1966	41575
2368	L	G	E	N	1965	54313
2369	S	D	E	N	1966	33970
2369	S	D	F	N	1964	36710
2369	S	D	F	N	1964	37148
2369	S	D	E	N	1964	37320
2369	S	D	R	N	1966	39993
2369	S	D	E	N	1966	39994
2369	S	D	E	N	1966	40528
2369	S	D	F	N	1966	42151
2369	S	D	E	N	1968	42270
2369	S	D	E	N	1969	49627
2369	S	D	E	N	1969	56122
2369	S	D	E	N	1968	56833
2369	S	D	E	N	1968	56848
2369	S	D	E	N	1969	57413
2369	S	C	E	N	1963	59542
2370	S	D	E	N	1965	40305
2370	S	D	E	N	1967	41136
2374	S	D	F	N	1964	37154
2374	S	D	E	N	1964	37501
2374	S	G	E	N	1964	40182
2374	S	D	E	N	1961	40793
2374	S	D	F	N	1966	41228
2374	S	D	E	N	1968	47949
2374	S	D	E	N	1968	51925
2375	S	D	E	N	1964	37501
2375	S	D	E	N	1964	37971
2375	S	G	E	N	1964	40182
2375	S	D	E	N	1961	40793
2375	S	D	F	N	1967	44045
2375	S	D	E	N	1968	48547
2376	S	D	E	N	1965	36348
2377	S	D	E	N	1965	36348
2378	S	D	E	N	1965	36348
2379	S	D	E	N	1965	36348
2380	S	D	E	N	1965	36348
2381	S	D	E	N	1965	36348
2382	S	D	E	N	1965	36348
2383	S	D	E	N	1965	36348
2384	S	D	E	N	1964	40756
2388	S	D	E	N	1965	35573

Phys. State: **G.** Gas; **L.** Liquid; **M.** Multiphase; **P.** Powder; **S.** Solid

Reflectance

B34

Substance Number	Phys. State	Subject	Language	Temperature	Year	TPRC Number	Substance Number	Phys. State	Subject	Language	Temperature	Year	TPRC Number	Substance Number	Phys. State	Subject	Language	Temperature	Year	TPRC Number
551-2391	S	D	E	N	1964	40756	551-2436	S	D	E	U	1967	36884	551-2530	S	D	E	N	1963	39496
2392	S	D	E	N	1960	40746	2436	S	D	E	N	1969	37877	2531	S	D	E	N	1963	39496
2394	S	D	E	N	1960	40746	2436	S	D	E	N	1966	39951	2533	S	D	E	N	1965	34041
2395	S	D	E	N	1965	34029	2436	S	D	E	N	1966	47098	2533	S	D	E	N	1965	39754
2395	S	C	E	N	1966	44307	2436	S	D	E	N	1967	47274	2534	S	D	E	N	1967	55869
2395	S	G	G	N	1967	51605	2436	S	D	E	N	1970	59654	2535	S	D	E	N	1966	34814
2396	S	D	E	N	1965	34029	2438	S	D	E	N	1931	47008	2535	S	D	E	N	1966	42978
2396	S	C	E	N	1966	44307	2438	S	D	E	N	1970	59141	2535	S	D	E	N	1963	54703
2397	S	D	E	N	1965	34029	2438	S	D	E	N	1970	60590	2537	S	D	E	N	1943	50820
2397	S	C	E	N	1966	44307	2439	S	D	E	N	1965	39086	2538	S	D	E	N	1966	34814
2397	S	D	E	N	1967	47244	2439	S	D	F	N	1966	39294	2541	S	D	E	N	1966	38732
2397	S	D	E	N	1967	49181	2439	S	D	F	N	1964	43112	2542	S	D	E	N	1960	38776
2397	S	G	G	N	1967	51605	2439	S	D	F	N	1969	45286	2543	S	D	E	N	1970	37675
2399	S	D	F	N	1970	57251	2441	S	D	E	N	1966	40351	2543	S	D	E	N	1964	43112
2401	S	S	G	U	1966	41418	2444	S	D	E	N	1966	40351	2544	S	D	E	N	1980	38776
2402	S	D	E	N	1965	44137	2445	S	D	F	L	1966	39959	2546	S	D	E	N	1967	55869
2402	S	D	E	N	1970	56051	2447	S	D	E	N	1966	40351	2550	S	D	E	N	1970	58264
2403	S	D	E	N	1965	44137	2449	S	D	E	N	1968	58406	2554	S	D	E	N	1965	33879
2403	S	D	E	N	1970	56051	2454	S	D	E	N	1966	40351	2554	S	D	E	N	1965	39092
2404	S	C	E	N	1966	34621	2455	S	D	E	N	1964	37320	2554	S	D	E	N	1966	52897
2404	S	T	G	U	1965	36681	2458	S	D	F	N	1964	37153	2555	S	D	E	N	1966	42978
2405	S	C	E	N	1966	34621	2458	S	D	F	N	1964	38139	2557	S	D	E	N	1966	34488
2406	S	D	E	N	1963	40467	2458	S	S	E	N	1967	45501	2566	S	S	E	N	1964	36423
2407	S	C	E	N	1966	34621	2459	S	C	E	N	1966	34625	2566	S	T	E	N	1965	36524
2409	S	D	E	N	1970	58524	2459	S	D	E	N	1965	41156	2566	S	S	E	U	1965	36532
2412	S	D	E	N	1963	40467	2460	S	D	E	N	1970	58297	2566	S	S	E	U	1965	36536
2413	S	C	E	N	1966	34625	2461	S	C	E	N	1966	34625	2566	S	D	E	N	1963	45545
2414	S	D	E	N	1963	40467	2462	S	C	E	N	1966	34625	2566	S	E	E	N	1966	48373
2415	S	D	E	N	1966	40351	2463	S	C	E	N	1966	34625	2566	S	S	E	U	1970	59975
2415	S	D	E	N	1963	40467	2470	S	D	E	N	1966	34548	2567	S	D	R	N	1966	45706
2415	S	D	E	N	1955	47062	2470	S	D	E	N	1965	39086	2567	S	E	E	N	1966	45707
2415	S	D	E	N	1967	52855	2470	S	D	E	N	1967	42092	2568	S	D	R	N	1966	45706
2416	S	D	E	N	1961	33844	2473	P	G	E	N	1966	34908	2568	S	D	E	N	1966	45707
2416	S	D	E	N	1966	33882	2474	S	G	E	N	1966	33961	2569	S	D	R	N	1966	45706
2416	S	G	E	N	1965	34683	2474	S	D	E	N	1964	33974	2569	S	D	E	N	1966	45707
2416	S	D	E	N	1950	35074	2474	S	D	E	N	1965	46366	2570	S	D	R	N	1966	45706
2416	S	D	F	N	1964	35321	2474	P	G	E	N	1966	33961	2570	S	D	E	N	1966	45707
2416	S	C	E	N	1963	36197	2476	P	G	E	N	1966	34908	2572	S	D	E	H	1931	47008
2416	S	D	E	N	1965	36504	2478	S	D	E	N	1964	33974	2572	S	D	E	N	1964	37399
2416	S	G	F	N	1964	36678	2478	S	D	E	N	1965	46366	2573	S	D	E	N	1964	37399
2416	S	D	E	N	1964	37483	2478	P	G	E	N	1966	34908	2575	S	D	E	N	1970	57802
2416	S	D	E	N	1968	37491	2480	S	D	E	N	1964	33974	2576	S	D	E	N	1966	33854
2416	S	D	E	N	1964	37501	2485	S	D	E	N	1960	40655	2577	S	D	E	N	1966	33854
2416	S	D	F	N	1964	37833	2485	S	G	E	N	1968	44882	2577	S	S	E	U	1946	39854
2416	S	D	E	N	1964	37970	2485	P	D	E	N	1966	39220	2577	S	D	E	N	1965	42685
2416	S	G	E	N	1965	38649	2487	S	D	E	N	1966	39215	2577	S	D	E	N	1970	57794
2416	S	D	F	N	1966	39288	2489	P	D	E	N	1966	39220	2577	S	C	E	N	1970	61485
2416	S	D	E	N	1965	39754	2491	S	D	E	N	1965	39754	2578	S	D	E	N	1966	33854
2416	S	D	E	N	1949	39837	2494	S	C	E	N	1966	39479	2579	S	D	E	N	1943	50820
2416	S	G	E	N	1964	40182	2495	S	C	E	N	1966	39479	2581	S	D	E	N	1966	33854
2416	S	D	E	N	1966	40343	2496	S	C	E	N	1966	39479	2583	S	D	E	N	1966	33854
2416	S	D	E	N	1966	40471	2497	S	C	E	N	1966	39479	2584	S	D	E	N	1966	33854
2416	S	G	E	N	1928	40591	2498	S	C	E	N	1966	39479	2586	S	D	E	N	1966	52897
2416	S	D	E	N	1960	40746	2499	S	C	E	N	1966	39479	2587	S	D	E	N	1966	33854
2416	S	D	E	L	1967	42889	2500	S	C	E	N	1966	39479	2589	S	E	E	L	1968	56536
2416	S	D	E	N	1965	43413	2501	S	C	E	N	1966	39479	2589	S	C	E	N	1965	58823
2416	S	D	E	N	1948	43655	2502	S	C	E	N	1966	39479	2589	G	E	E	N	1965	58823
2416	S	D	F	N	1967	44044	2503	S	C	E	N	1966	39479	2590	S	D	E	N	1968	52892
2416	S	C	E	N	1966	44304	2504	S	C	E	N	1966	39479	2593	S	D	E	N	1946	55801
2416	S	C	F	N	1966	44306	2505	S	C	E	N	1966	39479	2594	S	D	E	N	1970	57511
2416	S	C	E	N	1966	44314	2506	S	C	E	N	1966	39479	2594	S	D	E	N	1969	58322
2416	S	D	F	N	1948	44326	2507	S	D	E	N	1965	39754	2596	S	D	E	N	1970	56770
2416	S	D	F	N	1966	44924	2508	S	D	E	N	1965	39754	2599	M	D	E	N	1966	34050
2416	S	G	E	N	1966	44933	2509	S	D	E	N	1967	55869	2600	S	D	F	N	1965	35934
2416	S	D	R	N	1967	45014	2510	S	D	E	N	1965	39754	2600	S	D	E	N	1966	52897
2416	S	D	E	N	1967	45015	2511	S	D	E	N	1965	39754	2604	S	D	E	N	1969	55822
2416	S	G	E	N	1966	45458	2512	S	D	E	N	1967	39460	2607	S	D	R	N	1969	58573
2416	S	D	G	N	1963	45738	2513	S	D	E	N	1965	39754	2608	S	D	E	N	1969	58574
2416	S	D	G	N	1938	46997	2514	S	D	E	N	1965	39754	2611	S	D	E	N	1965	45840
2416	S	D	E	N	1955	47062	2515	S	D	E	N	1965	39754	2611	S	D	E	N	1962	33902
2416	S	D	E	N	1967	47275	2516	S	D	E	N	1965	39754	2611	S	D	E	N	1965	36525
2416	S	C	E	N	1968	47821	2517	S	D	E	N	1965	39754	2613	S	D	E	N	1966	36090
2416	S	D	E	N	1967	48032	2518	S	D	E	N	1965	39754	2613	S	D	E	N	1964	37390
2416	S	D	F	N	1960	48051	2519	S	D	E	N	1965	39754	2614	S	D	E	N	1964	34058
2416	S	D	F	E	1968	49177	2520	S	D	E	N	1963	39496	2614	S	D	E	N	1964	36090
2416	S	D	R	N	1968	50272	2521	S	D	E	N	1963	39496	2614	S	D	E	N	1964	37398
2416	S	D	E	N	1966	50297	2522	S	D	E	N	1963	39496	2614	S	D	E	N	1964	37399
2416	S	G	E	N	1968	50483	2524	S	D	E	N	1965	39239	2615	S	D	E	N	1964	36686
2416	S	D	E	U	1969	52328	2524	S	D	E	N	1966	39977	2618	L	D	E	N	1966	40529
2416	S	D	E	N	1968	53491	2524	S	D	E	N	1966	40471	2618	L	D	E	N	1969	57233
2416	S	C	R	N	1969	54211	2524	S	C	E	N	1966	44307	2619	S	D	E	N	1964	36090
2416	S	C	E	N	1969	54212	2524	S	C	E	N	1966	45458	2619	S	D	E	N	1964	37390
2416	S	D	E	N	1969	56122	2524	S	G	G	N	1967	51605	2620	S	D	E	N	1964	36090
2416	S	C	G	N	1969	56190	2525	S	D	E	N	1968	58406	2620	S	D	E	N	1964	37390
2416	S	G	E	N	1968	56538	2526	S	D	E	N	1966	34733	2621	S	D	E	N	1964	36090
2416	S	D	E	N	1968	56849	2526	S	D	E	N	1968	47309	2621	S	D	E	N	1964	37390
2416	S	D	E	N	1970	57015	2526	S	D	E	N	1970	57342	2623	S	D	E	N	1964	37390
2416	S	D	E	N	1970	57794	2527	S	D	R	N	1966	33747	2623	S	D	E	N	1964	37398
2416	S	D	E	N	1970	59319	2527	S	D	E	N	1965	34385	2623	S	C	E	N	1965	46109
2416	S	C	E	N	1963	59542	2528	S	D	E	N	1965	38531	2623	S	J	E	N	1968	53166
2416	S	D	G	N	1969	59668	2528	S	C	E	N	1966	39015	2623	S	D	E	N	1970	57784
2416	S	D	R	N	1969	61852	2528	S	D	R	N	1966	39609	2624	S	D	E	N	1965	46109
2416	S	D	E	N	1969	61853	2528	S	D	E	N	1966	39610	2625	S	D	E	N	1965	34394
2416	M	D	E	N	1966	41373	2528	S	S	E	N	1967	43517	2627	S	D	E	N	1965	33944
2421	S	D	E	N	1966	40351	2528	S	T	E	U	1969	53985	2627	S	D	E	N	1965	35221
2426	S	C	G	N	1964	37015	2528	S	D	E	N	1969	54014	2627	S	D	E	N	1957	41118
2426	S	S	E	N	1966	52912	2528	S	D	E	N	1968	54260	2627	S	D	E	N	1967	43686
2429	S	D	R	N	1970	60790	2529	S	D	E	N	1963	39496	2627	S	D	E	N	1967	44765
2429	S	D	E	N	1971	61959															
2430	M	D	E	N	1966	39631															
2436	S	G	E	N	1965	34683															

Substance Number	Phys. State	Sub-ject	Lan-guage	Temper-ature	Year	TPRC Number
551-2627	S	S	E	N	1966	52913
2628	S	D	E	N	1966	45921
2630	S	D	E	N	1963	36219
2630	S	D	E	N	1967	43447
2631	S	D	E	N	1965	34344
2631	S	D	E	N	1965	35808
2631	S	D	E	N	1965	39263
2632	S	D	E	N	1966	45921
2633	S	D	E	N	1965	34344
2633	S	D	E	N	1965	35808
2634	S	D	E	N	1964	37272
2638	S	D	E	N	1967	43447
2640	S	D	E	N	1955	47062
2642	S	D	E	N	1964	34680
2643	S	D	E	N	1964	34680
2644	S	G	E	N	1964	34682
2645	S	D	E	N	1964	34680
2646	S	D	E	N	1964	34680
2648	S	G	E	N	1964	34682
2649	S	D	E	N	1964	34680
2650	S	D	E	N	1964	34680
2651	S	D	E	N	1964	34680
2652	S	D	E	N	1964	34680
2653	S	D	E	N	1964	34680
2654	S	D	E	N	1964	34680
2655	S	D	E	N	1964	34680
2656	S	D	E	N	1964	34680
2657	S	D	E	N	1964	34680
2658	S	D	E	N	1964	34680
2659	S	G	E	N	1964	34682
2660	S	D	R	N	1969	58573
2660	S	D	E	N	1969	58574
2661	S	D	E	N	1964	34680
2662	S	D	E	N	1964	36090
2662	S	D	E	N	1964	37390
2662	S	D	R	N	1969	58573
2662	S	D	E	N	1969	58574
2663	S	G	E	N	1964	34682
2664	S	D	E	N	1964	34680
2665	S	D	E	N	1964	34680
2666	S	D	E	N	1964	34680
2667	S	D	E	N	1964	34680
2668	S	D	E	N	1964	34680
2670	S	D	E	N	1964	34680
2671	S	D	E	N	1964	34680
2672	S	D	E	N	1964	34680
2673	S	D	E	N	1964	46335
2674	S	D	E	N	1964	46335
2675	S	G	E	N	1964	34682
2676	S	G	E	N	1964	34682
2677	S	D	E	N	1964	34680
2677	S	G	E	N	1964	34682
2677	S	D	E	N	1965	39754
2677	S	D	E	N	1963	40420
2677	S	D	E	N	1968	51403
2677	S	D	E	N	1961	52153
2677	S	D	E	N	1967	52855
2677	S	D	R	N	1969	56239
2678	S	G	E	N	1964	34682
2679	S	G	E	N	1964	34682
2680	S	G	E	N	1964	34682
2681	S	G	E	N	1964	34682
2682	S	G	E	N	1964	34682
2683	S	G	E	N	1964	34682
2684	S	G	E	N	1964	34682
2684	S	D	E	N	1960	35546
2684	S	G	E	N	1967	45871
2685	S	G	E	N	1964	34682
2686	S	G	E	N	1964	34682
2687	S	G	E	N	1964	34682
2688	S	G	E	N	1964	34682
2689	S	G	E	N	1964	34682
2690	S	G	E	N	1964	34682
2691	S	G	E	N	1964	34682
2692	S	G	E	N	1964	34682
2693	S	G	E	N	1964	34682
2694	S	G	E	N	1964	34682
2695	S	D	R	N	1969	58573
2695	S	D	E	N	1969	58574
2696	S	G	E	N	1964	34682
2697	S	G	E	N	1964	34682
2698	S	G	E	N	1964	34682
2699	S	G	E	N	1964	34682
2700	S	G	E	N	1964	34682
2701	S	G	E	N	1964	34682
2706	S	G	E	N	1965	34683
2706	S	D	E	N	1966	47098
2707	S	D	E	N	1955	47062
2709	S	G	E	N	1965	34683
2710	S	G	E	N	1965	34683
2711	S	G	E	N	1965	34683
2712	S	G	E	N	1965	34683
2713	S	D	E	N	1955	47062
2714	S	G	E	N	1965	34683
2714	S	D	E	N	1964	37398
2715	S	G	E	N	1965	34683
2715	S	D	E	N	1966	40528
2717	S	G	E	N	1965	34683
2717	S	D	E	N	1965	36500
2717	S	D	E	N	1968	37491
2717	S	D	E	N	1967	39212

Substance Number	Phys. State	Sub-ject	Lan-guage	Temper-ature	Year	TPRC Number
551-2717	S	S	E	N	1960	44278
2717	S	D	E	N	1961	44317
2717	S	D	E	N	1948	44326
2717	S	D	E	N	1963	55806
2717	S	D	F	N	1970	60240
2717	S	G	E	N	1970	60616
2717	S	G	E	N	1968	63393
2720	S	G	E	N	1965	34683
2723	S	D	E	N	1955	47062
2724	S	D	R	N	1966	45706
2724	S	D	E	N	1966	45707
2725	S	D	R	N	1966	45706
2725	S	D	E	N	1966	45707
2726	S	D	R	N	1966	45706
2726	S	D	E	N	1966	45707
2727	S	D	R	N	1966	45706
2727	S	D	E	N	1966	45707
2728	S	D	R	N	1966	45706
2728	S	D	E	N	1966	45707
2729	S	D	E	N	1955	47062
2731	S	D	E	N	1955	47062
2732	S	D	E	N	1955	47062
2733	S	D	E	N	1955	47062
2734	S	D	E	N	1955	47062
2735	S	D	E	N	1955	47062
2736	S	D	E	N	1965	34045
2736	S	D	E	N	1968	53491
2738	S	D	E	N	1955	47062
2739	S	D	E	N	1965	39263
2740	S	D	R	N	1966	45706
2740	S	D	E	N	1966	45707
2741	S	D	E	N	1955	47062
2742	S	D	E	N	1955	47062
2743	S	D	R	N	1966	45706
2743	S	D	E	N	1966	45707
2744	S	D	R	N	1966	45706
2744	S	D	E	N	1966	45707
2745	S	D	R	N	1966	45706
2745	S	D	E	N	1966	45707
2746	S	D	R	N	1966	45706
2746	S	D	E	N	1966	45707
2747	S	D	R	N	1966	45706
2747	S	D	E	N	1966	45707
2748	S	D	R	N	1966	45706
2748	S	D	E	N	1966	45707
2750	S	D	E	N	1955	47062
2751	S	G	E	N	1966	34928
2751	S	D	E	N	1969	35545
2751	S	D	E	N	1960	35546
2751	S	D	F	N	1965	35934
2751	S	D	E	N	1968	36046
2751	S	D	E	U	1967	36884
2751	S	D	E	N	1965	39754
2751	S	D	E	N	1967	45529
2751	S	D	E	N	1967	47756
2751	S	D	E	N	1967	48234
2751	S	D	E	N	1966	48366
2751	S	E	E	N	1966	48374
2751	S	D	E	N	1968	49665
2751	S	D	E	N	1965	52152
2751	S	D	E	N	1968	52298
2751	S	D	E	N	1968	52906
2751	S	D	E	N	1969	55070
2751	S	D	E	N	1969	55221
2751	S	D	E	N	1969	58335
2751	S	D	E	N	1970	59648
2752	S	D	E	N	1955	47062
2753	S	D	E	N	1955	47062
2755	S	C	R	N	1964	34769
2755	S	C	E	N	1965	34770
2758	S	D	E	N	1965	34045
2758	S	D	E	N	1965	34817
2758	S	D	E	N	1969	37388
2758	S	D	R	N	1966	44625
2758	S	D	E	N	1968	53513
2759	S	D	F	L	1966	39959
2759	S	D	E	N	1965	42781
2759	S	D	E	N	1963	54703
2772	S	D	E	N	1970	59648
2774	S	D	R	N	1969	58573
2774	S	D	E	N	1969	58574
2775	S	D	E	N	1940	45700
2776	S	D	E	N	1940	45700
2777	S	D	E	N	1940	45700
2778	S	D	E	N	1940	45700
2779	S	D	E	N	1940	45700
2780	S	D	E	N	1940	45700
2781	S	D	E	N	1940	45700
2782	S	D	E	N	1940	45700
2784	S	D	E	N	1964	37398
2788	S	D	R	N	1969	58573
2788	S	D	E	N	1969	58574
2789	S	D	R	N	1969	58573
2789	S	D	E	N	1969	58574
2791	S	D	E	N	1964	36414
2792	S	D	R	N	1969	58573
2792	S	D	E	N	1969	58574
2793	M	D	E	N	1962	45667
2794	M	D	E	N	1962	45667
2795	M	D	E	N	1962	45667
2796	M	D	E	N	1962	45667

Substance Number	Phys. State	Sub-ject	Lan-guage	Temper-ature	Year	TPRC Number
551-2797	M	D	E	N	1962	45667
2800	S	D	R	N	1967	46082
2800	S	D	E	N	1967	46083
2810	S	D	E	N	1969	55810
2810	P	D	E	N	1969	55810
2815	S	D	E	N	1965	36544
2816	S	D	E	N	1965	36544
2817	S	D	R	N	1967	46082
2817	S	D	E	N	1967	46083
2818	P	D	R	N	1967	46082
2818	P	D	E	N	1967	46083
2819	S	D	R	N	1967	46082
2819	S	D	E	N	1967	46083
2820	S	D	R	N	1967	46082
2820	S	D	E	N	1967	46083
2822	S	D	E	N	1968	55175
2822	S	D	E	N	1969	55741
2823	S	D	E	N	1969	35545
2823	S	D	E	N	1969	55741
2824	S	C	E	N	1966	44307
2824	S	C	E	N	1966	44308
2824	S	D	E	N	1967	49397
2828	P	D	R	N	1967	46082
2828	P	D	E	N	1967	46083
2829	S	D	E	N	1967	41836
2829	S	D	E	N	1969	53180
2830	S	D	E	N	1965	35808
2832	S	D	E	N	1946	55801
2833	S	D	E	N	1965	35323
2833	S	D	E	N	1950	46403
2834	S	C	E	N	1963	36197
2836	S	D	E	N	1964	36126
2837	S	D	E	N	1963	36112
2838	S	D	E	N	1963	36112
2839	S	D	E	N	1955	40788
2839	S	D	E	N	1967	45741
2839	S	D	E	N	1955	47062
2839	S	D	E	N	1969	57187
2840	S	D	E	N	1965	39754
2840	S	D	E	N	1967	45741
2840	S	D	E	N	1966	52897
2840	S	D	E	N	1969	57187
2840	S	D	E	N	1970	60103
2841	S	D	E	N	1955	40788
2841	S	D	E	N	1967	45741
2841	S	D	E	N	1966	52897
2841	S	D	E	N	1969	57187
2842	S	D	E	N	1964	36090
2842	S	D	E	N	1967	40221
2842	S	D	E	N	1967	40233
2843	S	D	E	N	1946	55801
2844	S	D	E	N	1963	36112
2848	S	D	E	N	1967	49397
2849	S	D	E	N	1946	55801
2853	S	D	E	N	1964	33974
2855	S	D	E	N	1964	33974
2857	S	D	E	N	1946	55801
2858	S	D	E	N	1946	55801
2861	S	D	E	N	1964	36090
2861	S	D	E	N	1965	36533
2862	S	D	E	N	1946	55801
2864	S	D	E	N	1964	36090
2869	P	D	E	N	1970	61071
2870	S	D	F	N	1966	47170
2871	S	D	E	N	1965	36466
2873	S	D	E	N	1965	36466
2874	S	D	E	N	1964	33974
2874	S	D	E	N	1964	35819
2874	S	D	E	N	1965	40303
2875	S	D	E	N	1964	37272
2876	S	D	E	N	1964	37272
2877	S	D	E	N	1964	37272
2878	S	D	E	N	1964	37272
2880	S	D	E	N	1964	37272
2882	S	D	E	N	1964	37272
2885	S	D	E	N	1968	52892
2886	S	D	E	N	1968	52892
2887	S	G	E	N	1967	52883
2888	S	G	E	N	1967	52883
2890	S	D	E	N	1969	35545
2890	S	D	E	N	1968	52298
2891	S	D	E	N	1968	47309
2893	S	D	G	N	1934	50420
2894	S	D	E	N	1969	35545
2894	S	D	E	N	1969	41573
2894	S	D	E	N	1968	52298
2905	S	D	E	L	1970	60029
2908	S	D	E	N	1967	39169
2908	S	D	E	N	1968	47869
2908	S	S	E	N	1970	57511
2908	S	D	E	N	1970	59219
2909	S	D	E	N	1967	52151
2910	S	D	E	N	1967	33824
2910	S	D	E	N	1964	36467
2910	S	D	E	L	1967	39806
2913	S	D	F	N	1966	47170
2929	S	D	R	N	1967	44859
2929	S	D	E	N	1967	44860
2937	S	D	E	N	1967	52855
2943	S	D	E	N	1967	46261
2951	S	D	E	N	1964	37790

Phys. State: G. Gas; L. Liquid; M. Multiphase; P. Powder; S. Solid

Substance Number	Phys. State	Subject	Language	Temperature	Year	TPRC Number
551-2953	S	D	E	N	1964	37790
2954	S	D	E	N	1964	37790
2955	S	D	E	N	1964	37790
2956	S	D	E	N	1968	52892
2957	S	D	E	N	1968	56375
2958	S	D	R	N	1967	47189
2958	S	D	E	N	1967	48674
2959	S	D	R	N	1967	47189
2959	S	D	E	N	1967	48674
2960	S	D	R	N	1967	47189
2960	S	D	E	N	1967	48674
2961	S	D	R	N	1967	47189
2961	S	D	E	N	1967	48674
2962	S	D	E	N	1968	56375
2963	S	D	E	N	1968	47421
2964	S	D	E	N	1968	47421
2965	S	D	E	N	1958	56416
2967	S	D	E	N	1968	47421
2971	S	G	E	N	1968	47917
2971	S	D	E	N	1970	58988
2973	S	D	E	N	1969	51456
2979	S	D	F	N	1968	48131
2979	S	D	E	N	1953	55041
2980	S	C	E	U	1968	47918
2987	S	G	E	N	1968	47919
2988	S	D	E	N	1962	53496
2988	S	D	E	N	1961	53497
2990	S	D	F	N	1970	60692
2991	S	D	E	N	1964	35819
2991	S	D	E	N	1964	36090
2991	S	D	E	N	1965	36683
2991	S	D	E	N	1964	37390
2991	S	D	F	N	1964	37830
2991	S	D	F	N	1964	43131
2991	S	D	E	N	1970	60299
2995	P	D	E	N	1969	59686
2996	S	D	E	N	1969	59686
2996	S	D	E	N	1970	62587
2997	S	D	E	H	1965	36511
2997	S	D	E	H	1965	37899
2998	S	D	E	N	1970	59115
2999	S	C	E	U	1965	36491
3000	S	D	E	N	1969	35541
3001	S	D	E	N	1969	35541
3002	S	D	E	N	1968	56849
3003	S	D	E	N	1968	56848
3004	S	D	E	N	1970	56051
3005	S	D	E	N	1970	56051
3008	S	D	E	N	1968	56848
3011	S	G	E	N	1965	36506
3012	S	D	E	N	1968	49327
3014	S	D	E	N	1968	49327
3015	S	D	E	N	1968	49327
3035	S	D	E	N	1970	57800
3037	S	D	E	N	1960	35546
3037	S	D	E	N	1959	46025
3040	S	D	E	U	1967	36884
3040	S	D	E	N	1965	39036
3040	S	D	E	N	1966	40294
3040	S	D	E	N	1964	40756
3040	S	D	E	N	1969	41573
3040	S	D	E	N	1967	47274
3040	S	G	E	N	1966	48359
3042	S	D	E	N	1968	56848
3043	S	D	E	N	1968	56848
3045	P	D	E	N	1969	55600
3048	S	D	E	N	1968	47874
3049	S	C	E	N	1966	44307
3050	S	D	E	N	1968	47874
3051	S	C	E	N	1966	44307
3053	S	D	E	N	1964	33974
3054	S	D	E	N	1965	34040
3054	S	D	E	N	1965	34344
3054	S	G	E	N	1966	34928
3054	S	D	E	N	1966	40529
3054	S	D	E	H	1966	43494
3054	S	D	E	N	1965	43527
3054	S	D	E	N	1967	43801
3054	S	G	E	N	1966	48359
3054	S	D	E	N	1969	58335
3054	S	D	E	N	1970	61721
3054	M	D	E	N	1965	34345
3055	S	C	E	N	1966	44314
3057	S	D	E	N	1968	48720
3058	S	D	E	N	1964	37398
3058	S	D	E	N	1968	49482
3058	S	D	E	N	1968	49963
3065	S	D	E	N	1969	57187
3067	S	D	E	N	1969	57187
3068	S	D	R	N	1969	61199
3070	S	G	E	N	1966	34928
3070	S	D	E	N	1965	39754
3070	S	G	E	N	1966	48359
3074	S	G	E	N	1966	48365
3076	P	D	E	L	1966	48370
3078	S	C	R	N	1969	57363
3078	S	C	E	N	1969	57364
3081	S	E	E	N	1966	48374
3083	S	C	E	N	1970	35861
3086	S	D	E	N	1965	39263
3090	S	D	E	N	1965	39263
551-3091	S	D	E	N	1965	39263
3092	S	D	E	N	1966	48375
3093	S	D	E	N	1965	39263
3095	S	D	E	N	1965	39263
3099	S	E	E	N	1966	48374
3100	S	E	E	N	1966	48374
3101	S	D	E	N	1965	39092
3108	S	D	E	N	1968	36046
3108	S	D	E	N	1967	44986
3108	S	D	E	N	1966	48383
3110	S	D	E	N	1965	39872
3111	S	D	E	N	1965	39872
3112	S	D	E	N	1965	39872
3113	S	D	E	N	1964	34819
3113	S	D	E	N	1964	50030
3113	M	D	E	N	1964	49031
3114	M	D	E	N	1964	49031
3115	S	D	E	N	1964	49031
3116	S	D	E	N	1964	49031
3117	S	D	E	N	1964	49031
3117	S	D	E	N	1970	59986
3118	S	D	E	N	1964	49031
3119	S	D	E	N	1964	49031
3120	S	D	E	N	1964	49031
3121	S	D	E	N	1964	49031
3122	S	D	E	N	1968	36046
3122	S	D	E	N	1964	49031
3123	S	D	E	N	1964	49031
3124	S	D	E	N	1964	49031
3125	S	D	E	N	1964	49031
3126	S	D	E	N	1964	49031
3127	S	D	E	N	1964	49031
3128	S	D	E	N	1964	49031
3129	S	D	E	N	1964	49031
3130	S	D	E	N	1964	49031
3131	S	D	E	N	1964	49031
3132	S	D	E	N	1964	49031
3133	S	D	E	N	1964	49031
3134	S	D	E	N	1964	49031
3135	S	D	E	N	1964	49031
3136	S	D	E	N	1964	49031
3137	S	D	E	N	1964	49031
3138	S	D	E	N	1964	49031
3139	S	D	E	N	1965	39872
3140	M	D	E	N	1964	49031
3141	M	D	E	N	1964	49031
3142	M	D	E	N	1964	49031
3143	S	D	E	N	1965	39872
3144	S	D	E	N	1964	49031
3149	S	D	E	N	1964	40756
3151	S	D	E	N	1968	48926
3153	S	D	E	N	1966	40333
3154	S	D	E	N	1966	40333
3157	S	D	E	N	1966	34814
3157	S	D	E	U	1967	36884
3157	S	D	E	N	1967	47274
3158	S	D	E	N	1964	41019
3159	S	D	E	N	1964	41019
3160	S	D	E	N	1964	41019
3164	S	D	E	U	1967	36884
3164	S	D	E	N	1966	40294
3166	S	D	E	N	1968	53089
3167	S	D	E	N	1968	53089
3175	S	D	E	N	1968	40103
3175	S	D	E	N	1966	48231
3175	S	D	E	N	1966	48232
3175	S	D	E	N	1967	48234
3175	S	D	E	N	1970	59637
3176	P	D	E	N	1966	48232
3190	P	D	E	N	1966	48232
3191	P	D	E	N	1966	48232
3193	P	D	E	N	1967	45664
3193	P	D	E	N	1966	48232
3193	P	G	E	N	1967	48233
3194	P	D	E	N	1966	48232
3195	P	D	E	N	1966	48232
3196	S	D	E	N	1964	40756
3199	S	D	E	N	1968	55175
3200	S	G	E	N	1967	48233
3200	S	D	E	N	1967	48234
3201	S	D	E	H	1969	41934
3201	S	D	E	N	1967	48234
3203	S	D	E	N	1968	40103
3207	S	D	E	N	1966	48375
3211	S	D	E	N	1961	41149
3213	S	D	R	N	1968	48731
3213	S	D	E	N	1968	51640
3213	P	G	E	N	1966	34908
3214	S	D	E	N	1968	55175
3218	S	D	E	N	1966	52897
3219	S	D	E	N	1961	41149
3221	S	D	E	N	1964	51508
3222	S	D	E	N	1968	51403
3222	S	D	E	N	1970	57800
3223	S	D	E	N	1968	51403
3224	S	D	E	N	1953	36071
3224	S	G	E	N	1967	41159
3224	S	D	E	N	1968	51403
3224	S	D	G	N	1958	51618
3224	S	D	E	N	1969	52594
3225	S	D	E	N	1968	51403
551-3226	S	D	E	N	1968	51403
3227	S	D	E	N	1964	33974
3228	S	D	E	N	1960	35546
3228	S	D	F	N	1959	51245
3229	S	D	F	N	1959	51245
3230	S	D	E	N	1960	35546
3230	S	D	E	N	1965	39754
3230	S	D	F	N	1959	51245
3233	S	D	E	N	1961	41149
3234	S	D	E	N	1968	50539
3235	S	D	E	N	1968	50539
3236	S	D	E	N	1968	50539
3237	S	D	E	N	1968	50539
3238	S	D	E	N	1968	50539
3239	S	D	E	N	1968	50539
3240	S	D	E	N	1968	50539
3241	S	D	E	N	1961	41149
3245	S	D	E	N	1961	41149
3253	S	C	E	N	1964	35417
3253	S	G	E	N	1965	36534
3256	S	D	E	N	1968	49251
3258	S	D	E	N	1969	55080
3260	S	D	E	N	1968	49665
3261	S	D	E	N	1968	49665
3264	S	D	E	N	1964	33974
3267	S	D	E	N	1967	49233
3268	S	D	E	N	1969	55221
3270	S	D	F	N	1965	35934
3271	S	D	E	N	1967	47274
3272	S	D	E	N	1968	36046
3272	S	D	E	N	1969	55221
3272	S	D	E	N	1970	59648
3275	S	D	R	N	1966	41413
3275	S	D	R	N	1968	55318
3279	S	D	R	N	1967	48463
3282	S	D	E	N	1966	48843
3283	S	D	E	U	1958	48169
3283	S	D	E	U	1957	48171
3284	S	D	E	U	1958	48169
3285	S	D	E	U	1958	48169
3286	S	D	E	U	1957	48171
3287	S	D	E	N	1968	48805
3287	S	D	E	N	1969	59832
3296	S	D	E	N	1964	36113
3296	S	D	E	N	1964	37351
3296	S	D	E	N	1969	41334
3296	S	D	E	N	1967	48032
3296	S	D	E	N	1970	59649
3299	S	D	E	N	1967	47332
3301	S	D	R	N	1970	59389
3301	S	D	E	N	1970	59390
3309	S	D	E	N	1957	48106
3310	S	D	E	N	1957	48106
3311	S	D	E	N	1957	48106
3312	S	D	E	N	1957	48106
3313	S	D	E	N	1968	36046
3313	S	D	E	N	1969	41589
3313	S	D	E	N	1967	47756
3313	S	D	E	N	1968	54303
3314	S	D	E	N	1969	51937
3316	S	D	E	N	1969	41573
3318	P	G	E	N	1966	42783
3321	S	D	E	N	1970	37675
3323	S	D	E	N	1962	43493
3325	S	D	E	L	1970	59627
3326	S	D	E	N	1970	63400
3329	S	D	R	N	1968	50016
3330	S	D	R	N	1967	49823
3331	S	D	R	N	1967	49823
3332	S	D	R	N	1967	49823
3333	S	D	E	N	1967	52871
3336	S	D	E	N	1967	52871
3337	S	D	E	N	1967	52871
3338	S	D	E	N	1967	52871
3342	S	D	E	N	1970	60299
3351	S	D	E	N	1965	39872
3354	S	D	E	N	1965	39872
3356	S	D	E	N	1965	39872
3364	S	D	E	N	1965	39872
3365	S	D	E	N	1965	39872
3369	S	D	R	N	1969	56323
3370	S	D	R	N	1969	56323
3371	S	D	R	N	1969	56323
3374	S	D	R	N	1969	56239
3375	S	D	R	N	1969	56239
3376	S	D	R	N	1969	56239
3377	S	D	R	N	1969	61199
3380	S	D	E	N	1970	63134
3381	S	D	E	N	1970	62417
3382	S	C	E	N	1966	36893
3382	S	C	R	N	1966	42262
3383	S	C	E	N	1970	61485
3384	S	D	R	N	1967	46157
3392	S	D	E	N	1968	52643
3392	S	D	E	N	1968	52644
3395	S	C	E	N	1968	48832
3396	S	D	E	N	1969	53068
3397	L	D	E	N	1923	51606
3398	L	D	E	N	1923	51606
3409	S	D	E	N	1965	36466
3410	S	D	E	N	1965	36466

Phys. State: **G.** Gas; **L.** Liquid; **M.** Multiphase; **P.** Powder; **S.** Solid

Reflectance

Substance Number	Phys. State	Sub-ject	Lan-guage	Temper-ature	Year	TPRC Number
551-3412	S	S	E	H	1958	36385
3412	S	D	E	N	1947	44094
3415	S	D	R	N	1962	52540
3415	S	D	E	N	1962	52541
3416	S	D	F	N	1969	54712
3424	S	S	E	N	1964	36423
3446	S	D	E	N	1958	51592
3446	L	D	E	N	1959	51593
3447	S	D	E	N	1958	51592
3448	S	D	E	N	1958	51592
3449	S	G	E	N	1964	34682
3449	S	D	E	N	1958	51592
3450	S	D	E	N	1958	51592
3451	S	D	E	N	1966	37595
3452	S	D	E	N	1966	37595
3454	S	D	E	N	1966	37595
3455	S	D	E	N	1966	37595
3456	S	D	E	N	1966	37595
3457	S	D	E	N	1966	37595
3458	S	D	E	N	1966	37595
3459	S	D	E	N	1966	37595
3460	S	D	E	N	1966	37595
3461	S	D	E	N	1966	37595
3461	S	D	E	N	1955	47062
3462	S	D	E	N	1966	37595
3463	S	D	E	N	1966	37595
3463	S	D	E	N	1955	47062
3466	S	D	E	N	1966	37595
3468	S	D	R	N	1966	43360
3468	S	D	E	N	1965	45840
3401	C	D	E	N	1966	44271
3481	S	D	E	N	1970	57800
3493	S	D	E	N	1955	40788
3496	S	D	R	N	1967	45226
3506	P	D	E	N	1969	59686
3509	S	D	E	N	1969	52293
3511	S	D	E	N	1969	54714
3512	S	D	E	N	1969	54714
3513	S	D	E	N	1969	54714
3514	S	D	E	N	1969	54714
621-0004	S	D	E	N	1968	52892
0012	S	D	E	N	1955	47062
0014	S	D	E	N	1966	33879
0017	S	D	E	N	1966	33879
0023	S	D	E	N	1969	55534
0028	S	D	E	N	1955	47062
0040	S	D	E	N	1966	52897
0040	S	D	R	N	1958	63386
0040	S	G	E	N	1970	63387
0040	M	D	J	N	1966	51615
0041	S	D	E	N	1965	34262
0042	S	D	E	N	1965	35323
0062	S	D	E	N	1968	52892
0075	S	D	E	N	1966	33840
0075	S	D	E	N	1966	52897
0076	M	D	J	N	1966	51615
0078	M	D	J	N	1966	51615
0079	M	D	J	N	1966	51615
0080	M	D	J	N	1966	51615
0081	M	D	J	N	1966	51615
0082	M	D	J	N	1966	51615
0083	S	D	E	N	1970	63038
0088	S	D	E	N	1967	52899
0092	S	D	E	N	1955	47062
0094	S	D	E	N	1955	47062
0098	S	D	E	N	1964	36100
0098	S	D	E	N	1965	39092
0099	S	D	E	N	1970	36116
0099	S	D	E	N	1970	59614
0101	S	D	E	N	1966	52897
0101	S	D	E	N	1967	52899
0102	S	D	E	N	1967	52899
0103	S	D	E	N	1967	52899
0104	S	D	E	N	1968	52892
0105	S	D	E	N	1967	52899
0106	S	D	E	N	1967	52899
0107	S	D	E	N	1967	52899
0108	S	D	E	N	1968	52892
0110	S	D	E	N	1968	52892
0111	S	D	E	N	1968	52892
0112	S	D	E	N	1969	54291
0114	S	D	F	N	1969	54291
0115	S	D	E	N	1969	54291
0188	S	D	E	N	1970	59835
0214	S	D	E	N	1969	54898
0217	S	E	E	N	1964	38247
0218	S	D	E	N	1955	47062
0275	S	S	E	U	1970	59133
0282	S	D	E	N	1955	47062
0283	S	D	E	N	1955	47062
0292	S	D	E	N	1955	47082
0295	S	D	E	N	1955	47062
0352	S	D	E	N	1966	52897
0376	S	D	E	N	1955	47062
0377	S	D	E	N	1955	47062
0378	S	D	E	N	1955	47062
0383	S	D	E	N	1955	47062
0393	S	D	E	N	1955	47062
651-0011	P	C	G	N	1966	44476
651-0015	L	D	E	U	1968	48979
0038	L	D	E	U	1968	48979
0039	L	D	E	U	1968	48979
0053	P	D	E	N	1962	42292
0067	L	D	E	U	1968	48979
0341	P	D	E	N	1959	40630
661-0004	S	D	E	N	1966	52897
0004	S	D	E	N	1967	52899
0004	S	D	E	N	1969	57187
0009	S	G	E	N	1965	38431
0030	S	G	E	N	1965	38431
0037	S	D	E	N	1969	57187
0042	S	E	E	N	1931	40174
0042	S	S	E	N	1966	52909
0048	S	D	E	N	1965	35573
0048	S	D	E	N	1964	37351
0048	S	C	E	N	1966	39479
0048	S	D	E	N	1968	49251
0048	S	G	R	N	1958	63386
0048	S	G	E	N	1970	63387
0049	S	D	E	N	1955	47062
0049	S	D	E	N	1966	52897
0049	S	D	E	N	1967	52899
0050	S	C	E	N	1968	55094
0050	S	C	R	N	1968	55330
0050	S	S	G	N	1969	56621
0050	S	D	E	N	1969	61035
0060	S	D	E	N	1969	57187
0067	S	D	E	N	1965	35573
0067	S	C	E	N	1966	39479
0067	S	D	E	N	1931	40580
0143	S	D	E	N	1966	52897
0170	S	G	E	N	1966	33879
0187	S	D	E	N	1966	33879
0223	S	D	E	N	1966	33879
0232	S	D	E	N	1966	33879
0245	S	D	E	N	1966	33879
0315	S	D	E	N	1966	33879
0334	S	G	E	N	1928	40591
0334	S	D	E	N	1966	52897
0367	S	D	E	N	1969	37877
0380	S	D	R	N	1969	56239
0415	S	D	E	N	1955	47062
0435	S	D	E	N	1969	54711
0436	S	D	E	N	1969	54711
0440	S	D	E	N	1928	40591
0443	S	G	E	N	1928	40591
0446	S	D	E	N	1955	47062
0453	S	D	E	N	1964	36100
0454	S	D	E	N	1964	36100
0458	S	D	E	N	1967	46041
0465	S	D	E	N	1966	52897
0467	S	D	E	N	1965	35573
0470	S	D	E	N	1969	57187
0480	S	D	E	N	1970	58535

Phys. State: G. Gas; L. Liquid; M. Multiphase; P. Powder; S. Solid

Chapter 9 Absorptance

Substance Number	Phys. State	Subject	Language	Temperature	Year	TPRC Number
528-0186	S	D	F	N	1965	35934
0186	S	D	E	N	1968	49665
0186	S	G	F	U	1968	51637
0253	S	D	E	N	1960	40746
0256	S	T	G	U	1965	43024
531-0056	S	D	E	N	1967	45553
0122	S	S	E	H	1960	49983
0127	M	D	E	N	1969	59184
0301	S	D	E	N	1967	45553
0301	S	D	E	N	1968	46815
0328	S	D	E	N	1970	59835
0365	S	D	E	R	1968	53887
0365	S	C	E	U	1968	53888
0790	M	D	R	N	1967	46073
0790	M	D	E	N	1967	46074
0791	M	D	R	N	1967	46073
0791	M	D	E	N	1967	46074
0909	S	D	G	N	1952	43597
1160	S	D	E	N	1967	47038
1254	S	D	E	N	1966	40903
1254	S	D	E	N	1966	40904
1254	S	D	E	N	1965	40905
1254	S	D	E	N	1966	40908
1254	S	D	E	N	1966	40909
1356	S	D	E	N	1968	38611
1377	S	D	E	N	1966	45984
1559	S	D	E	N	1967	45553
1641	M	D	F	N	1968	55829
1642	M	D	E	N	1968	55829
541-0135	S	D	E	N	1967	45669
0450	S	D	E	N	1967	45669
0482	S	D	F	N	1965	35934
0482	S	D	E	N	1970	43918
0516	S	G	E	N	1964	34339
0517	S	G	E	N	1964	34339
0518	S	T	E	U	1964	34339
0651	S	D	E	N	1964	37040
0699	S	D	E	N	1967	47275
0707	S	D	E	N	1967	47275
0976	S	D	E	N	1964	37443
1050	S	D	E	N	1967	46040
1085	S	D	E	N	1967	45669
1093	S	D	E	N	1969	56056
1172	S	D	E	U	1968	53085
1238	S	D	E	N	1970	52895
1268	S	D	E	N	1964	40358
1413	S	D	R	N	1970	61438
1413	S	D	E	N	1970	61439
551-0129	S	D	E	N	1961	52095
0153	S	D	E	N	1961	52095
0154	S	D	E	N	1965	34736
0154	S	D	E	N	1960	40746
0154	S	D	E	N	1961	48258
0154	S	D	E	N	1962	49984
0154	S	D	E	N	1961	52095
0223	S	D	E	N	1965	34051
0223	S	D	E	N	1965	34736
0223	S	D	E	N	1963	34830
0223	S	D	E	U	1965	35265
0223	S	D	E	F	1965	35934
0223	S	D	E	N	1965	36504
0223	S	D	E	N	1965	36515
0223	S	D	E	N	1965	39754
0223	S	D	E	N	1963	40467
0223	S	D	E	N	1960	40746
0223	S	D	E	N	1966	42977
0223	S	C	G	N	1967	43384
0223	S	D	E	N	1966	43484
0223	S	D	E	N	1965	43527
0223	S	D	F	N	1953	44281
0223	S	D	E	N	1966	44771
0223	S	D	E	N	1965	47412
0223	S	D	E	N	1966	48372
0223	S	D	E	N	1962	49984
0223	S	D	E	N	1961	52095
0223	S	D	E	N	1968	53087
0223	S	D	E	N	1969	56125
0223	S	D	E	N	1969	57227
0223	S	D	E	N	1969	57597
0223	S	C	E	L	1968	58324
0223	S	D	E	N	1970	59278
0223	S	G	E	N	1970	61104
0253	S	D	E	N	1965	35840
0258	S	D	E	N	1965	35840
0260	S	D	E	N	1965	35840
0278	S	D	E	N	1965	34051
0278	S	D	E	N	1968	35552
0278	S	D	E	N	1966	36740
0278	S	D	E	N	1965	38109
0278	S	D	E	N	1965	38672
0278	S	D	E	N	1969	41590
551-0278	S	D	E	N	1964	42291
0278	S	D	E	N	1962	43493
0278	S	D	E	H	1965	44224
0278	S	D	E	N	1966	44772
0278	S	D	E	N	1965	46810
0278	S	D	E	N	1967	47142
0278	S	D	E	N	1967	47276
0278	S	D	E	N	1968	49665
0278	S	D	E	N	1968	52905
0278	S	D	E	N	1970	59649
0325	S	T	E	U	1969	52830
0325	S	T	R	N	1958	61716
0325	S	T	E	N	1970	61717
0347	S	D	E	N	1969	54052
0428	S	D	E	H	1962	49984
0428	S	D	E	N	1961	52095
0428	S	D	E	N	1961	52784
0435	S	D	E	N	1963	34830
0435	S	D	E	U	1965	35265
0435	S	D	E	N	1965	35934
0435	S	D	F	N	1965	36515
0435	S	D	E	U	1969	41919
0435	S	D	E	N	1966	44771
0435	S	D	E	N	1967	52871
0436	S	D	E	N	1965	39263
0436	S	G	E	U	1968	51607
0436	S	D	E	U	1970	56418
0436	S	D	E	N	1969	57227
0447	S	D	J	N	1968	54519
0447	P	D	E	N	1966	48232
0447	P	G	E	N	1967	48233
0460	S	D	E	N	1962	49984
0460	S	D	E	H	1961	53498
0466	S	D	E	N	1969	35545
0466	S	D	E	N	1966	44771
0466	S	D	E	N	1965	46810
0466	S	G	E	U	1968	51637
0466	S	D	E	N	1961	52095
0466	S	D	E	N	1961	52784
0468	S	D	E	N	1966	44771
0487	S	D	E	N	1969	35545
0487	S	D	E	N	1964	36090
0487	S	D	E	N	1965	36523
0487	S	D	E	N	1965	36683
0487	S	D	E	N	1969	41573
0487	S	D	E	N	1966	43484
0487	S	D	E	N	1967	47275
0487	S	D	E	N	1968	52905
0487	S	D	E	N	1969	57597
0487	S	D	E	N	1970	59649
0487	S	C	E	L	1970	61190
0487	P	D	E	H	1965	38992
0487	P	D	E	N	1966	48232
0500	S	D	E	N	1963	34830
0500	S	D	E	U	1965	35265
0500	S	D	E	N	1965	36470
0500	S	D	E	N	1965	36515
0500	S	D	E	N	1963	40467
0500	S	D	E	N	1960	40746
0500	S	D	E	N	1966	44771
0500	S	D	E	N	1961	52095
0501	S	D	E	N	1963	40467
0501	S	D	E	N	1966	44771
0501	S	D	E	N	1963	45698
0501	S	D	E	N	1961	52784
0501	S	D	E	U	1968	53085
0505	S	D	E	N	1966	44771
0507	S	D	E	N	1965	36470
0507	S	D	E	N	1966	37595
0507	S	D	E	N	1966	44771
0507	S	D	E	N	1961	52784
0507	S	D	E	N	1968	53087
0507	S	D	E	N	1968	55175
0533	S	D	E	N	1965	36523
0533	S	G	E	N	1965	36526
0533	S	D	E	N	1965	36527
0533	S	D	E	N	1966	48377
0533	S	D	E	N	1966	48377
0534	S	D	E	N	1965	36523
0539	S	D	E	N	1968	49251
0539	S	D	E	N	1968	52057
0570	S	D	E	N	1965	47145
0570	S	D	E	N	1961	52095
0570	S	D	E	N	1968	53087
0570	S	D	E	N	1965	36466
0577	S	D	E	N	1964	37040
0577	S	D	E	N	1966	40343
0577	S	D	E	N	1966	42914
0577	S	D	E	N	1966	43484
0577	S	D	E	N	1965	46810
0577	S	D	E	N	1967	47275
0577	S	D	R	N	1969	52920
0577	S	D	E	N	1969	56078
0577	S	D	E	N	1970	56079
551-0577	S	D	E	N	1970	59649
0577	S	C	E	N	1970	59985
0577	S	D	E	N	1970	63134
0582	S	D	F	L	1966	39181
0582	S	D	E	N	1968	52057
0582	S	C	E	L	1970	61190
0582	G	D	E	N	1959	51573
0601	S	D	E	N	1962	33902
0601	S	D	E	N	1965	36525
0601	S	D	E	N	1965	36527
0601	S	D	E	N	1965	36528
0601	S	D	E	N	1968	40103
0601	S	D	E	N	1964	40358
0601	S	D	E	N	1964	40415
0601	S	D	E	N	1964	40416
0601	S	D	E	N	1964	40417
0601	S	D	E	U	1962	41123
0601	S	D	E	U	1969	41919
0601	S	D	E	H	1969	41941
0601	S	D	E	U	1969	41945
0601	S	D	E	N	1965	42290
0601	S	D	E	N	1962	42292
0601	S	D	E	N	1966	42505
0601	S	G	E	N	1966	43481
0601	S	D	E	N	1966	43482
0601	S	D	E	N	1965	43527
0601	S	D	E	N	1966	43530
0601	S	D	E	N	1965	43840
0601	S	D	E	H	1965	44224
0601	S	D	E	N	1966	44272
0601	S	D	E	N	1966	44772
0601	S	D	E	N	1967	47142
0601	S	D	E	N	1967	47273
0601	S	D	E	N	1968	48230
0601	S	D	E	N	1966	48231
0601	S	D	E	N	1966	48232
0601	S	D	E	N	1966	48375
0601	S	D	E	N	1966	48377
0601	S	D	E	N	1966	48378
0601	S	D	E	N	1964	50030
0601	S	G	E	U	1968	51637
0601	S	D	E	N	1967	54978
0601	S	D	E	N	1969	55070
0601	S	D	E	N	1970	56418
0601	S	D	E	N	1969	57227
0601	S	D	G	N	1969	59668
0602	S	D	E	N	1965	36523
0605	S	D	E	N	1962	43493
0620	S	D	E	N	1960	38317
0620	S	D	E	U	1969	41919
0631	S	D	E	N	1959	51573
0631	S	D	E	N	1967	41161
0637	S	D	E	N	1964	40413
0637	S	D	E	N	1965	42351
0643	S	D	E	N	1964	36090
0648	S	D	E	N	1962	53496
0657	S	C	E	L	1969	52663
0657	S	D	E	N	1969	54056
0657	S	G	E	N	1968	56536
0690	S	T	E	U	1969	52929
0691	S	D	E	N	1963	45545
0692	S	D	E	N	1963	45545
0702	S	D	E	N	1962	54959
0702	S	D	E	N	1970	61104
0704	S	D	E	N	1962	54959
0705	S	D	E	N	1962	54959
0723	S	D	E	N	1962	54959
0751	S	D	E	N	1965	39754
0767	S	D	E	N	1965	34736
0799	S	D	E	N	1966	36802
0799	S	D	R	N	1966	42258
0799	S	D	E	N	1967	52871
0799	P	D	R	N	1965	36352
0799	P	D	E	N	1965	36353
0812	S	D	E	N	1965	38109
0812	S	D	E	N	1965	46810
0813	S	D	E	N	1965	38109
0813	S	D	E	N	1965	46810
0826	S	D	E	N	1965	38109
0826	S	D	E	N	1965	46810
0850	S	D	E	R	1967	47189
0850	S	D	E	N	1967	48674
0850	S	D	F	N	1968	49483
0854	S	D	E	N	1963	44419
0854	S	D	E	N	1970	56865
0888	S	D	E	U	1967	45665
0906	S	D	E	N	1963	45698
0906	S	D	E	N	1965	46109
0908	S	D	E	N	1964	40756
0924	S	D	E	N	1965	39754
0924	S	D	E	N	1960	40746
0924	S	D	E	N	1964	42291
0924	S	D	E	N	1962	43493
0949	S	D	E	N	1965	43527
0959	S	D	E	N	1970	62417

Phys. State: G. Gas; L. Liquid; M. Multiphase; P. Powder; S. Solid

Substance Number	Phys. State	Subject	Language	Temperature	Year	TPRC Number
551-0976	S	D	E	N	1966	44771
1064	S	D	E	N	1965	38109
1064	S	D	E	N	1967	52871
1064	S	D	E	N	1962	54959
1076	S	S	E	N	1965	39238
1076	S	D	E	N	1960	40746
1076	S	S	E	N	1967	46337
1076	S	S	E	N	1969	57666
1116	S	D	F	N	1965	35934
1125	S	D	E	N	1960	35546
1130	S	D	E	N	1960	35546
1135	S	D	E	U	1967	45665
1141	S	D	E	N	1964	34333
1141	S	D	E	N	1965	34736
1141	S	D	E	N	1965	36404
1141	S	D	E	N	1965	36523
1141	S	G	E	N	1965	36526
1141	S	D	E	N	1965	36527
1141	S	D	E	N	1965	36528
1141	S	D	E	N	1966	36740
1141	S	D	E	N	1964	37040
1141	S	D	E	N	1965	38109
1141	S	D	E	N	1965	39745
1141	S	D	E	N	1965	39754
1141	S	D	E	N	1964	40416
1141	S	D	E	N	1964	40417
1141	S	D	E	N	1965	47145
1141	S	D	E	N	1967	47273
1141	S	D	E	N	1966	48372
1141	S	D	E	N	1966	48377
1141	S	D	E	N	1962	49984
1144	S	D	E	N	1965	46109
1175	S	D	E	N	1965	34736
1175	S	D	E	N	1965	36523
1175	S	G	E	N	1965	36526
1175	S	D	E	N	1966	39363
1175	S	D	E	N	1966	44772
1175	S	D	E	N	1967	47142
1175	S	D	E	N	1965	47145
1175	S	D	E	N	1966	48372
1175	S	D	E	N	1962	49984
1179	S	D	E	N	1965	36525
1179	S	D	E	N	1966	40583
1179	S	D	E	N	1962	41123
1179	S	D	E	N	1969	41590
1179	S	D	E	N	1966	48231
1179	S	D	E	N	1966	48232
1181	S	D	E	N	1965	36523
1181	S	D	E	N	1965	42290
1184	S	D	E	N	1964	34819
1185	S	D	E	N	1965	36525
1186	S	D	E	N	1962	33902
1186	S	D	E	N	1965	36525
1186	S	D	E	N	1962	41123
1188	S	D	E	N	1962	41123
1188	S	D	E	N	1966	43530
1189	S	D	E	N	1962	33902
1189	S	D	E	N	1965	36525
1189	S	D	E	N	1962	41123
1190	S	D	E	N	1962	33902
1190	S	D	E	N	1965	36523
1190	S	D	E	N	1965	36525
1190	S	D	E	N	1962	42292
1191	S	D	E	N	1962	33902
1191	S	D	E	N	1965	36525
1191	S	D	E	N	1962	42292
1191	S	D	E	N	1964	50030
1192	S	D	E	N	1962	33902
1193	S	D	E	N	1962	33902
1193	S	D	E	N	1965	36525
1194	S	D	E	N	1962	33902
1197	S	D	E	N	1962	33902
1197	S	D	E	N	1964	34819
1197	S	D	E	N	1965	36525
1197	S	D	E	N	1962	41123
1197	S	D	E	N	1965	42290
1197	S	D	E	N	1962	42292
1197	S	D	E	N	1965	43840
1197	S	D	E	N	1965	48230
1197	S	D	E	N	1966	48377
1197	S	D	E	N	1966	48492
1197	S	D	E	N	1964	50030
1198	S	D	E	N	1962	41123
1202	S	D	E	N	1962	33902
1202	S	D	E	N	1965	36523
1202	S	D	E	N	1965	36525
1202	S	G	E	N	1965	36526
1202	S	D	E	N	1965	36527
1202	S	D	E	N	1962	42292
1202	S	D	E	N	1966	43482
1202	S	D	E	N	1966	44772
1202	S	D	E	N	1961	46640
1202	S	D	E	N	1967	47142
1202	S	D	E	N	1966	48372
1202	S	D	E	N	1966	48377
1202	S	D	E	N	1965	52152
1202	S	D	E	N	1967	52864
1202	S	D	E	N	1969	55070
1205	S	D	E	N	1962	33902
1205	S	D	E	N	1965	36523
1206	S	D	E	N	1964	34819

Substance Number	Phys. State	Subject	Language	Temperature	Year	TPRC Number
551-1207	S	D	E	N	1966	43530
1218	S	D	E	N	1964	40756
1226	S	D	E	U	1969	41919
1227	S	D	E	N	1965	36523
1227	S	D	E	N	1965	36525
1227	S	D	E	N	1965	36527
1230	S	D	E	N	1965	36525
1234	S	D	E	N	1965	36525
1258	S	D	E	N	1965	36527
1258	S	D	E	N	1963	37193
1258	S	D	E	N	1964	37443
1258	S	D	E	N	1965	40136
1258	S	G	E	N	1964	40414
1258	S	D	E	N	1963	42295
1266	S	D	E	N	1965	34736
1266	S	D	E	N	1962	49984
1289	S	D	E	N	1965	38109
1289	S	D	E	N	1965	46810
1290	S	D	E	N	1965	38109
1290	S	D	E	N	1965	46810
1319	S	D	E	N	1963	45545
1320	S	D	E	N	1963	45545
1321	S	D	E	N	1964	36090
1321	S	D	E	N	1965	36523
1321	S	D	E	N	1963	45545
1321	P	G	E	N	1967	48233
1323	S	D	E	N	1964	36090
1323	S	D	E	N	1965	36523
1323	S	D	E	N	1963	45545
1324	S	D	E	N	1965	36523
1324	S	D	E	N	1963	45545
1325	S	D	E	N	1964	36090
1325	S	D	E	N	1963	45545
1326	S	D	E	N	1965	36523
1326	S	D	E	N	1963	45545
1326	S	D	E	N	1967	47142
1326	S	D	E	N	1965	47145
1327	S	D	E	N	1965	36523
1327	S	D	E	N	1963	45545
1328	S	D	E	N	1965	36523
1328	S	D	E	N	1963	45545
1329	S	D	E	N	1963	45545
1330	S	D	E	N	1963	45545
1331	S	D	E	N	1965	38109
1331	S	D	E	N	1963	45545
1331	S	D	E	N	1965	46810
1333	S	D	E	N	1963	45545
1373	S	D	E	N	1965	35840
1377	S	D	E	N	1964	36090
1384	S	D	E	N	1970	63134
1394	S	D	F	N	1964	38258
1417	S	D	E	N	1962	43493
1421	S	D	E	N	1960	40746
1425	S	D	E	N	1964	36090
1430	S	G	E	N	1970	61104
1440	S	D	E	N	1961	52095
1460	S	D	F	N	1966	40124
1460	S	D	E	N	1968	56848
1462	S	D	E	N	1960	48099
1463	S	D	E	N	1960	48099
1471	S	D	E	N	1965	35840
1508	S	D	E	N	1965	46109
1514	S	D	E	N	1966	34622
1520	S	G	E	U	1968	51637
1521	S	D	E	N	1967	48234
1521	S	G	E	U	1968	51637
1521	S	C	E	N	1970	59985
1535	S	D	E	N	1966	37595
1545	S	D	E	N	1962	33902
1545	S	D	E	N	1965	36404
1545	S	D	E	N	1965	36521
1545	S	D	E	N	1965	36523
1545	S	D	E	N	1965	36525
1545	S	G	E	N	1965	36526
1545	S	D	E	N	1965	36527
1545	S	D	E	N	1965	39745
1545	S	D	E	N	1964	40416
1545	S	D	E	U	1969	41919
1545	S	D	E	H	1969	41934
1545	S	D	E	U	1969	41945
1545	S	D	E	N	1965	42290
1545	S	D	E	N	1966	42914
1545	S	G	E	N	1966	43481
1545	S	D	E	N	1966	43482
1545	S	D	E	N	1965	43527
1545	S	D	E	N	1966	43530
1545	S	D	E	H	1965	44224
1545	S	D	E	N	1966	44270
1545	S	D	E	N	1966	44772
1545	S	D	E	N	1968	46990
1545	S	D	E	N	1967	47142
1545	S	D	E	N	1967	47273
1545	S	D	E	N	1968	48228
1545	S	D	E	N	1965	48230
1545	S	D	E	N	1966	48231
1545	S	G	E	N	1967	48233
1545	S	D	E	N	1967	48234
1545	S	D	E	N	1966	48374
1545	S	D	E	N	1966	48375
1545	S	D	E	N	1966	48377
1545	S	D	E	N	1966	48492

Substance Number	Phys. State	Subject	Language	Temperature	Year	TPRC Number
551-1545	S	D	E	N	1965	49030
1545	S	D	E	N	1968	49665
1545	S	D	E	N	1964	50030
1545	S	D	E	N	1968	53089
1545	S	D	E	N	1969	55070
1545	S	D	E	N	1969	55085
1545	S	D	E	N	1969	57227
1571	S	S	E	N	1965	34736
1571	S	D	E	N	1965	36504
1571	S	E	E	N	1965	36810
1571	S	D	E	N	1966	48362
1571	S	S	E	H	1960	49983
1571	S	S	E	N	1959	52098
1571	S	T	E	U	1969	53985
1571	S	T	E	N	1969	54038
1571	S	S	E	U	1968	54883
1571	S	S	E	N	1969	57666
1571	S	D	E	N	1970	57796
1571	S	S	E	U	1970	59133
1571	S	T	E	N	1970	61718
1572	S	D	E	N	1968	49568
1574	S	D	E	N	1965	36504
1575	S	D	E	N	1965	36504
1575	S	D	E	N	1970	37023
1591	S	D	E	H	1962	49984
1601	S	D	E	N	1962	43493
1606	S	D	E	N	1965	46810
1631	S	D	E	N	1960	48099
1659	S	D	E	N	1966	48378
1661	S	D	E	N	1965	35840
1689	S	D	E	N	1969	41421
1689	S	D	E	U	1969	41919
1689	S	D	E	H	1969	41934
1689	S	D	E	U	1969	41945
1689	S	D	E	N	1967	52151
1689	S	D	E	N	1969	57227
1689	S	D	E	N	1969	57597
1689	S	D	E	N	1970	57854
1689	L	D	E	N	1965	42814
1716	S	D	R	N	1964	38805
1716	S	D	E	N	1964	38806
1730	S	D	E	N	1965	39263
1730	S	D	E	N	1964	44415
1730	S	D	E	N	1963	45698
1739	S	D	E	N	1966	44771
1768	S	D	E	N	1964	37040
1768	S	D	E	N	1962	54959
1789	S	D	E	N	1967	52151
1796	S	D	E	N	1969	41421
1796	S	D	E	N	1967	52151
1799	S	D	E	N	1963	36112
1799	S	D	E	N	1964	37489
1802	S	D	E	N	1963	36112
1805	S	D	E	N	1963	36112
1805	S	D	E	N	1964	37489
1805	S	D	E	N	1965	39745
1805	S	D	E	N	1966	44772
1805	S	D	E	N	1967	47142
1811	S	D	E	N	1965	36523
1811	S	D	E	N	1962	41123
1811	S	D	E	N	1969	41573
1811	S	D	E	U	1969	41945
1811	S	D	E	N	1966	48231
1811	S	D	E	N	1966	48232
1858	S	D	E	N	1966	34449
1858	S	D	E	N	1966	34454
1858	S	D	E	N	1963	34830
1858	S	D	E	U	1965	35265
1858	S	D	F	N	1965	35934
1858	S	D	E	N	1965	36504
1858	S	D	E	N	1965	36515
1858	S	D	E	N	1966	39013
1858	S	D	E	N	1965	39754
1858	S	D	E	N	1966	40343
1858	S	D	E	U	1969	41919
1858	S	D	E	N	1966	43484
1858	S	D	E	N	1965	43527
1858	S	D	E	H	1965	44224
1858	S	D	E	N	1966	44771
1858	S	D	E	N	1969	46420
1858	S	D	R	N	1968	50016
1858	S	D	E	N	1968	52057
1858	S	D	E	N	1967	52151
1858	S	D	E	U	1968	53085
1858	S	D	E	N	1968	53087
1858	S	D	E	N	1969	56347
1858	S	D	E	N	1969	57227
1858	S	C	E	N	1970	57793
1858	S	D	G	N	1969	59668
1858	L	D	E	N	1965	42814
1861	S	D	E	N	1951	55224
1884	S	D	E	N	1961	52095
1890	S	D	E	U	1969	41919
1908	S	D	E	N	1962	41123
1908	S	G	E	U	1968	51637
1917	S	D	E	N	1964	34333
1917	S	D	E	N	1965	36527
1917	S	D	E	N	1965	36528
1917	S	D	E	N	1966	36740
1917	S	D	E	N	1964	40416
1917	S	D	E	N	1964	40417

Phys. State: **G.** Gas; **L.** Liquid; **M.** Multiphase; **P.** Powder; **S.** Solid

B41 Absorptance

Substance Number	Phys. State	Subject	Language	Temperature	Year	TPRC Number
551-1917	S	D	E	N	1966	48377
1938	S	D	E	N	1969	35545
1938	S	D	E	N	1965	35818
1938	S	D	E	N	1964	35819
1938	S	D	E	N	1965	36683
1938	S	D	E	N	1965	40303
1938	S	D	E	N	1964	43423
1938	S	C	E	N	1970	57793
1939	S	D	E	N	1966	48492
1943	S	D	E	N	1965	34736
1943	S	D	F	N	1965	35934
1943	S	D	E	N	1964	37040
1943	S	D	E	N	1965	40136
1943	S	D	E	N	1966	42505
1943	S	D	E	N	1966	44771
1943	S	D	E	U	1967	45665
1943	S	D	E	N	1965	48230
1943	S	D	E	N	1962	49984
1943	S	D	E	U	1967	54978
1943	S	D	E	N	1969	57597
1944	S	D	E	N	1965	39263
1944	S	D	E	N	1959	46025
1953	S	D	E	N	1965	36504
1953	S	D	E	N	1965	39754
1953	S	C	G	N	1967	43384
1953	S	D	E	N	1965	43413
1953	S	D	E	H	1962	49984
2020	S	D	E	N	1966	44771
2020	S	D	E	N	1968	49665
2020	L	D	E	N	1965	42814
2031	S	D	E	N	1968	53491
2071	S	D	F	N	1961	52095
2071	S	D	E	N	1961	52784
2071	S	D	E	N	1969	57597
2129	S	D	E	N	1968	49568
2132	S	D	E	N	1966	44272
2132	S	D	E	N	1965	52152
2132	S	D	E	N	1969	55070
2147	S	D	E	N	1965	47145
2147	S	D	E	N	1969	55070
2147	S	D	G	N	1969	59668
2169	S	C	E	N	1966	36893
2169	S	C	R	N	1966	42262
2193	S	D	E	N	1966	44771
2217	S	D	E	L	1965	36535
2217	S	G	E	N	1968	56536
2240	S	D	E	N	1964	37840
2240	S	D	E	N	1964	42291
2247	S	D	E	N	1966	40159
2266	S	D	E	N	1969	37388
2266	S	D	E	N	1965	39263
2266	S	D	E	N	1967	49229
2266	S	D	E	N	1968	53087
2280	S	D	R	N	1966	39236
2280	S	D	R	N	1962	40744
2280	S	D	G	N	1965	40745
2280	S	D	E	N	1966	41003
2283	S	D	E	N	1966	40583
2284	S	D	E	N	1966	40583
2285	S	D	E	N	1966	40583
2288	S	D	E	N	1967	47142
2291	S	D	E	N	1967	47142
2298	S	D	E	N	1969	57597
2305	S	D	E	N	1961	52095
2315	S	D	E	N	1969	55085
2316	S	D	E	N	1963	45545
2322	S	D	E	N	1968	48228
2331	S	D	E	U	1967	45665
2332	S	D	E	U	1967	45665
2333	S	D	E	U	1967	45665
2334	S	D	E	U	1967	45665
2337	S	D	E	N	1965	36418
2337	S	D	E	N	1964	37489
2337	S	D	E	N	1966	44772
2337	S	D	E	N	1967	47142
2339	S	D	E	N	1964	37320
2345	S	D	E	N	1966	44771
2347	S	D	E	N	1966	44771
2347	S	D	E	N	1961	52095
2347	S	D	E	N	1961	52784
2369	S	D	E	N	1968	56848
2369	S	D	E	N	1969	57413
2372	S	D	E	N	1965	35946
2384	S	D	E	N	1964	40756
2387	S	D	E	N	1964	40756
2389	S	D	E	N	1964	40756
2390	S	D	E	N	1964	40756
2391	S	D	E	N	1964	40756
2392	S	D	E	N	1960	40746
2394	S	D	E	N	1960	40746
2406	S	D	E	N	1963	40467
2411	S	D	E	N	1966	34622
2412	S	D	E	N	1963	40467
2414	S	D	E	N	1963	40467
2415	S	D	E	N	1965	39754
2415	S	D	E	N	1963	40467
2416	S	D	E	N	1965	36504
2416	S	G	E	N	1965	38649
2416	S	D	E	N	1965	39754
2416	S	D	E	N	1966	40343
2416	S	D	E	N	1960	40746
551-2416	S	D	E	U	1969	41919
2416	S	C	G	N	1967	43384
2416	S	D	E	N	1965	43413
2416	S	D	E	N	1966	43484
2416	S	D	E	N	1966	44771
2416	S	D	E	N	1967	47275
2416	S	D	E	N	1965	47412
2416	S	D	E	N	1968	49666
2416	S	D	E	N	1968	53086
2416	S	D	G	N	1969	59668
2421	S	D	E	N	1966	40351
2436	S	D	E	U	1968	53085
2438	S	D	E	N	1966	33989
2441	S	D	E	N	1966	40351
2444	S	D	E	N	1966	40351
2445	S	D	E	N	1968	48933
2447	S	D	E	N	1966	40351
2454	S	D	E	N	1966	40351
2490	S	D	E	N	1966	34454
2491	S	D	E	N	1965	39754
2492	S	D	E	N	1965	39754
2492	S	D	E	N	1969	57597
2493	S	D	E	N	1965	39754
2507	S	D	E	N	1965	39754
2508	S	D	E	N	1967	52151
2524	S	G	E	N	1964	34339
2524	S	D	F	N	1967	42312
2547	S	D	E	N	1966	44771
2566	S	S	E	N	1964	36423
2566	S	S	E	U	1965	36522
2566	S	T	E	N	1965	36524
2566	S	S	E	U	1965	[illegible]
2566	S	S	E	U	1965	36536
2566	S	D	E	N	1963	41266
2566	S	S	E	N	1963	45545
2566	S	E	E	N	1966	48373
2566	S	S	E	N	1967	54175
2566	S	D	E	N	1962	55198
2566	S	S	E	U	1970	59975
2566	M	S	E	N	1967	43123
2574	S	D	E	N	1964	37489
2580	S	D	E	N	1964	37489
2582	S	D	E	N	1964	37489
2589	S	C	E	N	1968	56536
2589	G	T	E	U	1967	50321
2600	S	D	E	N	1965	34051
2600	S	D	F	N	1965	35934
2600	S	D	E	N	1968	53087
2609	S	D	E	N	1965	43840
2610	S	D	E	N	1965	43840
2611	S	D	E	N	1966	43530
2611	S	D	E	N	1966	44772
2611	S	D	E	N	1965	47145
2611	S	D	E	N	1966	48372
2611	S	D	E	N	1964	50030
2612	S	D	E	N	1965	43840
2613	S	D	E	N	1964	36090
2614	S	D	E	N	1963	37859
2614	S	D	R	N	1963	37860
2614	S	D	E	N	1970	57246
2621	S	D	E	N	1961	41149
2623	S	D	E	N	1965	46109
2624	S	D	E	N	1965	46109
2627	S	G	E	N	1964	34339
2630	S	D	E	N	1963	36219
2631	S	D	E	N	1965	39263
2640	S	D	E	N	1961	52095
2662	S	D	E	N	1964	36090
2677	S	D	E	N	1965	39754
2677	S	D	E	N	1965	40136
2677	S	D	E	N	1961	52095
2677	S	D	E	N	1961	52153
2677	S	D	E	N	1961	52784
2678	S	D	E	N	1961	52095
2684	S	D	E	N	1965	47145
2717	S	D	E	N	1965	36504
2717	S	D	E	N	1966	42505
2721	S	D	E	N	1965	36525
2736	S	D	E	N	1965	34736
2736	S	D	E	N	1962	49984
2739	S	D	E	N	1965	34736
2739	S	D	E	N	1965	39263
2739	S	D	E	N	1962	49984
2749	P	D	R	N	1967	43797
2749	P	D	E	N	1967	43798
2751	S	D	E	N	1965	34736
2751	S	D	E	N	1960	35546
2751	S	D	F	N	1965	35934
2751	S	D	E	N	1968	36046
2751	S	D	E	N	1965	36523
2751	S	G	E	N	1965	36526
2751	S	D	E	N	1965	36527
2751	S	D	E	N	1965	36528
2751	S	D	E	N	1964	37040
2751	S	D	E	N	1964	37489
2751	S	D	E	N	1966	39363
2751	S	D	E	N	1964	40358
2751	S	D	E	N	1964	40416
2751	S	D	E	N	1964	40417
2751	S	D	E	N	1966	42505
2751	S	D	E	N	1966	43482
551-2751	S	D	E	N	1966	43530
2751	S	D	E	N	1966	44772
2751	S	D	E	N	1967	47142
2751	S	D	E	N	1965	47145
2751	S	D	E	N	1967	47756
2751	S	D	E	N	1965	48230
2751	S	D	E	N	1967	48234
2751	S	D	E	N	1966	48372
2751	S	D	E	N	1968	49665
2751	S	D	E	N	1962	49984
2751	S	D	E	N	1965	52152
2751	S	D	E	N	1968	54303
2751	S	D	E	N	1969	55070
2751	S	D	E	N	1969	57227
2751	S	D	G	N	1969	59668
2754	S	D	E	N	1965	36418
2756	S	D	E	N	1965	36418
2757	S	D	E	N	1965	36418
2760	S	D	E	N	1965	36418
2764	S	D	E	N	1963	34830
2764	S	D	E	U	1965	35265
2764	S	D	E	N	1965	36515
2766	S	D	E	N	1964	34819
2767	S	D	E	N	1965	36418
2768	S	D	E	N	1964	34819
2768	S	D	E	N	1965	42290
2768	S	D	E	N	1966	48492
2769	S	D	E	N	1965	36418
2770	S	D	E	N	1965	36418
2772	S	D	E	N	1965	36470
2772	S	D	E	U	1968	53085
2772	S	D	E	N	1968	53087
2772	S	D	E	N	1969	57597
2786	S	D	E	N	1965	36470
2837	S	D	E	N	1963	36112
2838	S	D	E	N	1963	36112
2840	S	D	E	N	1966	44771
2841	S	D	E	N	1966	44771
2844	S	D	E	N	1963	36112
2845	S	D	E	N	1963	36112
2846	S	D	E	N	1963	36112
2863	S	D	E	N	1964	36090
2890	S	D	E	N	1969	35545
2894	S	D	E	N	1969	41573
2904	S	C	G	N	1967	43384
2909	S	D	E	N	1969	41421
2911	S	D	E	N	1967	52151
2912	S	D	E	N	1967	52151
2914	S	D	E	N	1967	52151
2915	S	D	E	N	1967	52151
2916	S	D	E	N	1967	52151
2917	S	D	E	N	1967	52151
2918	S	D	E	N	1967	52151
2919	S	D	E	N	1967	52151
2920	S	D	E	N	1967	52151
2921	S	D	E	N	1961	52095
2922	S	D	E	N	1961	52095
2952	S	C	E	N	1970	59985
2971	S	D	E	N	1965	34051
2971	S	D	E	N	1968	53087
2973	S	D	E	N	1969	51456
2977	S	D	F	N	1968	48122
2980	S	D	E	N	1968	48933
2988	S	D	E	N	1962	53496
2991	S	D	E	N	1964	36090
2991	S	D	E	N	1963	45545
2993	S	D	E	N	1968	53087
2998	S	D	E	N	1970	59115
3003	S	D	E	N	1968	56848
3013	S	D	E	N	1970	56865
3017	S	D	E	N	1965	36523
3018	S	D	E	N	1965	36523
3019	S	D	E	N	1965	36523
3020	S	D	E	N	1965	36523
3021	S	D	E	N	1965	36523
3022	S	D	E	N	1965	36523
3023	S	D	E	N	1965	36523
3024	S	D	E	N	1965	36523
3025	S	D	E	N	1965	36523
3027	S	D	E	N	1965	36523
3028	S	D	E	N	1965	36523
3029	S	D	E	N	1965	36523
3030	S	D	E	N	1965	36523
3031	S	D	E	N	1965	36523
3032	S	D	E	N	1965	36523
3033	S	D	E	N	1965	36523
3034	S	D	E	N	1965	36523
3035	S	D	E	N	1965	36523
3035	S	D	G	N	1969	59668
3036	S	D	E	N	1965	36523
3037	S	D	E	N	1960	35546
3037	S	D	E	N	1965	36523
3037	S	D	E	N	1959	46025
3038	S	D	E	N	1965	36523
3040	S	D	E	N	1969	35545
3040	S	D	E	N	1965	36523
3040	S	D	E	N	1966	39363
3040	S	D	E	N	1964	40358
3040	S	D	E	N	1964	40756
3040	S	D	E	N	1969	41573
3040	S	D	E	N	1966	48372

Phys. State: **G.** Gas; **L.** Liquid; **M.** Multiphase; **P.** Powder; **S.** Solid

Substance Number	Phys. State	Sub- ject	Lan- guage	Temper- ature	Year	TPRC Number
551-3041	S	D	E	N	1965	34736
3041	S	D	E	N	1961	46640
3041	S	D	E	N	1962	49984
3044	S	D	E	N	1968	48185
3054	S	D	E	N	1965	39754
3054	S	D	E	N	1966	42505
3054	S	D	E	N	1965	43527
3054	S	D	E	H	1965	47231
3054	S	G	E	U	1968	51637
3054	S	D	E	U	1967	54978
3054	S	D	E	N	1969	57227
3056	S	D	E	N	1965	39745
3075	S	D	E	N	1966	43530
3079	S	D	E	N	1966	39363
3079	S	D	E	N	1966	48372
3080	S	D	E	N	1966	43530
3080	S	D	E	N	1965	47145
3080	S	D	E	N	1966	48372
3080	S	D	E	N	1966	48377
3082	S	D	E	N	1968	48933
3084	S	D	E	N	1968	48933
3085	S	D	E	N	1968	48933
3086	S	D	E	N	1965	39263
3087	S	D	E	N	1968	48933
3088	S	D	E	N	1968	48933
3089	S	D	E	N	1968	48933
3090	S	D	E	N	1965	39263
3091	S	D	E	N	1965	39263
3092	S	D	E	N	1966	48375
3093	S	D	E	N	1965	39263
3094	S	D	E	N	1966	48377
3095	S	D	E	N	1965	39263
3096	S	D	E	N	1962	41123
3097	S	D	E	N	1965	39262
3107	S	D	E	N	1965	39745
3138	S	D	E	N	1964	50030
3147	S	D	E	N	1966	43530
3148	S	D	E	N	1964	50030
3149	S	D	E	N	1964	40756
3149	S	D	E	N	1965	49030
3151	S	D	E	N	1968	48926
3153	S	D	E	N	1966	40333
3155	S	D	E	N	1964	40358
3161	S	D	E	N	1962	41123
3163	S	D	E	N	1965	42290
3164	S	D	E	N	1965	42290
3165	S	D	E	N	1962	41123
3166	S	D	E	N	1965	42290
3166	S	D	E	N	1966	48492
3166	S	D	E	N	1968	53089
3167	S	D	E	N	1965	42290
3167	S	D	E	N	1968	53089
3168	S	D	E	N	1966	48231
3168	S	D	E	N	1966	48492
3169	S	D	E	N	1966	48231
3169	S	D	E	N	1966	48492
3170	S	D	E	N	1966	48231
3170	S	D	E	N	1966	48492
3171	S	D	E	N	1966	48492
3172	S	D	E	N	1962	41123
3173	S	D	E	N	1962	41123
3174	S	D	E	N	1962	41123
3175	S	D	E	N	1968	40103
3175	S	D	E	N	1967	48234
3176	P	D	E	N	1966	48232
3177	S	D	E	N	1962	41123
3178	S	D	E	N	1965	48230
3179	S	D	E	N	1965	48230
3180	S	D	E	N	1962	41123
3181	S	D	E	N	1965	48230
3182	S	D	E	N	1962	41123
3182	S	D	E	N	1961	41149
3184	S	D	E	N	1962	41123
3188	S	D	E	N	1962	41123
3189	S	D	E	N	1966	48492
3192	S	D	E	N	1962	41123
3193	P	G	E	N	1967	48233
3196	S	D	E	N	1964	40756
3196	P	D	E	N	1966	48232
3200	S	D	E	N	1965	42290
3200	S	D	E	N	1965	48230
3200	S	G	E	N	1967	48233
3200	S	D	E	N	1967	48234
3201	S	D	E	H	1969	41934
3201	S	D	E	U	1969	41945
3201	S	D	E	N	1967	48234
3202	S	D	E	N	1962	41123
3203	S	D	E	N	1968	40103
3204	S	D	E	N	1962	41123
3206	S	D	E	N	1962	41123
3208	S	D	E	N	1962	41123
3209	S	D	E	N	1962	41123
3209	S	D	F	N	1966	43530
3211	S	D	E	N	1961	41149
3212	S	D	E	N	1961	41149
3213	S	D	E	N	1966	44942
3216	S	D	E	N	1961	41149
3216	S	D	E	N	1965	47145
3217	S	D	E	N	1961	41149
3218	S	D	E	N	1968	55175
3219	S	D	E	N	1961	41149

Substance Number	Phys. State	Sub- ject	Lan- guage	Temper- ature	Year	TPRC Number
551-3231	S	D	E	N	1961	41149
3232	S	D	E	N	1961	41149
3233	S	D	E	N	1961	41149
3241	S	D	E	N	1961	41149
3242	S	D	E	N	1961	41149
3243	S	D	E	N	1961	41149
3244	S	D	E	N	1961	41149
3245	S	D	E	N	1961	41149
3247	S	D	E	N	1969	41421
3249	S	D	E	U	1967	45665
3250	S	D	E	U	1967	45665
3251	S	D	E	U	1967	45665
3253	S	D	E	U	1967	45665
3256	S	D	E	N	1968	49251
3258	S	D	E	N	1967	49229
3258	S	D	E	N	1969	55080
3260	S	D	E	N	1968	49665
3261	S	D	E	N	1968	49665
3266	S	D	E	L	1970	58803
3270	S	D	F	N	1965	35934
3270	P	D	E	N	1966	40583
3277	S	D	E	N	1969	57597
3281	S	D	E	N	1969	57597
3290	S	D	E	N	1969	57597
3296	S	D	E	N	1964	36113
3296	S	D	E	U	1965	36514
3296	S	D	E	N	1965	39754
3297	S	D	E	U	1970	37845
3316	S	D	E	N	1969	41573
3323	S	D	E	N	1962	43493
3324	S	D	E	N	1962	43493
3329	S	D	R	N	1968	50016
3334	S	D	E	N	1961	52784
3341	S	D	E	N	1970	60268
3348	S	C	E	N	1969	49590
3349	S	D	E	N	1962	54959
3350	S	D	E	N	1962	54959
3352	S	D	E	N	1962	47962
3353	S	D	E	N	1962	47962
3362	S	D	E	N	1961	52784
3363	S	D	E	N	1961	52784
3378	S	D	R	N	1970	62859
3380	S	D	E	N	1970	63134
3382	S	C	E	N	1966	36893
3382	S	C	R	N	1966	42262
3424	S	S	E	N	1964	36423
3451	S	D	E	N	1966	37595
3452	S	D	E	N	1966	37595
3454	S	D	E	N	1966	37595
3455	S	D	E	N	1966	37595
3456	S	D	E	N	1 66	37595
3457	S	D	E	N	1966	37595
3458	S	D	E	N	1966	37595
3459	S	D	E	N	1966	37595
3460	S	D	E	N	1966	37595
3461	S	D	E	N	1966	37595
3461	S	D	E	N	1966	44771
3462	S	D	E	N	1966	37595
3463	S	D	E	N	1966	37595
3466	S	D	E	N	1966	37595
3474	S	D	E	N	1960	38317
3509	S	D	E	N	1969	52293
621-0002	S	D	R	N	1966	41097
0002	S	D	E	N	1966	41098
0003	M	D	E	N	1970	57854
0028	S	D	E	N	1967	49229
0040	S	G	R	N	1958	63386
0040	S	G	E	N	1970	63887
0083	S	D	E	N	1970	63038
0188	S	D	E	N	1970	59835
0217	S	E	E	N	1964	38247
0275	S	S	E	U	1970	59133
651-0467	S	S	E	N	1969	56355
661-0042	S	S	E	N	1966	52909
0048	S	D	E	N	1968	49251
0048	S	G	R	N	1958	63386
0048	S	G	E	N	1970	63887
0075	S	D	R	N	1969	58510

Phys. State: G. Gas; L. Liquid; M. Multiphase; P. Powder; S. Solid

Chapter 10 Transmittance

Substance Number	Phys. State	Subject	Language	Temperature	Year	TPRC Number
528-0095	S	G	E	N	1967	40918
0106	S	T	F	N	1966	39539
0253	S	C	E	N	1967	49124
0253	S	C	E	N	1967	49418
0253	S	D	E	N	1970	60817
0253	S	D	E	N	1971	60818
0254	S	C	R	N	1967	49124
0254	S	C	E	N	1967	49418
0256	S	T	G	U	1965	43024
0345	S	D	E	N	1965	41122
0394	S	D	E	N	1965	39750
0395	S	D	E	N	1965	39750
0396	S	D	E	N	1965	39750
531-0122	S	D	E	N	1967	43820
0125	S	D	E	N	1966	40523
0135	S	D	E	N	1966	40530
0156	S	D	E	U	1970	40641
0312	M	D	E	N	1966	33925
0315	M	D	E	N	1966	33925
0320	M	D	E	N	1966	33925
0324	M	D	E	N	1966	33925
0328	S	G	E	N	1966	34627
0328	S	D	E	N	1964	37054
0328	S	D	E	N	1963	40283
0328	S	D	E	N	1970	44256
0328	S	D	R	N	1965	46685
0328	S	D	E	N	1965	46696
0328	S	D	E	N	1964	47973
0328	S	D	E	N	1970	58660
0328	S	D	E	N	1970	59835
0329	M	D	E	N	1966	33925
0344	M	D	E	N	1966	33925
0346	M	D	E	N	1966	33925
0438	P	T	E	N	1965	38827
0458	S	D	E	U	1966	39178
0461	S	D	E	N	1970	43840
0505	M	D	E	N	1966	33965
0512	S	D	E	N	1966	33965
0558	M	D	E	N	1965	43840
0564	M	D	E	N	1965	43840
0567	M	D	E	N	1965	43840
0590	S	G	E	N	1965	36758
0591	M	D	E	N	1965	43840
0629	S	C	R	U	1965	46683
0629	S	C	E	U	1966	46684
0633	S	T	E	N	1969	52925
0633	S	D	E	N	1969	52926
0633	S	T	R	U	1970	60823
0633	S	T	E	U	1970	60824
0696	S	D	E	N	1964	38213
0696	S	D	E	N	1962	55198
0714	S	T	E	U	1968	47049
0822	M	D	E	N	1967	43179
0823	M	D	E	N	1967	43179
0831	S	D	E	N	1965	34485
0843	S	D	E	N	1968	36811
0843	S	D	E	N	1966	52850
0843	S	D	E	N	1969	56009
0883	S	D	R	N	1967	39136
0883	S	D	E	N	1967	39137
0909	S	D	G	N	1952	43597
0909	S	D	E	N	1968	49289
0928	S	C	E	N	1965	43415
0991	S	D	E	N	1968	42204
1026	M	D	E	N	1968	49289
1027	S	C	E	N	1968	42185
1027	M	C	E	N	1968	50190
1030	S	D	E	N	1968	49882
1055	S	D	E	N	1968	39750
1055	S	D	E	N	1964	40324
1158	S	T	E	U	1959	47972
1160	S	D	E	N	1967	47038
1254	S	D	E	N	1966	40903
1254	S	D	E	N	1966	40904
1254	S	D	E	N	1965	40905
1254	S	D	E	N	1966	40908
1254	S	D	E	N	1966	40909
1259	S	D	E	N	1969	52319
1260	S	D	E	N	1969	52319
1261	S	D	E	N	1969	52319
1290	S	D	E	N	1966	40764
1314	M	D	G	N	1960	54085
1324	S	D	E	N	1960	40738
1325	S	D	E	N	1960	40738
1379	S	D	E	N	1969	55350
1380	S	D	E	N	1969	55550
1381	S	D	E	N	1964	36084
1414	S	D	E	N	1969	42114
1417	S	D	E	N	1969	42109
1418	S	D	E	N	1969	42109
1419	S	D	E	N	1969	42109
1594	S	D	E	N	1964	35892
1595	S	D	E	N	1964	35892
1596	S	D	E	N	1964	35892
531-1597	S	D	E	N	1964	35892
1598	S	D	E	N	1964	35892
1599	S	D	E	N	1964	35892
1600	S	D	E	N	1964	35892
1619	S	D	E	N	1963	39738
1619	S	D	R	N	1968	55471
1636	S	D	E	N	1970	59319
1649	M	C	E	N	1970	35888
1674	M	D	E	N	1970	59738
1674	M	D	E	N	1970	59739
1675	M	D	R	N	1970	59738
1675	M	D	E	N	1970	59739
1676	M	D	R	N	1970	59740
1676	M	D	E	N	1970	59741
1677	M	D	E	N	1970	59740
1677	M	D	R	N	1970	59741
1678	S	D	E	N	1970	59833
1678	S	C	R	N	1970	61323
1678	S	C	E	N	1970	61324
1797	S	D	E	N	1967	53123
1799	S	D	R	N	1970	58767
541-0209	S	G	E	N	1967	40918
0211	S	G	E	N	1967	40918
0449	S	D	E	N	1967	45666
0449	S	D	E	N	1969	53828
0620	S	D	E	N	1967	36371
0621	S	D	E	N	1967	36371
0622	S	D	E	N	1967	36371
0664	M	D	E	N	1968	47908
1011	S	D	E	N	1963	48776
1019	S	D	E	N	1969	52319
1072	S	G	E	N	1969	53829
1128	S	D	E	N	1969	38042
1128	S	D	E	N	1966	52850
1154	S	D	E	N	1969	37884
1159	S	D	E	N	1966	52900
1164	S	D	E	N	1968	53509
1165	S	D	E	N	1968	53509
1169	S	D	E	N	1968	53509
1216	M	E	E	N	1970	60173
1238	M	D	E	N	1970	60173
1266	M	D	E	N	1966	39620
1287	M	D	E	N	1970	58301
1288	S	D	E	N	1970	58301
1289	S	D	E	N	1970	58301
1311	S	D	E	N	1970	58562
1317	S	C	E	N	1970	59520
1317	S	C	E	N	1970	59521
1318	S	C	E	N	1970	59521
1319	S	C	E	N	1970	59521
1333	M	D	R	N	1970	59740
1333	M	D	E	N	1970	59741
1334	M	D	R	N	1970	59740
1334	M	D	E	N	1970	59741
1335	M	D	R	N	1970	59740
1335	M	D	R	N	1970	59741
1336	M	D	R	N	1970	59741
1354	S	D	E	N	1970	59085
1354	S	D	R	N	1970	59086
1355	S	D	E	N	1970	59085
1385	S	D	E	N	1970	59086
1413	S	D	R	N	1970	61438
1413	S	D	E	N	1970	61439
551-0126	S	D	E	L	1969	54848
0223	S	D	E	N	1964	34325
0223	S	D	F	N	1965	34981
0223	S	G	F	N	1964	35818
0223	S	D	F	N	1964	36878
0223	S	C	F	N	1966	36900
0223	S	C	F	N	1964	37150
0223	S	C	F	N	1969	38143
0223	S	D	F	N	1966	39016
0223	S	D	E	N	1966	39111
0223	S	D	F	N	1966	39279
0223	S	C	E	N	1957	39840
0223	S	D	E	N	1965	40303
0223	S	D	E	N	1966	42977
0223	S	D	E	N	1953	43647
0223	S	D	E	N	1953	44281
0223	S	G	F	N	1962	46817
0223	S	C	G	N	1951	46924
0223	S	D	G	N	1938	46997
0223	S	D	G	N	1965	47412
0223	S	D	E	N	1959	48098
0223	S	C	F	U	1968	51949
0223	S	D	E	N	1969	53318
0223	S	D	E	N	1969	56125
0223	S	D	E	N	1968	58406
0223	S	D	E	N	1970	59319
0223	P	D	E	N	1965	36813
0233	S	D	E	N	1965	35440
0233	S	D	E	N	1969	45437
551-0233	S	D	R	N	1968	49788
0233	S	D	E	N	1968	51769
0233	S	D	E	N	1969	62975
0233	S	D	E	N	1969	62977
0278	S	D	E	N	1968	48912
0278	S	D	E	N	1970	60756
0295	S	D	E	L	1969	54848
0298	S	D	E	N	1967	52871
0313	S	D	F	N	1964	37832
0313	S	D	F	N	1964	38140
0313	S	D	F	N	1967	41348
0325	S	T	E	U	1967	36912
0325	S	T	E	N	1967	44703
0325	S	T	E	U	1969	52830
0325	S	T	E	N	1970	59527
0325	S	T	R	N	1958	61716
0325	S	T	E	N	1970	61717
0325	S	E	F	N	1969	62979
0337	S	D	E	N	1965	35527
0337	S	D	E	N	1961	40793
0346	S	D	E	N	1969	45437
0347	S	C	E	N	1964	35417
0390	S	D	E	L	1969	54848
0436	S	D	E	N	1967	52871
0447	S	D	E	N	1968	07604
0447	S	D	R	N	1966	44373
0447	S	D	J	N	1968	54519
0447	S	D	E	N	1967	55869
0447	S	D	R	N	1966	59381
0447	S	D	E	N	1966	59382
0447	S	D	E	N	1970	59640
0463	S	D	E	L	1969	54848
0468	S	D	E	N	1963	44319
0485	S	C	G	N	1928	43153
0485	S	D	R	N	1969	58438
0487	S	D	E	N	1964	34325
0487	S	D	E	N	1965	35028
0487	S	D	E	N	1968	49666
0487	S	G	E	N	1968	59164
0487	P	G	E	N	1966	34908
0501	S	D	E	N	1967	36371
0528	L	D	E	N	1965	42976
0548	S	D	F	N	1969	38833
0548	S	D	E	N	1967	42885
0548	S	D	R	N	1968	48803
0548	S	D	E	N	1968	49363
0548	S	D	F	N	1969	53707
0577	S	D	E	N	1959	36386
0577	S	D	E	N	1965	36688
0577	S	D	E	N	1970	37043
0577	S	D	E	N	1965	39262
0577	S	D	E	N	1967	39452
0577	S	D	R	N	1966	39661
0577	S	D	E	N	1966	39662
0577	S	D	R	N	1966	40598
0577	S	D	E	N	1966	40599
0577	S	C	E	N	1966	41612
0577	S	D	E	N	1966	42737
0577	S	D	E	N	1965	43456
0577	S	D	E	N	1967	44710
0577	S	D	E	N	1961	46843
0577	S	D	E	N	1960	48007
0577	S	D	E	N	1958	48105
0577	S	D	E	N	1968	48919
0577	S	D	E	N	1968	49766
0577	S	D	E	N	1967	50856
0577	S	D	E	N	1968	50900
0577	S	D	E	N	1968	50966
0577	S	D	R	N	1968	51932
0577	S	D	E	N	1968	51944
0577	S	D	E	N	1968	51945
0577	S	D	E	N	1967	52872
0577	S	D	E	N	1969	52940
0577	S	D	E	N	1968	53509
0577	S	D	E	N	1968	54815
0577	S	D	R	N	1968	55464
0577	S	D	F	N	1968	55465
0577	S	D	R	N	1969	57171
0577	S	D	E	N	1969	57172
0577	S	D	E	N	1970	58708
0577	S	D	E	N	1970	62676
0582	S	D	E	N	1966	34487
0582	S	D	E	N	1965	35963
0582	S	D	E	N	1966	39391
0582	S	D	E	N	1968	52204
0612	S	D	E	L	1969	54848
0618	S	D	F	N	1961	38059
0620	S	C	E	N	1964	35417
0620	S	D	E	N	1966	41412
0623	S	E	F	N	1968	55320
0626	S	D	E	L	1969	54848
0631	S	D	E	N	1955	43603
0631	S	D	E	N	1953	43647
0637	S	C	E	N	1957	39840

Phys. State: G. Gas; L. Liquid; M. Multiphase; P. Powder; S. Solid

Transmittance

B44

Substance Number	Phys. State	Subject	Language	Temperature	Year	TPRC Number
551-0637	S	G	E	N	1966	39976
0643	S	D	E	N	1953	43647
0648	S	D	E	N	1959	38288
0648	S	C	E	L	1967	42867
0648	S	D	E	N	1968	48974
0648	S	D	E	L	1970	58255
0648	S	D	G	N	1932	59552
0690	S	T	E	U	1969	52929
0720	S	D	E	N	1969	54893
0723	S	D	F	N	1961	38059
0746	S	D	E	N	1969	52343
0799	S	D	E	N	1966	36802
0799	S	C	E	N	1953	41767
0799	S	D	R	N	1966	42258
0799	S	D	E	N	1967	52871
0799	P	D	R	N	1965	36352
0799	P	D	E	N	1965	36353
0826	S	D	E	H	1970	58670
0850	S	D	E	N	1965	42590
0850	S	D	R	N	1968	50444
0850	S	D	E	N	1968	50924
0850	S	G	E	N	1968	56550
0850	S	D	F	N	1970	58814
0850	S	D	R	N	1970	60598
0850	S	D	E	N	1970	60599
0850	S	D	R	N	1969	61986
0850	S	D	E	N	1969	61987
0850	P	D	R	N	1969	57365
0850	P	D	E	N	1969	57366
0854	S	D	E	N	1963	44419
0854	S	D	R	N	1968	51944
0854	S	D	E	N	1968	51945
0959	S	D	E	L	1969	52342
0988	S	D	E	N	1968	53509
0988	S	S	E	N	1970	57511
0989	S	D	R	N	1965	38765
0989	S	D	E	N	1965	38766
0989	S	D	E	N	1968	47874
0989	S	D	E	N	1968	50246
0989	S	D	E	N	1957	55043
0989	S	C	G	N	1966	57510
0989	S	D	E	N	1968	59010
1064	S	D	E	N	1967	52871
1064	S	D	E	N	1970	57578
1071	S	D	R	N	1969	58438
1076	S	D	E	N	1965	34327
1140	S	S	E	N	1967	43519
1142	S	G	E	N	1966	34928
1142	S	D	E	N	1965	35808
1142	S	G	E	N	1966	48359
1206	S	D	E	N	1967	45212
1237	S	D	E	N	1969	52343
1321	S	D	R	N	1967	45018
1321	S	D	E	N	1967	45019
1321	S	D	R	N	1967	48617
1321	S	D	E	N	1967	48618
1328	S	D	R	N	1970	61693
1328	S	D	E	N	1970	61694
1394	S	D	E	N	1968	37034
1394	S	D	E	L	1969	56458
1430	S	D	E	N	1965	36144
1430	S	D	E	N	1966	39638
1430	S	D	R	N	1966	39661
1430	S	D	E	N	1966	39662
1430	S	D	E	N	1967	40852
1430	S	D	R	N	1966	41413
1430	S	D	E	N	1953	43647
1430	S	D	F	N	1968	48179
1430	S	D	R	N	1968	55318
1430	S	D	R	N	1969	57110
1430	S	D	E	N	1969	57111
1430	S	D	E	N	1970	59641
1460	S	D	E	N	1965	36755
1460	S	D	F	N	1966	40124
1460	S	D	E	N	1968	56848
1460	S	D	E	N	1970	58623
1460	S	D	R	N	1970	60598
1460	S	D	E	N	1970	60599
1464	S	D	E	N	1966	44300
1464	S	D	E	N	1968	47416
1464	S	D	R	N	1968	50444
1464	S	D	E	N	1968	50924
1464	S	D	R	N	1970	60598
1464	S	D	E	N	1970	60599
1508	S	D	E	N	1958	38278
1508	S	G	E	N	1961	38456
1508	S	G	E	N	1950	38459
1508	S	D	E	N	1967	44931
1508	S	D	E	N	1931	47008
1508	S	G	E	N	1935	47009
1508	S	D	E	N	1969	54291
1508	S	D	E	N	1969	55534
1508	S	D	E	N	1953	60518
1511	S	G	E	N	1966	42783
1511	S	C	E	N	1968	50367
1511	S	D	E	N	1970	58659
1511	M	D	G	N	1970	59331
1514	S	D	E	N	1963	36800
1514	S	D	R	N	1965	38771
1514	S	D	E	N	1965	38772
1514	S	D	R	N	1968	52604

Substance Number	Phys. State	Subject	Language	Temperature	Year	TPRC Number
551-1514	S	D	E	N	1969	52605
1514	S	D	E	N	1961	55042
1571	S	D	E	N	1967	33824
1571	S	D	E	N	1965	34330
1571	S	T	E	U	1967	34450
1571	S	S	E	N	1965	34736
1571	S	E	E	N	1965	35527
1571	S	T	R	U	1965	36354
1571	S	T	E	U	1965	36355
1571	S	D	E	N	1967	36372
1571	S	T	F	N	1963	37633
1571	S	T	R	U	1965	38248
1571	S	T	E	U	1965	38924
1571	S	D	E	N	1966	39619
1571	S	T	R	U	1966	39653
1571	S	T	E	U	1966	39654
1571	S	D	E	N	1965	39750
1571	S	S	E	N	1955	40651
1571	S	E	E	N	1966	41612
1571	S	S	R	U	1966	41739
1571	S	T	F	U	1966	41989
1571	S	D	E	N	1967	47094
1571	S	D	E	N	1968	51492
1571	S	T	E	U	1969	53985
1571	S	C	G	N	1969	54627
1571	S	T	E	N	1970	61718
1621	S	D	F	N	1966	39717
1621	S	D	E	N	1969	55810
1666	S	D	E	N	1969	55634
1683	S	D	F	N	1967	34972
1683	S	D	E	N	1967	44016
1683	S	D	E	N	1967	44708
1683	S	D	E	N	1967	45033
1683	S	D	E	N	1968	48932
1683	S	G	E	N	1968	49973
1683	S	D	E	N	1970	59639
1684	S	D	E	N	1963	36800
1684	S	G	F	L	1968	42321
1684	S	D	E	N	1967	44016
1684	S	D	E	N	1968	50379
1689	S	D	E	N	1965	42290
1689	S	D	E	N	1967	45212
1689	S	D	E	N	1965	49030
1689	S	D	E	N	1964	49031
1716	S	D	E	N	1965	36342
1716	S	D	F	N	1966	39286
1716	S	D	R	N	1970	60598
1716	S	D	E	N	1970	60599
1722	S	D	F	N	1965	36963
1726	S	D	E	N	1954	40796
1728	S	D	E	N	1970	58664
1734	S	T	F	U	1965	36823
1767	S	G	F	N	1966	40114
1768	S	D	E	N	1965	35880
1768	S	G	F	N	1964	36678
1768	S	D	E	N	1963	36800
1768	S	D	E	N	1970	37677
1768	S	D	F	N	1963	38629
1768	S	G	E	N	1966	39439
1768	S	D	E	N	1949	47204
1768	S	D	E	N	1968	49482
1768	S	G	E	N	1968	49973
1768	S	D	R	N	1969	50054
1768	S	D	R	N	1968	50444
1768	S	D	E	N	1968	50924
1768	S	D	E	N	1968	53509
1768	S	D	E	N	1969	54327
1768	S	D	E	N	1961	55195
1768	S	D	E	N	1969	55821
1768	S	D	E	N	1970	58562
1768	S	D	R	N	1970	60598
1768	S	D	E	N	1970	60599
1802	S	D	E	N	1963	36112
1805	S	D	E	N	1963	36112
1810	S	D	E	N	1960	39894
1818	S	D	E	N	1969	56033
1858	S	D	E	N	1965	35116
1858	S	D	E	N	1965	35818
1858	S	D	E	N	1965	38512
1858	S	D	E	N	1966	39013
1858	S	D	E	N	1965	40303
1858	S	D	F	N	1965	43129
1858	S	D	G	N	1941	43657
1858	S	D	E	N	1968	47526
1858	S	D	R	N	1968	50016
1858	S	D	R	N	1968	50444
1858	S	D	E	N	1968	50924
1858	S	D	R	N	1970	60598
1858	S	D	E	N	1970	60599
1858	P	D	R	N	1969	57365
1858	P	D	E	N	1969	57366
1861	S	D	E	N	1951	55224
1890	S	D	E	N	1969	52343
1926	S	D	E	N	1964	35029
1926	S	D	E	N	1965	35311
1926	S	D	E	N	1964	39032
1926	S	D	E	L	1969	56458
1938	S	D	E	N	1966	39013
1938	S	D	E	N	1958	48105
1938	S	D	E	N	1957	48109
1953	S	D	F	N	1967	34971

Substance Number	Phys. State	Subject	Language	Temperature	Year	TPRC Number
551-1953	S	D	F	N	1965	34981
1953	S	D	E	N	1965	35963
1953	S	G	F	N	1964	36678
1953	S	D	F	N	1965	36819
1953	S	D	E	N	1966	39016
1953	S	G	E	N	1966	39976
1953	S	D	E	N	1965	43413
1953	S	D	F	N	1967	46482
1953	S	D	F	L	1965	49118
1953	S	D	E	N	1968	52204
1953	S	D	E	N	1969	62975
1990	S	D	E	L	1970	62676
2007	S	D	E	N	1967	49181
2020	S	D	E	N	1968	49665
2052	S	D	E	N	1970	62501
2130	S	S	E	N	1969	62978
2133	S	D	F	N	1969	49110
2133	S	D	F	N	1970	59143
2133	S	D	F	N	1969	62979
2136	S	D	E	N	1955	53231
2156	S	D	E	N	1963	36800
2156	S	D	F	N	1964	37153
2156	S	D	E	N	1965	38180
2156	S	D	F	N	1966	40957
2156	S	D	F	N	1965	43296
2156	S	D	E	N	1967	43439
2165	S	D	G	L	1965	36633
2165	S	C	F	N	1963	37633
2165	S	D	R	N	1966	42462
2168	S	D	E	N	1961	46843
2168	S	D	E	N	1968	56848
2170	S	D	R	N	1967	45016
2170	S	D	E	N	1967	45017
2170	S	D	R	N	1967	45018
2170	S	D	E	N	1967	45019
2207	M	D	E	N	1967	47761
2217	S	D	R	N	1970	59085
2217	S	D	E	N	1970	59086
2229	S	D	E	N	1963	36800
2247	S	D	R	N	1967	48617
2247	S	D	E	N	1967	48618
2266	S	D	E	N	1967	56447
2275	S	C	G	N	1970	59192
2281	S	C	F	N	1969	38143
2286	S	D	E	N	1965	35527
2286	S	G	E	N	1965	38673
2287	S	T	E	N	1967	44702
2293	S	D	R	N	1966	40893
2293	S	D	E	N	1966	40894
2294	S	D	R	N	1966	40893
2294	S	D	E	N	1966	40894
2295	S	D	R	N	1966	40893
2295	S	D	E	N	1966	40894
2299	S	D	R	N	1966	41095
2299	S	D	E	N	1966	41096
2300	S	D	R	N	1966	41095
2300	S	D	E	N	1966	41096
2302	S	D	R	N	1966	41095
2302	S	D	E	N	1966	41096
2308	S	D	E	N	1967	42237
2308	S	D	E	N	1970	58903
2312	S	D	R	N	1966	41075
2312	S	D	E	N	1966	41076
2313	S	D	E	N	1966	34623
2315	S	D	E	N	1965	35069
2315	S	D	E	N	1931	47008
2315	S	D	E	N	1970	60085
2315	P	G	E	N	1966	34908
2317	S	D	E	N	1965	35069
2318	S	D	E	N	1965	35069
2319	S	D	E	N	1965	35069
2320	S	D	E	N	1965	36325
2321	S	D	E	N	1965	38717
2321	S	D	E	H	1964	43741
2327	S	D	F	N	1965	36961
2330	S	D	E	N	1967	45873
2339	S	D	E	N	1969	35809
2348	S	G	E	N	1965	35113
2348	S	D	E	N	1965	36686
2348	S	D	E	N	1965	39314
2348	S	D	E	N	1965	39746
2348	S	D	E	N	1966	44340
2348	S	D	E	N	1966	46559
2351	S	D	E	L	1966	34542
2351	S	D	F	L	1965	36933
2353	S	D	E	L	1966	34542
2353	S	D	F	L	1965	36933
2353	P	D	F	N	1968	51129
2354	S	D	E	L	1966	34542
2354	S	D	F	L	1965	36933
2355	S	G	E	N	1965	35113
2355	S	D	E	N	1968	52870
2361	S	G	E	L	1967	42889
2362	S	D	O	N	1965	35099
2367	S	D	E	N	1966	34626
2367	S	D	R	N	1966	39777
2367	S	D	E	N	1966	39778
2367	S	D	E	N	1967	44016
2367	S	D	E	N	1961	46843
2367	S	G	E	N	1966	52103
2367	S	G	E	N	1967	52883

Phys. State: **G.** Gas; **L.** Liquid; **M.** Multiphase; **P.** Powder; **S.** Solid

Substance Number	Phys. State	Subject	Language	Temperature	Year	TPRC Number
551-2367	S	D	E	N	1968	53509
2367	S	C	E	N	1969	57070
2369	S	D	G	L	1965	36633
2369	S	D	E	N	1961	41599
2369	S	D	E	N	1968	56848
2369	S	C	E	N	1963	59542
2369	S	D	G	N	1932	59552
2371	S	D	G	L	1965	36633
2373	S	D	R	N	1968	34183
2373	S	D	E	N	1968	52757
2374	S	D	E	N	1965	36346
2374	S	D	R	N	1965	38765
2374	S	D	E	N	1965	38766
2374	S	D	E	N	1961	40793
2374	S	D	E	N	1968	47949
2374	S	D	R	N	1968	50444
2374	S	D	E	N	1968	50924
2374	S	D	E	N	1970	59632
2374	S	D	R	N	1970	60598
2374	S	D	E	N	1970	60599
2375	S	D	E	N	1965	36346
2375	S	D	E	N	1964	37971
2375	S	D	E	N	1961	40793
2385	S	D	F	N	1970	56733
2386	S	D	F	N	1970	54229
2393	S	D	R	N	1965	38771
2393	S	D	E	N	1965	38772
2395	S	G	G	N	1967	51605
2397	S	D	E	N	1967	47244
2397	S	D	E	N	1967	49181
2398	S	D	R	N	1965	33976
2398	S	D	E	N	1965	33977
2399	S	D	F	N	1970	57251
2400	P	D	E	N	1970	61715
2401	S	S	G	U	1966	41418
2408	S	C	E	N	1966	34621
2409	S	D	E	N	1966	34622
2409	S	D	E	N	1963	36800
2409	S	C	E	N	1968	47827
2410	S	D	E	N	1966	34622
2415	S	D	E	N	1966	49395
2416	S	D	F	N	1965	34981
2416	S	D	E	N	1950	35074
2416	S	D	E	N	1965	35818
2416	S	G	F	N	1964	36678
2416	S	D	E	N	1964	37970
2416	S	G	E	N	1965	38649
2416	S	D	E	N	1966	39016
2416	S	D	F	N	1966	39288
2416	S	D	F	N	1966	39299
2416	S	D	E	N	1963	39497
2416	S	D	E	N	1949	39837
2416	S	C	E	N	1957	39839
2416	S	D	E	N	1965	40303
2416	S	D	E	N	1966	42716
2416	S	G	E	L	1967	42889
2416	S	D	E	N	1965	43413
2416	S	D	F	N	1967	44044
2416	S	D	E	N	1948	44326
2416	S	G	E	N	1963	45735
2416	S	D	G	N	1938	46997
2416	S	D	E	N	1965	47412
2416	S	D	F	N	1960	48051
2416	S	D	E	N	1968	49666
2416	S	D	R	N	1968	50272
2416	S	D	E	N	1968	50483
2416	S	D	E	N	1969	52210
2416	S	D	E	U	1969	52328
2416	S	D	E	N	1969	53318
2416	S	D	E	N	1969	53852
2416	S	C	R	N	1969	54211
2416	S	C	E	N	1969	54212
2416	S	C	G	N	1969	54627
2416	S	C	G	N	1969	56190
2416	S	D	E	N	1970	57316
2416	S	D	E	N	1970	59319
2416	S	C	E	N	1963	59542
2416	P	D	R	N	1969	57365
2416	P	D	E	N	1969	57366
2426	S	C	G	N	1964	37015
2426	S	D	E	N	1966	39604
2426	S	D	E	N	1953	43647
2426	S	S	E	N	1966	52912
2426	S	D	E	N	1969	54041
2427	S	D	E	N	1966	39604
2429	S	D	E	N	1966	39604
2429	S	D	E	N	1953	43647
2429	S	D	R	N	1969	57110
2429	S	D	E	N	1969	57111
2438	S	D	E	L	1966	33775
2438	S	D	E	N	1966	39504
2438	S	D	F	N	1966	39942
2438	S	D	E	N	1931	47008
2439	S	D	E	N	1966	39447
2439	S	D	F	N	1967	44051
2445	S	D	E	N	1964	37736
2445	S	D	F	L	1966	39959
2448	S	D	E	L	1969	35583
2450	S	D	E	L	1966	34480
2451	S	D	E	L	1966	34480
2452	S	D	E	L	1966	34480
551-2453	S	D	E	L	1966	34480
2455	S	D	R	N	1966	36054
2455	S	D	E	N	1963	36800
2455	S	D	E	N	1964	37320
2455	S	D	E	N	1967	44016
2455	S	D	E	N	1968	50379
2455	S	D	E	N	1966	51630
2458	S	D	E	N	1963	36800
2458	S	D	F	N	1964	37153
2458	S	D	E	N	1967	44016
2458	S	S	E	N	1967	45501
2458	S	D	E	N	1967	47112
2458	S	G	E	N	1968	49973
2459	S	C	E	N	1966	34625
2463	S	D	E	N	1966	46559
2484	S	T	E	U	1966	34547
2508	S	D	E	N	1967	56447
2509	S	D	E	N	1967	55869
2523	S	D	E	N	1966	39638
2523	S	D	E	N	1969	52343
2526	S	D	E	N	1959	38288
2526	S	C	E	L	1967	42867
2526	S	D	E	L	1961	46830
2526	S	D	E	N	1968	47309
2526	S	D	E	N	1968	48974
2526	S	D	E	N	1968	53509
2526	S	D	E	N	1970	57342
2526	S	D	G	N	1932	59552
2528	S	D	E	N	1966	39481
2528	S	S	E	N	1967	43517
2528	S	D	E	N	1967	44016
2528	S	D	E	N	1968	53509
2528	S	T	E	U	1969	53985
2528	S	D	E	N	1969	55747
2528	S	C	E	N	1969	57070
2528	P	D	R	N	1969	46485
2528	P	D	E	N	1969	48513
2532	M	D	R	N	1966	39777
2532	M	D	E	N	1966	39778
2533	M	D	R	N	1966	39777
2533	M	D	E	N	1966	39778
2534	S	D	E	N	1967	55869
2536	S	D	E	N	1965	39084
2539	S	D	E	N	1966	34918
2540	S	D	E	N	1966	34918
2546	S	D	E	N	1967	55869
2551	P	D	R	N	1969	40867
2551	P	D	E	N	1969	40868
2552	S	G	E	L	1969	40362
2553	P	D	R	N	1969	40867
2553	P	D	E	N	1969	40868
2561	S	D	E	N	1966	39383
2564	S	D	R	N	1967	48617
2564	S	D	E	N	1967	48618
2565	S	D	R	N	1967	48617
2565	S	D	E	N	1967	48618
2566	S	T	E	N	1965	36524
2566	S	D	R	U	1968	49367
2566	S	D	E	U	1968	49368
2571	S	D	E	N	1966	39563
2571	S	D	E	N	1968	47767
2571	S	G	E	N	1968	49973
2571	S	D	E	N	1969	55735
2572	S	D	E	N	1966	39677
2585	S	C	E	N	1970	59711
2588	S	C	E	N	1970	59711
2594	S	S	E	N	1970	57511
2595	S	D	E	N	1966	39670
2596	S	D	E	N	1969	58322
2601	S	D	E	N	1967	56447
2614	S	D	E	N	1965	34058
2623	S	C	J	N	1968	53166
2627	S	D	E	N	1964	33944
2627	S	D	R	N	1965	34117
2627	S	D	E	N	1965	35221
2627	S	D	E	N	1967	41118
2627	S	D	E	N	1967	43686
2627	S	D	E	N	1967	44765
2627	S	S	E	N	1966	52913
2627	S	D	E	N	1966	60092
2628	S	D	E	N	1966	45921
2630	S	D	E	N	1963	36219
2630	S	D	E	N	1967	43447
2630	S	D	E	N	1969	54044
2631	S	D	E	N	1965	34344
2631	S	D	E	N	1965	35808
2631	S	D	E	N	1965	42975
2632	S	D	E	N	1966	45921
2635	S	D	E	N	1965	34344
2638	S	D	E	N	1967	43447
2638	S	D	E	N	1969	54044
2647	S	D	E	N	1967	43447
2647	S	D	E	N	1969	54044
2706	S	D	E	N	1967	38578
2715	S	D	R	N	1967	45016
2715	S	D	E	N	1967	45017
2758	S	D	E	N	1967	56447
2761	S	D	E	N	1965	35021
2762	S	D	E	N	1965	35021
2774	S	D	R	N	1969	58573
2774	S	D	E	N	1969	58574
551-2790	S	D	E	N	1967	44931
2790	S	D	F	N	1968	50554
2790	S	D	G	N	1932	59552
2801	S	D	E	L	1969	55814
2802	S	D	E	N	1959	38288
2802	S	D	E	L	[illegible]	[illegible]
2806	S	D	E	N	1970	58672
2807	S	D	G	N	1964	43767
2807	S	D	E	N	1966	43768
2808	S	D	G	N	1964	43767
2808	S	D	E	N	1966	43768
2810	S	D	E	N	1969	55810
2812	S	D	E	N	1970	58567
2824	S	C	E	N	1966	44307
2824	S	D	E	N	1969	52387
2826	S	D	E	N	1965	42590
2827	S	D	E	N	1965	42590
2829	S	D	E	N	1969	53180
2839	S	D	E	N	1967	45741
2840	S	D	E	N	1967	45741
2841	S	D	E	N	1967	45741
2842	S	D	E	N	1967	40233
2847	S	D	E	N	1969	53374
2859	S	D	E	N	1963	36800
2859	S	G	E	N	1968	49973
2884	S	D	F	N	1967	45943
2887	S	G	E	N	1967	52883
2888	S	G	E	N	1967	52883
2889	S	D	E	N	1967	52872
2891	S	D	E	N	1959	38288
2891	S	C	E	L	1967	42867
2891	S	D	E	N	1968	48974
2891	S	D	G	N	1932	59552
2892	S	D	E	N	1959	38288
2892	S	C	E	L	1967	42867
2892	S	D	E	N	1968	48974
2892	S	D	G	N	1932	59552
2893	S	D	E	N	1959	38288
2893	S	C	E	L	1967	42867
2893	S	D	E	N	1968	48974
2893	S	D	E	N	1968	53509
2893	S	D	G	N	1932	59552
2895	S	D	E	N	1959	38288
2895	S	C	E	L	1967	42867
2895	S	D	E	N	1968	48974
2895	S	D	G	N	1932	59552
2896	S	D	E	N	1959	38288
2896	S	C	E	L	1967	42867
2897	S	D	E	N	1959	38288
2897	S	C	E	L	1967	42867
2897	S	D	E	N	1968	48974
2897	S	D	G	N	1932	59552
2898	S	D	E	N	1959	38288
2898	S	D	E	N	1931	40580
2898	S	C	E	L	1967	42867
2898	S	D	E	N	1968	48974
2898	S	D	E	N	1968	48978
2898	S	D	G	N	1932	59552
2899	S	D	E	N	1967	52030
2900	S	D	E	N	1959	38288
2900	S	C	E	L	1967	42867
2901	S	D	E	N	1959	38288
2901	S	C	E	L	1967	42867
2901	S	D	E	L	1969	48001
2902	S	D	E	N	1959	38288
2902	S	C	E	L	1967	42867
2903	S	D	E	N	1959	38288
2903	S	C	E	L	1967	42867
2904	S	D	E	N	1959	38288
2904	S	C	E	L	1967	42867
2904	S	D	E	N	1969	54044
2904	S	D	G	N	1932	59552
2905	S	D	E	L	1966	33775
2905	S	D	E	N	1959	38288
2905	S	D	E	N	1970	40705
2905	S	C	E	L	1967	42867
2905	S	D	E	N	1968	48974
2905	S	C	F	L	1969	50060
2905	S	D	E	N	1968	53509
2905	S	D	E	L	1970	60029
2906	S	D	E	N	1959	38288
2906	S	C	E	L	1967	42867
2906	S	D	E	N	1968	48974
2906	S	C	F	L	1969	50060
2906	S	D	E	N	1968	53509
2906	S	D	G	N	1932	59552
2907	S	D	E	N	1959	38288
2907	S	C	E	L	1967	42867
2907	S	D	E	N	1968	48974
2907	S	C	F	L	1969	50060
2907	S	D	G	N	1932	59552
2908	S	D	E	N	1969	44528
2908	S	D	E	N	1968	47869
2908	S	S	E	N	1970	57511
2909	S	D	E	N	1969	41421
2909	S	D	E	N	1967	52151
2910	S	D	E	N	1959	38288
2910	S	C	E	L	1967	42867
2910	S	D	E	N	1968	48919
2910	S	D	E	N	1968	48974
2910	S	D	G	N	1932	59552

Phys. State: **G.** Gas; **L.** Liquid; **M.** Multiphase; **P.** Powder; **S.** Solid

Substance Number	Phys. State	Sub-ject	Lan-guage	Temper-ature	Year	TPRC Number
551-2923	S	D	E	N	1969	56041
2924	S	D	E	N	1967	35225
2925	S	D	E	N	1969	62975
2926	S	D	E	N	1967	42885
2927	S	D	E	N	1969	62976
2928	S	D	E	N	1969	56041
2930	S	D	E	N	1969	35809
2931	S	D	E	N	1969	35809
2932	S	D	E	N	1969	35809
2933	S	D	E	N	1969	35809
2934	S	D	E	N	1970	56163
2935	S	D	E	N	1970	56163
2936	S	D	E	N	1970	56163
2938	S	D	E	N	1969	56046
2939	S	D	E	N	1969	56046
2941	S	D	E	N	1967	45950
2942	S	D	E	N	1967	45950
2944	S	D	R	N	1967	43989
2944	S	D	E	N	1967	43990
2945	S	D	R	N	1967	43989
2945	S	D	E	N	1967	43990
2946	S	D	R	N	1967	43989
2946	S	D	E	N	1967	43990
2947	S	D	R	N	1967	43989
2947	S	D	E	N	1967	43990
2948	S	D	R	N	1967	43989
2948	S	D	E	N	1967	43990
2949	S	D	R	N	1967	43989
2949	S	D	E	N	1967	43990
2950	S	D	R	N	1967	43989
2950	S	D	E	N	1967	43990
2965	S	D	E	N	1958	56416
2968	S	D	E	N	1968	53509
2969	S	D	E	N	1968	53509
2970	S	D	E	N	1968	53509
2972	S	D	E	N	1968	53509
2973	S	D	G	N	1967	43383
2974	S	D	E	N	1968	53509
2975	S	D	E	N	1968	53509
2976	S	D	E	N	1968	53509
2978	S	D	E	N	1968	53509
2980	S	C	E	U	1968	47918
2980	S	D	E	N	1968	53509
2980	S	D	E	N	1970	56965
2981	S	D	E	N	1968	53509
2982	S	D	E	N	1968	53509
2983	S	D	E	N	1967	44231
2984	S	D	E	N	1968	53509
2985	S	D	E	N	1968	53509
2986	S	D	E	N	1968	48135
2986	S	D	E	N	1969	52343
2991	S	D	E	N	1965	35818
2991	S	D	F	N	1964	37830
2991	S	D	F	N	1964	43131
2994	S	D	E	N	1968	50375
2994	S	D	E	N	1969	62974
2996	S	D	E	N	1969	59686
2998	S	D	E	N	1970	59115
2999	S	D	E	N	1967	45212
3003	S	D	E	N	1968	56848
3007	S	D	E	N	1970	56864
3008	S	D	E	N	1968	56848
3013	S	D	E	N	1970	56865
3016	S	D	E	N	1970	54297
3026	S	D	E	N	1970	54297
3039	S	D	E	N	1970	54297
3048	S	D	E	N	1968	47874
3049	S	C	E	N	1966	44307
3050	S	C	E	N	1966	47874
3056	S	D	E	N	1970	54297
3057	S	D	E	N	1969	52343
3058	S	D	E	N	1968	49482
3058	S	D	E	N	1968	49963
3062	S	C	E	N	1969	57070
3068	S	D	E	N	1965	36172
3070	S	G	E	N	1966	34928
3070	S	G	E	N	1966	48359
3076	P	D	E	L	1966	48370
3077	S	D	E	N	1965	38697
3108	S	D	E	N	1967	44986
3108	S	D	E	N	1967	45212
3108	S	D	E	N	1967	56447
3145	S	D	E	N	1962	39896
3150	S	D	E	N	1968	48925
3151	S	D	E	N	1968	48926
3197	S	D	E	N	1953	43647
3205	S	D	F	N	1968	50554
3205	S	D	E	N	1968	53509
3213	S	D	E	N	1966	44942
3213	S	D	E	N	1967	45177
3213	S	D	E	N	1967	45954
3213	S	D	E	N	1968	48136
3213	S	D	E	N	1967	52872
3213	S	D	R	N	1970	60152
3213	S	D	E	N	1970	60153
3215	P	D	F	N	1968	51129
3258	S	D	E	N	1967	49229
3258	S	D	E	N	1969	55080
3259	S	D	E	N	1967	49229
3261	S	D	E	N	1968	49665
3273	S	D	E	N	1970	58969

Substance Number	Phys. State	Sub-ject	Lan-guage	Temper-ature	Year	TPRC Number
551-3273	S	D	G	N	1969	63268
3275	S	D	R	N	1966	41413
3275	S	D	R	N	1968	55318
3280	S	D	E	N	1967	49181
3288	S	D	R	N	1966	41095
3288	S	D	E	N	1966	41098
3288	S	D	E	N	1958	48105
3288	S	D	E	N	1958	48105
3289	S	D	E	N	1958	48105
3291	S	D	E	N	1958	48105
3293	S	D	E	N	1970	58903
3298	S	D	E	N	1968	51926
3300	S	D	R	N	1970	59387
3300	S	D	E	N	1970	59388
3302	S	D	E	N	1970	56965
3303	S	D	E	N	1970	56965
3304	S	D	E	N	1970	56965
3305	S	D	E	N	1970	56965
3306	S	D	E	N	1970	56965
3307	S	D	E	N	1970	56965
3307	S	D	E	N	1970	60495
3308	S	D	E	N	1970	56965
3315	S	D	G	N	1932	59552
3317	S	G	E	N	1966	42783
3319	L	D	E	N	1965	42976
3329	S	D	R	N	1968	50016
3335	S	D	E	N	1970	60031
3338	S	D	E	N	1967	52871
3343	S	D	E	N	1970	60495
3344	S	D	E	N	1970	60495
3345	S	D	E	N	1970	60495
3348	S	D	E	N	1970	60680
3355	S	D	E	N	1965	35909
3357	S	D	E	N	1969	47997
3358	S	D	E	N	1969	47997
3359	S	D	E	N	1969	47997
3360	S	D	E	N	1969	47997
3361	S	D	E	N	1969	47997
3368	S	D	E	N	1961	46843
3372	S	D	G	N	1969	63268
3373	S	D	G	N	1969	63268
3377	S	D	R	N	1969	61199
3379	S	D	R	N	1966	63236
3379	S	D	E	N	1966	63237
3382	S	C	E	N	1966	36893
3382	S	C	R	N	1966	42262
3392	S	D	R	N	1968	52643
3392	S	D	E	N	1968	52644
3393	S	D	R	N	1968	52643
3393	S	D	E	N	1968	52644
3394	S	D	R	N	1968	52643
3394	S	D	E	N	1968	52644
3412	S	S	E	H	1958	36385
3424	S	S	E	N	1964	36423
3481	S	D	E	N	1966	44271
3501	S	D	E	N	1969	52343
3503	S	D	E	N	1969	52343
3504	S	D	E	N	1969	52343
3505	S	D	E	N	1969	52343
3506	P	D	E	N	1969	59686
3507	S	D	E	N	1969	52343
3508	S	D	E	N	1969	52329
3510	S	D	E	N	1968	49882
621-0023	S	D	E	N	1969	55534
0029	S	D	E	N	1966	49395
0040	S	G	R	N	1958	63386
0040	S	G	E	N	1970	63887
0043	M	D	G	N	1960	54085
0083	S	D	E	N	1970	63038
0085	S	D	G	N	1898	46995
0096	S	D	E	N	1966	49395
0097	S	D	E	N	1966	49395
0112	S	D	E	N	1969	54291
0114	S	D	E	N	1969	54291
0115	S	D	E	N	1969	54291
0188	S	C	E	N	1970	59520
0188	S	C	E	N	1970	59521
0188	S	D	E	N	1970	59835
0214	S	D	E	N	1965	50496
0214	S	D	E	N	1969	54898
0217	S	E	E	N	1964	38247
0275	S	D	E	U	1970	59133
0308	S	D	E	N	1968	53086
651-0066	S	D	E	N	1931	47008
0068	S	D	E	N	1961	54099
0070	S	D	E	N	1940	40581
0071	S	D	E	N	1940	40581
0257	S	D	E	N	1966	41030
0272	S	D	E	N	1966	41030
0273	S	D	E	N	1966	41030
0274	S	D	E	N	1966	41030
0318	S	D	E	N	1962	56354
0322	S	D	E	N	1966	41030
0323	S	D	E	N	1966	41030
0324	S	D	E	N	1966	41030
0416	S	D	E	N	1966	41030
0476	S	D	E	N	1950	40784
0476	S	D	E	N	1955	40789
0476	S	D	E	N	1955	40790

Substance Number	Phys. State	Sub-ject	Lan-guage	Temper-ature	Year	TPRC Number
651-0476	S	D	G	N	1898	46995
661-0040	S	D	E	H	1959	56353
0048	S	G	R	N	1958	63386
0048	S	G	E	N	1970	63387
0050	S	D	E	N	1966	40530
0050	S	D	E	N	1955	40789
0050	S	D	E	N	1967	44551
0060	S	D	E	N	1955	40789
0067	S	D	E	N	1931	40580
0067	S	D	E	N	1932	59587
0171	S	D	G	N	1960	54085
0415	S	D	E	N	1967	44551
0426	S	D	E	N	1967	44551
0432	S	D	E	N	1967	44551
0438	S	D	E	N	1967	44551
0439	S	D	E	N	1967	44551
0440	S	D	E	N	1967	44551
0441	S	D	E	N	1967	44551

Phys. State: G. Gas; L. Liquid; M. Multiphase; P. Powder; S. Solid

Chapter 11 Absorptance To Emittance Ratio

Substance Number	Phys. State	Subject	Language	Temperature	Year	TPRC Number
531-0056	S	D	E	N	1967	45553
0301	S	D	E	N	1967	45553
1454	S	D	E	N	1968	48858
1454	S	D	E	N	1968	51985
1454	S	D	E	N	1970	56418
1559	S	D	E	N	1967	45553
541-0135	S	D	E	N	1967	45669
0516	S	D	E	N	1964	34339
0517	S	D	E	N	1964	34339
0651	S	D	E	N	1964	37040
1050	S	D	E	N	1968	53086
1268	S	D	E	N	1964	40358
551-0129	S	D	E	N	1961	52095
0153	S	D	E	N	1961	52095
0154	S	D	E	N	1965	34736
0154	S	D	E	N	1962	49984
0154	S	D	E	N	1961	52095
0223	S	D	E	N	1965	34736
0223	S	D	E	N	1966	42977
0223	S	D	E	N	1962	49984
0223	S	D	E	N	1969	57597
0278	S	D	E	N	1966	36740
0278	S	D	E	N	1965	38672
0278	S	D	E	N	1964	42291
0278	S	D	E	N	1963	54703
0428	S	D	E	N	1961	52095
0428	S	D	E	N	1961	52784
0435	S	D	E	N	1963	54703
0447	P	D	E	N	1963	54703
0460	S	D	E	N	1962	49984
0460	S	D	E	N	1961	53498
0466	S	D	E	N	1961	52095
0466	S	D	E	N	1961	52784
0466	S	D	E	N	1963	54703
0482	S	D	E	N	1963	54703
0487	S	D	E	N	1963	54703
0487	S	D	E	N	1969	57597
0487	P	D	E	H	1965	38992
0500	S	D	E	N	1961	52095
0500	S	D	E	N	1961	52784
0501	S	D	E	N	1961	52784
0507	S	D	E	N	1961	52095
0570	S	D	E	N	1961	52095
0577	S	D	E	N	1964	37040
0577	S	D	E	N	1966	43484
0577	S	D	E	N	1965	46810
0577	S	D	E	N	1969	52920
0577	S	D	E	N	1970	57797
0601	S	D	E	N	1964	40358
0601	S	D	E	H	1969	41934
0601	S	D	E	U	1969	41945
0601	S	D	E	N	1966	42505
0601	S	D	E	N	1966	43530
0601	S	D	E	U	1967	47280
0601	S	D	E	N	1966	48376
0601	S	D	E	N	1968	53088
0601	S	D	G	N	1969	59668
0637	S	D	E	N	1964	40413
0637	S	D	E	N	1965	42351
0648	S	D	E	N	1962	53496
0680	S	D	E	N	1966	50298
0767	S	D	E	N	1965	34736
0888	S	D	E	N	1966	50298
0924	S	D	E	N	1964	42291
0924	S	D	E	N	1963	54703
0926	S	D	E	N	1963	54703
1076	S	S	E	N	1965	39238
1076	S	D	E	N	1969	57666
1116	S	D	E	N	1963	54703
1135	S	D	E	N	1966	50298
1141	S	D	E	N	1965	34736
1141	S	D	E	N	1965	36404
1141	S	D	E	N	1966	36740
1141	S	D	E	N	1964	37040
1141	S	D	E	N	1965	39745
1141	S	D	E	N	1962	49984
1175	S	D	E	N	1965	34736
1175	S	D	E	N	1962	49984
1199	S	D	E	N	1966	43530
1202	S	D	E	N	1961	46640
1234	S	D	E	N	1965	36525
1258	S	D	E	N	1963	37193
1258	S	D	E	N	1964	37443
1258	S	D	E	N	1965	40136
1258	S	G	E	N	1964	40414
1258	S	D	E	N	1963	42295
1258	S	D	E	U	1967	47280
1266	S	D	E	N	1965	34736
1266	S	D	E	N	1962	49984
1440	S	D	E	N	1961	52095
1545	S	D	E	N	1965	36404
1545	S	D	E	N	1965	39745
1545	S	D	E	H	1969	41934
1545	S	D	E	U	1969	41945

Substance Number	Phys. State	Subject	Language	Temperature	Year	TPRC Number
551-1545	S	D	E	N	1966	42914
1545	S	D	E	N	1966	43530
1545	S	D	E	N	1968	46990
1545	S	D	E	U	1967	47280
1545	S	D	E	N	1968	60177
1571	S	D	E	N	1965	34736
1571	S	E	E	U	1964	36423
1571	S	S	E	U	1968	54883
1571	S	S	E	N	1969	57666
1571	S	S	E	U	1970	59133
1689	S	D	E	H	1969	41934
1689	S	D	E	U	1969	41945
1689	S	D	E	N	1969	57597
1768	S	D	E	N	1964	37040
1805	S	D	E	N	1965	39745
1811	S	D	E	H	1969	41934
1811	S	D	E	U	1969	41945
1858	S	D	G	N	1969	59668
1884	S	D	E	N	1961	52095
1917	S	D	E	N	1966	36740
1943	S	D	E	N	1965	34736
1943	S	D	E	N	1964	37040
1943	S	D	E	N	1965	40136
1943	S	D	E	N	1966	48376
1943	S	D	E	N	1962	49984
1943	S	D	C	N	1969	57597
1944	S	D	E	N	1959	46025
2052	S	C	E	N	1967	33870
2071	S	D	E	N	1961	52095
2071	S	D	E	N	1961	52784
2071	S	D	E	N	1969	57597
2147	S	D	G	N	1969	59668
2240	S	D	E	N	1964	37840
2240	S	D	E	N	1964	42291
2298	S	D	E	N	1969	57597
2305	S	D	E	N	1961	52095
2347	S	D	E	N	1961	52095
2347	S	D	E	N	1961	52784
2387	S	D	E	N	1964	40756
2389	S	D	E	N	1964	40756
2390	S	D	E	N	1964	40756
2416	S	D	E	N	1966	43484
2416	S	D	G	N	1969	59668
2492	S	D	E	N	1969	57597
2524	S	D	E	N	1964	34339
2566	S	S	E	N	1964	36423
2566	S	T	E	N	1965	36524
2566	S	E	E.	N	1963	45545
2566	S	E	E	N	1967	47281
2566	S	D	E	N	1962	55198
2572	S	D	E	N	1963	54703
2611	S	D	E	N	1966	43530
2627	S	D	E	N	1964	34339
2640	S	D	E	N	1961	52095
2677	S	D	E	N	1966	40136
2677	S	D	E	N	1961	52095
2677	S	D	E	N	1961	52153
2677	S	D	E	N	1961	52784
2678	S	D	E	N	1961	52095
2736	S	D	E	N	1965	34736
2736	S	D	E	N	1962	49984
2739	S	D	E	N	1965	34736
2739	S	D	E	N	1962	49984
2751	S	D	E	N	1965	34736
2751	S	D	E	N	1964	37040
2751	S	D	E	N	1964	40358
2751	S	D	E	N	1966	43530
2751	S	D	E	U	1967	47280
2751	S	D	E	N	1962	49984
2759	S	D	E	N	1963	54703
2772	S	D	E	N	1969	57597
2921	S	D	E	N	1961	52095
2922	S	D	E	N	1961	52095
2968	S	E	E	N	1967	47281
2988	S	D	E	N	1962	53496
3037	S	D	E	N	1959	46025
3040	S	D	E	N	1964	40358
3041	S	D	F	N	1965	46640
3041	S	D	E	N	1961	46640
3041	S	D	E	N	1962	49984
3054	S	D	E	N	1968	46990
3054	S	D	E	N	1966	48376
3054	S	D	E	N	1963	54703
3056	S	D	E	N	1965	39745
3069	S	G	E	N	1965	37369
3075	S	D	E	N	1966	43530
3080	S	D	E	N	1966	43530
3107	S	D	E	N	1965	39745
3155	S	D	E	N	1964	40358
3158	S	D	E	U	1967	47280
3201	S	D	E	H	1969	41934
3201	S	D	E	U	1969	41945
3249	S	D	E	N	1966	50298
3250	S	D	E	N	1966	50298
3251	S	D	E	N	1966	50298

Substance Number	Phys. State	Subject	Language	Temperature	Year	TPRC Number
551-3253	S	D	E	N	1966	50298
3254	S	D	E	N	1966	50298
3277	S	D	E	N	1969	57597
3281	S	D	E	N	1969	57597
3290	S	D	E	N	1969	57597
3334	S	D	E	N	1961	52784
3352	S	D	E	N	1962	47962
3353	S	D	E	N	1962	47962
3362	S	D	E	N	1961	52784
3363	S	D	E	N	1961	52784
3366	S	D	E	N	1963	54703
661-0040	S	D	E	H	1959	56353

Phys. State: G. Gas; L. Liquid; M. Multiphase; P. Powder; S. Solid

Chapter 12 Diffusion Coefficient

Substance Number	Phys. State	Sub- ject	Lan- guage	Temper- ature	Year	TPRC Number	Substance Number	Phys. State	Sub- ject	Lan- guage	Temper- ature	Year	TPRC Number	Substance Number	Phys. State	Sub- ject	Lan- guage	Temper- ature	Year	TPRC Number
528-0145	S	D	E	N	1967	44688														
0310	S	D	F	N	1962	49526														
531-0243	S	T	R	U	1965	33797														
0243	S	T	E	U	1965	33798														
0514	M	G	E	N	1966	39801														
0717	S	D	E	N	1965	36419														
0795	S	D	J	N	1967	39779														
0795	S	D	E	N	1967	39780														
0798	S	D	J	N	1967	39779														
0798	S	D	E	N	1967	39780														
0803	S	D	J	N	1967	39779														
0803	S	D	E	N	1967	39780														
1110	S	D	E	H	1965	49207														
1111	S	D	E	H	1965	49207														
1112	S	D	E	H	1965	49207														
1113	S	D	E	H	1965	49207														
1114	S	D	E	H	1965	49207														
551-0233	S	D	E	H	1966	51120														
1778	S	D	E	N	1965	34394														
2464	S	D	E	H	1967	45173														
2626	S	D	E	N	1965	34394														
2702	S	D	E	N	1965	34251														
2703	S	D	E	N·	1965	34251														
2704	S	D	E	N	1965	34251														
2705	S	D	E	N	1965	34251														
2809	S	G	G	N	1964	43767														
2809	S	G	E	N	1966	43768														
3213	S	D	R	N	1968	48731														
3213	S	D	E	N	1968	51640														
3246	S	D	R	N	1967	48743														

Phys. State: G. Gas; L. Liquid; M. Multiphase; P. Powder; S. Solid

Chapter 13 Thermal Linear Expansion Coefficient

Substance Number	Phys. State	Sub-ject	Lan-guage	Temper-ature	Year	TPRC Number
526-0004	S	D	E	N	1936	38383
0004	S	D	E	N	1953	43480
0004	S	D	E	N	1956	59993
0004	P	D	G	H	1961	62304
0004	P	D	G	H	1962	62305
0004	M	D	E	N	1939	38901
0005	S	D	G	N	1967	48442
0005	S	D	E	H	1959	52529
0005	S	D	E	H	1957	53168
0005	S	D	R	H	1969	58844
0005	S	D	E	H	1969	58845
0005	S	D	E	N	1950	59944
0006	S	D	E	N	1950	59944
0006	S	D	E	N	1956	59993
0009	S	D	E	N	1953	43480
0020	S	D	E	N	1950	59944
0059	S	D	J	N	1983	37596
0067	S	D	E	N	1953	43480
0067	S	S	F	H	1965	47334
0067	S	D	R	N	1968	48742
0067	S	D	E	N	1968	51954
0067	S	D	E	N	1968	55303
0067	S	D	R	N	1969	57094
0067	S	D	R	N	1969	57095
0067	S	D	E	N	1901	59874
0067	S	D	E	N	1956	59993
0067	S	D	R	H	1970	61174
0067	S	D	E	H	1970	61175
0068	S	D	E	N	1950	59944
0073	S	D	E	N	1950	59944
0074	S	D	E	N	1936	38383
0074	S	D	E	N	1953	43480
0074	M	D	E	N	1939	38901
0086	S	D	R	N	1969	57514
0099	S	D	E	N	1953	43480
0099	M	D	E	N	1936	38383
0130	S	D	E	N	1950	59944
0151	S	D	E	N	1950	59944
0152	S	D	E	N	1950	59944
0155	S	D	E	N	1950	59944
0164	S	D	E	N	1950	59944
0168	S	D	E	N	1950	59944
0182	S	D	E	N	1936	38383
0182	S	D	E	N	1950	59944
0187	S	D	E	N	1965	34838
0188	S	D	E	N	1936	38383
0188	S	D	E	N	1950	59944
0192	S	D	G	N	1967	48442
0192	S	D	E	H	1957	53168
0193	S	D	E	L	1968	49266
0194	M	D	E	N	1936	38383
0196	S	D	G	N	1967	48442
0197	S	D	G	N	1967	48442
0198	S	E	E	H	1911	48358
0199	S	D	E	N	1965	47549
0200	S	D	E	N	1930	38880
0200	S	D	E	N	1931	60443
0201	M	D	E	N	1939	38901
0202	S	D	E	N	1965	47549
0207	S	D	R	H	1967	45279
0207	S	D	E	H	1967	50880
0208	S	D	R	N	1968	49780
0210	S	D	R	N	1966	61240
0210	S	D	E	N	1966	61240
0211	S	D	R	N	1966	41240
0212	S	D	E	N	1969	38663
0215	S	D	R	N	1969	55263
0219	S	D	R	N	1968	50146
0220	S	D	E	N	1950	59944
0220	M	D	E	N	1936	38383
0227	S	D	R	N	1969	57389
0228	S	D	F	N	1968	57389
0229	S	D	F	N	1968	57389
0232	S	D	E	N	1946	50114
0233	S	D	R	N	1968	56709
0234	S	D	R	N	1800	56701
0235	S	D	E	N	1946	50114
0236	S	D	E	N	1950	59944
0237	S	D	E	N	1950	59944
0238	S	D	E	N	1950	59944
0239	S	D	E	N	1950	59944
0240	S	D	E	N	1950	59944
0241	S	D	E	N	1950	59944
0242	S	D	E	N	1950	59944
0243	S	D	E	N	1950	59944
0244	S	D	E	N	1950	59944
0245	S	D	E	N	1950	59944
0246	S	D	E	N	1950	59944
0247	S	D	E	N	1950	59944
0248	S	D	E	N	1950	59944
0249	S	D	E	N	1950	59944
0250	S	D	E	N	1950	59944
0251	S	D	E	N	1950	59944
0251	M	D	E	N	1936	38383
0252	S	D	E	N	1950	59944
526-0253	S	D	E	N	1950	59944
0254	S	D	E	N	1950	59944
0255	S	D	E	N	1950	59944
0256	S	D	E	N	1950	59944
0257	S	D	E	N	1950	59944
0258	S	D	E	N	1950	59944
0259	S	D	E	N	1950	59944
0260	S	D	E	N	1950	59944
0261	S	D	E	N	1950	59944
0262	S	D	E	H	1957	53168
0265	S	D	R	N	1967	63152
0265	S	D	E	N	1967	63153
0269	S	G	E	N	1968	44400
0269	M	G	E	N	1968	44400
528-0015	S	D	E	N	1965	37995
0319	S	D	E	N	1960	52654
0320	S	D	E	N	1960	52654
0321	S	D	E	N	1960	52654
0323	S	D	E	N	1960	52654
0330	S	D	E	J	1957	40706
0330	S	D	E	N	1966	40707
0330	S	D	E	N	1934	43056
0330	S	D	E	N	1941	60370
0338	S	D	R	N	1967	40356
0338	S	D	E	N	1967	49357
0339	S	D	E	H	1965	34059
0340	S	D	E	H	1965	34059
0344	S	D	F	N	1933	58015
0348	S	D	E	N	1965	37995
0349	S	D	E	N	1965	37995
0350	S	D	E	N	1965	37995
0375	S	D	E	H	1967	52873
0384	S	D	E	N	1967	52901
0391	S	D	E	N	1968	53091
0402	S	D	E	N	1969	58105
0411	S	D	E	N	1970	34631
0425	S	D	E	N	1968	54811
0425	S	D	E	N	1968	54817
0426	S	D	E	N	1968	54811
0426	S	D	E	N	1968	54817
0427	S	D	E	N	1968	54811
0427	S	D	E	N	1968	54817
0429	S	D	E	N	1963	52276
531-0023	S	D	E	L	1952	55206
0060	S	D	E	N	1965	35530
0060	S	D	E	N	1964	37499
0060	S	D	E	N	1967	44987
0060	S	D	E	N	1968	44806
0060	S	D	E	N	1968	52893
0060	S	D	E	N	1968	52896
0064	S	D	E	H	1953	55659
0066	S	D	E	N	1968	57225
0068	S	D	E	N	1965	35530
0092	S	D	E	N	1965	35530
0201	S	D	E	L	1966	42657
0301	S	D	E	L	1966	42657
0301	S	D	E	N	1969	50857
0303	S	D	E	N	1958	40795
0303	S	D	E	L	1966	42657
0303	S	G	E	N	1962	52660
0303	S	D	E	N	1968	52896
0303	S	D	E	N	1968	43742
0303	S	D	E	L	1968	42657
0303	L	D	R	N	1967	45230
0304	S	D	E	N	1964	37499
0304	S	D	E	N	1966	42657
0434	S	D	E	N	1960	52260
0463	M	D	E	N	1966	58327
0476	M	D	E	N	1959	52668
0476	M	D	E	N	1959	52669
0515	S	D	E	L	1966	35110
0515	S	D	E	L	1966	42657
0515	S	D	E	N	1970	52866
0515	S	D	E	L	1952	55206
0525	S	D	E	L	1966	42657
0537	S	D	E	L	1966	34659
0537	S	D	E	N	1966	46528
0537	S	D	E	N	1964	50019
0568	S	D	E	L	1966	35110
0568	S	D	E	N	1965	35899
0568	S	D	E	N	1964	36202
0568	S	D	E	N	1964	37499
0568	S	D	E	N	1966	39083
0568	S	D	E	L	1965	40148
0568	S	D	E	N	1966	42657
0568	S	G	E	N	1966	43742
0568	S	D	E	N	1966	44937
0568	S	D	E	N	1964	50019
0568	S	G	E	L	1968	50604
0568	S	D	E	L	1968	59952
0568	L	D	R	N	1967	45230
0569	S	D	E	N	1965	36419
0569	S	D	E	N	1964	37499
531-0569	S	D	E	N	1968	52893
0585	S	D	E	L	1966	42657
0617	S	D	E	N	1970	52866
0620	S	D	E	N	1964	37255
0620	S	D	E	N	1964	37499
0620	S	D	E	H	1965	40305
0620	S	D	E	H	1967	41136
0620	S	D	E	L	1966	42657
0620	S	G	E	N	1966	43742
0620	S	D	E	N	1968	48806
0620	S	D	E	N	1968	52893
0633	S	S	E	L	1967	56366
0679	S	D	E	L	1968	42657
0679	S	D	E	N	1968	52896
0708	S	D	J	H	1969	57565
0839	S	D	E	N	1966	43937
0839	S	D	E	L	1967	43938
0929	S	D	E	L	1966	42657
0957	S	D	E	L	1966	42657
0960	S	D	E	L	1966	42657
0960	S	D	E	N	1964	50019
0962	S	D	E	H	1965	46471
0962	S	D	E	H	1966	52891
0962	S	D	E	H	1963	58962
0963	S	D	E	L	1966	42657
0963	S	D	E	N	1964	50019
0966	S	D	E	L	1966	42657
0966	S	D	E	N	1964	50019
0974	S	S	E	N	1967	47070
0974	S	D	E	N	1960	52560
1054	S	D	E	N	1968	46814
1054	S	D	R	N	1969	58530
1055	S	D	R	N	1967	46367
1095	S	S	E	H	1967	47070
1095	S	D	E	N	1961	47215
1095	S	D	E	H	1967	50153
1095	S	D	E	N	1970	60164
1157	S	D	E	N	1968	50308
1159	L	D	R	N	1967	47327
1244	S	D	R	N	1966	43080
1246	S	D	J	U	1967	45441
1246	S	E	E	N	1960	52260
1246	S	D	E	L	1967	56366
1267	P	D	E	U	1961	52267
1268	P	D	E	U	1961	52267
1269	P	D	E	U	1961	52267
1287	S	D	G	N	1966	41631
1287	S	D	E	N	1970	52866
1288	S	D	E	N	1962	47240
1320	S	D	E	N	1963	52562
1332	S	D	E	N	1966	40343
1363	S	D	E	N	1964	37499
1381	S	D	E	N	1984	36084
1390	S	D	E	N	1965	36419
1391	S	D	E	N	1965	36419
1392	S	D	E	H	1967	47068
1420	S	D	E	N	1964	36285
1426	S	D	E	N	1965	47397
1427	S	D	E	N	1965	47397
1428	S	D	E	N	1965	47397
1429	S	D	E	N	1965	47397
1430	S	D	E	N	1965	47397
1433	S	D	E	N	1967	47565
1434	S	D	E	N	1967	47565
1435	S	D	E	N	1967	47565
1457	S	S	E	H	1965	35862
1457	S	S	E	H	1965	37590
1457	S	T	E	U	1969	49144
1457	S	T	E	U	1968	51629
1457	S	T	E	U	1970	59107
1457	S	T	E	U	1971	62683
1476	S	D	E	H	1966	52891
1501	S	D	E	N	1968	50857
1505	S	D	E	N	1968	53519
1506	S	D	E	N	1968	53519
1507	S	D	E	N	1968	53519
1508	S	D	E	N	1968	53519
1509	S	U	E	N	1969	53519
1513	S	D	E	N	1969	58878
1518	S	D	E	N	1970	34631
1518	S	D	E	N	1966	53098
1518	S	D	E	N	1940	60284
1550	S	D	E	N	1970	52866
1551	S	D	E	N	1970	52866
1584	S	D	E	N	1970	52866
1585	S	D	E	N	1970	52866
1586	S	D	E	N	1970	52866
1603	S	D	E	N	1964	36084
1624	M	G	E	N	1963	39738
1651	S	D	R	N	1969	58842
1651	S	D	E	N	1969	58842
1652	S	D	R	N	1969	58842
1652	S	D	E	N	1969	58843
1665	S	D	E	H	1966	42738
1665	S	D	E	H	1965	53111

Phys. State: G. Gas; L. Liquid; M. Multiphase; P. Powder; S. Solid

Thermal Linear Expansion Coefficient   B52

Substance Number	Phys. State	Sub- ject	Lan- guage	Temper- ature	Year	TPRC Number
531-1670	S	D	J	H	1969	57565
1671	S	D	O	H	1969	58147
1672	S	D	R	N	1969	58530
1673	S	D	R	N	1969	58261
1679	S	C	E	N	1969	54502
1679	S	C	E	N	1970	60282
1687	S	D	E	N	1965	47549
1688	S	D	E	N	1965	47549
1689	S	D	E	N	1965	47549
1690	S	D	E	N	1965	47549
1693	S	C	E	N	1969	54502
1700	S	D	E	N	1970	52866
1701	S	D	E	N	1970	52866
1704	S	D	E	H	1960	41864
1705	S	C	E	N	1970	58317
1749	S	C	E	N	1969	54502
1755	S	D	E	N	1967	60099
1765	S	D	E	N	1968	57225
1766	S	D	E	N	1968	57225
1767	S	D	E	N	1968	57225
1787	S	D	E	N	1968	55303
1800	M	D	E	N	1969	56642
541-0002	S	D	E	H	1966	40133
0019	S	D	E	N	1964	37626
0108	S	S	E	U	1961	52771
0411	S	D	R	N	1961	52569
0411	S	D	E	N	1961	52570
0416	S	D	E	N	1962	52661
0417	S	D	E	H	1965	46471
0426	S	D	E	H	1959	52421
0433	P	D	R	N	1963	58408
0433	P	D	E	N	1964	58409
0438	S	D	E	H	1960	52756
0616	M	D	E	N	1964	37044
0678	S	D	E	H	1964	37378
0694	S	D	E	N	1966	40343
0696	S	D	E	N	1966	40343
0907	S	D	E	N	1966	46055
0909	S	D	E	H	1966	46935
0910	S	S	E	H	1967	47070
0962	S	D	E	N	1968	50308
1002	S	D	E	H	1964	46645
1023	S	D	E	N	1960	52274
1024	S	D	E	N	1960	52274
1025	S	D	E	N	1960	52274
1027	S	D	E	N	1960	52282
1028	S	D	E	N	1960	52282
1029	S	D	E	N	1960	52282
1030	S	D	E	N	1957	43566
1030	S	D	E	N	1960	52282
1031	S	D	E	N	1960	52282
1032	S	D	E	N	1960	52282
1036	S	D	E	N	1958	52287
1037	S	D	E	N	1958	52287
1042	S	D	E	N	1961	52782
1042	S	D	E	N	1962	52787
1050	S	T	E	U	1960	52125
1060	S	D	E	N	1962	45217
1080	S	D	E	N	1966	35741
1080	S	D	E	N	1968	52904
1081	S	D	E	N	1966	35741
1130	S	D	E	N	1966	52891
1160	S	D	E	N	1968	52904
1161	S	D	E	H	1968	52904
1162	S	D	E	N	1968	52904
1163	S	D	E	N	1968	52904
1214	S	D	E	N	1969	55331
1253	S	S	E	N	1966	39080
1254	S	S	E	N	1966	39080
1286	P	D	R	N	1963	58408
1286	P	D	E	N	1964	58409
1295	S	S	E	N	1969	57666
1308	S	D	E	N	1970	34631
1308	S	G	E	N	1968	50604
1330	S	D	R	N	1969	58530
1332	S	D	R	N	1969	58286
1345	S	D	E	N	1970	60262
1346	S	D	E	N	1970	60262
1347	S	C	E	N	1970	60282
1349	S	G	E	N	1968	50604
1358	S	S	E	F	1966	42294
1358	S	D	E	H	1962	46907
1358	S	D	E	H	1962	48004
1359	S	D	E	H	1962	46907
1359	S	D	E	H	1962	48004
1360	S	D	E	N	1966	40776
1362	S	D	E	H	1964	37488
1429	S	D	R	N	1970	61567
551-0153	S	D	E	H	1959	47025
0153	S	S	E	H	1967	50724
0153	S	D	E	F	1955	80523
0154	S	D	E	H	1959	47025
0154	S	S	E	H	1967	50724
0154	S	D	E	F	1955	80523
0826	S	D	E	H	1970	58670
1289	S	S	E	N	1967	50724
1289	S	D	E	F	1955	60523
1290	S	S	E	H	1967	50724
2608	S	D	E	N	1965	39065

Substance Number	Phys. State	Sub- ject	Lan- guage	Temper- ature	Year	TPRC Number
551-2763	S	S	E	H	1967	50724
621-0028	S	D	E	N	1964	36084
0028	S	D	E	N	1964	37509
651-0025	M	D	E	N	1966	44981
0232	S	D	G	N	1941	41603
0286	S	D	G	N	1941	41603
0287	S	D	G	N	1941	41603
0288	S	D	G	N	1941	41603
0289	S	D	E	N	1944	57948
0314	S	D	G	N	1941	41603
0319	S	D	E	N	1962	56354
0351	S	D	G	N	1941	41603
0365	S	D	G	N	1941	41603
0476	S	D	F	N	1869	59537
0476	M	D	E	N	1966	44981
0478	M	D	E	N	1966	44981
0479	M	D	E	N	1966	44981
0544	S	S	E	N	1968	42148
661-0002	S	D	E	N	1969	58221
0008	S	D	E	N	1908	40328
0050	S	D	E	N	1969	58221
0433	M	E	R	H	1968	50878
0437	S	C	E	N	1946	45807
0481	S	D	E	N	1947	59935

Phys. State: G. Gas; L. Liquid; M. Multiphase; P. Powder; S. Solid

Chapter 14 Thermal Volumetric Expansion Coefficient

Substance Number	Phys. State	Subject	Language	Temperature	Year	TPRC Number	Substance Number	Phys. State	Subject	Language	Temperature	Year	TPRC Number	Substance Number	Phys. State	Subject	Language	Temperature	Year	TPRC Number
526-0199	S	D	E	N	1965	47549														
0201	S	T	E	U	1969	55130														
0201	S	E	E	N	1951	60526														
0202	S	D	E	N	1965	47549														
0205	S	D	E	N	1965	47549														
0206	S	G	E	N	1968	44400														
0206	S	D	E	N	1965	47549														
0206	L	G	E	N	1968	44400														
531-0568	S	D	E	N	1968	54248														
0620	S	D	E	N	1964	41910														
1095	S	D	E	H	1967	47532														
1457	S	T	E	U	1968	51629														
1457	S	T	E	U	1970	59107														
1457	S	T	E	U	1970	59599														
1651	S	D	R	N	1969	58842														
1651	S	D	E	N	1969	58843														
1652	S	D	R	N	1969	58842														
1652	S	D	E	N	1969	58843														
1687	S	D	E	N	1965	47549														
1688	S	D	E	N	1965	47549														
1690	S	D	E	N	1965	47549														
541-0002	S	D	F	H	1966	40133														
1030	S	C	E	N	1958	52783														
1043	S	C	E	N	1958	52783														
651-0476	S	D	E	N	1902	43063														
0476	S	D	E	N	1956	43309														
661-0481	S	D	E	N	1947	59935														

Phys. State: G. Gas; L. Liquid; M. Multiphase; P. Powder; S. Solid

Chapter 15 Surface Tension

Substance Number	Phys. State	Subject	Language	Temperature	Year	TPRC Number
528-0099	M	T	E	U	1967	43177
0181	M	G	E	H	1969	54683
0187	M	G	E	H	1969	54683
0255	M	D	E	H	1964	43408
0257	M	G	E	H	1969	54683
0336	M	D	E	H	1969	52841
0372	L	D	E	N	1961	52158
0373	L	D	E	N	1961	52158
0424	L	D	R	U	1968	56194
531-0169	L	D	E	N	1967	44385
0169	L	D	E	N	1967	47575
0169	L	C	R	N	1967	47599
0169	L	D	E	N	1969	55853
0169	M	D	E	N	1970	57052
0181	L	D	E	N	1967	46357
0182	L	D	E	N	1967	39612
0182	L	D	E	N	1967	44385
0182	L	D	E	N	1967	45975
0182	L	D	E	N	1967	46029
0182	L	D	E	N	1967	46357
0182	L	C	R	N	1967	47599
0182	L	D	E	N	1968	51016
0182	M	D	E	N	1970	57052
0186	L	D	E	N	1967	44385
0186	L	D	E	N	1967	45975
0186	L	D	E	N	1967	46357
0186	L	C	R	N	1967	47599
0186	L	D	E	N	1968	51016
0186	L	D	E	N	1969	55853
0186	M	D	E	N	1970	57052
0193	L	D	E	N	1967	46357
0194	L	D	E	N	1967	45975
0207	L	D	E	N	1967	46357
0207	L	C	R	N	1967	47599
0207	L	D	E	N	1968	51016
0207	M	D	E	N	1970	57052
0208	L	D	E	N	1967	40915
0209	L	C	E	N	1967	36997
0209	L	G	E	N	1965	40433
0209	L	D	E	N	1967	40915
0209	L	D	E	N	1967	46357
0209	L	D	E	N	1968	47509
0209	L	D	E	N	1967	47575
0209	L	C	R	N	1967	47599
0209	L	D	E	N	1968	47748
0209	L	D	E	N	1968	48408
0209	L	D	E	N	1967	50218
0209	L	D	E	N	1969	50252
0209	L	D	E	N	1968	51016
0209	L	D	J	N	1968	56971
0209	M	C	E	N	1964	35526
0209	M	D	E	N	1966	45911
0209	M	D	E	N	1970	57052
0211	L	D	E	N	1967	46357
0211	L	D	E	N	1967	40915
0222	L	D	E	N	1967	47599
0222	L	C	R	N	1969	50252
0222	L	D	J	N	1968	56971
0222	M	D	E	N	1966	45906
0224	L	D	E	N	1967	40915
0230	L	C	F	N	1968	50794
0230	L	D	J	N	1969	56189
0230	L	D	G	N	1967	56626
0230	M	E	F	N	1966	40612
0230	M	E	E	N	1966	40613
0238	L	D	E	N	1967	40915
0240	L	D	E	N	1967	40915
0241	L	D	E	N	1967	40915
0246	L	D	E	N	1967	40915
0265	L	D	E	N	1967	40915
0270	L	D	E	N	1967	40915
0286	L	D	E	N	1967	40915
0287	L	D	E	N	1967	40915
0289	L	D	E	N	1807	40915
0292	L	D	E	N	1969	58184
0292	L	D	E	N	1967	43078
0354	L	D	E	N	1967	43078
0356	L	D	E	N	1967	43078
0357	L	D	E	N	1967	43078
0360	L	D	F	N	1967	42876
0382	L	D	E	N	1966	46554
0379	M	T	E	U	1968	51088
0450	L	C	R	N	1966	35198
0460	L	D	E	N	1966	35199
0460	L	D	E	N	1967	38962
0488	L	D	E	N	1967	38962
0489	L	D	E	N	1968	53722
0489	M	D	E	N	1967	38962
0522	L	D	E	N	1967	46029
0593	L	D	E	N	1965	46357
0593	L	D	E	N	1967	46632
0593	L	G	R	N	1967	47599
531-0593	L	D	E	N	1968	47748
0593	L	D	E	N	1968	47905
0593	L	D	E	N	1969	50252
0593	L	D	E	N	1968	51016
0593	L	D	G	N	1968	51049
0593	M	D	E	N	1970	57052
0594	L	D	E	N	1966	46542
0601	L	D	E	N	1966	46542
0602	L	D	E	N	1966	46542
0603	L	D	E	N	1966	46542
0614	L	D	E	N	1966	46542
0616	L	D	E	N	1966	46542
0619	L	D	E	N	1966	46625
0621	L	C	E	U	1968	51088
0621	L	D	E	N	1966	46625
0622	L	C	E	U	1968	51088
0622	L	C	E	U	1966	46625
0623	L	D	E	N	1966	46625
0624	L	D	E	N	1968	47748
0624	L	C	E	U	1968	51088
0625	L	C	E	U	1966	46625
0625	L	C	E	U	1968	51088
0626	I	D	E	N	1966	46625
0626	L	C	E	U	1808	51088
0627	L	D	E	N	1966	46625
0628	L	D	E	N	1966	46625
0630	M	G	R	N	1965	46721
0630	M	G	R	N	1965	46722
0632	M	G	R	N	1965	46721
0632	M	G	R	N	1965	46722
0634	M	G	R	N	1965	46721
0634	M	G	R	N	1965	46722
0645	M	G	R	N	1965	46721
0645	M	G	R	N	1965	46722
0689	L	D	F	N	1969	47317
0714	L	T	E	N	1956	36085
0714	L	E	E	U	1965	40433
0714	L	T	R	U	1965	45977
0714	L	T	E	U	1965	46551
0714	L	T	R	U	1965	46713
0714	L	E	E	U	1965	46714
0714	L	T	E	U	1967	47298
0714	L	E	E	U	1967	47575
0714	L	T	E	U	1968	47738
0714	L	T	E	U	1968	48410
0714	L	E	E	U	1968	48566
0714	L	E	E	U	1967	51818
0714	L	C	E	N	1969	52213
0714	G	T	E	N	1969	56364
0714	G	T	E	U	1956	36085
0714	M	T	E	U	1965	42827
0714	M	T	R	E	1965	46713
0714	M	T	E	U	1965	46714
0714	M	M	E	E	1968	48006
0714	M	S	T	E	1965	51447
0721	L	T	D	E	1967	39612
0721	L	D	G	N	1965	40433
0721	L	D	E	R	1967	46357
0721	L	D	E	R	1967	46360
0721	L	C	E	R	1967	47599
0721	L	D	E	N	1968	51016
0721	L	D	J	N	1968	56971
0721	M	D	E	N	1967	39612
0727	L	D	E	N	1967	45911
0728	L	D	R	N	1967	39612
0732	L	D	E	N	1967	45077
0732	L	D	E	N	1967	45077
0751	L	D	R	N	1967	45077
0751	L	D	E	N	1967	45077
0752	L	D	R	N	1967	45077
0752	L	D	E	N	1967	45077
0753	L	D	R	N	1967	46077
0753	L	D	E	N	1967	45105
0760	L	D	R	N	1967	46077
0760	L	D	E	H	1967	45105
0764	L	D	E	N	1967	46077
0764	L	D	R	N	1967	45105
0765	L	D	R	N	1967	46077
0765	L	D	E	N	1967	45105
0771	L	D	R	N	1967	46077
0771	L	D	E	N	1967	45105
0804	L	D	E	N	1967	38962
0821	L	D	E	N	1967	38962
0881	L	C	E	N	1967	45834
0881	L	D	G	N	1968	50965
0881	L	G	F	U	1969	58185
0910	L	C	E	U	1967	51088
0914	L	D	E	N	1967	38962
0914	M	D	E	N	1968	53722
0915	L	C	C	N	1967	36997
0915	L	C	C	N	1967	36997
0916	L	C	C	N	1967	44385
0917	L	D	E	N	1967	46029
0920	L	D	E	N	1967	46029
0920	L	D	E	N	1967	46029
531-0921	L	D	E	N	1967	46029
0922	L	D	E	N	1967	46029
0924	L	D	F	N	1969	47317
0925	L	E	R	N	1967	44263
0925	M	G	E	N	1965	40433
0925	M	D	E	N	1967	44349
0925	M	D	F	N	1968	49574
0925	L	D	O	N	1969	55698
0926	L	C	E	U	1968	51088
0930	L	C	E	U	1968	51088
0932	L	D	R	N	1967	47688
0932	L	D	E	N	1967	47689
0933	L	D	E	N	1967	47689
0933	L	D	E	N	1967	47689
0933	L	D	R	N	1967	47690
0933	L	D	E	N	1967	47691
0934	L	D	R	N	1967	47688
0934	L	D	E	N	1967	47689
0934	L	D	E	N	1967	47690
0934	L	D	E	N	1967	47691
0935	L	D	R	N	1967	47688
0935	L	D	E	N	1967	47689
0935	L	D	R	N	1967	47690
0935	L	D	D	N	1967	47691
0936	L	D	J	N	1968	56971
0936	L	D	R	N	1067	47688
0937	L	D	E	N	1967	47689
0937	L	D	R	N	1967	47688
0938	L	D	E	N	1967	47689
0938	L	D	R	N	1967	47688
0939	L	D	R	N	1967	47688
0939	L	D	E	N	1967	47689
0939	L	D	R	N	1967	47691
0940	L	D	E	N	1967	47688
0940	L	D	R	N	1967	47689
0941	L	D	R	N	1967	47688
0941	L	D	E	N	1967	47689
0941	L	D	E	N	1967	47690
0941	L	D	R	N	1967	47691
0942	L	D	E	N	1967	47688
0942	L	D	R	N	1967	47689
0943	L	D	E	N	1967	47688
0943	L	D	E	N	1967	47689
0944	L	D	R	N	1967	47688
0944	L	D	E	N	1967	47689
0945	L	D	R	N	1967	47688
0945	L	D	E	N	1967	47689
0945	L	D	E	N	1967	47690
0945	L	D	E	N	1967	47691
0946	L	D	R	N	1967	47688
0946	L	D	E	N	1967	47689
0946	L	D	E	N	1967	47690
0946	L	D	E	N	1967	47691
0946	L	D	J	N	1968	56971
0947	L	D	E	N	1967	47688
0947	L	D	E	N	1967	47689
0948	L	D	R	N	1967	47688
0948	L	D	E	N	1967	47689
0949	L	D	E	N	1967	47689
0949	L	D	R	N	1967	47688
0950	L	D	E	N	1967	47689
0950	M	D	E	N	1966	45911
0950	M	D	E	N	1967	47688
0951	L	D	E	N	1967	47688
0951	L	D	E	N	1967	47689
0952	L	D	E	N	1967	47688
0953	L	D	E	N	1967	47689
0953	L	D	E	N	1967	47688
0954	L	D	E	N	1967	47689
0954	L	D	E	N	1967	47688
0956	L	D	F	N	1969	47317
0959	L	D	F	N	1968	50117
0965	M	D	E	U	1968	51088
0977	L	C	E	M	1968	50965
0980	L	D	E	N	1964	51273
0982	L	D	E	N	1964	51273
0983	L	D	E	N	1964	51273
0988	L	D	E	N	1968	51255
0989	L	D	E	N	1968	49842
0989	L	D	F	N	1968	48119
0990	L	D	F	N	1968	54449
0990	L	D	F	N	1968	48163
0992	M	D	F	N	1969	47317
0993	L	D	E	N	1966	45727
0994	L	D	E	N	1968	47748
0994	L	D	E	N	1968	47748
0995	L	D	E	N	1967	47599
0996	L	C	E	N	1968	47748
0996	L	D	E	N	1968	47748
0996	L	D	E	N	1968	50484
0996	L	D	E	N	1968	51016
0996	M	C	E	N	1970	56508

Phys. State: G. Gas; L. Liquid; M. Multiphase; P. Powder; S. Solid

Substance Number	Phys. State	Subject	Language	Temperature	Year	TPRC Number
531-0997	L	D	E	N	1968	47748
0998	L	D	E	N	1968	47748
0998	L	D	R	U	1968	56233
0998	M	D	E	N	1966	45906
0999	L	D	E	N	1968	47748
1000	L	D	E	N	1968	47750
1001	L	D	E	N	1968	49571
1002	L	D	E	N	1968	49571
1003	L	D	E	N	1968	49571
1004	L	D	E	N	1968	49571
1005	L	D	E	N	1968	49571
1006	L	D	E	N	1968	49571
1007	L	D	E	N	1968	49571
1008	L	D	E	N	1968	49571
1009	L	D	E	N	1968	49571
1010	L	D	F	N	1968	48119
1010	L	D	F	N	1968	54469
1011	L	D	F	N	1968	48119
1011	L	D	F	N	1968	54469
1012	L	D	F	N	1968	48119
1012	L	D	F	N	1968	54469
1013	L	D	F	N	1968	48119
1014	L	D	F	N	1968	48119
1014	L	D	F	N	1968	54469
1015	L	D	F	N	1968	48119
1015	L	D	F	N	1968	54469
1016	L	D	F	N	1968	48119
1017	L	D	F	N	1968	48119
1017	L	D	F	N	1968	54469
1018	L	D	F	N	1968	48119
1018	L	D	F	N	1968	54469
1019	L	D	F	N	1968	48119
1020	L	D	F	N	1968	48119
1021	L	D	F	N	1968	48119
1021	L	D	F	N	1968	54469
1022	L	D	F	N	1968	48119
1022	L	D	F	N	1968	54469
1023	L	D	F	N	1968	48119
1023	L	D	F	N	1968	54469
1024	L	D	F	N	1968	48119
1024	L	D	F	N	1968	54469
1025	L	D	F	N	1968	48119
1028	L	D	E	N	1968	50569
1029	L	D	E	N	1968	50569
1031	L	D	E	N	1968	47509
1032	L	D	E	N	1968	47509
1033	L	D	E	U	1967	47144
1033	L	D	E	N	1968	47509
1033	L	C	R	N	1967	47599
1034	L	D	E	N	1968	47509
1034	L	C	R	N	1967	47599
1035	L	D	E	N	1968	47509
1036	L	D	E	N	1968	47509
1037	L	D	E	N	1968	50484
1037	L	D	E	N	1968	51016
1041	L	D	E	N	1968	50484
1041	L	D	E	N	1968	51016
1044	L	D	E	N	1968	50484
1044	L	D	E	N	1968	51016
1045	L	D	E	N	1968	48950
1046	L	D	E	N	1968	48950
1047	L	D	E	N	1968	48950
1048	L	D	E	N	1968	48950
1049	L	D	E	N	1968	48950
1049	L	D	E	N	1977	56969
1050	L	D	E	N	1968	48950
1051	L	D	E	N	1968	48950
1052	L	D	E	N	1967	45003
1052	L	C	E	N	1968	48481
1058	L	D	F	N	1969	47317
1059	M	D	F	H	1968	49574
1060	M	D	F	H	1968	49574
1061	M	D	F	H	1968	49574
1062	M	D	F	H	1968	49574
1063	M	D	F	H	1968	49574
1064	M	D	F	H	1968	49574
1065	L	D	E	N	1968	51016
1066	L	D	E	N	1968	51016
1067	L	D	E	N	1968	51016
1067	M	C	E	N	1970	56508
1068	L	D	E	N	1968	51016
1069	L	D	E	N	1968	51016
1070	L	D	E	N	1968	51016
1071	L	D	R	N	1968	49842
1071	L	D	E	N	1968	51255
1072	L	D	R	N	1968	49842
1072	L	D	E	N	1968	51255
1073	L	D	R	N	1968	49842
1073	L	D	E	N	1968	51255
1074	L	D	R	N	1968	49842
1074	L	D	E	N	1968	51255
1076	L	D	E	N	1968	51016
1076	M	C	E	N	1970	56508
1081	L	D	E	N	1968	51016
1086	L	D	E	N	1968	51016
1087	L	D	R	N	1965	51617
1088	L	D	R	N	1965	51617
1089	L	D	R	N	1965	51617
1089	L	D	R	N	1969	56858
1089	M	D	E	N	1967	47999
1090	L	D	R	N	1965	51617
531-1096	L	D	E	N	1968	51016
1097	L	D	E	N	1968	51016
1098	L	D	E	N	1968	51016
1098	M	C	E	N	1970	56508
1119	L	D	E	N	1968	51016
1119	M	C	E	N	1970	56508
1121	L	D	E	N	1968	48408
1122	L	D	E	N	1968	48408
1123	L	D	E	N	1968	48408
1124	L	D	E	N	1968	48408
1125	L	D	E	N	1968	48408
1126	L	D	E	N	1968	48408
1127	L	D	O	N	1967	48421
1128	L	D	O	N	1967	48421
1129	L	D	O	N	1967	48421
1130	L	D	O	N	1967	48421
1131	L	D	O	N	1967	48421
1132	L	D	O	N	1967	48421
1138	L	D	E	N	1967	44385
1138	L	D	E	N	1967	49810
1139	L	D	E	N	1967	44385
1139	L	C	R	N	1967	47599
1139	M	D	E	N	1970	57052
1140	L	D	E	N	1964	44417
1141	L	D	E	N	1964	44417
1148	L	D	E	N	1968	51016
1150	L	C	E	U	1966	44534
1151	L	D	E	N	1968	51016
1155	M	D	E	N	1967	47999
1156	M	D	E	N	1967	47999
1166	L	D	E	N	1967	46357
1166	L	C	R	N	1967	47599
1166	L	D	E	N	1968	51016
1167	L	D	E	N	1967	46357
1167	L	C	R	N	1967	47599
1167	L	D	E	N	1968	51016
1168	S	D	E	N	1960	51290
1169	L	C	R	N	1967	47599
1169	M	D	E	N	1970	57052
1170	L	D	E	N	1967	46357
1170	L	C	R	N	1967	47599
1171	L	D	E	N	1967	46357
1171	L	C	R	N	1967	47599
1172	L	D	E	N	1967	46357
1172	L	C	R	N	1967	47599
1173	L	D	E	N	1967	46357
1173	L	C	R	N	1967	47599
1173	L	D	E	N	1968	51016
1174	L	D	E	N	1967	46357
1174	L	C	R	N	1967	47599
1175	L	C	R	N	1967	47599
1176	L	C	R	N	1967	47599
1177	L	C	R	N	1967	47599
1177	L	D	E	U	1967	56167
1178	L	C	R	N	1967	47599
1179	L	C	R	N	1967	47599
1180	L	C	R	N	1967	47599
1181	L	C	R	N	1967	47599
1182	L	C	R	N	1967	47599
1183	L	C	R	N	1967	47599
1183	L	D	E	N	1968	51016
1184	L	C	R	N	1967	47599
1185	L	C	R	N	1967	47599
1186	L	C	R	N	1967	47599
1187	L	C	R	N	1967	47599
1188	L	C	R	N	1967	47599
1189	L	C	R	N	1967	47599
1190	L	C	R	N	1967	47599
1191	L	C	R	N	1967	47599
1192	L	C	R	N	1967	47599
1193	L	C	R	N	1967	47599
1194	L	C	R	N	1967	47599
1195	L	C	R	N	1967	47599
1196	L	C	R	N	1967	47599
1196	L	D	E	N	1968	51016
1196	M	C	E	N	1970	56508
1197	L	C	R	N	1967	47599
1197	L	D	E	N	1968	50484
1197	L	D	E	N	1968	51016
1197	M	C	E	N	1970	56508
1198	L	C	R	N	1967	47599
1198	L	D	E	N	1968	51016
1198	M	C	E	N	1970	56508
1199	L	C	R	N	1967	47599
1199	L	D	E	N	1968	50484
1199	L	D	E	N	1968	51016
1199	M	C	E	N	1970	56508
1200	L	C	R	N	1967	47599
1200	L	D	E	N	1968	50484
1200	L	D	E	N	1968	51016
1200	M	C	E	N	1970	56508
1201	L	C	R	N	1967	47599
1201	L	D	E	N	1968	50484
1201	L	D	E	N	1968	51016
1201	M	C	E	N	1970	56508
1202	L	C	R	N	1967	47599
1202	M	C	E	N	1970	56508
1203	L	C	R	N	1967	47599
1204	L	C	R	N	1967	47599
1205	L	C	R	N	1967	47599
1206	L	C	R	N	1967	47599
531-1206	M	C	E	N	1970	56508
1207	L	C	R	N	1967	47599
1208	L	C	R	N	1967	47599
1208	M	D	E	N	1970	57052
1209	L	C	R	N	1967	47599
1209	L	D	E	N	1968	51016
1209	M	C	E	N	1970	56508
1210	L	C	R	N	1967	47599
1211	L	C	R	N	1967	47599
1215	L	D	E	N	1968	51016
1215	M	C	E	N	1970	56508
1216	L	D	E	N	1968	51016
1216	M	C	E	N	1970	56508
1220	L	D	R	N	1967	46360
1221	L	D	R	N	1967	46360
1222	L	D	R	N	1967	46360
1223	L	D	R	N	1967	46360
1224	L	D	R	N	1967	46360
1225	L	D	R	N	1967	46360
1226	L	D	R	N	1967	46360
1231	L	C	E	N	1968	48654
1232	L	C	E	N	1968	48654
1233	L	C	E	N	1968	48654
1234	L	C	E	N	1968	48654
1235	L	C	E	N	1968	48654
1236	L	C	E	N	1968	48654
1237	L	D	E	N	1967	46357
1238	L	D	E	N	1967	46357
1239	L	D	E	N	1967	46357
1241	L	D	E	N	1967	46357
1242	L	D	E	N	1967	46357
1243	L	D	E	N	1967	46357
1255	L	D	E	N	1967	46357
1271	M	D	E	U	1969	52588
1272	M	D	E	U	1969	52588
1273	L	D	E	N	1967	46357
1279	L	D	F	N	1969	47317
1282	L	D	F	N	1969	47317
1283	L	C	E	L	1969	53332
1285	L	D	E	N	1938	54104
1286	L	D	E	N	1938	54104
1291	L	D	E	N	1967	46357
1292	L	D	E	N	1967	46357
1293	L	D	E	N	1967	46357
1295	L	D	E	N	1967	46357
1296	L	D	E	N	1967	46357
1297	L	D	E	N	1967	46357
1298	L	D	E	N	1967	46357
1299	L	D	E	N	1967	46357
1300	L	D	E	N	1967	46357
1301	L	D	E	N	1967	46357
1302	L	D	E	N	1967	46357
1303	L	D	E	N	1967	46357
1304	L	D	E	N	1967	46357
1305	L	D	E	N	1967	46357
1306	L	D	E	N	1967	46357
1307	L	D	E	N	1967	46357
1308	L	D	E	N	1967	46357
1310	L	D	E	N	1967	46357
1311	L	D	E	N	1967	46357
1311	M	D	E	N	1970	57052
1313	M	T	E	N	1966	45237
1316	L	D	E	N	1967	46357
1317	L	D	E	N	1968	51016
1326	L	D	E	N	1958	40940
1326	M	D	E	H	1968	52999
1327	L	D	E	N	1958	40940
1328	L	D	E	N	1958	40940
1329	L	D	E	N	1958	40940
1333	L	D	E	N	1968	53136
1334	L	D	E	N	1967	45003
1335	L	D	E	N	1967	45003
1338	L	D	E	N	1968	53136
1339	L	D	E	N	1968	53136
1341	L	D	G	N	1968	51049
1342	L	D	G	N	1968	51049
1343	L	D	G	N	1968	51049
1344	L	D	G	N	1968	51049
1345	L	D	G	N	1968	51049
1346	L	D	G	N	1968	51049
1347	L	D	G	N	1968	51049
1348	L	D	G	N	1968	51049
1349	L	D	G	N	1968	51049
1350	L	D	G	N	1968	51049
1351	L	D	G	N	1968	51049
1352	L	D	G	N	1968	51049
1353	L	D	G	N	1968	51049
1354	L	D	G	N	1968	51049
1355	L	D	G	N	1968	51049
1362	L	D	O	U	1966	41280
1365	L	D	E	N	1966	45727
1366	M	D	E	N	1966	45906
1366	M	C	E	N	1970	56508
1370	M	D	E	N	1966	45906
1371	M	D	E	N	1966	45906
1372	M	D	E	N	1966	45911
1373	M	D	E	N	1966	45911
1374	M	D	E	N	1966	45911
1375	M	D	E	N	1966	45911
1386	L	D	J	N	1968	56971
1387	L	D	J	N	1968	56971

Phys. State: G. Gas; L. Liquid; M. Multiphase; P. Powder; S. Solid

Substance Number	Phys. State	Sub-ject	Lan-guage	Temper-ature	Year	TPRC Number
531-1388	L	D	J	N	1968	56971
1393	L	D	J	N	1968	56971
1397	L	D	J	N	1968	56971
1398	L	D	J	N	1968	56971
1403	L	D	J	N	1968	56971
1404	L	D	J	N	1968	56971
1406	L	D	J	N	1968	56971
1411	L	D	J	N	1968	56971
1412	L	D	J	N	1968	56971
1413	L	D	J	N	1968	56971
1415	L	D	E	N	1969	38408
1415	L	D	R	N	1969	55149
1416	L	D	E	N	1969	38408
1416	L	D	R	N	1969	55149
1432	L	D	J	N	1968	56971
1437	L	D	E	N	1970	57957
1446	L	D	J	N	1968	50787
1470	M	D	R	H	1969	53681
1478	L	D	E	N	1967	50218
1479	L	D	E	N	1967	50218
1481	L	D	E	N	1970	35802
1482	L	D	E	N	1970	35802
1483	L	D	E	N	1970	35802
1484	L	D	E	N	1970	35802
1485	L	D	E	N	1970	35802
1486	L	D	E	N	1970	35802
1488	L	C	E	N	1957	56360
1489	L	C	E	N	1957	56360
1490	L	C	E	N	1957	56360
1491	L	C	E	N	1957	56360
1494	M	D	E	N	1968	53722
1495	M	D	E	N	1968	53722
1496	M	D	E	N	1968	53722
1500	L	D	I	N	1968	52998
1502	L	T	E	U	1969	49815
1502	M	S	E	N	1969	54996
1502	M	T	E	U	1970	56508
1504	M	T	E	U	1969	49612
1504	M	S	E	N	1969	54542
1504	M	D	G	N	1967	56852
1510	M	S	E	N	1969	54542
1511	M	T	E	U	1969	54237
1514	L	D	I	N	1968	52998
1521	L	D	R	N	1969	56515
1521	L	D	E	N	1969	56516
1523	L	D	E	U	1967	47144
1523	L	D	E	N	1969	53733
1524	L	D	E	U	1967	47144
1524	L	D	E	N	1969	53733
1525	L	D	E	U	1967	47144
1526	M	D	E	U	1967	48529
1527	L	D	E	U	1967	49010
1528	L	D	E	U	1967	49010
1529	L	D	E	U	1967	49010
1530	L	D	E	U	1967	49010
1531	L	D	E	U	1967	49010
1532	L	D	E	U	1967	49010
1533	L	D	E	U	1967	49010
1534	L	D	E	U	1967	49010
1535	L	D	E	U	1967	49010
1536	L	D	E	U	1967	49010
1537	L	D	R	U	1967	50463
1538	L	D	E	N	1969	55679
1543	M	C	E	N	1970	56508
1544	M	D	E	N	1970	54349
1545	M	D	E	N	1970	54349
1546	M	D	E	N	1970	54349
1547	M	D	E	N	1970	54349
1548	M	D	E	N	1970	54349
1549	L	D	E	N	1967	53250
1552	M	C	E	N	1964	35526
1553	M	C	E	N	1964	35526
1554	M	C	E	N	1964	35526
1555	M	C	E	N	1964	35526
1556	M	C	E	N	1964	35526
1557	M	C	E	N	1964	35526
1558	M	C	E	N	1964	35526
1560	M	D	E	N	1970	57052
1561	M	D	E	N	1970	57052
1562	M	D	E	N	1970	57052
1563	M	D	E	N	1970	57052
1564	M	D	E	N	1970	57052
1565	M	D	E	N	1969	55663
1566	L	D	J	U	1967	50422
1566	M	D	E	N	1969	55660
1566	M	D	E	N	1969	55663
1567	M	D	E	N	1969	55663
1568	M	D	E	N	1969	55663
1569	M	D	E	N	1969	55663
1570	M	D	E	N	1969	55663
1571	M	D	E	N	1969	55663
1572	M	D	E	N	1969	55663
1573	M	D	E	N	1969	55663
1574	M	D	E	N	1969	55663
1575	M	D	E	N	1969	55663
1576	M	D	E	N	1969	55663
1577	M	D	E	N	1969	55663
1578	M	D	E	N	1969	55660
1589	M	D	E	N	1969	55566
1590	M	D	E	N	1969	55566
1591	M	D	E	N	1969	55566

Substance Number	Phys. State	Sub-ject	Lan-guage	Temper-ature	Year	TPRC Number
531-1593	L	D	O	U	1966	41280
1606	L	D	E	N	1970	51438
1607	L	D	E	N	1970	51438
1613	M	D	E	U	1967	48529
1613	M	D	E	N	1970	57378
1614	L	D	E	N	1969	55679
1614	L	D	E	N	1970	57378
1615	L	D	E	N	1969	55679
1615	L	D	E	N	1970	57378
1637	L	D	J	N	1969	56189
1637	L	D	G	N	1967	56441
1638	L	D	J	N	1969	56189
1648	M	G	R	N	1965	46721
1648	M	G	E	N	1965	46722
1655	L	D	R	H	1969	37711
1662	L	D	R	H	1966	41986
1706	L	D	E	N	1969	55679
1707	L	D	E	N	1969	55679
1708	L	D	E	N	1969	55679
1709	L	D	E	N	1969	55679
1710	L	D	R	U	1968	55859
1711	L	D	R	U	1968	55859
1712	L	D	R	U	1968	55859
1713	L	D	R	U	1968	55859
1714	L	D	R	U	1968	55859
1715	L	D	R	U	1968	55859
1716	L	D	R	U	1968	55859
1718	L	D	E	N	1969	56184
1720	L	D	R	U	1968	56233
1721	M	D	E	N	1968	51385
1722	L	D	R	N	1966	48692
1722	L	D	E	N	1967	50701
1722	L	D	E	N	1969	56670
1723	L	D	R	N	1969	56746
1725	L	D	E	N	1970	56988
1726	L	D	E	N	1970	56988
1727	L	D	E	N	1970	56988
1728	L	D	E	N	1970	56988
1729	L	D	E	N	1970	56988
1730	L	D	E	N	1970	56988
1731	L	D	E	N	1970	56988
1732	L	D	E	N	1970	56988
1733	L	D	D	N	1965	48544
1734	L	D	R	N	1969	56174
1736	L	D	E	N	1967	56969
1737	L	D	E	N	1967	56969
1738	L	D	E	N	1967	56969
541-0049	L	D	R	H	1958	33893
0049	L	D	R	H	1960	40985
0049	L	D	E	H	1960	43057
0049	L	D	R	H	1965	43310
0049	L	D	E	H	1965	43842
0049	M	D	E	H	1957	34615
0049	M	D	E	H	1967	40653
0049	M	D	R	H	1959	41114
0049	M	D	R	H	1959	41117
0085	L	D	R	H	1960	33892
0085	L	D	R	H	1958	33893
0085	L	D	R	H	1959	41117
0085	L	D	E	H	1960	43057
0085	M	D	E	H	1967	40653
0093	M	D	E	H	1967	40653
0101	M	D	E	H	1967	40653
0102	M	D	E	H	1967	40653
0103	M	D	E	H	1967	40653
0115	L	D	R	N	1968	49641
0115	L	D	E	N	1968	59361
0126	M	D	E	H	1967	40653
0127	M	D	E	H	1967	40653
0130	M	D	E	H	1967	40653
0132	M	D	E	H	1967	40653
0147	M	D	E	H	1967	40653
0148	M	D	E	H	1967	40653
0149	M	D	E	H	1967	40653
0150	M	D	E	H	1967	40653
0159	L	D	R	N	1968	49641
0159	L	D	E	N	1968	59361
0171	L	D	R	N	1968	49641
0171	L	D	E	N	1968	59361
0172	L	D	R	N	1968	49641
0172	L	D	E	N	1968	59361
0173	L	D	R	N	1968	49641
0173	L	D	E	N	1968	59361
0179	L	D	R	N	1968	49641
0179	L	D	E	N	1968	59361
0228	L	D	R	N	1966	40575
0228	L	D	E	N	1966	40576
0276	L	D	R	N	1966	40575
0276	L	D	E	N	1966	40576
0407	L	D	R	H	1965	35050
0407	L	D	R	H	1966	46667
0407	L	D	R	H	1966	46668
0407	L	D	E	H	1968	50036
0407	L	D	E	H	1968	55245
0408	L	D	R	H	1965	35050
0408	L	D	R	H	1966	46668
0420	L	D	F	H	1961	39958
0420	L	D	R	H	1960	40985
0420	L	D	E	H	1960	43057
0420	L	D	R	H	1966	46668

Substance Number	Phys. State	Sub-ject	Lan-guage	Temper-ature	Year	TPRC Number
541-0420	L	D	R	H	1968	53225
0420	L	D	E	H	1968	53226
0420	M	D	E	H	1957	34615
0420	M	D	R	H	1966	39946
0420	M	D	R	H	1959	41114
0420	M	D	R	H	1959	41117
0421	M	D	R	H	1966	39946
0422	M	D	R	H	1966	39946
0423	M	D	R	H	1966	39946
0467	L	D	R	H	1960	33892
0467	L	D	R	H	1958	33893
0467	L	D	R	H	1959	41117
0470	M	D	R	H	1960	33892
0470	M	D	E	H	1954	39695
0471	M	D	R	H	1960	33892
0472	M	D	R	H	1960	33892
0473	M	D	R	H	1960	33892
0475	L	D	R	H	1958	33893
0475	L	D	R	H	1960	40985
0475	L	D	E	H	1960	43057
0475	M	D	E	H	1957	34615
0475	M	D	R	H	1959	41114
0475	M	D	R	H	1959	41117
0480	L	D	R	H	1960	33892
0481	L	D	R	H	1960	33892
0510	L	D	O	H	1965	34146
0522	M	D	E	H	1956	34611
0523	M	D	E	H	1956	34611
0524	M	D	E	H	1956	34611
0525	M	D	E	H	1956	34611
0527	L	D	R	H	1968	53225
0527	L	D	E	H	1968	53226
0527	M	D	R	H	1959	34610
0528	L	E	E	N	1962	55223
0529	L	D	E	N	1964	37391
0529	L	D	R	N	1960	40985
0529	M	D	E	N	1967	38575
0529	M	D	E	N	1968	48401
0548	M	C	E	N	1966	46543
0549	M	C	E	N	1966	46543
0550	M	C	E	N	1966	46543
0551	L	T	R	N	1965	46641
0551	L	T	E	N	1965	46642
0552	L	C	R	N	1965	46641
0552	L	C	E	N	1965	46642
0552	L	G	E	N	1967	49172
0553	L	C	R	N	1965	46641
0553	L	C	E	N	1965	46642
0554	L	C	R	N	1965	46641
0554	L	C	E	N	1965	46642
0555	L	C	R	N	1965	46641
0555	L	C	E	N	1965	46642
0556	L	C	R	N	1965	46641
0556	L	C	E	N	1965	46642
0557	L	C	R	N	1965	46641
0557	L	C	E	N	1965	46642
0558	L	C	R	N	1965	46641
0558	L	C	E	N	1965	46642
0559	L	C	R	N	1965	46641
0559	L	C	E	N	1965	46642
0560	L	C	R	N	1965	46641
0560	L	C	E	N	1965	46642
0561	L	C	R	N	1965	46641
0561	L	C	E	N	1965	46642
0562	L	C	R	N	1965	46641
0562	L	C	E	N	1965	46642
0563	L	C	R	N	1965	46641
0563	L	C	E	N	1965	46642
0564	L	C	R	N	1965	46641
0564	L	C	E	N	1965	46642
0568	L	C	R	N	1965	46641
0568	L	C	E	N	1965	46642
0569	L	C	R	N	1965	46641
0569	L	C	E	N	1965	46642
0569	L	C	E	U	1967	48534
0569	L	D	E	N	1968	49061
0569	L	G	E	N	1967	49172
0573	L	E	E	N	1962	55223
0579	M	E	E	N	1962	55223
0580	L	G	R	H	1965	46748
0580	L	G	E	H	1965	46749
0581	L	G	R	H	1965	46748
0581	L	G	E	H	1965	46749
0582	L	G	R	H	1965	46748
0582	L	G	E	H	1965	46749
0589	L	D	R	H	1966	46667
0591	L	D	R	H	1966	46667
0592	L	D	R	H	1966	46667
0597	L	D	R	H	1966	46667
0598	L	D	R	H	1966	46668
0599	L	D	R	H	1966	46668
0601	L	D	R	H	1966	46668
0602	L	D	R	H	1966	46668
0603	L	D	R	H	1966	46668
0604	L	D	R	H	1966	46668
0605	L	D	E	N	1960	43057
0605	L	D	R	N	1966	46671
0605	M	D	R	N	1959	44084
0605	M	D	E	N	1959	44085
0606	L	D	R	H	1966	46668
0607	L	D	R	H	1966	46668

Phys. State: **G.** Gas; **L.** Liquid; **M.** Multiphase; **P.** Powder; **S.** Solid

Substance Number	Phys. State	Subject	Language	Temperature	Year	TPRC Number
541-0608	L	D	R	H	1966	46668
0613	L	D	R	H	1965	43310
0613	L	D	E	H	1965	43842
0614	L	D	G	N	1948	40939
0614	L	D	R	N	1960	40985
0614	L	D	E	N	1960	43057
0614	M	D	E	N	1970	35560
0614	M	D	E	N	1952	35937
0614	M	D	R	N	1959	44084
0614	M	D	E	N	1959	44085
0615	L	D	E	N	1960	43057
0615	M	D	R	N	1959	44084
0615	M	D	E	N	1959	44085
0615	M	D	E	N	1966	55189
0618	M	D	R	N	1959	44084
0618	M	D	E	N	1959	44085
0639	L	D	R	N	1967	45820
0639	L	D	E	N	1967	47687
0640	L	D	R	N	1967	45820
0640	L	D	E	N	1967	47687
0641	L	D	R	N	1967	45820
0641	L	D	E	N	1967	47687
0642	L	D	E	N	1960	43057
0642	L	D	G	U	1968	48712
0642	L	D	E	N	1968	49394
0643	L	D	E	H	1963	35938
0643	L	D	R	H	1964	40968
0643	L	D	R	H	1960	40985
0643	L	D	E	H	1960	43057
0645	L	D	E	N	1960	43057
0646	L	D	G	N	1948	40939
0646	L	D	E	H	1960	43057
0647	L	D	E	N	1960	43057
0648	L	D	G	N	1948	40939
0648	L	D	E	N	1960	43057
0648	M	D	R	N	1959	41117
0649	L	D	E	N	1960	43057
0654	L	D	R	H	1965	42042
0655	L	D	R	H	1965	42042
0659	M	D	E	N	1968	48475
0660	L	C	R	H	1967	45773
0660	L	T	R	H	1967	48523
0662	M	D	E	N	1970	35560
0662	M	D	E	N	1968	48475
0665	L	D	E	N	1968	48786
0673	L	D	E	U	1968	42354
0674	L	D	E	U	1968	42354
0674	L	D	E	N	1968	49813
0674	L	D	E	N	1968	53036
0675	L	D	E	U	1968	42354
0676	L	D	E	U	1968	42354
0676	L	D	E	N	1968	49813
0680	L	D	E	N	1967	42893
0681	L	D	E	N	1969	54476
0686	L	D	E	H	1958	40934
0687	L	D	E	H	1958	40934
0688	L	D	E	H	1958	40934
0698	L	D	F	H	1961	39958
0709	L	D	E	N	1927	39700
0711	L	D	E	N	1927	39700
0714	M	D	E	N	1927	39700
0719	M	D	E	N	1927	39700
0740	L	D	R	N	1967	47472
0740	L	D	E	N	1967	47521
0741	L	D	R	N	1967	47472
0741	L	D	E	N	1967	47521
0742	L	D	R	N	1967	47472
0742	L	D	E	N	1967	47521
0743	L	D	R	N	1967	47472
0743	L	D	E	N	1967	47521
0744	L	D	R	N	1967	47472
0744	L	D	E	N	1967	47521
0745	L	D	R	N	1967	47472
0745	L	D	E	N	1967	47521
0746	L	D	R	N	1967	47472
0746	L	D	E	N	1967	47521
0747	L	D	R	N	1967	47472
0747	L	D	E	N	1967	47521
0748	L	D	R	N	1967	47472
0748	L	D	E	N	1967	47521
0749	L	D	R	N	1967	47472
0749	L	D	E	N	1967	47521
0750	L	D	R	N	1967	47472
0750	L	D	E	N	1967	47521
0751	L	D	R	N	1967	47472
0751	L	D	E	N	1967	47521
0752	L	D	R	N	1967	47472
0752	L	D	E	N	1967	47521
0755	L	D	E	H	1969	55489
0755	M	D	E	H	1968	51218
0763	M	D	E	H	1954	39695
0764	M	D	E	H	1954	39695
0765	L	D	E	U	1961	40006
0775	L	D	E	H	1955	40992
0786	L	D	G	H	1935	40926
0787	L	D	G	H	1935	40926
0788	L	D	G	N	1935	40926
0791	L	D	E	N	1968	49061
0793	L	D	E	N	1968	49061
0822	L	D	G	N	1935	40926
0824	L	D	G	N	1935	40926

Substance Number	Phys. State	Subject	Language	Temperature	Year	TPRC Number
541-0826	L	D	G	N	1948	40939
0827	L	D	F	N	1969	54974
0851	M	D	R	H	1969	37734
0852	M	D	R	H	1969	37734
0854	L	D	E	N	1952	35937
0854	L	D	E	N	1927	39700
0854	L	D	E	N	1927	44091
0854	L	D	R	N	1968	51806
0855	L	D	E	N	1927	44091
0855	M	D	E	N	1966	55189
0856	L	D	E	N	1927	44091
0857	L	D	E	N	1927	39700
0857	L	D	E	N	1927	44091
0858	L	D	E	N	1927	44091
0859	L	D	E	N	1927	44091
0864	M	D	E	N	1968	48401
0871	L	G	E	N	1967	49172
0872	L	G	E	N	1967	49172
0873	L	G	E	N	1967	49172
0874	L	G	E	N	1967	49172
0875	L	G	E	N	1967	49172
0876	L	G	E	N	1967	49172
0877	L	G	E	N	1967	49172
0878	L	G	E	N	1967	49172
0879	L	D	O	H	1967	48399
0884	L	D	E	N	1964	44417
0885	L	D	E	N	1964	44417
0886	L	D	E	N	1964	44417
0887	L	D	E	N	1964	44417
0888	L	D	E	N	1964	44417
0908	M	D	R	H	1969	37734
0911	L	D	R	H	1969	37711
0916	L	D	R	H	1969	37711
0917	L	D	R	H	1969	37711
0921	M	D	R	H	1962	55081
0922	L	D	R	N	1968	54118
0923	L	D	R	N	1968	50741
0924	L	D	R	H	1969	37711
0968	M	D	J	H	1967	46441
0969	L	D	O	H	1967	46438
0970	L	D	O	H	1967	46438
0972	M	D	J	H	1967	46441
0973	M	D	J	H	1967	46441
0974	M	D	J	H	1967	46441
0991	L	D	R	H	1969	37711
0996	L	D	R	H	1969	37711
1001	L	D	R	N	1959	41117
1014	L	G	E	N	1965	40433
1041	M	C	R	N	1968	52629
1041	M	C	E	N	1968	52630
1046	L	D	R	H	1968	53225
1046	L	D	E	H	1968	53226
1048	L	D	R	H	1968	53225
1048	L	D	E	H	1968	53226
1049	L	D	R	H	1968	53225
1049	L	D	E	H	1968	53226
1052	L	D	R	H	1968	53225
1052	L	D	E	H	1968	53226
1055	L	D	R	H	1968	53225
1055	L	D	E	H	1968	53226
1063	L	C	R	H	1968	53753
1063	L	C	E	H	1968	53871
1065	L	D	E	N	1969	54060
1066	L	D	E	N	1969	54060
1067	L	D	E	N	1969	54060
1068	L	D	E	N	1969	54060
1069	L	D	E	N	1969	54060
1070	L	D	E	N	1969	54060
1071	L	D	E	N	1969	54060
1073	L	D	E	H	1968	50036
1073	L	D	E	H	1968	55245
1074	L	D	E	H	1968	50036
1074	L	D	E	H	1968	55245
1075	L	D	E	H	1968	50036
1075	L	D	E	H	1968	55245
1076	L	D	E	H	1968	50036
1076	L	D	R	H	1968	53017
1076	L	D	E	H	1968	55245
1076	L	D	E	H	1968	63448
1098	L	D	R	N	1968	51381
1099	L	D	R	N	1968	51381
1100	L	D	R	N	1968	51381
1101	L	D	R	N	1968	51381
1111	L	D	R	N	1968	51806
1112	L	D	R	N	1968	51806
1138	L	C	E	N	1970	56291
1180	L	D	E	N	1968	53036
1181	L	D	R	N	1965	41872
1182	L	E	R	F	1965	41870
1183	M	D	R	U	1965	41879
1184	L	D	J	U	1966	44482
1185	L	D	J	U	1966	44482
1186	L	D	J	U	1966	44482
1187	L	G	G	U	1967	46328
1188	L	G	G	U	1967	46328
1189	L	G	G	U	1967	46328
1190	L	C	E	U	1967	48534
1191	L	D	G	U	1968	48712
1192	L	D	G	U	1968	48712
1193	L	D	G	U	1968	48712
1194	L	D	G	U	1968	48712

Substance Number	Phys. State	Subject	Language	Temperature	Year	TPRC Number
541-1195	L	D	G	U	1968	48712
1196	L	D	G	U	1968	48712
1197	L	D	G	U	1968	48712
1198	L	D	R	U	1967	48778
1198	L	T	R	U	1969	53436
1198	L	T	E	U	1969	63449
1199	L	D	E	N	1968	53036
1200	L	D	E	N	1968	53036
1202	L	D	E	N	1968	53036
1203	M	D	E	N	1970	35560
1204	M	D	E	N	1970	35560
1205	M	D	E	N	1970	35560
1206	M	D	E	N	1970	35560
1207	M	D	E	N	1970	35560
1208	M	D	E	N	1970	35560
1209	M	D	E	N	1970	35560
1210	M	D	E	N	1970	35560
1211	M	D	E	N	1970	35560
1212	M	D	E	N	1970	35560
1247	L	D	E	N	1969	55599
1248	L	D	E	N	1969	55599
1249	L	D	E	N	1969	55599
1250	L	D	E	N	1969	55599
1251	L	D	E	N	1969	55599
1256	L	D	E	N	1969	40356
1260	L	D	E	N	1969	40356
1261	L	D	E	N	1969	40356
1262	L	D	E	N	1969	40356
1263	L	D	E	N	1969	40356
1278	L	D	R	N	1968	51006
1284	L	D	F	N	1969	56060
1285	L	D	F	N	1969	56060
1305	L	D	R	H	1969	37701
1306	L	D	R	H	1969	37701
1337	L	D	O	H	1968	49557
1338	L	D	O	H	1968	49557
1367	L	D	E	N	1969	40356
1368	L	D	E	N	1969	40356
1369	L	D	E	H	1969	55489
1370	L	D	E	H	1969	55489
1371	L	D	E	H	1969	55489
1372	L	D	E	H	1969	55489
1373	L	D	R	U	1968	55512
1374	L	D	R	U	1968	55512
1375	L	D	R	U	1968	55859
1376	L	D	R	U	1968	55859
1377	L	D	R	U	1968	55859
1378	L	D	R	U	1968	55859
1379	L	D	R	U	1968	55859
1380	L	D	R	U	1968	55859
1381	L	D	E	U	1970	56990
1382	L	D	E	U	1970	56990
1383	M	D	E	N	1968	51385
1423	L	D	R	H	1968	53017
1423	L	D	E	H	1968	63448
1424	L	D	R	H	1968	53017
1424	L	D	E	H	1968	63448
1425	L	D	R	H	1968	53017
1425	L	D	E	H	1968	63448
1426	L	D	R	H	1968	53017
1426	L	D	E	H	1968	63448
1427	L	D	R	H	1968	53017
1427	L	D	E	H	1968	63448
1428	L	T	R	U	1969	53436
1428	L	T	E	U	1969	63449
1430	L	D	R	H	1968	49800
1430	L	D	E	H	1968	50636
1431	L	D	R	H	1968	49800
1431	L	D	E	H	1968	50636
1432	L	D	R	H	1968	49800
1432	L	D	E	H	1968	50636
1433	L	D	R	H	1968	49800
1433	L	D	E	H	1968	50636
1434	L	D	R	H	1968	49800
1434	L	D	E	H	1968	50636
1435	L	D	R	H	1968	49800
1435	L	D	E	H	1968	50636
1436	L	D	R	H	1968	50073
1436	L	D	E	H	1968	63445
1437	L	D	R	H	1968	50073
1437	L	D	E	H	1968	63445
1438	L	D	R	H	1968	50073
1438	L	D	E	H	1968	63445
1439	L	D	R	H	1968	50073
1439	L	D	E	H	1968	63445
1440	L	D	R	H	1968	50073
1440	L	D	E	H	1968	63445
1441	L	D	R	H	1968	50073
1441	L	D	E	H	1968	63445
1442	L	D	R	H	1968	50073
1442	L	D	E	H	1968	63445
1443	L	D	R	H	1968	50073
1443	L	D	E	H	1968	63445
1444	L	D	R	H	1968	50073
1444	L	D	E	H	1968	63445
1445	L	D	R	H	1968	50073
1445	L	D	E	H	1968	63445
651-0066	L	D	G	U	1966	44338
0074	L	D	E	N	1948	36924
0237	M	D	O	N	1964	44490

Phys. State: **G.** Gas; **L.** Liquid; **M.** Multiphase; **P.** Powder; **S.** Solid

Substance Number	Phys. State	Sub-ject	Lan-guage	Temper-ature	Year	TPRC Number	Substance Number	Phys. State	Sub-ject	Lan-guage	Temper-ature	Year	TPRC Number	Substance Number	Phys. State	Sub-ject	Lan-guage	Temper-ature	Year	TPRC Number
651-0476	L	D	E	N	1967	45271														
0476	L	D	R	U	1967	47608														
0544	L	D	E	U	1966	51158														
0564	L	D	E	N	1966	46545														

Phys. State: G. Gas; L. Liquid; M. Multiphase; P. Powder; S. Solid

Part C
BIBLIOGRAPHY

USE OF THE BIBLIOGRAPHY

A. GENERAL CONSIDERATIONS

Because of the wide variety of literature sources cited (i.e., serial publications, dissertations, government and industrial reports, books, etc.), three specific formats for bibliographic citations are used in the *Bibliography*. In connection with the establishment of bibliographic format, a number of problems of general character are encountered in reporting bibliographic information on a broad scope, TPRC's procedures in coping with these problems are fully described below:

1. Titles reported in the *Bibliography* are taken either from an abstract or from the original work. In the case of translated titles, wide discrepancies often exist between various sources. In general, TPRC makes no special effort to check the accuracy of titles.

2. The names of authors reported in abstracts, and even in original publications, are at times misspelled and/or incomplete. TPRC attempts to report the correct names of authors in the *Author Index*. However, the *Bibliography* cites author names as found in the original work or its abstract.

3. The names of scientific and technical journals are normally abbreviated according to the notations used by the abstracting journals. In those cases where a journal name is not applicable, the name of the publisher, symposium, or disseminating agency is entered in place of the journal name, depending upon the reference work.

4. Keypunching format limitations necessitated the adoption of substitute representations for some of the symbols and alphabetic and numeric arrangements. The following are examples of substitute representations used in the *Bibliography*:

   a. Brackets [ ] are shown as // //.
   b. Parentheses ( ) are shown as / /.
   c. Subscripts are either spelled out or written on the line. Examples:
   C_p ≡ specific heat at constant pressure; $(C_{12}H_{26})$ ≡ /C12H26/; etc.

d. Superscripts are spelled out. Examples: V^2 ≡ V squared; $\sqrt{2}$ ≡ square root of two; etc.

e. Diacritical marks used with proper names or words have been omitted, and no attempt has been made to insert vowels or other speech sounds.

B. ABBREVIATIONS AND ACRONYMS USED IN BIBLIOGRAPHIC CITATIONS

AAAS........American Association for the Advancement of Science
AAIE........ American Association of Industrial Engineers
AD.......... Prefix Catalog Codes for Defense Documentation Center (DDC) (Formerly ASTIA)
AEC......... Atomic Energy Commission
AEDC........Arnold Engineering and Development Center
AF...........Air Force
AFB......... Air Force Base
AFML.......Air Force Materials Laboratory
AFWL.......Air Force Weapons Laboratory
AGARD......Advisory Group for Aeronautical Research and Development
AIAA........ American Institute of Aeronautics and Astronautics
AIAE........ Associate Institute of Automobile Engineering
AICE........American Institute of Chemical Engineering
AIEE........American Institute of Electrical Engineers (Now IEEE)
AIIE......... American Institute of Industrial Engineers
AIME........ American Institute of Mining and Metallurgical Engineers
AIP......... American Institute of Physics
AISL........American Iron and Steel Institute
AM......... Applied Mechanics Reviews
AMRA....... Army Materials Research Agency
ANL........ Argonne National Laboratory
AOA........American Ordnance Association
APDA.......Atomic Power Development Association, Inc.
API......... American Petroleum Institute
ARB........ ASTIA Report Bibliography
ARC........ Aeronautical Research Council
ARDE.......Armament Research and Development Establishment
ARF........ Armour Research Foundation
ARL........ Aeronautical Research Laboratories
ARS........ American Rocket Society (Now AIAA)
ASHRAE..... American Society of Heating, Refrigeration, Air Conditioning Engineers
ASM......... American Society for Metals
ASME.......American Society for Mechanical Engineers
ASR........ American Society of Rocketry
ASSE........ American Society of Safety Engineers
ASTE........ American Society of Tool Engineers

ASTIA....... Armed Services Technical Information
Agency (Now DDC)
ASTM........ American Society for Testing and Materials
ASTME....... American Society of Tool and Manufacturing
Engineers

BC........... British Ceramic Society
BISI......... British Iron and Steel Industry
BR........... Battelle Technical Review
BTJ.......... ASTIA Bibliographies
BU ORD..... Bureau of Ordnance
BU WEPS.... Bureau of Weapons

CA.......... Chemical Abstracts
CFSTI........ Clearinghouse for Federal Scientific and
Technical Information (Now NTIS)
CNEN........ Comitato Nazionale per L'Energia Nucleare
(National Nuclear Energy Committee)
CNDR........ Consiglio Nazionale Delle Ricerche
(National Research Council of Italy)
CNES........ Centre National D'Etudes Spatiales
(National Center for Space Studies)
COSATI...... Committee on Scientific and Technical
Information (Formerly COSI)
COSI......... Committee on Scientific Information
(Now COSATI)
CPIA......... Chemical Propulsion Information Agency
CRREL....... Cold Regions Research and Engineering
Laboratory
CSTAR....... Confidential Scientific and Technical
Aerospace Reports (Now STAR)

DASA........ Defense Atomic Support Agency
DDC......... Defense Documentation Center
DDR&E...... Directorate of Defense Research and
Engineering
DMIC........ Defense Metals Information Center
DOD......... U. S. Department of Defense
DOFL........ Diamond Ordnance Fuze Laboratory (HDL)
DPGR....... Dugway Proving Ground

ECOM....... Electronics Command, U. S. Army
EPIC........ Electronic Properties Information Center
EURATOM... European Atomic Energy Community

FAI.......... Federation Aeronautique Internationale
FID.......... International Federation of Documentation
FPL.......... Forest Production Laboratory
FTD......... Foreign Technology Division

GRA........ Government Research Announcements
(Formerly UGAR)

HDL......... Harry Diamond Laboratory

IAA......... International Academy of Astronautics
IAS.......... Institute of Aero/Space Science
(Now AIAA)
IEE......... Institute of Electrical and Electronic
Engineers (Formerly AIEE & IRE)
IEEE........ Institute of Electrical and
Electronic Engineers
IIT.......... Illinois Institute of Technology Research
Institute
IPST........ Israel Program for Scientific
Translation, Ltd.
IRE......... Institute of Radio Engineers (Now IEEE)
ISA.......... Instrument Society of America

JA.......... Ceramic Abstracts (in Journal of the
American Ceramic Society)
JAERI....... Japan Atomic Energy Research Institute

JANAF....... Joint Army–Navy–Air Force
JPL......... Jet Propulsion Laboratory

KAPL....... Knolls Atomic Power Plant

MA......... Metallurgical Abstracts, Series II
MR......... ASM Review of Current Metal Literature
(in Metals Review)

NADC....... Naval Air Development Center
NAS........ National Academy of Science
NASA....... National Aeronautics and Space
Administration
NASC....... National Aeronautics and Space Council
NAS/NRC.... National Academy of Science—National
Research Council
NATO....... North Atlantic Treaty Organization
NBS......... National Bureau of Standards
NDAC....... National Defense Advisory Committee
NDRC....... National Defense Research Council
NOTS....... Naval Ordnance Test Station
NPL....... National Physical Laboratory (England)
NRC........ National Research Council (Canada)
NRD........ Naval Research and Development
(Now ONR)
NRL........ Naval Research Laboratory
NS......... Nuclear Science Abstracts
NSF......... National Science Foundation
NTIS........ National Technical Information Service
(Formerly CFSTI)

OAR........ Office Aerospace Research
ONERA..... Office National D'Etudes et de Recherches
Aerospatiales (France)
ONR........ Office of Naval Research
ORNL....... Oak Ridge National Laboratory
OSI......... Office of Scientific Information
OSR........ Office of Scientific Research
OTS......... Office of Technological Service

RA.......... Refrigeration Abstracts
REIC........ Radiation Effects Information Center
RIA......... Rock Island Arsenal
RM.......... ASM Review of Metal Literature
RPL......... Rocket Propulsion Laboratory
RR.......... U. S. Government Research Reports (OTS)

SA.......... Physics Abstracts (Section A of Science
Abstracts)
SIPRE....... Snow, Ice and Permafrost Research
Establishment
STAR....... Scientific and Technical Aerospace Reports
(Formerly CSTAR)

TA.......... Technical Abstracts Bulletin (AD)
TML........ Titanium Metallurgical Laboratory
TPRC....... Thermophysical Properties Research Center
TT.......... Technical Translations (OTS)

UCRL....... University of California Radiation
Laboratory
UGAR....... United States Government Announcements
Reports (Now GRA)
USAEC...... U. S. Atomic Energy Commission
USBM....... U. S. Bureau of Mines
USDA....... U. S. Department of Agriculture
USJPRS...... U. S. Joint Publication Research Service
USL......... Underwater Sound Laboratory

WADC....... Wright Air Development Center
WADD...... Wright Air Development Division
WRAC....... Willow Run Aeronautical Center

BIBLIOGRAPHY

TPRC Number	Bibliographic Citation

33747 AUGMENTATION OF THE INTENSITY OF RADIATION EMITTED BY LOW-VOLTAGE GALLIUM ARSENIDE DIODES.
SHIROKSHINA Z V   KONSTANTINOVA E N
OPT I SPEKTROSKOPIYA
20   1   173-5   1966

33748 AUGMENTATION OF THE INTENSITY OF RADIATION EMITTED BY LOW-VOLTAGE GALLIUM-ARSENIDE DIODES. //ENGLISH TRANSLATION OF OPT. I SPEKTROSKOPIYA 20 /1/ 173-5, 1966.//
SHIROKSHINA Z V   KONSTANTINOVA E N
OPT SPECTRY
20   1   92-3   1966

33749 THE TOTAL HEMISPHERICAL THERMAL EMITTANCE OF NICKEL AS A FUNCTION OF OXIDE THICKNESS IN THE TEMPERATURE RANGE 400-900 DEGREES C. PH.D. THESIS.   /SEE ALSO TPRC NO. 48367./
SHELTON J L   RICE UNIV   HOUSTON TEXAS
UNIVERSITY MICROFILMS INC   ANN ARBOR MICH
65-10351   1-146   1965

33775 SPECTROSCOPY IN THE 5 TO 400 WAVENUMBER REGION WITH THE GRUBB PARSONS INTERFEROMETRIC SPECTROMETER.
WHEELER R G   HILL J C
J OPT SOC AM
56   5   657-65   1966

33797 DIFFUSION KINETICS IN BINARY SYSTEMS IN THE PRESENCE OF CERTAIN PHASES.
BORISOV V I   GOLIKOV V M   DUBININ G N
FIZ METAL I METALLOVED
20   1   69-77   1965

33798 DIFFUSION KINETICS IN BINARY SYSTEMS IN THE PRESENCE OF CERTAIN PHASES. //ENGLISH TRANSLATION OF FIZ. METAL. I METALLOVED. 20 /1/ 69-77, 1965.//
BORISOV V I   GOLIKOV V M   DUBININ G N
PHYS METALS METALLOG /USSR/
20   1   63-70   1965

33799 OPTICAL PROPERTIES OF VACUUM-EVAPORATED WHITE TIN.
MAC RAE R A   ARAKAWA E T   WILLIAMS M W
PHYS REV
162   3   615-20   1967

33800 THERMAL RESISTANCE OF PRESSED CONTACTS.
THOMAS T R   PROBERT S D   WELSH COLLEGE OF ADVANCED TECHNOLOGY   CARDIFF WALES   UKAEA
HMSO AND CFSTI
TRG REP   T 1013//R/X//
1-22   1965

33811 A CELL MODEL FOR CONDUCTION PROCESSES IN POWDERS AT LOW TEMPERATURES.
BRODIE D E   MATE C F
BULL INST INTERN FROID ANNEXE
2   133-8   1965   CA   65   11821

33813 CALCULATING THERMAL CONTACT RESISTANCE OF MACHINED METAL SURFACES. //ENGLISH TRANSLATION OF TEPLOENERGETIKA 12 /10/ 79-83, 1965.//
SHLYKOV YU P
THERMAL ENG /USSR/
12   10   102-8   1965

33814 CALCULATING THERMAL CONTACT RESISTANCE OF MACHINED METAL SURFACES.
SHLYKOV YU P
TEPLOENERGETIKA
12   10   79-83   1965

33824 OPTICAL TUNNELING AND ITS APPLICATIONS TO OPTICAL FILTERS.
BAUMEISTER P W
APPL OPT
6   5   897-905   1967

33840 UNUSUAL RECONNAISSANCE CONCEPTS VOL. II.   SOURCES OF EXPERIMENTAL ERRORS IN SPECTROPHOTOMETRIC MEASUREMENTS.
GOERGE D   LIMPERIS T   WILLOW RUN LABORATORIES
UNIVERSITY OF MICHIGAN   ANN ARBOR MICHIGAN   USAF
DDC
AFAL-TR-65-331 AD 481796
1-13   1966

33844 ELLIPSOMETER STUDIES OF SILVER SINGLE CRYSTALS. PH.D. THESIS.
FROMHOLD A T JR   CORNELL UNIVERSITY
UNIV MICROFILMS PUBL
62-169   1-187   1961
NUOVO CIMENTO
28   1127-   1963

33854 PHYSICAL AND OPTICAL PROPERTIES OF PROJECTION SCREENS.
KLAIBER R J   NAVAL TRAINING DEVICE CENTER
ORLANDO FLORIDA
DDC

33858 HEAT TRANSFER IN STRUCTURAL HONEYCOMB COMPOSITES AT HIGH TEMPERATURE.
MINGES M L   AIR FORCE MATERIALS LAB
WRIGHT-PATTERSON AIR FORCE BASE   OHIO
DDC
AFML-TR-65-233 AD 481418
1-53   1966

33864 MEASUREMENT OF THE THERMAL DIFFUSIVITIES OF SOME SINGLE-LAYER WALLS IN BUILDINGS.
PRATT A W   LACY R E
INTERN J HEAT MASS TRANSFER
9   4   345-53   1966   CA   64   11072

33865 RECOMMENDED VALUES OF THE THERMOPHYSICAL PROPERTIES OF EIGHT ALLOYS, MAJOR CONSTITUENTS AND THEIR OXIDES.
TOULOUKIAN Y S   DIRECTOR AND EDITOR FOR STAFF OF THERMOPHYSICAL PROPERTIES RESEARCH CENTER PURDUE UNIV
LAFAYETTE INDIANA   NBS   AND
TPRC
TPRC   REPT 16 NASA CR-71699 N66-23802
1-540   1966

33870 METHOD FOR ESTIMATING RATIO OF ABSORPTANCE TO EMITTANCE.
WIEBELT J A   OKLAHOMA STATE UNIV RESEARCH FOUNDATION   COLLEGE
NASA
NASA   CR-87322 HT 3-67   N67-33459
1-18   1967

33879 SKY LUMINANCES AND THE DIRECTIONAL LUMINOUS REFLECTANCES OF OBJECTS AND BACKGROUNDS FOR A MODERATELY HIGH SUN.
GORDON J I   CHURCH P V
APPL OPT
5   5   793-801   1966

33880 ATMOSPHERIC PROPERTIES AND REFLECTANCES OF OCEAN WATER AND OTHER SURFACES FOR A LOW SUN.
BOILEAU A R   GORDON J I
APPL OPT
5   5   803-13   1966

33882 NOTE ON REFLECTANCE MEASUREMENTS ON METALS.
MUELLER W E
APPL OPT
5   5   876-7   1966

33884 THE PHOTOEMISSION METHOD FOR OBTAINING MIRROR AND DIFFUSE REFLECTION SPECTRA IN THE VACUUM ULTRAVIOLET REGION OF THE SPECTRUM.
VILESOV F I   AZGRUBSKII A A   KIRILLOVA M M
OPTIKA I SPEKTROSKOPIYA
23   1   153-7   1967

33885 THE PHOTOEMISSION METHOD FOR OBTAINING MIRROR AND DIFFUSE REFLECTION SPECTRA IN THE VACUUM ULTRAVIOLET REGION OF THE SPECTRUM. //ENGLISH TRANSLATION OF OPTIKA I SPEKTROSKOPIYA 23 /1/ 153-7, 1967.//
VILESOV F I   AZGRUBSKII A A   KIRILLOVA M M
OPT SPECTRY
23   1   79-82   1967

33892 EFFECT OF CARBON ON THE SURFACE TENSION OF LIQUID COBALT AND NICKEL AND THEIR INTERFACIAL TENSIONS AT THE SURFACE OF AL2O3.
EREMENKO V N   NIZHENKO V I
UKRAIN KHIM ZHUR
26   423-8   1960

33893 SURFACE TENSION OF METALS OF THE IRON GROUPS.
NIZHENKO YU N   EREMENKO V N   IVASHCHENKO YU N
NIZHENKO V I   FESENKO V V
IZV AKAD NAUK SSSR OTD TEKH NAUK
7   144-6   1958

33896 OFF-SPECULAR PEAKS IN THE DIRECTIONAL DISTRIBUTION OF REFLECTED THERMAL RADIATION.
TORRANCE K E   SPARROW E M
TRANS ASME J HEAT TRANSFER
88   C   2   223-30   1966

33902 STABLE WHITE COATINGS.   INTERIM REPORT.   APRIL 1 TO OCTOBER 1, 1962.
ZERLAUT G A   HARADA Y   ARMOUR RESEARCH FOUNDATION ILL INST OF TECH   CHGO ILL
NASA
NASA-CR-51002 ARF 3207-14 X-63-15369
N65-10374   1-94   1962

33905 EXPERIMENTAL STUDY OF THE INFLUENCE OF THE INTERSTITIAL FLUID UPON THERMAL CONTACT RESISTANCE.
BARDON J-P   CORDIER H
COMPT REND
261   23   5013-16   1965

33923 ON THE TRANSFER OF ENERGY BETWEEN A GAS AND A SOLID.
SHIN H
J PHYS CHEM

TPRC Number	Bibliographic Citation

33924 COLOR CENTERS AND HYDROGEN-DEUTERIUM EXCHANGE IN GAMMA-IRRADIATED SILICA-ALUMINA CATALYSTS.
SHIPMAN G F
J PHYS CHEM
70   4   1120-5   1966

33925 THE ADSORPTION OF ETHYLENE ON A SERIES OF NEAR-FAUJASITE ZEOLITES STUDIED BY INFRARED SPECTROSCOPY AND CALORIMETRY.
CARTER J L   YATES D J C   LUCCHESI P J
ELLIOTT J J   KEVORKIAN V
J PHYS CHEM
70   4   1126-36   1966

33944 FOUNDATIONAL RESEARCH 1964. PART I.
SLAWSKY Z I   NAVAL ORDNACE LAB   WHITE OAK   MD
DDC
AD 469433L NOLR-1259 /PT 1/
    1-98   1964   TA   65-20 A-10

33961 THERMAL RADIATION PROPERTIES OF NONOXIDE COMPOUNDS.
SCHATZ E A
J OPT SOC AM
56   4   465-9   1966   CA   64   18450

33962 INFRARED SPECTRAL REFLECTANCE OF FROST.
KEEGAN H J   WEIDNER V R
J OPT SOC AM
56   4   523-4   1966   CA   65   3181

33965 D-XYLOSE. AN ULTRAVIOLET-TRANSMITTING CEMENT.
LAULAINEN N S   MC DERMOTT M N
J OPT SOC AM
56   4   528   1966

33966 STRUCTURAL AND OPTICAL CHARACTERISTICS OF THIN GALLIUM ARSENIDE FILMS. II.
PANKEY T JR   DAVEY J E
J APPL PHYS
37   4   1507-15   1966

33970 EFFECT OF CRYSTALLITE SIZE ON THE INFRARED DISPERSION OF LITHIUM FLUORIDE FILMS.
MARTIN T P   TURNER A F
J APPL PHYS
37   4   1749-54   1966

33971 RADIANT HEAT TRANSFER IN FIBROUS THERMAL INSULATION.
HAGER N E JR   STEERE R C
J APPL PHYS
38   12   4663-8   1967

33974 HIGH TEMPERATURE, HIGH EMITTANCE INTERMETALLIC COATINGS. PART II EMITTANCE AND REFLECTANCE OF INTERMETALLIC COATINGS.
SCHATZ E A   ALVAREZ G H   BURKS T L
COUNTS C R III   DUNDERLEY F J   AMERICAN
MACHINE AND FOUNDRY CO   ALEXANDRIA VA   USAF
DDC
ML-TDR-64-179
AD-472439   1-85   1964

33976 THE PHOTOCHEMICAL REACTION IN A SOLID LAYER OF POLY/VINYL CINNAMATE/.
KIRSH YU E   LYALIKOV K S   KALNINSH K K
ZH FIZ KHIM
39   8   1886-   1965

33977 THE PHOTOCHEMICAL REACTION IN A SOLID LAYER OF POLY/VINYL CINNAMATE/. //ENGLISH TRANSLATION OF ZH. FIZ. KHIM. 39 /8/ 1886- , 1965.//
KIRSH YU E   LYALIKOV K S   KALNINSH K K
RUSS J PHYS CHEM
39   8   1002-4   1965

33988 THERMAL INSULATION SYSTEMS. A SURVEY.
GLASER P E   BLACK I A   LINDSTROM R S
RUCCIA F E   WECHSLER A E   TECHNOLOGY
UTILIZATION DIVISION   NASA
NASA
NASA-SP-5027
    1-148   1967

33989 EXPERIMENTAL MEASUREMENTS OF NONEQUILIBRIUM AND EQUILIBRIUM RADIATION FROM PLANETARY ATMOSPHERES.
THOMAS G M   MENARD W A
AIAA JOURNAL
4   2   227-37   1966

34005 THEORETICAL MODEL FOR THE MOTION OF KRYPTON IN A HYDROQUINONE CLATHRATE.
BARNETT B   HAZONY Y
J CHEM PHYS
43   10   3462-7   1965

34029 OPTICAL PROPERTIES OF BIVALENT RARE EARTH METALS AND ALKALINE EARTH METALS.
MUELLER W E
PHYS LETTERS
17   2   82-3   1965   CA   63   9235

34038 SELECTIVE SPECTRAL CHARACTERISTICS AS AN IMPORTANT FACTOR IN THE EFFICIENCY OF SOLAR COLLECTORS. FROM CONFERENCE ON SOLAR ENERGY. THE SCIENTIFIC BASIS.
GIER J T   DUNKLE R V   UNIVERSITY OF CALIFORNIA
BERKELEY CALIFORNIA
LC
    41-56   1955

34040 THERMAL RADIATIVE PROPERTIES OF CARBON CRYO-DEPOSITS FROM 0.5 TO 1.1 MICRONS. /SEE ALSO TPRC NO. 34345./
MC CULLOUGH B A   WOOD B E   DAWSON J P   ARO
INC ARNOLD AIR FORCE STATION   TENN   USAF
DDC
AEDC-TR-65-94 AD 468632
    1-84   1965   CA   67   69105

34041 VACUUM INTEGRATING SPHERES FOR MEASURING CRYODEPOSIT REFLECTANCES FROM 0.35 TO 15 MICRONS.
WOOD B E   MC CULLOUGH B A   DAWSON J P
BIRKEBAK R C   ARO INC ARNOLD AIR FORCE
STATION TENN   USAF
DDC
AEDC-TR-65-178 AD 468609 N65-33338
    1-24   1965

34045 INSTRUMENTATION FOR MEASURING THERMAL CHARACTERISTICS OF SURFACES. PART 2. AN INTEGRATING SPHERE TO MEASURE THE VARIATION OF REFLECTANCE WITH ANGLE OF INCIDENCE.
PORTER J   BUTLER E A W   ROYAL AIRCRAFT
ESTABLISHMENT   FARNBOROUGH ENGLAND
DDC
RAE-TR-65155 AD 470387
    1-23   1965   TA   65-21 A-139

34048 THERMAL CONDUCTIVITY-TEMPERATURE RELATIONSHIP FOR NINE GLASS AND ASBESTOS FIBER-REINFORCED AIRCRAFT PLASTICS.
LEWIS W   FOREST PRODUCTS LAB   MADISON WIS
DDC
RN-FPL-36   AD 470821
    1-15   1965   CA   63   16535

34050 A METHOD TO IMPROVE THE OPTICAL CHARACTERISTICS OF THE BLACK POLYETHYLENE SKIN SIMULANT. FINAL RESEARCH REPT.
MURTHA T D   DERKSEN W L   MONAHAN T I   NAVAL
APPLIED SCIENCE LAB   BROOKLYN N Y
DDC
AFSWP 845   AD 471034
    1-7   1955

34051 THERMAL SPACE SIMULATION TESTING. COMPARISON OF RESULTS OBTAINED FROM SOLAR AND HEAT FLUX IRRADIATED SURFACES.
LATTURE N C   ARO INC   ARNOLD
STATION TENN   USAF
DDC
AEDC-TR-65-107 AD 468409
    1-38   1965   PA   N65-3-21 3719

34058 SILICON CARBIDE DATA SHEETS.
NEUBERGER M   ELECTRONIC PROPERTIES INFORMATION
CENTER   CULVER CITY CALIF   USAF
DDC
AD 465161   1-110   1965

34059 MECHANICAL AND PHYSICAL PROPERTIES OF SUPER ALLOY AND COATED REFRACTORY ALLOY FOILS.
LEGGETT H   COOK J L   SCHWAB D E   POWERS C T
DOUGLAS AIRCRAFT COMPANY INC   SANTA MONICA CALIF
USAF
DDC
AFML-TR-65-147 AD 468607
    1-954   1965

34064 INTERCHANGE OF ENERGY BETWEEN GASEOUS MOLECULES AND A SOLID SURFACE. ACCOMMODATION COEFFICIENT.
PEREZ-MASIA A
ANALES REAL SOC ESPAN FIS QUIM
61   B   1   93-107   1965

34072 STUDY OF SPECTRAL REFLECTION COEFFICIENTS OF THERMAL RADIATION RECEIVER COATINGS FOR THE WAVELENGTH RANGE FROM 10 TO 200 A. //ENGLISH TRANSLATION OF OPTIKA I SPEKTROSKOPIYA 10 /5/ 657-62, 1961.//
KOZYREV B P   KROPOTKIN M A
OPTICS AND SPECTROCOPY
10   345-8   1961

34076 EFFECTIVE THERMAL CONDUCTIVITY OF GRANULAR BEDS.
BRETSZNAJDER S   ZIOLKOWSKI D
CHEM STOSOWANA
2   B   2   129-53   1965

34115 HEAT TRANSFER IN POROUS MEDIA WITH VARIOUS MOISTURE CONTENTS.
PALOSAARI S   NORDEN H V
KEM TEOLLISUUS
22   5   359-64   1964   CA   63   15873

TPRC Number	Bibliographic Citation

34117 OPTICAL PROPERTIES OF SINGLE-CRYSTAL FILMS OF LEAD SULFIDE, LEAD SELENIDE, AND LEAD TELLURIDE.
SEMILETOV S A   VORONINA I P   KORTUKOVA E I
KRISTALLOGRAFIYA
10   4   515-19   1965   CA   63   12425

34121 DETERMINATION OF MECHANICAL AND THERMOPHYSICAL PROPERTIES OF REFRACTORY METALS.
BECK E J   MARTIN CO   DENVER COLO   USAF
DDC
RTD-CR-65-1 ML-TR-65-247
AD 471505   1-192   1965   TA   65-22 A-54

34137 EMISSIVITY OF GRANULAR SURFACES AT RESONANCE FREQUENCIES. FROM PROC. OF THE THIRD SYMP. ON REMOTE SENSING OF ENVIRONMENT.
BLOCK M J   BLOCK ENGINEERING INC   CAMBRIDGE MASS
CFSTI
AD 614032   N65-33578
435-40   1965   CA   67   17678

34138 THE CONSEQUENCES OF TERRESTRIAL SURFACE INFRARED EMISSIVITY. FROM PROC. OF THE THIRD SYMP. ON REMOTE SENSING OF ENVIRONMENT.
BUETTNER K J K   KERN C D   CRONIN J F
WASHINGTON UNIV   SEATTLE   USAFCRL
CFSTI
AD 614032   N65-33586
549-61   1965   PA   N65-3-22 3909

34146 PHYSICAL PROPERTIES OF HIGH-MANGANESE SLAG AND SEMIPRODUCT.
RASHEVA I   BOBKOVA O   RASHEV T   OIKS G
PETUKHOV V
RUDODOBIV MET /SOFIA/
20   7   19-23   1965   CA   63   15931

34183 INFRARED-SPECTROSCOPIC INVESTIGATION OF THE VITRIFICATION AND CRYSTALLIZATION OF LEAD GLASS WITH A COMPOSITION CORRESPONDING TO THE COMPOUND PB O.SI O2.
SMIRNOVA E V
IZV AKAD NAUK SSSR NEORG MATER
4   7   1124-8   1968

34195 DETERMINATION OF THE THERMAL CONDUCTIVITY OF PARAFFIN AT LOW TEMPERATURES.
KRUPSKII I N   DOLGOPOLOV D G   MANZHELII V G
KOLOSKOVA L A
INZHENER-FIZ ZH
8   1   11-15   1965

34196 DETERMINATION OF THE THERMAL CONDUCTIVITY OF PARAFFIN AT LOW TEMPERATURES. //ENGLISH TRANSLATION OF INZHENER.-FIZ. ZH. 8 /1/ 11-5, 1965.//
KRUPSKII I N   DOLGOPOLOV D G   MANZHELII V G
KOLOSKOVA L A
J ENG PHYS /BSSR/
8   1   7-10   1965

34197 PROPERTIES OF PNEUMATICALLY PLACED REFRACTORY CONCRETES.
LIVOVICH A F
BULL AM CERAM SOC
45   1   11-15   1966

34226 SPECTRAL EMISSIVITY AND ELECTRON EMISSION CONSTANTS OF THORIA CATHODES FORMED BY CATAPHORESIS.
HANLEY T E   NAVAL RESEARCH LAB   WASHINGTON D C
CFSTI AND DDC
NRL-3157 AD 620580
    1-17   1947
PB 129171   1-17   1960

34251 TIME - TEMPERATURE EMITTANCE STUDY OF GOLD-PLATED STAINLESS STEEL. M.S. THESIS.
DEITCH M E   UNIV OF DENVER   DENVER COLO
UNIV OF DENVER
    1-117   1965

34258 THE TOTAL HEMISPHERE EMITTANCE OF POLISHED AND OF OXIDIZED ALPHA PLUTONIUM. PRELIMINARY REPORT.
KARLSSON R H   ROCKY FLATS DIV DOW CHEMICAL CO
GOLDEN COLORADO   USAEC
NASA AND CFSTI
RFP-616 N65-30024
    1-11   1965

34262 A STUDY OF THE REFLECTION AND POLARIZATION CHARACTERISTICS OF SELECTED NATURAL AND ARTIFICIAL SURFACES.
COULSON K L   GRAY E L   BOURICIUS G M B
GENERAL ELECTRIC CO SPACE SCIENCES LAB
PHILADELPHIA PA
DDC
R65SD4   AD   619032 N65-32209
    1-150   1965

34264 THERMAL STUDIES HASTELLOY X METALLIC INSULATION.
HICKERSON J P   BATTELLE-NORTHWEST   RICHLAND WASH
USAEC
CFSTI AND NASA
BNWL-102 N65-31564

34297 AN APPARATUS FOR THE CALORIMETRIC MEASUREMENT OF THE TOTAL HEMISPHERICAL EMITTANCE OF SOLID SURFACES. M.S. THESIS, MECH. ENG.
RAMSEY J W   U OF MINNESOTA   MINNEAPOLIS
U OF MINNESOTA   MINNEAPOLIS
    1-136   1965

34309 THERMAL RESISTIVITY AT PB-CU AND SN-CU INTERFACES BETWEEN 1.3 AND 2.1 K.
BARNES L J   DILLINGER J R
PHYS REV
141   2   615-20   1966

34312 SUMMARY OF THE NINTH MEETING OF THE REFRACTORY COMPOSITES WORKING GROUP.
BARTLETT E S   OGDEN H R   BATTELLE MEMORIAL INSTITUTE   COLUMBUS OHIO   USAF
DDC AND DMIC
DMIC-MEMO-200
    1-14   1965

34325 POLARIZERS FOR THE EXTREME ULTRAVIOLET.
HUNTER W R   US NAVAL RESEARCH LAB   WASHINGTON D C
DDC
AD 609535   N65-19059 N65-19061
    1-7   1964

34327 INVESTIGATION OF LIGHT SCATTERING IN HIGH REFLECTING PIGMENTED COATINGS. QUARTERLY REPORT, 1 MAY - 1 AUG. 1965.
ZERLAUT G A   KAYE B H   KATZ S   RAZIUNAS V
RESEARCH INST   CHICAGO ILL   TECHNOLOGY CENTER
NASA AND CFSTI
NASA-CR-64619 IITRI-C6018-15
N65-33255   1-56   1965

34329 THEORETICAL AND EXPERIMENTAL STUDY OF THERMAL CONDUCTANCE OF WAVY SURFACES. SEMIANNUAL STATUS REPORT, NOV. 1964 - JUN. 1965.
YOVANOVICH M M   MASSACHUSETTS INST OF TECH
CAMBRIDGE
NASA AND CFSTI
NASA-CR-64808 N65-33279
    1-33   1965

34330 COMBINED SPACE ENVIRONMENT EFFECTS ON TYPICAL SPACECRAFT WINDOW MATERIALS. FINAL REPORT, JUN. 1964 TO JAN. 1965.
JONES R H   AVCO CORP   TULSA OKLA
NASA AND CFSTI
NASA-CR-65142 TR-65-351-F
N65-33370   1-55   1965

34333 ULTRAVIOLET STABILITY TESTS FOR TWO WHITE PAINTS.
CLAUSEN W   MEYER K H   PERCY J L   GENERAL DYNAMICS/ASTRONAUTICS   SAN DIEGO CALIF
NASA
GDA-ERR-AN-64-487
N64-28053   1-43   1964   PA   N64-2 2836

34339 STUDIES OF ABSORPTANCE AND EMITTANCE OF LUMINESCENT MATERIALS FOR CONTROLLING TEMPERATURE OF A SPACECRAFT SURFACE. SUMMARY TECHNICAL REPORT.
GEORGETOWN UNIV DEPT OF PHYSICS
WASHINGTON D C
NASA AND CFSTI
NASA-CR-58215 N65-29462
    1-74   1964

34341 THE TOTAL NORMAL ABSORPTANCE OF A LAMPBLACK COATING ON COPPER. FINAL REPORT NO. 12.
SOUTHERN RESEARCH CENTER   BIRMINGHAM ALA
NASA AND CFSTI
NASA-CR-64360 N65-31014
    1-31   1965

34344 FAR INFRARED OPTICAL PROPERTIES OF SOME SPACECRAFT PAINTS. FROM SPACE PROGRAMS SUMMARY NO. 37-33, VOL. IV.
HALL W M   CALIF INST OF TECH JET PROPULSION LAB
PASADENA
CFSTI AND NASA
NASA-CR-64605 N65-32410 N65-32428
    81-3   1965

34345 THERMAL RADIATIVE PROPERTIES OF CARBON DIOXIDE CRYODEPOSITS. FROM 6TH ANN. SYMP. ON SPACE ENVIRON. SIMULATION. /SEE ALSO TPRC NO. 34040/.
DAWSON J P   MC CULLOUGH B A   WOOD B E
BIRKEBAK R C   ARO INC ARNOLD AIR FORCE STATION
TENN AEROSPACE ENVIRONMENTAL FACILITY   USAF
DDC
AD 465701   N65-34034 N65-34025 22-11
    9.10-9.19   1965

34354 EFFECT OF SURFACE IMPURITIES ON THE THERMAL ACCOMMODATION COEFFICIENT.
ALLEN R T   FEUER P
J CHEM PHYS
43   12   4500-5   1965   CA   64   4293

TPRC Number	Bibliographic Citation

34367 PREPARATION OF TWO-LAYER ACHROMATIC ANTIREFLECTION COATINGS.
UMEROV R I   SHKLYAREVSKII I N   PONAMAREVA G I
OPTIKA I SPEKTROSKOPIYA
26   6   1027-30   1969

34368 PREPARATION OF TWO-LAYER ACHROMATIC ANTIREFLECTION COATINGS. //ENGLISH TRANSLATION OF OPTIKA I SPEKTROSKOPIYA 26 /6/ 1027-30, 1969.//
UMEROV R I   SHKLYAREVSKII I N   PONAMAREVA G I
OPT SPECTROSC
26   6   556-8   1969

34385 DEVELOPMENT OF OPTICAL COATINGS FOR CADMIUM SULFIDE THIN FILM SOLAR CELLS. SECOND QUARTERLY REPORT, 1 MAR. - 1 1965.
SCHAEFER J C   HILL E R   HARSHAW CHEMICAL CO CLEVELAND OHIO
NASA
NASA-CR-54452 N65-29293
1-25   1965

34391 PROBLEMS OF INFRARED ATMOSPHERIC SPECTROSCOPY RELATED TO THE SATELLITE DETERMINATION OF TEMPERATURE OF UNDERLYING SURFACE. //ENGLISH TRANSLATION OF FIZ. ATM. I OKEANA 5 /6/ 616-30, 1969.//
KONDRATYEV K YA
NASA AND CFSTI
NASA-TT-F-12540
N69-35644   1-19   1969

34394 THE DISSOLUTION AND DIFFUSION OF OXYGEN IN ZIRCONIUM.
DOERFFLER W W   ATOMIC ENERGY OF CANADA LTD   CHALK RIVER ONTARIO
CFSTI
AECL-2268   N65-35001
1-29   1965   CA   63   12343

34401 TRANSIENT METHODS OF MEASURING THERMAL PROPERTIES OF SOLIDS.
DAS M B   HOSSAIN M A
BRIT J APPL PHYS
17   1   87-97   1966   CA   64   7396

34406 INFRARED STUDY OF ADSORBED MOLECULES ON METAL SURFACES BY REFLECTION TECHNIQUES.
GREENLER R J
J CHEM PHYS
44   1   310-15   1966

34428 THE INDUSTRIAL GRAPHITE ENGINEERING HANDBOOK.
UNION CARBIDE CORPORATION CARBON PRODUCTS DIV   NEW YORK N Y
UNION CARBIDE CORP   NEW YORK   N Y
4TH EDITION/REV/
1 VOLUME   1965

34449 EMISSIVITY PHYSICS. FROM RESEARCH ACHIEVEMENTS REVIEW, VOLUME 11.
MILLER E R   NATIONAL AERONAUTICS AND SPACE ADMINISTRATION MARSHALL SPACE FLIGHT CENTER HUNTSVILLE ALABAMA
NASA AND CFSTI
NASA-TM-X-53557 N67-24644
N64-24641   11-17   1966

34450 TRIPLE LAYER INFRARED FILTER.
MATHUR K C
INDIAN J PURE APPL PHYS
5   6   240-1   1967

34453 ABSOLUTE RADIATION STANDARD IN THE FAR INFRARED.
LICHTENBERG A J   SESNIC S
J OPT SOC AM
56   1   75-9   1966

34454 HIGH-PRECISION METHOD FOR MEASURING THE ABSORPTANCE OF EVAPORATED METALS.
BRANDENBERG W M   CLAUSEN O W   MC KEOWN D
J OPT SOC AM
56   1   80-6   1966   CA   64   5937

34455 UNIDIRECTIONAL REFLECTANCE OF IMPERFECTLY DIFFUSE SURFACES.
BRANDENBERG W M   NEU J T
J OPT SOC AM
56   1   97-103   1966

34456 REFLECTANCES OF CONCAVE DIFFRACTION GRATINGS FOR POLARIZED VACUUM ULTRAVIOLET.
HANSON W F   ARAKAWA E T
J OPT SOC AM
56   1   124-5   1966

34480 INFRARED AND RAMAN STUDIES OF CRYSTALLINE HCL, DCL, HBR AND DBR.
SAVOIE R   ANDERSON A
J CHEM PHYS
44   2   548-56   1966

34485 THE THERMAL PROPERTIES OF SELECTED SPACE SUIT MATERIALS.
BEVANS J T   TRW SYSTEMS   REDONDO BEACH CALIF
NASA
NASA-CR-65678
N67-34338   1-172   1965   PA   N67-5-20 3582

34487 OPTICAL PROPERTIES OF COPPER OXIDE FILMS.
WIEDER H   CZANDERNA A W
J APPL PHYS
37   1   184-7   1966

34488 ALLOY FILMS OF PB TE/X/ SE/1-X/.
BIS R F   ZEMEL J N
J APPL PHYS
37   1   228-30   1966

34514 THE MECHANISM OF CONTACT BETWEEN METAL SURFACES- THE PENETRATING DEPTH AND THE AVERAGE CLEARANCE.
TSUKIZOE T   HISAKADO T
J BASIC ENG
87   3   666-74   1965   CA   63   17493

34541 APPARATUS TO MEASURE MID-INFRARED SPECTRAL EMITTANCE OF COLD POWDERS IN A VACUUM.
GOETZ A F H   BAUMAN C A
REV SCI INSTRUM
38   6   775-8   1967   CA   67   16298

34542 FAR INFRARED STUDY OF THE COPPER HALIDES AT LOW TEMPERATURES.
PLENDL J N   HADNI A   CLAUDEL J   HENNINGER Y
MORLOT G   STRIMER P   MANSUR L C
APPL OPT
5   3   397-401   1966

34545 ELECTROREFLECTION BY GAS ADSORBTION ON ZINC OXIDE.
HOFFMANN B
Z PHYSIK
206   3   293-308   1967

34547 OPTICAL CONSTANTS OF THIN FILMS BY A KRAMERS-KRONIG METHOD.
KOZIMA K   SUETAKA W   SCHATZ P N
J OPT SOC AM
56   2   181-4   1966

34548 OPTICAL CONSTANTS OF VACUUM EVAPORATED FILMS OF CADMIUM AND THALLIUM IN THE VACUUM ULTRAVIOLET.
JELINEK T M   HAMM R N   ARAKAWA E T   HUEBNER R H
J OPT SOC AM
56   2   185-8   1966   CA   64   9091

34550 EVALUATION OF ABSOLUTE REFLECTANCE FOR STANDARDIZATION PURPOSES.
VAN DEN AKKER J A   DEARTH L R   SHILLCOX W M
J OPT SOC AM
56   2   250-2   1966

34599 HEAT TRANSFER IN A DISPERSE HEAT-INSULATING SYSTEM.
KOSTYLEV V M   NABATOV V G
INZH FIZ ZH AKAD NAUK BELORUSSK SSR
9   3   877-83   1965   CA   64   1676

34610 MEASUREMENT OF THE SURFACE TENSION AND DENSITY OF LIQUID CHROMIUM.
EREMENKO V N   MAIDICH YU V
IZVEST AKAD NAUK SSSR OTDEL TEKH NAUK MET I TOPLIVO
2   111-12   1959

34611 SURFACE TENSION AT ELEVATED TEMPERATURES. III. EFFECT OF CR, IN, SN AND TI ON LIQUID NICKEL SURFACE TENSION AND INTERFACIAL ENERGY WITH AL2 O3.
KURKJIAN C R   KINGERY W D
J PHYS CHEM
60   961-3   1956

34615 SURFACE TENSION OF PURE LIQUID IRON, COBALT, AND NICKEL AT 1550 C.
KOZAKEVITCH P   URBAIN G
J IRON AND STEEL INST /LONDON/
186   167-73   1957

34621 EVAPORATED INHOMOGENEOUS THIN FILMS.
JACOBSSON R   MARTENSSON J O
APPL OPT
5   1   29-34   1966

34622 OPTICAL PROPERTIES OF MULTISOURCE THERMALLY EVAPORATED III-V SEMICONDUCTOR COMPOUNDS.
POTTER R F
APPL OPT
5   1   35-40   1966

34623 COMPUTATIONAL METHOD FOR DETERMINING N AND K FOR A THIN FILM FROM THE MEASURED REFLECTANCE, TRANSMITTANCE, AND FILM THICKNESS.
BENNETT J M   BOOTY M J
APPL OPT
5   1   41-3   1966

TPRC Number	Bibliographic Citation
34824	FURTHER STUDIES ON MGF2-OVERCOATED ALUMINUM MIRRORS WITH HIGHEST REFLECTANCE IN THE VACUUM ULTRAVIOLET. CANFIELD L R HASS G WAYLONIS J E APPL OPT 5 1 45-50 1966
34825	MULTILAYER MIRRORS WITH HIGH REFLECTANCE OVER AN EXTENDED SPECTRAL REGION. TURNER A F BAUMEISTER P W APPL OPT 5 1 69-76 1966
34626	HIGH PERFORMANCE BLOCKING FILTERS FOR THE REGION 1 TO 20 MICRONS. SEELEY J S SMITH S D APPL OPT 5 1 81-5 1966
34627	FACTORS AFFECTING THE PERFORMANCE OF COMMERCIAL INTERFERENCE FILTERS. BLIFFORD I H JR APPL OPT 5 1 105-11 1966
34628	ZERO REFLECTION FROM THIN ABSORBING FILMS ON A METAL BASE. WEINBERGER H APPL OPT 5 1 165-6 1966 CA 64 9079
34631	ADVANCED COMPOSITE MATERIALS IN SPACECRAFT. FOREST J D MODERN PLASTICS 47 9 136-40 1970
34640	THERMAL CONDUCTIVITY OF FREEZE-DRIED MODEL FOOD GELS. SARAVACOS G PILSWORTH M N JR J FOOD SCI 30 5 773-8 1965 CA 64 2657
34659	THERMOPHYSICAL PROPERTIES OF PLASTIC MATERIALS AND COMPOSITES TO LIQUID HYDROGEN TEMPERATURE /-423 F/. CAMPBELL M D O BARR G L HASKINS J F HERTZ J GENERAL DYNAMICS CONVAIR SAN DIEGO CALIF USAF GENERAL DYNAMICS/CONVAIR SAN DIEGO CALIF AF-TDR-64-33 /PT 3/ AD 468155 1-88 1965
34666	TEMPERATURE DEPENDENCE OF THERMAL CONDUCTIVITY AND THERMAL DIFFUSIVITY FOR SOME POLYMERIC MATERIALS. BIL V S AVTOKRATOVA N D PLASTICHESKIE MASSY 10 37-9 1965 CA 64 2230
34669	THERMAL DIFFUSIVITY, THERMAL CONDUCTIVITY, AND SPECIFIC HEAT OF CONCRETE. TOKUDA H ITO T PROC JAPAN CONGR TESTING MATER 8 112-14 1965 CA 63 17673
34680	SPECULAR SPECTRAL REFLECTANCE OF PAINTS FROM 0.4 TO 40.0 MICRONS. RAMSEY W Y METEOROLOGICAL SATELLITE LAB WEATHER BUREAU WASHINGTON D C CFSTI REPT NO MSL-31 PB 167179 1-36 1964
34682	REFLECTIVITY OF AIRFIELD MARKING PAINTS. DRISKO R W NAVAL CIVIL ENGINEERING LAB PORT HUENEME CALIF CFSTI AND DDC NCEL-TR-R-323 PB-167322 AD-446327 1-24 1964 RR 40 S-32
34683	PROCEDURES FOR THE PRECISE DETERMINATION OF THERMAL RADIATION PROPERTIES. RICHMOND J C DUNN S T DE WITT D P HAYES W D JR NATIONAL BUREAU OF STANDARDS NBS AND SUPERINTENDENT OF DOCUMENTS U S GOVERNMENT PRINTING OFFICE WASHINGTON D C NBS-IN-267 ML-TDR-04-257/PT II/ AD 478229 AD 628586 N66-22418 1-62 1965
34706	TRANSFER OF HEAT BELOW 0.2 K. A COMPARISON OF BONDING AGENTS. CONNOLLY J I ROACH W R SARWINSKI R J REV SCI INSTR 36 9 1370-1 1965 CA 64 1869
34724	INFRARED SIGNATURE CHARACTERISTICS. DURAND J L HOUSTON C K DIRECTORATE OF ARMAMENT DEVELOPMENT RESEARCH AND TECHNOLOGY DIVISION AIR FORCE SYSTEM COMMAND EGLIN AIR FORCE BASE FLORIDA THE MARTIN COMPANY DIV OF MARTIN-MARIETTA CORP ORLANDO FLORIDA USAF ATL-TR-66-8 OR-6820 302 1-174 1966
34726	RESEARCH AND DEVELOPMENT OF VACUUM FOIL-TYPE INSULATION FOR RADIOISOTOPE POWER SYSTEMS. QUARTERLY PROGRESS REPORT 4, NOV. 1966-1967. DUNLAY J B FRONDUTO J PAQUIN M L POIRIER V L THERMO ELECTRON ENGINEERING CORP WALTHAM MASS OFSTI ALO-3634-04 UC-33-4 N67-38546 1-116 1967 PA N67-5-23 4363
34733	REFLECTION MEASUREMENTS OF RUBIDIUM IODIDE FILMS IN THE ULTRAVIOLET. WATANABE M KATO R NAKAI Y J PHYS SOC JAPAN 21 1 191 1966
34736	SPACE MATERIALS HANDBOOK /SECOND EDITION/. LOCKHEED AIRCRAFT CORP LOCKHEED MISSILES AND SPACE CO SUNNYVALE CALIF NASA AND DDC ML-TDR-64-40 WITH SUPPL. NASA-SP-3025 WITH SUPPL. AD 460399 /X65-14878/ AD 629720 1-712,S1-2631965-6
34740	EXPERIMENTAL EVALUATION OF EXPANDED PYROLYTIC GRAPHITE FOR USE IN SPACE RADIATORS. QUARTERLY PROGRESS REPT. NO. 3, 1 SEP. 30 NOV. 66. MADSEN J KING P P HILTZ E F ALLEN R D DOUGLAS AIRCRAFT CO INC MISSILE AND SPACE SYSTEMS DIV SANTA MONICA CALIF USAF DDC DAO-50081 AD 805434 1-37 1966
34751	THERMAL CONDUCTIVITY OF HYDROGEN FROM 2000 TO 4700 F. ISRAEL S L HAWKINS T D HYMAN S C UNITED NUCLEAR CORP WHITE PLAINS N Y NASA NASA-CR-403 N66-19386 1-60 1966 PA N66-4-9 1562
34752	THE OPTIMIZATION OF THERMAL COMPOSITES. RYAN J M CROSS R I PAULSEN N J BLACK W E GENERAL DYNAMICS/CONVAIR SAN DIEGO CALIF DDC AFML-TR-65-244 AD 479531L 1-115 1965 TA 66-9 A-44
34753	SURVEY OF THERMAL PROPERTIES OF SELECTED MATERIALS. HERTZ J KNOWLES D GENERAL DYNAMICS/CONVAIR SAN DIEGO CALIF GENERAL DYNAMICS/CONVAIR ZZL-65-008 AR-504-1-553 N65-31775 1-172 1965 CA 67 15485
34756	THERMAL RESISTIVITY AT INTERFACES BETWEEN METAL AND DIELECTRIC FILMS AT 1.5 AND 4.2 K. HOLT V E J APPL PHYS 37 2 798-802 1966
34762	SHIELDED CERAMIC COMPOSITE STRUCTURE. FINAL REPT. 1 JUNE 63-18 JUNE 65. KUMMER D L ROSENTHAL J J LUM D W MC DONNELL AIRCRAFT CORP ST LOUIS MO USAF DDC MAC A131 AFML-TR-65-331 AD 475002 1-405 1965 TA 66-2 A31
34769	DETERMINING THE EMISSIVITY OF MATERIALS ON THE BASIS OF THEIR INFRARED REFLECTION SPECTRA. KROPOTKIN M A KOZYREV B P INZHEN-FIZ ZHUR /USSR/ 7 9 108-12 1964
34770	DETERMINING THE EMISSIVITY OF MATERIALS ON THE BASIS OF THEIR INFRARED REFLECTION SPECTRA. //ENGLISH TRANSLATION OF INZHEN-FIZ. ZHUR. /USSR/ 7, /9/ 108-12 1964.// KROPOTKIN M A KOZYREV B P CFSTI OR ETC JPRS-R-5536-D TT-65-63682 /FOREIGN TEXT INCL/ 1-14 1965 TT 14-9 75
34814	OPTICAL REFLECTORS FOR USE IN INTERNAL SAMPLE AQUEOUS CHERENKOV COUNTERS. STRINDEHAG O M REV SCI INSTR 37 3 344-9 1966
34817	DEVELOPMENT OF OPTICAL COATINGS FOR CADMIUM SULFIDE THIN FILM SOLAR CELLS. THIRD QUARTERLY REPORT, JUNE 1-AUG. 1. 1965. SCHAEFER J C HILL E R HARSHAW CHEMICAL CO CLEVELAND OHIO CRYSTAL-SOLID STATE DIV NASA NASA-CR-54750 N65-35072 1-21 1965 PA N65-3-23 3929

TPRC Number	Bibliographic Citation

34819 DEVELOPMENT OF SPACE-STABLE THERMAL-CONTROL COATINGS /PAINTS WITH LOW SOLAR ABSORPTANCE/EMITTANCE RATIOS/. TRIANNUAL REPORT, 20 JUNE-20 OCT. 1964.
ZERLAUT G A   GILLIGAN J E   HARADA Y   IIT
RESEARCH INST   CHICAGO ILL
NASA AND CFSTI
NASA-CR-67295 IITRI-C6014-18
N65-35122   1-72   1964   PA   N65-3-23 4018

34820 HIGH-PERFORMANCE THERMIONIC CONVERTER.   QUARTERLY PROGRESS REPORT, 13 MAY-13 AUG. 1965.
CAMPBELL A E   POLLOCK D H   ELECTRO-OPTICAL
SYSTEMS INC   PASADENA CALIF
NASA AND CFSTI
NASA-CR-67299 EOS-6952-Q-1
N65-35354   1-91   1965   PA   N65-3-23 4002

34829 ON THERMAL CONDUCTIVITY OF POROUS MATERIALS.
SUGAWARA A
J APPL PHYS /JAPAN/
30   17-23   1961

34830 A THERMAL VACUUM TECHNIQUE FOR MEASURING THE SOLAR ABSORPTANCE OF SATELLITE COATINGS AS A FUNCTION OF ANGLE OF INCIDENCE.
HOKE M G   NATIONAL AERONAUTICS AND SPACE
ADMINISTRATION GODDARD SPACE FLIGHT CENTER
GREENBELT MD
NASA AND CFSTI
NASA-TM-X-51837
N65-35221   1-31   1963   PA   N65-3-23 4079

34832 A COMPARISON OF TWO EMITTANCE MEASUREMENT TECHNIQUES.
HEANEY J B   NATIONAL AERONAUTICS AND SPACE
ADMINISTRATION GODDARD SPACE FLIGHT CENTER
GREENBELT MD
NASA AND CFSTI
NASA-TM-X-55294 X-713-65-354
N66-10687   1-20   1965   PA   66-4-1 147

34834 A CORRECTION FOR THE NET RADIOMETER REFLECTION ERROR. ANALYSIS OF MEASURED NET RADIATION VALUES FOR CANADA.
FINKE D D   WENDLAND W M   WISCONSIN UNIV MADISON
DEPT OF METEOROLOGY   ONR
DDC AND CFSTI
AD 618721   N65-35196 TR 19
                             1-86   1965   PA   N65-3-23 4060

34835 THERMAL RADIATION OF COMPLEX CERAMIC SOLIDS.
GRENIS A F   ARMY MATERIALS RESEARCH AGENCY
MATERIALS ENGINEERING DIV   WATERTOWN MASS
DDC AND CFSTI
AMRA-TR-65-14 AD 620004 N66-10374
                             1-19   1965   PA   66-4-1 146

34838 HIGH-COMPRESSIVE-STRENGTH CONCRETE.   FINAL REPT. NO. 3 FOR JUL 61 - JAN 65.
SAUCIER K L   TYNES W O   SMITH E F   ARMY
ENGINEER WATERWAYS EXPERIMENT STATION   VICKSBURG MISS
DDC AND CFSTI
AFWL-TR-65-16 AD 622445
                             1-91   1965   TA   65-23 59

34851 DEVELOPMENT AND TESTING OF A TEMPERATURE TRANSIENT METHOD FOR PERFORMING IN-PILE THERMAL CONDUCTANCE TESTING.   FINAL SUMMARY REPORT, MARCH 1, 1964-NOV. 30, 1965.
BURDG C E   PARRETTE J R   BROCKETT R I
CHERNOCK W P   COMBUSTION ENG INC   WINDSOR CONN
CFSTI
CEND-3336-260 EURAEC-1567
                             1-98   1966   CA   67   69742

34908 EFFECT OF PRESSURE ON THE REFLECTANCE OF COMPACTED POWDERS.
SCHATZ E A
J OPT SOC AM
56   3   389-94   1966   CA   64   12050

34918 LOW PASS FILTERS FOR FAR INFRARED SPECTROSCOPY.
BERREMAN D W
REV SCI INSTR
37   4   513-14   1966

34925 UNIFORM VACUUM ULTRAVIOLET REFLECTING COATINGS ON LARGE SURFACES.
HERZIG H   GODDARD SPACE FLIGHT CENTER   GREENBELT MD
NASA AND CFSTI
NASA-TN-D-3357 N66-19759
                             1-6   1966   CA   66   120357

34928 FAR-INFRARED REFLECTANCE OF SPACECRAFT COATINGS.
EDWARDS D K   HALL W M   CALIF INSTITUTE OF
TECHNOLOGY   JET PROPULSION LAB   PASADENA CALIF
NASA
JPL-TR-32-873 NASA-CR-71187
N66-19633   1-11   1966

34936 THE MEASUREMENT OF THE EXCHANGE OF INTERNAL ENERGY BETWEEN PROPANE AND A METAL SURFACE.
EHRHARDT H   EINHAUS R   ENGELKE H
Z PHYSIK

34971 OPTICAL PROPERTIES OF COPPER IN THIN FILMS IN THE SPECTRAL DOMAIN GOING FROM 1.5 TO 7 MICRONS.
VARENNE S
COMPT REND
265   B   20   1115-18   1967

34972 PREPARATION OF STOICHIOMETRIC CDSE IN THIN FILMS.
FINCK C   PAPARODITIS C
COMPT REND
265   B   20   1119-22   1967

34978 HIGH-TEMPERATURE MATERIALS PROGRAM.   PROGRESS REPORT NO. 50, PT. A.
GENERAL ELECTRIC CO ATOMIC PRODUCTS
DIV   CINCINNATI OHIO   AT
CFSTI
GEMP-50A N66-10207
                             1-81   1965   PA   66-4-1 71

34979 PERFORMANCE EVALUATION OF THERM-LAG MATERIAL FOR ENTRY HEAT PROTECTION OF ADVANCED MANNED SPACECRAFT. MATERIAL DEVELOPMENT REPORT, 1 OCT. 1962-1 APR. 1963.
EMERSON ELECTRIC CO ELECTRONIC AND
SPACE DIV   ST LOUIS MO
NASA AND CFSTI
NASA-CR-65166 REPT 1558
N66-10712   1-205   1963   PA   66-4-1 85

34981 REMARKS ON THE OPTICAL ABSORPTION OF GROUP IB METALS IN THE FORM OF VERY THIN FILMS FOR DIFFERENT VELOCITIES OF FORMATION.
EMERIC N   EMERIC A   PHILIP R
J PHYSIQUE
26   12   769-75   1965   CA   65   1609

35021 EFFECT OF ULTRAVIOLET IRRADIATION ON SELECTED PLASTIC FILMS IN VACUUM.
ANAGNOSTOU E   LEWIS RESEARCH CENTER   CLEVELAND OHIO   NASA
NASA AND CFSTI
NASA-TM-X-1124
                             1-14   1965   PA   N65-3 2961

35023 SNAP 19 TEST REPORT THERMAL PERFORMANCE OF MATERIALS AND MECHANICAL JOINTS.
SHIH C   MARTIN NUCLEAR CO   BALTIMORE MD
CFSTI
MND-3607-112 N67-23693
                             1-52   1966   PA   N67-5 2146

35027 THE INFLUENCE OF LONGWAVE REFLECTIVITY OF NATURAL SURFACES ON THE MEASUREMENTS OF SURFACE TEMPERATURE USING RADIOMETERS.
LORENZ D   JOHANN-WOLFGANG-GOETHE UNIV INST FOR
METEOROLOGY AND GEOPHYSICS   FRANKFURT AM MAIN GERMANY
U S DEPT OF ARMY
DDC
AD 464604   1-51   1965   TA   65-14 A-187

35028 OPTICAL PROPERTIES AND THERMAL CONDUCTIVITY OF ALUMINUM OXIDE.
NEUBERGER M   ELECTRONIC PROPERTIES INFORMATION
CENTER HUGHES AIRCRAFT CO   CULVER CITY CALIF
DDC
EPIC-S-6 AD 464823
N65-34038   1-20   1965   PA   N65-3-22 3875

35029 THEORY OF THE OPTICAL PROPERTIES OF THIN POLYCRYSTALLINE METAL LAYERS.
DRUMHELLER C E   GENERAL DYNAMICS/ELECTRONICS
ROCHESTER N Y   AIR FORCE AVIONICS LAB
WRIGHT-PATTERSON AFB   OHIO
DDC
ASD-TDR-63-519 AD 461605
                             1-8   1964   TA   65-11 234

35050 VISCOSITY AND SURFACE PROPERTIES OF SYNTHETIC WHITE SLAGS WITH ADDITIVES AL2O3, CAF2 AND NA3ALF6.
NA3 AL F6.
SMOLYARENKO V D   YAKUSHEV A M   EDNERAL F P
IZV VYSSHIKH UCHEBN ZAVEDENII CHERNAYA MET
8   6   72-7   1965   CA   63   15915

35069 COLOR SENSITIZATION OF ZINC OXIDE WITH CYANINE DYES.
NAMBA S   HISHIKI Y
J PHYS CHEM
69   3   774-9   1965

35074 XCIII SOME OPTICAL PROPERTIES OF EVAPORATED LAYERS OF SILVER, COPPER, AND TIN.
AVERY D G
PHIL MAG
41   1018-31   1950

35077 MULTILAYER COATING WITH A SELECTIVE REFLECTANCE.
SHKLYAREVSKII I N   LUPASHKO E A
OPT I SPEKTROSKOPIYA
18   4   661-7   1965   CA   63   3734

35078 MULTILAYER COATING WITH A SELECTIVE REFLECTANCE. //ENGLISH TRANSLATION OF OPT. I SPEKTROSKOPIYA 18 /4/ 661-7, 1965.//
SHKLYAREVSKII I N   LUPASHKO E A

TPRC Number	Bibliographic Citation

35099 STUDY OF THE OPTICAL PROPERTIES OF THIN MERCURY SULFIDE FILMS.
BILENKY B F   MILIANSHUK M V   PASHKOVSKY M V
UKR FIZ ZHUR

| 10 | | 6 | 687-9 | 1965 | CA | 63 | 9247 |

35110 PROGRAM FOR THE EVALUATION OF STRUCTURAL REINFORCED PLASTIC MATERIALS AT CRYOGENIC TEMPERATURES.   FINAL REPT. JUL. 1, 1963 TO JUL. 1, 1966.
TOTH L W   BOLLER T J   BUTCHER I R   KARIOTIS A H   YODER F D
NASA
NASA-CR-80061 GER 12792

| N67-12051 | | | 1-239 | 1966 |

35113 BANDPASS FILTERS FOR THE ULTRAVIOLET.
BAUMEISTER P W   COSTICH V R   PIEPER S C
APPL OPT

| 4 | | 8 | 911-14 | 1965 |

35116 WIRE GRID POLARIZER.
YOUNG J B   GRAHAM H A   PETERSON E W
APPL OPT

| 4 | | 8 | 1023-6 | 1965 | CA | 63 | 6523 |

35119 OPTICAL COATINGS FOR LASER USE.
LAIKIN M
APPL OPT

| 4 | | 8 | 1032-3 | 1965 | CA | 63 | 12547 |

35120 INVESTIGATION INTO THE MECHANISM OF FREEZE DRYING.   TECHN. REPT.
SUNDERLAND J E   NORTHWESTERN UNIV TECHN INST   EVANSTON ILL
CFSTI AND DDC
AD 615890

| | | | 1-45 | 1961 | TA | 65-14 30 |

35129 ON THE THEORY OF ACCOMMODATION COEFFICIENTS-IV.   SIMPLE DISTRIBUTION FUNCTION THEORY OF GAS-SOLID INTERACTION SYSTEMS.
GOODMAN F O
J PHYS CHEM SOLIDS

| 26 | | 1 | 85-105 | 1965 |

35151 SUMMARY OF THE EIGHTH REFRACTORY COMPOSITES WORKING GROUP MEETING.
JAMES D R   HJELM L N   AIR FORCE MATERIALS LAB
AIR FORCE SYSTEMS COMMAND
DDC
ML-TDR-64-233 /VOLUME 3/
AD 470695

| | | | 632-950 | 1964 |

35198 MEASUREMENT OF THE STATIC SURFACE TENSION BY THE SESSILE BUBBLE METHOD DURING A LONG PERIOD OF ADSORPTION LAYER FORMATION WITHIN A WIDE TEMPERATURE RANGE.
KULMAN R A
DOKL AKAD NAUK SSSR

| 168 | | 1 | 149-51 | 1966 |

35199 MEASUREMENT OF THE STATIC SURFACE TENSION BY THE SESSILE BUBBLE METHOD DURING A LONG PERIOD OF ADSORPTION LAYER FORMATION WITHIN A WIDE TEMPERATURE RANGE. //ENGLISH TRANSLATION OF DOKL. AKAD. NAUK SSSR 168 /1/ 149-51, 1966.//
KULMAN R A
DOKL PHYS CHEM

| 168 | | 1 | 304-6 | 1966 |

35200 LAW OF COMPOSITION OF THERMAL RESISTANCES IN CONTACT BETWEEN TWO SUBSTANCES.
FOUCHE F
COMPT REND

| 265 | B | 1 | 9-11 | 1967 | IAA 7-19 3347 |

35221 OPTICAL CONSTANTS OF LEAD SULFIDE IN THE FUNDAMENTAL ABSORPTION EDGE REGION.
SCHOOLAR R B   DIXON J R
PHYS REV

| 137 | A | 2 | 667-70 | 1965 |

35225 MAGNETO-OPTICAL PROPERTIES OF GARNET FILMS.
MAC DONALD R E   VOEGELI O   MEE C D
J APPL PHYS

| 38 | | 10 | 4101-2 | 1967 |

35265 A THERMAL VACUUM TECHNIQUE FOR MEASURING THE SOLAR ABSORPTANCE OF SATELLITE COATINGS AS A FUNCTION OF ANGLE OF INCIDENCE.
HOKE M G
AIAA J

| 3 | | 5 | 947-51 | 1965 | AM | 18 | 4992 |

35268 ACCOMMODATION AND HEAT-TRANSFER COEFFICIENTS IN GAS-CONTAINING VOIDS.
MUSTACCHI C
ANALES REAL SOC ESPAN FIS QUIM

| 60 | B | 2 | 267-70 | 1964 |

35311 PROPERTIES AND PREPARATION OF THIN BISMUTH FILMS.
CONDAS G A
REV SCI INSTR

| 36 | | 8 | 1252-4 | 1965 |

35321 OPTICAL CONSTANTS IN THE FAR ULTRAVIOLET OF THICK SILVER FILMS UNEXPOSED TO AIR.
PRIOL M   LARVOR M   ROBIN S
C R ACAD SCI /FRANCE/

| 259 | | 22 | 3983-6 | 1964 | SA | 68 | 15573 |

35323 DESIGN AND ANALYSIS OF AN ELLIPSOIDAL MIRROR REFLECTOMETER.   PH.D. THESIS.
DUNN S T   OKLAHOMA STATE UNIV   STILLWATER
UNIV MICROFILMS PUBL

| 65-11677 | | | 1-185 | 1965 |

35338 THERMAL CONDUCTIVITY OF COMPOSITES.
D ANDREA G
UNIV MICROFILMS PUBL

| UM70-2614 | | | 1-174 | 1969 | CA | 73 | 102701 |

35374 A NEW METHOD FOR RAPID MEASUREMENT OF THERMAL DIFFUSIVITY AT ELEVATED TEMPERATURES WITH AN APPLICATION TO A SANDWICH STRUCTURE.   TECH. NOTE.
SMITH W K   NAVAL ORDNANCE TEST STATION   CHINA LAKE CALIF
DDC /FOR REF ONLY/ NAVAL ORDNANCE TEST STAT
TN 4061-31 AD 614034

| | | | 1-7 | 1959 | TA | 65-11 141 |

35414 THE DETERMINATION OF INFRARED EMISSIVITIES OF TERRESTRIAL SURFACES.
BUETTNERK J K   KERN C D
J GEOPHYS RES /USA/

| 70 | | 6 | 1329-38 | 1965 | SA | 68 | 18354 |

35417 RESEARCH ON OPTICS OF SELECTIVE SURFACES.   FINAL REPT. FOR 1 JUN 63-31 OCT 64.
TABOR H   HEBREW UNIV   JERUSALEM /ISRAEL/   USAF
CFSTI AND DDC
AFCRL-65-625 N66-24577
AD-620506

| | | | 1-74 | 1964 | TA | 65-20 86 |

35434 LEAD OXIDE.
NEUBERGER M   HUGHES AIRCRAFT CO ELECTRONIC PROPERTIES INFORMATION CENTER   CULVER CITY CALIFORNIA
DDC
EPIC-DS-155 AD-814147

| | | | 1-84 | 1967 | TA | 67-13 A-23 |

35440 CURIE TEMPERATURE AND LATTICE CONSTANT OF EVAPORATED NICKEL FILMS.
MORITA N   INOUE N
J PHYS SOC JAPAN

| 20 | | 5 | 694-9 | 1965 |

35455 OPTICAL POLARIZATION REFLECTION BEHAVIOUR OF THIN, SCATTERING, ABSORBING, SLIGHTLY DICHROIC, BIREFRINGENT AND INHOMOGENEOUS SINGLE FILMS ON GLASS.
RASSOW J
OPTIK /GERMANY/

| 22 | | 5 | 369-87 | 1965 | SA | 68 | 16326 |

35458 OPTICAL COEFFICIENTS OF THIN CHROMIUM FILMS PREPARED BY EVAPORATION IN A VACUUM.   //ENGLISH TRANSLATION OF OPTIKA I SPEKTROSK. 17 /6/ 923-6, 1964.//
IDCHAK E F
OPTICS AND SPECTROSC /USA/

| 17 | | 6 | 501-3 | 1964 | SA | 68 | 18232 |

35523 REFLECTION STUDIES OF EXCITIONS IN LIQUID AND SOLID XENON.
BEAGLEHOLE D
DDC
AFOSR-66-0310 AD 629813

| | | | 1-5 | 1965 | TA | 66-9 31 |

PHYS REV LETTERS

| 15 | | 13 | 551-3 |

CORRECTION.
REV SCI INSTR

| 39 | | 6 | 921-2 |
| 15 | | 13 | 551-3 | 1965 |

35526 INTERFACE AND TRANSFERRING SPECIES IN AMINE EXTRACTION OF URANIUM.
MC DOWELL W J   COLEMAN C F   OAK RIDGE NATL LAB   OAK RIDGE TENN
CFSTI
ORNL-P-2359 CONF-660805-4

| | | | 1-15 | 1964 | CA | 66 | 98770 |

35527 SIMULTANEOUS MEASUREMENTS OF OPTICAL TRANSMISSION AND REFLECTION IN THIN FILMS.
VARADI P F   SUFFREDINI J R
REV SCI INSTR

| 36 | | 9 | 1331-3 | 1965 |

35530 THERMOPHYSICAL PROPERTIES OF SIX CHARRING ABLATORS FROM 140 TO 700 K. AND TWO CHARS FROM 800 TO 3000 K.
WILSON R G   COMPILER   LANGLEY RESEARCH CENTER   LANGLEY STATION   HAMPTON VA   NASA
NASA AND CFSTI
NASA-TN-D-2991

| N65-34502 | | | 1-172 | 1965 | CA | 67 | 15535 |

TPRC Number	Bibliographic Citation

35535 ON THE THERMAL CONDUCTIVITY OF POWDER INSULATIONS. FROM PROC. INTERN. CONGRESS OF REFRIGERATION, 11TH, MUNICH 1963.
EVEREST A   GLASER P E   WECHSLER A E
PROGRESS IN REFRIGERATION SCIENCE AND TECHNOLOGY
1   255-63   1965

35536 THE INFLUENCE OF GAS-FILLED CELLS ON THERMAL CONDUCTIVITY OF RIGID POLYURETHANE FOAM. FROM PROC. INTERN. CONGRESS OF REGRIGERATION, 11TH, MUNICH 1963.
KAHLENBERG F
PROGRESS IN REFRIGERATION SCIENCE AND TECHNOLOGY
1   265-9   1965

35537 EFFECTIVENESS OF EVACUATED MULTIPLE-LAYER INSULATIONS. FROM PROC. INTERN. CONGRESS OF REFRIGERATION, 11TH, MUNICH, 1963.
BLACK I A   DOHERTY P   GLASER P E   MELLNER M
PROGRESS IN REFRIGERATION SCIENCE AND TECHNOLOGY
1   283-92   1965

35539 TRANSIENT TEMPERATURES IN A THERMOELECTRIC REFRIGERATOR FOLLOWING A STEP CHANGE IN CURRENT. FROM PROC. INTERN. CONGRESS OF REFRIGERATION, 11TH, MUNICH, 1963.
STOECKER W F   CHADDOCK J B
PROGRESS IN REFRIGERATION SCIENCE AND TECHNOLOGY
1   631-41   1965

35541 THE REFLECTANCE SPECTRA OF SOME HETEROGENEOUS MATERIALS. M.S. THESIS.
RICHARDSON J S   TEXAS TECHNICAL UNIV   LUBBOCK TEXAS
TEXAS TECHNICAL UNIV
1-42   1969

35545 OVI-10 THERMAL CONTROL COATING ORBITAL EXPERIMENT.
BOEBEL C P   AIR FORCE MATERIALS LAB
WRIGHT-PATTERSON AIR FORCE BASE OHIO
DDC
AFML-TR-68-392 PT. 1
1-109   1969

35546 UTILIZATION OF PIGMENTED COATINGS FOR THE CONTROL OF EQUILIBRIUM SKIN TEMPERATURES OF SPACE VEHICLES.
ZERLAUT GENE A   ARMY MISSILE COMMAND REDSTONE ARSENAL ALA
DDC
ABMA-MISC-32 AD 463738
1-46   1960   TA   65-13 A-269

35552 EFFECTS OF VACUUM-ULTRAVIOLET ENVIRONMENT ON OPTICAL PROPERTIES OF BRIGHT ANODIZED ALUMINUM TEMPERATURE CONTROL COATINGS.
WEAVER J H   AIR FORCE MATERIALS LAB
WRIGHT-PATTERSON AIR FORCE BASE   OHIO
DDC
AFML-TR-67-421
1-23   1968

35556 EVALUATION OF TITANIUM CARBIDE EMITTANCE IMPREEOVEMENT TOPCOAT ON DISIL COATED TZM /MO-0.5 TI-0.1ZR/.
GUNDERSON J W   LINDH D V   STRATTON W K
BOEING CO   SEATTLE
DDC
D2-36145-1 AD-462018
1-78   1964   TA   65-11 A-157

35560 WETTING OF CERAMIC OXIDES BY MOLTEN METALS UNDER ULTRAHIGH VACUUM.
HARDING F L   ROSSINGTON D R
J AM CERAM SOC
53   2   87-90   1970

35561 DEVELOPMENT OF OPTICAL COATINGS FOR CADMIUM SULFIDE THIN FILM SOLAR CELLS.
SCHAEFER J C   HILL E R   HARSHAW CHEMICAL CO
CLEVELAND   NASA
NASA AND CFSTI
NASA-CR-54336
N65-23844   1-20   1965   PA   N65-3 2127

35573 A SIMPLE SEMI-MICRO CELL FOR THE MEASUREMENT OF SPECTRAL REFLECTANCE.
FREI R W   FRODYMA M M
ANAL CHIM ACTA
32   6   501-7   1965   CA   63   4919

35580 THE PERFORMANCE OF DIELECTRIC THIN FILM BEAM-DIVIDING SYSTEMS.
CATALAN L A   PUTNER T
BRIT J APPL PHYS
12   9   499-502   1961   CA   63   7768

35583 INFRARED SPECTRUM OF MATRIX-ISOLATED DIMERIC NITROGEN MONOXIDE.
HUNTER C E
UNIV MICROFILMS PUBL
UM-70-1999   1-72   1969   CA   73   125228

35591 THIRD-SOUND RESONANCE.
RATMAN B   MOCHEL J
PHYS REV LETTERS

35620 SHORT-CIRCUIT CURRENT CARRYING CAPACITY OF ALUMINUM-CLAD STEEL WIRE.
TANAKA A   NUMAJIRI F
HITACHI REV
S. I.   12   69-79   1965

35659 DETERMINATION OF THERMAL ACCOMMODATION OF TRANSLATIONAL ENERGY OF VAPORS AT A GLASS SURFACE.
MORRISON J A   TUZI Y
J VACUUM SCI TECHNOL
2   3   109-12   1965   JA   50   17

35723 AN INFRARED REFLECTOMETER WITH A SPHEROIDAL MIRROR.
BLEVIN W R   BROWN W J
J SCI INSTR
42   6   385-9   1965   CA   63   2536

35739 THERMAL CONDUCTIVITIES OF FUELED GRAPHITE SPHERES FROM IRRADIATION TEST DATA.
PRADOS J W   DE CARLO V A   SCOTT J L
TRANS AM NUCL SOC
8   37-8   1965   NS   19   28900

35741 INVESTIGATION OF METAL MATRIX COMPOSITES WITH HIGH MODULUS, LOW DENSITY CONTINUOUS FILAMENT REINFORCEMENTS. ANNUAL SUMMARY TECH. REPT. 1 JULY 65 30 JUNE 66.
CHUANG K C   GENERAL TECHNOLOGIES CORP ALEXANDRIA VA   USAF
DDC
AFML-TR-66-330
AD-802423   1-59   1966   TA   67-1 A-51

35796 BASIC INVESTIGATION OF MULTI-LAYER INSULATION SYSTEMS. FINAL REPT.
LITTLE /ARTHUR D/ INC   CAMBRIDGE MASS   NASA
NASA AND CFSTI
ADL-65958-00-04 NASA-CR-54191
N65-23738   1-298   1964   PA   N65-3 2271

35797 THERMAL DIFFUSIVITY OF SOLID PROPELLANTS—DEVELOPMENT OF APPARATUS AND INITIAL TEST RESULTS. PROGRESS REPORT NO. 8, NOV 1963 - FEB 1964.
TANGER G E   NIX G H   AUBURN UNIV   ALA   DEPT OF MECH ENG
DDC
AD-457934   N65-22828
1-40   1964   PA   N65-3 2085

35802 SURFACE AND INTERFACIAL TENSIONS OF POLYMER MELTS. II. POLY/METHYL METHACRYLATE/, POLY/N-BUTYL METHACRYLATE/, AND POLYSTYRENE.
WU S
J PHYS CHEM
74   3   632-8   1970

35806 STUDY OF MICROMETEOROID DAMAGE TO THERMAL CONTROL MATERIALS. FINAL TECHNICAL REPT. 7 FEB - 3 NOV 1964.
FRIICHTENICHT J F   SPACE TECHNOLOGY LABS INC REDONDO BEACH CALIF   NASA
NASA AND CFSTI
STL-4146-6009-SU-000 NASA-CR-62810
N65-24293   1-60   1965   PA   N65-3 2269

35808 FAR INFRARED OPTICAL PROPERTIES OF SOME SPACECRAFT PAINTS. FROM JPL SPACE PROGRAMS SUMMARY NO. 37-33, VOL. 4, APRIL 1 - MAY 31, 1965.
HALL W M   JET PROPULSION LAB CALIF INST OF TECH PASADENA
CFSTI
JPL-RS-37-33-4
81-3   1965

35809 OPTICAL ABSORPTION AND PHOTOCONDUCTIVITY IN UNSTABLE AZIDES. PART 2. LEAD AZIDE.
DEB S K
TRANS FARADAY SOC
65   12   3187-94   1969

35810 HIGH THERMAL CONDUCTANCE DEVICES UTILIZING THE BOILING OF LITHIUM OR SILVER.
DEVERALL J E   KEMME J E   LOS ALAMOS SCIENTIFIC LAB   N MEX   USAEC
CFSTI
LA-3211   1-33   1964   NS   19   24667

35812 CARBIDE FUEL DEVELOPMENT. PHASE V REPORT FOR PERIOD OF OCT 1, 1962 - SEP 30, 1963.
STRASSER A   STAHL D   UNITED NUCLEAR CORP   WHITE PLAINS N Y   USAEC
TAYLOR K   ANDERSEN J   FALLS N Y   USAEC
CFSTI
UNC-5081   1-97   1964   NS   19   24923

35815 IRRADIATION TESTING OF CERAMIC FUELS. INTERIM REPORT, OCT 1, 1963 THROUGH NOV 30, 1964.
ZUROMSKY G   KOZIOL J J   CHERNOCK W P
COMBUSTION ENGINEERING INC   WINDSOR CONN   NUCLEAR DIV USAEC
CFSTI

TPRC Number	Bibliographic Citation

35818 OPTICAL COATINGS FOR ABSORPTANCE OF SOLAR ENERGY.
QUARTERLY PROGRESS REPT. NO. 6, 1 OCT. 64-1 JAN 65.
SCHMIDT R N   PARK K C   JANSSEN J E   HONEYWELL
RESEARCH CENTER   HOPKINS MINN   USAF
DDC
AD-466933   1-38   1965

35819 OPTICAL COATINGS FOR ABSORPTANCE OF SOLAR ENERGY.
QUARTERLY PROGRESS REPT. NO. 5, 1 JULY - 1 OCT. 64.
SCHMIDT R N   PARK K C   JANSSEN J E   HONEYWELL
RESEARCH CENTER   HOPKINS MINN   USAF
DDC
AD-466932   1-32   1964   TA   65-17 A-92

35821 TOTAL HEMISPHERICAL EMITTANCE OF OXIDIZED HEAT
RESISTANT ALLOYS.
DESANTIS V J   GENERAL ELECTRIC CO   PHILADELPHIA
PA   MISSILE AND SPACE DIV
DDC
R64SD60 AD-466356
                      1-30   1964   TA   65-16 A-87

35826 LIGHTWEIGHT THERMAL PROTECTION SYSTEM DEVELOPMENT.
VOL 2.   INSULATION MATERIAL DATA AND TEST APPARATUS.
TECH. REPT.
RYAN J M   GENERAL DYNAMICS/CONVAIR   SAN DIEGO
CALIF   USAF
DDC
GD/A-DCB-64-104 VOL 2 ML-TR-65-26 VOL 2
AD-467868   1-120   1965   TA   65-18 A-111

35832 THERMAL-PHYSICAL PARAMETERS OF MATERIALS.   ANNUAL
PROGRESS REPT. 21 MAY 62 - 20 MAY 63.
STEINBERG S   LARSON R E   KYDD A R   LITTON
SYSTEMS INC   ST PAUL MINN   DEPT OF
DDC
REPT   2409   USA-NLABS-TPMR-63-2
AD-465672   1 VOLUME   1963   TA   65-15 A-90

35840 HIGH TEMPERATURE, HIGH EMITTANCE INTERMETALLIC
COATINGS.
SCHATZ E A   ALVAREZ G H   COUNTS C R III
HOPPKE M A   AMERICAN MACHINE AND FOUNDRY CO
ALEXANDRIA VA   USAF
DDC
AFML-TR-65-217 /PT III/
AD-468059   1-100   1965   TA   65-18 A-113

35851 EXPERIMENTAL STUDY OF HEAT CONDUCTION THROUGH
RAREFIED GASES CONTAINED BETWEEN CONCENTRIC
CYLINDERS.
DYBBS A   SPRINGER G S   FLUID MECHANICS LAB
MASS INST OF TECH   CAMBRIDGE   ONR
DDC AND CFSTI
AD-618219   N65-32051
                      1-29   1965   TA   65-17 143

35856 EVALUATION OF INFRARED EMISSION OF CLOUDS AND GROUND
AS MEASURED BY WEATHER SATELLITES.   PH.D. THESIS.
KERN C D   WASHINGTON UNIV   SEATTLE   USAF
DDC AND CFSTI
AD-617417   1-157   1964   PA   N65-3-21 3610

35861 INFRARED REFLECTION SPECTRA OF GA/1-X/ AL/X/ AS MIXED
CRYSTALS.
ILEGEMS M   PEARSON G L
PHYS REV
1   B   4   1576-82   1970

35862 REFRACTORY MATERIALS SUITABLE FOR USE IN GUIDED
MISSILE PROPULSION SYSTEMS.   QUARTERLY PROGRESS REPT.
NO. 6.
MICCIOLI G R   JULIEN H P   CARBORUNDUM CO
NIAGARA FALLS N Y   USN
DDC
AD-465428   1-27   1965   TA   65-15 A-88

35880 THE OPTICAL PROPERTIES OF THIN GERMANIUM FILMS.
GRANT P M   HARVARD UNIV DIV OF ENGINEERING AND
APPLIED PHYSICS   HARVARD UNIV   ONR
APPLIED PHYSICS   CAMBRIDGE MASS
IIP-14   N65-35348
AD-619071   1-213   1965   TA   65-18 144

35886 STUDY COMPARING THE INFLUENCE OF THE SURFACE STATE OF
A METAL ON RADIATION AND THERMAL ACCOMMODATION.
PAULMIER D
PUBL SCI TECH MIN AIR /FRANCE/
413   1-80   1965   CA   63   15970

35888 INTERACTIONS BETWEEN SURFACE HYDROXYL GROUPS AND
ADSORBED MOLECULES. II.   INFRARED SPECTROSCOPIC
STUDY OF BENZENE ADSORPTION.
CUSUMANO J A   LOW M J D
J PHYS CHEM
74   9   1950-6   1970

35889 DETERMINATION OF THE EMISSIVITY OF MATERIALS.
SEMIANNUAL PROGRESS REPT., NOV. 15, 1964 - MAY 14,
1965.
LUOMA W   EMANUELSON R C   PRATT AND WHITNEY
AIRCRAFT   EAST HARTFORD CONN   NASA
NASA AND CFSTI
PWA-2608 NASA-CR-54444
N65-28951   1-56   1965   PA   N65-3 3033

35892 EVALUATION OF OPTICAL PROPERTIES AND ENVIRONMENTAL
STABILITY OF SOLAR CELL ADHESIVES.
MAURI R E   LOCKHEED MISSILES AND SPACE CO   PALO
ALTO CALIF   NASA
NASA AND CFSTI
LMSC-AO-34229
N64-28643   1-22   1964

35893 MOLECULAR BEAM VELOCITY DISTRIBUTION MEASUREMENTS.
REPORT 65-1.
SCOTT P B   MASSACHUSETTS INST OF TECH FLUID
DYNAMICS RESEARCH LAB   CAMBRIDGE   ONR
USAF
DDC AND CFSTI
AFOSR-65-0192 N65-27582
AD-614230   1-183   1965   PA   N65-3-16 2700

35894 EXPERIMENTAL VERIFICATION OF AN ANALYTICAL
DETERMINATION OF OVERALL THERMAL CONDUCTIVITY OF
HONEYCOMB-CORE PANELS.
STROUD C W   LANGLEY RESEARCH CENTER LANGLEY
STATION   VA   NASA
NASA AND CFSTI
NASA-TN-D-2866
N65-28646   1-15   1965   CA   66   106220

35899 PROPERTIES OF GLASS-REINFORCED EPOXY THROUGH THE
20 K. RANGE.
TOTH L W
MOD PLASTICS
42   12   123-30   1965   CA   63   11781

35902 INFRARED RADIATION OF SOLIDS.   REFRACTORY
MATERIALS.
GRENIS A F   LEVITT A P
AM CERAM SOC BULL
44   11   901-6   1965

35907 IMPROVED RADIATOR COATINGS. PART II
SCHATZ E A   COUNTS C R III   ALVAREZ G H
HOPPKE M A   AMERICAN MACHINE AND FOUNDRY CO
ALEXANDRIA VIRGINIA   USAF
DDC
ML-TDR-64-146 /PT 2/
AD-468576   1-99   1965

35909 ULTRA THIN GAUGE POLYMERIC FILMS FOR SPACE
APPLICATIONS.
COX D W JR   SEA-SPACE SYSTEMS INC   TORRANCE CALIF
NASA
NASA AND CFSTI
NASA-CR-274
                      1-45   1965

35922 A STUDY OF THE EFFECTIVE THERMAL DIFFUSIVITY OF A
FLUIDIZED BED.
BORODULYA V A   TAMARIN A I
INZHENER FIZ ZH
                      12   8-12   1964

35923 A STUDY OF THE EFFECTIVE THERMAL DIFFUSIVITY OF A
FLUIDIZED BED.   /ENGLISH TRANSLATION OF INZHENER.
FIZ. ZH. NO. 12, 8-12, 1964./
BORODULYA V A   TAMARIN A I
INTERN CHEM ENG
5   3   432-5   1965

35934 SPECTRAL MEASUREMENT OF SOLAR ABSORPTANCE.
FAUGERE J F
NASA
CNES-NT-2   N65-28516
                      1-20   1965   PA   N65-3 3029

35937 SIMULTANEOUS DETERMINATION OF THE SURFACE TENSION OF
TIN AND ITS CONTACT ANGLE WITH SILICA BY THE USE OF
CONICAL CAPILLARIES.
ATTERTON D V   HOAR T P
J INST METALS
81   541-51   1952-3

35938 THE SURFACE TENSION OF COPPER BY OPTICAL
MEASUREMENTS.
BELFORTI D A   LEPIE M P
MET SOC OF AIME TRANS
227   80-3   1963

35944 APPARATUS FOR THE DETERMINATION OF THERMAL
CONDUCTIVITY OF CONSTRUCTION MATERIALS.
KUBIS L P   BABANINA T I
STROITEL MATERIALY
9   5   37-8   1963

TPRC Number	Bibliographic Citation

35945 APPARATUS FOR THE DETERMINATION OF THERMAL CONDUCTIVITY OF CONSTRUCTION MATERIALS. //FRENCH TRANSLATION OF STROITEL. MATERIALY 9 /5/ 37-8, 1963.//
KUBIS L P   BABANINA T I
CNRS
CNRS   H/R-170 TT-65-28221
                                          1-3   1963   TT   15-2 67

35946 BAND STRUCTURE OF BISMUTH TELLURIDE, BISMUTH SELENIDE AND THEIR RESPECTIVE ALLOYS.
GREENAWAY D L   HARBEKE G
J PHYS CHEM SOLIDS
26   10   1585-1604   1965

35963 THE REDUCTION OF CU O/0.67/ IN HYDROGEN.
CZANDERNA A W
J PHYS CHEM
69   10   3607-10   1965

35988 OPTICAL PROPERTIES OF COLD- AND HOT-DEPOSITED GOLD FILMS.
DAVEY J E   PANKEY T
J APPL PHYS
36   8   2571-6   1965   CA   63   9243

36021 COMPILATION OF THERMAL CONDUCTIVITY OF FOODS. M.S. THESIS.
QASHOU M S   AUBURN UNIVERSITY   AUBURN ALABAMA
AUBURN UNIVERSITY
                        1-221   1970

36038 REFLECTANCE OF EVAPORATED GERMANIUM FILMS.
DONOVAN T M   ASHLEY E J
J OPT SOC AM
54   9   1141-4   1964   CA   61   13998

36046 IN SITU ELECTRON, PROTON, AND ULTRAVIOLET RADIATION EFFECTS ON THERMAL CONTROL COATINGS.
FOGDALL L B   CANNADAY S S   BROWN R R   BOEING COMP   SEATTLE WASHINGTON
NASA AND CFSTI
NASA-CR-100840
N69-24925   1-133   1968   CA   71   75766

36054 OPTICAL ABSORPTION OF POLYCRYSTALLINE FILMS OF ZINC TELLURIDE, CUBIC MODIFICATION.
SHALIMOVA K V   SPINULESCU-CARNARU I   PIROGOVA N V
IZV VUZ FIZIKA
9   4   12-17   1966

36058 METHODS FOR MEASURING THERMAL CONDUCTIVITY. APPLICATION TO FIBROUS INSULATING MATERIALS.
GASQUET R
REV THERM
2   23   1245-58   1963   CA   61   13900

36066 OPTICAL PROPERTIES OF GERMANIUM WITH CLEAN AND OXIDIZED SURFACES. //ENGLISH TRANSLATION OF FIZ. TEKH. POLUPROV. 4 /9/ 1770-4, 1970.//
ROZUMNYUK V T   NESTERENKO V A   TSEBULYA G G
LISITSA M P   SNITKO O V
SOVIET PHYSICS-SEMICONDUCTORS
4   9   1517-20   1971

36071 AN IMPROVED SPHERE PAINT.
MIDDLETON W E K   SANDERS C L
ILLUM ENGINEERING
48   254-6   1953

36080 CALORIMETER FOR MEASURING THE HEAT CAPACITY OF DISPERSED BODIES AND ADSORPTION SYSTEMS FROM 120 TO 300 K.
BEREZIN G I   KISELEV A V   KOZLOV A A
ZH FIZ KHIM
38   8   2106-10   1964   CA   61   14198

36084 DEVELOPMENT OF A 1200 F. RADOME.   SUMMARY ENGINEERING REPT. NO. 1, 1 APR 63 - 31 OCT 64.
CHASE V A   COPELAND R L   BRUNSWICK CORP
MARION VA   USAF
DDC
AD-454643   1-155   1964   TA   U65-4 A-138

36085 THERMODYNAMICAL FUNDAMENTAL EQUATION FOR SPHERICAL INTERFACE.
KONDO S
J CHEM PHYSICS
25   4   662-9   1956

36088 THERMAL-CONTACT RESISTANCE IN FINNED TUBING.
GARDNER K A   CARNAVOS T C
J HEAT TRANSFER TRANS ASME
82   279-93   1960

36090 HIGH TEMPERATURE SOLAR ABSORBER COATINGS.   PART II. REPT. FOR 1 JUN 63 - 1 JUL 64.
SCHMIDT R N   PARK K C   JANSSEN J E   HONEYWELL RESEARCH CENTER   HOPKINS MINN   USAF
DDC
ML-TDR-64-250/PT 2/
AD-455068   1-89   1964   TA   65-5 A-126

36100 DIRECTIONAL REFLECTANCE OF CERTAIN MATERIALS IN THE NEAR INFRARED.
MARY D J   HARRY DIAMOND LABS U S ARMY MATERIEL COMMAND   WASHINGTON D C
DDC
HDL-TM-64-29
AD-454083   1-29   1964   TA   U65-4 A-227

36109 BERYLLIUM COMPOSITE STRUCTURES.   VOLUME II. MATERIALS AND PROCESSES.
KRUSOS J N   KJELBY A S   BOROSIC J   BYRNE T J
AERONCA MANUFACTURING CORP   MIDDLETOWN OHIO   USAF
DDC AND CFSTI
ASD-TR-61-706 VOL. 2
AD-278526   1 VOLUME   1962

36112 PIGMENTED SURFACE COATINGS FOR USE IN THE SPACE ENVIRONMENT.
SEARLE N Z   HIRT R C   SCHMITT R G   AMERICAN CYANAMID CO CENTRAL RESEARCH DIV   STAMFORD CONN
USAF
DDC AND CFSTI
ASD-TDR-62-840
AD-297802   1-54   1963

36113 SOLAR REFLECTOR FOAMING TECHNOLOGY DEVELOPMENT.
SWANSON P   GOODYEAR AEOSPACE CORP   AKRON OHIO
USAF
DDC AND CFSTI
GER-11755   APL-TR-64-128
AD-609223   1-157   1964   TA   U65-3 9

36116 MEASUREMENT OF BIDIRECTIONAL REFLECTANCE USING A PHOTOGRAPHIC TECHNIQUE.   M.S. THESIS.
LOEHRLEIN J E   PURDUE UNIVERSITY   WEST LAFAYETTE INDIANA
PURDUE UNIVERSITY
                        1-65   1970

36118 A STUDY OF THERMAL RADIATION FROM A UNIT ELEMENT.
HALL F F   AXTELLE G E   ITT FEDERAL LABS SAN FERNANDO CALIF   USAF
DDC
SSD-TDR-62-86
AD-454090   1-45   1962   TA   U65-4 A-227

36123 RESEARCH ON LOW DENSITY THERMAL INSULATION MATERIALS FOR USE ABOVE 3000 F.   THIRD QUARTERLY STATUS REPORT, 1 JULY - 30 SEPTEMBER 1964.
STYHR K H   NATIONAL BERYLLIA CORP   HASKELL N Y
NASA
NASA AND CFSTI
NASA-CR-60542 N65-16258
                        1-18   1964   PA   N65-3 997

36126 RADIATION PROPERTIES OF MATERIALS DETAILED REFINED MEASUREMENTS.   REPT. FOR MAY 62 - SEP 64.
CAMUS J   GUNDERSON R B   HALL F F   SPADE G L
ITT FEDERAL LABS   SAN FERNANDO CALIF   USAF
DDC
SSD-TDR-64-224
AD-450295   1 VOLUME   1964   TA   U64-24 A-1

36128 CESIUM NEUTRAL AND ION EMISSION FROM CARBURIZED AND OXYGENATED POROUS TUNGSTEN.
CHO A Y   SHELTON H
AIAA J
2   12   2135-7   1964   CA   62   8481

36144 ADVANCES IN THE MATERIALS TECHNOLOGY RESULTING FROM THE X-20 PROGRAM.
STRATTON W K   STACY J T   GUNDERSON R D
HONEBRINK D A   TRIPP J T   BRESLICH F N
PERKOWSKI W S   OTTESTAD D J   ARMSTRONG C S
LEGAN D J   TREPUS G E JR   THE BOEING CO
SEATTLE WASH   USAF
DDC
AFML-TR-64-396
                        1-162   1965

36145 THE THERMOPHYSICAL PROPERTIES OF PLASTIC MATERIALS FROM MINUS 50 TO OVER 700 F.   REPT PART II.   REPT FOR 15 JAN. 64 TO 22 JAN. 65.
PEARS C D   SOUTHERN RESEARCH INSTITUTE BIRMINGTON ALA   USAF
DDC
ML-TDR-64-87 PT. 2
AD-462523   1-87   1965   TA   65-12 A-144

36146 THE OPTICAL PROPERTIES OF EVAPORATED GOLD IN THE VACUUM ULTRAVIOLET FROM 300 TO 2000 ANGSTROMS.
CANFIELD L R   HASS G   HUNTER W R   ARMY ENGINEER RESEARCH AND DEVELOPMENT LABS   FORT BELVOIR VA AND NAVAL RESEARCH LAB   WASHINGTON D C
NASA
NASA AND CFSTI
NASA-CR-53161
N65-15365   1-21   1963   PA   N65-5 822

TPRC Number	Bibliographic Citation

36154 EFFECT OF LOW TEMPERATURE ON THE THERMAL EMITTANCE OF
THREE BLACK PAINTS. COMPARISON OF NORMAL AND
HEMISPHERICAL EMITTANCES. FROM JPL RESEARCH SUMMARY
NO. 37 21, VOL. IV, 1965.
HALL W M   JET PROPULSION LAB   CALIF INST OF TECH
PASADENA
CFSTI
JPL-RS-37-31 /VOL 4/
                              108-9   1965

36167 EMISSIVITY MEASUREMENTS ON SOLAR CELLS.
SCHOCKEN K   THERMODYNAMICS SECTION RESEARCH
PROJECTS LAB DEVELOPMENT OPERATIONS DIVISION ARMY
BALLISTIC MISSILE AGENCY REDSTONE ARSENAL   ALABAMA
NASA
ABMA-DV-TN-16-59
N-82955   1-14   1959

36172 GROWTH AND PROPERTIES OF THIN SINGLE-CRYSTAL FILMS
OF YTTRIUM IRON GARNET.
LINARES R C   PERKIN-ELMER CORP   NORWALK CONN
USAF
DDC AND CFSTI
AFCRL-65-128 AD-615016
                    1-31   1965   CA   63   15629

36178 STUDY OF THE OPTICAL PROPERTIES OF CERTAIN METALS IN
THE INFRARED.
BURTIN R
REV OPT /FRANCE/
43   9   463-87   1964   CA   65   3183

36197 OPTICAL CONSTANTS OF METALS IN THE FAR ULTRAVIOLET
AND THEIR RELATION TO ENERGY BAND STRUCTURE.
SCIENTIFIC REPORT NO. 1, JUN 1962 - MAY 1963.
GOURNAY L S   COLORADO U ATOMOSPHERIC AND SPACE
PHYSICS LAB   BOULDER   NASA
NASA
NASA-CR-50680
N65-16327   1-111   1963   PA   N65-3 975

36199 A REVIEW OF DATA ON THERMAL CONDUCTIVITY.
RATCLIFFE E H
WOOD
JULY/SEPT   1-12   1964

36202 HOLLOW GLASS FIBER REINFORCED LAMINATES. FINAL REPT.
15 JUNE 1963 - 15 AUG. 1964.
GENERAL ELECTRIC CO SPACE SCIENCES
LAB   PHILADELPHIA PA   BUWEPS
LAB   BUWEPS
DDC
AD-451684   N65-15830
                        1-142   1964   PA   N65-3 937

36209 ON THE USE OF MAGNETOHYDRODYNAMICS DURING HIGH SPEED
RE-ENTRY.
JARVINEN P O   AVCO-EVERETT RESEARCH LAB   EVERETT
MASS   NASA
NASA AND CFSTI
NASA-CR-206
                              1-62   1965

36212 EFFECTS OF SURFACE EMITTANCE ON TURBULENT SKIN
FRICTION AT SUPERSONIC AND LOW HYPERSONIC SPEEDS.
ALLEN J M   CZARNECKI K R   LANGLEY RESEARCH
CENTER LANGLEY STATION   HAMPTON VA   NASA
NASA AND CFSTI
NASA-TN-D-2706
                              1-25   1965

36213 RADIOMETRIC OBSERVATIONS OF THE EARTHS HORIZON FROM
ALTITUDES BETWEEN 300 AND 600 KILOMETERS.
MC KEE T B   WHITMAN R I   ENGLE C D   LANGLEY
RESEARCH CENTER LANGLEY STATION   HAMPTON VA   NASA
NASA AND CFSTI
NASA-TN-D-2528
                              1-24   1964

36219 INVESTIGATION OF LIGHT SCATTERING IN HIGHLY
REFLECTING PIGMENTED COATINGS.
ZERLAUT G A
NASA AND CFSTI
IITRI-C6018-4 NASA-CR-55196
N65-16838   1-11   1963   PA   N65-7 1117

36229 EFFECTS OF ROUGHNESS OF METAL SURFACES ON ANGULAR
DISTRIBUTION OF MONOCHROMATIC REFLECTED RADIATION.
BIRKEBAK R C   ECKERT E R G
TRANS ASME J HEAT TRANSFER
87   C   1   85-94   1965   AM   18   432

36246 PHYSICAL PROPERTIES OF POROUS MATERIALS AND THEIR
APPLICATIONS. III.
WAKASHIMA H   KUWANA K
KANAZAWA DAIGAKU KOGAKUBU KIYO
3   2   172-6   1963   CA   62   9808

36268 THE NATURE OF LIQUID FILM EVAPORATION DURING NUCLEATE
BOILING.
SHARP R R   LEWIS RESEARCH CENTER   CLEVELAND OHIO
NASA
NASA AND CFSTI

36269 COMPARISON OF MEASUREMENTS OF INTERNAL TEMPERATURES
IN ABLATION MATERIAL BY VARIOUS THERMOCOUPLE
CONFIGURATIONS.
DOW M B   LANGLEY RESEARCH CENTER LANGLEY STATION
HAMPTON VA   NASA
NASA AND CFSTI
NASA-TN-D-2165
                              1-19   1964

36270 THE EFFECT OF SURFACE ROUGHNESS AND WAVINESS UPON THE
OVERALL THERMAL CONTACT RESISTANCE.
MIKIC B B   YOVANOVICH M M   ROHSENOW W M
DEPARTMENT OF MECHANICAL ENGINEERING MASSACHUSETTS
INST TECHNOLOGY
MIT
MIT-TR-76361-43 NASA-CR-81158
N67-15385   1-36   1966

36285 TESTING OF ABLATION MATERIAL. PRELIMINARY DATA
CONCERNING NONDEGRADED ABLATION MATERIALS, NOS. 1, 2,
3, AND 4. TECHNICAL PROGRESS LETTER NO. 4, 10
JAN.- 10 FEB. 1964.
ELIASON L K   RICE D H   SANFORD W L   POE T L
MELPAR INC   FALLS CHURCH VA   NASA
NASA AND CFSTI
NASA-CR-53658
N65-16831   1-62   1964   PA   N65-7 1100

36290 THE MEASUREMENT OF THE THERMAL CONDUCTIVITY OF UO2
UNDER IRRADIATION IN THE TEMPERATURE RANGE
150 - 1600 C.
CLOUGH D J   SAYERS J B   UNITED KINGDOM ATOMIC
ENERGY AUTHORITY METALLURGY DIV   HARWELL BERKSHIRE
AERE AND CFSTI
AERE-R-4690
N65-25179   1-55   1964   SA   68   9403

36320 AN INTEGRATING SPHERE SYSTEM FOR MEASURING AVERAGE
REFLECTANCE AND TRANSMITTANCE.
DAVIES J M   ZAGIEBOYLO W
APPL OPT
4   2   167-74   1965   CA   62   14040

36325 OPTICAL INTERFERENCE FILTER FOR THE SPECTRAL LINE OF
MERCURY 1849 A.
BAUMEISTER P W   COSTICH V R
APPL OPT
4   3   364   1965

36327 THERMAL CONDUCTANCE TESTS ON CABIN WALL INSULATION
ASSEMBLIES FOR A SUPERSONIC TRANSPORT AIRCRAFT.
SUPERSEDES ARC-27497.
MC NAUGHTAN I I   KEENE P A   AERONAUTICAL
RESEARCH COUNCIL   LONDON
DDC
RAE-TN-MECH-ENG-405 ARC-CP-910 N67-24037
AD-809896   1-79   1964   TA   67-10 A-2

36337 DIGITAL EVALUATION OF THE COMPLEX INDEX OF REFRACTION
FROM REFLECTANCE DATA.
JUENKER D W
J OPT SOC AM
55   3   295-9   1965

36342 OPTICAL AND PHOTOELECTRIC PROPERTIES OF THIN METALLIC
FILMS IN THE VACUUM ULTRAVIOLET.
RUSTGI OM P   WEISSLER G L
J OPT SOC AM
55   4   456   1965   CA   65   19443

36346 TRANSMITTANCE OF THIN METALLIC FILMS IN THE VACUUM-
UNTRAVIOLET REGION BELOW 1000 ANGSTROMS.
RUSTGI OM P
J OPT SOC AM
55   6   630-4   1965   CA   62   15587

36348 PRECISION OF COLOR MEASUREMENT WITH THE GE
SPECTROPHOTOMETER. I. ROUTINE INDUSTRIAL
PERFORMANCE.
BILLMEYER F W JR
J OPT SOC AM
55   6   707-17   1965

36352 A BOLOMETER FOR THE FAR INFRARED.
MARKOV M N
OPT I SPEKTROSKOPIYA
18   1   119-22   1965

36353 A BOLOMETER FOR THE FAR INFRARED. //ENGLISH
TRANSLATION OF OPT. I SPEKTROSKOPIYA 18 /1/ 119-22,
1965.//
MARKOV M N
OPT SPECTRY /USSR/
18   1   60-2   1965

36354 THE OPTICAL PROPERTIES OF MAGNESIUM OXIDE.
SAKHNOVSKII M YU
OPT I SPEKTROSKOPIYA
18   1   179-82   1965   CA   62   14041

TPRC Number	Bibliographic Citation

36355 THE OPTICAL PROPERTIES OF MAGNESIUM OXIDE. //ENGLISH
TRANSLATION OF OPT. I SPEKTROSKOPIYA 18 /1/ 179-82,
1965.//
SAKHNOVSKII M YU
OPT SPECTRY /USSR/
18   1   100-2   1965

36356 REFLECTION AT GRAZING INCIDENCE OF OPTICAL WAVES FROM
MIRRORS WITH A PROTECTIVE COATING.
KAZNACHEEV YU I   GORSHKOVA N K   KOLESNIKOVA N A
OPT I SPEKTROSKOPIYA
18   2   295-9   1965

36357 REFLECTION AT GRAZING INCIDENCE OF OPTICAL WAVES FROM
MIRRORS WITH A PROTECTIVE COATING. //ENGLISH
TRANSLATION OF OPT. I SPEKTROSKOPIYA 18 /2/ 295-9,
1965.//
KAZNACHEEV YU I   GORSHKOVA N K   KOLESNIKOVA N A
OPT SPECTRY /USSR/
18   2   163-5   1965

36360 HEAT TRANSFER CHARACTERISTICS OF CONCURRENT
GAS-LIQUID FLOW IN PACKED BEDS.
WEEKMAN V W JR   MYERS J E
AICHE JOURNAL
11   1   13-17   1965

36367 PART II. HEAT TRANSFER.
GABOR J D   STANGELAND B   MECHAM W J
AICHE JOURNAL
11   1   130-2   1965

36371 MINIATURE OPTICALLY IMMERSED THERMISTOR BOLOMETER
ARRAYS.
DE WAARD R   WEINER S
APPL OPT
6   8   1327-31   1967

36372 EFFECTS OF THE VARIATION OF ANGLE OF INCIDENCE AND
TEMPERATURE ON INFRARED FILTER CHARACTERISTICS.
BAKER M L   YEN VICTOR L
APPL OPT
6   8   1343-51   1967

36377 HEAT-TRANSFER CHARACTERISTICS OF PACKED BEDS WITH
STAGNANT FLUIDS.
OFUCHI K   KUNII D
INTERN J HEAT MASS TRANSFER
8   5   749-57   1965

36385 THE DESIGN OF HEAT REFLECTIVE PAINTS.
TOMPKINS L M   TOMPKINS E H
J OIL AND COLOUR CHEMISTS ASSOC
44   1   98-108   1958

36386 TEMPERATURE STABILIZATION OF HIGHLY REFLECTING
SPHERICAL SATELLITES.
HASS G   DRUMMETER L F JR   SCHACH M
J OPT SOC AM
49   9   918-24   1959

36404 PROPERTIES OF TWO WHITE PAINTS FOR APPLICATION TO
INFLATABLE SPACECRAFT. TITANIUM-DIOXIDE-PIGMENTED
EPOXY AND ZINC-OXIDE-PIGMENTED METHYL SILICONE
ELASTOMER.
WOERNER C V   LANGLEY RESEARCH CENTER LANGLEY
STATION   HAMPTON VA   NASA
NASA AND CFSTI
NASA-TN-D-2834
N65-25273   1-21   1965   CA   66   116746

36413 THERMAL CONTACT RESISTANCE. VOLUME I-A REVIEW OF THE
LITERATURE.
MINGES M L   WRIGHT-PATTERSON AFB   OHIO
DDC
AFML-TR-65-375
AD-482633   1-70   1966

36414 NEW METHOD FOR MEASURING DIFFUSE REFLECTANCE IN THE
INFRARED.
WHITE J U
J OPT SOC AM
54   11   1332-7   1964   CA   65   6517

36416 THERMOPHYSICAL PROPERTIES OF PLASTIC MATERIALS AND
COMPOSITES TO LIQUID HYDROGEN TEMPERATURE %-423 F.
CAMPBELL M D   HERTZ J   O BARR G L   HASKINS J F
GENERAL DYNAMICS/ASTRONAUTICS   SAN DIEGO CALIF
USAF
DDC AND CFSTI
ML-TDR-64-33/PT 2/ X65-16852 X65-18921
AD-464555   1-37   1965   TA   65-14 A-93

36418 PIGMENTED SURFACE COATINGS FOR USE IN THE SPACE
ENVIRONMENT.
SEARLE N Z   DANIEL J H JR   HIRT R C   MULLEN P A
STEHMAN W J   AMERICAN CYANAMID CO STAMFORD RES
LAB
DDC
ML-TDR-64-319/PT II/
AD-464802   1-54   1965

36419 THE THERMAL AND MECHANICAL PROPERTIES OF FIVE
ABLATIVE REINFORCED PLASTICS FROM ROOM TEMPERATURE TO
750 F.
PEARS C D   ENGELKE W T   THORNBURGH J D
SOUTHERN RESEARCH INST   BIRMINGHAM ALA   USAF
DDC
AFML-TR-65-133 X-65-17697
AD-465175   1-188   1965   TA   65-15 A-84

36420 INVESTIGATION AND DEVELOPMENT OF HIGH TEMPERATURE
INSULATION SYSTEMS.
WECHSLER A E   KRITZ M A   ARTHUR D LITTLE INC
CAMBRIDGE MASS   USAF
DDC
AFML-TR-65-138
          1-110   1965

36423 NASA CONTRIBUTIONS TO THE TECHNOLOGY OF INORGANIC
COATINGS.
PLUNKETT J D   DENVER UNIV DENVER RESEARCH
INSTITUTE COLO   NASA
NASA AND CFSTI
NASA-SP-5014
          1-260   1964   RM   22   M12-20581

36425 FAR INFRARED REFLECTOMETER FOR IMPERFECTLY DIFFUSE
SPECIMENS.
NEHER R T   EDWARDS D K
APPL OPT
4   7   775-80   1965   CA   63   3808

36440 METHOD FOR DETERMINING ACCOMMODATION COEFFICIENTS
FROM DATA IN THE TEMPERATURE JUMP RANGE WITHOUT
APPLYING TEMPERATURE-JUMP THEORY.
WACHMAN H Y
J CHEM PHYS
42   5   1850-1   1965   CA   62   13878

36460 ON THE EFFECT OF ADSORBED PARTICLES ON THE
ACCOMMODATION COEFFICIENTS.
SHIN H
J CHEM PHYS
42   10   3442-5   1965   SA   68   19783

36466 IMPROVED DIELECTRIC FILMS FOR OPTICAL AND SPACE
APPLICATIONS. FROM ARMY SCIENCE CONFERENCE
PROCEEDINGS, 17-19 JUNE 1964, VOL. I.
COX J T   HASS G   RAMSEY J B   U S ARMY
ENGINEER R AND D LABS   FT BELVOIR VA
DDC AND CFSTI
AD-612134   193-205   1964
AD-611432   193-205   1965

36467 INFRARED COATING STUDIES.   FIRST QUARTER REPORT 1964.
MARTIN T P   BAUSCH AND LOMB INC   ROCHESTER N Y
DEPT OF THE ARMY
DDC
AD-446686   N65-11272
          1-8   1964   RR   68-15 130

36469 SPACE TECHNOLOGY.   VOL. I.   SPACECRAFT SYSTEMS.
ABRAHAM L H   DOUGLAS AIRCRAFT CO INC   SANTA
MONICA CALIF   NASA
NASA AND CFSTI
NASA-SP-65   1-85   1965

36470 SPACE TECHNOLOGY.   VOL. II.   SPACECRAFT MECHANICAL
ENGINEERING.
ADAMS J L   JET PROPULSION LAB CALIF INST OF TECH
PASADENA CALIF   NASA
NASA AND CFSTI
NASA-SP-66   1-166   1965

36471 DEVELOPMENT OF A PREDISTRIBUTED AZIDE BASE
POLYURETHANE FOAM FOR RIGIDIZATION OF SOLAR
CONCENTRATORS IN SPACE.
JOURILES N   WELLING C E   GOODYEAR AEROSPACE
CORP   AKRON OHIO   NASA
NASA AND CFSTI
NASA-CR-235
          1 VOLUME   1965

36477 REFLECTANCE OF ANODIZED TITANIUM AND BERYLLIUM.
JANSSEN J   LUCK J   TORBORG R
J ELECTROCHEM SOC
109   3   1-31   1962

36480 CORRELATION OF THE HEAT TRANSFER PROPERTIES AND
ULTRASONIC TRANSMISSION PROPERTIES OF BONDS.
DI NOVI R A   ARGONNE NATL LAB   ARGONNE ILL

CFSTI AND AEC
ANL-7074 N66-35715
          1-21   1966   PA   N66-4-21 4267

36491 FUNDAMENTALS OF THERMAL RADIATION IN CERAMIC
MATERIALS. FROM SYMP. ON THERMAL RADIATION OF
SOLIDS, SAN FRANCISCO, CALIF., MARCH 4,5,6 1964.
COX R L   LING-TEMCO-VOUGHT ASTRONAUTICS DIV
DALLAS TEX   USAF NBS
NASA AND CFSTI
NASA-SP-55 ML-TDR-64-159 N65-26863
AD 629980   83-101   1965

TPRC Number	Bibliographic Citation

36493 A TEST OF ANALYTICAL EXPRESSIONS FOR THE THERMAL EMITTANCE OF SHALLOW CYLINDRICAL CAVITIES. FROM SYMP. ON THERMAL RADIATION OF SOLIDS, SAN FRANCISCO, CALIF., MARCH 4,5,6, 1964.
KELLY F J   MOORE D G   NATIONAL BUREAU OF STANDARDS WASHINGTON D C   USAF
NASA
NASA AND CFSTI
NASA-SP-55 ML-TDR-64-159 N65-26865
AD 629980   117-31   1965
APPL OPT
4   1   31-40   1965

36496 INFLUENCE OF SURFACE ROUGHNESS, SURFACE DAMAGE, AND OXIDE FILMS ON EMITTANCE. FROM SYMP. ON THERMAL RADIATION OF SOLIDS, SAN FRANCISCO, CALIF., MARCH 4,5,6, 1964.
BENNETT H E   MICHELSON LAB   CHINA LAKE CALIF
USAF NBS AND NASA
NASA AND CFSTI
NASA-SP-55 ML-TDR-64-159 N65-26868
AD 629980   145-52   1965

36499 SURFACE PROPERTIES OF METALS. FROM SYMP. ON THERMAL RADIATION OF SOLIDS, SAN FRANCISCO, CALIF., MARCH 4,5,6, 1964.
BLAU H H JR   FRANCIS H A   LITTLE /ARTHUR D/ INC
CAMBRIDGE MASS   USAF
NASA AND CFSTI
NASA-SP-55 ML-TDR-64-159 N65-26871
AD 629980   159-63   1965

36500 EFFECT OF SURFACE TEXTURE ON DIFFUSE SPECTRAL REFLECTANCE. A. DIFFUSE SPECTRAL REFLECTANCE OF METAL SURFACES. FROM SYMP. ON THERMAL RADIATION OF SOLIDS, SAN FRANCISCO, CALIF., MARCH 4,5,6, 1964.
KEEGAN H J   SCHLETER J C   WEIDNER V R
NATIONAL BUREAU OF STANDARDS WASHINGTON D C   USAF
NBS   NASA
NASA AND CFSTI
NASA-SP-55 ML-TDR-64-159 N65-26872
AD 629980   165-9   1965

36504 SOLAR ABSORPTANCE AND THERMAL EMITTANCE OF EVAPORATED METAL FILMS WITH AND WITHOUT SURFACE COATINGS. FROM SYMP. ON THERMAL RADIATION OF SOLIDS, SAN FRANCISCO, CALIF., MARCH 4,5,6, 1964.
HASS G   U S ARMY ENGINEER R AND D LABS   FORT BELVOIR VA   USAF
NASA AND CFSTI
NASA-SP-55 ML-TDR-64-159 N65-26876
AD 629980   189-95   1965

36506 LIGHT-SCATTERING BEHAVIOR OF PIGMENTED COATINGS. FROM SYMP. ON THERMAL RADIATION OF SOLIDS, SAN FRANCISCO, CALIF., MARCH 4,5,6, 1964.
COX R L   LING-TEMCO-VOUGHT ASTRONAUTICS DIV
DALLAS TEX   USAF NBS
NASA AND CFSTI
NASA-SP-55 ML-TDR-64-159 N65-26878
AD 629980   205-9   1965

36511 HEMISPHERICAL SPECTRAL EMITTANCE OF ABLATION CHARS, CARBON, AND ZIRCONIA. FROM SYMP. ON THERMAL RADIATION OF SOLIDS, SAN FRANCISCO, CALIF., MARCH 4,5,6, 1964.
WILSON R G   NASA LANGLEY RESEARCH CENTER LANGLEY AFB VA   USAF
NASA AND CFSTI
NASA-SP-55 ML-TDR-64-159 N65-26883
AD 629980   259-75   1965

36514 DIRECTIONAL SOLAR ABSORPTANCE MEASUREMENTS. FROM SYMP. ON THERMAL RADIATION OF SOLIDS, SAN FRANCISCO, CALIF., MARCH 4,5,6, 1964.
DOUGLAS N J   LOCKHEED MISSILES AND SPACE CO   PALO ALTO CALIF   USAF NBS
NASA AND CFSTI
NASA-SP-55 ML-TDR-64-159 N65-26886
AD 629980   293-301   1965

36515 A THERMAL VACUUM TECHNIQUE FOR MEASURING SOLAR ABSORPTANCE OF SATELLITE COATINGS AS A FUNCTION OF ANGLE OF INCIDENCE. FROM SYMP. ON THERMAL RADIATION OF SOLIDS, SAN FRANCISCO, CALIF., MARCH 4,5,6, 1964.
HOKE M G   NASA GODDARD SPACE FLIGHT CENTER
GREENBELT MD   USAF
NASA AND CFSTI
NASA-SP-55 ML-TDR-64-159 N65-26887
AD 629980   303-11   1965

36516 THE DIRECTIONAL SPECTRAL EMITTANCE OF SURFACES BETWEEN 200 AND 600 C. FROM SYMP. ON THERMAL RADIATION OF SOLIDS, SAN FRANCISCO, CALIF., MARCH 4,5,6, 1964.
BRANDENBERG W M   CLAUSEN O W   GENERAL DYNAMICS/ASTRONAUTICS SAN DIEGO CALIF   USAF NBS
NASA AND CFSTI
NASA-SP-55 ML-TDR-64-159 N65-26888
AD 629980   313-19   1965

36517 MEASUREMENT OF EMITTANCE OF ORGANIC MATERIALS /ELECTRIC WIRE INSULATION/. FROM SYMP. ON THERMAL RADIATION OF SOLIDS, SAN FRANCISCO, CALIF., MARCH 4,5,6, 1964.
MC INTYRE G W   THE BOEING CO   SEATTLE WASH
USAF   NBS
NASA AND CFSTI
NASA-SP-55 ML-TDR-64-159 N65-26889
AD 629980   321-4   1965

36519 THE APPLICATION OF TEMPERATURE RATE MEASUREMENTS TO THE DETERMINATION OF THERMAL EMITTANCES. FROM SYMP. ON THERMAL RADIATION OF SOLIDS, SAN FRANCISCO, CALIF., MARCH 4,5,6, 1964.
HALL W M   CALIF INST OF TECH JET PROPULSION LAB
PASADENA   USAF
NASA AND CFSTI
NASA-SP-55 ML-TDR-64-159 N65-26891
AD 629980   331-6   1965

36521 SOLAR-WIND BOMBARDMENT OF A SURFACE IN SPACE. FROM SYMP. ON THERMAL RADIATION IN SOLIDS, SAN FRANCISCO, CALIF., MARCH 4,5,6, 1964.
WEHNER G K   LITTON SYSTEMS INC   MINNEAPOLIS MINN
USAF NBS AND NASA
NASA AND CFSTI
NASA-SP-55 ML-TDR-64-159 N65-26893
AD 629980   345-9   1965

36522 SOME FUNDAMENTAL ASPECTS OF NUCLEAR RADIATION EFFECTS IN SPACECRAFT THERMAL CONTROL MATERIALS. FROM SYMP. ON THERMAL RADIATION OF SOLIDS, SAN FRANCISCO, CALIF., MARCH 4,5,6, 1964.
GILLIGAN J E   CAREN R P   LOCKHEED MISSILES AND SPACE CO   PALO ALTO CALIF   USAF NBS
NASA AND CFSTI
NASA-SP-55 ML-TDR-64-159 N65-26894
AD 629980   351-64   1965

36523 NUCLEAR ENVIRONMENTAL EFFECTS ON SPACECRAFT THERMAL CONTROL COATINGS. FROM SYMP. ON THERMAL RADIATION IN SOLIDS, SAN FRANCISCO, CALIF., MARCH 4,5,6, 1964.
BREUCH R A   POLLARD H E   LOCKHEED MISSILES AND SPACE CO   PALO ALTO CALIF   USAF NBS
NASA AND CFSTI
NASA-SP-55 ML-TDR-64-159 N65-26895
AD 629980   365-79   1965

36524 DEVELOPMENT OF A TECHNIQUE FOR THE CORRELATION OF FLIGHT- AND GROUND-BASED STUDIES OF THE ULTRAVIOLET DEGRADATION OF POLYMER FILMS. FROM SYMP. ON THERMAL RADIATION OF SOLIDS, SAN FRANCISCO, CALIF., MARCH 4,5,6, 1964.
PARKER J A   NEEL C B   GOLUB M A   NASA AMES RESEARCH CENTER AND STANFORD RESEARCH INSTITUTE MENLO PARK CALIF   USAF
NASA AND CFSTI
NASA-SP-55 ML-TDR-64-159 N65-26896
AD 629980   381-9   1965

36525 ULTRAVIOLET IRRADIATION OF WHITE SPACECRAFT COATINGS IN VACUUM. FROM SYMP. ON THERMAL RADIATION OF SOLIDS, SAN FRANCISCO, CALIF., MARCH 4,5,6, 1964.
ZERLAUT G A   HARADA Y   TOMPKINS E H   IIT RESEARCH INSTITUTE CHICAGO ILL   USAF NBS
NASA AND CFSTI
NASA-SP-55 ML-TDR-64-159 N65-26897
AD 629980   391-420   1965

36526 THE EFFECTS OF ULTRAVIOLET RADIATION ON LOW ALPHA SUB S/EPSILON SURFACES. FROM SYMP. ON THERMAL RADIATION OF SOLIDS, SAN FRANCISCO, CALIF., MARCH 4,5,6, 1964.
OLSON R L   MC KELLAR L A   STEWART J V
LOCKHEED MISSILES AND SPACE CO   PALO ALTO CALIF
USAF   NBS
NASA AND CFSTI
NASA-SP-55 ML-TDR-64-159 N65-26898
AD 629980   421-32   1965

36527 A STUDY OF THE PHOTODEGRADATION OF SELECTED THERMAL-CONTROL SURFACES. FROM SYMP. ON THERMAL RADIATION OF SOLIDS, SAN FRANCISCO, CALIF., MARCH 4,5,6, 1964.
PEZDIRTZ G F   JEWELL R A   NASA LANGLEY AFB   VA
USAF   NBS
NASA AND CFSTI
NASA-SP-55 ML-TDR-64-159 N65-26899
AD 629980   433-41   1965

36528 PRELIMINARY RESULTS FROM A ROUND-ROBIN STUDY OF ULTRAVIOLET DEGRADATION OF SPACECRAFT THERMAL-CONTROL COATINGS. FROM SYMP. ON THERMAL RADIATION OF SOLIDS, SAN FRANCISCO, CALIF., MARCH 4,5,6, 1964.
ARVESEN J C   NEEL C B   SHAW C C   NASA AMES RESEARCH CENTER MOFFETT FIELD CALIF   USAF NBS AND NASA
NASA AND CFSTI
NASA-SP-55 ML-TDR-64-159 N65-26900
AD 629980   443-52   1965

TPRC Number	Bibliographic Citation
36530	ALTERATION OF SURFACE OPTICAL PROPERTIES BY HIGH-SPEED MICRON-SIZE PARTICLES. FROM SYMP. ON THERMAL RADIATION OF SOLIDS, SAN FRANCISCO, CALIF., MARCH 4,5,6, 1964. MIRTICH M J MARK H NASA LEWIS RESEARCH CENTER CLEVELAND OHIO USAF NASA AND CFSTI NASA-SP-55 ML-TDR-64-159 N65-26902 AD 629980 473-81 1965
36532	AREAS OF RESEARCH ON SURFACES FOR THERMAL CONTROL. FROM SYMP. ON THERMAL RADIATION OF SOLIDS, SAN FRANCISCO, CALIF., MARCH 4,5,6, 1964. SNODDY W MILLER E NASA MARSHALL SPACE FLIGHT CENTER HUNTSVILLE ALA USAF NASA NASA AND CFSTI NASA-SP-55 ML-TDR-64-159 N65-26904 AD 629980 495-508 1965
36533	SELECTIVE COATINGS FOR VACUUM-STABLE HIGH-TEMPERATURE SOLAR ABSORBERS. FROM SYMP. ON THERMAL RADIATION OF SOLIDS, SAN FRANCISCO, CALIF., MARCH 4,5,6, 1964. SCHMIDT R N JANSSEN J E HONEYWELL RESEARCH CENTER HOPKINS MINN USAF NBS NASA AND CFSTI NASA-SP-55 ML-TDR-64-159 N65-26905 AD 629980 509-24 1965
36534	SURFACES OF CONTROLLED SPECTRAL ABSORPTANCE. FROM SYMP. ON THERMAL RADIATION OF SOLIDS, SAN FRANCISCO, CALIF., MARCH 4,5,6, 1964. TABOR H WEINBERGER H HARRIS J NATIONAL PHYSICAL LAB OF ISRAEL JERUSALEM USAF NBS NASA NASA AND CFSTI NASA-SP-55 ML-TDR-64-159 N65-26906 AD 629980 525-30 1965
36535	AN EXPERIMENTAL DETERMINATION OF THE ABSORPTANCES OF CRYODEPOSITED FILMS USING CALORIMETRIC TECHNIQUES. FROM SYMP. ON THERMAL RADIATION OF SOLIDS, SAN FRANCISCO, CALIF., MARCH 4,5,6, 1964. CAREN R P GILCREST A S ZIERMAN C A LOCKHEED MISSILES AND SPACE CO PALO ALTO CALIF USAF NBS NASA AND CFSTI NASA-SP-55 ML-TDR-64-159 N65-26907 AD 629980 531-4 1965
36536	THE STUDY OF LOW SOLAR ABSORPTANCE COATINGS FOR A SOLAR PROBE MISSION. FROM SYMP. ON THERMAL RADIATION OF SOLIDS, SAN FRANCISCO, CALIF., MARCH 4,5,6, 1964. STREED E R BEVERIDGE C M PHILCO CORP PALO ALTO CALIF USAF NBS NASA AND CFSTI NASA-SP-55 ML-TDR-64-159 N65-26908 AD 629980 535-48 1965
36538	STUDIES OF ABLATIVE MATERIAL PERFORMANCE FOR SOLID ROCKET NOZZLE APPLICATIONS. FINAL REPT. SCHAEFER J W DAHM T J RODRIGUEZ D A REESE J J JR WOOL M R AEROTHERM CORP PALO ALTO CALIF NASA AND CFSTI NASA-CR-72429 N68-31937 1-150 1968 PA N68-8-19 3360
36539	THERMAL SENSOR DESIGN FOR GLIDE REENTRY VEHICLES. FROM SYMP. ON THERMAL RADIATION OF SOLIDS, SAN FRANCISCO, CALIF., MARCH 4,5,6, 1964. BRUNSCHWIG F S KOCH G E WILHELM J K THE BOEING CO SEATTLE WASH USAF NBS NASA AND CFSTI NASA-SP-55 ML-TDR-64-159 N65-26911 AD 629980 567-9 1965
36544	COLOR AND STRUCTURAL CHARACTER OF CDS-CDSE PIGMENTS. EROLES A J FRIEDBERG A L J AM CERAM SOC 48 5 223-7 1965
36550	A THERMAL INSULATION AND STRUCTURAL SUPPORT FOR THERMOELECTRIC DEVICES. KETCHMAN J J WITTMAN R H PROCEEDINGS IEEE/AIAA THERMOELECTRIC SPECIALISTS CONFERENCE MAY 17-19 1966 19.1—19.20 1966
36571	EXPERIMENTAL STUDIES ON AL-FOIL THERMAL INSULATIONS FOR NUCLEAR REACTOR VESSELS. ISHIKAWAJIMA-HARIMA ENG REV 4 549-55 1964
36607	MOLECULAR ACCOMMODATION COEFFICIENTS OF EIGHT HALOGENATED DERIVATIVES OF METHANE ON PT. II. PEREZ-MASIA A VALLE BRACERO A RIENDA J M B ANALES REAL SOC ESPAN FIS QUIM /MADRID/ 60 A 5 101-8 1964

TPRC Number	Bibliographic Citation						
36612	DETERMINATION OF COEFFICIENT OF THERMAL CONDUCTIVITY OF GRAPHITOPLASTS BASED ON EPOXY RESINS. GRABOI L P LENSKAYA L P CHUDNOVSKII A R PLASTICHESKIE MASSY	3	41-3	1965	CA	62	13322
36633	STRUCTURE AND INFRARED LATTICE VIBRATION OF CALCIUM FLUORIDE AND LITHIUM FLUORIDE. BERTHOLD G Z PHYSIK	185	400-6	1965			
36634	MEASUREMENT OF THERMAL RADIATION CHARACTERISTICS. REPLOGLE B T MARTIN CO BALTIMORE DDC IDEP-428-90-40-60F2-01 ER-13638 AD-458082		1-25	1964			
36678	CONTRIBUTION TO THE STUDY OF THE ABNORMAL OPTICAL PROPERTIES OF CERTAIN METALS IN THE FORM OF THIN FILMS. RICHARD J ANN PHYS /FRANCE/ 9	11	697-728	1964	CA	62	14028
36681	OPTICAL PROPERTIES AND PRACTICAL USES OF INHOMOGENEOUS THIN FILMS. ANDERS H EICHINGER R APPL OPT 4	8	899-905	1965	CA	63	6486
36683	HIGH-TEMPERATURE SPACE-STABLE SELECTIVE SOLAR ABSORBER COATINGS. SCHMIDT R N PARK K C APPL OPT 4	8	917-25	1965	CA	63	6524
36684	ENHANCED REFLECTANCE OF RESTSTRAHLEN REFLECTION FILTERS. TURNER A F CHANG L MARTIN T P APPL OPT 4	8	927-33	1965			
36686	USE OF AN EVAPORATED DIELECTRIC FILM FOR DETERMINING THE OPTICAL CONSTANTS OF A METAL. I. BENNETT J M ASHLEY E J BENNETT H E APPL OPT 4	8	961-6	1965	CA	63	6523
36688	EFFECT OF ULTRAVIOLET IRRADIATION ON THE OPTICAL PROPERTIES OF SILICON OXIDE FILMS. BRADFORD A P HASS G MC FARLAND M RITTER E APPL OPT 4	8	971-8	1965	CA	63	6486
36700	THE OPTICAL CONSTANTS OF ALUMINIUM OXIDE FROM 12.0 TO 24.4 EV. FREEMAN G H C BRIT J APPL PHYS 16	7	927-31	1965			
36706	DETERMINATION AND VARIATION OF THE ACCOMMODATION COEFFICIENTS OF HIGH-ENERGY ARGON ATOMS, WITH REFERENCE TO DIFFERENT METALS. DEVIENNE F M COMPT REND 259 CORRECTION. 260	25 9	4575-8 2481	1964 1965	CA	62	13858
36710	OPTICAL CONSTANTS OF COPPER IN THE FAR-ULTRA-VIOLET, OBTAINED FROM THIN /VACUUM-DEPOSITED/ FILMS COVERED WITH A LITHIUM FLUORIDE LAYER. SEIGNAC A PRIOL M ROBIN S COMPT REND 258	20	4948-51	1964	MA	32	883
36740	SOLAR ABSORPTANCE AND TOTAL HEMISPHERICAL EMITTANCE AT CRYOGENIC TEMPERATURES. SUMMARY TECHNICAL REPORT. JUL. 1, 1964- SEPT. 30, 1965. WEBB L A U S RADIOLOGICAL DEFENSE LAB AIR FORCE MATERIALS LAB WRIGHT-PATTERSON AFB OHIO ATTN MANE AFML-TR-66-90 AD-489177L		1-26	1966	TA	86-21 A-43	
36751	P-POLARIZED REFLECTANCES FOR TRANSPARENT THIN FILMS ON TRANSPARENT SUBSTRATES. CATALAN L A J OPT SOC AM 55	7	857-9	1965	BR	14	5893
36754	DETERMINATION OF OPTICAL CONSTANTS OF METALS BY REFLECTIVITY MEASUREMENTS. COLLINS J R BOCK R O REV SCI INSTR 14	5	135-41	1943			
36755	LITHIUM FLUORIDE COLOR-CENTER FORMATION AND UV TRANSMISSION LOSSES FROM ARGON AND HYDROGEN DISCHARGES. WARNECK P J OPT SOC AM						

TPRC Number	Bibliographic Citation

36758 FIBER OPTICS. XI. PERFORMANCE IN THE INFRARED REGION.
KAPANY N S   SIMMS R J
J OPT SOC AM
55   8   963-8   1965   CA   63   6480

36797 AN APPARATUS TO MEASURE THE THERMAL CONDUCTIVITY OF POOR CONDUCTORS TO 1000 C.
ENGEL N N   MC ELROY D L   OAK RIDGE NATL LAB
TENN   USAEC
CFSTI
CONF-764-9 AEC NO. 2797
N66-10449   1-11   1964   CA   62   12473

36800 ABSORPTION SPECTRUM OF GERMANIUM AND ZINC-BLEND-TYPE MATERIALS AT ENERGIES HIGHER THAN THE FUNDAMENTAL ABSORPTION EDGE.
CARDONA M   HARBEKE G
J APPL PHYS /PT 1/
34   4   813-18   1963

36802 A STUDY OF THE PROPERTIES OF GOLD BLACK. //ENGLISH TRANSLATION OF ZH. PRIKL. SPEKTROSKOPII, AKAD. NAUK BELORUSSK. SSR 4 /6/ 503-8, 1966.//
SINTSOV V N
J APPLIED SPECTROSCOPY
4   6   362-5   1966

36807 AN ACCURATE METHOD OF MEASURING TOTAL EMITTANCE.
RUSSELL M W   GENERAL ELECTRIC   RECEIVING TUBE
DEPT   OWENSBORO KENTUCKY
GENERAL ELECTRIC CO
R59-ETR-1   1-19   1959

36809 THERMAL CONDUCTIVITY OF KERAMZIT CONCRETES. /WITH GERMAN TRANSLATION OF UNKNOWN ORIGIN./
SPIVAK N YA
BETON I ZHELEZOBETON
9   3   137-40   1963

36810 MULTI-REFLEXION ATTENUATED TOTAL REFLECTANCE INFRARED SPECTROSCOPY.
BAXTER B H   PUTTNAM N A
NATURE
207   288   1965

36811 COASTAL WATER PENETRATION USING MULTISPECTRAL PHOTOGRAPHIC TECHNIQUES. FROM PROCEEDINGS OF THE 5TH SYMPOSIUM ON REMOTE SENSING OF ENVIRONMENT. SEPT. 1968, UNIV. OF MICHIGAN.
YOST E   WENDEROTH S
DDC
AD-876327   571-86   1968

36813 SPECTROSCOPIC EVIDENCE OF CHEMICAL INTERACTIONS AT THE GOLD FILM-SILICA SUBSTRATE INTERFACE.
GUERRA C R   HEALY T W   FUERSTENAU D W
NATURE
207   518-20   1965   CA   63   14233

36819 INFLUENCE OF THE SPEED OF FORMATION OF VERY THIN LAYERS OF COPPER UPON THEIR TRANSPARENCY.
EMERIC N   EMERIC A
COMPT REND
260   18   4703-6   1965

36823 REFLECTION AND TRANSMISSION AT THE SURFACE OF AN ABSORBANT MEDIUM. APPLICATION TO THE CRITICAL STUDY OF THE ABBE REFRACTOMETRIC METHOD FOR THE DETERMINATION OF THE INDICES OF REFRACTION.
VINCENT-GEISSE J   DAYET J
J PHYSIQUE
26   2   66-72   1965

36824 APPARATUS FOR MEASURING THE THERMAL CONDUCTIVITY FOR THERMOELEMENTS.
GIRAUDIER L
J PHYSIQUE
26   4   129A-136A   1965   CA   63   1247

36876 MATERIAL PROPERTY DATA FOR UO2.
KOENIG N R   ATOMICS INTERNATIONAL
AEC
NAA-SR-MEMO   2529
                    1-7   1958

36884 REFLECTIVITY MEASUREMENTS OF SELECTED THERMAL CONTROL COATINGS IRRADIATED IN HIGH VACUUM. TECH. REPT. 1 OCT. 1964-1 DEC. 1966.
MC DANIEL R H   BELL J R   WATTIER J B   NUCLEAR AEROSPACE RESEARCH FACILITY GENERAL DYNAMICS   FORT WORTH TEXAS
DDC
AFWL-TR-67-22
AD-815911   1-124   1967   TA   67-16 A-41

36893 A STUDY OF MULTILAYERED DIELECTRIC LEAD OXIDE AND CRYOLITE COATINGS. //ENGLISH TRANSLATION OF ZH. PRIKL. SPEKTROSKOPII, AKAD. NAUK BELORUSSK. SSR 5 /2/ 153-7, 1966.//
NABOIKIN YU V   KRAMARENKO N L   AKOPOV V M
J APPLIED SPECTROSCOPY
5   2   114-17   1966

36903 EFFECTIVE THERMAL CONDUCTIVITY AND THE INTEGRATED EMISSIVITY FOR HEAT-RESISTANT CERAMIC COATINGS PREPARED FROM REFRACTORY OXIDES BY GAS-FLAME SPUTTERING.
BOGANOV A G   PIROGOV YU A   MAKAROV L P
TEPLOFIZ VYSOKIKH TEMPERATUR AKAD NAUK SSSR
3   1   64-9   1965   CA   63   332

36912 EFFECTIVE REFRACTIVE INDEXES OF METAL-DIELECTRIC INTERFERENCE FILTERS.
HEMINGWAY D J   LISSBERGER P H
APPL OPT
6   3   471-6   1967   CA   66   89797

36924 DETERMINATION OF SURFACE TENSION OF MOLTEN MATERIALS
DAVIS J K   BARTELL F E
ANAL CHEM
20   12   1182-5   1948

36933 FUNDAMENTAL FREQUENCIES OF ABSORPTION OF HALOGENATED COPPERS IN THE FAR INFRARED, BETWEEN ORDINARY TEMPERATURE AND THAT OF LIQUID HELIUM.
HENNINGER Y   MORLOT G   HADNI A
J PHYSIQUE
26   3   143-6   1965

36943 THERMAL CONTACT RESISTANCE IN A VACUUM ENVIRONMENT.
CLAUSING A M   CHAO B T
J HEAT TRANSFER TRANS ASME
87   C   2   243-51   1965   CA   63   52

36946 THE THERMAL RESISTANCE OF ADHESIVE BONDS.
LEWIS D M   SAUER H J JR
J HEAT TRANSFER TRANS ASME
87   C   2   310-11   1965

36947 OPTICAL PROPERTIES OF PALLADIUM IN THE FAR ULTRAVIOLET.
SEIGNAC A   STEPHAN D   ROBIN S
COMPT REND
260   13   3587-90   1965   RM   22M15 25647

36960 VARIATIONS OF THE TRANSMISSION FACTOR OF VERY THIN LAYERS OF GOLD, AS A FUNCTION OF THE VELOCITY OF FORMATION.
EMERIC N   EMERIC A
COMPT REND
260   3   845-8   1965   SA   68   15553

36961 PREPARATION, ELECTRICAL AND OPTICAL PROPERTIES OF THIN LAYERS OF GALLIUM INDIUM ARSENIDE GA AS IN.
MARTINUZZI S
COMPT REND
260   5   1379-82   1965

36963 ABSORPTION /OF LIGHT/ BY VERY THIN LAYERS OF SODIUM IN ULTRA-VACUUM.
BLANC R   PAYAN R   RIVOIRA R
COMPT REND
260   21   5504-7   1965   CA   63   10870

36997 INTERFACIAL TENSION OF POLAR MIXTURES.
O TOOLE J T   SHENDALMAN L H
TRANS FARADAY SOC
63   7   1584-95   1967

37011 RE-ENTRY HEAT CONDUCTION OF A FINITE SLAB WITH A NONCONSTANT THERMAL CONDUCTIVITY.
WELLS W R
AIAA J
2   2   379-81   1964   AM   17   4053
NASA
NASA-RP-167   N64-20200   1964   PA   N64-2   1587

37014 LOW-TEMPERATURE THERMAL CONDUCTIVITY OF A SUSPENSION OF COPPER PARTICLES.
ANDERSON A C   RAUCH R B
J APPL PHYS
41   9   3648-51   1970   CA   73   81448

37015 THE OPTICAL CONSTANTS OF CADMIUM OXIDE IN THE REGION OF INTRINSIC ABSORPTION.
FINKENRATH H
Z ANGEW PHYS
16   6   503-10   1964   CA   61   1405

37016 INVESTIGATION OF CRYSTAL STRUCTURE OF SILICATE MATERIALS IN REFLECTED LIGHT.
MINKO N I   MINAKOV V A
ZAVODSK LAB
30   4   465   1964   CA   61   108

37022 AN ANALYTICAL STUDY OF THERMAL CONTACT CONDUCTANCE FOR ROUGH WAVY SURFACES IN CONTACT. M.S. THESIS.
YEDDANAPUDI K M   SOUTH DAKOTA SCHOOL OF MINES AND TECHNOLOGY   RAPID CITY SOUTH DAKOTA
SOUTH DAKOTA SCHOOL OF MINES AND TECHNOLOGY
                    1-84   1969

37023 STRUCTURE ANALYSIS OF THERMAL OXIDE FILMS OF SILICON BY ELECTRON DIFFRACTION AND INFRARED ABSORPTION.
NAGASIMA N
JAPANESE J APPLIED PHYS

TPRC Number	Bibliographic Citation						

37034 OPTICAL PROPERTIES OF THIN GALLIUM FILMS IN THE WAVELENGTH RANGE FROM 4000 TO 10,000 A.
WESOLOWSKA C
ACTA PHYS POLON
25 — 3 — 323-36 — 1964 — CA — 62 — 1179

37035 ON A METHOD OF MEASURING THE THERMAL CONDUCTIVITY OF SEMICONDUCTORS IN THE HELIUM II TEMPERATURE RANGE, AS APPLIED TO GRAPHITE.
RAFALOWICZ J
ACTA PHYSICA POLONICA
25 — 3 — 427-36 — 1964 — BR — 13 — 5690

37040 MEASUREMENT OF THERMAL-RADIATION CHARACTERISTICS OF TEMPERATURE-CONTROL SURFACES DURING FLIGHT IN SPACE.
NEEL C B
ISA TRANSACTIONS
3 — 2 — 108-22 — 1964 — RM — 21 — 02632P

37043 SOME PROPERTIES OF SILICA FILM MADE BY RF GLOW DISCHARGE SPUTTERING.
KOZUMA T   KOBAYASHI T   HAMAGUCHI C   NAKAI J
FUJIOKA T   MATSUZAWA A
JAPANESE J APPLIED PHYS
9 — 8 — 983-91 — 1970

37044 A NEW ALUMINUM-CERAMIC COMPOSITE.
PALMISANO R R   DRAGER J
METAL PROGRESS
85 — 6 — 113-16 — 1964 — RM — 21 — 01871P

37053 SIMULTANEOUS MEASUREMENT OF SIX THERMAL PROPERTIES OF A CHARRING PLASTIC.
PFAHL R C JR   MITCHEL B J
INTERN J HEAT MASS TRANSFER
13 — 2 — 275-86 — 1970 — CA — 72 — 122265

37054 USE OF A RADIOACTIVATED LIGHT SOURCE FOR THE ABSOLUTE CALIBRATION OF TWO-COLOUR NIGHT AIRGLOW PHOTOMETER.
KULKARNI P V   SANDERS C L
PLANETARY AND SPACE SCIENCE
12 — 189-94 — 1964 — TA — U64-13 315

37058 THERMOFLASH APPARATUS AND COEFFICIENTS OF HEAT CONDUCTION OF INSULATING AND OTHER BUILDING MATERIALS.
SAXENA B K
INDIAN J TECHNOL
2 — 1 — 27-30 — 1964 — JA — 47 — 198

37066 THE TEMPERATURE WAVE METHOD OF MEASURING THERMAL DIFFUSIVITY AND THERMAL CONTACT CONDUCTANCE. M.S. THESIS.
TOMSIC M   KANSAS STATE UNIVERSITY   MANHATTAN KANSAS
KANSAS STATE UNIVERSITY
1-170 — 1969

37122 EXPERIMENTAL DETERMINATION OF THE DIRECTIONAL REFLECTANCE OF DIELECTRIC COATINGS ON METALLIC SUBSTRATES. M.S. THESIS.
PIEROWAY C S   UNIVERSITY OF OKLAHOMA   NORMAN OKLAHOMA
UNIVERSITY OF OKLAHOMA
1-45 — 1969

37146 SIMULTANEOUS MEASUREMENT OF REFLECTION AND TRANSMISSION FACTORS OF THIN FILMS AND THEIR CORRESPONDING PHASE CHANGES.
BOUSQUET P   DELEUIL R   GASTAUD A
J PHYS /PARIS/
25 — 31-8 — 1964 — CA — 61 — 137

37148 THE OPTICAL CONSTANTS AND REFLECTING POWER OF THIN FLUORIDE FILMS IN THE FAR ULTRA-VIOLET.
FABRE D   ROMAND J   VODAR B
J PHYS /FRANCE/
25 — 1 — 55-9 — 1964 — SA — 67 — 22820

37150 THE OPTICAL PROPERTIES OF EVAPORATED GOLD IN THE VACUUM ULTRAVIOLET FROM 300 TO 2000 A.
CANFIELD L R   HASS G   HUNTER W R
J PHYS /PARIS/
25 1/2 — 124-9 — 1964 — CA — 61 — 1400

37153 THE PREPARATION AND OPTICAL PROPERTIES OF THIN FILMS OF GALLIUM ARSENIDE AND CADMIUM TELLURIDE.
MARTINUZZI S   PERROT M   FOURNY J
J PHYS /FRANCE/
25 — 1 — 203-8 — 1964 — SA — 67 — 22857

37154 OPTICAL PROPERTIES OF THIN FILMS OF TELLURIUM UNEXPOSED TO AIR.
MERDY H   ROBIN-KANDARE S   ROBIN J
J PHYS /FRANCE/
25 — 1 — 223-8 — 1964 — SA — 67 — 22897

37173 SIMPLIFIED METHOD FOR CALCULATING THERMAL CONDUCTANCE OF ROUGH, NOMINALLY FLAT SURFACES IN HIGH VACUUM.
MC KINZIE D J JR   LEWIS RESEARCH CENTER
CLEVELAND OHIO
NASA

37193 ULTRAVIOLET STABILITY OF SOME MODIFIED METAL PHOSPHATES FOR THERMAL-CONTROL SURFACES.
PEZDIRTZ G F   WAKELYN N T
SYMP MATER SPACE VEHICLE USE 6TH SEATTLE
3 — 1 — 1-20 — 1963 — CA — 61 — 148

37194 HIGH EMITTANCE FOR REFRACTORY ALLOYS.
KERLEE C
SYMP MATER SPACE VEHICLE USE
3 — 11 — 1-19 — 1963 — CA — 61 — 386

37200 THE THERMAL CONDUCTIVITY OF DISPERSED MATERIALS AT DIFFERENT ATMOSPHERIC PRESSURES.
KOSTYLEV V M
TEPLOFIZ VYSOKIKH TEMP /USSR/
2 — 1 — 21-8 — 1964 — SA — 68 — 4834

37242 THE THERMAL CONDUCTIVITY OF A MULTILAYER SAMPLE OF UNDERWEAR MATERIAL UNDER A VARIETY OF EXPERIMENTAL CONDITIONS.
HOGE H J   FONSECA G F
TEXTILE RES J
34 — 5 — 401-10 — 1964 — CA — 61 — 3248

37255 THE THERMOPHYSICAL PROPERTIES OF ABLATIVE PLASTIC MATERIALS FROM ROOM TEMPERATURE TO DECOMPOSITION. PROGRESS REPORT NO. 3 OVER THE PERIOD APRIL 8 TO JULY 10, 1964.
SOUTHERN RESEARCH INST RESEARCH AND TECHNOLOGY DIV   AF SYSTEMS COMMAND   BIRMINGHAM ALA
DDC AND CFSTI
6868-1399-1-XII
1-46 — 1964

37272 PIGMENT OPTICS.
DE VORE J R
OFFIC DIG FEDERATION SOC PAINT TECHNOL /USA/
36 471 — 336-42 — 1964 — ZINC 22-7 280

37274 CHEMICAL ENGINEERING DIVISION RESEARCH HIGHLIGHTS, MAY 1963 - APRIL 1964.
ARGONNE NATIONAL LAB   ILL   ANL
USAEC
CFSTI
ANL-6875 — 1-191 — 1964

37275 THERMAL EMITTANCE OF MATERIALS FOR SPACE-CRAFT RADIATOR COATINGS.
HAYES R J   ATKINSON W H
AM CERAM SOC BULL
43 — 9 — 616-21 — 1964 — CA — 61 — 15805

37283 THERMAL RESISTANCE OF METALLIC CONTACTS.
SHLYKOV YU P   GANIN E A
INTERN J HEAT MASS TRANSFER
7 — 921-9 — 1964 — AM — 18 — 3666

37320 INFRARED COATING STUDIES. FINAL QUARTERLY REPT., 15 DEC 63 - 15 MAR 64.
BAUSCH AND LOMB INC   ROCHESTER N Y
U S ARMY
DDC AND CFSTI
AD-600264 — 1-23 — 1964 — TA — U64-14 142

37326 ON THE THEORY OF ACCOMMODATION COEFFICIENTS. V. CLASSICAL THEORY OF THERMAL ACCOMMODATION AND TRAPPING.
GOODMAN F O   ROYAL AIRCRAFT ESTABLISHMENT FARNBOROUGH ENGLAND
M O A AND DDC
RAE-TN-CPM-45
AD-438800 — 1-33 — 1963 — TA — U64-13 307

37327 THERMAL EMISSIVITIES OF SOME NON-METALLIC SURFACES AT LOW TEMPERATURES.
SEWELL J H   ROYAL AIRCRAFT ESTABLISHMENT FARNBOROUGH ENGLAND
M O A AND DDC
RAE-TN-CPM-80 AD-438802
1-16 — 1964 — TA — U64-13 367

37332 RESEARCH ON THE CHARACTERIZATION AND ANALYSIS OF NEW PLASTIC AND COMPOSITE MATERIALS IN ADVANCED RE-ENTRY ENVIRONMENTS. TEST REPT. NO. 2, 15 NOV - 15 DEC 61.
HOERCHER H E   RECESSO J V   BURKHARD K   AVCO CORP   WILMINGTON MASS   USAF
DDC
RAD-SR-62-4 AD-438219
1-65 — 1962 — TA — U64-13 232

37338 RESEARCH ON LOW DENSITY THERMAL INSULATION MATERIALS FOR USE ABOVE 3000 F. SIXTH QUARTERLY STATUS REPORT, 1 JUL - 30 SEP 1963.
STYHR K H   NATIONAL BERYLLIA CORP   HASKELL N J
NASA
NASA AND CFSTI
NASA-CR-55559
N64-15533 — 1-32 — 1963 — PA — N64-2 868

37342 DEVELOPMENT OF OXIDATION RESISTANT COATINGS FOR MOLYBDENUM.
GUNDERSON J M   BOEING CO   SEATTLE WASH   USAF
DDC

TPRC Number	Bibliographic Citation

37351 THE TIROS LOW RESOLUTION RADIOMETER.
BARTKO F   KUNDE V   CATOE C   HALEV M
GODDARD SPACE FLIGHT CENTER   GREENBELT MD   NASA
NASA AND CFSTI
NASA-TN-U-614
                                 1-34   1964

37355 VECTORIAL REFLECTANCE OF THE EXPLORER IX SATELLITE
MATERIAL.
KEATING G M   MULLINS J A   LANGLEY RESEARCH
CENTER LANGLEY STATION   HAMPTON VA   NASA
NASA AND CFSTI
NASA-TN-D-2388
N64-26270   1-45   1964   PA   N64-2 2553

37369 THEORETICAL AND EXPERIMENTAL STUDIES OF RADIATION
ABSORPTANCE AND EMITTANCE CHARACTERISTICS AND THEIR
CONTROL FOR POTENTIAL SPACE VEHICLE SURFACE
MATERIALS. TECHNICAL REPORT.
GEORGETOWN UNIV DEPT OF PHYS
WASHINGTON D C
NASA
NASA-CR-67168
X65-20133   1-105   1965

37370 VITRIFICATION OF THE SOLID FLUORIDE RESIDUE ARISING
FROM THE REACTION OF FLUORINE ON IRRADIATED FUEL.
BONNIAUD R   LABE P   COMMISSARIAT A LENERGIE
ATOMIQUE   FONTENAY-AUX-ROSES /FRANCE/
NASA AND CFSTI
CEA-CONF-1196
N69-29117   1-19   1968

37375 RESEARCH AND DEVELOPMENT OF VACUUM FOIL-TYPE
INSULATION FOR RADIOISOTOPE POWER SYSTEMS.
CARVALHO J   DUNLAY J B   FRONDUTO J   PAQUIN M L
POIRIER V L
NASA
ALO-3634-12 N69-34477
                                 1-59   1969

37378 DEVELOPMENT OF FRONTAL SECTION FOR SUPER-ORBITAL,
LIFTING, RE-ENTRY VEHICLE. VOL. II. MATERIALS AND
COMPOSITE STRUCTURE DEVELOPMENT.
LICCIARDELLO M R   OHNYSTY B   STETSON A R
SOLAR DIV OF INTERNATIONAAL HARVESTER CO   SAN DIEGO
CALIF   USAF
DDC
ER-1115-30 FDL-TDR-64-59/VOL 2/
AD-442590   1-248   1964

37381 AN INVESTIGATION OF THE THERMAL CONDUCTANCE OF BOLTED
JOINTS.
ANDREW I D C   ROYAL AIRCRAFT ESTABLISHMENT
FARNBOROUGH   GT BRIT
MOA AND NASA
RAE-TN-WE-46 N64-22371
AD-442270   1-38   1964

37388 DEVELOPMENT OF SERIES EMITTANCE THERMAL CONTROL
COATINGS. FINAL REPT. SEP. 1968-JUNE 1969.
LINDER B   GRIFFIN R N
NASA AND CFSTI
NASA-CR-66820
N69-35480   1-35   1969

37390 OPTICAL COATINGS FOR ABSORPTANCE OF SOLAR ENERGY.
QUARTERLY PROGRESS REPT. NO. 3, 1 DEC 63 - 1 MAR 64.
SCHMIDT R N   PARK K C   JANSSEN J E   TORBORG R H
HONEYWELL RESEARCH CENTER   HOPKINS MINN   USAF
DDC
HR-63-288   AD-442744
                                 1-27   1964   TA   U64-17 A-187

37391 THERMOPHYSICAL AND TRANSPORT PROPERTIES OF LIQUID
METALS. QUARTERLY PROGRESS REPT. NO. 1, JAN-MAR 64.
TEPPER F   MURCHISON A   ZELENAK J   ROEHLICH F
MSA RESEARCH CORP   CALLERY PA   USAF
DDC
MSAR-64-36 N64-27702
AD-441064   1-18   1964   PA   N64-2 2654

37393 RESEARCH ON LOW DENOITY THERMAL INSULATION MATERIALS
FOR USE ABOVE 3000 F. QUARTERLY STATUS REPORT, JAN.
1 - MAR. 31, 1964.
STYHR K H   NATIONAL BERYLLIA CORP   HASKELL N J
NASA
NASA AND CFSTI
NASA-CR-56546 QSR-1
N64-23530   1-25   1964   PA   N64-2 2064

37394 GAS-COOLED REACTOR PROGRAM SEMIANNUAL PROGRESS REPORT
FOR PERIOD ENDING SEPTEMBER 30, 1963.
OAK RIDGE NATIONAL LAB   TENN
USAEC
CFSTI
ORNL-3523   1-462   1964

37398 IMPROVED RADIATOR COATINGS. PART I. REPT. FOR 1 APR
63-1 APR 64.
SCHATZ E A   COUNTS C R III   BURKS T L
AMERICAN MACHINE AND FOUNDRY CO   ALEXANDRIA VA
USAF
DDC
ML-TDR-64-146
AD-442286   1-82   1964

37399 COATED REFRACTORY METAL EMITTANCE SPECIMENS.
LEAVENWORTH H W JR   SCHATZ E A   BROWNING M E
DUNKERLEY F J   AMERICAN MACHINE AND FOUNDRY CO
ALEXANDRIA VA   USAF
DDC AND CFSTI
ML-TDR-64-148 N65-10488
AD-607530   1-57   1964   PA   N65-3 83

37409 ELECTRICAL RESISTIVITY, THERMAL CONDUCTIVITY AND
THERMOPOWER OF GOLD AT LOW TEMPERATURES.
ANDERSEN H H   NIELSEN M   DANISH ATOMIC ENERGY
COMM   ROSKILDE   DAEC
DDC AND DAEC
RISO-REPT NO. 77
AD-443218   1-22   1964   NS   18   32146

37438 IN-PILE EFFECTIVE THERMAL CONDUCTIVITY OF OXIDE FUEL
ELEMENTS TO HIGH FISSION DEPLETIONS.
DANIEL R C   COHEN I   WESTINGHOUSE ELECTRIC CORP
BETTIS ATOMIC POWER LAB   PITTSBURGH   USAEC
CFSTI
WAPD-246   1-151   1964   NS   18   26073

37443 AMORPHOUS PHOSPHATE COATINGS FOR THERMAL CONTROL OF
ECHO II.
CLEMENS D L JR   CAMP J D
ELECTROCHEM TECHNOL
2 7-8   221-32   1964
NTIS
NASA-RP-303 N64-28752   1964

37446 TOTAL EMISSIVITY OF SILVER FILMS.
YODA E
JAPAN J APPL PHYS
8   11   1355   1969

37447 THERMAL CONDUCTIVITIES OF GAS-FILLED POROUS SOLIDS.
HARPER J C   EL-SAHRIGI A F
IND ENG CHEM FUNDAMENTALS
3   4   318-24   1964   CA   61   15255

37450 THE THERMAL CONDUCTIVITY OF DISPERSED MATERIALS AT
DIFFERENT ATMOSPHERIC PRESSURES. //ENGLISH
TRANSLATION OF TEPLOFIZ. VYSOKIKH TEMPERATUR, AKAD.
NAUK SSSR 2 /1/ 21-8, 1964.//
KOSTYLEV V M
HIGH TEMPERATURE
2   1   15-21   1964

37477 EMITTANCE MEASUREMENTS OF DISILICIDE-TYPE COATINGS AT
THE U. S. NAVAL RADIOLOGICAL DEFENSE LABORATORY.
FROM SUMMARY OF THE SEVENTH REFRACTORY COMPOSITES
WORKING GROUP MEETING, VOLUME 2.
ALVARES N   U S NAVAL RADIOLOGICAL DEFENSE LAB
SAN FRANCISCO CALIF   USAF
DDC
RTD-TDR-63-4131/VOL II/   N64-27020
AD-601265   341-51   1963   PA   N64-2 2603

37479 REFRACTORY COATING RESEARCH AND DEVELOPMENT AT U. S.
ARMY MATERIALS RESEARCH AGENCY /AMRA/. FROM SUMMARY
OF THE SEVENTH REFRACTORY COMPOSITES WORKING GROUP
MEETING, VOLUME 2.
LEVY M   U S ARMY MATERIALS RESEARCH AGENCY
WATERTOWN MASS   USAF
DDC
RTD-TDR-63-4131/VOL II/   N64-27020
AD-601265   450-87   1963

37480 COATED REFRACTORY ALLOYS FOR THE X-20 VEHICLE. FROM
SUMMARY OF THE SEVENTH REFRACTORY COMPOSITES WORKING
GROUP MEETING, VOLUME 2.
KUSHNER M   THE BOEING CO   SEATTLE WASH   USAF
DDC
RTD-TDR 63 4131/VOL II/   N64-27020
AD-601265   483-508   1963

37483 OPTICAL MEASUREMENTS ON CONES AND CYLINDERS.
BARTLE R C   BEARD C I   BURKE J E   SYLVANIA
ELECTRIC PRODUCTS INC   MOUNTAIN VIEW CALIF
ELECTRONIC DEFENSE LABS   DEPT   OF
DDC AND CFSTI
EDL-M614 N64-24758
AD-601477   1-50   1964   PA   N64-2 2363

37488 APPLICATION AND EVALUATION OF REINFORCED REFRACTORY
CERAMIC COATINGS. TECHNICAL DOCUMENTARY REPORT FOR
APR 61 - APR 62.
KALLUP C JR   SKLAREW S   CASTNER S V   USAF
DDC AND CFSTI
ML-TDR-64-81 AD-604423
                                 1-173   1964   NS   19   9608

TPRC Number	Bibliographic Citation						
37489	PIGMENTED SURFACE COATINGS FOR USE IN THE SPACE ENVIRONMENT. SEARLE N Z DANIEL J H JR HIRT R C MULLEN P A STEHMAN W J AMERICAN CYANAMID CO STAMFORD CONN USAF DDC ML-TDR-64-319 AD-458540	1-84	1964	TA	65-8 A-153		
37491	SOLID-CRYOGEN COOLER DESIGN STUDIES AND DEVELOPMENT OF AN EXPERIMENTAL COOLER. GROSS U E JR MANDAL R P LAWSON T W AEROJET-GENERAL CORP AZUSA CALIF DDC AFFDL-TR-68-1 AGC-3097 AD-834944	1-211	1968				
37493	WEAR RESISTANCE OF SPRAYED METALLIC AND CERAMIC COATINGS. CROUCHER T R KECK R R NORTRONICS DIV NORTHROP CORP HAWTHORNE CALIF DDC NOR-63-105 AD-445190	1-20	1963	TA	U64-20 A-360		
37499	THE THERMOPHYSICAL PROPERTIES OF PLASTIC MATERIALS FROM MINUS 50 TO OVER 700 F. PART I. REPT. FOR 1 MAY 62 - 15 JAN 64. PEARS C D ENGELKE W T THORNBURGH SOUTHERN RESEARCH INST BIRMINGHAM ALA USAF DDC ML-TDR-64-87/PT 1/ AD-446250	1-260	1964	TA	U64-20 A-188		
37501	STUDIES, RESEARCH AND INVESTIGATIONS OF THE OPTICAL PROPERTIES OF THIN FILMS OF METALS, SEMI-CONDUCTORS AND DIELECTRICS. SEMIANNUAL REPT. NO. 10, 16 FEB - 15 AUG 64. SCHEDDY C H COLORADO STATE UNIV FORT COLLINS DEPT OF THE ARMY DDC AD-446632	1-76	1964	TA	U64-20 A-290		
37504	DETERMINATION OF THE OPTICAL CHARACTERISTICS OF THIN FILM COATINGS. II. //ENGLISH TRANSLATION OF IZV. VUZ, FIZIKA 9 /4/ 122-9, 1966.// TEKUCHEVA I A SOVIET PHYSICS J 9	4	79-83	1966			
37509	DEVELOPMENT OF A 1200 F RADOME. INTERIM ENGINEERING REPORT NO. 5, 1 APR - 30 JUNE 1964. CHASE V A COPELAND R L BRUNSWICK CORP MARION VA USAF DDC AND CFSTI AD-603730	N64-28569 1-41	1964	PA	N64-2 2840		
37546	OPTICAL CONSTANTS OF METALS AT WAVELENGTHS SHORTER THAN THEIR CRITICAL WAVELENGTHS. HUNTER W R J PHYS /PARIS/ 25	154-60	1964	CA	61	3825	
37570	DEVELOPMENT OF 400 TO 1800 F. FIBROUS-TYPE INSULATION FOR RADIOISOTOPE POWER SYSTEMS. COLLINS J O JAUNARAJS K L REID D R JOHNS-MANVILLE RESEARCH AND ENGINEERING CENTER MANVILLE N J AEC AND CFSTI ALO-2661-12 N70-12852	1-236	1969	NSA 23-20 4267			
37580	THERMAL RESISTANCE AT INDIUM-SAPPHIRE BOUNDARIES BETWEEN 1.1 AND 2.1 K. NEEPER D A DILLINGER J R PHYS REV 135	A 4	A1028-33	1964			
37583	STATISTICAL STUDY OF THERMAL CONTACT CONDUCTANCE. M.S. THESIS. BHANDARI N K UNIVERSITY OF MIAMI CORAL GABLES FLORIDA UNIVERSITY OF MIAMI	1-96	1969				
37590	SUMMARY OF THE TENTH REFRACTORY COMPOSITES WORKING GROUP MEETING. HJELM L N JAMES D R BEARDSLEE E H AIR FORCE MATERIALS LAB AIR FORCE SYSTEMS COMMAND DDC AFML-TR-65-207 AD-472867	1-918	1965				
37595	THERMAL RADIATION ABSORPTANCE AND VACUUM OUTGASSING CHARACTERISTICS OF SEVERAL METALLIC AND COATED SURFACES. MIMURA T ANAGNOSTOU E COLARUSSO P E LEWIS RESEARCH CENTER CLEVELAND OHIO NASA NASA AND CFSTI NASA-TN-D-3234 N66-15791	1-56	1966				

TPRC Number	Bibliographic Citation						
37596	CLINKER BRICK. MATSUO C KATAOKA Y SEMENTO GIJUTSU NENPO 17	124-8	1963	JA	47	244	
37597	INFRARED SPECTRAL EMISSIVITIES OF COBALT OXIDE AND NICKEL OXIDE FILMS. KOKOROPOULOS P EVANS M V SOLAR ENERGY 8	2	69-73	1964	JA	47	235
37626	GAS-COOLED REACTOR PROGRAM. SEMIANNUAL PROGRESS REPORT FOR PERIOD ENDING MARCH 31, 1964. OAK RIDGE NATIONAL LAB TENN USAEC CFSTI ORNL-3619	1-368	1964	NS	18	30116	
37631	UO2 PELLET THERMAL CONDUCTIVITY FROM IRRADIATIONS WITH CENTRAL MELTING. LYONS M F COPLIN D H PASHOS T J WEIDENBAUM B GENERAL ELECTRIC CO ATOMIC POWER EQUIPMENT DEPT SAN JOSE CALIF USAEC CFSTI GEAP-4624	EURAEC-1142 1-38	1964	NS	18	34215	
37633	THE INFLUENCE OF SURFACE IRREGULARITIES ON THE OPTICAL PROPERTIES OF THIN FILMS. BOUSQUET P J PHYS /FRANCE/ 25	1	50-4	1964	SA	67	22819
37648	ESTIMATION OF THE EFFECT OF MOISTURE ON THE THERMAL CONDUCTIVITY OF STRUCTURE AND INSULATING MATERIALS. CAMMERER C ACHTZIGER J CHEM-INGR TECH 36	5	493-6	1964			
	ZIEGELINDUSTRIE 18	2	22-5	1965	JA	50	80
37675	INELASTIC TUNNELING DUE TO VIBRATIONAL MODES OF YTTRIUM AND CHROMIUM OXIDES. JAKLEVIC R C LAMBE J PHYS REV 2	B 4	808-12	1970			
37677	OPTICAL PROPERTIES OF AMORPHOUS GERMANIUM FILMS. DONOVAN T M SPICER W E BENNETT J M ASHLEY E J PHYS REV 2	B 2	397-413	1970			
37689	PROBABLE INTERFACE TEMPERATURES OF SOLIDS IN SLIDING CONTACT. LING F F PU S L AIR FORCE MATERIALS LAB WRIGHT-PATTERSON AIR FORCE BASE OHIO DDC RTD-TDR-63-4184 AD-433587	1-23	1964				
37701	EFFECT OF IMPURITIES ON THE SURFACE TENSION OF IRON MELTS. LEVI L I GLADYSHEV S A IZV VYSSH UCHEB ZAVED CHERN MET 12	7	151-4	1969	CA	71	127577
37711	PHYSICOCHEMICAL PROPERTIES OF WHEEL STEEL AND PRODUCTS OF ITS DEOXIDATION. POPEL S I DERYABIN A A SABUROV L N ISAEV N I VASILENKO G N IZV VYSSH UCHEB ZAVED CHERN MET 12	8	5-9	1969	CA	72	46425
37726	EFFECTIVE RADIAL THERMAL CONDUCTIVITIES IN PACKED BEDS. YAGI S KUNII D WAKAO N PROC HEAT TRANSFER CONF 1961-62 UNIV COLO 1961 /AND/ LONDON 1962	742-9	1963	CA	61	11633	
37734	OPERATIVE PRINCIPLES OF THE ADHESION AND WETTING OF OXIDES WITH IRON CARBIDE MELTS. KUPRIYANOV A A FILIPPOV S I IZV VYSSH UCHEB ZAVED CHERN MET 12	9	14-17	1969	CA	72	25265
37736	THE OPTICAL PROPERTIES AND CHEMICAL DECOMPOSITION OF LEAD IODIDE. TUBBS M R PROC ROY SOC A /GB/ 280	566-85	1964	SA	67	26202	
37790	EXPLORATORY PREPARATION OF FOILS STRONGLY COLORED IN THE 1-6 MICRON REGION OF THE INFRARED. PIKE E W LINCOLN LAB MASSACHUSETTS INST OF TECH LEXINGTON MASS USAF DDC AND CFSTI ESD-TDR-64-357 N65-10327 AD-604009	1-87	1964	PA	N65-3 119		

TPRC Number	Bibliographic Citation

37792 TRUDY TSENTRALNOY AEROLOGICHESKOY OBSERVATORII
//PROCEEDINGS OF THE CENTRAL AEROLOGICAL OBSERVATORY/
NO. 42.
KOKIN G A   LIVSHITS N S   EDITORS
GIDROMETEOROLOGICHESKOYE IZDATELSTOVO   MOSCOW
1-194   1962

37793 ACCOMMODATION OF AIR ON TUNGSTEN FILAMENTS AND THE
RADIATION COEFFICIENT OF TUNGSTEN. //FROM ENGLISH
TRANSLATION OF THE COLLECTION OF ARTICLES TRUDY
TSENTRALNOY AEROLOGICHESKOY OBSERVATORII NO. 42,
MOSCOW, 1962.//
ZAGORUYKO N V   KOKIN G A   KOKIN G A AND
LIVSHITS N S   EDITORS FOR THE STAFF
NASA AND CFSTI
NASA-TT-F-141
156-75   1964

37795 A CALORIMETRIC DEVICE FOR THE MEASUREMENT OF TOTAL
HEMISPHERICAL EMITTANCE. M.S. THESIS.
PUTNAM B R   AIR FORCE INST OF TECH
WRIGHT-PATTERSON AFB   OHIO
DDC AND CFSTI
AFIT-GAM/ME/64-15
AD-604834   1-156   1964   CA   62   4920

37815 EVALUATION OF THE MEAN COEFFICIENT OF HEAT
CONDUCTIVITY OF LAYERED WALLS BY THE METHOD OF
THERMAL FIELDS.
FOCSA V   RADU A
ARCH INZYN LADOWEJ
9   4   407-18   1963   AM   17   5340

37830 SOME OPTICAL PROPERTIES OF THIN MOLYBDENUM FILMS.
MOUTTET C
COMPT REND
258   19   4694-7   1964   MA   32   640

37831 OPTICAL CONSTANTS OF THIN, VACUUM-DEPOSITED COPPER
FILMS, UNEXPOSED TO AIR, IN THE FAR-ULTRA-VIOLET.
PRIOL M   SEIGNAC A   ROBIN S
COMPT REND
258   22   5398-401   1964   MA   32   883

37832 OPTICAL PROPERTIES OF THIN CALCIUM FILMS.
ROBRIEUX B.   CARLAN A
COMPT REND
258   22   5406-9   1964   MA   32   883

37833 OPTICAL PROPERTIES OF THIN SILVER LAYERS BETWEEN 35
AND 14 E.V.
ROBIN-KANDARE S   KANDARE S   ROBIN J
COMPT REND
259   4   765-8   1964   CA   61   15507

37840 PREPARATION OF TEMPERATURE CONTROL-SURFACES BY
ANODIZATION OF ALUMINUM. I.—THE EMATAL
PROCESS-EFFECTS OF PROCESS PARAMETERS. II.—OPTICAL
CHARACTERISTICS OF EMATAL ANODIC FILM.
HULQUIST A E   SIBERT M E
ELECTROCHEM TECHNOL
2 1/2   26-34   1964   MA   22   141

37845 COLOR CENTERS AND POINT DEFECTS IN IRRADIATED THORIA.
CHILDS B G   HARVEY P J   HALLETT J B
J AMERICAN CERAMIC SOC
53   8   431-5   1970

37859 DEVICE FOR THE SIMULTANEOUS DETERMINATION OF THE
DEGREE OF BLACKNESS OF THE TOTAL NORMAL RADIATION OF
MATERIALS FOR SIX SPECIMENS. //ENGLISH TRANSLATION
OF ZAVOD. LAB. 29, 4, 490-494, 1963.//
ZHOROV G A
INDUSTR LAB
29   4   503-7   1963   AM   17   6044

37860 DEVICE FOR THE SIMULTANEOUS DETERMINATION OF THE
DEGREE OF BLACKNESS OF THE TOTAL NORMAL RADIATION OF
MATERIALS FOR SIX SPECIMENS.
ZHOROV G A
ZAVODSKAYA LAB
29   4   490-4   1963

37877 POLARIZATION CHARACTERISTICS OF MAGNESIUM OXIDE AND
OTHER DIFFUSE REFLECTORS.
CARMER D C   BAIR M E
APPL OPT
8   8   1597-605   1969   CA   71   107506

37884 FAR INFRARED INTERFERENCE FILTERS.
VARMA S P   MOELLER K D
APPL OPT
8   8   1663-6   1969

37886 THERMAL RADIATIVE COATINGS FOR SNAP-8 FLIGHT SHIELD.
MAISEL L   ATOMICS INTERNATIONAL DIV OF NORTH
AMERICAN AVIATION INC   CANOGA PARK CALIF   USAEC
CFSTI
NAA-SR-MEMO-9749
1-17   1964   NS   18   39823

37889 INFORMATION ON PROPERTIES OF NAA-85 PROTECTIVE
COATING SYSTEM.
NORTH AMERICAN AVIATION INC   LOS
ANGELES CALIF
NASA AND CFSTI
TFD-63-522 N64-25998
1-9   1963   PA   N64-2 2526

37899 HEMISPHERICAL SPECTRAL EMITTANCE OF ABLATION CHARS,
CARBON, AND ZIRCONIA TO 3700 K.
WILSON R G   LANGLEY RESEARCH CENTER LANGLEY
STATION   HAMPTON VA   NASA
NASA AND CFSTI
NASA-TN-D-2704
N65-18607   1-28   1965   PA   N65-3 1357

37915 SEALED FOAM, CONSTRICTIVE-WRAPPED, EXTERNAL
INSULATION SYSTEM FOR LIQUID-HYDROGEN TANKS OF BOOST
VEHICLES.
LEWIS RESEARCH CENTER STAFF LEWIS
RESEARCH CENTER   CLEVELAND OHIO
NASA
NASA AND CFSTI
NASA-TN-D-2685
N65-19904   1-157   1965

37935 CHARACTERISTICS OF INTERFERENCE IN THIN GRANULAR
SILVER FILMS COVERED WITH ALUMINUM.
SHKLYAREVSKII I N   KORNEEVA T I   USOSKIN A I
SHKLYAREVSKII O I
OPT SPEKTROSK
27   5   840-4   1969

37970 OPTICAL CONSTANTS OF VACUUM-EVAPORATED SILVER FILMS.
HUEBNER R H   ARAKAWA E T   MAC RAE R A   HAMM R N
J OPT SOC AM
54   12   1434-7   1964   CA   62   1195

37971 OPTICAL AND ELECTRICAL PROPERTIES OF ANTIMONY
DEPOSITS.
HARRIS L   CORRIGAN F R   MASSACHUSETTS INST OF
TECHNOL   CAMBRIDGE
J OPT SOC AM
54   12   1437-41   1964   CA   62   1150

37995 NEW VALUES FOR THERMAL COEFFICIENTS.
VALENTICH J
PRODUCT ENGINEERING
63-71   1965

38028 PHYSICO-CHEMICAL BEHAVIOR OF SOME REFRACTORY
REINFORCED PHENOLIC.
MELNICK A M   FLORENTINE R A   TANZILLI R A
COHEN L   GENERAL ELECTRIC CO DEPT OF MISSILES AND
SPACE DIVISION   PHILADELPHIA PA
GENERAL ELECTRIC COMPANY
65-SD2048   1-6   1965

38030 PRESSURE DEPENDENCE OF ACCOMMODATION COEFFICIENTS.
NASR A A   SHERIF I I   AMMAR A S
PHYS LETTERS   NETHERLANDS
10   3   283   1964   SA   67   27218

38036 SPECTRAL EMISSION BY POLARITONS IN LITHIUM FLUORIDE.
HISANO K   OKAMOTO Y   MATUMURA O
J PHYS SOC JAPAN
28   2   425-9   1970

38042 FAR INFRARED BAND-PASS FILETERS IN THE 400 - 16
CM /-1/ SPECTRAL REGION.
VARMA S P   MOELLER K D
APPL OPT
8   10   2151-2   1969

38059 INVESTIGATION OF THE OPTICAL PROPERTIES OF VARIOUS
LAYERS IN VACUUM WHICH CHANGE ON EXPOSURE TO AIR.
RASIGNI G   RIVOIRA R
REV OPT
40   341-6   1961   CA   61   15452

38104 ROTATING CYLINDER APPARATUS FOR RAREFIED GAS FLOW
STUDIES.
ALOFS D J   SPRINGER G S
REV SCI INSTRUM
41   8   1161-3   1970

38109 APPARATUS FOR MEASURING EMITTANCE AND ABSORPTANCE AND
RESULTS FOR SELECTED MATERIALS.
CURTIS H B   NYLAND T W   LEWIS RESEARCH CENTER
CLEVELAND OHIO   NASA
NASA AND CFSTI
NASA-TN-D-2583
N65-13826   1-22   1965

38116 CONCERNING THE POSSIBILITY OF OBTAINING AN IDEAL
SPECTRAL DEPENDENCE IN THE REFLECTION COEFFICIENT OF
SILICON PHOTOVOLTAIC CELLS.
KOLTUN M M   LANDSMAN A P
OPTIKA I SPEKTROSKOPIYA
26   4   618-21   1969

TPRC Number	Bibliographic Citation

38117 CONCERNING THE POSSIBILITY OF OBTAINING AN IDEAL SPECTRAL DEPENDENCE IN THE REFLECTION COEFFICIENT OF SILICON PHOTOVOLTAIC CELLS. //ENGLISH TRANSLATION OF OPTIKA I SPEKTROSKOPIYA 26 /4/ 618-21, 1969.//
KOLTUN M M   LANDSMAN A P
OPT SPECTROSC
26   4   338-40   1969

38139 ABSORPTION AND REFLECTIVITY OF THIN FILMS OF CADMIUM TELLURIDE BETWEEN 3 AND 15 EV.
MERDY H   ROBIN-KANDARE S   ROBIN J
COMPT REND
259   5   1078-80   1964   SA   68   4301

38140 EFFECT OF THE PARTIAL PRESSURE OF WATER VAPOR ON THE OPTICAL PROPERTIES OF THIN CALCIUM LAYERS.
ROBRIEUX B
COMPT REND
259   21   3744-7   1964   CA   62   5990

38143 APPLICATION OF THE RELATIONS OF DISPERSION TO THE SIMULTANEOUS DETERMINATION OF THE THICKNESS AND THE COMPLEX INDEX OF A THIN METALLIC FILM.
FROISSART C
REVUE DE PHYSIQUE APPLIQUEE
4   3   397-403   1969

38175 A NEW METHOD FOR SIMULTANEOUS ESTIMATION OF THE EFFECTIVE THERMAL CONDUCTIVITY AND THE WALL HEAT TRANSFER COEFFICIENT IN A TUBE PACKED WITH GRANULAR MATERIAL.
ZIOLKOWSKI D
BULL ACAD POLON SCI SER SCI CHIM
18   4   221-6   1970   CA   73   57498

38180 INFRARED FILTERS OF EVAPORATED GALLIUM ARSENIDE.
HOWSON R P
J OPT SOC AM
55   3   271-5   1965

38189 ANOMALOUS THERMAL CONDUCTIVITY OF A FILM OF WATER ON MICA CRYSTALS.
METSIK M S   AIDANOVA O S
ISSLED V OBL POVERKHN SIL AKAD NAUK SSSR INST FIZ KHIM SB DOKL NA VTOROI KONF MOSCOE 1962
            188-95   1964   CA   62   2262

38190 THE USE OF THE FLAT PLATE METHOD TO INVESTIGATE THE THERMOPHYSICAL PROPERTIES OF HIGHLY EFFICIENT HEAT INSULATION AT LOW TEMPERATURES. //ENGLISH TRANSLATION OF INZH.-FIZ. ZH. 11 /5/ 634-8, 1964.//
MIKHALCHENKO R S   GERZHIN A G   PERSHIN N P
J ENGINEERING PHYS
11   5   356-8   1966

38247 THEORETICAL METHOD FOR DETERMINING THE APPARENT RADIATION PROPERTIES FOR MATERIALS IN SINUSOIDAL CONFIGURATION.
FARBER E A   VALANDANI P
J ENG POWER
86   4   472-4   1964   CA   62   4921

38248 OPTICAL CHARACTERISTICS OF THIN FILMS.
TEKUCHEVA I A
IZV VUZ FIZIKA
8   1   116-21   1965

38249 FLAME-CONTACT STUDIES.
STOLL A M   CHIANTA M A   MUNROE L R
J HEAT TRANSFER
86   3   449-56   1964   CA   62   2868

38251 A THEORY OF ENERGY ACCOMMODATION.
TRILLING L
J MECAN /FRANCE/
3   2   215-34   1964   SA   67   29674

38253 EFFECT OF A THIN SURFACE FILM ON THE ELLIPSOMETRIC DETERMINATION OF OPTICAL CONSTANTS.
BURGE D K   BENNETT H E
J OPT SOC AMER
54   12   1428-33   1964   SA   68   4277

38258 STUDY OF THE OPTICAL PROPERTIES OF GALLIUM THIN FILMS.
WESOLOWSKA C   RICHARD M J
J PHYS /FRANCE/
25   7   737-40   1964   SA   67   30680

38272 RESEARCHES ON HEAT AND MASS TRANSFER BY SUBLIMATION OF ICE.
KATAYAMA K   HAYASHI Y   KIMURA T
BULL JSME /JAPAN SOC MECH ENGR/
12   50   257-64   1969

38278 FUNDAMENTAL OPTICAL ABSORPTION IN MAGNESIUM OXIDE.
REILING G H   HENSLEY E B
PHYS REV
112   4   1106-11   1958

38288 ULTRAVIOLET ABSORPTION OF ALKALI HALIDES.
EBY J E   TEEGARDEN K J   DUTTON D B
PHYS REV

38296 THE CALCULATION OF THE REFLECTIVITY OF AN ABSORBING FILM ON AN ABSORBING SUBSTRATE.
FREEMAN G H C
OPTICA ACTA /INTERNAT/
11   3   219-22   1964   SA   68   5076

38299 OPTICAL COEFFICIENTS OF THIN CHROMIUM FILMS OBTAINED BY THERMAL SUBLIMATION IN VACUO.
IDCHAK E F
OPTIKA I SPEKTROSKOPIYA
17   6   923-6   1964   CA   62   4761

38309 MEASUREMENT OF THE COEFFICIENT OF THERMAL ACCOMMODATION OF NEON AS A MEANS OF DETERMINING THE CONTAMINATION OF METALLIC SURFACES. I. TUNGSTEN FILAMENTS.
FARON M J   TEICHNER S J
REV HAUTES TEMP REFRACTAIRES
1   2   111-20   1964   CA   61   15476

38310 MEASUREMENT OF THE COEFFICIENT OF THERMAL ACCOMMODATION OF NEON AS A MEANS OF DETERMINING THE CONTAMINATION OF METALLIC SURFACES. II. INFLUENCE OF THE SURFACE ROUGHNESS OF THE TUNGSTEN AND PALLADIUM-GOLD FILAMENTS.
REV HAUTES TEMP REFRACTAIRES
1   3   201-10   1964   CA   62   7463

38317 SELECTIVE-EMISSION SUBSTANCES FOR SOLAR-HEAT ABSORBERS.
GILLETTE R B
SOLAR ENERGY
4   4   24-32   1960   CA   68   3412

38320 THERMAL CONDUCTIVITY OF COMPRESSIBLE POROUS MATERIALS.
MC MASTER D G
TAPPI
47   12   796-801   1964   CA   63   260

38383 ELASTIC AND THERMAL EXPANSION PROPERTIES OF CONCRETE AS AFFECTED BY SIMILAR PROPERTIES OF THE AGGREGATE.
KOENITZER L H
PROC AM SOC TESTING MATERIALS
36   2   393-406   1936

38391 RADIATION CHARACTERISTICS OF ROUGH AND OXIDIZED METALS. FROM ADVANCES IN THERMOPHYSICAL PROPERTIES AT EXTREME TEMPERATURES AND PRESSURES.
EDWARDS D K   CATTON I
3RD ASME SYMP ON THERMOPHYSICAL PROPERTIES PURDUE UNIV LAFAYETTE IND MARCH 22-25
            189-99   1965   CA   63   9587

38392 EFFECT OF VACUUM-UV ON REFLECTANCE AND EMITTANCE OF SOLAR REFLECTORS. FROM ADVANCES IN THERMOPHYSICAL PROPERTIES AT EXTREME TEMPERATURES AND PRESSURES.
FINCH H L   RHODES B L   KRAUSE A J
3RD ASME SYMP ON THERMOPHYSICAL PROPERTIES PURDUE UNIV LAFAYETTE IND MARCH 22-25
            200-6   1965   CA   63   9189

38402 A METHOD FOR THE SIMULTANEOUS DETERMINATION OF ALL THERMAL PROPERTIES OF POOR HEAT CONDUCTORS OVER THE TEMPERATURE RANGE 80 TO 500 K. FROM ADVANCES IN THERMOPHYSICAL PROPERTIES AT EXTREME TEMPERATURES AND PRESSURES.
LUIKOV A V   VASILEV L L   SHASHKOV A G
3RD ASME SYMP ON THERMOPHYSICAL PROPERTIES PURDUE UNIV LAFAYETTE IND MARCH 22-25
            314-19   1965   CA   63   9085

38405 THERMAL RESISTANCE OF ADHESIVE LAYERS. FROM ADVANCES IN THERMOPHYSICAL PROPERTIES AT EXTREME TEMPERATURES AND PRESSURES.
SCHWALLER D L   SAUER H J JR   REMINGTON C R
3RD ASME SYMP ON THERMOPHYSICAL PROPERTIES PURDUE UNIV LAFAYETTE IND MARCH 22-25
            336-40   1965   CA   63   11034

38408 ADSORPTION OF CERTAIN DIAMINES AT WATER-AIR AND WATER-CHLOROFORM INTERFACES. //ENGLISH TRANSLATION OF ZH. FIZ. KHIM. 43 /4/ 1039-41, 1969.//
NIKONOV V Z   SOKOLOV L B
RUSS J PHYS CHEM
43   4   581-3   1969

38417 ZIG-ZAG EFFECT FOR THERMAL CONDUCTIVITY IN A TRANSVERSE MAGNETIC FIELD.
THORN J   WYDER P
PHYS LETTERS /NETHERLANDS/
13   1   11-12   1964   SA   68   6576

38431 DIFFUSE SCATTERED RADIATION THEORIES OF DUNTLEY AND OF KUBELKA-MUNK.
LATHROP A L
J OPT SOC AM
55   9   1097-1104 1965

38456 THE VIBRATIONS OF THE MGO CRYSTAL STRUCTURE AND ITS INFRA-RED ABSORPTION SPECTRUM.
RAMAN C V
PROC INDIAN ACAD SCI

TPRC Number	Bibliographic Citation

38459 THE INFRA-RED SPECTRUM OF MAGNESIUM OXIDE.
WILLMOTT J C
PROC PHYS SOC
63   A   389-402   1950

38460 CRYOGENICS, HELIUM THREE.
DAUNT J G   EDWARDS D O
ANN REV PHYS CHEM
15   83-108   1964   N9   19   1486

38505 ON THE PYROELECTRIC PROPERTIES OF SOME MATERIALS AND
THEIR APPLICATION TO THE DETECTION OF INFRARED.
HADNI A   HENNINGER Y   THOSAS R   VERGNAT P
WYNCKE B
J PHYSIQUE
26   6   345-60   1965

38512 INFRARED TRANSMITTANCE OF THIN ALUMINUM FILMS ON
COLLODION SUBSTRATES.
BRETT D A   SULLIVAN E J
J OPT SOC AM
55   11   1556-8   1965   CA   64   4453

38525 THERMAL CONDUCTIVITY   FROM   THE AMERICAN INSTITUTE OF
PHYSICS HANDBOOK /MC GRAW-HILL BOOK CO., INC. 1963/
POWELL R L   CRYOGENIC ENGINEERING LAB NATIONAL
BUREAU OF STANDARDS
NBS
NBS-R-307   4-77—4-101 1963

38531 ULTRAVIOLET REFLECTION SPECTRUM OF CUBIC CADMIUM
SULFIDE.
CARDONA M   WEINSTEIN M   WOLFE G A
PHYS REV
140   A   2   633-7   1965

38575 OPTICAL CONSTANTS OF A CLEAN MERCURY SURFACE AS A
FUNCTION OF TEMPERATURE.
SMITH T
J OPT SOC AM
57   10   1207-10   1967

38578 VERSATILE SPECTRORADIOMETER AND ITS APPLICATIONS.
SANDERS C L   GAW W
APPL OPT
6   10   1639-47   1967

38586 THERMAL CONDUCTIVITY OF PELLETIZED CERAMIC PACKINGS.
//ENGLISH TRANSLATION OF INZHEN.-FIZ. ZH. 9 /1/ 48-53
1965.//
KHARLAMOV A G
J ENG PHYS BSSR
9   1   36-40   1965

38609 A STUDY OF MULTI-LAYER INSULATION.
POVOLOTSKII L V   ARKADEV B A
TEPLOENERGETIKA
11   1   36-40   1964

38610 A STUDY OF MULTI-LAYER INSULATION.   //ENGLISH
TRANSLATION OF TEPLOENERGETIKA 11 /1/ 36-40, 1964.//
POVOLOTSKII L V   ARKADEV B A
THERMAL ENGINEERING
11   1   44-8   1964

38611 DEVELOPMENT OF AN ANALYTICAL TECHNIQUE FOR PREDICTING
PROPERTIES OF COMPOSITE MATERIALS.   PART II.   FINAL
REPT.   JUL 1, 1964- FEB 1, 1966.
SILBERNAGEL R E   WISE R A   STUDEBAKER CORP
CINCINNATI   OHIO   CTL DIVISION
NASA AND CFSTI
NASA-CR-82485 N67-19155
                    1-135   1966   PA   N67-5-8 1286

38615 HEAT TRANSFER BETWEEN SOLIDS.
KOLB R P   MARINE ENGINEERING LAB   ANNAPOLIS MD
DDC
MEL-157/64 AD-467809
                    1-11   1965   TA   65-18 A-234

38618 DETERMINATION OF THE EMISSIVITY OF MATERIALS.
SEMI-ANNUAL PROGRESS REPORT, MAY 14 - NOVEMBER 15,
1964.
EMANUELSON R C   PRATT AND WHITNEY AIRCRAFT   EAST
HARTFORD CONN
NASA
PWA-2518 NASA-CR-54268
N65-16746   1-52   1964   NS   19   18465

38627 INVESTIGATION OF THE EFFECTIVE THERMAL CONDUCTIVITY
AND TOTAL EMISSIVITY OF HEAT-RESISTANT CERAMIC
COATINGS OF REFRACTORY OXIDES PRODUCED BY GAS-FLAME
SPRAYING.   //ENGLISH TRANSLATION OF TEPLOFIZ.
VYSOKIKH TEMPERATUR 3 /1/ 64-9, 1965.//
BOGANOV A G   PIROGOV YU A   MAKAROV L P
HIGH TEMPERATURE /USSR/
3   1   53-8   1965

38629 EXPERIMENTAL STUDIES OF OPTICAL PROPERTIES OF THIN
FILMS OF GERMANIUM IN THE VISIBLE AND INFRARED
REGION.
RICHARD J
COMPT REND

38630 DETERMINATION OF THE EMISSIVITY OF MATERIALS.
PROGRESS REPORT.
PRATT AND WHITNEY AIRCRAFT
EAST HARTFORD CONN
S T I FACILITY
PWA-1994 NASA-CR-51003
X-64-80069   1-55   1961

38646 STUDIES OF THE CHARACTERISTICS OF PROBABLE LUNAR
SURFACE MATERIALS.   PART I.   SPECIAL REPORT NO. 20.
SALISBURY J W   GLASER P E   ARTHUR D LITTLE INC
CAMBRIDGE MASS
DDC AND CFSTI
AFCRL-64-970 AD-613018//PT 1/
N65-26232   1-309   1964

38649 OPTICAL CONSTANTS OF EVAPORATED METAL FILMS OF
SILVER AND INDIUM.   M.S. THESIS.
HEUBNER R H   VANDERBILT UNIV   NASHVILLE TENN
CFSTI
TID-21850   1-61   1965   CA   63   17287

38663 EFFECT OF CURING TEMPERATURE ON INITIAL THERMAL
EXPANSION OF CALCIUM ALUMINATE CEMENT.
DAVIDO K W   WHITTEMORE O J JR
AM CERAM SOC BULL
48   12   1137-8   1969

38672 EFFECTS OF VACUUM-ULTRAVIOLET ENVIRONMENT ON THE
OPTICAL PROPERTIES OF BRIGHT ANODIZED ALUMINUM.
TECH. REPT. 1 JAN. 1963—1 AUG. 1964.
WEAVER J H   AIR FORCE SYSTEMS COMMAND
WRIGHT-PATTERSON AFB AIR FORCE MATERIALS LAB   OHIO
DDC
AFML-TR-64-355
AD-612774   1-22   1965   PA   N65-3 2595

38673 TRANSMISSION CHARACTERISTICS OF FABRY-PEROT
INTERFEROMETERS AND A RELATED ELECTROOPTIC MODULATOR.
DEL PIANO V N JR   QUESADA A F
APPL OPT
4   11   1386-90   1965

38697 INFRARED-ABSORPTION SPECTRUM AND FERROELECTRIC
BEHAVIOR OF SODIUM TRIHYDROSELENITE.
KHANNA R K   DECIUS J C   LIPPINCOTT E R
J CHEM PHYS
43   9   2974-9   1965

38707 THERMAL DIFFUSIVITY OF MTR-ETR TYPE FUEL PLATES.
FROM   CERAMICS RESEARCH AND DEVELOPMENT REPORT.
OCT. - DEC., 1965.
BATES J L   BATTELLE-NORTHWEST LAB   RICHLAND WASH
CFSTI
BNWL-269 N67-19331
                    1.1-1.6   1965   PA   N67-5-8 1324

38716 HEAT CONDUCTION EXPERIMENTS IN RAREFIED GASES BETWEEN
CONCENTRIC CYLINDERS.
DYBBS A   SPRINGER G S
PHYS FLUIDS
8   11   1946-50   1965

38717 STRUCTURE AND CHEMISTRY OF OXIDE FILMS THERMALLY
GROWN ON MOLYBDENUM SILICIDES.
BARTLETT R W   MC CAMONT J W   GAGE P R
J AM CERAM SOC
48   11   551-8   1965

38727 CONDUCTIVITY MEASUREMENTS ON HEAT DEVELOPABLE
PHOTOGRAPHIC FILMS.
SOGIN H H   BORDELON F M   DEPT OF MECH ENGRG
TULANE UNIV   NEW ORLEANS   LOUISIANA
KALVAR CORPORATION   909 S BROAD STREET NEW ORLEANS
LOUISIANA
RD-12   1-83   1965

38731 LABORATORY REFLECTION MEASUREMENTS.   FROM   INFRARED
BALLON ASTRONOMY.
ZANDER R   THE JOHNS HOPKINS UNIVERSITY   BALTIMORE
MARYLAND
DDC
AFCRL-66-616 AD-643567
                    1-33   1966

38732 AMMONIA REFLECTANCE MEASUREMENTS.   FROM   INFRARED
BALLON ASTRONOMY.
ZANDER R   THE JOHNS HOPKINS UNIVERSITY   BALTIMORE
MARYLAND
DDC
AFCRL-66-616 AD-643567
                    1-5   1966

38737 NONREFLECTIVE COATINGS.
LEWIN G
APPL OPT
4   1   146-7   1965
CORRECTION.
4   9   1203   1965

TPRC Number	Bibliographic Citation					
38743	THERMODYNAMIC PROPERTIES OF UREA-HYDROCARBON ADDUCTS. HEAT CAPACITIES OF THE ADDUCTS OF N-C10H22, N-C12H26, N-C16H34, AND N-C20H42 FROM 12 TO 300 K. PEMBERTON R C PARSONAGE N G TRANS FARADAY SOC					
	61	2112-21	1965	CA	64	15076
38765	OPTICAL PROPERTIES OF THERMALLY DEPOSITED ANTIMONY TRISULFIDE AND TELLURIUM LAYERS IN THE INFRARED SPECTRAL REGION. VALEEV A S GISIN M A OPTIKA I SPEKTROSKOPIYA					
	19	1 121-7	1965	CA	64	167
38766	OPTICAL PROPERTIES OF THERMALLY DEPOSITED ANTIMONY TRISULFIDE AND TELLURIUM LAYERS IN THE INFRARED SPECTRAL REGION. //ENGLISH TRANSLATION OF OPTIKA I SPEKTROSKOPIYA 19 /1/ 121-7, 1965.// VALEEV A S GISIN M A OPT SPECTROSY /USSR/					
	19	1 62-5	1965			
38771	OPTICAL PROPERTIES OF POLYCRYSTALLINE INDIUM ANTIMONIDE FILMS. SEMILETOV S A AGALARZADE P S KORTUKOVA E M OPTIKA I SPEKTROSKOPIYA					
	19	2 252-4	1965	CA	63	17320
38772	OPTICAL PROPERTIES OF POLYCRYSTALLINE INDIUM ANTIMONIDE FILMS. //ENGLISH TRANSLATION OF OPT. I SPEKTR. 19 /2/ 252-4, 1965.// SEMILETOV S A AGALARZADE P S KORTUKOVA E M OPT SPECTROSCY /USSR/					
	19	2 142-3	1965			
38778	A NOVEL INTEGRATING SPHERE REFLECTOMETER FOR THE DETERMINATION OF ABSOLUTE HEMISPHERICAL SPECTRAL REFLECTANCE BETWEEN 0.2 AND 2.5 MICRONS /THE SOLAR SPECTRUM/. ZERLAUT G A KRUPNIK A C ENGRG MATERIALS BRANCH STRUCTURES AND MECHANICS DIVISION G C MARSHALL SPACE FLIGHT CENTER NASA NASA NASA-TM-X-50722 N63-84859					
		1-38	1960			
38805	OPTICAL AND ELECTRICAL PROPERTIES OF TIN. GOLOVASHKIN A I MOTULEVICH G P ZH EKSPERIM I TEOR FIZ					
	46	460-70	1964			
38806	OPTICAL AND ELECTRICAL PROPERTIES OF TIN. //ENGLISH TRANSLATION OF ZH. EKSPERIM. I TEOR. FIZ. 46, 460-70, 1964.// GOLOVASHKIN A I MOTULEVICH G P SOVIET PHYSICS-JETP					
	19	2 310-17	1964			
38827	INFRARED SPECTRA OF QUATERNARY AMMONIUM COMPOUNDS ADSORBED ON SILICA-GEL AEROSIL. BLACKMAN L C F HARROP R NATURE /LONDON/					
	208	777-8	1965			
38833	STUDY OF THE ELECTRICAL AND OPTICAL PROPERTIES OF CU1.8S FILMS. GUASTAVINO F LUQUET H BOUGNOT J COMPT REND					
	269	B 17 831-4	1969			
38834	HEAT TRANSFER IN GAS-FILLED POWDERS AT LOW TEMPERATURES. BRODIE D E MATE C F CAN J PHYS					
	43	12 2344-60	1965	CA	64	4293
38880	THERMAL EXPANSION OF SILICA BRICK AND MORTARS. COLE S S J AM CERAMIC SOC					
	13	7 437-46	1930			
38901	THERMAL VOLUME CHANGE AND ELASTICITY OF AGGREGATES AND THEIR EFFECT ON CONCRETE. WILLIS T F DE REUS M E PROC AMERICAN SOC TESTING MATERIALS					
	39	919-28	1939			
38913	DETERMINATION OF THE COEFFICIENT OF THERMAL CONDUCTIVITY IN AN ASYMMETRICAL HEATING MODE. //ENGLISH TRANSLATION OF ZAVODSKAYA LAB. 36 /9/ 1095-96, 1970.// KAGANER M G IND LAB USSR					
	36	9 1396-7	1970			
38924	OPTICAL CHARACTERISTICS OF THIN FILMS. //ENGLISH TRANSLATION OF IZV. VUZ, FIZIKA 8, NO. 1, 116-21, 1965.// TEKUCHEVA I A SOVIET PHYS J					
		1 84-7	1965			

TPRC Number	Bibliographic Citation					
38962	SOME PHYSICAL PROPERTIES OF BIMOLECULAR LIPID MEMBRANES PRODUCED FROM NEW LIPID SOLUTIONS. TIEN H TI DIANA A L NATURE /LONDON/					
	215	1199-200	1967			
38990	MODIFICATION OF HEMISPHERICAL EMISSOMETER FOR LOW TEMPERATURE AND AUTOMATIC DATA ACQUISITION. FROM JPL SPACE PROGRAMS SUMMARY NO. 37-35, VOL. IV. HALL W M JET PROPULSION LAB PASADENA CALIF NASA JPL-SPS-37-35					
		64-5	1965			
38992	SPECTRAL EMITTANCE OF ALUMINUM OXIDE AND ZINC OXIDE ON OPAQUE SUBSTRATES. LIEBERT C H LEWIS RESEARCH CENTER CLEVELAND OHIO NASA NASA AND CFSTI NASA-TN-D-3115 N66-10339					
		1-22	1965	CA	66	109918
38996	OPTICAL PROPERTIES OF ALUMINUM EVAPORATED IN ULTRAVIOLET BETWEEN 500 AND 1400 ANGSTROMS. DAUDE A SAVARY A JEZEQUEL G ROBIN S COMPT REND					
	269	B 18 901-4	1969			
39008	OVERCAST SKY LUMINANCES AND DIRECTIONAL LUMINOUS REFLECTANCES OF OBJECTS AND BACKGROUNDS UNDER OVERCAST SKIES. GORDON J I CHURCH P V APPL OPT					
	5	6 919-23	1966			
39013	INTERFERENCE FILTERS FOR THE FAR ULTRAVIOLET /1700 ANGSTROMS TO 2400 ANGSTROMS/. BATES B BRADLEY D J APPL OPT					
	5	6 971-5	1966			
39014	INTEGRATING SPHERE FOR THE INFRARED. MORRIS J C APPL OPT					
	5	6 1035-7	1966			
39015	PERFECT MATCH IN ANTIREFLECTION SYSTEMS. PARK K C APPL OPT					
	5	6 1082-3	1966			
39016	FAR INFRARED TRANSMISSION THROUGH METAL LIGHT PIPES WITH LOW THERMAL CONDUCTANCE. HARRIS R E CAPPELLETTI R L GINSBERG D M APPL OPT					
	5	6 1083-4	1966			
39032	THE PROPERTIES AND THE PREPARATION OF THIN BISMUTH FILMS. CONDAS G A LAWRENCE RADIATION LAB UNIV OF CALIFORNIA LIVERMORE AEC AND CFSTI UCRL-12001					
		1-15	1964	NSA 20-7	1394	
39036	THE EFFECTS OF EXTREME ULTRAVIOLET RADIATION ON THE REFLECTANCE OF THERMAL CONTROL SURFACE COATINGS. M.S. THESIS COWIE J M AIR FORCE INST OF TECH SCHOOL OF ENGINEERING WRIGHT-PATTERSON AFB OHIO DDC GSF/MECH/65-32 N66-19549 AD-625442					
		1-66	1965	PA	N66-4-9 1564	
39040	RADIATION BLISTERING. INTERFEROMETRIC AND MICROSCOPIC OBSERVATIONS OF OXIDES, SILICON, AND METALS. PRIMAK W LUTHRA J J APPL PHYS					
	37	6 2287-94	1966			
39065	PROPERTIES OF THERMO-LAG T-500 EX 167 SUBLIMING COMPOUND. - SUPERSEDES REPORT 1139. EMERSON ELECTRIC MFG CO ST LOUIS MO NASA REPT 1701 N66-16949					
		1-181	1965			
39075	ON THE NATURE OF THERMAL CONTACT CONDUCTANCE. M.S. THESIS. HSIEH C K PURDUE UNIVERSITY LAFAYETTE IND PURDUE UNIVERSITY 19363					
		1-93	1964			
39080	COATINGS OF HIGH-TEMPERATURE MATERIALS. PART I. //ENGLISH TRANSLATION OF POLRYTIYA IZ TUGOPLAVKIKH SOEDIENII, 1964.// SAMSONOV G V EPIK A P PLENUM PRESS NEW YORK N Y					
		1-111	1966			
39081	COATINGS OF HIGH-TEMPERATURE MATERIALS. PART II. PROPERTIES OF COATED REFRACTORY METALS. GIREAULT W A BARTLETT E S					

TPRC Number	Bibliographic Citation

39083 EARTH LOADING CHARACTERISTICS OF GLASS/EPOXY PRESSURIZED PIPE.
JONES H L JR
MODERN PLASTICS
43   11   125-30 192   1966

39084 ELECTRICAL AND OPTICAL PROPERTIES OF TANTALUM THIN FILMS SPUTTERED IN NITROGEN.
BLASINGAME J M   YOUNG C F   AIR FORCE AVIONICS LAB   RESEARCH AND TECHNOLOGY DIV   WRIGHT-PATTERSON AFB   OHIO   USAF
DDC
AFAL-TR-65-191
AD-473490   1-12   1965   TA   65-24 A-106

39086 OPTICAL PROPERTIES OF VACUUM-EVAPORATED FILMS OF CADMIUM, THALLIUM, AND ZINC IN THE VACUUM ULTRAVIOLET. JELLINEKS M. S. THESIS.
JELINEK T M   HAMM R N   ARAKAWA E T
CFSTI
ORNL-TM-1164 N66-20281
    1-55   1965   CA   64   5950

39092 SPECTRA NOTEBOOK. VOLUME I. MATERIAL, TARGET AND BACKGROUND DATA.
WILBURN D K   ARMY TANK AUTOMOTIVE CENTER COMPONENTS RESEARCH AND DEVELOPMENT LABS   WARREN MICH
DDC
TR-8863   VOL. 1
AD-475817   1-66   1965   TA   66-3 A-77

39094 BURNUP RATES OF POWDERS BEHIND A NORMAL SHOCK WAVE. RESEARCH NOTE NO. 25.
HOOKER W J   MORSELL A L   HELIODYNE CORP   LOS ANGELES CALIF   DA
DDC
AD-476227   1-24   1965   TA   66-4 A-132

39111 OPTICAL PROPERTIES OF THIN METALLIC FILMS IN ISLAND FORM.
DOREMUS R H
J APPL PHYS
37   7   2775-81   1966

39120 ELECTROREFLECTANCE IN METALS.
FEINLEIB J
PHYS REV LETTERS
16   26   1200-2   1966

39136 POLARIZATION TEXTURES FOR THE NEAR INFRARED RADIATION.
DISTLER G I   KORTUKOVA E I   KOTOV A V   LEBEDEVA V N   CHUDAKOV V S
OPTIKA I SPEKTROSKOPIYA
23   1   137-42   1967

39137 POLARIZATION TEXTURES FOR THE NEAR INFRARED RADIATION. //ENGLISH TRANSLATION OF OPTIKA I SPEKTROSKOPIYA 23 /1/ 137-42, 1967.//
DISTLER G I   KORTUKOVA E I   KOTOV A V   LEBEDEVA V N   CHUDAKOV V S
OPT SPECTRY
23   1   71-4   1967

39169 REFLECTIVITY OF TIN TELLURIDE IN THE INFRARED.
RIEDL H R   DIXON J R   SCHOOLAR R B
PHYS REV
162   3   692-700   1967

39178 AURORAL HYDROGEN EMISSION.
EATHER R H   JACKA F
AUSTRALIAN J PHYS
19   2   241-74   1966

39181 SPECTROPHOTOMETRIC STUDY OF THE BLUE AND VIOLET ABSORPTION OF CUPROUS OXIDE.
DAUNOIS A   DEISS J L   MEYER B
J PHYSIQUE
27 3/4/   142-6   1966

39185 DOUBLE BEAM SPECTROPHOTOMETRY IN THE FAR ULTRAVIOLET. I. 1150 ANGSTROMS TO 3600 ANGSTROMS.
SCHMITT R G   BREHM R K
APPL OPT
5   7   1111-1116   1966

39212 THERMAL RADIATION PROPERTY MEASUREMENT TECHNIQUES.
DUNN S T   GEIST J C   MOORE D G   CLARK H E   RICHMOND J C   NATL BUREAU OF STDS   WASHINGTON D C
CFSTI AND NASA
NASA-CR-66127 NBS-TN-415
N67-31130   1-83   1967   PA   N67-5-17 3069

39215 OPTICAL CONSTANTS OF EVAPORATED GOLD AND PLATINUM FILMS ON POTASSIUM TANTALATE.
RIDEOUT V L   WEMPLE S H
J OPT SOC AM
56   6   749-51   1966   CA   65   3194

39220 USE OF AN AUXILIARY SPHERE WITH A SPECTROFLECTOMETER TO OBTAIN ABSOLUTE REFLECTANCE.
GOEBEL D G   CALDWELL B P   HAMMOND H K III
J OPT SOC AM

39236 A STUDY OF THE EMISSIVITY OF SOLID BODIES.
MITOR V V   KONOPELKO I N
TEPLOENERGETIKA
13   7   67-71   1966

39238 EMISSIVITY COATINGS FOR LOW-TEMPERATURE SPACE RADIATORS. QUARTERLY PROGRESS REPORT NO. 1, 1 JUL.-30 SEP. 1965.
SMITH F J   OLSON R L   LOCKHEED MISSILES AND SPACE CO   SUNNYVALE CALIF   AEROSPACE SCIENCES LAB   NASA
NASA AND CFSTI
NASA-CR-54807
N66-15372   1-68   1965   PA   N66-4-6 969

39239 OPTICAL CONSTANTS OF BARIUM IN THE VACUUM ULTRAVIOLET REGION.
FISHER E I   FUJITA I   WEISSLER G L   UNIV OF SOUTHERN CALIF   LOS ANGELES   DEPT OF PHYSICS
NASA AND CFSTI
USC-VACUV-104 NASA-CR-69335
N66-15339   1-28   1965   PA   N66-4-6 919

39242 A FEASIBILITY STUDY ON THE USE OF POWDERS AS HEAT TRANSFER MEDIA IN IRRADIATION CAPSULES.
EVANS J L   INTERNATIONAL RESEARCH AND DEVELOPMENT CO LTD   NEWCASTLE ENGLAND
USAEC AND HMSO
TRG-920//D,X//
N66-14174   1-20   1965

39257 THEORETICAL EVALUATION OF THE HEAT CAPACITY OF THE GAS MIXTURE, THE EFFECTIVE THERMAL CONDUCTIVITY OF THE SYSTEM AND THE WALL HEAT TRANSFER COEFFICIENT.
ZIOLKOWSKI D
BULL ACAD POLON SCI SER SCI CHIM
18   4   227-33   1970   CA   73   57499

39262 ULTRAVIOLET DEGRADATION STUDY.
BOTTOMS W T   LILLYWHITE M A   WEBB J J   GODDARD SPACE FLIGHT CENTER   GREENBELT MD   NASA
NASA
NASA-TM-X-55331
N66-11226   1-18   1965   PA   N66-4-2 324

BALDRIGE J H   IIT RESEARCH INST   CHICAGO ILL
DEVELOPMENT OF THERMAL RADIATIVE SURFACE COATINGS FOR NASA-LEWIS SPACE SIMULATION CHAMBER. FINAL REPORT, 29 JUN.-30 MAY, 1965.
BALDRIGE J H   IIT RESEARCH INST   CHICAGO ILL
NASA
NASA AND CFSTI
IITRI-C6042-10 NASA-CR-54274
N65-36780   1-101   1965   CA   66   86708

39264 IRRADIATION EVALUATION OF FUEL ELEMENTS FOR THE PBRE AND AVR REACTORS.
SCOTT J L   MORGAN J G   DE CARLO V A   OAK RIDGE NATL LAB   OAK RIDGE TENN   USAEC
USAEC
ORNL-P-1507
    1-4   1965   CA   64   10706

39279 OPTICAL OBSERVATION OF OSCILLATIONS OF PLASMA IN GRANULAR LAYERS OF GOLD.
EMERIC A
COMPT REND
262   B   4   292-5   1966

39285 ON THE ANALOGICAL CALCULATION OF THE THERMAL CONTACT RESISTIVITIES. APPLICATION TO THE CALCULATION OF THE THERMAL CONDUCTIVITY OF THE INTERSTITIAL GAS.
BARDON J-P
COMPT REND
262   B   10   660-3   1966

39286 EXPERIMENTAL STUDY OF THE OPTICAL PROPERTIES OF VERY THIN LAYERS OF TIN.
RASIGNI G   CODACCIONI J P   MICHAUD-BONNET J   ABBA F   PETRAKIAN J-P
COMPT REND
262   B   11   772-5   1966

39288 VARIATIONS OF THE REFLECTION FACTORS AND TRANSMISSION OF VERY THIN FILMS OF SILVER AS A FUNCTION OF THE TEMPERATURE.
BARRAS H   GASPARINI J-P   PHILIP R
COMPT REND
262   B   13   889-91   1966

39294 OPTICAL PROPERTIES OF THIN LAYERS OF ZINC IN THE FAR ULTRAVIOLET.
KANDARE S   ROBIN-KANDARE S   ROBIN J
COMPT REND
262   B   19   1302-5   1966

39296 THERMAL CONTACT RESISTANCE BY PARALLEL BANDS.
FOUCHE F   CORDIER H
COMPT REND
262   B   21   1367-9   1966

TPRC Number	Bibliographic Citation

39299 DISPLACEMENT OF THE MAXIMUMS OF ABNORMAL ABSORPTION OBSERVED AT OBLIQUE INCIDENCE IN THIN LAYERS OF SILVER WHEN THEY PASS FROM VACUUM TO AIR.
EMERIC N   EMERIC A
COMPT REND
262   B   26   1699-702   1966

39303 A USEFUL INFRARED SOURCE.
CARLON H R
APPL OPT
5   8   1281-3   1966

39314 USE OF AN EVAPORATED DIELECTRIC FILM FOR DETERMINING THE OPTICAL CONSTANTS OF A METAL.   I AND II.
BENNETT J M   ASHLEY E J   BENNETT H E   BURGE D K
NAVAL ORDNANCE TEST STATION   CHINA LAKE CALIF
DDC
NOTS-TP-3989 NAVWEPS-9010
AD-624999   1-13   1965   TA   66-3 86

39338 SUPPLEMENTAL INFORMATION ON HIGH TEMPERATURE COATING AND MATERIAL PROGRAMS AT AMF.
BROWNING M E   SCHATZ E A   MC CANDLESS C
PEARSON E G   AMERICAN MACHINE AND FOUNDRY COMPANY
ALEXANDRIA VIRGINIA
NASA
AMF-AR63-502A NASA-CR-53234
N64-17588   1-4   1965

39347 SINTERED CERAMIC COATINGS FOR THERMAL CONTROL. PRESENTED AT THE 3D AM. ELECTROPLATERS SOC. AEROSPACE FINISHING SYMP., DALLAS, 16-17 JAN. 1964.
COX R L   LING-TEMCO-VOUGHT INC   DALLAS TEX
ASTRONAUTICS DIV
LING-TEMCO-VOUGHT INC   DALLAS TEXAS
REPT   NO   .00.369
N65-36213   1-23   1964   CA   67   14531

39352 HEAT TRANSFER ACROSS SURFACES IN CONTACT.   PRACTICAL EFFECTS OF TRANSIENT TEMPERATURE AND PRESSURE ENVIRONMENTS.   SEMIANNUAL REPT. APR. 1-OCT. 1, 1965.
BLUM H A   SOUTHERN METHODIST UNIV   DALLAS TEX
NASA AND CFSTI
NASA-CR-69698 N66-16072
        1-42   1965   PA   N66-4-6 973

39353 THERMAL AND MECHANICAL STUDIES OF SOLID-SOLID CONTACTS.
MUSTACCHI C   GIULIANI S   EUROPEAN ATOMIC ENERGY COMMUNITY   BRUSSELS   /BELGIUM/
USAEC
EUR-2486.E N66-14078
        1-53   1965   PA   N66-4-4 624

39355 PROBE CONSTANT FOR IN-SITU DETERMINATION OF THERMAL CONDUCTIVITIES OF SOILS. U S - A S M E - NO. 122.
HUS S T   LEE W W
PROC 3RD INTERN HEAT TRANSFER CONF
4   81-8   1966

39356 HEAT TRANSFER IN STRUCTURAL HONEYCOMB COMPOSITES AT HIGH TEMPERATURES. U S - A S M E - NO. 123.
MINGES M L
PROC 3RD INTERN HEAT TRANSFER CONF
4   89-99   1966

39357 CONTACT CONDUCTANCE MEASUREMENTS DURING TRANSIENT HEATING. U S - A S M E - NO. 124.
SCHAUER D A
PROC 3RD INTERN HEAT TRANSFER CONF
4   100-8   1966

39358 THERMAL CONDUCTIVITY MEASUREMENTS OF FIBROUS INSULATIONS UP TO 2500 F. U S - A I CH E - NO. 128.   ;
ROLINSKI E J   PURCELL G V
PROC 3RD INTERN HEAT TRANSFER CONF
4   133-40   1966

39362 AN INVESTIGATION OF RADIANT HEAT TRANSFER IN CLUSTERS OF PARALLEL RODS. U K - I MECH E - NO. 169.
FISHER S A   COWIN M
PROC 3RD INTERN HEAT TRANSFER CONF
5   174-83   1966

39363 SPACE RADIATION DAMAGE TO THE THERMAL RADIATIVE PROPERTIES OF MATERIALS. U S - A S M E - NO. 177.
BREUCH R A   GREENBERG S A
PROC 3RD INTERN HEAT TRANSFER CONF
5   246-56   1966

39383 PHOTOCONDUCTIVITY AND FADING MECHNISMS OF DYES.
PATTERSON D   PILLING B
TRANS FARADAY SOC
62   7   1976-84   1966

39391 ON THE BAND STRUCTURE AND THE ABSORPTION SPECTRUM OF CUPROUS OXIDE.
BRAHMS S   NIKITINE S   DAHL J P
PHYS LETTERS
22   1   31-3   1966

39439 OPTICAL PROPERTIES OF THIN GERMANIUM FILMS IN THE WAVELENGTH RANGE 2000-6000 ANGSTROMS.
GRANT P M   PAUL W
J APPL PHYS
37   8   3110-20   1966

39442 THERMAL PROPERTIES OF THIN-FILM POLYMERS BY TRANSIENT HEATING.
STEERE R B
J APPL PHYS
37   9   3338-44   1966

39447 PREPARATION AND PROPERTIES OF NONCRYSTALLINE ZINC OXIDE FILMS.
MICKELSEN R A   KINGERY W D
J APPL PHYS
37   9   3541-4   1966

39452 SOME PROPERTIES OF SI O2 FILMS DEPOSITED BY THE REACTION OF SI H4 WITH WATER VAPOR.
HANETA Y   NAKANUMA S
JAPAN J APPL PHYS
6   10   1176-83   1967

39460 SPECULAR REFLECTANCE CURVES OF WAX-POLISHED PAINTED SURFACE.   DETERMINATION OF WAX-FILM THICKNESS AND RMS ROUGHNESS.
NAGATA K-I
JAPAN J APPL PHYS
6   10   1198-202   1967

39479 COLOR PREDICTION USING THE TWO-CONSTANT TURBID-MEDIA THEORY.
DAVIDSON H R   HEMMENDINGER H
J OPT SOC AM
56   8   1102-9   1966

39481 DOUBLE-LAYER INTERFERENCE IN AIR-CDS FILMS.
GOTTLING J G   NICOL W S
J OPT SOC AM
56   9   1227-31   1966

39483 THEORY OF THERMAL CONDUCTANCE IN A COLLISIONLESS GAS. REPORT 746.
WU Y   PRINCETON UNIV DEPT OF AERONAUTICAL ENGINEERING   N J   USAF
DDC AND CFSTI
AD-631737   N66-27711
        1-31   1965   CA   66   119542

39494 RESTSTRAHL REFLECTION CHARACTERISTICS OF AMORPHOUS SILICA.
SATO K   SHIBATA M
J PHYS SOC JAPAN
21   6   1088-96   1966
CORRECTION.
MILER M
21   12   2737   1966

39496 RARE EARTH OXIDES FOR HIGH TEMPERATURE REFLECTIVE PIGMENTS. FROM SYMPOSIUM ON DYES AND PIGMENTS IN COATINGS AND PLASTICS.
CUTRIGHT R C   AERONAUTICAL SYSTEMS DIVISION
WRIGHT-PATTERSON AFB   OHIO
AM CHEM SOC DIV ORG COATINGS AND PLASTIC CHEMISTRY
        474-85   1963

39497 PLASMA RESONANCE ABSORPTION IN THIN METAL FILMS.
MC ALISTER A J   STERN E A
PHYS REV
132   4   1599-602   1963

39504 INFRARED STUDY OF CARBON-OXYGEN SURFACE COMPLEXES.
SMITH R N   YOUNG D A   SMITH R A
TRANS FARADAY SOC
62   8   2280-6   1966

39516 THERMAL PROPERTIES OF A SIMULATED LUNAR MATERIAL IN AIR AND IN VACUUM.
BERNETT E C   WOOD H L   JAFFE L D   MARTENS H E
JET PROPULSION LAB   CALIF INST OF TECH   PASADENA
NASA
JPL-TR-32-369
N63-13767   1-23   1962   PA   N63-1 552

39530 THERMAL ACCOMMODATION COEFFICIENTS OF HELIUM ON TUNGSTEN AND HYDROGEN ON TUNGSTEN-COVERED TUNGSTEN AT 325, 403, AND 473 K.
WACHMAN H Y
J CHEM PHYS
45   5   1532-8   1966   CA   65   12877

39539 DETERMINATION OF THE OPTICAL CONSTANTS OF A THIN ABSORBANT SHEET, TRIPLE AND SYMMETRICAL.
CASSET J
COMPT REND
263   B   4   299-302   1966

39563 OPTICAL PROPERTIES OF EPITAXIAL FILMS OF CDXHG1-XTE.
CD/X/ HG/1-X/ TE.
LUDEKE R   PAUL W
J APPL PHYS
37   9   3499-501   1966

TPRC Number	Bibliographic Citation

39580 REFLECTANCE AND STRUCTURE OF THIN LAYERS OF ELECTRO-DEPOSITED NICKEL.
VAGRAMYAN A T   BARABOSHKINA N K
ZHUR FIZ KHIM
40   1   63-   1966

39581 REFLECTANCE AND STRUCTURE OF THIN LAYERS OF ELECTRO-DEPOSITED NICKEL. //ENGLISH TRANSLATION OF ZHUR.
FIZ. KHIM. 40 /1/ 63- , 1966.//
VAGRAMYAN A T   BARABOSHKINA N K
RUSS J PHYS CHEM
40   1   31-4   1966

39602 OXIDIZED NICKEL AS A HEATING ELEMENT IN VACUUM.
WONG H Y
BRIT J APPL PHYS
17   10   1329-37   1966

39604 DIGITAL LIGHT DEFLECTORS.
KULCKE W   KOSANKE K   MAX E   HABEGGER M A
HARRIS T J   FLEISHER H
APPL OPT
5   10   1657-87   1966

39609 SPECTRAL DISTRIBUTION OF THE RELATIVE QUANTUM YIELD
OF PHOTOLUMINESCENCE OF POLYCRYSTALLINE CADMIUM
SULFIDE FILMS AT 77 K.
SHALIMOVA K V   KHIRIN V N   KOROLEV O I
OPT I SPEKTROSKOPIYA
20   6   1063-5   1966

39610 SPECTRAL DISTRIBUTION OF THE RELATIVE QUANTUM YIELD
OF PHOTOLUMINESCENCE OF POLYCRYSTALLINE CADMIUM
SULFIDE FILMS AT 77 K.   //ENGLISH TRANSLATION OF OPT.
I SPEKTROSKOPIYA 20 /6/ 1063-5, 1966.//
SHALIMOVA K V   KHIRIN V N   KOROLEV O I
OPT SPECTRY
20   6   587-8   1966

39612 DROP SIZE DISTRIBUTION IN AGITATED LIQUID-LIQUID
SYSTEMS.
CHEN H T   MIDDLEMAN S
AICHE JOURNAL
13   5   989-95   1967

39619 SINGLE AND MULTIPLE REFLEXION ATTENUATED TOTAL
REFLECTANCE INFRA-RED SPECTROSCOPY.
FORD C G
NATURE /LONDON/
212   72   1966

39620 ADSORPTION OF POLAR COMPOUNDS ON AMORPHOUS BORON.
GILLESPIE J S   HOBSON M C   GAGER H M
NATURE /LONDON/
212   137-9   1966

39622 THE TOTAL HEMISPHERICAL EMITTANCE OF COATED WIRES.
BRADLEY D   ENTWISTLE A G
BRIT J APPL PHYS
17   9   1155-64   1966   CA   65   14640

39631 EFFECTIVE THICKNESS OF BULK MATERIALS AND OF THIN
FILMS FOR INTERNAL REFLECTION SPECTROSCOPY.
HARRICK N J   DU PRE F K
APPL OPT
5   11   1739-43   1966

39638 ELLIPSOMETER STUDY OF ANOMALOUS ABSORPTION IN VERY
THIN DIELECTRIC FILMS ON EVAPORATED METALS.
BASHARA N M   PETERSON D W
J OPT SOC AM
56   10   1320-31   1966

39653 SYNTHESIS OF NEUTRAL NONREFLECTING COATINGS. 1.
FUNDAMENTALS OF THE THEORY OF THE SYNTHESIS OF
SPECTRAL CHARACTERISTICS.
FURMAN SH A
OPTIKA I SPEKTROSKOPIYA
21   1   82-90   1966

39654 SYNTHESIS OF NEUTRAL NONREFLECTING COATINGS. 1.
FUNDAMENTALS OF THE THEORY OF THE SYNTHESIS OF
SPECTRAL CHARACTERISTICS. //ENGLISH TRANSLATION OF
OPTIKA I SPEKTROSKOPIYA 21 /1/ 82-90, 1966.//
FURMAN SH A
OPT SPECTRY
21   1   44-8   1966

39661 GALLIUM ARSENIDE OPTICAL FILTERS.
KOLTUN M M   KAGAN M B
OPTIKA I SPEKTROSKOPIYA
21   1   116-18   1966

39662 GALLIUM ARSENIDE OPTICAL FILTERS. //ENGLISH
TRANSLATION OF OPTIKA I SPEKTROSKOPIYA 21 /1/ 116-8,
1966.//
KOLTUN M M   KAGAN M B
OPT SPECTRY
21   1   65   1966

39670 PHOTOGENERATION OF CHARGE CARRIERS IN TETRACENE.
GEACINTOV N   POPE M   KALLMANN H
J CHEM PHYS

39675 TRANSIENT DETERMINATIONS OF THERMAL DIFFUSIVITIES
AND DISSIPATIONS OF METAL FOILS.
JACOVELLI P B   ZINKE O H
J APPL PHYS
37   11   4117-20   1966

39677 OPTICAL DENSITY AND THICKNESS OF GRAPHITE LAMELLAE.
MYERS G E   MONTET G L
J APPL PHYS
37   11   4195-6   1966

39695 METAL-CERAMIC INTERACTIONS. IV. ABSOLUTE MEASUREMENT
OF METAL-CERAMIC INTERFACIAL ENERGY AND THE
INTERFACIAL ADSORPTION OF SILICON FROM IRON-SILICON
ALLOYS.
KINGERY W D
J AM CERAM SOC
37   42-5   1954

39700 SURFACE TENSION OF METALS WITH REFERENCE TO SOLDERING
CONDITIONS.
COFFMAN A W   PARR S W
IND ENG CHEM
19   12   1308-11   1927

39717 OPTICAL ABSORPTION OF VERY THIN FILMS OF GROUP IVA
ELEMENTS OF THE PERIODIC SYSTEM.
ABBA F   CODACCIONI J P   MICHAUD-BONNET J
PETRAKIAN J P   RASIGNI G
COMPT REND SER A B
262   B   14   954-7   1966

39721 EXPERIMENTAL DETERMINATION OF THE THERMAL
RESISTANCE OF AIR SPACES. M.S. THESIS.
VEST E W   UNIVERSITY OF MAINE   ORONO
UNIVERSITY OF MAINE
                          1-101   1962

39734 MEASUREMENT OF HEAT-INSULATING PROPERTIES OF
MATERIALS.
FEREBAUER R
VEDA A VYZKUM V PRUMYSLU KOZEDELNEM
2   27-44   1957   CA   52   16774

39737 THERMAL PROPERTIES OF WEST VIRGINIA HIGHWAY
MATERIALS. M.S. THESIS.
OBLENIS J D   WEST VIRGINIA UNIVERSITY   MORGANTOWN
WEST VIRGINIA
WEST VIRGINIA UNIVERSITY
                          1-172   1964

39738 ANISOTROPY OF THE OPTICAL CONSTANTS OF SINGLE
CRYSTAL GRAPHITE IN THE ULTRAVIOLET-VISIBLE SPECTRUM.
M.S. THESIS.
YASINSKY J B   UNIVERSITY OF PITTSBURGH
PITTSBURGH PENNSYLVANIA
UNIVERSITY OF PITTSBURGH
                          1-49   1963

39745 STUDY AND DEVELOPMENT OF MATERIALS AND TECHNIQUES FOR
PASSIVE THERMAL CONTROL OF FLEXIBLE EXTRA-VEHICULAR
SPACE GARMENTS.
RICHARDSON D L   LITTLE /ARTHUR/ INC   CAMBRIDGE
MASS
DDC AND CFSTI
AMRL-TR-65-156 N66-16743
AD-624886   1-101   1965   PA   N66-4-7 1009

39746 NEW HIGH TEMPERATURE INFRARED TRANSMITTING GLASSES.
FINAL TECHNICAL SUMMARY REPORT FOR 1 MAY 62-31
JUL 65.
HILTON A R   TEXAS INSTRUMENTS INC   DALLAS   ONR
ONR
DDC AND CFSTI
TT-08-65-121 AD-623262
                          1-168   1965   TA   65-24 47

39750 LONG WAVELENGTH INFRARED FIBER OPTICS. FINAL
TECHNICAL REPT. 20 MAY 64-20 SEPT. 65.
KAPANY N S   SIMMS R J   OPTICS TECHNOLOGY INC
PALO ALTO CALIF   USAF
DDC
OTI-9223-R AFAL-TR-65-313
AD-475508   1-48   1965   TA   66-3 A-100

39754 DEVELOPMENT OF MATERIAL RESISTANT TO HIGH-INTENSITY
THERMAL RADIATION.
ANDERSON R B   DOUGLAS AIRCRAFT CO INC   LONG BEACH
CALIF   USAF
DDC
AFML-TR-65-438
                          1-245   1965

39773 OPTICAL PROPERTIES OF TIN IN THE EXTREME ULTRAVIOLET.
LEMONNIER J-C   ROBIN S
COMPT REND
265   B   11   661-4   1967

39777 THE ABSORPTION OF ZINC SULFIDE.
MOROZOVA N K   SHALIMOVA K V
OPTIKA I SPEKTROSKOPIYA
21   2   192-6   1966

TPRC Number	Bibliographic Citation

39778 THE ABSORPTION OF ZINC SULFIDE. //ENGLISH TRANSLATION OF OPTIKA I SPEKTROSKOPIYA 21 /2/ 192-6, 1966.//
MOROZOVA N K   SHALIMOVA K V
OPT SPECTRY
21   2   112-14   1966

39779 DETERIORATION OF CATALYSTS FOR THE DEHYDROGENATION OF N-BUTANE DUE TO DIFFUSION IN PARTICLES.
SUGA K   MORITA Y   KUNUGITA E   OTAKE T
KOGYO KAGAKU ZASSHI
     2   136-41   1967

39780 DETERIORATION OF CATALYSTS FOR THE DEHYDROGENATION OF N-BUTANE DUE TO DIFFUSION IN PARTICLES. //ENGLISH TRANSLATION OF KOGYO KAGAKU ZASSHI NO. 2, 136-41, 1967.//
SUGA K   MORITA Y   KUNUGITA E   OTAKE T
INTERN CHEM ENG
7   4   742-8   1967

39801 NUCLEAR-MAGNETIC-RESONANCE STUDY OF SELF-DIFFUSION IN A BOUNDED MEDIUM.
WAYNE R C   COTTS R M
PHYS REV
151   1   264-72   1966

39806 OPTICAL CONSTANTS OF POTASSIUM BROMIDE IN THE FAR INFRARED.
HADNI A   CLAUDEL J   CHANAL D   STRIMER P
VERGNAT P
PHYS REV
163   3   836-43   1967

39833 HEAT TRANSMISSION THROUGH INSULATION AS AFFECTED BY ORIENTATION OF WALL.
ROWLEY F B   LUND C E   UNIV OF MINNESOTA INST OF TECH ENG EXPERIMENT STATION
UNIVERSITY OF MINNESOTA
TECH   PAPER NO. 44
     1-4   1943

39837 TRANSMISSION AND REFLECTION COEFFICIENTS OF ALUMINIUM FILMS FOR INTERFEROMETRY.
CRAWFORD M F   GRAY W M   SCHAWLOW A L
KELLY F M
J OPT SOC AM
39   10   888   1949

39839 INDUCED TRANSMISSION IN ABSORBING FILMS APPLIED TO BAND PASS FILTER DESIGN.
BERNING P H   TURNER A F
J OPT SOC AM
47   3   230-9   1957

39840 OPTICAL PROPERTIES OF THIN ABSORBING FILMS.
ABELES F
J OPT SOC AM
47   6   473-82   1957

39842 COLORS PRODUCED BY REFLECTION AT GRAZING INCIDENCE FROM ROUGH SURFACES.
MIDDLETON W E K   WYSZECKI G
J OPT SOC AM
47   11   1020-3   1957

39843 ALUMINOUS INSULATING MATERIALS RESIST HIGH TEMPERATURES.
BARNITT J B   HEILMAN R H
CHEM MET ENG
38   390-3   1931

39844 HEAT INSULATION DEVELOPED FOR EVERY PURPOSE.
TOWNSHEND B   WILLIAMS E R
CHEM MET ENG
39   4   219-22   1932

39850 SOME MATERIALS OF LOW THERMAL CONDUCTIVITY.
GRIFFITHS E
FARADAY SOC TRANS
18   252-8   1922

39852 INVESTIGATION OF INSULATED WALLS.
ROWLEY F B
TRANS AM SOC MECH ENGRS
49/50   49-55   1927-8

39854 HIGH-REFLEXION FILMS.
GREENLAND K M
J SCI INSTR
23   48-50   1946

39858 THERMAL CONDUCTIVITIES OF POROUS ROCKS FILLED WITH STAGNANT FLUID.
KUNII D   SMITH J M
J SOC PETROL ENGRS
1   1   37-42   1961

39860 DEVELOPMENT OF THE THERMAL CONDUCTIVITY PROBE.
HOOPER F C   CHANG S C
AM SOC HEATING VENTILATING ENGRS RES BULL
59   463-72   1953

39866 THERMAL INSULATION WITH ALUMINUM FOIL.
MASON R B
IND ENG CHEM
25   3   245-55   1933

39869 THE THERMAL CONDUCTIVITY OF SPHERICAL METAL POWDERS INCLUDING THE EFFECT OF AN OXIDE COATING.
SWIFT D L
INTERN J HEAT MASS TRANSFER
9   10   1061-74   1966   CA   66-2 6202

39872 DEVELOPMENT OF AN OLIVE DRAB SOLAR HEAT REFLECTING AND LOW VISIBILITY ENAMEL.
SANDLER M H   COATING AND CHEMICAL LAB ABERDEEN PROVING GROUND   MD
DDC
CCL-188 N66-13437
AD-473571   1-32   1965   TA   65-24 A-34

39873 RESEARCH ON REFURBISHABLE THERMOSTRUCTURAL PANELS FOR MANNED LIFTING ENTRY VEHICLES.
LA PORTE A H   MARTIN COMPANY   BALTIMORE MD
NASA
NASA-CR-638
     1-217   1966

39881 FORMULA FOR THERMAL ACCOMMODATION COEFFICIENT. /SEE ALSO TPRC NO. 42096./
GOODMAN F O   WACHMAN H Y   MASS INST OF TECH CAMBRIDGE   FLUID DYNAMICS RES LAB   USAF
DDC AND CFSTI
AFOSR-66-0295 N66-26422
AD-631007   1-44   1966   TA   66-10 86

39894 OPTICAL PROPERTIES OF PB SE FILMS IN THE ULTRAVIOLET. M.S. THESIS.
COULTER J K   UNIVERSITY OF ROCHESTER   ROCHESTER NEW YORK
UNIVERSITY OF ROCHESTER
     1-50   1960

39896 THE EFFECT OF COMPOSITION AND TEMPERATURE ON THE ULTRAVIOLET ABSORPTION OF GLASS. M.S. THESIS.
MC SWAIN B D   UNIVERSITY OF ROCHESTER   ROCHESTER NEW YORK
UNIVERSITY OF ROCHESTER
     1-116   1962

39919 HEAT FLOW BETWEEN NORMAL AND SUPERCONDUCTING REGIONS.
GRIFFIN A   MAKI K
PHYS LETTERS
23   7   429-30   1966

39942 OPTICAL PROPERTIES OF THIN FILMS OF CARBON IN THE NEAR INFRARED.
LEVY-MANNHEIM C   MERING J
COMPT REND
263   B   18   1033-6   1966

39944 OPTICAL PROPERTIES IN THE FAR ULTRAVIOLET OF ALUMINUM FILMS EVAPORATED IN HIGH VACUUM AND UNEXPOSED TO AIR.
DAUDE A   PRIOL M   ROBIN S
COMPT REND
263   B   21   1178-81   1966

39946 EFFECT OF NITROGEN ON THE SURFACE TENSION OF LIQUID IRON AND ITS ALLOYS.
BORODULIN E K   KUROCHKIN K T   UMRIKHIN P V
IZV VYSSHIKH UCHEBN ZAVEDENII CHERN MET
     10   5-11   1966

39951 OPTICAL CHARACTERISTICS OF A PROPOSED REFLECTANCE STANDARD.
TRYTTEN G   FLOWERS W
APPL OPTICS
5   12   1895-7   1966   CA   66-6 24113
CORRECTION.
6   5   979   1967

39952 INFRARED SPECTRAL EMITTANCE MEASUREMENTS OF OPTICAL MATERIALS.
STIERWALT D L
APPL OPTICS
5   12   1911-15   1966   CA   66-6 24080

39958 SURFACE TENSION OF MOLTEN IRON AND ITS ALLOYS.
KOZAKEVITCH P   URBAIN G
MEM SCI REV MET
58   6   401-13   1961

39959 NEW OBSERVATIONS IN SPECTRAL ABSORPTION AND EXCITON REFLECTION OF THIN MONOCRYSTALLINE FILMS OF LEAD IODIDE. FROM COLLOQUE SUR LA SPECTROMETRIE DU SOLIDE MONTPELLIER, FRANCE, 11-14 NOVEMBER, 1965.
NIKITINE S   BIELLMANN J
J PHYSIQUE SUPPL
5/6   C2-95-99   1966

39976 OPTICAL PROPERTIES OF EVAPORATED FILMS OF CHROMIUM AND COPPER.
HENDERSON G   WEAVER C
J OPT SOC AM
56   11   1551-3   1966   CA   66-6 23789

TPRC Number	Bibliographic Citation

39977 OPTICAL CONSTANTS OF EVAPORATED BARIUM IN THE VACUUM ULTRAVIOLET.
FISHER E I   FUJITA I   WEISSLER G L
J OPT SOC AM
56   11   1560-4   1966   CA   66-2 6710

39987 ACCELERATED THERMOCOUPLE METHOD OF MEASURING THERMAL CONDUCTIVITY.
YANKELEV L F   ROIFE V S
INZH FIZ ZH
8   4   511-15   1965

39988 ACCELERATED THERMOCOUPLE METHOD OF MEASURING THERMAL CONDUCTIVITY.   //ENGLISH TRANSLATION OF INZH. FIZ. ZH. 8 /4/ 511-5, 1965.//
YANKELEV L F   ROIFE V S
J ENG PHYS /BSSR/
8   4   355-8   1965

39993 SPECTRAL DEPENDENCE OF THE ABSORPTION, REFLECTION, AND PHOTOEMISSION COEFFICIENTS OF LIF IN THE REGION FROM 60 TO 120 EV.
LUKIRSKII A P   ERSHOV O A   ZIMKINA T M
SAVINOV E P
FIZ TVERDOGO TELA
8   6   1787-90   1966

39994 SPECTRAL DEPENDENCES OF THE ABSORPTION, REFLECTION, AND PHOTOEMISSION COEFFICIENTS OF LIF IN THE REGION FROM 60 TO 120 EV.   //ENGLISH TRANSLATION OF FIZ. TVERD. TELA 8 /6/ 1787-90. 1966.//
LUKIRSKII A P   ERSHOV O A   ZIMKINA T M
SAVINOV E P
SOVIET PHYSICS-SOLID STATE
8   6   1422-4   1966

40006 DEVELOPMENT OF LOW TEMPERATURE BRAZING ALLOYS FOR TITANIUM HONEYCOMB SANDWICH MATERIALS.
ELSNER N B   MACK E B   METCALF A G   TROY W C
SOLAR AIRCRAFT CO   USAF
CFSTI AND DDC
ASD-TR-61-313
AD-272147   1-61   1961

40017 THERMAL CONTACT RESISTANCE IN A VACUUM ENVIRONMENT. PH. D. THESIS.
CLAUSING A M   UNIVERSITY OF ILLINOIS   URBANA ILLINOIS
UNIVERSITY OF ILLINOIS   URBANA ILLINOIS
1-156   1963

40018 THE CONTACT THERMAL CONDUCTANCE OF AN ALUMINUM SHEATHED NICKEL OR TIN PLATED URANIUM ROD.
DUGEON E H   PRIOR B W   NATIONAL RESEARCH LABORATORIES   OTTAWA CANADA
NATIONAL RESEARCH COUNCIL
NRC MT-29   1-17   1955

40019 A STUDY OF THERMAL RESISTANCE ACROSS ALUMINUM JOINTS IN A HIGH VACUUM.
ELLIOTT D H   DOUGLAS AIRCRAFT CO INC
SANTA MONICA CALIFORNIA
DDC
SM-46759 AD 458087   1-136   1964

40020 CONTACT HEAT TRANSFER BETWEEN PLANE METAL SURFACES. //ENGLISH TRANSLATION OF INZHEN. FIZ. ZH. 7 /3/ 3-9, 1964.//
DYBAN E F   SHVETS I T
INTERNATIONAL CHEMICAL ENGINEERING
4   4   621-4   1964
CORRECTION.
TSAPLIN M I
J ENG PHYS
9   4   357-8   1965
REPLY
9   4   359   1965

40025 THE THERMAL CONDUCTIVTY OF SOME HEAT INSULATORS. PH.D. THESIS.
VAN DUSEN M S   JOHNS HOPKINS UNIVERSITY
BALTIMORE MARYLAND
JOHNS HOPKINS UNIVERSITY
1-27   1921

40056 HEAT FLOW IN A SUPERCONDUCTOR IN THE VORTEX STATE.
PARKS R D   ZUMSTEG F C   MOCHEL J M
PHYS REV LETTERS
18   2   47-9   1967   CA   66   41617

40082 THE THERMAL CONDUCTIVITY OF BEDS OF SPHERICAL PARTICLES.   M.S. THESIS.
WILSON W G
CLEMSON COLLEGE   CLEMSON SOUTH CAROLINA
1-48   1964

40083 AN EXPERIMENTAL INVESTIGATION OF THE HEAT TRANSFER MECHANISM ACROSS AN AIRCRAFT STRUCTURAL JOINT.   M.S. THESIS.
COLE R L
SOUTHERN METHODIST UNIVERSITY   DALLAS TEXAS
1-51   1961

40099 SYNTHESIS OF NEUTRAL ANTIREFLECTION COATINGS.   II. CALCULATION OF COATINGS WHICH, OVER A BROAD SPECTRAL RANGE, PREVENT REFLECTION FROM COMPONENTS HAVING REFRACTIVE INDICES N EQUAL TO OR GREATER THAN 2.
FURMAN SH A
OPTIKA I SPEKTROSKOPIYA
21   3   357-64   1966

40100 SYNTHESIS OF NEUTRAL ANTIREFLECTION COATINGS.   II. CALCULATION OF COATINGS WHICH, OVER A BROAD SPECTRAL RANGE, PREVENT REFLECTION FROM COMPONENTS HAVING REFRACTIVE INDICES N EQUAL OR GREATER THAN 2.
//ENGLISH TRANSLATION OF OPTIKA I SPEKTROSKOPIYA 21 /3/ 357-64, 1966.//
FURMAN SH A
OPT SPECTRY
21   3   201-4   1966

40103 DEVELOPMENT OF SPACE-STABLE THERMAL-CONTROL COATINGS. /TRIANNUAL REPORT./
ZERLAUT G A   NOBLE G   IIT RESEARCH INSTITUTE
TECHNOLOGY CENTER   CHICAGO ILL
NASA
IITRI-U6002-63   1-36   1968

40114 ON THE IRRADIATION WITH ELECTIONS OF THIN FILMS OF LANTHANUM FLUORIDE.
PORRECA FLAVIO   CARRELLI ANTONIO   BOURG MARCEL
COMPT REND
263   8   23   1334-6   1966

40124 OPTICAL PROPERTIES OF THIN INDIUM FILMS.
VAN DE VOORDE M   JONES A
J PHYSIQUE
27   9   543-8   1966

40133 THE EFFECT OF DENSITY AND GRAIN SIZE ON THE THERMAL CONDUCTIVITY OF UO2 DURING IRRADIATION.
ARAGONES M   GUERRERO H   ATOMIC ENERGY OF CANADA LTD   CHALK RIVER /ONTARIO/
AEC AND CFSTI
AECL-2564   N66-31977
1-52   1966   CA   65   8281

40135 THE ACCOMMODATION COEFFICIENTS OF GASES AND VAPORS ON THE SURFACE OF METAL.
SPIVAK G V
UCHENYE ZAPISKI LENINGRAD GOSUDARST UNIV SER FIZ NAUK
38   7-13   1939

40136 DEVELOPEMENT AND QUALIFICATION OF THERMAL CONTROL COATINGS FOR SNAP SYSTEMS.
CROSBY J R   AT INTERN   CANOGA PARK CALIF
AEC AND CFSTI
NAA-SR-9908 N65-34983
1-69   1965   CA   65   6654

40148 PHYSICAL PROPERTIES OF FILAMENT WOUND GLASS EPOXY STRUCTURES AS APPLIED TO POSSIBLE USE IN LIQUID HYDROGEN BUBBLE CHAMBERS.
BRECHNA H   HALDEMANN W   STANFORD UNIV   STANFORD CALIF
AEC
SLAC-PUB-121 CONF-650802-14 N66-25109
1-47   1965   CA   65   13161

40159 OPTICAL PROPERTIES AND COLOR-CENTER FORMATION IN THIN FILMS OF MOLYBDENUM TRIOXIDE.
DEB S K   CHOPOORIAN J A
J APPL PHYS
37   13   4818-25   1966   CA   66   41680

40164 OPTICAL PROPERTIES OF TIO2 AS A REFLECTOR FOR PLASTIC SCINTILLATORS.
MAYER R   WENSEL A   UNIV FRANKFURT/M   GERMANY
AEC
IKF-12   1-12   1965   CA   65   11711

40167 CONSTRUCTION OF APPARATUS FOR DETERMINATION OF THERMAL CONDUCTIVITY OF COAL STONES IN GAS AT PRESSURES IN THE 0-15 ATMOSPHERE RANGE AND AT TEMPERATURES UP TO 250 C.
LANGEN J   HECKER R   KERNFORSCHUNGSANLAGE JUELICH /WEST GERMANY/
AEC
THTR-5   EUR-2775-D N66-32378
1-25   1966   PA   N66-4-18 3672

40173 DETERMINATION OF THE EXISTENCE OF CERTAIN PROPERTIES PECULIAR TO MOLECULAR TRIATOMIC JETS OF HYDROGEN.
DEVIENNE F M   ROUSTAN J-C
COMPT REND
263   8   25   1389-92   1966

40174 THE REFLECTIVITIES OF BUILDING MATERIALS FOR SOLAR RADIATION.   FROM   THE EXCLUSION OF SOLAR HEAT.
BECKETT H E
J INST HEATING AND VENTILATING ENGRS
3   26   84-8   1931

TPRC Number	Bibliographic Citation
40182	THE DETERMINATION OF OPTICAL PROPERTIES OF OPAQUE METAL FILMS BY ELLIPSOMETRIC MEASUREMENTS ON THE SUBSTRATE BACK SURFACE. M.S. THESIS. SHEDDY C H COLORADO STATE UNIVERSITY FORT COLLINS COLORADO 1-76 1964
40193	THERMAL CONDUCTIVITY OF COMPACTED POWDER CARBONS AND MIXTURES OF CARBON POWDERS WITH URANIUM CARBIDE PARTICLES. TYE R P WOODMAN M J CARBON 4 2 167-76 1966 CA 65 19329
40194	THERMAL RESISTANCE OF METALLIC CONTACTS. GENERAL PRINCIPLES AND NEW EXPERIMENTAL TECHNIQUES. PIAZZESI G DASSU G TERMOTECNICA /MILAN/ 19 1 31-46 1965
40197	THERMAL PROPERTIES OF FUR. HAMMEL H T AM J PHYSIOL 369-76 1955
40221	OPTICAL PROPERTIES OF EVAPORATED IRIDIUM IN THE VACUUM ULTRAVIOLET FROM 500 ANGSTROMS TO 2000 ANGSTROMS. HASS G JACOBUS G F HUNTER W R J OPT SOC AM 57 6 758-62 1967 IAA 7-16 2733
40226	THERMAL CONDUCTIVITIES OF BUILDING MATERIALS IN DWELLING CONSTRUCTION. TUOMOLA T T RUSO R R THE STATE INSTITUTE FOR TECHNICAL RESEARCH LABORATORY OF TECHNOLOGY HELSINKI FINLAND STATE INSTITUTE FOR TECHNICAL RESEARCH HELSINKI FINLAND LABORATORY REPORT C-1 1-7 1954
40228	MODERN HEAT INSULATING AND DECKING MATERIALS. GRIFFITHS E HICKMAN M J TRANSACTIONS NORTH EAST COAST INSTITUTION OF ENGINEERS AND SHIPBUILDERS IN NEWCASTLE UPON TYNE ON THE 26TH MARCH 1943 59 207-30 1943
40233	REFLECTANCE AND RELATIVE TRANSMITTANCE OF LASER-DEPOSITED IRIDIUM IN THE VACUUM ULTRAVIOLET. SAMSON J A R PADUR J P SHARMA A J OPT SOC AM 57 7 966-7 1967
40268	THE THERMAL AND ELECTRICAL CONDUCTIVITIES OF METALS AND ALLOYS. PH.D. THESIS. POWELL R W LONDON UNIVERSITY ENGLAND LONDON UNIVERSITY 1-79 1938
40269	PHASE-SHIFT-CORRECTED THICKNESS DETERMINATION OF SILICON DIOXIDE ON SILICON BY ULTRAVIOLET INTERFERENCE. WESSON R A PHILLIPS R P PLISKIN W A J APPL PHYS 38 6 2455-60 1967
40280	OPTICAL PROPERTIES AND PHOTOELECTRIC EMISSION OF SINGLE CRYSTALS AND THIN FILMS OF BAF2 IN THE FAR ULTRAVIOLET, BETWEEN 1800 AND 950 A. /7 TO 13.5 EV./ ROBIN-KANDARE S ROBIN J J PHYS /PARIS/ 26 2 85-8 1965 CA 64 7545
40283	ULTRA-HIGH VACUUM MEASUREMENT OF THE OPTICAL PROPERTIES OF COPPER. PH.D. THESIS. SPENCER W T UNIVERSITY OF ROCHESTER ROCHESTER NEW YORK UNIV MICROFILMS PUBL 63-7783 1-113 1963
40294	MEASUREMENT OF THE SPECTRAL REFLECTANCE OF REACTOR-IRRADIATED THERMAL-CONTROL COATINGS. REPORT FZK-256. LEWIS J H MC DANIEL R H BELL J R GENERAL DYNAMICS NUCLEAR AEROSPACE RESEARCH FACILITY FORT WORTH TEXAS DDC AFWL-TR-66-66 AD-801516 1-114 1966 TA 66-24 A-83
40298	CARBONIZED PLASTIC COMPOSITES FOR HYPERTHERMAL ENVIRONMENTS. CARLSON R K FORCHT B A MEDFORD J A MC KINNEY A R MC QUISTON F C SCOTT R O CHANCE VOUGHT ASTRONAUTICS DIVISION DALLAS TEXAS USAF DDC ASD-TDR-62-352 AD 277376 1-265 1962
40299	A PRELIMINARY EVALUATION OF THE MECHANICAL, PHYSICAL, AND THERMAL PROPERTIES OF THIRTEEN REINFORCED PLASTIC MATERIALS. AEROJET-GENERAL CORPORATION SOLID ROCKET PLANT SACRAMENTO CALIFORNIA USAF DDC AD-286219 1-10 1962
40300	A SURVEY OF INSULATION MATERIALS. 1 OCTOBER 1962 THROUGH 31 DECEMBER 1962. AEROJET-GENERAL CORPORATION SOLID ROCKET PLANT SACRAMENTO CALIFORNIA USAF DDC AD-295682 1-57 1963
40301	CARBONIZED PLASTICS COMPOSITES FOR HYPERTHERMAL ENVIRONMENTS. PT II. SYNTHESIS OF IMPROVED THERMALLY PROTECTIVE PLASTICS AND COMPOSITES. FORCHT B A HAVILAND J K MC KINNEY A R LING-TEMCO-VOUGHT ASTRONAUTICS DIVISION DALLAS TEXAS DDC ASD-TDR-62-352 /PT 2/ AD 403363 1-138 1963
40303	HIGH TEMPERATURE SOLAR ABSORBER COATINGS, PART III. SCHMIDT R N PARK K C JANSSEN J E HONEYWELL RESEARCH CENTER HOPKINS MINN USAF DDC AFML-TR-65-317 AD-475639L 1-55 1965 TA 66-3 A-42
40304	THE ROLE OF EMITTANCE IN REFRACTORY METAL COATING PERFORMANCE. PART I-REVIEW AND ANALYSIS. BARTSCH K O HUEBNER A NORTH AMERICAN AVIATION INC LOS ANGELES CALIF DDC NA66-760/PT 1/ AFML-TR-66-55/PT 1/ AD-801274 1-153 1966 TA 66-24 A36
40305	EMERGING AEROSPACE MATERIALS. TECHNICAL STAFF AIR FORCE MATERIALS LAB OHIO USAF DDC AFML-TR-65-114 1-105 1965
40314	DEVELOPMENT, EVALUATION AND APPLICATION OF HIGH TEMPERATURE INDICATING PAINTS FOR ASSET ASV-4. RUSERT E L MC DONNELL AIRCRAFT CORP ST LOUIS MO DDC DDC AD-478295 1-77 1965 TA 66-7 A-44
40318	THERMAL CONDUCTANCE OF CONTACTS AND JOINTS. FONTENOT J E JR BOEING CO AEROSPACE GROUP SEATTLE WASH DDC DDC AD 479008 D5-12206 1-166 1964 TA 66-8 A-122
40323	A MODEL FOR THE WALL BOUNDARY CONDITION IN KINETIC THEORY. EPSTEIN M AEROSPACE CORP LAB OPERATIONS EL SEGUNDO CALIF DDC DDC SSD-TR-66-49 AD-480131 1-29 1966 TA 66-10 A-106
40324	A MULTI-CHANNEL OPTICAL PYROMETER FOR THE 1500 TO 3000 K RANGE. PIERSON A H ZEIGLER G WEISS M BARNES ENGINEERING CO STAMFORD CONN DDC DDC AFFDL-TR-65-133 AD-480387 1-69 1966 TA 66-10 A-56
40327	THERMAL PROPERTY DATA UTILIZED FOR ASSET MATERIALS. SCHATTYN J M MC DONNELL AIRCRAFT CORP ST LOUIS MO DDC DDC A-656 AD-480414 1-45 1964 TA 66-10 A-43
40328	FURTHER MEASUREMENTS OF THE COEFFICIENT OF LINEAR EXPANSION AT LOW TEMPERATURES. DORSEY H G PHYS REV 27 1 1-10 1908
40333	PHASE CHANGE IN POLYMERIC SYSTEMS FOR ACTIVE THERMAL CONTROL. KELLIHER W C PEZDIRTZ G F YOUNG P R NASA-LANGLEY RESEARCH CENTER NASA N66-32952 1-6 1966

TPRC Number	Bibliographic Citation

40343 PROGRAM ASTEC /ADVANCED SOLAR TURBO ELECTRIC CONCEPT/. PART I. CANDIDATE MATERIALS LAB TESTS.
HURTT W W   BLAKNEY T L   CUNNINGTON G R JR
BRADSHAW W G   POLLARD H E   LOCKHEED MISSILES AND SPACE COMP   SUNNYVALE CALIF
DDC
AFAPL-TR-85-53 PT-1
AD-482282   1-318   1966   TA   66-13 A-45

40347 VACUUM ULTRAVIOLET REFLECTIVITY OF INDIUM BISMUTHIDE. M.S. THESIS.
HANNING W A   CALIFORNIA STATE COLLEGE   LONG BEACH CALIFORNIA
CALIFORNIA STATE COLLEGE
                                    1-64   1969

40351 IMPROVED RADIATOR COATINGS.
SCHATZ E A   COUNTS C R III   AMERICAN MACHINE AND FOUNDRY CO   ALEXANDRIA VIRGINIA   USAF
DDC
AFML-TDR-64-146 /PT III/
AD-486446L   1-95   1966   TA   66-17 A-48

40356 THE ELECTRICAL DOUBLE LAYER IN AMIDE SOLUTIONS.
PAYNE R
J PHYS CHEM
73   11   3598-608   1969

40358 SPACECRAFT COATINGS, BEHAVIOR IN THE SPACE ENVIRONMENT.
NEVILLE T   HACKWORTH J   BOEBEL C   GENERAL ELECTRIC CO MISSILE AND SPACE DIVISION
NASA
64-SD-264   N66-24685
                                    1-16   1964

40362 INFRARED SPECTRA AND BONDING IN THE SODIUM SUPEROXIDE AND SODIUM PEROXIDE MOLECULES.
ANDREWS L
J PHYS CHEM
73   11   3922-8   1969

40368 UO2 POWDER AND PELLET THERMAL CONDUCTIVITY DURING IRRADIATION.
LYONS M F   COPLIN D H   HAUSNER H   WEIDENBAUM B
PASHOS T J   GENERAL ELECTRIC CO   PLEASANTON CALIF
USAEC AND EURATOM
AEC AND CFSTI
GEAP-5100-1 EURAEC-1626
                                    1-76   1966

40396 HYDROGEN THERMAL CONDUCTIVITY AT TEMPERATURES FROM 2000 TO 4000 F.
ISRAEL S L   HAWKINS T D   SALTER R T   HYMAN S C
UNITED NUCLEAR CORP   WHITE PLAINS N Y
NASA AND CFSTI
NASA-CR-78167 UNC-5082
N66-37325   1-132   1964   PA   N66-4-22 4454

40398 WALL TO FLUID HEAT TRANSFER IN LIQUID FLUIDIZED BEDS.
WASMUND B   SMITH J W
CAN J CHEM ENG
45   3   156-65   1967

40399 DETERMINATION OF THERMOPHYSICAL PROPERTIES OF ABLATIVE MATERIALS. PHASE I. LABORATORY DETERMINATIONS. PART A. HEAT CAPACITY, DENSITY, DECOMPOSITION AND MELT TEMPERATURE.
BLUMENTHAL J L   LARSON R D   NORDBERG R C   TRW SYSTEMS   REDONDO BEACH CALIF
NASA AND CFSTI
NASA-CR-65486 TRW-04812-6001-R000
N66-35661   1-79   1966   PA   N66-4-21 4266

40400 DETERMINATION OF THE EMISSIVITY OF MATERIALS. SEMIANNUAL PROGRESS REPORT, 15 NOV. 1965-14 MAY 1966.
WALEK W J   LUOMA W L   EMANUELSON R C
PRATT AND WHITNEY AIRCRAFT   EAST HARTFORD CONN
NASA AND CFSTI
NASA-CR-72058 PWA-2877 N66-37039
                                    1-88   1966   PA   N66-4-22 4379

40404 EFFECT OF PARTICLE SHAPE ON THE AXIAL EFFECTIVE THERMAL CONDUCTIVITY OF POROUS MEDIA. M.S. THESIS
REYMOND B   UNIVERSITY OF AKLAHOMA   NORMAN OKLAHOMA
UNIVERSITY OF OKLAHOMA
                                    1-52   1964

40413 SPECTRAL EMISSIVITY OF METALS AFTER DAMAGE BY PARTICLE IMPACT. FROM PROC. OF CONF. ON SPACECRAFT COATINGS DEVELOP. 1964.
SCHOCKEN K   FOUNTAIN J A
NASA AND CFSTI
NASA-TM-X-56167 N66-37820
                                    1-20   1964   PA   N66-4-23 4632

40414 SOME POLY-BASIC PHOSPHATE CONVERSION COATINGS FOR THERMAL CONTROL. FROM PROC. OF CONF. ON SPACECRAFT COATINGS DEVELOP. 1964.
WAKELYN N T   PEZDIRTZ G F
NASA AND CFSTI
NASA-TM-X-56167 N66-37821

40415 THE INFLUENCE OF TEMPERATURE ON THE STABILITY OF LOW SOLAR ABSORPTANCE AND THERMAL COATINGS. FROM PROC. OF CONF. ON SPACECRAFT COATINGS DEVELOP. 1964.
STREED E R   AMES RESEARCH CENTER   MOFFETT FIELD CALIFORNIA
NASA AND CFSTI
N66-37822   1-10   1964

40416 THE EFFECTS OF ULTRAVIOLET AND GAMMA RAYS ON THERMAL CONTROL COATINGS. FROM PROC. OF CONF. ON SPACECRAFT COATINGS DEVELOP. 1964.
JEWELL R A   PEZDIRTZ G F   BURKS H D   LANGLEY RESEARCH CENTER
NASA AND CFSTI
N66-37823   1-16   1964

40417 PRELIMINARY RESULTS FROM A ROUND-ROBIN STUDY OF ULTRAVIOLET DEGRADATION OF SPACECRAFT THERMAL-CONTROL COATINGS. FROM PROC. OF CONF. ON SPACECRAFT COATINGS DEVELOP. 1964.
ARVESEN J C   NEEL C B   SHAW C C   AMES RESEARCH CENTER   MOFFETT FIELD CALIF
NASA AND CFSTI
N66-37824   1-21   1964

40418 LOW DENSITY ABLATION MATERIALS SURVEY.
WELSH W E   STARNER K E   LEEDS D H
SLAUGHTER J I   AEROSPACE CORP   EL SEGUNDO CALIF
LAB OPERATIONS   USAF
DDC
SSD-TR-66-35 AD 480846
                                    1-76   1966   TA   66-11 A-123

40420 MONOCHROMATIC REFLECTANCE TESTS.
WETMORE R A   BOEING CO   SEATTLE WASH   USAF
DDC
BSD-TR-66-140 AD 483037
                                    1-116   1963   TA   66-14 A-107

40433 CAPILLARY PROPERTIES OF ALKALI METAL SYSTEMS. SURFACE TENSION OF POTASSIUM UNDER ARGON. INTERFACIAL TENSION OF LITHIUM AND LITHIUM CHLORIDE. M.S. THESIS.
CHUNG J-W   COLUMBIA UNIVERSITY
AEC AND ORINS
CU-2660-24   1-97   1965   CA   65   17728

40447 THE MEASUREMENT OF THE THERMAL PROPERTIES OF BUILDING MATERIALS - THE THERMAL CONDUCTIVITY OF THIN 12 INCHES SQUARE SAMPLES.
ROUX A J A   RICHARDS S J   RENNHACKKAMP W M H
SOUTH AFRICAN COUNCIL FOR SCIENTIFIC AND INDUSTRIAL RESEARCH
SOUTH AFRICAN COUNCIL FOR SCIENTIFIC AND INDUSTRIAL RESEARCH
DR-6   1-28   1950

40448 ADHESION AND THERMAL PROPERTIES OF REFRACTORY COATING METAL SUBSTRATE SYSTEMS.
LEVY M   SKLOVER G N   SELLERS D J   ARMY MATERIALS RESEARCH AGENCY MATERIALS ENGINEERING DIV WATERTOWN MASS
DDC
AMRA-TR-66-01 AD 482760
                                    1-19   1966   TA   66-13 A-49

40467 TESTING OF REFLECTIVE COATINGS - HSM 80C.
WETMORE R A   BOEING CO   SEATTLE WASH   USAF
DDC
D2-30484-1 BSD-TR-66-218 AD-443339
AD-800433   1-16   1963   TA   66-23 A-40

40471 OPTICAL CONSTANTS OF SILVER AND BARIUM IN THE VACUUM ULTRAVIOLET SPECTRAL REGION.
FISHER E I   FUJITA I   WEISSLER G L
UNIVERSITY OF SOUTHERN CALIF   LOS ANGELES CALIF
NASA AND CFSTI
NASA-CR-76364 USC-VACUV-108
N66-30847   1-107   1966   PA   N66-4-17 3425

40474 HEAT TRANSFER ACROSS SURFACES IN CONTACT. PRACTICAL EFFECTS OF TRANSIENT TEMPERATURE AND PRESSURE ENVIRONMENTS.
BLUM H A   MOORE C J JR   SOUTHERN METHODIST UNIV DALLAS TEXAS
NASA AND CFSTI
NASA-CR-76878
N66-32762   1-40   1966   PA   N66-4-19 3888

40476 ANALYTICAL DETERMINATION OF THE EFFECT OF THERMAL PROPERTY VARIATIONS ON THE PERFORMANCE OF A CHARRING ABLATOR.
PITTMAN C M   BREWER W D   LANGLEY RESEARCH CENTER   LANGLEY STATION VA
NASA AND CFSTI
NASA-TN-D-3486
N66-30176   1-28   1966   CA   66   119520

TPRC Number	Bibliographic Citation
40514	STUDIES, RESEARCH AND INVESTIGATIONS OF THE OPTICAL PROPERTIES OF THIN FILMS OF METALS, SEMICONDUCTORS AND DIELECTRICS. HADLEY L N SHAKLEE K COLORADO STATE UNIV DEPT OF PHYSICS FORT COLLINS DA DDC AND CFSTI SAR-11 AD 634034 N66-33580 1-12 1965 TA 66-14 87
40523	INFRARED COATING STUDIES. TURNER A F BAUSCH AND LOMB INC ROCHESTER N Y RESEARCH AND DEVELOPMENT DIV DDC AD 635629 N66-37228 1-13 1966 TA 66-17 92
40526	THERMAL ACCOMMODATION COEFFICIENT FOR AIR ON TUNGSTEN. M.S. THESIS. LAU C A AIR FORCE INST OF TECH SCHOOL OF ENGINEERING WRIGHT-PATTERSON AFB OHIO DDC GAM/ME/66B-14 N66-34973 AD-635231 1-76 1966 CA 67 25952
40528	INFRARED COATING STUDIES. SULZBACH F TURNER A F BAUSCH AND LOMB INC ROCHESTER NY RESEARCH AND DEVELOPMENT DIV DDC AD 635670 1-33 1966 TA 66-17 92
40529	EFFECTS OF DIFFUSION PUMP OIL CONTAMINATION ON THE REFLECTANCE CHARACTERISTICS OF VARIOUS SURFACES. PINION E C ARNOLD ENGINEERING DEV CENTER ARNOLD AIR FORCE STATION TENN USAF DDC AEDC-TR-66-53 AD-635897 1-41 1966 TA 66-17 80
40530	MEASUREMENTS OF THE INFRARED SPECTRAL TRANSMITTANCE OF OPTICALLY THICK SAMPLES. LOW M J D ABRAMS L RUTGERS UNIV NEW BRUNSWICK N J SCHOOL OF CHEMISTRY DDC CFSTI AND DDC AD 636035 N66-37254 TR-2 1-23 1966 TA 66-17 92
40565	EFFECT OF SUPPORT ON THE QUENCHING OF THE PHOSPHORESCENCE OF TRYPAFLAVINE BY OXYGEN. ZAKHAROV I A ALESKOVSKII V B ZH FIZ KHIM 40 5 985- 1966
40566	EFFECT OF SUPPORT ON THE QUENCHING OF THE PHOSPHORESCENCE OF TRYPAFLAVINE BY OXYGEN. //ENGLISH TRANSLATION OF ZH. FIZ. KHIM. 40 /5/ 985- , 1966.// ZAKHAROV I A ALESKOVSKII V B RUSS J PHYS CHEM 40 5 530-4 1966
40575	DYNAMIC SURFACE EFFECT IN A CATALYTIC PROCESS. GRIGORYAN V A ZUKHOVITSKII A A SHVINDLERMAN L S CHIKOMASOVA M I ZH FIZ KHIM 40 5 1144- 1966
40576	DYNAMIC SURFACE EFFECT IN A CATALYTIC PROCESS. //ENGLISH TRANSLATION OF ZH. FIZ. KHIM. 40 /5/ 1144- 1966.// GRIGORYAN V A ZUKHOVITSKII A A SHVINDLERMAN L S CHIKOMASOVA M I RUSS J PHYS CHEM 40 5 615-18 1966
40580	INVESTIGATION IN THE FAR INFRARED. STRONG J PHYS REV 38 1818-26 1931
40581	THE INFRA-RED TRANSMISSION OF THIN FILMS OF VARIOUS ORGANIC MATERIALS. WELLS A J J APP PHYS 11 137-40 1940
40583	HIGH REFLECTANCE COATINGS. DUBS C W IEEE TRANS AND NUCLEAR SCI 13 1 729-34 1966 CA 65 279 DDC AFCRL-IP-102 AFCRL-66-414 AD 637893 1-10 1966 TA 66-20 73
40587	THE EVALUATION OF THE ACCOMMODATION COEFFICIENT OF AIR. M.S. THESIS. BURCH B A UNIV OF OKLAHOMA NORMAN OKLAHOMA NORMAN OKLAHOMA UNIVERSITY OF OKLAHOMA 1-47 1964
40591	ABSOLUTE METHODS IN REFLECTOMETRY. RESEARCH PAPER NO. 3. MC NICHOLAS H J J RESEARCH OF THE NAT BUR STANDARDS

TPRC Number	Bibliographic Citation
40598	TRANSMITTANCE OF ANTIREFLECTION-COATED SILICON IN THE FAR INFRARED. RUDYAVSKAYA I G KUDRYAVTSEVA A G KISLOVSKII L D OPTIKA I SPEKTROSKOPIYA 21 4 476-81 1966 CA 66-6 24065
40599	TRANSMITTANCE OF ANTIREFLECTION-COATED SILICON IN THE FAR INFRARED. //ENGLISH TRANSLATION OF OPTIKA I SPEKTROSKOPIYA 21 /4/ 476-81, 1966.// RUDYAVSKAYA I G KUDRYAVTSEVA A G KISLOVSKII L D OPT SPECTRY 21 4 266-9 1966
40600	SYNTHESIS OF NEUTRAL ANTIREFLECTION COATINGS. III. CALCULATION OF COATINGS WHICH, OVER A BROAD SPECTRAL RANGE, PREVENT REFLECTION FROM OPTICAL COMPONENTS HAVING REFRACTIVE INDICES N LESS THAN OR EQUAL TO 2. FURMAN SH A OPTIKA I SPEKTROSKOPIYA 21 4 503-8 1966
40601	SYNTHESIS OF NEUTRAL ANTIREFLECTION COATINGS. III. CALCULATION OF COATINGS WHICH, OVER A BROAD SPECTRAL RANGE, PREVENT REFLECTION FROM OPTICAL COMPONENTS HAVING REFRACTIVE INDICES N LESS THAN OR EQUAL TO 2. //ENGLISH TRANSLATION OF OPTIKA I SPEKTROSKOPIYA 21 /4/ 503-8, 1966.// FURMAN SH A OPT SPECTRY 21 4 280-3 1966
40608	REFLECTANCE OF EVAPORATED ALUMINUM IN THE VACUUM ULTRAVIOLET. HASS G HUNTER W R TOUSEY R J OPT SOC AM 46 12 1009-10 1956
40611	THE THERMAL CONDUCTIVITIES OF PLASTICS WITH GLASS, ASBESTOS, AND CELLULOSIC FIBRE REINFORCEMENT. RATCLIFFE E H APPL MATERIALS RESEARCH 5 200-1 1966 CA 66 56064
40612	INTERFACIAL SURFACE TENSION MEASUREMENTS BY MEANS OF THE PENDANT DROP METHOD AND APPLICATIONS OF THE SAME WITH RESPECT TO HYDROCARBON FLUIDS WITHIN RESERVOIRS. GAULIER C PACSIRSZKY J REV LINST FRANC PETROLE ET ANN COMBUSTIBLES LIQUIDES 21 2 227-38 1966
40613	INTERFACIAL SURFACE TENSION MEASUREMENTS BY MEANS OF THE PENDANT DROP METHOD AND APPLICATIONS OF THE SAME WITH RESPECT TO HYDROCARBON FLUIDS WITHIN RESERVOIRS. //ENGLISH TRANSLATION OF REV. LINST. FRANC. PETROLE ET ANN. COMBUSTIBLES LIQUIDES. 21 /2/ 227-38, 1966.// GAULIER C PACSIRSZKY J SLA TT-66-13938 1-26 1966 TT 16-11 54
40619	VACUUM AS AN INSULATOR. DUEVEL C O JR REFRIG ENG 20 223-8 1930
40620	REFLECTANCE AND PHASE ENVELOPES OF AN ITERATED MULTILAYER. ARNDT J BAUMEISTER P W J OPT SOC AM 56 12 1760-2 1966
40630	STUDY OF DIFFUSE REFLECTION FROM POWDERS UNDER DIFFUSE ILLUMINATION. TOPORETS A S OPTICS AND SPECTROSCOPY 7 471-3 1959
40641	ANALYSIS OF A MODIFIED FRUSTRATED TOTAL REFLECTION FILTER. NOYES G R BAUMEISTER P W APPL OPT 6 2 355-6 1967
40651	FILMED SURFACES FOR REFLECTING OPTICS. HASS G J OPT SOC AM 45 11 945-52 1955
40653	ADHERENCE AND WETTABILITY OF NICKEL, NICKEL-TITANIUM ALLOYS, AND NICKEL-CHROMIUM ALLOYS TO SAPPHIRE. RITTER J E JR BURTON M S TRANS MET SOC AIME 239 1 21-6 1967
40655	STANDARDS OF REFLECTANCE. BUDDE W J OPT SOC AM 50 3 217-20 1960

TPRC Number	Bibliographic Citation

40663 AN APPARATUS FOR THE MEASUREMENT OF CLOTHING INSULATION.
NELMS J D   FLYING PERSONNEL RESEARCH COMMITTEE
LONDON /ENGLAND/
CFSTI
FPRC/MEMO-206 N66-19722
N66-24829   1-18   1964   CA   66   96617

40664 THERMAL CONTACT CONDUCTANCE IN A VACUUM. M.S. THESIS.
YOVANOVICH M M   MASS INST TECH   CAMBRIDGE
TECHNOLOGY   CAMBRIDGE MASS
CFSTI NASA AND MIT
NASA-CR-74619 DSR-4542-39 N66-24593
NP-15897   1-132   1966   CA   65   17723

40669 AN INVESTIGATION OF THE INFLUENCE OF MOISTURE PENETRATION ON THE THERMAL CONDUCTIVITY OF A LOW TEMPERATURE INSULATING MATERIAL. PH.D. THESIS.
ZAKI M A   VICTORIA UNIVERSITY OF MANCHESTER
LONDON
VICTORIA UNIVERSITY OF MANCHESTER
     1-137   1953

40670 UNSTEADY STATE METHODS FOR THE DETERMINATION OF THE THERMAL CONSTANTS OF INSULATING MATERIALS IN SLAB FORM. PH.D. THESIS.
HATTON A P   VICTORIA UNIVERSITY OF MANCHESTER
LONDON
VICTORIA UNIVERSITY OF MANCHESTER   LONDON
     1-85   1956

40671 AN INVESTIGATION OF THE EFFECT OF DENSITY AND WATER VAPOUR CONDENSATION ON THE THERMAL CONDUCTIVITY OF LOW TEMPERATURE INSULATING MATERIALS. PH.D. THESIS.
AFIFY M Y M   VICTORIA UNIVERSITY OF MANCHESTER
LONDON
VICTORIA UNIVERSITY OF MANCHESTER
     1-116   1955

40677 A CRITICAL ANALYSIS OF THE HEAT CONDUCTION THROUGH PARTIALLY RAREFIED GASES. PH.D. THESIS.
PURSLOW B W   UNIVERSITY OF LONDON   LONDON ENGLAND
UNIVERSITY OF LONDON
     1-221   1952

40678 THERMAL RESISTANCE OF METAL CONTACTS. PH.D. THESIS.
POTTER J H   THE JOHN HOPKINS UNIVERSITY
BALTIMORE MARYLAND
THE JOHN HOPKINS UNIVERSITY
     1-164   1948

40681 THERMAL RESISTANCE OF BONDS AND CONTACTS. M.S. THESIS.
LEWIS D M   UNIVERSITY OF MISSOURI   ROLLA MISSOURI
UNIVERSITY OF MISSOURI
     1-48   1964

40683 THE REDUCTION OF THERMAL CONTACT RESISTANCE BY USE OF INTERFACIAL METALLIC FOILS. M.S. THESIS.
KOH B   UNIVERSITY OF MARYLAND   BALTIMORE MARYLAND
UNIVERSITY OF MARYLAND
     1-48   1964

40684 EFFECTIVE THERMAL CONDUCTIVITY IN A PACKED BED. M.S. THESIS.
PARK C M   UNIVERSITY OF MARYLAND   BALTIMORE
MARYLAND
UNIVERSITY OF MARYLAND
     1-89   1964

40689 THERMAL CONTACT CONDUCTANCE. M.S. THESIS.
DEAN R A   UNIVERSITY OF PITTSBURGH   PITTSBURGH
PENNSYLVANIA
UNIVERSITY OF PITTSBURGH
     1-52   1963

40699 INVESTIGATION OF THE EFFECT OF A PHASE CHANGE ON THE TOTAL EMISSIVITY OF ALUMINIUM. M.S. THESIS.
KEAGY B J   WEST VIRGINIA UNIVERSITY   MORGANTOWN
WEST VIRGINIA
WEST VIRGINIA UNIVERSITY
     1-44   1963

40705 MULTIPLET ABSORPTION BAND OF CESIUM IODIDE IN THE EXTREME ULTRAVIOLET REGION.
SAITO H   ONAKA R
J PHYS SOC JAPAN
28   5   1380   1970

40706 THERMAL EXPANSION OF DUMET WIRES.
KISHII T
TOSHIBA REV INTERN ED
12   3   265-72   1957

40707 THERMAL EXPANSION OF DUMET WIRES. //ENGLISH TRANSLATION OF TOSHIBA REV. INTERN. ED. 12, /3/ 265-72, 1957.//
KISHII T
SLA
TT-66-12506
     1-39   1966   TT   14-8 72

40709 HEAT TRANSFER ACROSS THE FUEL-CLAD INTERFACE IN SHRINK FIT HYDRIDE - STAINLESS STEEL FUEL ELEMENTS.
ACCOMAZZO M A   MILLER J   ATOMICS INTERNATIONAL
DIV OF NORTH AMERICAN AVIATION INC   CANOGA PARK
CALIFORNIA   UGAEC
AEC AND CFSTI
NAA-SR-MEMO-9184
     1-18   1963

40712 THE EFFECT OF MOISTURE ON THE THERMAL CONDUCTIVITY IN THE CASE OF FIBROUS SLAB MATERIAL. M.S. THESIS, CHEM. ENG.
BECK E R   MICHIGAN TECHNOLOGICAL UNIV   HOUGHTON
MICHIGAN
MICHIGAN TECHNOLOGICAL UNIV
     1-56   1947

40713 THE EFFECT OF MOISTURE CONTENT ON THE THERMAL CONDUCTIVITY IN THE CASE OF FIBROUS SLAB MATERIAL. M.S. THESIS, CHEM. ENG.
WEBER F J JR   MICHIGAN TECHNOLOGICAL UNIV
HOUGHTON MICHIGAN
MICHIGAN TECHNOLOGICAL UNIV
     1-88   1948

40724 METHOD FOR CALCULATING THERMAL RESISTANCE OF CONDUCTIVELY COOLED ELECTRONIC PARTS.
PRASINOS T   GRUMMAN AIRCRAFT ENGINEERING CORP
BETHPAGE NEW YORK
DDC
RC-R-10-9.0 IDEP 347.40.00.00-K4-04
AD-483519L   1-14   1965   TA   66-14 A-29

40738 SPECTRAL EMISSION OF RADIATION BY GLASS.
BEATTIE J R   COEN E
BRIT J APPL PHYS
11   4   151-7   1960

40744 METHODS FOR THE DETERMINATION OF THE SURFACE TEMPERATURE, THE ABSORPTION NUMBER AND THE THERMAL CONDUCTIVITY OF LOOSE THIN COATINGS.
PRASOLOV R S
IZV VYSSHIKH UCHEB ZAVEDENII PRIBOROSTROENIE USSR
5   3   122-31   1962

40745 METHODS FOR THE DETERMINATION OF THE SURFACE TEMPERATURE, THE ABSORPTION NUMBER AND THE THERMAL CONDUCTIVITY OF LOOSE THIN COATINGS. //GERMAN TRANSLATION OF IZV. VYSSHIKH UCHEB. ZAVEDENII PRIBOROSTROENIE USSR, 5, /3/ 122-31, 1962.//
PRASOLOV R S
SLA
TT-65-28395
     1-15   1965   TT   14-6 59

40746 RADIATIVE PROPERTIES OF SURFACES CONSIDERED FOR USE ON THE EXPLORER SATELLITES AND PIONEER SPACE PROBES.
SHIPLEY W S   THOSTESEN T O   JET PROPULSION
LABORATORY CALIFORNIA INSTITUTE OF TECHNOLOGY
PASADENA CALIFORNIA
DDC
JPL MEM   O NO. 20-194
     1-26   1960

40756 IMPROVED ORGANIC COATINGS FOR TEMPERATURE CONTROL IN A SPACE ENVIRONMENT.
HORMANN H H   WRIGHT-PATTERSON AIR FORCE BASE
OHIO
DDC
ML-TDR-64-177
AD-453437   1-31   1964

40761 THERMAL CONDUCTIVITY AND THERMAL SHOCK QUALITIES OF ZIRCONIA COATINGS ON THIN GAGE NI-MO-C METAL.
BUCKLEY J D
AM CERAM SOC BULL
49   6   588-91   1970

40764 PROCEDURES FOR THE PRECISE DETERMINATION OF THERMAL RADIATION PROPERTIES. NOVEMBER 1964-OCTOBER 1965.
RICHMOND J C   KNEISSL G J   KELLEY D L
KELLY F J   NATIONAL BUREAU STANDARDS
CFSTI
NBS-TN-292 AFML-TR-66-302
AD-649694   1-80   1966   TA   67-11 91

40769 THERMAL CONDUCTIVITY MEASUREMENTS.
BERMAN R
EXPTL CRYOPHYS
     327-36   1961   CA   58   11976

40776 EQUILIBRIUM DEFECT CONCENTRATION IN CRYSTALLINE SODIUM.
FEDER R   CHARBNAU H P
PHYS REV
149   2   464-71   1966

40783 THREE-LAYERED REFLECTION-REDUCING COATINGS.
LICKHART L B JR   KING P
J OPT SOC AM
37   9   689-94   1947

TPRC Number	Bibliographic Citation

40784 RECENT INVESTIGATIONS IN THE FAR INFRA-RED.
MC CUBBIN T K   SINTON W M
J OPT SOC AM
40   8   537-9   1950

40788 INTEGRATING SPHERE FOR THE MEASUREMENT OF REFLECTANCE
WITH THE BECKMAN MODEL DR RECORDING
SPECTROPHOTOMETER.
JACQUEZ J A   MC KEEHAN W   DIMITROFF J M
KUPPENHEIM H F
J OPT SOC AM
45   11   971-5   1955

40789 OBSERVATIONS OF SOLAR AND LUNAR RADIATION AT 1.5
MILLIMETERS.
SINTON W M
J OPT SOC AM
45   11   975-9   1955

40790 SAMPLING TECHNIQUE FOR TRANSMISSION MEASUREMENTS IN
THE FAR INFRARED REGION.
YOSHINAGA H   OETJEN R A
J OPT SOC AM
45   1085   1955

40793 OPTICAL PROPERTIES OF SB, TE, AND TI FILMS IN THE
VACUUM ULTRAVIOLET.
RUSTGI OM P   WALKER W C   WEISSLER G L
J OPT SOC AM
51   12   1357-9   1961

40795 MEASURED VALUES FOR THE COEFFICIENTS OF LINEAR
EXPANSION OF POLYCEL 420 AND CONOLON 506 AT LOW
TEMPERATURES.
HASKINS J F   GENERAL DYNAMICS ASTRONAUTICS
MATERIALS RESEARCH GROUP
GENERAL DYNAMICS ASTRONAUTICS
MRG-194   1-10   1960

40796 ON THE OPTICAL ABSORPTION OF THIN BARIUM OXIDE FILM
IN THE ULTRAVIOLET REGION.
TAKAZAWA K   TOMITIKA T
J PHYS SOC JAPAN
9   6   996-1000   1954

40798 INFRARED EMISSION SPECTRUM OF SILICON CARBIDE HEATING
ELEMENTS.   /RESEARCH PAPER NO. 2810./
STEWART J E   RICHMOND J C
J RESEARCH NATL BUR STANDARDS
59   6   405-9   1957

40852 OPTICAL AND SPECTRAL PROPERTIES OF S-20
PHOTOCATHODES ON NESA SUBSTRATES.
MORRISON C W
APPL OPT
6   3   573-4   1967

40863 A STUDY OF THE STRUCTURE OF THE ADSORPTION CENTERS ON
ALUMINUM-CHROMIUM CATALYSTS THROUGH OPTICAL SPECTRA.
SHVETS V A   KAZANSKII V B
DOKLADY AKAD NAUK SSSR
167   6   1331-4   1966

40864 A STUDY OF THE STRUCTURE OF THE ADSORPTION CENTERS ON
ALUMINUM-CHROMIUM CATALYSTS THROUGH OPTICAL SPECTRA.
//ENGLISH TRANSLATION OF DOKLADY AKAD. NAUK SSSR 167
/6/ 1331-4, 1966.//
SHVETS V A   KAZANSKII V B
DOKLADY PHYS CHEM
167   6   268-71   1966

40867 PHOTOCHEMICAL TRANSFORMATIONS OF AURAMINE BASE IN
CARBON TETRACHLORIDE.
KORSUNOVSKII G A
ZH FIZ KHIM
43   10   2537-   1969

40868 PHOTOCHEMICAL TRANSFORMATIONS OF AURAMINE BASE IN
CARBON TETRACHLORIDE. //ENGLISH TRANSLATION OF ZH.
FIZ. KHIM. 43 /10/ 2537- , 1969.//
KORSUNOVSKII G A
RUSS J PHYS CHEM
43   10   1422-5   1969

40893 NEW LUSTER COLORS FOR GLASS AND PORCELAIN.
LEVIN P A
STEKLO I KERAMIKA
23   1   26-9   1966

40894 NEW LUSTER COLORS FOR GLASS AND PORCELAIN. //ENGLISH
TRANSLATION OF STEKLO I KERAMIKA 23 /1/ 26-9, 1966.//
LEVIN P A
GLASS AND CERAMICS
23   1   31-3   1966

40903 SUNPATH DIAGRAMS AND OVERLAYS FOR SOLAR HEAT GAIN
CALCULATIONS.
PETHERBRIDGE P   BUILDING RESEARCH STATION
GARSTON WATFORD HERTS ENGLAND
MINISTRY OF TECHNOLOGY BUILDING RESEARCH STATION
GARSTON WATFORD HERTS ENGLAND
BRS CURRENT   PAPERS RESEARCH SERIES NO. 39
                     1-12   1966

40904 SUNPATH DIAGRAMS AND OVERLAYS FOR SOLAR HEAT GAIN
CALCULATIONS.   SUPPLEMENT 1.
PETHERBRIDGE P   BUILDING RESEARCH STATION
GARSTON WATFORD HERTS ENGLAND
MINISTRY OF TECHNOLOGY BUILDING RESEARCH STATION
GARSTON WATFORD HERTS ENGLAND
BRS CURRENT   PAPERS RESEARCH SERIES NO. 39
SUPPL. 1   1-12   1966

40905 TRANSMISSION CHARACTERISTICS OF WINDOW-GLASSES AND
SUN-CONTROLS.
PETHERBRIDGE P   BUILDING RESEARCH STATION
GARSTON WATFORD HERTS ENGLAND
MINISTRY OF TECHNOLOGY BUILDING RESEARCH STATION
GARSTON HERTS ENGLAND
BRS-NOTE-EN-53-65
                     1-21   1965

40908 INVESTIGATIONS OF SUMMER OVERHEATING AT THE BUILDING
RESEARCH STATION, ENGLAND.
LOUDON A G   DANTER E
BUILDING SCI
1   1   89-94   1966

40909 RADIATION TRANSMISSION CHARACTERISTICS OF LOUVER
SYSTEMS.   BUILDING RESEARCH SERIES /CURRENT PAPERS/
NO. 53.
NICOL J F
BUILDING SCI
1   167-82   1966

40915 ADSORPTION OF SODIUM DODECYL SULFATE AT VARIOUS
HYDROCARBON-WATER INTERFACES.
REHFELD S J
J PHYS CHEM
71   3   738-45   1967

40918 RADIANT TRANSFER THROUGH SPECULAR TUBES.
O BRIEN P F   SOWELL E F
J OPT SOC AM
57   1   28-34   1967

40922 REFLECTION OF LIGHT FROM FILMED ROUGH SURFACE.
DETERMINATION OF FILM THICKNESS AND RMS ROUGHNESS.
NAGATA K-I   NISHIWAKI J
JAPANESE J APPL PHYS
6   2   251-7   1967

40926 SURFACE TENSION OF MOLTEN METALS AND ALLOYS.
SAUERWALD F
Z ANORG CHEM
223   84-91   1935

40934 SURFACE TENSION OF TITANIUM, ZIRCONIUM, AND HAFNIUM.
PETERSON A W   KEDESDY H   KECK P H   SCHWARZ E
J APPL PHYS
29   2   213-16   1958

40939 THE SURFACE TENSION OF LIQUID METALS AND ALLOYS.
PELZEL E
BERG HUTTENMAENN MONATSH MONTAN HOCHSCHULE LEOBEN
93   12   248-54   1948

40940 EFFECT OF ATMOSPHERE ON SURFACE TENSION OF GLASS.
PARIKH N M
J AM CERAM SOC
41   1   18-22   1958

40952 REFLECTANCE OF A SINGLE-LAYER DIELECTRIC COATING FOR
LIGHT INCIDENT AT AN ANGLE EQUAL TO OR LARGER THAN
THE CRITICAL ANGLE FOR TOTAL REFLECTION.
MINLOV I M
OPTIKA I SPEKTROSKOPIYA
21   5   624-9   1966   CA   66   41904

40953 REFLECTANCE OF A SINGLE-LAYER DIELECTRIC COATING FOR
LIGHT INCIDENT AT AN ANGLE EQUAL TO OR LARGER THAN
THE CRITICAL ANGLE FOR TOTAL REFLECTION. //ENGLISH
TRANSLATION OF OPTIKA I SPEKTROSKOPIYA 21/5/ 624-9,
1966.//
MINLOV I M
OPT SPECTRY
21   5   344-6   1966

40954 ANTIREFLECTION COATINGS FOR SILICON PHOTOCELLS.
KOLTUN M M   GOLOVNER T M
OPTIKA I SPEKTROSKOPIYA
21   5   630-7   1966   CA   66   42058

40955 ANTIREFLECTION COATINGS FOR SILICON PHOTOCELLS.
//ENGLISH TRANSLATION OF OPTIKA I SPEKTROSKOPIYA
21/5/ 630-7, 1966.//
KOLTUN M M   GOLOVNER T M
OPT SPECTRY
21   5   347-50   1966

40957 CONTRIBUTION TO THE STUDY OF THIN FILMS OF
SEMICONDUCTING MATERIALS USED IN THE PREPARATION OF
PHOTOCELLS.
PERROT M   DAVID J P   MARTINUZZI S
REVUE DE PHYSIQUE APPLIQUEE
1   3   164-72   1966

TPRC Number	Bibliographic Citation

40968 MEASURING THE SURFACE TENSION OF ELECTROLYTIC COPPER
BY THE METHOD OF MAXIMUM GAS BUBBLE PRESSURE.
YASHKICHEV V I   LAZAREV V B
IZV AKAD NAUK SSSR SER KHIM
1   170-2   1964

40985 APPARATUS FOR MEASURING THE SURFACE TENSION OF METALS
AT HIGH TEMPERATURES BY THE MAXIMUM PRESSURE GAS
BUBBLE METHOD.
FESENKO V V   EREMENKO V N
UKRAIN KHIM ZHUR
26   198-200   1960

40992 SURFACE TENSION AT ELEVATED TEMPERATURES. II.
EFFECT OF C, N, O AND S ON LIQUID IRON SURFACE
TENSION AND INTERFACIAL ENERGY WITH AL2O3.
HALDEN F A   KINGERY W D
J PHYS CHEM
59   557-9   1955

40996 INFLUENCE OF TEMPERATURE AND VIBRATION ON THE
VISCOSITY OF FREE FILMS AND SURFACE LAYERS IN
SOLUTIONS OF SODIUM DODECYL SULPHATE CONTAINING
ADDED DODECANOL.
TRAPEZNIKOV A A   DOKUKINA E S
ZH FIZ KHIM
40   7   1641-   1966

40997 INFLUENCE OF TEMPERATURE AND VIBRATION ON THE
VISCOSITY OF FREE FILMS AND SURFACE LAYERS IN
SOLUTIONS OF SODIUM DODECYL SULPHATE CONTAINING
ADDED DODECANOL. //ENGLISH TRANSLATION OF
ZH. FIZ. KHIM. 40 /7/ 1641-, 1966.//
TRAPEZNIKOV A A   DOKUKINA E S
RUSS J PHYS CHEM
40   7   888-90   1966

41003 A STUDY OF THE EMISSIVITY OF SOLID BODIES.
//ENGLISH TRANSLATION OF TEPLOENERGETIKA 13 /7/
67-71, 1966.//
MITOR V V   KONOPELKO I N
THERMAL ENGINEERING
13   7   92-7   1966

41019 DEVELOPMENT OF COATINGS TO PREVENT OVERHEATING OF
UNDERLYING SUBSTRATES ON EXPOSURE TO INTENSE THERMAL
RADIATION FOR SHORT PERIODS.
NAVAL AIR ENGINEERING CTR
PHILADELPHIA PA
DDC
NAEC-AML-1872
AD-433744   1-3   1964

41030 EFFECT OF PHOTODEGRADATION ON ATTENUATED TOTAL
REFLECTANCE SPECTRA OF ORGANIC COATINGS.
HEARST P J   U S NAVAL CIVIL ENG LAB   PORT HUENEME
CALIFORNIA
DDC AND CFSTI
AD-640733   1-26   1966

41075 INVESTIGATION OF THE PRODUCTS OF INTERACTION OF
ACTIVE SILICA WITH METAL IONS IN AMMONIACAL SOLUTION.
BULATOV M I   ALESKOVSKII V B
ZH PRIKL KHIM
39   2   284-8   1966

41076 INVESTIGATION OF THE PRODUCTS OF INTERACTION OF
ACTIVE SILICA WITH METAL IONS IN AMMONIACAL SOLUTION.
//ENGLISH TRANSLATION OF ZH. PRIKL. KHIM. 39 /2/
284-8, 1966.//
BULATOV M I   ALESKOVSKII V B
J APPL CHEM /USSR/
39   2   258-61   1966

41087 CAUSES OF THE COLORATION OF RED COPPER-CONTAINING
GLASSES.
AVGUSTINIK A I   ZHURAVLEV G I   VIGDERGAUZ V S
ZH PRIKL KHIM
39   4   934-5   1966

41088 CAUSES OF THE COLORATION OF RED COPPER-CONTAINING
GLASSES. //ENGLISH TRANSLATION OF ZH. PRIKL. KHIM.
39 /4/ 934-5, 1966.//
AVGUSTINIK A I   ZHURAVLEV G I   VIGDERGAUZ V S
J APPL CHEM /USSR/
39   4   876-7   1966

41095 INVESTIGATION OF THE CHEMICAL STABILITY OF
WATER-REPELLENT ORGANOSILICON COATINGS ON GLASS.
VORONKOV M G   PASHCHENKO A A   LASSKAYA E A
KARIBAEV K K
ZH PRIKL KHIM
39   6   1345-51   1966

41096 INVESTIGATION OF THE CHEMICAL STABILITY OF WATER-
REPELLENT ORGANOSILICON COATINGS ON GLASS. //ENGLISH
TRANSLATION OF ZH. PRIKL. KHIM. 39 /6/ 1345-51,
1966.//
VORONKOV M G   PASHCHENKO A A   LASSKAYA E A
KARIBAEV K K
J APPL CHEM /USSR/
39   6   1256-61   1966

41097 CHANGES IN THE INFRARED SPECTRA OF CERTAIN MODIFIED
CELLULOSE MATERIALS AFTER ULTRAVIOLET IRRADIATION AND
AFTER ISOLATION UNDER NATURAL CONDITIONS.
GUSEV S S   MAKHKAMOV K   ERMOLENKO I N
VIRNIK A D   ROGOVIN Z A
ZH PRIKL KHIM
39   6   1356-60   1966

41098 CHANGES IN THE INFRARED SPECTRA OF CERTAIN MODIFIED
CELLULOSE MATERIALS AFTER ULTRAVIOLET IRRADIATION AND
AFTER ISOLATION UNDER NATURAL CONDITIONS. //ENGLISH
TRANSLATION OF ZH. PRIKL. KHIM. 39 /6/ 1356-60,
1966.//
GUSEV S S   MAKHKAMOV K   ERMOLENKO I N
VIRNIK A D   ROGOVIN Z A
J APPL CHEM /USSR/
39   6   1266-9   1966

41113 SINGLE-PHONON ENERGY TRANSFER BETWEEN MOLECULAR BEAMS
AND SOLID SURFACES.
GADZUK J W
PHYS REV
153   3   759-63   1967

41114 MAXIMUM BUBBLE-PRESSURE METHOD FOR DETERMINING
SURFACE TENSION OF METALS OF THE IRON FAMILY.
FESENKO V V   EREMENKO V N
BYUL INST METALLOKERAM I SPETS SPLAVOV AKAD NAUK UKR
SSR
4   52-64   1959

41117 SESSILE DROP METHOD FOR MEASURING SURFACE TENSION OF
METALS OF THE IRON FAMILY.
EREMENKO V V   NIZHENKO V I   IVASHCHENKO YU N
BYUL INST METALLOKERAM I SPETS SPLAVOV AKAD NAUK UKR
SSR
4   65-71   1959

41118 USE OF THIN FILMS IN DETERMINING THE OPTICAL
CONSTANTS OF PBS FROM 1 TO 5 ELECTRON VOLTS.
WESSEL P R
PHYS REV
153   3   836-40   1967

41122 POLARIZATION OF INFRARED RADIATION BY BREWSTER ANGLE
REFLECTION FROM A GERMANIUM SURFACE. M.S. THESIS.
SCHROPP G E   AIR FORCE INST OF TECH AIR UNIV
WRIGHT-PATTERSON AFB   OHIO   USAF
DDC AND CFSTI
GE/EE/65-23
AD-623141   1-84   1965

41123 FURTHER DEVELOPMENT OF A THERMAL RADIATION CONTROL
COATING FOR USE ON SNAP VEHICLES.
KREDER K   MILLER E   HAGAN M   NORTH AMERICAN
AVIATION INC LOS ANGELES DIVISION
AEC
NA-62-622   1-31   1962

41128 TEMPERATURE-MODULATED REFLECTANCE OF GOLD FROM 2 TO
10 ELECTRON VOLTS.
SCOULER W J
PHYS REV LETTERS
18   12   445-8   1967

41136 EMERGING AEROSPACE MATERIALS AND FABRICATION
TECHNIQUES.
OLEVITCH A   WRIGHT-PATTERSON AFB   OHIO   USAF
DDC
AFML-TR-67-1   1-190   1967

41146 INTEGRATING-SPHERE RELECTOMETER FOR THE DETERMINATION
OF ABSOLUTE HEMISPHERICAL SPECTRAL REFLECTANCE.
ZERLAUT G A   KRUPNICK A C
AIAA J
4   7   1227-32   1966   JA   49-11 310

41149 PROGRAM FOR THE DEVELOPMENT OF SPACE RADIATOR
SURFACES /SNAP 2 APU/ - PART 1. DEVELOPMENT OF
PROTOTYPE SYSTEM.
MILLER E   CARROLL W   KLEMM R   NORTH AMERICAN
AVIATION INC LOS ANGELES DIVISION
AEC
NA-61-802   1-56   1961

41156 INFRARED COATING STUDIES. FINAL REPORT 18 FEB.-18
MAY 1965.
CHANG L   BAUSCH AND LOMB INC   ROCHESTER N Y
USDA
DDC
AD-464854   1-23   1965   TA   65-14 A227

41159 A 5-M INTEGRATING SPHERE.
NONAKA M   KASHIMA T   KONDO Y
APPL OPT
4   757-64   1967

41161 PREPARATION AND PROPERTIES OF SPUTTERED BISMUTH OXIDE
FILMS.
CLAPHAM P B
BRIT J APPL PHYS
18   3   363-6   1967

TPRC Number	Bibliographic Citation

41165 A FLAT-PLATE THERMAL-CONDUCTIVITY APPARATUS FOR MEASURING LOW-CONDUCTIVITY MATERIALS AT CRYOGENIC TEMPERATURES. FROM THERMAL CONDUCTIVITY MEASUREMENTS OF INSULATING MATERIALS AT CRYOGENIC TEMPERATURES.
HASKINS J F
ASTM
ASTM-STP-411
3-12   1966   IAA 7-22 3802

41168 A GUARDED-HOT-PLATE THERMAL-CONDUCTIVITY APPARATUS FOR MULTILAYER CRYOGENIC INSULATION. FROM THERMAL CONDUCTIVITY MEASUREMENTS OF INSULATING MATERIALS AT CRYOGENIC TEMPERATURES.
KARP G S   LANKTON C S
ASTM
ASTM-STP-411
13-24   1966   IAA 7-22 3802

41169 CRYOGENIC THERMAL CONDUCTIVITY MEASUREMENT OF INSULATING MATERIALS. FROM THERMAL CONDUCTIVITY MEASUREMENTS OF INSULATING MATERIALS AT CRYOGENIC TEMPERATURES.
COSTON R M   ZIERMAN C A
ASTM
ASTM-STP-411
25-42   1966   IAA 7-22 3802

41182 EFFECT OF CONVECTION IN HELIUM-CHARGED, PARTIAL FOAM INSULATIONS FOR LIQUID-HYDROGEN PROPELLANT TANKS.
TAYLOR B N   MACK F E
ADVAN CRYOG ENG 1965
11   65-76   1966   CA   65   12351

41210 A NONSTEADY-STATE THERMAL-CONDUCTIVITY TESTER FOR CRYOGENIC INSULATIONS. FROM THERMAL CONDUCTIVITY MEASUREMENTS OF INSULATING MATERIALS AT CRYOGENIC TEMPERATURES.
GIBBON N C   MATSCH L C   WANG D I J
ASTM
ASTM-STP-411
61-73   1966   IAA 7-22 3803

41217 AN IMPROVED GUARDED-COLD-PLATE THERMAL-CONDUCTIVITY APPARATUS. FROM THERMAL CONDUCTIVITY MEASUREMENTS OF INSULATING MATERIALS AT CRYOGENIC TEMPERATURES.
BLACK I A   WECHSLER A E   GLASER P E
FOUNTAIN J A
ASTM
ASTM-STP-411
74-94   1966   IAA 7-22 3803

41218 A SIMPLE MULTILAYER INSULATION CALORIMETER. FROM THERMAL CONDUCTIVITY MEASUREMENTS OF INSULATING MATERIALS AT CRYOGENIC TEMPERATURES.
DEHAAN J R
ASTM
ASTM-STP-411
95-109   1966   IAA 7-22 3803

41228 OPTICAL AND PHOTOELECTRIC PROPERTIES OF TELLURIUM IN THE FUNDAMENTAL ABSORPTION RANGE.
MERDY H
ANN PHYS /PARIS/
1 5/6   289-325   1966   CA   66   23421

41240 INFLUENCE OF NEUTRON RADIATION ON SOME PROPERTIES OF REFRACTORY CONCRETES.
DUBROVSKII V B   IBRAGIMOV SH SH   LADYGIN A YA
PERGAMENSHCHIK B K
AT ENERG /USSR/
21   2   108-12   1966   CA   65   16451

41250 THERMAL RESISTANCE OF GALLIUM ARSENIDE LASER DIODES.
CAPLAN S   GONDA T   LAMORTE M F   NYUL P
TRANS MET SOC AIME
239   3   403-8   1967

41266 AIR DRY THERMAL CONTROL COATINGS.
HOCKRIDGE R R   FOSTER C F
SYMPOSIUM MATERIALS SPACE VEHICLE USE
1-27   1963

41280 PHYSICOCHEMICAL PHENOMENA ACCOMPANYING DECONTAMINATION. II. MOISTENING OF YPERITE BY SOLUTIONS OF SURFACE-ACTIVE AGENTS.
SZCZUCKI E   DURKA K
BIUL WOJSK AKAD TECH IMENI JAROSLAWA DABROWSKIEGO
15 168   113-18   1966   CA   66   39068

41292 AUTOMATIC APPARATUS WITH SQUARE PLATE HEATER FOR DETERMINATION OF THE THERMAL CONDUCTIVITY OF INSULATING MATERIALS USED UP TO 1000 DEGREES.
FERRO V   SACCHI A
CALORE
37   4   135-49   1966   CA   65   13242

41302 THERMAL CONDUCTIVITY OF PACKED BEDS AND POWDER BEDS.
CHENG S C   VACHON R I
INTERN J HEAT MASS TRANSFER
12   9   1201-8   1969   CA   71   103552

41319 ANALYSIS OF THERMAL CONDUCTIVITY IN GRANULAR MATERIALS.
KRUPICZKA R
CHEMIA STOSOWANA
3   B   2   183-226   1966   CA   66-2 620

41334 DETERMINATION OF OPTICAL AND PHYSICAL PROPERTIES OF ARTIFICIAL SATELLITES BY PASSIVE GROUND-BASED PHOTOMETRY. FROM THERMAL DESIGN PRINCIPLES OF SPACECRAFT AND ENTRY BODIES. PROGRESS IN ASTRONAUTICS AND AERONAUTICS. VOL. 21.
PRESKI R J   GOODYEAR AEROSPACE CORP   AKRON OHIO
ACADEMIC PRESS   NEW YORK
447-68   1969
AIAA 3RD THERMOPHYSICS CONFERENCE
PAPER 68-742   1969

41336 A HELIUM DEWAR VESSEL WITHOUT NITROGEN COOLING.
ANASHKIN O P   DANILOV I B   KRIVENKO V G
CRYOGENICS
6   2   106-7   1966   CA   65   8367

41348 OPTICAL PROPERTIES OF THIN FILMS OF CALCIUM BETWEEN 2,300 AND 6,900 ANGSTROMS.
BLANC R   RIVOIRA R   ROUARD P
COMPT REND
264   B   8   634-7   1967

41373 REFLECTIVITY MEASUREMENTS ON MOLYBDENITE.
LEOW J H
ECON GEOL
61   3   598-612   1966   CA   65   6924
CORRECTION.
EALES HUGH V
62   1   151-3   1967   CA   67   23867

41375 EMISSION, TOTAL INTERNAL REFLECTION, AND TUNNELING OF THERMAL RADIATION IN METALS. FROM THERMAL DESIGN PRINCIPLES OF SPACECRAFT AND ENTRY BODIES. PROGRESS IN ASTRONAUTICS AND AERONATUTICS. VOL. 21.
CAREN R P   LIU C K   LOCKHEED PALO ALTO RESEARCH LAB   PALO ALTO CALIF
ACADEMIC PRESS   NEW YORK
509-30   1969
AIAA 3RD THERMOPHYSICS CONFERENCE
PAPER 68-773   1969

41386 CALORIMETER FOR MEASURING HEAT CAPACITIES OF DISPERSE MATERIALS AND ADSORPTION SYSTEMS BETWEEN 120 AND 300 K.   //ENGLISH TRANSLATION OF ZH. FIZ. KHIM. 38 /8/ 2106-10, 1964.//
BEREZIN G I   KISELEV A V   KOZLOV A A
RUSSIAN J PHYS CHEM
38   8   1145-8   1964

41389 CONTACT THERMAL RESISTANCE.
FOUCHE F
ENTROPIE
10   37-46   1966   CA   66   49808

41397 THE THERMAL CONTACT RESISTANCE AT GOLD FOIL SURFACES.
MOLGAARD J   SMELTZER W W
INT J HEAT MASS TRANSFER
13   7   1153-62   1970   CA   73   70530

41412 OPTICAL AND ENERGY PROPERTIES OF SUNLIGHT PROTECTING GLASSES WITH COBALT OXIDE COATING.
REKANT N B   BORISOVA I I
GELIOTEKHNIKA AKAD NAUK UZ SSR
3   42-7   1966   CA   65   15020

41413 SOME OPTICAL CHARACTERISTICS OF ELECTRICALLY CONDUCTIVE TIN OXIDE FILMS ON GLASS.
SHEKLEIN A V   REKANT N B   ZHUKOVSKAYA E A
YURKOVA S V   BAULINA M A
GELIOTEKH AKAD NAUK UZB SSR
4   57-63   1966   CA   66   33207

41418 GLASSES WITH SELECTIVE REFLECTION PROPERTIES.
SCHROEDER H
GLAS-EMAIL-KERAMO-TECH
17   5   161-5   1966   JA   49   265

41421 DEVELOPMENT OF PHASE-CHANGE COATINGS. FROM THERMAL DESIGN PRINCIPLES OF SPACECRAFT AND ENTRY BODIES. PROGRESS IN ASTRONAUTICS AND AERONAUTICS. VOL. 21.
GRIFFIN R N   LINDER B   GENERAL ELECTRIC COMP
KING OF PRUSSIA   PA
ACADEMIC PRESS   NEW YORK
559-74   1969
AIAA 3RD THERMOPHYSICS CONFERENCE
PAPER 68-776   1969

41427 EFFECTS OF TRANSIENT PRESSURE ENVIRONMENTS ON HEAT TRANSFER IN ONE-DIMENSIONAL COMPOSITE SLABS WITH CONTACT RESISTANCE. FROM THERMAL DESIGN PRINCIPLES OF SPACECRAFT AND ENTRY BODIES. PROGRESS IN ASTRONAUTICS AND AERONAUTICS. VOL. 21.
MOORE C J JR   BLUM H A
ACADEMIC PRESS   NEW YORK
621-36   1969
AIAA 3RD THERMOPHYSICS CONFERENCE
PAPER 68-780   1969

TPRC Number	Bibliographic Citation

41450 THE DETERMINATION OF THERMAL CONDUCTIVITY BY MEANS OF MELTING PHENOMENA.
BISHOP P H H   ROGERS K F   ROYAL AIRCRAFT ESTABLISHMENT
AEC AND CFSTI
RAE-TR-66328 N67-28806
AD-649698   1-18   1966   TA   67-11 130

41466 MATHEMATICAL MODEL FOR ESTIMATION OF EFFECTIVE DIFFUSIVITIES FOR HEAT AND MASS TRANSFER IN PACKED BEDS.
RAJAGOPALAN R   LADDHA G S
INDIAN CHEM ENGR
8   2   43-6   1966   CA   65   14866

41480 DEPENDENCE OF CONTACT THERMAL CONDUCTIVITY OF GRANULAR SYSTEMS ON EXTERNAL LOADING.
DULNEV G N   ZARICHNYAK YU P   MURATOVA B L   SIGALOVA Z V
INZH-FIZ ZH
11   2   202-6   1966   CA   66   30422

41482 APPLICATION OF A FLAT-PLATE CALORIMETER TO THE INVESTIGATION OF THERMAL PROPERTIES OF INSULATORS AT LOW TEMPERATURES.
MIKHALCHENKO R S   GERZHIN A G   PERSHIN N P
INZH-FIZ ZH
11   5   634-8   1966   CA   66   39101

41506 A BICALORIMETER IN THE SHAPE OF A PARALLELEPIPED WITH A THIN HEAT-INSULATING ENVELOPE.
AKHUNDOV S K   CULMANOV A A
IZV VYSSH UCHEB ZAVED PRIBOROSTR
5   126-8   1967

41516 DESIGN AND CONSTRUCTION OF AN INSTRUMENT FOR MEASURING THERMAL CONDUCTIVITY BY A STEADY-STATE METHOD. M.S. THESIS.
TURNER J H   UNIVERSITY OF MAINE   ORONO MAINE
UNIVERSITY OF MAINE
1-61   1964

41573 EFFECTS OF EXTREME ULTRAVIOLET ON THE OPTICAL PROPERTIES OF THERMAL CONTROL COATINGS. FROM THERMAL DESIGN PRINCIPLES OF SPACECRAFT AND ENTRY BODIES. PROGRESS IN ASTRONAUTICS AND AERONAUTICS. VOL. 21.
SWOFFORD D D   MANGOLD V L   JOHNSON S W
ACADEMIC PRESS   NEW YORK
667-95   1969   CA   72   113265
AIAA 3RD THERMOPHYSICS CONFERENCE   LOS ANGELES CALIF
PAPER 68-783   1968

41575 REFLECTANCE OF FILM-COVERED WATER SURFACES AS RELATED TO EVAPORATION SUPPRESSION.
BEARD J T   WIEBELT J A
J GEOPHYS RES
71   16   3835-41   1966   CA   65   8562

41589 ELECTRON-ULTRAVIOLET RADIATION EFFECTS ON THERMAL CONTROL COATINGS. FROM THERMAL DESIGN PRINCIPLES OF SPACECRAFT AND ENTRY BODIES. PROGRESS IN ASTRONAUTICS AND AERONAUTICS. VOL. 21.
BROWN R R   FOGDALL L B   CANNADAY S S   THE BOEING COMP   SEATTLE WASHINGTON
ACADEMIC PRESS   NEW YORK
697-724   1969   CA   72   113264
AIAA 3RD THERMOPHYSICS CONFERENCE   LOS ANGELES CALIF
PAPER 68-779   1968

41590 EFFECTS OF COMBINED ELECTRON-ULTRAVIOLET IRRADIATION ON THERMAL CONTROL COATINGS IN VACUO AT 77 K. FROM THERMAL DESIGN PRINCIPLES OF SPACECRAFT AND ENTRY BODIES. PROGRESS IN ASTRONAUTICS AND AERONAUTICS. VOL. 21.
MILES J K   CHEEVER P R   ROMANKO J   GENERAL DYNAMICS   FORTH WORTH TEXAS
ACADEMIC PRESS   NEW YORK
725-40   1969   CA   72   113263
AIAA 3RD THERMOPHYSICS CONFERENCE   LOS ANGELES CALIF
PAPER 68-781   1968

41593 THERMAL CONDUCTIVITIES OF PORTLAND CEMENT PASTE, AGGREGATE, AND CONCRETE DOWN TO VERY LOW TEMPERATURES.
LENTZ A E   MONFORE G E
J PCA RES DEVELOP LAB
8   3   27-33   1966   CA   66   31774

41599 OPTICAL PROPERTIES OF LITHIUM FLUORIDE IN THE EXTREME ULTRAVIOLET.
KATO R
J PHYS SOC JAPAN
16   12   2525-33   1961   JA   49   289

41603 THERMAL EXPANSION OF PAINT FILMS.
KONIG W
KUNSTSTOFF-TECHNIK
11   6   165-70   1941

41610 THE DEVELOPMENT OF S-13G-TYPE THERMAL-CONTROL COATINGS. FROM THERMAL DESIGN PRINCIPLES OF SPACECRAFT AND ENTRY BODIES. PROGRESS IN ASTRONAUTICS AND AERONAUTICS. VOL. 21.
ZERLAUT G A   ROGERS F O   NOBLE G   IIT RESEARCH INSTITUTE   CHICAGO ILLINOIS
ACADEMIC PRESS   NEW YORK
741-66   1969
AIAA 3RD THERMOPHYSICS CONFERENCE   LOS ANGELES CALIF
PAPER 68-790   1968

41612 A SIMPLE SPECULAR REFLECTANCE ATTACHMENT FOR INFRARED SPECTROMETERS.
ALLAM D S
J SCI INSTRUM
43   11   834-5   1966   CA   65   17904

41631 GLASS FIBER-REINFORCED POLYCARBONATES. II. ELECTRICAL AND THERMAL PROPERTIES.
STREIB H   OBERBACH K
KUNSTSTOFFE
56   2   100-4   1966   CA   65   7369

41664 SUPERCONDUCTIVITY IN METALS AND ALLOYS.
CHERRY W H   CODY G D   COOPER J L   GITTLEMAN J I
HANAK J J   MC CONVILLE G T   RAYL M   ROSI F D
RCA LABORATORIES   PRINCETON NEW JERSEY   USAF
DDC
ASD-TDR-62-269
AD-286456   1-86   1962

41679 HEAT TRANSMISSION THROUGH WALLS—II.
SHEARD H
J INST HEATING VENTILATING ENGRS
5   388-90   1937

41684 ON THE HEAT TRANSMISSION OF TEXTILE FABRICS.
NIVEN C D   BABBITT J D
J TEXTILE INST TRANSACTIONS
29   161-72   1938

41719 DETERMINATION OF THERMAL CONDUCTIVITY OF GRAPHITE-FILLED PLASTICS BASED ON EPOXY RESINS.
//ENGLISH TRANSLATION OF PLASTICHESKIE MASSY /3/ 41-3, 1965.//
GRABOI L P   LENSKAYA L P   CHUDNOVSKII A R
SOVIET PLASTICS
3   45-6   1965

41739 OPTICAL PROPERTIES OF THIN LAYERS OF VARIOUS MATERIALS IN THE VACUUM ULTRAVIOLET SPECTRAL REGION.
NOVIKOV V M
OPTIKA-MEKHAN PROM
33   3   38-45   1966   CA   65   6523

41744 RADIATION CAPACITY OF MOLYBDENUM DISILICIDE COATINGS.
ZHOROV G A   SIVAKOVA E V
TEPLOFIZ VYSOKIKH TEMPERATUR AKAD NAUK SSSR
4   2   182-8   1966   CA   65   1930

41745 EMISSIVITY OF MOLYBDENUM DISILICIDE COATINGS.
//ENGLISH TRANSLATION OF TEPLOFIZ. VYSOKIKH TEMPERATUR 4 /2/ 182-8, 1966.//
ZHOROV G A   SIVAKOVA E V
HIGH TEMPERATURE
4   2   180-6   1966

41764 DISPERSION OF THE PHASE JUMP DUE TO THE REFLECTION OF LIGHT FROM MULTILAYER DIELECTRIC COATINGS.
SHKLYAREVSKII I N   LUPASHKO E A
OPT SPEKTROSK
21   4   482-6   1966   CA   66-6 24115

41765 DISPERSION OF THE PHASE SHIFT DUE TO THE REFLECTION OF LIGHT FROM MULTILAYER DIELECTRIC COATINGS.
//ENGLISH TRANSLATION OF OPTIKA I SPEKTROSKOPIYA 21/4/ 482-6, 1966.//
SHKLYAREVSKII I N   LUPASHKO E A
OPT SPECTRY
21   4   269-71   1966

41767 CONDUCTANCE AND RELAXATION TIME OF ELECTRONS IN GOLD BLACKS FROM TRANSMISSION AND REFLECTION MEASUREMENTS IN THE FAR INFRARED.
HARRIS L   LOEB A L
J OPT SOC AM
43   11   1114-18   1953

41768 TOTAL EMITTANCE MEASUREMENTS OF REFRACTORY MATERIALS.
WADE W R   SLEMP W S
SPACE/AERONAUTICS
38   9   133-9   1962

41776 A PHASE TRANSITION AND DIELECTRIC ABSORPTION IN THE QUINOL METHANOL CLATHRATE COMPOUND.
MATSUO T   SUGA H   SEKI S
J PHYS SOC JAPAN
22   2   677-8   1967

41836 INFRARED REFLECTION SPECTRUM OF THIN-FILM SOLID BENZENE.
HOLLENBERG J L   GLOVER D E

TPRC Number	Bibliographic Citation

41839 EFFECTIVE THERMAL CONDUCTIVITY OF THE COPOLYMER MSN DISPERSED IN COMPRESSED GASES.
BRAGER N N   GOLUBEV I F
PLASTICHESKIE MASSY
7   49-51   1966   CA   65   15587

41849 EFFECT OF SURFACE ROUGHNESS ON THE REFLECTANCE OF REFRACTORY METALS.
STEVISON D F   MATERIALS ENGINEERING BRANCH
MATERIALS APPLICATIONS DIV AF MATERIALS LAB
WRIGHT-PATTERSON AFB   OHIO
DDC
AFML-TR-66-232
AD-812801   1-94   1966

41864 SYSTEM 400 COATING FOR THE PROTECTION OF COLUMBIUM.
HALL W B   APPLIED RESEARCH OPERATION FLIGHT
PROPULSION LAB
FLIGHT PROPULSION LABORATORY
DM-60-97   1-29   1960

41870 APPARATUS FOR DETERMINATION OF SURFACE TENSION OF LIQUID METALS.
BLIZNYUKOV S A   VISHKAREV A F   YAVOISKII V I
POVERKH YAVLENIYA RASPLAVAKH VOZNIKAYUSHCHIKH NIKH
TVERD FAZAKH NALCHIK
223-9   1965   CA   66   40946

41872 TEMPERATURE DEPENDENCE OF THE FREE SURFACE ENERGY OF LIQUID MAGNESIUM.
IVASHCHENKO YU N   EREMENKO V N   BOGATYRENKO B B
KHILYA G P
POVERKH YAVLENIYA RASPLAVAKH VOZNIKAYUSHCHIKH NIKH
TVERD FAZAKH NALCHIK
281-6   1965   CA   66   40250

41879 SURFACE TENSION OF IRON-PHOSPHORUS-OXYGEN MELTS.
VOLKOV S E   LEVENETS N P   SAMARIN A M
POVERKH YAVLENIYA RASPLAVAKH VOZNIKAYUSHCHIKH NIKH
TVERD FAZAKH NALCHIK
411-15   1965   CA   66   40029

41910 RESEARCH AND DEVELOPMENT ON ADVANCED GRAPHITE MATERIALS.   VOLUME IX.   FABRICATION AND PROPERTIES OF CARBONIZED CLOTH COMPOSITES.
BEASLEY W C   MC HENRY E R   PIPER E L
NATIONAL CARBON COMPANY UNION CARBIDE CORP
LAWRENCEBURG TENNESSEE   USAF
DDC
WADD-TR-61-72/9/
AD-802735   1-95   1964

41919 RESULTS FROM THE THERMAL CONTROL COATINGS EXPERIMENT ON OSO-III.   FROM   THERMAL DESIGN PRINCIPLES OF SPACECRAFT AND ENTRY BODIES.   PROGRESS IN ASTRONAUTICS AND AERONAUTICS.   VOL. 21.
MILLARD J P   NASA AMES RESEARCH CENTER   MOFFETT FIELD CALIF
ACADEMIC PRESS   NEW YORK
769-95   1969
AIAA 3RD THERMOPHYSICS CONFERENCE   LOS ANGELES CALIF
PAPER 68-794

41927 EFFECT OF THE PRESSURE, TEMPERATURE, AND INITIAL TREATMENT OF THE SEPARATING SURFACE ON THE THERMAL RESISTANCE OF CONTACT BETWEEN STAINLESS STEEL AND A LIQUID METAL.
BLUENVEN J   HUETZ J   VAUTREY L
REV GEN THERMIQUE
5   53   437-44   1966   CA   65   8453

41932 THERMAL CONDUCTANCE OF LAP-JOINTS IN VACUUM.
OSBORN A B   MAIR W N   ROYAL AIRCRAFT
ESTABLISHMENT MINISTRY OF AVIATION
DDC AND AEC
RAE-TR-66034
AD-637770   1-45   1966

41934 ENVIRONMENTAL STUDIES OF THERMAL CONTROL COATINGS FOR LUNAR ORBITER.   FROM   THERMAL DESIGN PRINCIPLES OF SPACECRAFT AND ENTRY BODIES.   PROGRESS IN ASTRONAUTICS AND AERONAUTICS.   VOL. 21.
SLEMP W S   HANKINSON T W E   NASA LANGLEY
RESEARCH CENTER   HAMPTON VA
ACADEMIC PRESS   NEW YORK
797-817   1969
AIAA 3RD THERMOPHYSICS CONFERENCE   LOS ANGELES CALIF
PAPER 68-792   1968

41945 THERMAL CONTROL EXPERIMENTS ON THE LUNAR ORBITER SPACECRAFT.   FROM   THERMAL DESIGN PRINCIPLES OF SPACECRAFT AND ENTRY BODIES.   PROGRESS IN ASTRONAUTICS AND AERONAUTICS.   VOL. 21.
CALDWELL C R   NELSON P A   THE BOEING COMPANY
SEATTLE WASHINGTON
ACADEMIC PRESS   NEW YORK
819-52   1969
AIAA 3RD THERMOPHYSICS CONFERENCE   LOS ANGELES CALIF
PAPER 68-793   1968

41953 VARIATION OF MERCURY VAPOR PRESSURE IN DISCHARGE TUBE WITH IGNITOR AFTER INITIATION OF CATHODE SPOT.
GOSHO Y

41986 THE PHYSICOCHEMICAL PROPERTIES OF MOLTEN SLAGS IN ELECTRO-SMELTING OF COPPER.
BAKYRDZHIEV P N   VANYUKOV A V
SB MOSK INST STALI I SPLAVOV
41   407-19   1966   CA   65   5088

41988 THERMAL ACCOMMODATION COEFFICIENTS FROM HETEROGENEOUS VAPOR-SOLID NUCLEATION DATA.
GRETZ R D
SURFACE SCI
5   2   261-2   1966   CA   66   14266

41989 METHOD OF CALCULATING THE OPTICAL CONSTANTS OF ABSORBING THIN FILMS FROM MEASURED VALUES OF REFLECTANCE AND TRANSMITTANCE.
ABELES F   THEYE M L
SURFACE SCI
5   3   325-31   1966   CA   66   33292

42004 INVESTIGATION OF THE THERMAL INSULATING PROPERTIES OF KAOLIN FIBERS.
KRZHIZHANOVSKII R E   CHUDNOVSKAYA I II
TEPLOFIZIKA VYSOKIKH TEMPERATUR
4   3   355-9   1966   CA   65   10301

42012 THERMAL CONDUCTIVITY AND EMISSIVITY OF ALUMINUM OXIDE COATINGS AT HIGH TEMPERATURES.
ZHOROV G A   KOVALEV A I   SIVAKOVA E V
TEPLOFIZ VYSOKIKH TEMPERATUR
4   5   643-8   1966
CORRECTION.
PETROV V A
6   3   573-4   1968

42017 HEAT TRANSFER THROUGH VACUUM-MULTILAYER INSULATION.
KAGANER M G   VELIKANOVA M G   FETISOVA L I
TEPLO- I MASSOPERENOS
4   143-53   1966   CA   66   20415

42042 INTERFACIAL TENSION AT THE INTERFACE OF FE-P ALLOYS WITH MNO-SIO2-P2O5 MELTS.
GOGIBERIDZE YU M   KEKELIDZE M A   MIKIASHVILI SH M
TR GRUZ INST MET
14   97-100   1965   CA   66   13267

42071 HEAT TRANSPORT THROUGH CARBON RADIO RESISTORS AT LOW TEMPERATURES AND THEIR USE IN CONJUCTION WITH SUPERCONDUCTING THERMAL VALVES.
PANDORF R C   CHEN C Y   DAUNT J G
CRYOGENICS
2   4   239-42   1962

42092 THE OPTICAL PROPERTIES OF INDIUM, GALLIUM AND THALLIUM.
BOR J   BARTHOLOMEW C
PROC PHYS SOC /LONDON/
90 /PT 4/   1153-7   1967

42096 FORMULA FOR THERMAL ACCOMMODATION COEFFICIENTS.
GOODMAN F O   WACHMAN H Y
J CHEM PHYS
46   6   2376-86   1967

42109 INFRARED SPECTRA AND SPECTRAL SHIFTS OF CARBON MONOXIDE ADSORBED ON EVAPORATED ALKALI-HALIDES. PART I.   FILMS OF SODIUM SALTS.
GEVIRZMAN R   KOZIROVSKI Y   FOLMAN M
TRANS FARADAY SOC
65   8   2206-14   1969

42114 MECHANISM OF ALCOHOL DECOMPOSTION OVER ALUMINA.   A DYNAMIC TREATMENT OF CHEMISORBED SPECIES DURING THE COURSE OF THE REACTION.
SOMA Y   ONISHI T   TAMARU K
TRANS FARADAY SOC
65   8   2215-23   1969

42148 LIMITATIONS IN THERMAL SIMILITUDE.   MONTHLY PROG REPT NO. 13.   JULY 1-AUG. 1, 1968.
LOCKHEED MISSILES AND SPACE COMPANY
PALO ALTO CALIFORNIA
SUNNYVALE CALIF
NASA AND LMSC
LMSC-681555 TP-2699
1-27   1968

42151 OPTICAL PROPERTIES AND STRUCTURE OF THIN FILMS OF LITHIUM FLUORIDE.
BARBAROUX N   BOURG A   BOURG M   PORRECA F
VIDE
21   75-9   1966   CA   65   4847

42169 COMPARATIVE STUDY OF CHARACTERISTICS OF INDIGENOUS AND IMPORTED TRANSPARENT CELLULOSE FILMS.
SHAH C J   SRIVASTAVA S K
SILK RAYON IND INDIA
9   1   12-15   1966   CA   65   847

42185 INTERACTION OF LIGHT WITH A PLANT CANOPY.
ALLEN W A   RICHARDSON A J
J OPT SOC AM
58   8   1023-8   1968

TPRC Number	Bibliographic Citation

42204 PLASTIC FIBER OPTICS. II. LOSS MEASUREMENTS AND LOSS MECHANISMS.
BROWN R G   DERICK B N
APPL OPT
7   8   1565-9   1968

42237 EFFECT OF ORDERING ON THE INFRARED SPECTRUM OF AMMONIUM CHLORIDE.
GARLAND C W   SCHUMAKER N E
J PHYS CHEM SOLIDS
28   .   5   799-803   1967

42258 PROPERTIES OF GOLD LAYERS.
SINTSOV V N
ZH PRIKL SPEKTROSKOPII AKAD NAUK BELORUSSK SSR
4   6   503-8   1966   CA   65   14650

42262 MULTILAYER DIELECTRIC LEAD OXIDE AND CRYOLITE COATINGS.
NABOIKIN YU V   KRAMERENKO N L   AKOPOV V M
ZH PRIKL SPEKTROSKOPII AKAD NAUK BELORUSSK SSR
5   2   153-7   1968   CA   66-6 21842

42270 FURTHER STUDIES ON LITHIUM FLUORIDE-OVERCOATED ALUMINUM MIRRORS WITH HIGHEST REFLECTANCE IN THE VACUUM ULTRAVIOLET.
COX J T   HASS G   WAYLONIS J E
APPL OPT
7   8   1535-9   1968

42284 HANDBOOK OF PHYSICAL CONSTANTS. THERMAL EXPANSION. VISCOSITY. THERMAL CONDUCTIVITY.
SKINNER B J   CLARK S P JR
GEOL SOC AM MEM
97   75-96
                                              1966   CA   65   4087

42290 DEVELOPMENT OF SPACE-STABLE THERMAL-CONTROL COATINGS.
ZERLAUT G A   FIRESTONE R F   JAMISON W E
GEORGE C MARSHALL SPACE FLIGHT CENTER   HUNTSVILLE ALABAMA
NASA
IITRI-C6014-26 NASA-CR-67250
                                    1-53   1965

42291 ELECTROCHEMICAL CONVERSION COATINGS ON ALUMINUM.
HULTQUIST A E   LOCKHEED MISSILES AND SPACE COMPANY   SUNNYVALE CALIF
NASA
NASA-CR-74534 X66-16931
                                    1-55   1964

42292 STABLE WHITE COATINGS. INTERIM REPT. NO. 1 SEPT. 22, 1961 - APR. 1, 1962.
TOMPKINS E H   ARMOUR RESEARCH FOUNDATION ILLINOIS INST OF TECH   CHICAGO ILL
NASA AND CFSTI
ARF-3207-5 NASA-CR-50814
N63-19854   1-98   1962

42294 ALUMINA CERAMICS.
GITZEN W H   OHIO STATE UNIV RESEARCH FOUNDATION COLUMBUS   USAF
DDC AND CFSTI
AFML-TR-66-13
AD-480064   1-821   1966   RR   69-10 102

42295 ULTRAVIOLET STABILITY OF SOME MODIFIED METAL PHOSPHATES FOR THERMAL-CONTROL SURFACES. PRESENTED AT THE SOCIETY AERO-SPACE MATERIALS PROCESSES ENGINEERS, SEATTLE, WASHINGTON, NOVEMBER 18-20, 1963.
PEZDIRTZ G F   WAKELYN N T
NASA
N65-88851   1-20   1963

42306 THE EFFECT OF MOISTURE ON THE THERMAL CONDUCTIVITY OF INSULATING MATERIALS. M.S. THESIS.
BEYERSDORFF L E   MICHIGAN COLLEGE OF MINING AND TECHNOLOGY   HOUGHTON MICHIGAN
MICHIGAN COLLEGE OF MINING AND TECHNOLOGY
                                    1-39   1949

42312 OPTICAL PROPERTIES OF THIN FILMS OF BARIUM BETWEEN 2,300 AND 6,000 ANGSTROMS.
BLANC R   RIVOIRA R
COMPT REND
264   B   13   983-5   1967

42321 ABSORPTION AND REFLECTION SPECTRA, IN THE FAR INFRARED, OF ZN SE, ZN TE, AND CD SE AT LOW TEMPERATURE.
HADNI A   CLAUDEL J   STRIMER P
PHYS STAT SOLIDI
26   1   241-52   1968   CA   68   91541

42325 OPTICAL CONSTANTS OF MAGNESIUM FOR THE REGION 1100 -2500 ANGSTROMS.
PRIOL M   DAUDE A   ROBIN S
COMPT REND
264   B   12   935-8   1967

42332 EXPERIMENTAL AND THEORETICAL INVESTIGATION OF THE INFLUENCE OF NATURAL CONVECTION IN WALLS WITH SLAB-TYPE INSULATION.
LORENTZEN G   NESJE R
BULL INST INT FROID ANNEXE-2.
                                  115-25   1900

42351 EFFECT OF MICROPARTICLE IMPACT ON THE OPTICAL PROPERTIES OF METALS.
GANNON R E   LASZLO T S   LEIGH C H   WOLNIK S J
AIAA J
3   11   2096-103   1965   CA   64   10581

42354 ELLIPSOMETRY MEASUREMENTS AT AND BELOW MONOLAYER COVERAGE.
SMITH T
J OPT SOC AM
58   8   1069-79   1968   CA   69   69920

42359 PHYSICOCHEMICAL BEHAVIOR OF SOME REFRACTORY REINFORCED PHENOLICS.
MELNICK A M   FLORENTINE R A   TANZILLI R A   COHEN L
AM CHEM SOC DIV ORG COATINGS PLASTICS CHEM PREPRINTS
24   2   170-3   1964   CA   64   12895

42368 DETERMINATION OF EMISSIVITIES WITH A DIFFERENTIAL SCANNING CALORIMETER.
ROGERS R N   MORRIS E D JR
ANAL CHEM
38   3   410-12   1966   CA   64   18520

42373 THEORETICAL AND EXPERIMENTAL RESEARCHES ON THERMAL CONDUCTIVITY OF CONCRETES.
MISSENARD F A
ANN INST TECH BATIMENT TRAV PUBL
18   949-68   1965   CA   64   15557

42374 THERMAL PROPERTIES OF POSTULATED LUNAR SURFACE MATERIALS.
GLASER P E   WECHSLER A E   GERMELES A E
ANN N Y ACAD SCI
123   2   856-70   1965   CA   64   5782

42392 RELATION BETWEEN THERMAL CONDUCTIVITY AND OXIDE CONCENTRATIONS IN SODIUM.
KOZLOV F A   ANTONOV I N
AT ENERG
19   4   391-2   1965   CA   64   13420

42424 INFLUENCE OF POROSITIES AND AIR SPACES ON THE THERMAL CONDUCTIVITIES OF ENGINEERING MATERIALS.
RYUTANI M
BULL TOKYO INST TECHNOL
66   95-107   1965   CA   64   4621

42440 THERMAL CONDUCTIVITY OF WET-WALL LIQUID HYDROGEN STORAGE TANK INSULATIONS FOR SPACE APPLICATIONS.
RAWUKA A C   YUNDT C G
CHEM ENG PROGR SYMP SER
62   61   219-24   1966   CA   65   565

42451 OPTICAL PROPERTIES OF POTASSIUM FOR PHOTONS OF ENERGY 3.96 TO 9.69 EV.
SUTHERLAND J C   ARAKAWA E T
J OPT SOC AM
58   .   8   1080-3   1968

42482 LIGHT FILTERS FOR THE ULTRAVIOLET REGION OF THE SPECTRUM, MADE FROM THIN LAYERS OF ALKALI METALS.
GURZADYAN G A   NOVIKOV V M
DOKL AKAD NAUK ARM SSR
42   1   15-18   1966   CA   65   3191

42504 HEAT CONDUCTIVITY OF INSULATION MATERIALS IN VACUO.
KAGANER M G   GLEBOVA L I
KISLOROD
12   1   13-18   1959

42505 DIRECTIONAL SOLAR ABSORPTANCE OF TEMPERATURE-CONTROL SURFACE FINISHES. FROM JPL SPACE PROGRAMS SUMMARY NO. 37-37, VOL. IV.
HALL W M   JET PROPULSION LAB   PASADENA CALIF
NASA
JPL-SPS-37-37
4   81-3   1966

42529 THERMAL ACCOMMODATION OF THE NOBLE GASES ON BARE SURFACES OF LITHIUM AND POTASSIUM. PH.D. THESIS.
DELMORE J E   UNIV OF MISSOURI   COLUMBIA MO
UNIV MICROFILMS PUBL
67-13847   1-91   1967   CA   68   72603

42531 THERMAL CONDUCTIVITY OF PELLETIZED CERAMIC PACKINGS.
KHARLAMOV A G
INZHEN-FIZ ZH
9   1   48-53   1965   CA   64   3178

42535 THERMAL PROPERTIES OF CHAMOTTE CERAMICS WITHIN THE TEMPERATURE RANGE 80 - 1200 K.
VASILEV L L   FRAIMAN YU E
INZH FIZ ZH AKAD NAUK BELORUSSK SSR
9   6   762-7   1965   CA   64   15540

TPRC Number	Bibliographic Citation						
42558	INTEGRAL METHOD OF THERMAL CONDUCTIVITY MEASUREMENT OF CYLINDRICAL SEMICONDUCTOR SPECIMENS IMMERSED IN A HELIUM BATH. RAFALOWICZ J ACTA PHYS POL						
	30	2	205-22	1966	CA	67	37511
42590	OPTICAL MEASUREMENT OF SEVERAL MATERIALS IN THE FAR INFRARED REGION. MITSUISHI A YOSHINAGA H YATA K MANABE A JAPAN J APPL PHYS SUPPL /1964/						
	4	1	581-7	1965	CA	64	4448
42593	MEASUREMENTS REPORT. THERMAL PROPERTY MEASUREMENTS OF MANNED SPACECRAFT CENTER SPACESUIT MATERIALS. TURNBOW F J TRW SYSTEMS REDONDO BEACH CALIFORNIA NASA AND CFSTI NASA-CR-92133						
	N68-25365		1-3	1968	RR	68-16 54	
42596	HEAT TRANSFER ACROSS SURFACES IN CONTACT. TRANSIENT EFFECTS OF AMBIENT TEMPERATURES AND PRESSURES. BLUM H A MOORE C J JR SOUTHERN METHODIST UNIV DALLAS TEXAS NASA NASA-CR-57137						
	N65-18445		1-11	1964			
42599	THE REFLECTION AND THERMAL ACCOMMODATION OF HELIUM BEAMS ON PLATINUM. MOORE G E DATZ S TAYLOR E H J CATALYSIS						
	5	12	218-23	1966	CA	64	16689
42612	MEASUREMENT OF THE REFLECTANCE OF AN ELECTRODE SURFACE. KOCH D F A SCAIFE D E J ELECTROCHEM SOC						
	113	3	302-5	1966	CA	64	15367
42628	THERMAL CONDUCTIVITY OF BRAZED EXPANDED PYROLYTIC GRAPHITE TO 700 C. ALLEN R D AM CERAM SOC BULL						
	47	8	717-21	1968			
42635	THE TOTAL HEMISPHERICAL EMITTANCE OF PLUTONIUM AT 89 C. KARLSSON R H J NUCL MATER						
	19	1	79-80	1966	CA	67	84979
42639	THE THERMAL CONDUCTIVITY OF PITCH-BONDED GRAPHITES. KRSTIC DJ PERME T NUKLEARNI INSTITUT JOZEF STEFAN LJUBLJANA /YUGOSLAVIA/ AEC AND CFSTI NIJS-R-517 N88-25094						
			1-20	1968	RR	68-16 177	
42657	THERMOPHYSICAL PROPERTIES OF REINFORCED PLASTICS AT CRYOGENIC TEMPERATURES. CAMPBELL M D HASKINS J F O BARR G L HERTZ J J SPACECRAFT ROCKETS						
	3	4	596-9	1966	CA	64	19886
42684	CHARACTERISTICS OF INTERFERENCE IN THIN GRANULAR SILVER FILMS COVERED WITH ALUMINUM. //ENGLISH TRANSLATION OF OPT. SPEKTROSK. 27 /5/ 840-4, 1969.// SHKLYAREVSKII I N KORNEEVA T I USOSKIN A I SHKLYAREVSKII O I OPT SPECTRY						
	27	5	458-60	1969			
42685	HIGH-REFLECTIVITY PAINTS. CHOWDHRY K K LABDEV /KANPUR INDIA/						
	3	4	219-22	1965	CA	64	11438
42688	THERMAL CONSTRICTION RESISTANCE DUE TO NON-UNIFORM SURFACE CONDITIONS. CONTACT RESISTANCE AT NON-UNIFORM INTERFACE PRESSURE. MIKIC B B INTERN J HEAT MASS TRANSFER						
	13	9	1497-500	1970			
42690	SOME DATA ON THE EFFECT OF ADDED OXYGEN ON HEAT EMISSION AS SODIUM FLOWS THROUGH A COOLED PIPE. ANDREEV A S FEDOROVICH E D SHCHEDRIN A V ZHIDKIE METALLY SBORNIK STATEI N P						
			109-13	1963			
42691	SOME DATA ON THE EFFECT OF ADDED OXYGEN ON HEAT EMISSION AS SODIUM FLOWS THROUGH A COOLED PIPE. //ENGLISH TRANSLATION OF ZHIDKIE METALLY SBORNIK STATEI, N. P., P. 109-13, 1963.// ANDREEV A S FEDOROVICH E D SHCHEDRIN A V DDC AND CFSTI FTD-TT-65-1511 TT-66-61103 AD-631581						
			1-12	1966			

TPRC Number	Bibliographic Citation						
42695	THE DEVELOPMENT, FABRICATION, AND EVALUATION OF A DEVICE FOR THE MEASUREMENT OF THERMAL PROPERTIES OF SOIL. HSU S T KAO H S VIRGINIA POLYTECHNIC INST BLACKSBURG DDC AND CFSTI AD-632080						
			1-250	1966	TA	66-12 60	
42708	DEVELOPMENT OF THERMAL CONDUCTIVITY PROBES FOR SOILS AND INSULATIONS. WECHSLER A E LITTLE /ARTHUR D/ INC CAMBRIDGE MASS DA DDC AND CFSTI CRREL-TR-182 N67-22531 AD-645337						
			1-113	1966	TA	67-5 49	
42716	STUDIES, RESEARCH AND INVESTIGATIONS OF THE OPTICAL PROPERTIES OF THIN FILMS OF METALS SEMI-CONDUCTORS AND DIELECTRICS. HADLEY L N COLORADO STATE UNIV FORT COLLINS DA DDC AND CFSTI AD-648429						
			1-5	1966	TA	67-9 107	
42737	THE THERMAL EQUILIBRIUM OF WIRES IN HOT GAS STREAMS. BASCOMBE K N EXPLOSIVES RES AND DEVELOPMENT ESTABLISHMENT WALTHAM ABBEY /ENGLAND/ DDC ERDE-11/M/66 N68-17305 AD-804163						
			1-28	1966	TA	67-3 A-72	
42738	THE HIGH TEMPERATURE EVALUATION OF AEROSPACE MATERIALS. WURST J C CHERRY J A GERDEMAN D A HECHT N L DAYTON UNIV OHIO RES INST USAF DDC AFML-TR-66-308 AD-804192						
			1-158	1966	TA	67-3 A-33	
42740	TARGET SIGNATURE MEASUREMENTS LABORATORY STUDY. ANDRYCHUK D TEXAS INSTRUMENTS INC DALLAS SCIENCE SERVICES DIV USAF DDC AD-805346						
			1-24	1966	TA	67-5 A-85	
42759	EFFECTS OF VARIOUS GASES ON HANDGEAR INSULATION. HALL J F JR BUEHRING W J STROBL W W AEROSPACE MEDICAL DIV AEROSPACE MEDICAL RESEARCH LABS /657TH/ WRIGHT-PATTERSON AFB OHIO DDC AND CFSTI AMRL-TR-65-4 N66-22473 AD-628367						
			1-18	1965	TA	66-7 31	
42764	EVALUATION OF INFRARED EMISSION OF CLOUDS AND GROUND AS MEASURED BY WEATHER SATELLITES. INTERIM REPORT. KERN C D AIR FORCE CAMBRIDGE RESEARCH LABS AND TERRESTRIAL SCIENCES LAB BEDFORD MASS AEC DDC AND CFSTI AFCRL-65-840 N66-22701 AD-629081						
			1-111	1965	TA	66-8 6	
42780	DETERMINATION OF THE EMISSIVITY OF MATERIALS. SEMIANNUAL PROGRESS REPORT, MAY 15-NOV. 14, 1965. LUOMA W EMANUELSON R C PRATT AND WHITNEY AIRCRAFT EAST HARTFORD CONN NASA AND CFSTI NASA-CR-54891 N66-17436 PWA-2750						
			1-46	1965	PA	N66-4-8 1238	
42781	DEVELOPMENT OF OPTICAL COATINGS FOR CDS THIN FILM SOLAR CELLS. FINAL REPORT. SCHAEFER J C HILL E R HARSHAW CHEMICAL CO CLEVELAND OHIO NASA AND CFSTI NASA-CR-54965						
	N66-25373		1-63	1965	CA	66	109746
42783	INVESTIGATION OF LIGHT SCATTERING IN HIGHLY REFLECTING PIGMENTED COATINGS. VOL. 2. CLASSICAL INVESTIGATIONS, THEORETICAL AND EXPERIMENTAL FINAL REPORT, 1 MAY-30 SEP. 1966. ZERLAUT G A KATZ S RAZIUNAS V JACKSON M IIT RES INST CHICAGO ILL NASA AND CFSTI NASA-CR-61166 IITRI-U6003-19 VOL 2 N67-19075						
	N67-19075		1-107	1966	PA	N67-5-8 1327	
42814	THERMAL RADIATION ABSORPTION OF METALS AND LAYERED INSULATIONS AT 80 K. COLYER B NATL INST FOR RES IN NUCLEAR SCIENCE HARWELL /ENGLAND/ RUTHERFORD HIGH ENERGY LAB AEC NIRL-R-83						
		N66-18788	1-16	1965	CA	68	6255
42824	FLAT PANEL VACUUM THERMAL INSULATION. STRONG H M BUNDY F P BOVENKERK H P J APPLIED PHYSICS						
	31	1	39-50	1960			
42827	SURFACE TENSION AND MOLECULAR CORRELATIONS NEAR THE CRITICAL POINT. WIDOM B						

TPRC Number	Bibliographic Citation

42840 MEASUREMENT OF THE ACCOMMODATION COEFFICIENTS OF MOLECULES OF VERY HIGH ENERGY AND RESULTS.
DEVIENNE F M
INTERN J HEAT MASS TRANSFER
10   8   1109-19   1967

42849 THERMAL CONTACT RESISTANCE IN VACUUM.
JENG D R
J HEAT TRANSFER TRANS ASME
89   C   3   275-6   1967

42851 CONTRAST ENHANCEMENT OF THE TRANSVERSE KERR EFFECT.
HUNT R P
J APPL PHYS
38   3   1215-16   1967

42858 PLANE LAYER TYPE APPARATUS FOR GAS THERMAL CONDUCTIVITY MEASUREMENTS.
TEAGAN W P   SPRINGER G S
REV SCI INSTR
38   3   335-9   1967

42867 OPTICAL ABSORPTION SPECTRA OF THE ALKALI HALIDES AT 10 K.
TEAGARDEN K   BALDINI G
PHYS REV
155   3   896-907   1967

42876 ON THE ELECTROADSORPTION OF PICRATE OCTADECYLTRI-METHYLAMMONIUM.
DUPEYRAT M   MICHEL J
COMPT REND
264   C   15   1240-3   1967

42885 FLASH EVAPORATION AND THIN FILMS OF CUPROUS SULFIDE, SELENIDE, AND TELLURIDE.
ELLIS S G
J APPL PHYS
38   7   2906-12   1967

42889 OPTICAL PROPERTIES OF DILUTE SILVER-INDIUM ALLOYS IN THE ULTRAVIOLET. PH.D. THESIS.
MORGAN R M   AMES LABORATORY IOWA STATE UNIVERSITY AMES IOWA
AEC
IS-T-187   1-103   1967
UNIV MICROFILMS PUBL
68-5967   1-108   1967

42890 REFLECTANCE AND TRANSMITTANCE OF EVAPORATED ALUMINUM AND ALUMINUM-MAGNESIUM FLUORIDE FILMS IN THE ULTRAVIOLET /GREATER THAN 1800 ANGSTROMS/.
BATES B   BRADLEY D J
J OPT SOC AM
57   4   481-5   1967   CA   67   37992

42893 A SESSILE DROP TEST TO STUDY WETTING BY LIQUID METALS IN METALLIC SYSTEMS. M.S. THESIS.
MAZE R C   AMES LABORATORY IOWA STATE UNIVERSITY AMES IOWA
AEC AND CFSTI
IS-T-205   1-60   1967   NSA 22-3 490

42894 NORMAL-INCIDENCE RELECTANCE OF ALUMINUM FILMS IN THE WAVELENGTH REGION 800-2000 ANGSTROMS.
VEHSE R C   ARAKAWA E T   STANFORD J L
J OPT SOC AM
57   4   551-2   1967   CA   67   86254

42907 A SIMPLE TRANSIENT-FLOW METHOD OF MEASURING THERMAL CONDUCTIVITY AND DIFFUSIVITY.
BALL E F
J REFRIGERATION
10   3   65-71   1967

42908 THE THERMAL AND ACOUSTIC PROPERTIES OF LIGHT-WEIGHT CONCRETES.
LOUDON A G   STACY E F
STRUCTURAL CONCRETE
3   58-76   1966

42914 EVALUATION OF LENTICULAR SPIRAL ARRAY.
STIMLER F J   GOODYEAR AEROSPACE CORPORATION AKRON OHIO   USAF
DDC
RADC-TR-66-61 GER-12369
AD-480596   1-202   1966

42916 AN INVESTIGATION OF THERMAL INSULATING MATERIALS FOR UNDERSEA HABITATS. INTERIM REPT. 1 JUL-15 DEC. 65.
TAYLOR L B   NAVY MINE DEFENSE LAB   PANAMA CITY FLA
DDC AND CFSTI
AD-630936   1-20   1966   TA   66-10 58

42921 MEASUREMENT OF HEMISPHERICAL TOTAL EMITTANCE OF A COATING.
FAUGERE J F   DELPONT J P   CENTRE SPATIAL DE BRETIGNY   BRETIGNY-SUR-ORGE /ESSONNE/
CNES
CNES-NT-4   N66-30190
                   1-23   1966

42939 EMISSIVITY AND ELECTRICAL RESISTIVITY OF FLAME-SPRAYED AL2O3 AND ZRO2 COATINGS AT HIGH TEMPERATURE.
REISS F
BER DEUT KERAM GES
43   7   477-8   1966   CA   67   26935

42955 THERMOPHYSICAL PROPERTIES OF A TWO-PHASE DISPERSED SYSTEM SATURATED WITH DIFFERENT LIQUIDS. //ENGLISH TRANSLATION OF INZH. FIZ. ZH. 12 /1/ 38-42, 1967.//
VERZHINSKAYA A B   VAINBERG V SH
J ENGINEERING PHYS
12   1   20-2   1967

42966 THE OPTICS OF COVELLINE.
VON GEHLEN K   PILLER H
MINING MAG
113   6   438-45   1965   CA   64   12046

42974 DESIGN AND DEVELOPMENT OF AN EM WINDOW FOR AIR LIFT REENTRY VEHICLES.
POULOS N E   MURPHY C A   HARRIS J N   WOLF J M
ENGINEERING EXPERIMENT STATION GEORGIA INSTITUTE OF TECHNOLOGY   ATLANTA GEORGIA   USAF
DDC
AD-465541   1-35   1965

42975 TEST OF APCO 1260.
GARZA J J   AUTONETICS   DOWNEY CALIFORNIA
DDC
AD-470031   1-7   1965

42976 EXPERIMENTS IN THE IDENTIFICATION OF PAINTS BY ATTENUATED TOTAL REFLECTANCE.
MC GOWAN R J   U S NAVAL CIVIL ENGINEERING LAB PORT HUENEME CALIF
DDC
AD-615951   1-7   1965

42977 THE LOCKSPRAY-GOLD PROCESS.
LEVY D J   LOCKHEED MISSILES AND SPACE CO   PALO ALTO CALIF
DDC
AD 632177   1-38   1966

42978 RECENT DEVELOPMENTS IN THE FIELD OF HIGH-GLOSS ALUMINUM.
TRAGNER E   KAPPEL G   INST OF MODERN LANGUAGES INC   WASHINGTON D C
DDC AND CFSTI
AERDL-T-1852 AD 635649
                   1-17   1966

42999 REFLECTANCES OF FREE AND BONDED WHITE PAINT FILMS.
KUDVA A K   WILLIAMS G C
OFFIC DIG J PAINT TECHNOL ENG
38 494   156-62   1966   CA   64   17860

43021 OPTICAL PROPERTIES AND STRUCTURE OF THIN LAYERS OF LANTHANUM FLUORIDE.
BOURG A   BARBAROUX N   BOURG M
OPT ACTA
12   2   151-60   1965   CA   64   5926

43022 THE RELATION BETWEEN THE THERMAL CONDUCTIVITY AND THE STRUCTURE OF ICE.
SHULEIKIN V V   RUSANOV N I   RYABCHIKOV V A
ZH GEOFIZIKI
1 1/2   179-86   1931

43023 DEPENDENCE OF THE THERMAL CONDUCTIVITY OF PURE ICE AND POROUS ICE, CONTAINING CARBON-DIOXIDE GAS, FROM THE PRESSURE.
VLASSOV L J   USPENSKI P N
ZH GEOFIZIKI
1 1/2   187-92   1931

43024 INFLUENCE OF THE OPTICAL PROPERTIES ON THE QUALITY OF METALLIC INTERFERENCE FILTERS.
ANDERS H   EICHINGER R
OPTIK
23   4   350-61   1965   CA   64   16821

43043 OPTICAL PROPERTIES OF NONCRYSTALLINE SEMICONDUCTORS.
TAUC J   ABRAHAM A   PAJASOVA L   GRIGOROVICI R   VANCU A
PHYS NON-CRYSTALLINE SOLIDS PROC INTERN CONF DELFT NETH 1964
                   606-15   1965   CA   64   7838

43052 A COMPARISON OF THE THERMAL CONDUCTIVITY OF MOIST CAPILLARY-POROUS MATERIALS AT TEMPERATURES BELOW AND ABOVE 0 C. //ENGLISH TRANSLATION OF INZH. FIZ. ZH. 12 /1/ 68-71, 1967.//
PAK N V
J ENGINEERING PHYS
12   1   36-7   1967

43056 GLASS-TO-METAL SEALS.
HULL A W   BURGER E E
PHYSICS
5   384-405   1934

TPRC Number	Bibliographic Citation						

43057 EXPERIMENTAL STUDIES ON THE SURFACE TENSION OF MOLTEN METALS AND ALLOYS.
MONMA K   SUTO H
TRANS JAPAN INST OF METALS

| | 1 | | 69-76 | 1960 | | | |

43063 COEFFICIENTS OF THE CUBICAL EXPANSION OF ICE, HYDRATED SALTS, SOLID CARBONIC ACID, AND OTHER SUBSTANCES AT LOW TEMPERATURES.
DEWAR J
PROC ROYAL SOC /LONDON/

| | 70 | | 237-46 | 1902 | | | |

43078 TEMPERATURE-INTERFACIAL TENSION STUDIES OF SOME 1-ALKENES AGAINST WATER.
JASPER J J   DUNCAN J C
J CHEM ENG DATA

| | 12 | 2 | 257-9 | 1967 | | | |

43079 DETERMINATION OF THE HEAT CONDUCTIVITY OF GLASS REINFORCED PLASTICS ACCORDING TO THE STRUCTURE AND PROPERTIES OF THEIR COMPONENTS.
SHLENSKII O F
PLASTICHESKIE MASSY

| | 12 | 33-7 | 1965 | CA | 64 | 8388 |

43080 CONTACT GLASS FABRIC-REINFORCED MATERIAL VPS-4.
KRAVCHENKO L I   AVRASIN YA D
PLASTICHESKIE MASSY

| | 1 | 52-6 | 1966 | CA | 64 | 11384 |

43092 HEATING OF THE COVER LAYER DURING SINTERING.
KOSTORNOV A G
POROSHKOVAYA MET AKAD NAUK UKR SSR

| | 6 | 1 | 72-5 | 1966 | CA | 64 | 13827 |

43097 THERMOPHYSICAL PROPERTIES OF BUILDING MATERIALS AT LOW TEMPERATURES.
VASILEV L L
PROBL STROIT TEPLOFIZ /MINSK VYSSH SHKOLA/ SB

| | | 407-13 | 1965 | CA | 64 | 15557 |

43098 QUASI-STATIONARY METHOD FOR COMPLEX DETERMINATION OF THE THERMOPHYSICAL CHARACTERISTICS OF BUILDING MATERIALS.
FRAIMAN YU E
PROBL STROIT TEPLOFIZ /MINSK VYSSH SHKOLA/ SB

| | | 419-26 | 1965 | CA | 64 | 15557 |

43100 STRUCTURE AND PROPERTIES OF PYROLYTIC CARBONS PREPARED IN A FLUIDIZED BED BETWEEN 1900 AND 2400 C.
PRICE R J   BOKROS J C   KOYAMA K   CHIN J
CARBON

| | 4 | | 263-72 | 1966 | | | |

43102 DETERMINING THE SPECTRAL RADIATION PROPERTIES /REFLECTING AND ABSORPTIVE POWER/ OF MATERIALS AS A FUNCTION OF TEMPERATURE.
SCHADACH P
ELEKTROWARME

| | 23 | 1 | 8-15 | 1965 | | | |

43112 SPECTRAL AND DIRECTIONAL THERMAL RADIATION CHARACTERISTICS OF SELECTIVE SURFACES FOR SOLAR COLLECTORS.
EDWARDS D K   GIER J T   NELSON K E   RODDICK R D
PROC U N CONF NEW SOURCES ENERGY ROME 1961

| | 4 | | 536-53 | 1964 | CA | 65 | 1810 |

43123 EXPERIMENTERS DESIGN HANDBOOK FOR THE MANNED LUNAR SURFACE PROGRAM.
FUCHS R A   HUGHES AIRCRAFT COMPANY   CULVER CITY CALIF
NASA
CHAPTER   7   7-1 - 7-114
CHAPTER   10   10-1 - 10-36 1967

43124 HEAT TRANSFER BETWEEN A LIQUID METAL AND THE SOLID METAL CRUST FORMED ON A WATER-COOLED COPPER WALL.
SOKOLOV L A
INZH FIZ ZH

| | 12 | 1 | 92-8 | 1967 | | | |

43128 OPTICAL PROPERTIES OF THINLY SLICED SOLIDS IN THE FAR ULTRAVIOLET REGION.
FABRE D
REV OPT

| | 43 | | 393-424 | 1964 | CA | 64 | 16841 |

43129 THE ANOMALOUS OPTICAL PROPERTIES OF THIN ALUMINUM FILMS IN ULTRA-HIGH VACUUM.
BLANC R   RIVOIRA R
REV OPT

| | 44 | 4 | 170-4 | 1965 | CA | 64 | 5953 |

43131 OPTICAL PROPERTIES OF THIN LAYERS OF HIGH-MELTING METALS UNDER VACUUM.
MOUTTET C   GRAVIER P
REV OPT THEOR INSTR

| | 43 | | 245-54 | 1964 | CA | 64 | 15195 |

43140 HEAT TRANSFER BETWEEN A LIQUID METAL AND THE SOLID METAL CRUST FORMED ON A WATER-COOLED COPPER WALL. //ENGLISH TRANSLATION OF INZH. FIZ. ZH. 12 /1/ 92-98, 1967.//
SOKOLOV L A
J ENGINEERING PHYS

| | 12 | 1 | 51-4 | 1967 | | | |

43152 ABSOLUTE MEASUREMENT OF THE THERMAL ACCOMMODATION OF GAS MOLECULES ON SOLID SURFACES AND DETERMINATION OF EVAPORATION VELOCITIES.
NIESLER R A   STRANSKI I N
Z PHYSIK CHEM

| | 27 | | 357-71 | 1961 | | | |

43153 INVESTIGATION OF THE INFRARED NORMAL MODES OF BINARY OXIDES /BEO, MGO, CAO, ZNO/.
TOLKSDORF S
ZEIT PHYSIK CHEM

| | 132 | | 161-84 | 1928 | | | |

43177 WETTING OF SOLID-METAL SURFACES BY MOLTEN METALS.
WASSINK R J K
J INST METALS

| | 95 | 2 | 38-43 | 1967 | | | |

43179 HIGH-TEMPERATURE INFRARED SPECTROSCOPY OF OLEFINS ADSORBED ON FAUJASITES.
EBERLY P E JR
J PHYS CHEM

| | 71 | 6 | 1717-22 | 1967 | | | |

43185 OPTICAL PROPERTIES OF SODIUM IN THE VACUUM ULTRAVIOLET.
SUTHERLAND J C   ARAKAWA E T   HAMM R N
J OPT SOC AM

| | 57 | 5 | 645-50 | 1967 | | | |

43207 HIGHER DIMENSIONAL CRYSTAL MODELS THEORY OF THERMAL ACCOMMODATION COEFFICIENTS.
CHAMBERS C M   KINZER E T
SURFACE SCI

| | 4 | 1 | 33-47 | 1966 | CA | 65 | 111 |

43295 MEASUREMENT OF THERMAL DIFFUSIVITY, HEAT CAPACITY, AND THERMAL CONDUCTIVITY IN TWO-LAYER COMPOSITE SAMPLES BY THE FLASH METHOD. FROM   PROCEEDINGS OF THE 5TH CONFERENCE ON THERMAL CONDUCTIVITY, DENVER HILTON HOTEL.   DENVER, COLORADO, OCTOBER 20-22, 1965. VOLUME 1.
LARSON K B   KOYAMA K   GENERAL ATOMIC DIVISION GENERAL DYNAMICS CORPORATION   SAN DIEGO CALIFORNIA UNIVERSITY OF DENVER

| | | 5 | 1-23 | 1965 | | | |

43296 PREPARATION AND STUDY OF THE FUNDAMENTAL PROPERTIES OF THIN FILMS OF GALLIUM ARSENIDE.
BOURGEOIS P   MOCH P
VIDE
20 119

| | | | 376-87 | 1965 | CA | 64 | 9031 |

43297 AN EXPERIMENTAL INVESTIGATION OF THE DIRECTIONAL EMISSIVITY CHARACTERISTICS OF FERROUS METALS.   M.S. THESIS.
TALBERT S G   UNIVERSITY OF MINNESOTA
UNIVERSITY OF MINNESOTA

| | | | 1-62 | 1959 | | | |

43309 COEFFICIENT OF VOLUME EXPANSION FOR PETROLEUM WAXES AND PURE N-PARAFFINS.
TEMPLIN P R
IND ENG CHEM

| | 48 | 1 | 154-61 | 1956 | | | |

43310 THE SURFACE TENSION OF LIQUID NICKEL-SILICON ALLOYS.
VASILIU M I   EREMENKO V N
POROSHKOVAYA MET AKAD NAUK UKR SSR

| | | 3 | 80-2 | 1965 | | | |

43349 TRUE ADSORPTION AND ACCOMMODATION OF SEVERAL GASES ON GLASS SURFACES.
GERSTACKER H   SCHAFER K
Z ELEKTROCHEM

| | 59 | 10 | 1023-9 | 1955 | | | |

43357 ADSORPTION, PARTICAL THERMAL ACCOMMODATION OF GASES ON SURFACES AND THEIR RELATION TO CATALYTIC ACTIVITY.
SCHAFER K   GERSTACKER H
Z ELEKTROCHEM

| | 60 | 8 | 874-87 | 1956 | | | |

43360 DETERMINATION OF OPTICAL PROPERTIES OF OBJECTS WITH ARBITRARY SCATTERING DIAGRAMS ON AN INTEGRAL PHOTOMETER.
RVACHEV V P   SAKHNOVSKII M YU
ZH PRIKL SPEKTROSKOPII AKAD NAUK BELORUSSK SSR

| | 4 | 2 | 172-4 | 1966 | CA | 65 | 1608 |

43368 THERMAL CONDUCTIVITY OF PLASTIC FOAMS -423 TO 75 F.
HASKINS J F   HERTZ J
ADV IN CRYOGENIC ENG

| | 7 | | 353-9 | 1961 | | | |

TPRC Number	Bibliographic Citation			
43383	PLASMA RESONANCE ABSORPTION OF POTASSIUM. BRAMBRING J Z PHYSIK			
	200	2	186-93	1967
43384	OPTICAL AND ELECTRICAL BEHAVIOR OF METALS ON THE BASIS OF MEASUREMENTS ON VERY THIN LAYERS, OBTAINED BY EVAPORATION. V. COMPARISON OF THEORY AND MEASUREMENT IN THE CASE OF OPTICAL ABSORPTION OF THIN DROPLET-SHAPED LAYERS OF SILVER. INFLUENCE OF THE CHANGE OF THE CRITICAL VALUES CAUSED BY GEOMETRICAL LIMITING OF THE CARRIER PATH-LENGTHS AND AT COOLING. FLECHSIG W PHYSIKALISCH TECHNISCHEN BUNDESANSTALT			
	19	2	1-	1967
	Z PHYSIK			
	200	3	304-31	
43388	THE EFFECTS OF PREOXIDATION TREATMENTS ON THE SPECTRAL NORMAL AND TOTAL NORMAL EMITTANCE OF INCONEL, INCONEL-X, AND TYPE 347 STAINLESS STEEL AT TEMPERATURES OF 900, 1200, 1500, AND 1800 F. SLEMP W S NASA LANGLEY RESEARCH CENTER LANGLEY STATION HAMPTON VA NASA NASA-TM-X-51016			
	N65-88898		1-30	1964
43408	SURFACE TENSION BY PENDANT DROP TECHNIQUE. FLINT O U K AT ENERGY AUTHORITY HARWELL ENGL UKAEA AND CFSTI AERE-R-4782 N65-21952			
			1-26	1964 CA 65 10270
43413	OPTICAL PROPERTIES AND SURFACE ACTIVITIES OF THIN METALLIC FILMS DEPOSITED IN ULTRA HIGH VACUA. M.S. THESIS. KHIM J H VIRGINIA POLYTECHNIC INST BLACKSBURG VA VIRGINIA POLYTECHNIC INST			
			1-67	1965
43414	SOME PRINCIPLES UNDERLYING THE SUCCESSFUL USE OF METALS AT HIGH TEMPERATURES. FAHRENWALD F A AMERICAN SOC TESTING MATERIALS PROC			
	24		310-47	1924
43415	THEORY AND APPLICATIONS OF DIFFUSE REFLECTANCE SPECTROSCOPY. COMPANION A L DEVELOP APPL SPECTRY			
	4		221-34	1965
43421	OXIDATION RESISTANT COATING PROGRAM FOR COLUMBIUM IN APOLLO AND TRANSTAGE EXIT CONES. PETERSON L E DDC AD-439102			
			1-64	1963
43422	SHIELDED CERAMIC COMPOSITE STRUCTURE. INTERIM ENGR. PROGRESS REPT. JUNE 1964-AUG. 1964. KUMMER D L ADVANCED FABRICATION TECHNIQUES BRANCH RES AND TECH DIVISION AIR FORCE SYSTEMS COMMAND WRIGHT-PATTERSON AIR FORCE BASE OHIO USAF DDC AD-447207			
			1-42	1964
43423	NEW SELECTIVELY ABSORBING COATINGS FOR SOLAR POWERED THERMOELECTRIC DEVICES. LONG R L 1ST LT/USAF AF AND TECH DIV AIR FORCE SYSTEMS COMMAND WRIGHT-PATTERSON AIR FORCE BASE OHIO USAF DDC ML-TDR-64-302 AD 455304			
			1-12	1964
43439	CURRENT SATURATION IN THE EVAPORATED GA AS FILMS. YAMASHITA A TUZAKI T YAMADA T YAMAUCHI I J APPL PHYS			
	38	5	2359-61	1967
43447	INFRARED LATTICE VIBRATIONAL SPECTRA OF AGCL, AGBR, AND AGI. BOTTGER G L GEDDES A L J CHEM PHYS			
	46	8	3000-4	1967
43452	ON THE DETERMINATION OF THERMAL ACCOMMODATION COEFFICIENTS IN THE TEMPERATURE-JUMP REGION. HARRIS R E J CHEM PHYS			
	46	8	3217-20	1967
43456	STUDY OF A THERMOPHOTOVOLTAIC CONVERTER. GM DEFENSE RES LABS SANTA BARBARA CALIF DA DDC GM-DRL-TR-65-03 AD 454940			
			1-34	1965

TPRC Number	Bibliographic Citation			
43457	STUDY OF A THERMOPHOTOVOLTAIC CONVERTER. FINAL REPORT III. GM DEFENSE RES LABS SANTA BARBARA CALIF DDC GM-DRL-TR-65-23 AD-462029			
			1-39	1965
43458	EFFECT OF ENVIRONMENTAL EXPOSURE ON MECHANICAL PROPERTIES OF SEVERAL FOIL GAGE REFRACTORY ALLOYS AND SUPERALLOYS. KERR J R COX J D AIR FORCE MATERIALS LAB RES AND TECH DIV AIR FORCE SYSTEMS COMMAND WRIGHT-PATTERSON AIR FORCE BASE OHIO DDC AFML-TR-65-92 AD 464905			
			1-146	1965
43464	ACCOMMODATION COEFFICIENT, COEFFICIENT OF CAPTURE AND VELOCITY OF DESORPTION OF AN ATOM ON A METALLIC LATTICE /ONE-DIMENSIONAL/. ARMAND G COMPT REND			
	264	B 17	1221-4	1967
43468	INVESTIGATION OF THE THERMOPHYSICAL PROPERTIES OF CHAMOTTE CERAMICS IN THE 80-1200 K RANGE. //ENGLISH TRANSLATION OF INZHEN.-FIZ. ZH. 9 /6/ 762-7, 1965.// VASILEV L L FRAIMAN YU E J ENG PHYS			
	9	6	467-9	1965
	CFSTI OR ETC TT-66-61303 N66-19881			
			1-6	1966
43470	THE SOLAR REFLECTANCE AND/OR HEMISPHERICAL EMITTANCE OF SELECTED SPACE SUIT MATERIALS. BEVANS J T T R W SYSTEMS GROUP REDONDO BEACH CALIF NASA AND CFSTI TRW-5324-6001-SU-000 NASA-CR-65244			
	N66-21000		1-232	1965 CA 66 66556
43471	INTEGRAL GLASS COATINGS FOR SOLAR CELLS. FROM PROC. 5TH PHOTOVOLTAIC SPECIALISTS CONF., VOL. 1, DEC. 1965. ILES P A HOFFMAN ELECTRONICS CORP EL MONTE CALIF NASA NASA AND CFSTI PIC-SOL-209/6 NASA-CR-70168			
	N66-17317	B-4 1-19		1965 PA N66-4-8 11
43475	INVESTIGATION OF THE EFFECT OF MATERIAL PROPERTIES ON COMPOSITE ABLATIVE MATERIAL BEHAVIOR. THIRD QUARTERLY REPORT, DEC. 11, 1965-MAR. 10, 1966. SCHULTZ F E GEN ELEC CO PHIL PA MISSILE AND SPACE DIV NASA AND CFSTI NASA-CR-71295			
	N66-20897		1-30	1966 PA N66-4-10 1689
43479	THERMAL RESISTANCE MEASUREMENT TECHNIQUES STUDY PROGRAM. FINAL REPORT. GREER P H DAVIDSON K W MOTOROLA INC SEMICONDUCTOR PRODUCTS DIV PHOENIX ARIZ NASA AND CFSTI NASA-CR-71745			
	N66-24586		1-197	1966 CA 67 5950
43480	THERMAL EXPANSION TESTS ON AGGREGATES, NEAT CEMENTS, AND CONCRETES. MITCHELL L J AMERICAN SOC TESTING MATERIALS PROC			
	53		963-75	1953
43481	EMISSIVITY COATINGS FOR LOW-TEMPERATURE SPACE RADIATORS. QUARTERLY PROGRESS REPORT, 1 JAN.-31 MAR. 1966. SMITH F J GRAMMER J G LOCKHEED MISSILES AND SPACE CO PALO ALTO CALIF AEROSPACE SCIENCES LAB NASA AND CFSTI NASA-CR-72060			
	N67-14914		1-37	1966 PA N67-5-5 726
43482	EMISSIVITY COATINGS FOR LOW-TEMPERATURE SPACE RADIATORS. QUARTERLY PROGRESS REPT. 5, 1 JULY-30 SEP. 1966. GRAMMER J R LOCKHEED MISSILES AND SPACE CO SUNNYVALE CALIF AEROSPACE SCIENCES LAB NASA AND CFSTI NASA-CR-72130			
	N67-16568		1-41	1966 PA N67-5-6 982
43483	DETERMINATION OF THE EMISSIVITY OF MATERIALS. SEMIANNUAL PROGRESS REPT., 15 MAY-14 NOV. 1966. LUOMA W L EMANUELSON R C PRATT AIRCRAFT EAST HARTFORD CONN NASA AND CFSTI PWA-2979 NASA-CR-72149			
	N67-16569		1-28	1966 PA N67-5-6 982

TPRC Number	Bibliographic Citation			
43484	LOW SOLAR ABSORPTANCE AND EMITTANCE SURFACES UTILIZING VACUUM DEPOSITED TECHNIQUES. FINAL REPT., 29 JUN. 1965-28 SEP. 1966. LOCKHEED MISSILES AND SPACE CO RES LAB PALO ALTO CALIF NASA AND CFSTI NASA-CR-73039			
	N67-17182	1-82	1966	PA	N67-5-7 1090

| 43489 | RESEARCH ON THERMAL TRANSFER PHENOMENA. FINAL REPT., JAN. 1-DEC. 31, 1965. ALLEN W C NATL BERYLLIA CORP HASKELL N J NASA AND CFSTI NASA-CR-74409 | | | |
| | N66-23512 | 1-82 | 1965 | CA | 66 107874 |

| 43491 | SOME INFLUENCES OF MACROSCOPIC CONSTRICTIONS ON THE THERMAL CONTACT RESISTANCE. CLAUSING A M ILLINOIS UNIV URBANA ENG EXPERIMENT STATION NASA AND CFSTI MF-TN-242-2 NASA-CR-74622 | | | |
| | N66-24595 | 1-53 | 1965 | PA | N66-4-13 2477 |

| 43493 | AN INVESTIGATION OF THE THERMAL RADIATION PROPERTIES OF CERTAIN SPACECRAFT MATERIALS. FINAL REPT. BEVANS J T BROWN G L LUEDKE E E MILLER W D NELSON K E RUSSELL D A SPACE TECH LABS INC REDONDO BEACH CALIF NASA AND CFSTI STL-8633-6014-SU-000 NASA-CR-74772 | | | |
| | N66-24938 | 1-104 | 1962 | PA | N66-4-13 2480 |

| 43494 | MEASURED DIRECTIONAL REFLECTANCES OF CHARS IN THE UV, VISIBLE, AND NEAR INFRARED. LITTLE /ARTHUR D/ INC CAMBRIDGE MASS NASA AND CFSTI NASA-CR-75128 C-66658 | | | |
| | N66-26660 | 1-79 | 1966 | PA | N66-4-14 2649 |

| 43495 | HIGH-PERFORMANCE THERMIONIC CONVERTERS. FINAL REPT., 19 MAY 1965-19 JUL. 1966. CAMPBELL A E ELECTRO-OPTICAL SYSTEMS INC PASADENA CALIF NASA AND CFSTI NASA-CR-79388 EOS-6952 | | | |
| | N67-11843 | 1-181 | 1966 | PA | N66-5-2 176 |

| 43517 | CADMIUM SULFIDE. /DATA SHEETS./ NEUBERGER M HUGHES AIRCRAFT CO ELECTRONIC PROPERTIES INFORMATION CENTER CULVER CITY CALIF DDC EPIC-DS-124/2E AD-810354 | | | |
| | | 1-256 | 1967 | TA | 67-10 A-165 |

| 43519 | BORON. /DATA SHEET./ MILEK J T WELLES S H HUGHES AIRCRAFT CO ELECTRONIC PROPERTIES INFORMATION CENTER CULVER CITY CALIF USAF DDC AD-810423 | | | |
| | | 1-251 | 1967 | TA | 67-10 A-15 |

| 43525 | STABLE WHITE COATINGS. SUMMARY TECHNICAL REPORT, 3 JUL. 1964-1 DEC. 1965. GILLIGAN J E IIT RESEARCH INST CHICAGO ILL NASA AND CFSTI NASA-CR-70780 N66-20053 IITRI-U6004-21 | | | |
| | | 1-69 | 1966 | PAN 66-4-10 1734 |

| 43527 | STUDY OF THE EFFECTS OF JPL STERILIZATION TECHNIQUES ON THERMAL CONTROL SURFACES. BLAIR P M JR HUGHES AIRCRAFT CO CULVER CITY CALIF NASA AND CFSTI NASA-CR-70838 N66-18443 | | | |
| | | 1-25 | 1965 | PAN 66-4-8 1246 |

| 43528 | INFRARED TESTING OF ELECTRONIC COMPONENTS, PHASE II, 8 JULY 1965-14 JANUARY 1966. CHADDERDON G HARTMAN T MARTIN CO ORLANDO FLA NASA AND CFSTI NASA-CR-71014 OR-8031 | | | |
| | N66-20139 | 1-69 | 1966 | PAN 66-4-10 1632 |

| 43530 | EXPERIMENTAL INVESTIGATION OF TOTAL EMITTANCE AND SOLAR ABSORPTANCE OF SEVERAL COATINGS BETWEEN 300 AND 575 K. DIEDRICH J H CURTIS H B NATL AERONAUTICS AND SPACE ADMINISTRATION LEWIS RES CTR CLEVELAND OHIO NASA AND CFSTI NASA-TN-D-3381 | | | |
| | N66-21038 | 1-41 | 1966 | CA | 66 77048 |

| 43546 | HEAT TRANSFER FROM THE URANIUM FUEL TO THE MAGNOX CAN IN A GAS-COOLED REACTOR. SANDERSON P D INTERN DEVELOP HEAT TRANSFER ASME | | | |
| | | 53-64 | 1961 |

| 43547 | THERMAL CONTACT BELOW 1 K. HART H R JR WHEATLEY J C | | | |

43548	CONTACT HEAT TRANSFER BETWEEN PLANE METAL SURFACES. DYBAN E F SHVETS I T INZHEN FIZ ZH			
	7	3	3-9	1964
	CORRECTION. TSAPLIN M I			
	9	4	539-41	1965
	REPLY SHVETS I T DYBAN E F			
	9	4	542-3	1965

| 43552 | HEAT EXCHANGE BETWEEN CONTACTING PARTS. DYBAN E P KONDAK N M SHVEP I T IZV AKAD NAUK SSSR OTDEL TEKH NAUK | | | |
| | 9 | | 63-70 | 1954 |

| 43566 | INVESTIGATION OF GLASS-METAL COMPOSITE MATERIALS. SECOND ANNUAL PROGRESS REPT. MARCH 1956-MARCH 1957. AILES H B GLASS-METALS RESEARCH LAB OWENS-CORNING FIBERGLAS CORP NEWARK OHIO DDC AD-153297 | | | |
| | | 1-34 | 1957 |

| 43583 | CRYOSTAT FOR MEASUREMENT OF HEAT CONDUCTION UNDER MECHANICAL LOADS. THOMAS T R PROBERT S D WARMAN D J SCI INSTR | | | |
| | 41 | | 88-91 | 1964 |

| 43591 | THERMAL CONDUCTIVITY MEASUREMENTS ON INERT MEDIA USING INTERNAL CALORIMETRY. HAIGH A L NATURE /LONDON/ | | | |
| | 173 | | 493-4 | 1954 |

| 43597 | ON THE MEASUREMENT OF REFLECTION, PERMEABILITY, AND ABSORPTION ON TEST SUBSTANCES OF OPTIONAL FORM IN THE ULBRICHT SPHERE-TYPE PHOTOMETER. TINGWALDT C P OPTIK | | | |
| | 9 | 7 | 323-32 | 1952 |

| 43599 | STUDY OF SPECTRAL REFLECTION COEFFICIENTS OF THERMAL RADIATION RECEIVING COATINGS FOR THE WAVELENGTH RANGE FROM 10-200 MICRONS. KOZYREV B P KROPOTKIN M A OPTIKA I SPEKTROSKOPIYA | | | |
| | 10 | 5 | 657-62 | 1961 |

| 43603 | A STUDY OF TRANSPARENT HIGHLY CONDUCTING GOLD FILMS. GILLHAM E J PRESTON J S WILLIAMS B E PHIL MAG | | | |
| | 46 | | 1051-68 | 1955 |

| 43605 | HEAT LOSS FROM THE NUDE BODY AND PERIPHERAL BLOOD FLOW AT TEMPERATURES OF 22 TO 35 C. HARDY J D SODERSTROM G F J NUTRITION | | | |
| | 16 | | 493-510 | 1938 |

| 43611 | THE REFLECTION POWER OF BLACK SURFACES. ROYDS T PHYSIK Z | | | |
| | 11 | | 316-19 | 1910 |

| 43616 | THERMAL BOUNDARY RESISTANCE BETWEEN SOME SUPERCONDUCTING AND NORMAL METALS. BARNES L J DILLINGER J R PHYS REV LETT | | | |
| | 10 | 7 | 287-9 | 1963 |

| 43634 | THERMAL CONTACT RESISTANCE. //ENGLISH TRANSLATION OF ATOMNAYA ENERGIYA 9 /6/ 496-8, 1960.// SHLYKOV YU P GANIN E A SOVIET J AT ENERGY | | | |
| | 9 | 6 | 1041-3 | 1961 |

| 43636 | RAPID METHOD FOR THE DETERMINATION OF THE THERMAL CONDUCTIVITY OF INSULATING MATERIALS. BETTANINI E CAVALLINI A DI PHILIPPO P TERMOTECNICA | | | |
| | 19 | 10 | 566-71 | 1965 |

| 43637 | THE THERMAL CONDUCTIVITY OF INSULATING AND OTHER MATERIALS. TAYLOR T S TRANS ASME | | | |
| | 41 | | 605-21 | 1919 |

| 43646 | THE REFLECTION FACTOR OF MAGNESIUM OXIDE. PRESTON J S TRANS OPT SOC / LONDON/ | | | |
| | 31 | | 15-35 | 1929 |

| 43647 | THE PROPERTIES OF SOME REACTIVELY SPUTTERED METAL OXIDE FILMS. HOLLAND L SIDDALL G VACUUM | | | |
| | 3 | 4 | 375-91 | 1953 |

| 43655 | THE OPTICAL CONSTANTS AND ELECTRICAL RESISTANCE OF THICK METAL FILMS. WEISS K | | | |

TPRC Number	Bibliographic Citation

43657 OPTICAL PROPERTIES OF ALUMINUM FILMS, VAPOR DEPOSITED IN HIGH VACUUM.
WALKENHORST W
Z TECH PHYSIK
22   1   14-21   1941

43670 REACTION BETWEEN STEARIC ACID AND CALCIUM IONS AT THE AIR-WATER INTERFACE USING SURFACE VISCOMETRY.
ENEVER R P   PILPEL N
TRANS FARADAY SOC
63   3   781-92   1967

43684 OPTICAL PROPERTIES OF MAGNESIUM FLUORIDE IN THE VACUUM ULTRAVIOLET.
WILLIAMS M W   MAC RAE R A   ARAKAWA E T
J APPL PHYS
38   4   1701-5   1967

43686 OPTICAL PROPERTIES OF EPITAXIAL PBS FILMS IN THE ENERGY RANGE 2-6 EV.
ROSSI C E   PAUL W
J APPL PHYS
38   4   1803-8   1967

43687 AN UNCERTAINTY ANALYSIS FOR SATELLITE CALORIMETRIC MEASUREMENTS.
MILLARD J P   AMES RESEARCH CENTER   MOFFETT FIELD CALIFORNIA
NASA AND CFSTI
NASA-TN-D-4384
N68-16529   1-64   1968   PA   N68-8-7 997

43690 THE RELATIONSHIP BETWEEN TRANSPORT PROPERTIES AND RATES OF FREEZE-DRYING OF POULTRY MEAT.
SANDALL O C   KING C J   WILKE C R
AICHE JOURNAL
13   3   428-38   1967

43692 THIRTEEN-MOMENT THEORY OF THE THERMAL FORCE ON A SPHERICAL PARTICLE.
DWYER H A
PHYS FLUIDS
10   5   976-84   1967   CA   67   34402

43733 SOME PROBLEMS OF THE FLUIDIZED-BED DRYING OF HEAT-SENSITIVE MATERIALS IN A CYCLIC REGIME.   FROM NON-STATIONARY HEAT AND MASS TRANSFER.
PIKUS I F   ISRAEL PROGRAM FOR SCIENTIFIC TRANSLATIONS LTD   JERUSALEM
NASA AND CFSTI
NASA-TT-F-432 TT-67-51368
N67-22054   98-105   1967

43735 DESIGN REQUIREMENTS FOR COATED REFRACTORY ALLOYS.
DREISBACH G W   BOEING CO   SEATTLE WASHINGTON
USAF
DDC
D2-81113-1 AD-433991
1-58   1963

43736 SOURCE AND DETECTOR OF RADIATION IN THE WAVELENGTH REGION 1500-50 ANGSTROMS SUITABLE FOR RADIATION EFFECTS STUDIES ON MATERIALS IN VACUO.   II.   VACUUM AND ULTRAVIOLET RADIATION EFFECTS ON MATERIALS AT ABOUT 10/TO THE MINUS 9 POWER/ MM. OF MERCURY.
MOORE H R   BERNSTEIN R   REYNOLDS R S
ELECTRO-OPTICAL SYSTEMS INC   PASADENA CALIF   USAF
DDC
WADD-TR-60-371 AD 295895
PART   II   1-79   1962

43741 INVESTIGATION OF MECHANISMS FOR OXIDATION PROTECTION AND FAILURE OF INTERMETALLIC COATINGS FOR REFRACTORY METALS.
BARTLETT R W   GAGE P R   PHILCO CORP RES LABS
NEWPORT BEACH CALIF   USAF
DDC AND CFSTI
ASD-TDR-63-753 /PT II/
AD 609167   1-127   1964

43742 THERMAL EXPANSION OF REINFORCED COMPOSITES-THERMAL HYSTERESIS EFFECTS.
DENMAN G L   MATERIALS LAB RES AND TECH DIV
WRIGHT PATTERSON AFB OHIO   USAF
DDC
AFML-TR-65-279
AD 484094   1-41   1966

43744 OPTIMIZATION OF INSULATION AND MECHANICAL SUPPORTS FOR HYPERSONIC AND ENTRY VEHICLES.
CROSS R I   BLACK W E   CONVAIR DIV OF GEN DYNAMICS   SAN DIEGO CALIF   USAF
DDC
AFML-TR-66-414
1-208   1967

43767 INVESTIGATION OF THE DIFFUSION IN SILICON AND IN THE DIFFUSION INHIBITING FILMS OF SILICON DIOXIDE WITH RADIOACTIVE METHODS.
FRAENZ I   LANGHEINRICH W   LOECHERER K H
TELEFUNKEN-ZEITUNG
37 3/4   194-209   1964

43768 INVESTIGATION OF THE DIFFUSION IN SILICON AND IN THE DIFFUSION INHIBITING FILMS OF SILICON DIOXIDE WITH RADIOACTIVE METHODS.   //ENGLISH TRANSLATION OF TELEFUNKEN-ZEITUNG 37 /3/-/4/ 194-209, 1964.//
FRAENZ I   LANGHEINRICH W   LOECHERER K H
SLA
TT66-11027   1-45   1966   TT   15-10 68

43774 THERMAL CONDUCTIVITY OF ROCKS BY A RING SOURCE DEVICE.   /M. S. THESIS, MECH. ENG./
MOSSAHEBI M   U OF CALIFORNIA   BERKELEY
U OF CALIFORNIA BERKELEY
1-77   1966

43777 ON THE THEORY OF THE ACCOMMODATION COEFFICIENT.
LANDAU L
PHYSIK Z SOWJETUNIOUN /USSR/
8   5   489-500   1935

43778 ON THE THEORY OF THE ACCOMMODATION COEFFICIENT.
//ENGLISH TRANSLATION OF PHYSIK. Z. SOWJETUNION /USSR/ 8 /5/ 489-500, 1935.//
LANDAU L
SLA
TT66-10712   1-13   1966   TT   15-8 36

43783 THERMAL CONDUCTIVITY OF WOOD-BASE FIBER AND PARTICLE PANEL MATERIALS.
LEWIS W C   FOREST PRODUCTS LAB   MADISON WISCONSIN
CFSTI
PB-175520   FPL-77
1-16   1967   PA   67-18 85

43788 THERMAL CONTACT CONDUCTANCE IN A VACUUM AND RELATED PARAMETER STUDY.   FROM   THE PROCEEDINGS OF CONFERENCE ON THERMAL JOINT CONDUCTANCE.
ATKINS H L   NASA   MARSHALL SPACE FLIGHT CTR
NASA AND CFSTI
NASA-TM-X-56300
N66-37807   7-60   1964   CA   66   116969

43789 MEASUREMENT OF THERMAL CONTACT CONDUCTANCE IN A VACUUM.   FROM   THE PROCEEDINGS OF CONFERENCE ON THERMAL JOINT CONDUCTANCE.
KASPARECK W E   DAILEY R M   NASA   MARSHALL SPACE FLIGHT CTR
NASA AND CFSTI
NASA-TM-X-56300
N66-37808   61-96   1964   CA   66   116970

43790 THERMAL JOINT CONDUCTANCE.   FROM   THE PROCEEDINGS OF CONFERENCE ON THERMAL JOINT CONDUCTANCE.
VICKERS J M F   ENG RES SECTION JET PROPULSION LAB
UNIV OF ILL
NASA AND CFSTI
NASA-TM-X-56300
N66-37809   97-118   1964   PAN 66-4-23 4631

43791 THERMAL CONDUCTANCE OF MOLYBDENUM AND STAINLESS STEEL INTERFACES IN A VACUUM ENVIRONMENT.   FROM   THE PROCEEDINGS OF CONFERENCE ON THERMAL JOINT CONDUCTANCE.
SOMMERS R D   COLES W D   LEWIS RES CTR
CLEVELAND OHIO   NASA
NASA AND CFSTI
NASA-TM-X-56300
N66-37810   119-36   1964   PAN 66-4-23 4631

43794 SOLAR CELL COATINGS.
MARKS B S   LOCKHEED MISSILES AND SPACE CO   PALO ALTO CALIF RES LABS
NASA AND CFSTI
NASA-CR-70168 N66-17318
PIC-SOL-20916   1-20   1965   PAN 66-4-8 1162

43797 KINETICS AND MECHANISM OF THE REACTION OF MANGANESE SULPHIDE WITH OXYGEN.
BATSANOV S S   KAZAKOV V P   DERBENEVA S S
ZH NEORG KHIM
12   6   1417-   1967

43798 KINETICS AND MECHANISM OF THE REACTION OF MANGANESE SULPHIDE WITH OXYGEN.   //ENGLISH TRANSLATION OF ZH. NEORG. KHIM. 12 /6/ 1417-, 1967.//
BATSANOV S S   KAZAKOV V P   DERBENEVA S S
RUSS J INORG CHEM
12   6   749-52   1967

43801 A VACUUM INTEGRATING SPHERE FOR IN SITU REFLECTANCE MEASUREMENTS AT 77K FROM 0.5 TO 10 MICRONS.
MC CULLOUGH B A   WOOD B E   SMITH A M   ARNOLD ENG DEV CTR ARO INC   OPERATING CONTRACTOR   ARNOLD AIR FORCE STATION   TENN   USAF
DDC AND CFSTI
AEDC-TR-67-10 AD 650072
1-27   1967   TA   67-11 117

43811 SUMMARY OF INVESTIGATIONS OF LIGHT SCATTERING IN HIGHLY REFLECTING PIGMENTED COATINGS.
ZERLAUT G A   KAYE B H   IIT RESEARCH INSTITUTE CHICAGO ILL
NASA AND CFSTI

TPRC Number	Bibliographic Citation

43820 SPACE ENVIRONMENTAL EFFECTS ON EXPANDABLE STRUCTURES MATERIALS.
SOUTHERLAN R E   ARNOLD ENG DEV CTR ARO INC
OPERATING CONTRACTOR ARNOLD AIR FORCE STATION   TENN
USAF
DDC
AEDC-TR-67-57 AD 811678
1-26   1967   TA   67-11 A-91

43837 ON THE GRAVIMETRIC DETERMINATION OF THERMAL CHARACTERISTICS OF PLASTIC-MATERIALS.
PLATO G
Z ANGEW PHYS
3   7   263-7   1951

43840 STABLE WHITE COATINGS.
ZERLAUT G A   GILLIGAN J E   HARADA Y   IIT RES
INST TECH CTR   CHICAGO ILL
NASA
IITRI-C-6027 N65-33883
1-110   1965

43842 THE SURFACE TENSION OF LIQUID NICKEL-SILICON ALLOYS. //ENGLISH TRANSLATION OF POROSHKOVAYA MET. AKAD. NAUK UKR. SSR /3/ 80-2, 1965.//
VASILIU M I   EREMENKO V N
SOVIET POWDER METALLURGY AND METAL CERAMICS
3   238-40   1965

43894 HEAT TRANSFER PROPERTIES OF LOOSE-FIBER MATERIALS IN VACUUM.
KOSTYLEV V M
TEPLOFIZIKA VYSOKIKH TEMPERATUR
4   3   351-4   1966

43895 HEAT TRANSFER PROPERTIES OF LOOSE-FIBER MATERIALS IN VACUUM. //ENGLISH TRANSLATION OF TEPLOFIZIKA VYSOKIKH TEMPERATUR 4 /3/, 351-4, 1966.//
KOSTYLEV V M
HIGH TEMPERATURE
4   3   341-3   1966

43896 INVESTIGATION OF THE THERMAL INSULATING PROPERTIES OF KAOLIN FIBERS. //ENGLISH TRANSLATION OF TEPLOFIZIKA VYSOKIKH TEMPERATUR 4 /3/ 355-9, 1966.//
KRZHIZHANOVSKII R E   CHUDNOVSKAYA I I
HIGH TEMPERATURE
4   3   344-7   1966

43918 THERMAL RADIATIVE CHARACTERISTICS OF SOLAR ARRAYS DETERMINED BY CALORIMETRIC TECHNIQUES.
WEN L
NASA
JPL-SPS-37-65
137-41   1970

43937 ADVANCED GRAPHITE FUEL SYSTEMS.
BOKROS J C   ENGLE G B   PRICE R J   WHITE J L
GENERAL DYNAMICS CORP GENERAL ATOMIC DIV   SAN DIEGO CALIF
AEC AND CFSTI
GA-7390   1-39   1966   NSA 22-6 1124

43938 ADVANCED GRAPHITE FUEL SYSTEMS.
BOKROS J C   ENGLE G P   GOEDDEL W V   PRICE R J
WHITE J L   GENERAL DYNAMICS CORP GENERAL ATOMIC
DIV   SAN DIEGO CALIF
AEC AND CFSTI
GA-7100   1-93   1967   NSA 22-6 1151

43989 LONG-WAVELENGTH INFRARED ABSORPTION SPECTRA OF PHTHALOCYANINES.
ALEKSANDROV A N   SIDOROV A N   YAROSLAVSKII N G
OPTIKA I SPEKTROSKOPIYA
22   4   560-5   1967

43990 LONG-WAVELENGTH INFRARED ABSORPTION SPECTRA OF PHTHALOCYANINES. //ENGLISH TRANSLATION OF OPTIKA I SPEKTROSKOPIYA 22 /4/ 560-5, 1967.//
ALEKSANDROV A N   SIDOROV A N   YAROSLAVSKII N G
OPT SPECTRY
22   4   307-9   1967

43996 THERMAL CONDUCTIVITY AND EMISSIVITY OF ALUMINUM OXIDE COATINGS AT HIGH TEMPERATURES. //ENGLISH TRANSLATION OF TEPLOFIZ. VYSOKIKH TEMPERATUR 4 /5/ 643-8, 1966.//
ZHOROV G A   KOVALEV A I   SIVAKOVA E V
HIGH TEMPERATURE USSR
4   5   603-8   1966
CORRECTION.
PETROV V A
6   3   556-7   1968

43997 HEAT EXCHANGE BETWEEN CONTACTING PARTS.
DYBAN E F   KONDAK N M   SHVETS I T
IZVEST AKAD NAUK UZBEKSKOI SSR
9   63-79   1954

43999 APPLICATION OF THE RITZ AND TREFTTS VARIATIONAL METHODS IN THE CALCULATION OF THERMAL CONDUCTIVITY IN DEFECTIVE NUCLEAR-REACTOR FUEL ELEMENTS.
TURILINA E S   VOSKRESENSKII K D

44000 APPLICATION OF THE RITZ AND TREFTTS VARIATIONAL METHODS IN THE CALCULATION OF THERMAL CONDUCTIVITY IN DEFECTIVE NUCLEAR-REACTOR FUEL ELEMENTS. //ENGLISH TRANSLATION OF TEPLOFIZ. VYSOKIKH TEMPERATUR 4 /5/ 660-9, 1966.//
TURILINA E S   VOSKRESENSKII K D
HIGH TEMPERATURE USSR
4   5   619-27   1966

44016 INFRARED LATTICE REFLECTION SPECTRA OF II-VI COMPOUNDS.
MANABE A   MITSUISHI A   YOSHINAGA H
JAPANESE J APPL PHYS
6   5   593-600   1967

44041 EXCITATION BY PHOTONS OF PLASMONS OF VOLUME AND SURFACE BY MAGNESIUM FILMS EVAPORATED AND STUDIED IN THE ULTRAVIOLET.
DAUDE A   PRIOL M   ROBIN S
COMPT REND
264   B   22   1489-92   1967

44044 OPTICAL PROPERTIES OF VERY THIN FILMS OF SILVER UNDER VACUUM AND UNDER AN ARGON ATMOSPHERE.
CARLAN A
COMPT REND
264   B   23   1595-8   1967

44045 OPTICAL PROPERTIES OF EVAPORATED FILMS OF ANTIMONY IN THE EXTREME ULTRAVIOLET.
LEMONNIER J-C   LE CALVEZ Y   STEPHAN G   ROBIN S
COMPT REND
264   B   23   1599-602   1967

44048 OPTICAL PROPERTIES OF MAGNESIUM FLUORIDE IN THE EXTREME ULTRAVIOLET.
STEPHAN G   LE CALVEZ Y   LEMONNIER J-C   ROBIN S
COMPT REND
264   B   24   1667-70   1967

44051 ON THE OPTICAL ABSORPTION OF VERY THIN FILMS OF ZINC PREPARED AT LOW TEMPERATURE.
BARRAS H   MICHAUD-BONNET J
COMPT REND
265   B   2   152-4   1967

44080 OPTICAL PROPERTIES OF GERMANIUM IN THE FAR ULTRAVIOLET.
MARTON L   TOOTS J
PHYS REV
160   3   602-8   1967

44084 USE OF GAMMA RAYS FOR DETERMINATION OF SURFACE TENSION AND CONTACT ANGLE AT HIGH TEMPERATURE.
ZIV D M   SHESTAKOV B I
ZHUR PRIKLAD KHIM
32   1767-70   1959

44085 USE OF GAMMA RAYS FOR DETERMINATION OF SURFACE TENSION AND CONTACT ANGLE AT HIGH TEMPERATURE. //ENGLISH TRANSLATION OF ZHUR. PRIKLAD. KHIM. 32, 1767-70, 1959.//
ZIV D M   SHESTAKOV B I
J APPL CHEM USSR
32   1804-7   1959

44091 ON THE SURFACE TENSION OF MOLTEN METALS AND ALLOYS.
MATUYAMA Y
SCI REP TOHOKU IMP UNIV
16   555-62   1927

44094 THE USE OF HEAT INSULATION IN BUILDING DESIGN AND CONSTRUCTION.
ALLCUT E A
J ROYAL ARCHITECT INST OF CANADA
24   90-100   1947

44096 EFFECTS OF PROTONS AND ALPHA PARTICLES ON THERMAL PROPERTIES OF SPACECRAFT AND SOLOR CONCENTRATOR COATINGS. FROM PROC. CONF. AIAA THERMOPHYSICS SPECIALISTS HELD AT MONTEREY, CALIF. SEPT. 13-15, 1965.
GILLETTE R B   BROWN R R   SEILER R F
SHELDON W R
ACADEMIC PRESS   NEW YORK AND LONDON
413-40   1965

44137 PRECISION IN MEASUREMENT OF SMALL COLOR DIFFERENCES.
ILLING A M   BALINKIN I
BULL AM CERAM SOC
44   12   956-62   1965

44155 THE THICKNESS AND OPTICAL PROPERTIES OF FILMS OF ANODIC ALUMINUM OXIDE.
KHAN I H   LEACH J S LL   WILKINS N J M
CORROS SCI
6   11   483-97   1966   CA   66   110877

44158 LOW-TEMPERATURE THERMAL CONDUCTIVITIES OF TWO HIGH COMPRESSIVE STRENGTH MATERIALS.
LYON D N   PARRISH W R
CRYOGENICS
7   1   21-5   1967   CA   86   97714

TPRC Number	Bibliographic Citation

44182 RELATION BETWEEN THERMAL CONDUCTIVITY AND OXIDE CONCENTRATIONS IN SODIUM. //ENGLISH TRANSLATION OF AT. ENERG. 19 /4/ PP. 391-92, 1965.//
KOZLOV F A   ANTONOV I N
SOVIET J ATOMIC ENERGY
19   4   1333-4   1965

44183 ANALYSIS OF THERMAL CONDUCTIVITY IN GRANULAR MATERIALS. //ENGLISH TRANSLATION OF CHEMIA STOSOWANA 2B, 183-226, 1966.//
KRUPICZKA R
INTERN CHEM ENG
7   1   122-44   1967

44224 EFFECTS OF PROTONS AND ALPHA-PARTICLES ON THERMAL PROPERTIES OF SPACECRAFT AND SOLAR CONCENTRATOR COATINGS. FROM AIAA THERMOPHYS. SPECIALIST CONF. MONTEREY, CALIF. SEPT. 13-15, 1965.
GILETTE R B   BROWN R R   SEILER R F   SHELDON W R
BOEING CO   SEATTLE WASH
BOEING CO   SEATTLE EASH
A-66-10230   1-38   1965   CA   64   16984

44226 INVESTIGATION OF NON-DESTRUCTIVE METHODS FOR THE EVALUATION OF GRAPHITE MATERIALS.
LOCKYER G E   SHULTZ A W   SERABIAN S   CARTER S W
AIR FORCE MATERIALS LAB   WRIGHT-PATTERSON AIR FORCE BASE   OHIO
DDC
AFML-TR-67-128 AD-816960
          1-170   1967

44231 SOME OPTICAL PROPERTIES OF THIN EVAPORATED CADMIUM ARSENIDE FILMS.
ZDANOWICZ L
PHYS STAT SOL
20   2   473-80   1967

44256 OPTICAL INSTRUMENTATION FOR TRACKING HIGH ALTITUDE VAPOR RELEASES BY DAY.
BEST G T
APPLIED OPTICS
9   12   2666-72   1970

44263 SURFACE TENSION AND ELECTRICAL CONDUCTIVITY OF MAGNETIZED WATER.
KUSHCHENKO A D   BOGUSLAVSKII L I
ELEKTROKHIMIYA
3   1   123-6   1967   CA   66   108518

44270 EMISSIVITY COATINGS FOR LOW-TEMPERATURE SPACE RADIATORS. QUARTERLY PROGRESS REPORT NO. 4, JUNE 30, 1966.
SMITH F J   GRAMMER J G   AEROSPACE SCIENCES LABORATORY LOCKHEED PALO ALTO RESEARCH LAB LOCKHEED MISSILES AND SPACE CO   SUNNYVALE CALIF
NASA AND CFSTI
NASA-CR-72089 N67-10862
          1-26   1966

44271 AN INTEGRATING SPHERE COATING FOR THE VACUUM ULTRAVIOLET SPECTRAL REGION.
HEANEY J B   GODDARD SPACE FLIGHT CENTER GREENBELT MARYLAND
NASA AND CFSTI
NASA-TM-X-55645
N67-16547   1-16   1966   PA   N67-5-6 919

44272 EMISSIVITY COATINGS FOR LOW-TEMPERATURE SPACE RADIATORS. QUARTERLY PROGRESS REPORT OCT. 10 DEC. 31, 1966.
GRAMMER J R
NASA AND CFSTI
NASA-CR-72161 N67-19871
          1-26   1966

44273 A STUDY ON EMISSIVITIES.
SCHOCKEN KLAUS   THERMODYNAMICS SECTION   RESEARCH PROJECTS LAB   DEVELOP   OPERATIONS DIVISION
ARMY BALLISTIC MISSILE AGENCY   REDSTONE ARSENAL ALA
NASA
DV-TN-64-58 N-83592
          1-23   1958

44278 REFLECTANCE-INCREASING COATINGS FOR THE VACUUM ULTRAVIOLET AND THEIR APPLICATIONS.
BERNING P H   HASS G   MADDEN R P
J OPT SOC AM
50   6   586-97   1960

44281 DETERMINATION OF THE REFLECTIVITY, TRANSMISSIVITY, AND ABSORBTIVITY OF THIN FILMS OF GOLD BY EVAPORATION.
ROUARD P   MALE D   TROMPETTE J
J PHYS RADIUM
14   11   587-90   1953

44283 THERMAL CONDUCTIVITY, LONGITUDINAL AND TRANSVERSE, OF STACKED TRANSFORMER SHEETS.
SCHUEMICHEN M
ELEKTRIE
20   12   457-60   1966   CA   66   87874

44300 OBSERVATION OF ABSORPTION EDGES IN THE EXTREME ULTRAVIOLET BY TRANSMITTANCE MEASUREMENTS THROUGH THIN UNBACKED METAL FILMS. FROM OPTICAL PROPERTIES AND ELECTRONIC STRUCTURE OF METALS AND ALLOYS.
HUNTER W R   HULBURT E O   CENTER FOR SPACE RESEARCH US NAVAL RESEARCH LAB   WASHINGTON D C
NORTH-HOLLAND PUB CO AMSTERDAM
          136-46   1966

44304 VALIDITY OF THE DRUDE THEORY FOR SILVER, GOLD, AND ALUMINUM IN THE INFRARED. FROM OPTICAL PROPERTIES AND ELECTRONIC STRUCTURE OF METALS.
BENNETT H E   BENNETT J M   MICHELSON LAB U S NAVAL ORDNANCE TEST STATION   CHINA LAKE CALIFORNIA
NORTH-HOLLAND PUB CO AMSTERDAM
          175-88   1966

44305 HIGH SENSITIVITY PIEZO-REFLECTIVITY MEASUREMENTS USING A STRESS-MODULATION TECHNIQUE. FROM OPTICAL PROPERTIES AND ELECTRONIC STRUCTURE OF METALS AND ALLOYS.
ENGELER W E   GARFINKEL M   TIEMANN J J   GENL ELEC RES LAB   SCHENECTADY N Y AND H FRITZSCHE INST FOR STUDY OF METALS AND SEPT PHYS UNIV CHGO   CHGO ILL
NORTH-HOLLAND PUB CO AMSTERDAM
          189-95   1966

44306 OPTICAL PROPERTIES OF SILVER AND PALLADIUM IN THE FAR ULTRAVIOLET. FROM OPTICAL PROPERTIES AND ELECTRONIC STRUCTURE OF METALS AND ALLOYS.
ROBIN S   LABORATOIRE DE SPECTROSCOPIE FACULTE DES SCIENCES   RENNES FRANCE
NORTH-HOLLAND PUB CO AMSTERDAM
          202-9   1966

44307 RECENT STUDIES ON THE OPTICAL PROPERTIES OF RARE-EARTH METALS. FROM OPTICAL PROPERTIES AND ELECTRONIC STRUCTURE OF METALS AND ALLOYS.
SCHULER C C   IBM ZURICH RESEARCH LAB   RUSCHLIKON ZURICH SWITZERLAND
NORTH-HOLLAND PUB CO AMSTERDAM
          221-36   1966

44308 THE BAND STRUCTURE OF GASOLINIUM, PHOTOEMISSION AND OPTICAL STUDIES. FROM OPTICAL PROPERTIES AND ELECTRONIC STRUCTURE OF METALS AND ALLOYS.
BLODGETT A J JR   SPICER W E   YU A Y C
STANFORD ELECTRONICS LABS STANFORD UNIV   STANFORD CALIF
NORTH-HOLLAND PUB CO AMSTERDAM
          246-56   1966

44314 OPTICAL PROPERTIES OF AG-AU ALLOYS. FROM OPTICAL PROPERTIES AND ELECTRONIC STRUCTURE OF METALS AND ALLOYS.
FUKUTANI H   SUEOKA O   INST PHYSICS COLLEGE OF GENL EDUC UNIV TOKYO   MEGURO-KU TOKYO JAPAN
NORTH-HOLLAND PUB CO AMSTERDAM
          565-73   1966   CA   68   55148

44317 APPARATUS FOR THE MEASUREMENT OF VACUUM ULTRAVIOLET OPTICAL PROPERTIES OF FRESHLY EVAPORATED FILMS BEFORE EXPOSURE TO AIR.
MADDEN R P   CANFIELD L R
J OPT SOC AM
51   8   838-45   1961

44319 ON THE VACUUM-ULTRAVIOLET REFLECTANCE OF EVAPORATED ALUMINUM BEFORE AND DURING OXIDATION.
MADDEN R P   CANFIELD L R   HASS G
J OPT SOC AM
53   5   620-5   1963

44326 REFLECTION AND TRANSMISSION INTERFERENCE FILTERS. II. EXPERIMENTAL-COMPARISON WITH THEORY, RESULTS.
HADLEY L N   DENNISON D M
J OPT SOC AM
38   6   483-96   1948

44338 ELECTROSTATIC DISPERSION OF LACQUERS. I.
SCHENE H
IND LACKIER-BETR
34   10   431-7   1966   CA   66   56604

44339 THE THERMAL CONDUCTIVITY OF POROUS MEDIA IV SANDSTONES. THE EFFECT OF TEMPERATURE AND SATURATION. FROM PROCEEDINGS OF THE 5TH CONFERENCE ON THERMAL CONDUCTIVITY. DENVER HILTON HOTEL. DENVER, COLORADO, OCTOBER 20-22, 1965. VOLUME 1.
MESSMER J H   GULF RESEARCH AND DEVELOPMENT COMPANY   PITTSBURGH PENNSYLVANIA
UNIVERSITY OF DENVER
5   1-29   1965

44340 HIGH-TEMPERATURE INFRARED TRANSMITTING GLASSES. III.
HILTON A R   JONES C E   BRAU M
INFRARED PHYS
6   4   183-94   1966   CA   66   68567

44342 DETERMINATION OF THE THERMAL RESISTANCE IN VACUUM OF CONTACTS BETWEEN METALLIC SURFACES OF VARIOUS DEGREES OF ROUGHNESS.
KAGANER M G   ZHUKOVA R I

TPRC Number	Bibliographic Citation						

44349 THE INFLUENCE OF MASS TRANSFER ON LIQUID FILM
BREAKDOWN.
PONTER A B   DAVIES G A   ROSS T K   THORNLEY P G
INT J HEAT MASS TRANSFER
10   3   349-59   1967   CA   66   96887

44373 DETERMINATION OF OPTICAL CHARACTERISTICS OF
THIN-LAYERED COATINGS. II.
TEKUCHEVA I A
IZV VYSSHIKH UCHEBN ZAVEDENII FIZ
9   4   122-9   1966   CA   66   89794

44376 OPTICAL CONSTANTS OF CESIUM FOR PHOTONS OF ENERGY 5
TO 9.6 EV. M.S. THESIS
WHANG U S   CALLCOTT T A   ARAKAWA E T   UNIV OF
TENNESSEE   OAK RIDGE NATL LAB
AEC AND CFSTI
ORNL-TM-2622 N70-29073
1-79   1970

44385 INTERFACIAL PROPERTIES OF HYDROCARBONS.
GILLAP W R   WEINER N D   GIBALDI M
J AMER OIL CHEM SOC
44   2   71-3   1967   CA   66   108526

44400 THERMAL EXPANSION-CONTRACTION OF ASPHALTIC CONCRETE.
M.S. THESIS.
DARTER M I   UNIVERSITY OF UTAH   SALT LAKE CITY
UTAH
UNIVERSITY OF UTAH
1-95   1968

44415 IMPROVED BLACK RADIATION DETECTOR.
EISENMAN W L   BATES R L
J OPT SOC AMER
54   10   1280-1   1964   CA   66   109846

44417 CRITICAL SURFACE TENSION OF GLASS.
OLSEN D A   OSTERAAS A J
J PHYS CHEM
68   9   2730-2   1964   JA   50   175

44419 OPTICAL ABSORPTION OF THERMALLY OXIDIZED SILICON
OXIDE FILM.
EDAGAWA H   MORITA Y   INUISHI Y
J PHYS SOC JAPAN
18   2   314-15   1963   JA   50   137

44428 THERMAL CONTACT RESISTANCE ACROSS ELASTICALLY
DEFORMED SPHERES.
YOVANOVICH M M
J SPACECRAFT ROCKETS
4   1   119-22   1967   CA   66   77618

44435 THE EFFECT ON AN INTERFACE ON THE TRANSIENT
DISTRIBUTION IN COMPOSITE AIRCRAFT JOINTS. M.S.
THESIS.
HOLLOWAY G F   SYRACUSE UNIVERSITY   SYRACUSE NEW
YORK
SYRACUSE UNIVERSITY
1-94   1954

44453 FUEL-CLADDING THERMAL CONDUCTANCE. FROM METALS AND
CERAMICS DIVISION ANNUAL PROGRESS REPORT FOR PERIOD
ENDING JUNE 30, 1968.
WILLIAMS R K   BANKS T E   MC ELROY D L
AEC
ORNL-4370
90-1   1968

44457 THERMOPHYSICAL CHARACTERISTICS OF PAINT AND VARNISH
COATINGS.
NOVICHENOK L N   SMEKHOV F M   NITSBERG L V
LAKOKRASOCH MATER IKH PRIMEN
6   30-2   1966   CA   66   86673

44476 REFLECTION SPECTRA OF SOLIDS. HYPOTHESIS AND LIMITS
OF QUANTITATIVE EVALUATION OF THE KUBELKA-MUNK
FUNCTION.
KORTUM G   OELKRUG D
NATURWISSENSCHAFTEN
53   23   600-9   1966   CA   66   89901

44482 INTERFACIAL PROPERTIES BETWEEN MOLTEN SILVER, COPPER,
NICKEL, AND THESE BINARY ALLOYS AND MOLTEN SLAG.
OGINO K   ADACHI A
NIPPON KINZOKU GAKKAISHI
30   10   965-70   1966   CA   66   78948

44490 THIXOTROPIC PROPERTIES OF PRINTING INKS.
GARA M
PAPIRIPAR MAGY GRAF
8   6   372-80   1964   CA   66   86717

44514 ACCOMMODATION COEFFICIENTS OF NITROGEN AND HELIUM ON
NITROGEN COVERED TUNGSTEN BETWEEN 325-496 K.
WACHMAN H Y
PROC INT SYMP RAREFIED GAS DYN
1   173-86   1967   CA   66   108539

44524 MEASUREMENTS OF COLOR AND OTHER APPEARANCE ATTRIBUTES
IN THE PLASTICS INDUSTRY.
HUNTER R S
SPE J

44528 TEMPERATURE DEPENDENCE OF THE OPTICAL ABSORPTION EDGE
OF P-TYPE SNTE.
BURKE J R JR   RIEDL H R
PHYS REV
184   3   830-6   1969

44534 CORRELATION OF INTERFACIAL TENSION OF HYDROCARBONS.
HOUGH E W   WARREN H G
SOC PETROL ENG J
6   4   345-9   1966   CA   66   87263

44551 INSTRUMENTATION STUDIES. LXXXVIII. A STUDY OF
PHOTOELECTRIC INSTRUMENTS FOR THE MEASUREMENT OF
COLOR, REFLECTANCE, AND TRANSMITTANCE. 16.
AUTOMATIC COLOR-BRIGHTNESS TESTER.
DEARTH L R   SHILLCOX W M   VAN DEN AKKER J A
TAPPI
50   2   51-8A   1967   CA   66   96387

44552 SPECTRAL EMISSIVITY OF MATERIALS IN THE TEMPERATURE
RANGE 100 TO 1100 C.
SMIRNOV E V   KONDRASHOV YU A
TEPLOFIZ VYSOKIKH TEMPERATUR
5   1   44-7   1967   CA   66   107457

44554 PROPERTIES AND USES OF BORON NITRIDE FIBERS.
ECONOMY J   ANDERSON R V
TEXTILE RES J
36   11   994-1003   1966   CA   66   66594

44560 HIGH TEMPERATURE COMPOSITE STRUCTURE. TECH.
DOCUMENTARY REPORT.
DAVIS R M   MILEWSKI C   WRIGHT-PATTERSON AIR
FORCE BASE   OHIO
DDC
ASD-TDR-62-418
AD-289092   1-270   1962

44594 EFFECT OF CROSS-LINKING NUCLEI ON THE MECHANICAL AND
THERMOPHYSICAL PROPERTIES OF POLYMERIC COATINGS.
SUKHAREVA L A   VORONKOV V A   ZUBOV P I
VYSOKOMOL SOEDIN
8   11   1857-9   1966   CA   66   19952

44625 REFLECTIVITY OF HIGHLY DOPED GALLIUM ARSENIDE IN A
WIDE SPECTRAL RANGE.
KAGAN M B   KOLTUN M M   LANDSMAN A P
ZH PRIKL SPEKTROSK
5   6   770-3   1966   CA   66   109831

44681 MEASUREMENT OF THERMAL CONDUCTANCE OF MULTILAYER AND
OTHER INSULATION MATERIALS. FINAL REPORT. JAN.,
1966.
KARP G S   FRIED E   GENERAL ELECTRIC SPACECRAFT
DEPT VALLEY FORGE SPACE TECHNOLOGY CENTER
PHILADELPHIA PENN
NASA
NASA-CR-78979 N66-39708
1-245   1966

44688 X-RAY DIFFRACTION BY MULTILAYERED THIN-FILM
STRUCTURES AND THEIR DIFFUSION.
DINKLAGE J B
J APPL PHYS
38   9   3781-5   1967

44702 OPTICAL TRANSMISSION THROUGH WINDOWS SET NEAR THE
BREWSTER ANGLE.
SPORTON T M
J SCI INSTR
44   9   720-4   1967

44703 OPTICAL TRANSMISSION OF THIN ABSORBING FILMS
CONTAINING PINHOLES.
BROOM R F
J SCI INSTR
44   9   805   1967

44708 CHARACTERISTICS OF THE NEW VIDICON-TYPE CAMERA TUBE
USING CADMIUM SELENIDE AS A TARGET MATERIAL.
SHIMIZU K   KIUCHI Y
JAPAN J APPL PHYS
6   9   1089-95   1967

44710 INFRARED ABSORPTION OF SILICON OXIDE FILMS MADE BY
OXIDATION OF SILANE.
SHOHNO K
JAPAN J APPL PHYS
6   9   1136   1967

44718 THE USE OF LIQUID HELIUM AND POWDERED CERIUM
MAGNESIUM NITRATE IN VERY LOW TEMPERATURE
MEASUREMENTS.
WHEATLEY J C
ANN ACAD SCI FENN SER A II
210   15-30   1966   CA   66   116990

44754 HEMISPHERICAL REFLECTANCE OF METAL SURFACES AS A
FUNCTION OF WAVELENGTH AND SURFACE ROUGHNESS.
BIRKEBAK R C   DAWSON J P   MC CULLOUGH B A
WOOD B E
INTERN J HEAT MASS TRANSFER
10   9   1225-32   1967

TPRC Number	Bibliographic Citation

44765 OPTICAL PROPERTIES OF EPITAXIAL PBS FILMS IN THE
ENERGY RANGE 2-8 EV.
ROSS C E   GORDON MC KAY LABORATORY HARVARD
UNIVERSITY   CAMBRIDGE MASS
CFSTI
ARPA-TR-28 AD 651215 HP-17
N67-28616   1-35   1967   PA   N67-5-15 2727

44767 REFLECTANCE OF PIGMENTED COATINGS.
DEDRICK R L   VAN VLIET R M   WRIGHT AIR
DEVELOPMENT CENTER MATERIALS LABORATORY DIRECTORATE
OF LABORATORIES WRIGHT-PATTERSON AFB   OHIO
DDC
WCLT-TM-59-56
AD-491057   1-16   1959

44771 EMISSIVITY AND ABSORPTIVITY OF MATERIALS SUBJECTED TO
HIGH VACUUM AND SOLAR RADIATION.
YOUNG J F   DOUGLAS AIRCRAFT CO INC MISSILES AND
SPACE SYSTEMS DIVISION   SANTA MONICA CALIF
DDC
IDEP-347.60.00.00-D7-02
AD-807182L   1-22   1966   TA   67-7 A-64

44772 TEMPERATURE CONTROL COATINGS FOR CRYOGENIC
TEMPERATURE SUBSTRATES.
SCHMIDT W F   OLSON R L   GREENBERG S A
MC KELLAR L A   OLENICK R W   STARKEY R E
DDC
AFML-TR-66-10/PT I/
AD-483227   1-182   1966

44813 VISCOSITY OF A FLUIDIZED SYSTEM.
MURRAY J D
RHEOL ACTA
6   1   27-30   1967   CA   67   4276

44854 POSITION DEPENDENT SUPERCONDUCTIVITY BY THERMAL AND
ELECTRICAL CONDUCTIVITY IN SUPERCONDUCTING FILMS.
PH.D. THESIS.
SEIDEL T   STEVENS INST OF TECHNOL   HOBOKEN N J
UNIV MICROFILMS PUBL
UM65-12585   1-120   1965   CA   64   9050

44859 DETERMINATION OF EFFECTIVE MASS IN LANTHANUM
ANTIMONIDE FROM THE INFRARED REFLECTION SPECTRUM.
GONCHAROVA E V   KUKHARSKII A A
FIZ TVERD TELA
9   5   1543-5   1967

44860 DETERMINATION OF EFFECTIVE MASS IN LANTHANUM
ANTIMONIDE FROM THE INFRARED REFLECTION SPECTRUM.
//ENGLISH TRANSLATION OF FIZ. TVERD. TELA 9 /5/
1543-5, 1967.//
GONCHAROVA E V   KUKHARSKII A A
SOVIET PHYSICS-SOLID STATE
9   5   1214-15   1967

44867 GEOTHERMAL MEASUREMENTS IN AUSTRALIA.   PH.D. THESIS.
SASS J H   GEOPHYSICS DEPT AUSTRALIAN NATIONAL
UNIVERSITY
AUSTRALIAN NATIONAL UNIVERSITY   CANBERRA
                         1-122   1964

44873 MEASUREMENT OF THE THERMAL CONDUCTIVITY OF
INSULATING MATERIALS BY A STATIONARY METHOD.
MACHALICKY J
ELECTROTECH OBZOR
49   12   648-9   1960

44874 MEASUREMENT OF THE THERMAL CONDUCTIVITY OF
INSULATING MATERIALS BY A STATIONARY METHOD.
//ENGLISH TRANSLATION OF ELECTROTECH. OBZOR. 49 /12/
648-9, 1960.//
MACHALICKY J
CEGB
CEGB   TP.2968
TT66-26903   1-4   1961

44875 EVALUATION OF MULTILAYER ANTI-REFLECTION COATINGS FOR
INSTRUMENT WEDGES AND COVERGLASSES.   INTERIM REPT.
PARKER J W III   U S NAVAL AIR DEVELOPMENT CENTER
JOHNSVILLE PENNSYLVANIA
DDC
NADC-AM-6625
AD-483616L   1-22   1966

44882 RELATIVE BIDIRECTIONAL REFLECTANCE OF
THREE-DIMENSIONAL SURFACES.   M.S. THESIS.
GREENE R F JR   UNIVERSITY OF NEBRASKA   LINCOLN
NEBRASKA
UNIVERSITY OF NEBRASKA
                         1-130   1968

44924 OPTICAL PROPERTIES OF VARIOUS MATERIALS IN THE FAR
ULTRAVIOLET.
SIMONE R
METHOD PHYS ANAL
                         135-43   1966

44931 INFRARED COATING STUDIES.
MAIER R L   BAUSCH AND LOMB INC   ROCHESTER N Y
RES AND DEVELOPMENT DIV   DA
DDC AND CFSTI
AD 653850   N67-04125
                         1-17   1967   PA   N67-5-20 3686

44933 EFFECTS OF FRONT SURFACE ROUGHENING ON SOLAR
ABSORPTIVITY OF QUARTZ REAR SURFACE MIRROR SATELLITE
COATINGS.
STARNER K E   STARK R L   AEROSPACE CORP
LABS DIV   EL SEGUNDO   USAF
DDC
SSD-TR-67-69 TR-1001-/2240/-10
AD 815036   1-31   1967

44937 INVESTIGATION OF 3-D FABRICATION OF ABLATIVE
MATERIALS.   FINAL REPT.-SEPT. 8-OCT. 18, 1966.
AVCO CORP SPACE SYSTEMS DIVISION
WILMINGTON MASSACHUSETTS
DDC
NASA-CR-65560
N67-14920   1-265   1966   PAN 67-5-5 712

44938 ENERGY TRANSFER THROUGH EVACUATED NONMETALLIC
MATERIALS.
TARBELL D W   BROWN ENG CO   HUNTSVILLE ALA
NASA AND CFSTI
NASA-CR-77390 N66-34695
                         1-37   1966   CA   67   4217

44942 DEVELOPMENT OF HIGH TEMPERATURE INSULATION MATERIALS.
BERG D I   LEWIS D W   WESTINGHOUSE RESEARCH LABS
PITTSBURGH PA
DDC
AD-801194   1-107   1966

44944 CORRELATION OF MECHANICAL AND THERMAL PROPERTIES OF
EXTRATERRESTRIAL MATERIALS.   FINAL REPT.
HALAJIAN J D   REICHMAN J   KARAFIATH L L
GRUMMAN AIRCRAFT ENG CORP GEO-ASTROPHYSICS SECTION
BETHPAGE N Y
NASA AND CFSTI
NASA-CR-83895
N67-25958   1-193   1967   PAN 67-5-13 2402

44949 DEVELOPMENT OF THERMAL CONTROL COATINGS FOR SPACE
VEHICLES.
MOOK C P   NASA   WASHINGTON D C
NASA AND CFSTI
NASA-TM-X-54906
N66-33378   1-7   1964   CA   66   116745

44971 OPTICAL PROPERTIES OF THE SILVER AND CUPROUS HALIDES.
CARDONA M
PHYS REV
129   1   69-78   1963

44981 MELTING, FLOW, AND THERMAL EXPANSION CHARACTERISTICS
OF SOME DENTAL AND COMMERCIAL WAXES.
OHASHI M   PAFFENBARGER G C
J AM DENTAL ASSOC
72   1141-50   1966   CA   65   7483

44986 PERFORMANCE OF MULTILAYER INSULATION SYSTEMS FOR
TEMPERATURES TO 700 K.
CUNNINGTON G R JR   ZIERMAN C A   FUNAI A I
LINDAHN A
NASA
NASA-CR-907
                         1-106   1967

44987 THERMAL AND MECHANICAL PROPERTIES OF A NONDEGRADED
AND THERMALLY DEGRADED PHENOLIC-CARBON COMPOSITE.
ENGELKE W T   PYRON C M JR   PEARS C D
SOUTHERN RESEARCH INSTITUTE   BIRMINGHAM ALABAMA
NASA
NASA-CR-896 N67-38017
                         1-55   1967   CA   69   11008

44990 AN AUTOMATIC APPARATUS WITH A FLAT PLATE FOR THE
MEASUREMENT OF THERMAL CONDUCTIVITY OF INSULATING
MATERIALS AT HIGH TEMPERATURE /LIMIT AT 1000 0./.
FERRO V   SACCHI A
IL CALORE
                         4   2-16   1966

44995 SYSTEMATICAL TEST OF THE THERMAL CONDUCTANCE EXECUTED
ON PREFABRICATED BUILDING WALLS.
BONDI P   FERRO V   LOMBARDI C   SACCHI A
ATTI E RASSEGNA TECNICA DELLA SOCIETA DEGLI INGEGN
ERI E DEGLI ARCHITETTI IN TORINO NUOVA SERIE A
21   6   3-11   1967

45003 SPREADING PRESSURE, INTERFACIAL TENSION AND
ADHESIONAL ENERGY OF THE LOWER ALKANES, ALKENES, AND
ALKYL BENZENES ON WATER.
POMERANTZ P   CLINTON W C   ZISMAN W A   NAVAL
RESEARCH LAB   WASHINGTON D C
CFSTI
NRL-6495 AD-646995
N67-24747   1-30   1967   PA   N67-5 2046

TPRC Number	Bibliographic Citation

45014 OPTICAL PROPERTIES OF DOUBLE METALLIC FILMS.
IDCHAK E F
OPTIKA I SPEKTROSKOPIYA
22   6   935-9   1967

45015 OPTICAL PROPERTIES OF DOUBLE METALLIC FILMS.
//ENGLISH TRANSLATION OF OPTIKA I SPEKTROSKOPIYA 22
/6/ 935-9, 1967.//
IDCHAK E F
OPT SPECTRY
22   6   507-9   1967

45016 TRANSPARENCE LIMITS OF INTERFERENCE FILMS OF HAFNIUM
AND THORIUM OXIDES IN THE ULTRAVIOLET REGION OF THE
SPECTRUM.
SVIRIDOVA A A   SUIKOVSKAYA N V
OPTIKA I SPEKTROSKOPIYA
22   6   940-5   1967

45017 TRANSPARENCE LIMITS OF INTERFERENCE FILMS OF HAFNIUM
AND THORIUM OXIDES IN THE ULTRAVIOLET REGION OF THE
SPECTRUM. //ENGLISH TRANSLATION OF OPTIKA I
SPEKTROSKOPIYA 22 /6/ 940-5, 1967.//
SVIRIDOVA A A   SUIKOVSKAYA N V
OPT SPECTRY
22   6   509-12   1967

45018 NARROW-BAND INTERFERENCE FILTERS FOR THE ULTRAVIOLET
REGION OF THE SPECTRUM.
MONTOVILOV O A
OPTIKA I SPEKTROSKOPIYA
22   6   986-9   1967

45019 NARROW-BAND INTERFERENCE FILTERS FOR THE ULTRAVIOLET
REGION OF THE SPECTRUM. //ENGLISH TRANSLATION OF
OPTIKA I SPEKTROSKOPIYA 22 /6/ 986-8, 1967.//
MONTOVILOV O A
OPT SPECTRY
22   6   537-8   1967

45033 GROWTH AND OPTICAL PROPERTIES OF WURTZITE AND
SPHALERITE CADMIUM SELENIDE EPITAXIAL THIN FILMS.
LUDEKE R   PAUL W
PHYS STAT SOL
23   1   413-18   1967

45042 EFFECTIVE THERMAL CONDUCTIVITY AND MULTILAYERED
INSULATION.
ADELBERG M
ADVAN CRYOG ENG
12   252-4   1967 CA 67 26565

45044 MEASURING THE THERMAL CONDUCTIVITY OF IRRADIATED
FOAM-TYPE INSULATION MATERIALS. FROM CRYOGENIC
ENGINEERING CONFERENCE BOULDER COLO. JUNE 13-15,
1966.
SMITH E T   MILLER R E
ADVAN CRYOG ENG
12   315-21   1967 CA 67 22517

45080 HEAT AND MASS TRANSFER IN HETEROGENOUS CATALYSIS.
II. THE EFFECT OF CATALYST CONDUCTIVITY ON THE
APPARENT CONDUCTIVITY OF THE CATALYST BED AND THE
HEAT TRANSFER COEFFICIENT FROM THE BED TO REACTOR
BATH.
HORAK J   MAMULA P
CHEM PRUM
17   4   188-91   1967 CA 67 55555

45087 APPARENT THERMAL CONDUCTIVITY OF POROUS CONCRETES IN
CONTACT WITH LIQUEFIED NATURAL GAS.
ROUX C   POTTIER M   MORDCHELLES G   JANNOT M
COLLOQ INT CENTRE NAT RECH SCI 1966
160   137-47   1967 CA 67 45725

45105 SURFACE ACTIVITY OF A BIOLOGICALLY ACTIVE
SEMICOLLOID-SODIUM CHOLATE.
MARKINA Z N   TSIRKURINA N N   BOVKUN O P
KOPEINA A D   REBINDER P A
DOKL AKAD NAUK SSSR
173   5   1132-5   1967 CA 67 36611

45153 THERMAL PROPERTIES OF TWO-PHASE DISPERSE SYSTEMS
SATURATED WITH DIFFERENT LIQUIDS.
VERZHINSKAYA A B   VAINBERG V SH
INZH-FIZ ZH
12   1   38-42   1967 CA 67 57809

45155 TEMPERATURE FIELD IN MULTILAYER SYSTEMS WITH VARIABLE
THERMAL PROPERTIES.
MEEROVICH I G
INZH-FIZ ZH
12   4   484-90   1967 CA 67 57812

45173 MEASUREMENT OF THE DISTRIBUTION OF PHOSPHORUS
DIFFUSED IN SILICON DIOXIDE FILM USING P-32 AS A
TRACER.
WATANABE Y   YOSHIDA M
JAPAN J APPL PHYS
6   3   410   1967 CA 67 25837

45177 SOME PROPERTIES OF VAPOR-DEPOSITED SILICON NITRIDE
FILMS USING THE SIH4-NH3-H2 SYSTEM.
BEAN K E   GLEIM P S   YEAKLEY R L   RUNYAN W R
J ELECTROCHEM SOC
114   7   733-7   1967 CA 67 68288

45209 HEAT CONDUCTIVITY MEASUREMENTS ON FOAM PLASTICS AT
LOW TEMPERATURES.
ZEHNDNER H
KALTETECHNIK
19   1   2-8   1967 CA 67 117616

45212 INFRARED DIFFUSE REFLECTOR COATING. MONTHLY PROGRESS
REPORT. MAY 1967.
SCHMIDT R N   HONEYWELL INC SYSTEMS AND RESEARCH
CENTER   ST PAUL MINNESOTA
DDC
AD-815634   1-33   1967

45217 BOND STRENGTH AND ELASTIC PROPERTIES OF CERAMIC
ADHESIVES.
THORNTON H R
J AMERICAN CERAMIC SOC
45   5   201-9   1962

45226 PROPERTIES OF BLACK CHROMIUM ELECTRODEPOSITS.
ANDREEVA N V   SAMARTSEV A G
OPT-MEKH PROM
34   2   40-4   1967 CA 67 69993

45230 SELECTION OF BINDERS FOR GLASS-REINFORCED PLASTICS
WITH GLASS-FILM FILLERS.
TROSTYANSKAYA E B   GUNYAEV G M
PLAST MASSY
      5   30-4   1967 CA 67 54713

45237 INTERFACE TENSION THEORY OF MUSCULAR CONTRACTION.
METHERELL A F   DOUGLAS AIRCRAFT COMPANY INC
HUNTINGTON BEACH   CALIFORNIA
DDC
AD-644981   1-31   1966

45252 THERMAL ACCOMMODATION OF RARE GASES ON CLEAN METAL
SURFACES. COMPARISON OF A SIMPLIFIED CLASSICAL
LATTICE THEORY WITH EXPERIMENTS.
TRILLING L
PROC INT SYMP RAREFIED GAS DYN
1   139-54   1967 CA 67 36624

45253 MOMENTUM ACCOMMODATION OF ARGON IN THE 0.06 TO 5
EV. RANGE.
ABAUF N   MARSDEN D G H
PROC INT SYMP RAREFIED GAS DYN
1   199-210   1967 CA 67 25730

45257 THERMAL CONDUCTIVITY OF FINELY DIVIDED AND GRANULAR
MATERIALS AND METHODS OF MEASUREMENT IN STABLE
CONDITIONS.
DUMEZ P
REV GEN THERMIQUE
5   54   561-74   1966
5   55   673-86   1966 CA 67 26554

45271 PREDICTING THE ADHESION OF HOT-MELT BLENDS ON
PAPER-BOARD.
GLOSSMAN N
TAPPI
50   5   224-6   1967 CA 67 34012

45275 EXPERIMENTAL DETERMINATION OF HEAT AND TEMPERATURE
CONDUCTIVITIES OF ELECTRODE GRAPHITES.
BOLOTIN N K   VENERAKI I E   ROMANKO K S
TOPOLNITSKII G G   CHEBOTAREV V A
TEPLOFIZ SVOISTVA VESHCHESTV AKAD NAUK UKR SSR RESPUB
MEZHVEDOM SB
          52-60   1966 CA 67 26552

45279 SOME PECULIARITIES OF THE ELECTRIC INSULATING
CHARACTERISTICS OF HEAT-RESISTANT CONCRETE.
SHAKHTAKHTINSKII T I   ROMANOV A I   SMIRNOVA L G
TEPLOFIZ VYS TEMP
5   1   155-60   1967 CA 67 25143

45286 OPTICAL AND ELECTRONIC PROPERTIES OF ZINC IN THE
EXTREME ULTRAVIOLET.
LEMONNIER J C   GIRAULT P   ROBIN S
COMPT REND
269   B 8   329-32   1969

45307 TIME OF FLIGHT MEASUREMENTS IN AN ARGON BEAM
DEFLECTED BY A HEATED PLATINUM TARGET. FROM
FUNDAMENTALS OF GAS-SURFACE INTERACTIONS. PROC. OF
THE SYMPOSIUM, SAN DIEGO, CALIF., DEC. 14-16, 1966.
MORAN J P   WACHMAN H Y   TRILLING L
ACADEMIC PRESS   NEW YORK AND LONDON
          461-79   1967

45308 VELOCITY OF MOLECULAR BEAM MOLECULES SCATTERED BY
PLATINUM SURFACES. FROM FUNDAMENTALS OF GAS-SURFACE
INTERACTIONS. PROC. OF THE SYMPOSIUM SAN DIEGO,
CALIF., DEC. 14-16, 1966.
HINCHEN J J   MALLOY E S
ACADEMIC PRESS   NEW YORK AND LONDON

TPRC Number	Bibliographic Citation

45309 THEORY OF GAS-SURFACE COLLISIONS. FROM FUNDAMENTALS OF GAS-SURFACE INTERACTIONS. PROC. OF THE SYMPOSIUM, SAN DIEGO, CALIF. DEC. 14-16, 1966.
TRILLING L
ACADEMIC PRESS   NEW YORK AND LONDON
392-421   1967

45325 INSTRUMENT FOR DETERMINING THE THERMAL CONDUCTIVITY OF POROUS AND FIBROUS INSULATORS.
DULNEV G N   PLATUNOV E S   MURATOVA B L
SIGALOVA Z V
INZH FIZ ZH
9   6   751-6   1965

45327 SPECTRAL EMISSIVITY OF MATERIALS IN THE TEMPERATURE RANGE 100 TO 1100 C.   //ENGLISH TRANSLATION OF TEPLOFIZ. VYSOKIKH TEMPERATUR 5 /1/ 44-7, 1967.//
SMIRNOV E V   KONDRASHOV YU A
HIGH TEMPERATURE USSR
5   1   37-40   1967

45382 EMISSIVITY OF BARIUM TITANATE IN THE TEMPERATURE RANGE OF 40 TO 80 DEGREES.
VAFIADI V G   GANSIEVSKAYA YA I
ZH PRIKL SPEKTROSK
6   1   110   1967   CA   67   59058

45389 THE SURFACE EMITTANCE OF VACUUM-METALLIZED POLYESTER FILM.
RUCCIA F E   HINCKLEY R B
ADVAN CRYOG ENG
12   300-7   1967   CA   67   74041

45418 COMPARISON OF THE THERMAL CONDUCTIVITIES OF MOIST CAPILLARY-POROUS MATERIALS AT TEMPERATURES ABOVE AND BELOW 0 DEGREES.
PAK N V
INZH-FIZ ZH
12   1   68-71   1967   CA   67   68255

45437 THE PREPARATION OF NICKEL MONOXIDE THIN FILMS AND THEIR USE IN OPTICAL MEASUREMENTS IN THE VISIBLE AND ULTRAVIOLET.
ROSSI C E   PAUL W
J PHYS CHEM SOLIDS
30   9   2295-305   1969

45441 MOLDING TECHNIQUES AND PRODUCTION CONTROL OF REINFORCED PLASTICS.   PROPERTIES OF FRP.
TAKAGI S   HATOGAI Y   HIRANO H   IWAI T
PLAST AGE /OSAKA/
13   2   73-80   1967   CA   67   74215

45455 HEAT TRANSFER AND VERTICAL HEAT CONDUCTIVITY OF A FLUIDIZED LAYER OF FINE-GRAINED MATERIAL IN A PACKING.
BASKAKOV A P   VERSHININA V S   LUMMI A P
PAKHALUEV V M
TEPLO MASSOPERENOS
5   172-81   1966   CA   67   83284

45458 OPTICAL CONSTANTS OF SILVER AND BARIUM IN THE VACUUM ULTRAVIOLET SPECTRAL REGION.   PH.D. THESIS.
FISHER E I   UNIV OF SOUTHERN CALIFORNIA   LOS ANGELES
UNIV MICROFILMS PUBL
66-11571   1-98   1966   CA   66   70536

45459 THE TOTAL HEMISPHERICAL THERMAL EMITTANCE OF /100/, /110/, AND /111/ SINGLE CRYSTALS OF NICKEL AS A FUNCTION OF OXIDE THICKNESS IN THE TEMPERATURE RANGE 300-900 DEGREES.   PH.D. THESIS.
OUBRE C L   RICE UNIV   HOUSTON TEX
UNIV MICROFILMS PUBL
66-10368   1-262   1966   CA   66   80368

45460 THE LOW-TEMPERATURE HEAT CAPACITIES AND THERMODYNAMIC PROPERTIES OF THE BETA-QUINOL CLATHRATES OF NITROGEN AND HYDROGEN CHLORIDE.   PH.D. THESIS.
ROPER G C   BOSTON UNIV   BOSTON MASS
UNIV MICROFILMS PUBL
66-11269   1-152   1966   CA   66   69474

45475 OUT-OF-PILE THERMAL TESTING OF UO2 FUEL ELEMENTS.
ANTHONY A J   BURDG C E   SANDERSON R J
TRANS AM NUCL SOC
236-7   1962

45477 IN-PILE EFFECTIVE THERMAL CONDUCTIVITY OF OXIDE FUEL ELEMENTS TO HIGH FISSION DEPLETION.
DANIEL R C   COHEN I
TRANS AM NUCL SOC
6   2   332-3   1963

45482 A THERMAL INSULATION STUDY.   QUARTERLY PROGRESS REPT. OCT.-DEC. 1966.
DE WITT W D   LINDE DIV UNION CARBIDE CORP   IND
AEC AND CFSTI
ALO-3632-11 N67-28418
1-56   1967   PA   N67-5-15 2804

45491 THERMAL CONDUCTIVITIES OF UNIDIRECTIONAL MATERIALS.
SPRINGER G S   TSAI S W

45492 THERMAL CONDUCTANCES IN A COLLISIONLESS GAS BETWEEN COAXIAL CYLINDERS AND CONCENTRIC SPHERES.
WU Y
J PLASMA PHYS
1   2   209-17   1967   CA   67   85614

45494 EXPERIMENTAL DETERMINATION OF THE THERMAL CONDUCTIVITY OF REFRACTORY HIGH-ALUMINA CONCRETE IN THE 200-1500 C. RANGE.   FROM INTERNATIONAL SYMPOSIUM ON PRODUCTION OF ELECTRIC POWER BY MEANS OF MHD GENERATORS. SALZBURG, AUSTRIA. JULY 4-8, 1966.
CHEKHOVSKOY V YA   KAULENAS A A
DDC
FTD-HT-67-195 SM-74-23
AD-674611   312-6   1967   PAN 69-7-3 7494

45498 MEASUREMENTS OF THERMAL CONTACT CONDUCTANCE IN A VACUUM.
KASPARECK W E   DAILEY R M   MARSHALL SPACE FLIGHT CENTER   HUNTSVILLE ALABAMA
DDC
AD-439477   1-35   1964

45500 MEASUREMENT OF THERMAL CONDUCTIVITY AND MOISTURE CONTENT IN EXISTING WALLS. FROM STUDIES ON HEAT TRANSFER IN REFRIGERATION.
VOS B H   VAN MINNEN J
INST INTERN DU FROID COMMISSION 2 TRONDHEIM NORWAY
191-211   1966

45501 CADMIUM TELLURIDE AND THE CADMIUM TELLURIDE-MERCURY TELLURIDE SYSTEM.
NEUBERGER M   HUGHES AIRCRAFT CO   CULVER CITY CALIFORNIA
DDC
DS-157   AD-819287
1-200   1967

45517 NON-LINEAR I-V CHARACTERISTICS OF CARBON FILMS AT LOW TEMPERATURE.
BROWN C R   MATTHEWS P W
REV SCI INSTR
39   4   616-17   1968   CA   69   31494

45529 COATINGS DEGRADATION. SECTION VIII.   FROM RADIOMETRY PROGRESS.   FIRST TRIMESTER-APRIL 1967.
NICOLETTA C A   GODDARD SPACE FLIGHT CENTER   GREENBELT MD
NASA
NASA-TM-X-55827
N67-32601   1-20   1967

45543 CORRELATION OF MECHANICAL AND THERMAL PROPERTIES OF EXTRATERRESTRIAL MATERIALS.
HALAJIAN J D   REICHMAN J   GEO-ASTROPHYSICS SECTION GRUMMAN RESEARCH DEPARTMENT
CFSTI
NASA-CR-71750
N66-24605   1-41   1966

45545 PASSIVE THERMAL CONTROL COATINGS.   PRESENTED AT SEVENTH MEETING OF THE REFRACTORY COMPOSITES WORKING GROUP, PALO ALTO, CALIF. MARCH 12-14, 1963.
GILLIGAN J E   SIBERT M E   GREENING T A
LOCKHEED MISSILES AND SPACE CO   PALO ALTO CALIF
DDC AND CFSTI
AD-602894   1-40   1963   TA   U64-17 67

45552 HIGH-TEMPERATURE EMITTANCE OF COATED REFRACTORY METAL.
ALLEN T H   JOHNSON C R   RUSERT E L
EFF SPACE ENVIRON MATER SOC AEROSP MATER PROCESS ENG
NAT SYMP EXHIB 11TH ST LOUIS
111-23   1967   CA   69   13509

45553 EVALUATION OF ASBESTOS-FILLED PLASTICS AND ELASTOMERS FOR SPACE APPLICATIONS.
SHUMAN W P JR   WRONSKI J P
EFF SPACE ENVIRON MATER SOC AEROSP MATER PROCESS ENG
NAT SYMP EXHIB 11TH ST LOUIS
159-68   1967   CA   68   105829

45570 THERMAL PROPERTIES OF MULTILAYER INSULATION.
MIKHALCHENKO R S   GERZHIN A G   PERSHIN N P
KLIPACH I V
INZH-FIZ ZH
12   4   426-32   1967   CA   67   94608

45579 EFFECTIVE THERMAL CONDUCTIVITY AND STRUCTURAL CHANGE OF UO2 POWDER COMPACTED FUEL IN THERMAL SIMULATION EXPERIMENT.
AOKI S   KOBAYASHI Y   SATO K
J NUCL SCI TECHNOL
4   8   408-14   1967   CA   69   95950

45639 THERMAL CONDUCTIVITY OF HIGH TEMPERATURE HETEROGENEOUS INSULATIONS.   FROM PROCEEDINGS OF THE 4TH CONFERENCE ON THERMAL CONDUCTIVITY. SAN FRANCISCO, CALIFORNIA. OCTOBER 13-16, 1964.
WECHSLER A E   KRITZ M A   GLASER P E   ARTHUR D LITTLE INC   CAMBRIDGE MASSACHUSETTS
U S NAVAL RADIOLOGICAL DEFENSE LABORATORY
4   1-22   1964

TPRC Number	Bibliographic Citation						
45653	THERMAL CONDUCTIVITY OF POROUS MEDIA. PACKINGS OF PARTICLES. FROM PROCEEDINGS OF THE 4TH CONFERENCE ON THERMAL CONDUCTIVITY. SAN FRANCISCO, CALIFORNIA. OCTOBER 13-16, 1964. MESSMER J H GULF RESEARCH AND DEVELOPMENT COMPANY PITTSBURGH PENNSYLVANIA U S NAVAL RADIOLOGICAL DEFENSE LABORATORY						
	4	1-12	1964				
45664	DEVELOPMENT OF SPACE-STABLE THERMAL-CONTROL COATINGS. TRIANNUAL REPORT, OCT. 20, 1966-FEB. 20, 1967. ZERLAUT G A IIT RES INST CHICAGO ILL NASA AND CFSTI IITRI-U6002-51 NASA-CR-84062						
	N67-26323	1-18	1967				
45665	EFFECTS OF A SIMULATED SPACE ENVIRONMENT ON THERMAL RADIATION CHARACTERISTICS OF SELECTED BLACK COATINGS. WADE W R PROGAR D J NATL AERONAUTICS AND SPACE ADMINISTRATION LANGLEY RES CTR LANGLEY STATION VA NASA AND CFSTI NASA-TN-D-4116						
	N67-35729	1-64	1967	CA	68	31137	
45666	SPECTRAL TRANSMISSION OF TWO OPTICAL SYSTEMS. SMITH A E ROYAL AIRCRAFT ESTABLISHMENT FARNBOROUGH /ENGLAND/ AEC AND CFSTI RAE-TR-67010 N67-27800						
	AD-657593	1-10	1967	PA	N67-5-15 2707		
45667	THE MEASUREMENT OF ABSORPTIVITY AND REFLECTIVITY. DE LA PERRELLE E T HERBERT H AERONAUTICAL RES COUNCIL /GT BRIT/ HMSO RAE-TN-RAD-ARC-20879 ARC-CP-601 N62-14110						
	N64-28030	1-22	1962	PA	N64-2 2867		
45669	PLANETARY VEHICLE THERMAL INSULATION SYSTEMS, PHASE I. SUMMARY REPT. BABJACK S J CARR R W COHEN A D DUDLEY L V LANKTON C S MORGAN H RUBENSTEIN S TWEEDIE A GENERAL ELECTRIC CO MISSILE AND SPACE DIV PHIL PA NASA AND CFSTI NASA-CR-88009 67SD4289						
	N67-35552	1-181	1967	RR	67-22 79		
45678	ABSORPTION AND REFLECTIVITY OF THIN FILMS OF SELENIUM NOT EXPOSED TO AIR IN THE FUNDAMENTAL ABSORPTION DOMAIN. MERDY H BALDI J COMPT REND						
	265	B	17	936-9	1967		
45698	MEASUREMENT OF THE INFRARED SPECTRAL ABSORPTANCE OF OPTICAL MATERIALS. STIERWALT D L BERNSTEIN J B KIRK D D APPL OPT						
	2	1169-73	1963				
45699	PROPERTIES OF MATERIALS AT LOW TEMPERATURES. CORRUCCINI R J CHEM ENG PROGRESS						
	53	8	397-402	1957			
45700	PAINTS TO REFLECT ULTRAVIOLET LIGHT. WILCOCK D F SOLLER W IND ENG CHEM						
	32	1446-51	1940				
45706	ANALYTICAL CHEMISTRY. A COMPARISON OF REAGENTS FOR THE DIRECT PHOTOMETRIC DETERMINATION OF RARE EARTH ELEMENTS ON CHROMATOGRAMS. BABKO A K VDOVENKO M E UKRAIN KHIM ZH						
	32	2	209-12	1966			
45707	ANALYTICAL CHEMISTRY. A COMPARISON OF REAGENTS FOR THE DIRECT PHOTOMETRIC DETERMINATION OF RARE EARTH ELEMENTS ON CHROMATOGRAMS. //ENGLISH TRANSLATION OF UKRAIN, KHIM. ZH. 32 /2/ 209-12, 1966.// BABKO A K VDOVENKO M E SOVIET PROGRESS IN CHEMISTRY						
	32	2	166-8	1966			
45714	THE EFFECT OF SURFACE ROUGHNESS, CURE PRESSURE, AND TEMPERATURE LEVEL ON THE THERMAL RESISTANCE OF BONDS AND CONTACTS. M.S. THESIS. SCHWALLER D L UNIVERSITY OF MISSOURI ROLLA MISSOURI UNIVERSITY OF MISSOURI ROLLA MISSOURI						
		1-42	1964				
45727	SURFACE TENSIONS OF POLYMER LIQUIDS BY A ROTATING BUBBLE METHOD. M.S. THESIS, CHEMISTRY. LUISE R R DARTMOUTH COLLEGE HANOVER N H DARTMOUTH COLLEGE HANOVER N H						
		1-57	1966				
45735	OPTICAL PROPERTIES OF SILVER FILMS-DIRECT OBSERVATION OF THE IM/E-SPECTRUM. YAMAGUCHI S						

TPRC Number	Bibliographic Citation						
45738	CALCULATED REFLECTANCE OF ALUMINUM-OVERCOATED IRIDIUM IN THE VACUUM ULTRAVIOLET FROM 500 ANGSTROMS TO 2000 ANGSTROMS. HASS G HUNTER W R APPL OPT						
	6	12	2097-100	1967	CA	68	34467
45741	PRESERVATION OF COLOR INTEGRITY IN MULTISTAGE PHOTOGRAPHY. PERRY B L APPL OPT						
	6	12	2158-62	1967			
45745	ACCOMMODATION OF NOBLE GASES ON PURE TUNGSTEN AND MOLYBDENUM SURFACES AT ROOM TEMPERATURE. KOUPTSIDIS J MENZEL D BER BUNSENGES PHYS CHEM						
	71	7	720-30	1967	CA	67	103106
45752	OPTICAL PROPERTIES OF GERMANIUM WITH CLEAN AND OXIDIZED SURFACES. ROZUMNYUK V T NESTERENKO V A TSEBULYA G G LISITSA M P SNITKO O V FIZ TEKH POLUPROV						
	4	9	1770-4	1970			
45753	PRESSURE EFFECTS ON POSTULATED LUNAR MATERIALS. FROM PROCEEDINGS OF THE 4TH CONFERENCE ON THERMAL CONDUCTIVITY. SAN FRANCISCO, CALIFORNIA. OCTOBER 13-16, 1964. WECHSLER A E GLASER P E ARTHUR D LITTLE INC CAMBRIDGE MASSACHUSETTS U S NAVAL RADIOLOGICAL DEFENSE LABORATORY						
	4	1-31	1964				
45766	DETERMINATION OF CONTACT THERMAL RESISTANCE DURING UNSTEADY HEAT TRANSFER BETWEEN CONNECTED BODIES. ZGURA A A TAITS N YU INZH-FIZ ZH						
	12	6	731-5	1967	CA	67	101420
45768	METALLIC INTERFACE THERMAL CONDUCTANCE-A NOTE ON MEASUREMENT TECHNIQUES. FROM PROCEEDINGS OF THE 4TH CONFERENCE ON THERMAL CONDUCTIVITY. SAN FRANCISCO, CALIFORNIA. OCTOBER 13-16, 1964. FRIED E SPACECRAFT DEPARTMENT GENERAL ELECTRIC COMPANY VALLEY FORGE PENNSYLVANIA U S NAVAL RADIOLOGICAL DEFENSE LABORATORY						
	4	1-15	1964				
45773	DETERMINATION OF INTERFACIAL TENSION OF A MOLTEN CAST IRON AT THE BOUNDARY WITH COKE. VYKHODETS A M SMIRNOV A I IZV VYSSH UCHEB ZAVED CHERN MET						
	10	7	149-53	1967	CA	67	102215
45790	TEMPERATURE DEPENDENCE OF THERMOPHYSICAL PROPERTIES OF SOME POLYMERIC MATERIALS. BARANOVSKII V M DUSHCHENKO V P SHUT N I KRASNOBOKII YU N PLAST MASSY						
	9	66-7	1967	CA	67	109083	
45796	PREDICTION OF THE MACROSCOPIC CONTACT AREA IN THERMAL CONTACT RESISTANCE. PH.D. THESIS. MC NARY R O UNIVERSITY OF ILLINOIS URBANA ILLINOIS UNIV MICROFILMS PUBL						
	68-8165	1-101	1967	PA	N69-7-5 912		
45800	THERMOPHYSICAL CHARACTERISTICS AND HEAT OF MELTING OF BITUMEN-RUBBER COMPOSITION. FOGEL V O ZANEMONETS N A ALEKSEEV P G KAPLUNOV YA N STROIT TRUBOPROVODOV						
	12	2	13-14	1967	CA	67	100829
45807	THERMAL-EXPANSION STRESSES IN REINFORCED PLASTICS. RESEARCH PAPER NO. 1745. TURNER P S J RESEARCH NATL BUREAU STANDARDS						
	37	4	239-50	1946			
45808	VACUUM DEPOSITION OF DIELECTRIC AND SEMICONDUCTOR FILMS BY A CO2 LASER. HASS G RAMSEY J B APPL OPT						
	8	6	1115-18	1969			
45820	INVESTIGATION OF INTERFACIAL TENSION AT THE BOUNDARY BETWEEN LEAD-SODIUM ALLOYS AND SODIUM CHLORIDE MELTS. NARYSHKIN I I MORACHEVSKII A G PATROV B V ZHUR PRIKLAD KHIM						
	40	6	1315-18	1967			
45825	THERMAL CONDUCTIVITY OF BEDS OF COATED FUEL PARTICLES. REPT. GA-7241. STEVENS D W NUCL APPL						
	3	10	626-34	1967	CA	67	104286

TPRC Number	Bibliographic Citation

45829 INTEGRAL SHORT-CIRCUIT CURRENT OF SOLAR CELLS FROM THEIR SPECTRAL CHARACTERISTICS.
KOLTUN M M   GOLOVNER T M   KHOLEVA M N
GELIOTEKHNIKA
         6   3-7   1966

45830 INTEGRAL SHORT-CIRCUIT CURRENT OF SOLAR CELLS FROM THEIR SPECTRAL CHARACTERISTICS. FROM RESEARCH IN SOLAR ENGINEERING 1967. //ENGLISH TRANSLATION OF GELIOTEKHNIKA /6/ 3-7, 1966.//
KOLTUN M M   GOLOVNER T M   KHOLEVA M N
NASA
JPRS-41144 TT-67-31785
N67-35471   1-7   1967

45834 ROLE OF DIFFUSION AND ADSORPTION-DESORPTION ANTAGONISM IN THE DEVELOPMENT OF LIQUID-LIQUID INTERFACES.
BARET J F   MERIGOUX R
COMPT REND
264   C   23   1785-8   1967

45840 MEASUREMENT OF THE SPECTRAL REFLECTANCE OF SEVERAL AEROSPACE MATERIALS. M.S. THESIS, MECH. ENG.
GRIMM T C   WASHINGTON U   ST LOUIS MO
WASHINGTON U   ST LOUIS MO
      1-96   1965

45847 THERMAL RADIATION PROPERTIES OF CERAMIC MATERIALS. FROM PROCEEDINGS OF A SYMPOSIUM ON MECHANICAL AND THERMAL PROPERTIES OF CERAMICS. GAITHERSBURG, MARYLAND. APRIL 1-2, 1968.
RICHMOND J C   NATIONAL BUREAU OF STANDARDS
WASHINGTON D C
USGPO
NBS-SP-303 N69-28431
      125-37   1969   CA   71   53012

45871 REFLECTANCE AT PAINT-SUBSTRATE BOUNDARIES.
TUNSTALL D F
J OIL COLOUR CHEM ASSOC
50   11   989-1007   1967   CA   67   11813

45873 HIGH-REFLECTANCE MULTILAYER DIELECTRIC MIRRORS.
BEHRNDT K H   DOUGHTY D W
J VAC SCI TECHNOL
4   4   199-202   1967   CA   67   121249

45906 MEASUREMENT OF INTERFACIAL TENSIONS BY THE PENDENT DROP METHOD. SOLUTIONS OF SODIUM DODECYL SULFATE VERSUS AIR AND NUJOL. M.S. THESIS.
HALL B T   PENNSYLVANIA STATE UNIV
PENN STATE UNIV   UNIVERSITY PARK
      1-68   1966

45911 A STUDY OF INTERFACIAL TENSION IN TERNARY SYSTEMS. M.S. THESIS, CHEM. ENG.
PROCHASKA F O JR   TEXAS TECHNOLOGICAL COLLEGE
LUBBOCK
TEXAS TECHNOLOGICAL COLLEGE   LUBBOCK
      1-81   1966

45921 LIGHT TRANSMITTANCE AND REFLECTANCE OF COLORED PLANTS. M.S. THESIS, CHEM. ENG.
YEN W H   U OF LOUISVILLE   KENTUCKY
U OF LOUISVILLE   KENTUCKY
      1-96   1966

45925 AN INVESTIGATION OF THE MECHANISMS OF HEAT TRANSFER IN LOW-DENSITY PHENOLIC-NYLON CHARS.
SMYLY E D   PYRON C M JR   PEARS C D   SOUTHERN RESEARCH INSTITUTE
NASA
NASA-CR-966 N68-11978
      1-97   1967   TT   68-4 69

45943 INFRARED SPECTRAL ABSORPTION OF THE INTERNAL VIBRATIONS OF MONOPOTASSIUM PHOSPHATE.
RATAJCZAK H
COMPT REND
265   B   15   807-10   1967

45950 ELECTRONIC STRUCTURE AND STABILITY OF THE INORGANIC FULMINATES.
IQBAL Z   YOFFE A D
PROC ROY SOC /LONDON/
302   A   35-49   1967

45954 VAPOR DEPOSITION OF SILICON NITRIDE ON GALLIUM ARSENIDE BY SI CL4-NH3-N2 SYSTEM.
SEKI H   MORIYAMA K
JAPAN J APPL PHYS
6   11   1345-6   1967

45972 EXACT FINITE METHOD OF LATTICE STATISTICS. II. HONEYCOMB-LATTICE GAS OF HARD MOLECULES.
RUNNELS L K   COMBS L L   SALVANT J P
J CHEM PHYS
47   10   4015-20   1967

45975 EXPERIMENTAL ANALYSIS OF INTERFACIAL FORCES AT THE PLANE SURFACE OF SOLIDS.
TAMAI Y   MAKUUCHI M   SUZUKI M

45977 INTERFACIAL TENSION FROM LATTICE MODELS OF TERNARY RANDOM MIXTURES.
SHENDALMAN L H   O TOOLE J T
J PHYS CHEM
71   13   4222-7   1967

45984 THE DETERMINATION BY SPECTRAL REFLECTANCE OF COPPER CONCENTRATED ON CHROMATOGRAPHIC COLUMNS. M.S. THESIS, CHEMISTRY.
LABINOWICH E P   U OF HAWAII   HONOLULU
U OF HAWAII   HONOLULU
      1-49   1966

45987 SPECTRAL REFLECTANCE MEASUREMENTS OF TEFLON AND TEFLON-COATED COPPER AND ALUMINUM. M.S. THESIS, MECH. ENG.
GILPIN T M   U OF WASHINGTON   SEATTLE
U OF WASHINGTON   SEATTLE
      1-59   1966

45991 STUDIES ON EFFECTIVE THERMAL CONDUCTIVITY IN PACKED BED. M.S. THESIS, CHEM. ENG.
DANDH K V   U OF NEW HAMPSHIRE   DURHAM
U OF NEW HAMPSHIRE   DURHAM
      1-78   1965

46004 OPTICAL PROPERTIES OF SODIUM IN THE VACUUM ULTRAVIOLET. PH.D. THESIS OF J. C. SUTHERLAND.
SUTHERLAND J C   HAMM R N   STEVENSON J R
ARAKAWA E T   OAK RIDGE NATIONAL LAB   TENN
AEC
ORNL-TM-1776 N67-34986
      1-126   1967   PA   N67-5-20 3715

46008 PROBLEMS OF INFRARED ATMOSPHERIC SPECTROSCOPY RELATED TO THE SATELLITE DETERMINATION OF TEMPERATURE OF UNDERLYING SURFACE.
KONDRATYEV K YA
FIZ ATM I OKEANA /MOSCOW/
5   6   616-30   1969

46011 A NEW METHOD FOR THE DETERMINATION OF PARTIAL THERMAL ACCOMMODATION COEFFICIENTS.
SCHAFER K   RIGGERT K H
Z ELEKTROCHEM /WEST GERMANY/
57   8   751-7   1953

46012 A NEW METHOD FOR THE DETERMINATION OF PARTIAL THERMAL ACCOMMODATION COEFFICIENTS. //ENGLISH TRANSLATION OF Z. ELEKTROCHEM. /WEST GERMANY/ 57 /8/ 751-7, 1953.//
SCHAFER K   RIGGERT K H
TC
TC 3243   1-14   1967   TT   17-11 60

46022 THE HEAT CONDUCTIVITY OF COMPOSITE PROPELLANTS. FROM SESSION REPORT OF THE NGLR ENGINE FUEL COMMITTEE. IN GERMAN SYMP. HELD IN TRAUEN, WEST GERMANY.
SELZER H
CFSTI
DLR-MITT-66-14 N67-24394
N67-24381   171-81   1966   PAN 67-5 2196

46025 A PIGMENTED COATING FOR THE CONTROL OF RADIANT ENERGY TRANSFER ACROSS THE SURFACE OF THE AM-16 SATELLITE.
ZERLAUT G A   ARMY BALLISTIC MISSILE AGENCY
REDSTONE ARSENAL   ALABAMA
NASA
DSN-TN-18-59
N-95341   1-31   1959

46029 ADSORPTION FROM SOME N-ALKANE MIXTURES AT THE LIQUID/ VAPOUR, LIQUID/WATER AND LIQUID/SOLID INTERFACES.
AVEYARD R
TRANS FARADAY SOC
63   11   2778-88   1967

46040 THERMOPHYSICS RESEARCH AT MARSHALL SPACE FLIGHT CENTER. FROM RESEARCH ACHIEVEMENTS REVIEWS. VOL. 1. SERIES 1-22.
HELLER G B   NATIONAL AERONAUTICS AND SPACE ADMINISTRATION   HUNTSVILLE ALABAMA
NASA
NASA-TM-X-53620
N67-30556   1-30   1967   PA   N67-5-17 3155

46041 MATERIALS RESEARCH AT MARSHALL SPACE FLIGHT CENTER. FROM RESEARCH ACHIEVEMENTS REV. VOL. 1. SERIES 1-22.
LUCAS W R   NATIONAL AERONAUTICS AND SPACE ADMINISTRATION   HUNTSVILLE ALABAMA
NASA
NASA-TM-X-53620
N67-30565   1-18   1967

46042 LUNAR AND METEOROID PHYSICS RESEARCH AT MARSHALL SPACE FLIGHT CENTER. FROM RESEARCH ACHIEVEMENTS REVIEWS. VOL. 1. SERIES 1-22.
HELLER G B   DOZIER J B JR   NATIONAL AERONAUTICS AND SPACE ADMINISTRATION   HUNTSVILLE ALABAMA
NASA
NASA-TM-X-53620
N67-30611   1-18   1967

TPRC Number	Bibliographic Citation

46055 THERMAL EXPANSION OF CEMENTED CASING. M.S. THESIS, PET. ENG.
SMITH F M   UNIV OF PITTSBURGH PA
U OF PITTSBURGH PA
1-19   1966

46059 MEASUREMENT OF THERMAL CONDUCTIVITY DURING FREEZE-DRYING OF BEEF. M.S. THESIS, MECH. ENG.
MASSEY W M JR   GEORGIA INST OF TECH ATLANTA
GEORGIA INST OF TECH ATLANTA
1-74   1966

46073 INVESTIGATION OF THE ADSORPTION OF TRIMETHYLCARBINOL ON MONTMORILLONITE BY THE METHOD OF INFRARED SPECTROSCOPY.
TARASEVICH YU I   RADUL N M   OVCHARENKO F D
DOKL AKAD NAUK SSSR
173   3   615-17   1967

46074 INVESTIGATION OF THE ADSORPTION OF TRIMETHYLCARBINOL ON MONTMORILLONITE BY THE METHOD OF INFRARED SPECTROSCOPY. //ENGLISH TRANSLATION OF DOKL. AKAD. NAUK SSSR 173 /3/ 615-7, 1967.//
TARASEVICH YU I   RADUL N M   OVCHARENKO F D
DOKL PHYS CHEM
173   3   230-2   1967

46077 SURFACE ACTIVITY OF A BIOLOGICALLY ACTIVE SEMICOLLOID-SODIUM CHOLATE. //ENGLISH TRANSLATION OF DOKL. AKAD. NAUK SSSR 173 /5/ 1132-5, 1967.//
MARKINA Z N   TSIRKURINA N N   BOVKUN O P
KOPEINA A D   REBINDER P A
DOKL PHYS CHEM USSR
173   5   275-8   1967

46082 COMPOSITION AND PROPERTIES OF MANGANIC AMMONIUM PHOSPHATE.
RISKIN I V   KALINSKAYA T V
ZH PRIKLAD KHIM
40   1   19-30   1967

46083 COMPOSITION AND PROPERTIES OF MANGANIC AMMONIUM PHOSPHATE. //ENGLISH TRANSLATION OF ZH. PRIKLAD. KHIM. 40 /1/ 19-30, 1967.//
RISKIN I V   KALINSKAYA T V
J APPL CHEM USSR
40   1   16-24   1967

46106 MEASUREMENT OF THE COEFFICIENT OF THERMAL CONDUCTIVITY OF FELT-LIKE INSULATING MATERIAL. DEUTSCHE LUFT-UND RAUMFAHRT FORSCHUNGS.
STIEGLITZ C B V   INST FUER FLUGZEUGBAU BRUNSWICK
WEST GERMANY
CFSTI
N67-35065   DLR-FB-67-29
1-27   1967   RR   67-21 107

46109 MEASUREMENT OF SOLAR ABSORPTANCE AND TOTAL NORMAL EMITTANCE OF MATERIALS. M.S. THESIS, MECH. ENG.
BARRY D W   U OF HOUSTON TEXAS
U OF HOUSTON TEXAS
1-24   1965

46119 THE THERMAL ACCOMMODATION COEFFICIENT-A LITERATURE SURVEY. M.S. THESIS.
GILBERT J   RENSSELAER POLYTECHNIC INSTITUTE TROY
NEW YORK
RENSSELAER POLYTECHNIC INSTITUTE
1-88   1967

46135 THERMAL EMISSION DUE TO THE VIRTUAL MODE IN LITHIUM FLUORIDE.
HISANO K   OKAMOTO Y   KUBOTA T   MATUMURA O
J PHYS SOC JAPAN
23   6   1422   1967

46157 OPTICAL AND PHYSICAL PROPERTIES OF SYNTHETIC URANINITE.
OSIPOV B S
DOKL AKAD NAUK SSSR
176   3   672-5   1967   CA   68   7165

46167 MEASUREMENT OF TOTAL AND SPECTRAL EMISSIVITY OF A TITANIUM SURFACE. M.S. THESIS.
CONRAD A G   NORTHWESTERN UNIVERSITY EVANSTON
ILLINOIS
NORTHWESTERN UNIVERSITY
1-31   1960

46170 PLANE TRICALORIMETRIC METHOD FOR MEASURING THE THERMAL CONDUCTIVITY OF POOR HEAT CONDUCTORS.
NAZIEV YA M
IZV VYSSH UCHEB ZAVED ENERG
10   7   76-81   1967   CA   68   7057

46172 APPARATUS FOR STUDYING THE EFFECT OF LOAD ON THE THERMAL CONDUCTIVITY OF GRANULAR SYSTEMS IN VACUO.
ZARICHNYAK YU P   MURATOVA B L   PLATUNOV E S
IZV VYSSH UCHEB ZAVED PRIBOROSTR
10   2   105-9   1967   CA   68   7046

TPRC Number	Bibliographic Citation

46184 METHOD AND EQUIPMENT FOR MEASURING THE THERMAL CONDUCTIVITY OF INSULATING MATERIALS.
SZCZENIOSKI B
REV GEN THERM
6   68   1047-72   1967   CA   68   7059

46217 STORAGE AND HANDLING OF CRYOGENIC FLUIDS. FROM CONFERENCE ON SELECTED TECHNOLOGY FOR THE PETROLEUM INDUSTRY.
NORED D L   HENNINGS G   SINCLAIR D H   SMITH G T
SMOLAK G R   STOFAN A J   LEWIS RESEARCH CENTER
NASA
NASA   SP   5053
125-53   1966

46221 A RADICAL HEAT FLOW APPARATUS FOR THERMAL CONDUCTIVITY MEASUREMENTS OF FLEXIBLE MATERIALS TO 800 C. M.S. THESIS.
HANSEN A J   U OF TOLEDO OHIO
U OF TOLEDO OHIO
1-37   1965

46250 THERMAL CONDUCTIVITY AND DIELECTRIC CONSTANT OF SILICATE MATERIALS. PROGRESS REPT. APRIL 1965-AUGUST 1966.
WECHSLER A E   SIMON I   LITTLE /ARTHUR D/ INC
CAMBRIDGE MASS
NASA
NASA-CR-81495
N68-15849   1-133   1966   RR   68-7 94

46251 INFRARED RADIATION METHOD OF DETERMINING THERMAL DIFFUSIVITY, HEAT CAPACITY AND THERMAL CONDUCTIVITY OF SOLID MATERIALS.
OGAWA K   NATIONAL AEROSPACE LABORATORY TOKYO
NATIONAL AEROSPACE LABORATORY /TOKYO/
NAL-TR-128   1-11   1967
CFSTI
N67-34640   1-15   1967   PA   67-5-20 37

46252 INFRARED RADIATION METHOD OF DETERMINING THERMAL DIFFUSIVITY, HEAT CAPACITY AND THERMAL CONDUCTIVITY OF SOLID MATERIALS. //ENGLISH TRANSLATION OF NAL-TR-128, 1-11, 1967.//
OGAWA K   AZTEC SCHOOL OF LANGUAGES INC RESEARCH TRANSLATION DIV ACTON MASS
NASA AND CFSTI
NASA-TT-F-11453
N68-17901   1-21   1968   RR   68-9 154

46261 OPTIMIZATION OF LEAD STEARATE CRYSTALS FOR THE DIFFRACTION OF ULTRA SOFT X-RAYS.
CHARLES M W   COOKE B A
J SCI INSTR
44   12   976-82   1967

46277 THERMAL RADIATION OF MICROROUGH OR DISPERSED METALS.
PY B
INTERN J HEAT MASS TRANSFER
10   12   1735-42   1967

46313 DETERMINATION OF THE COMPLEX INDEX AND OPTICAL THICKNESS OF THIN FILMS OF GERMANIUM FROM MEASUREMENTS OF REFLECTIVITY BY MEANS OF OBLIQUE INCIDENCE.
PAUTY M   GOUDONNET J-P
COMPT REND
265   B   18   997-1000   1967

46328 BOUNDARY SURFACE ACTIVITY AT INTERFACES OF MERCURY AND ORGANIC LIQUIDS.
STEUDEL TH   NEUMANN A W
ABH DEUT AKAD WISS BERLIN KL CHEM GEOL BIOL 1966
6   861-7   1967   CA   68   16377

46334 THERMAL CONDUCTIVITY. SUMMARY OF THE LINE-SOURCE METHOD AND THEORY OF HETEROGENEOUS MIXTURES. SUMMARY REPT. NO. 10 JULY-SEPT. 1964.
TANGER G E   NIX G H   CHENG S C   LOWERY G W
SCHOOL OF ENGINEERING AUBURN UNIVERSITY ALABAMA
NASA
NASA-CR-59821
N65-13874   1-43   1964   PA   N65-4 698

46335 PHASE CHANGE IN POLYMERIC SYSTEMS FOR ACTIVE THERMAL CONTROL. FROM PROCEEDINGS OF CONFERENCE ON ACTIVE TEMPERATURE CONTROL
KELLIHER W C   PEZDIRTZ G F   YOUNG P   LANGLEY RESEARCH CENTER
NASA AND CFSTI
NASA-TM-X-56165 N66-32946
N66-32952   1-11   1964

46336 DETERMINATION OF THE EMISSIVITY OF MATERIALS. SEMIANNUAL PROGRESS REPT., NOV. 15, 1966-MAY 14, 1967.
WALEK W J   LUOMA W L   EMANUELSON R C   PRATT AND WHITNEY AIRCRAFT EAST HARTFORD CONNECTICUT
CFSTI AND NASA
NASA-CR-72294 PWA-3129
N67-37318   1-58   1967   PA   N67-5-22 4162

TPRC Number	Bibliographic Citation

46337 HANDBOOK OF OPTICAL PROPERTIES FOR THERMAL CONTROL
SURFACES, VOLUME III. FINAL REPT.
BREUCH R   LOCKHEED MISSILES AND SPACE CO
SUNNYVALE CALIF
CFSTI AND NASA
NASA-CR-87484 LMSC-A847882 VOL III
N67-34625   1-81   1967   PA   67-5-20 3746

46357 SPREADING PRESSURES AND COEFFICIENTS, INTERFACIAL
TENSIONS, AND ADHESION ENERGIES OF THE LOWER ALKANES,
ALKENES, AND ALKYL BENZENES ON WATER.
POMERANTZ P   CLINTON W C   ZISMAN W A
J COLLOID INTERFACE SCI
24   1   16-28   1967   CA   68   16383

46360 SURFACE ACTIVITY AND STABILITY OF STYRENE EMULSIONS
IN AQUEOUS SOLUTIONS OF POLY/VINYL ALCOHOL/. EFFECT
OF POLY/VINYL ALCOHOL/ ON INTERFACIAL TENSION OF THE
STYRENE-WATER BOUNDARY.
GROMOV E V   KREMNEV L YA   EGOROVA E I
BARENBAUM R KH   ABRAMZON A A   DOKUKINA L F
OSTROVSKII M V
KOLLOID ZH
29   4   484-8   1967   CA   68   16457

46366 ACTIVITY REPORT ON HIGH TEMPERATURE MATERIALS AND
COATING WORK AT AMF. FROM SUMMARY OF THE NINTH
REFRACTORY COMPOSITES WORKING GROUP MEETING.
BROWNING M E   SCHATZ E A   LEAVENWORTH H W JR
DUNKERLEY F J   AMERICAN MACHINE AND FOUNDRY CO
ALEXANDRIA VA
DDC
AFML-TR-64-398
AD-477414   979-1009   1965

46367 FEATURES OF THE THERMAL EXPANSION OF SINTERED OPTICAL
GLASS FIBERS.
GOMELSKII M S   POLUKHIN V N   KHABAROVA E A
BABKINA V A
OPT-MEKH PROM
34   5   41-7   1967   CA   68   15628

46368 VISCOUS FLOW OF GLASS DURING ELONGATION OF FILAMENTS
WITHOUT A SHELL AND IN A SHELL MADE FROM ANOTHER
GLASS.
ALEKSEENKO M P   KHABAROV YU M
OPT-MEKH PROM
34   7   35-9   1967   CA   68   15610

46369 THERMAL CONDUCTIVITY OF SOME PLASTICS.
ANDREEVA E A   BURAVOI S E   KUREPIN V V
PLAST MASSY
               7   61   1967   CA   68   13629

46396 HEAT TRANSFER ACROSS SURFACES IN CONTACT. PRACTICAL
EFFECTS OF TRANSIENT TEMPERATURE AND PRESSURE
ENVIRONMENTS. SEMIANNUAL REPT. 1 APR.-1 OCT. 1966.
BLUM H A   SOUTHERN METHODIST UNIV   DALLAS TEXAS
NASA
NASA-CR-82396
N67-18979   1-13   1966

46397 RESEARCH AND EVELOPMENT IN A THERMAL INSULATION
STUDY. A QUARTERLY PROGRESS REPT. JAN.-MAR. 1967.
DE WITT W D   LINDE DIV UNION CARBIDE CORP
INDIANAPOLIS IND
CFSTI
ALO-3632-14 N67-33700
                      1-48   1967   PA   N67-5-19 3555

46400 THERMAL ACCOMMODATION OF GASES ON SOLIDS. FROM
FUNDAMENTALS OF GAS-SURFACE INTERATIONS. PROC. OF
THE SYPOSIUM, SAN DIEGO, CALIF., DEC. 14-16, 1966.
THOMAS L B
ACADEMIC PRESS   NEW YORK AND LONDON
                      346-69   1967

46403 MEASUREMENT OF SIXTY-DEGREE SPECULAR GLOSS. /NBS RP
2105/
HAMMOND H K III   NIMEROFF I
J OF RESEARCH NATIONAL BUREAU OF STANDARDS
44   585-9   1950

46420 RADIATION PROPERTIES OF THERMISTOR BEADS.
TOSCANO W M   MASSACHUSETTS INSTITUTE OF
TECHNOLOGY   CAMBRIDGE
DDC
AFCRL-68-0448 N69-37867
AD-689805   1-78   1969   PA   N69-7-22 4154

46438 SURFACE TENSION OF YTTRIUM-MODIFIED CAST IRON.
COSNEANU C   MOISEEV I
METALLURGIA /BUCHAREST/
'19   6   311-15   1967   CA   68   23889

46441 THE WETTING OF AL2O3 AND SIO2 BY MOLTEN IRON AND
IRON-CARBON ALLOYS.
MUKAI K   SAKAO H   SANO K
NIPPON KINZOKU GAKKAISHI
31   8   923-8   1967   CA   68   23898

46471 X-20 NOSE CAP MATERIALS SUMMARY REPORT. FROM
SUMMARY OF THE NINTH REFRACTORY COMPOSITES WORKING
GROUP MEETING.
SEEGER J W   LOWRANCE D T   ROGERS D C   LTV
AETRONAUTICS DIVISION LING-TEMCO-VOUGHT INC   DALLAS
TEXAS
DDC
AFML-TR-64-398
AD-477414   1021-151   1965

46479 LOCAL STUDY OF THE TEMPERATURE FIELD IN THE
NEIGHBORHOOD OF THE CONTACT BETWEEN A SOLID AND A
LIQUID.
PADET J-P
COMPT REND
265   B   25   1365-8   1967

46480 THE BRIEF SIGNAL METHOD APPLIED TO THE MEASUREMENT OF
THE THERMAL CONTACT RESISTANCE.
LAURENT M   MACQUERON J L   GERY A   SINICKI G
COMPT REND
265   B   25   1369-71   1967

46482 OPTICAL PROPERTIES OF THIN COPPER FILM IN THE
SPECTRAL REGION BETWEEN 0.8 AND 4 MICRONS.
VARENNE S
COMPT REND
265   B   25   1467-70   1967

46485 FUNDAMENTAL ABSORPTION AND EXCITATION IN A SERIES OF
IIA-VIB COMPOUNDS.
LEVSHIN V L   MIKHAILIN V V   SAULEVICH L K
IZV AKAD NAUK SSSR SER FIZ
33   6   974   1969

46496 A STUDY OF GAS-SURFACE INTERACTIONS EMPLOYING A
CONTINUUM MODEL OF THE SOLID. PH.D. THESIS.
SCOTT L B JR   STANFORD UNIVERSITY   PALO ALTO
CALIFORNIA
UNIV MICROFILM PUBL
67-17501   1-161   1967   PA   N69-7-1 124

46504 MEASUREMENT OF THE IN-PILE CORE TEMPERATURE OF AN
EL-4 PENCIL ELEMENT, FIRST CHARGE /CAN OF STAINLESS
STEEL, TYPE 347-0.4 MM. THICK—UO2 FUEL—11 MM.
DIAMETER/. DETERMINATION OF THE APPARENT THERMAL
CONDUCTIVITY INTEGRAL OF IN-PILE UO2.
LAVAUD B   RINGOT C   VIGNESOULT N   CENTRE
D ETUDES NUCLEAIRES   SACLAY FRANCE
AEC
CEA-R 3041   1-23   1966

46508 CORROSION AND HEAT RESISTANCE OF ALUMINUM-COATED
STEEL.
SCHMITT R J   RIGO J H
ALLUM NUOVA MET
36   2   73-80   1967   CA   68   24051

46516 A STUDY OF TRANSIENT THERMAL CONTACT CONDUCTANCE.
M.S. THESIS.
WALTER J D   SOUTHERN METHODIST UNIVERSITY   DALLAS
UNIVERSITY   DALLAS TEXAS
SOUTHERN METHODIST UNIVERSITY AND DDC
AD-297995   1-76   1963

46528 EVALUATION OF POLYBENZIMIDAZOLE GLASS-FABRIC
LAMINATES.
MAC KAY H A
MODERN PLASTICS
43   5   149-52 247 1966

46542 SURFACE TENSION OF SLICK PATCHES NEAR KELP BEDS.
STURDY G   FISCHER W H
NATURE
211   951-2   1966

46543 ESTIMATION OF SURFACE ENERGIES FROM CONTACT ANGLES.
GOOD R J
NATURE
212   276-7   1966

46545 THE EXPANSION/CONTRACTION BEHAVIOR OF LAMINAR LIQUID
JETS
OLIVER D R
CAN J CHEM ENG
44   2   100-7   1966

46551 INTERFACIAL TENSION EFFECTS IN FINITE, PERIODIC,
TWO-DIMENSIONAL SYSTEMS.
MAYER J E   WOOD W W
J CHEM PHYSICS
42   12   4268-74   1965

46554 EFFECT OF SURFACE TENSION ON MULTILAYER GAS
ADSORPTION AT MODERATE PRESSURES.
BAKRI M M
J CHEM PHYS
44   6   2488-95   1966

TPRC Number	Bibliographic Citation

46559 INTERFERENCE FILTERS FOR THE NEAR ULTRA-VIOLET. FROM COLLOQUE SUR LES METHODES NOUVELLES DE SPECTROSCOPIE INSTRUMENTALE, 2ND, UNIVERSITE DE PARIS, APR 25-9, 1968.
NEILSON R G T   RING J
J DE PHYSIQUE SUPPL
28 3/4   C2- 270-5   1967

46589 MEASURING THERMAL CONDUCTIVITY OF MATERIALS WITH HIGH LAMBDA VALUES.
VISHNEVSKII I I   SKRIPAK V N
OGNEUPORY
9   9-11   1966

46590 MEASURING THERMAL CONDUCTIVITY OF MATERIALS WITH HIGH LAMBDA VALUES. //ENGLISH TRANSLATION OF OGNEUPORY /9/ 9-11, 1966.//
VISHNEVSKII I I   SKRIPAK V N
REFRACTORIES
9   499-500   1966

46606 QUARTZ PRODUCTS MADE BY SLIP CASTING.
VORONIN N I   CHURAKOVA R S
OGNEUPORY
1   47-9   1967

46607 QUARTZ PRODUCTS MADE BY SLIP CASTING. //ENGLISH TRANSLATION OF OGNEUPORY /1/ 47-9, 1967.//
VORONIN N I   CHURAKOVA R S
REFRACTORIES
1   50-2   1967

46625 INTERFACIAL TENSIONS IN TWO-LIQUID-PHASE TERNARY SYSTEMS.
PLISKIN I   TREYBAL R E
J CHEM ENG DATA
11   1   49-52   1966

46632 THE MEASUREMENT OF SURFACE TENSION BY THE PENDANT DROP TECHNIQUE.
STAUFFER C E
J PHYS CHEM
69   6   1933-8   1965

46639 MEASUREMENT OF THERMAL CONDUCTANCE OF MULTILAYER AND OTHER INSULATION MATERIALS. FINAL REPORT AND SUPPL.
FRIED E   KARP G S   HOBBS R B   HIESER G
GENERAL ELECTRIC CO   PHILADELPHIA PA
CFSTI AND NASA
67-SD-4388 NASA-CR-65797
N68-10401   1-52   1967   RR   68-2 52
SUPPLEMENT - NASA-CR-92329
N68-37110   1-15   1968   RR   69-1 164

46640 INORGANIC SURFACE COATINGS FOR SPACE APPLICATIONS.
SIBERT M E   LOCKHEED AIRCRAFT CORP MISSILES AND SPACE DIVISION   VAN NUYS CALIFORNIA
CFSTI
LMSC-3-77-61-12
1-36   1961

46641 SURFACE ENERGY AT METAL/DIELECTRIC LIQUID INTERFACES.
ZADUMKIN S N   KARASHAEV A A
FIZ KHIM MEKH MATERIALOV
1   2   139-41   1965

46642 SURFACE ENERGY AT METAL/DIELECTRIC LIQUID INTERFACES. //ENGLISH TRANSLATION OF FIZ.-KHIM. MEKH. MATERIALOV 1 /2/ 139-41, 1965.//
ZADUMKIN S N   KARASHAEV A A
SOVIET MATERIALS SCIENCE
1   2   86-8   1965

46645 THE LONGITUDINAL AND DIAMETRAL EXPANSIONS OF UO2 FUEL ELEMENTS.
NOTLEY M J F   BAIN A S   ROBERTSON J A L
CHALK RIVER ONTARIO
AEC
AECL-2143   1-14   1964

46646 EXPERIMENTAL DETERMINATION OF THE THERMAL CONDUCTIVITY OF HEAT-RESISTING HIGH-ALUMINA CONCRETES AT TEMPERATURES BETWEEN 200 AND 1500 C. FROM MEZHDYNARODNYI SYMP. PO PROIZVODSTVU ELECTROENERGII S POMOSH YU O GENERATOROV. 1966.
CHEKHOVSKOI V YA   KAULENAS A A   MOSCOW STATE UNIV
AEC
CONF-660704-47
1-4   1966

46647 THE DEVELOPMENT AND TESTING OF THE UO2 FUEL ELEMENT SYSTEM. SUMMARY REPORT.
MURTHA B E   CHERNOCK W P   NUCLEAR DIVISION COMBUSTION ENG INC   WINDSOR CONNECTICUT
AEC
CEND-141   1-134   1961

46648 INVESTIGATION OF RADIATION AND CONDUCTION HEAT TRANSFER IN FIBROUS HIGH TEMPERATURE INSULATIONS. TECHNICAL REPT.
ROLINSKI E J   PURCELL G V   WRIGHT-PATTERSON AIR FORCE BASE OHIO
DDC
AFML-TR-67-251
1-139   1967

46667 ADHESION OF IRON ALLOYS WITH CHROMIUM, MOLYBDENUM AND TUNGSTEN WITH WHITE ELECTRIC STEEL ALLOY SLAGS.
POPEL S I   DERYABIN A A   ZUPNIK A E
IZV VYSSHIKH UCHEBN ZAVEDENI CHERNAYA MET
1   21-4   1966

46668 INTERFACE TENSION OF LIQUID STEEL ON THE BOUNDARY WITH THE WHITE SLAG OF ELECTRIC STEEL.
YAKUSHEV A M   SMOLYARENKO V D   EDNERAL F P
IZV VYSSHIKH UCHEBN ZAVEDENI CHERNAYA MET
11   35-8   1966

46671 DETERMINATION OF THE SURFACE ENERGY OF SILICON IRON BY THE METHOD OF THE NEUTRAL DROP.
AVRAAMOV YU S   SEMENOV V M   LEVIN I YA
IZV VYSSHIKH UCHEBN ZAVEDENI CHERNAYA MET
7   129-32   1966

46683 PHASE-DISPERSION POLARIZATION INTERFERENCE FILTER.
IOFFE S B   DRICHKO N M
DOKLADY AKAD NAUK SSSR
164   4   793-5   1965

46684 PHASE-DISPERSION POLARIZATION INTERFERENCE FILTER. //ENGLISH TRANSLATION OF DOKLADY AKAD. NAUK SSSR 164 /4/ 793-95, 1965.//
IOFFE S B   DRICHKO N M
SOVIET PHYSICS-DOKLADY
10   10   971-3   1966

46695 EXPERIMENTAL DETERMINATION OF TEMPERATURE IN SHOCK-COMPRESSED NACL AND KCL AND OF THEIR MELTING CURVES AT PRESSURES UP TO 700 KBAR.
KORMER S B   SINITSYN M V   KIRILLOV G A   URLIN V D
J EXPTL THEORET PHYS /USSR/
48   1033-49   1965

46696 EXPERIMENTAL DETERMINATION OF TEMPERATURE IN SHOCK-COMPRESSED NACL AND KCL AND OF THEIR MELTING CURVES AT PRESSURES UP TO 700 KBAR. //ENGLISH TRANSLATION OF J. EXPTL. THEORET. PHYS. /USSR/ 48, 1033-49, 1965.//
KORMER S B   SINITSYN M V   KIRILLOV G A   URLIN V D
SOVIET PHYS JETP
21   4   689-700   1965

46713 CALCULATION OF THE INTERFACIAL TENSION AND ADSORPTION AT THE BOUNDARY BETWEEN CONDENSED PHASES.
PAVLOV V V   POPEL S I   ESIN O A
ZH FIZ KHIM
39   1   214-   1965

46714 CALCULATION OF THE INTERFACIAL TENSION AND ADSORPTION AT THE BOUNDARY BETWEEN CONDENSED PHASES. //ENGLISH TRANSLATION OF ZH. FIZ. KHIM. 39 /1/ 214-, 1965.//
PAVLOV V V   POPEL S I   ESIN O A
RUSS J PHYS CHEM
39   1   114-16   1965

46721 EXPERIMENTAL VERIFICATION OF THE EQUATION DESCRIBING THE SURFACE ACTIVITY OF A PROCESS.
ROI D L   MIKHALIK E
ZH FIZ KHIM
39   2   510-   1965

46722 EXPERIMENTAL VERIFICATION OF THE EQUATION DESCRIBING THE SURFACE ACTIVITY OF A PROCESS. //ENGLISH TRANSLATION OF ZH. FIZ. KHIM. 39 /2/ 510-, 1965.//
ROI D L   MIKHALIK E
RUSS J PHYS CHEM
39   2   274-7   1965

46748 ELECTROCAPILLARY CURVES IN OXIDE MELTS.
DERYABIN A A   ESIN O A   POPEL S I
ZH FIZ KHIM
39   4   966-   1965

46749 ELECTROCAPILLARY CURVES IN OXIDE MELTS. //ENGLISH TRANSLATION OF ZH. FIZ. KHIM. 39 /4/ 966-, 1965.//
DERYABIN A A   ESIN O A   POPEL S I
RUSS J PHYS CHEM
39   4   508-11   1965

46790 DETERMINATION OF OPTICAL CONSTANTS OF PALLADIUM BY ELLIPSOMETRY. M.S. THESIS.
RIEDINGER M S C   VANDERBILT UNIVERSITY   NASHVILLE TENNESSEE
VANDERBILT UNIVERSITY
1-41   1968

TPRC Number	Bibliographic Citation

46797 EMITTANCE OF HAYNES C8752 AND FANSTEEL ALLOYS.
STRUCTURAL MATERIALS DIVISION
AEROJET-GENERAL CORPORATION   AZUSA
CALIFORNIA
DDC
AGC REP   T M-2119
AD-297685   1-10   1982

46810 MEASUREMENT OF HEMISPHERICAL TOTAL EMITTANCE AND
NORMAL SOLAR ABSORPTANCE OF SELECTED MATERIALS IN THE
TEMPERATURE RANGE 280 TO 600 K.
CURTIS H B   LEWIS RESEARCH CENTER   CLEVELAND OHIO
NASA
NASA AND CFSTI
NASA-TM-X-56167 N66-37816
                         1-18   1964
NASA-TM-X-56273 N66-18345
                         1-20   1965
J SPACECRAFT ROCKETS
3   3   383-7   1966   CA   64   18692

46814 NEW CONCENTRATE REINFORCES POLYETHYLENE.
DE VENUTO G   STOTZ D S   WIGGILL J B
MODERN PLASTICS
45   7   97-106   1968

46815 ANALYSIS OF PLASTICS BY ATR SPECTROSCOPY.
KRAFT E A
MODERN PLASTICS
45   7   123-6   1968

46817 STUDY OF OPTICAL PROPERTIES OF THIN LAYERS IN THE
NEAR INFRARED.
RICHARD J   ROUX D
REV OPT
41   9   469-78   1962

46830 SPECTROSCOPY AT EXTREME INFRA-RED WAVELENGTHS.   II.
THE LATTICE RESONANCES OF IONIC CRYSTALS.
JONES G O   MARTIN D H   MAWER P A   PERRY C H
PROC ROY SOC
261   10-27   1961

46842 EXPERIMENTAL STUDY OF THE OPTICAL PROPERTIES OF
LIQUID HG AND LIQUID GA IN THE WAVELENGTH RANGE OF
0.23 TO 13 MICRONS.
SCHULZ L G
J OPT SOC AM
47   1   64-9   1957

46843 INFRARED FILTERS OF ANTIREFLECTED SI, GE, INDIUM
ARSENIDE, AND INDIUM ANTIMONIDE.
COX J T   HASS G   JACOBUS G F
J OPT SOC AM
51   7   714-18   1961

46856 DETERMINATION OF OPTICAL PARAMETERS OF A OPAQUE
METALLIC FILM BY MEANS OF GONIOPHOTOMETRIC
MEASUREMENTS.
ABELES F
J PHYS
15   4   303-1   1954

46860 THERMAL DIFFUSIVITY OF CARBONS AND GRAPHITES IN THE
TEMPERATURE RANGE FROM 1800 TO 3300 K.
KASPAR J   ZEHMS E H   AEROSPACE CORP   EL SEGUNDO
CALIF
CFSTI AND DDC
ESD-TR-67-116
AD-655786   1-33   1967   RR   67-18 80

46863 SINGULARITIES OF MEASURING THE COEFFICIENTS OF
THERMAL CONDUCTIVITY OF MULTILAYER INSULATIONS.
GOLOVANOV L B   JOINT INST FOR NUCLEAR RESEARCH
DUBNA /USSR/ LAB OF HIGH ENERGY
NASA
JINR-P-8-3236
N67-33879   1-21   1967

46864 THERMOPHYSICAL PROPERTIES OF SOME MULTILAYER
INSULATIONS IN CRYOGENIC TEMPERATURES.
GOLOVANOV L B   JOINT INST FOR NUCLEAR RESEARCH
DUBNA /USSR/ LAB OF HIGH ENERGY
NASA
JINR-237   3-20   1967
JINR-P-8-3237
N67-33878   1-21   1967

46879 A COLLECTION OF SOME CONTROLLED SURFACE THERMAL
ACCOMMODATION COEFFICIENT MEASUREMENTS.   FROM
RAREFIED GAS DYNAMICS 5TH SYMPOSIUM 1 PP. 155-82.
1967.
THOMAS L B   PHYSICAL CHEMICAL LAB MISSOURI UNIV
COLUMBIA
DDC
AFOSR-67-1921 AD-657242
                        1-10   1967

46882 FUEL-CLAD CONTACT CONDUCTANCE.
ARRIGHI J   MUSTACCHI C   ZANELLA S   JOINT
NUCLEAR RESEARCH CENTER   ISPRA /ITALY/
AEC
EUR-3155.E   1-116   1966

46907 REINFORCED INORGANIC REFRACTORY CERAMIC COATINGS.
FROM SUMMARY OF THE SIXTH REFRACTORY COMPOSITES
WORKING GROUP MEETING. VOL. 1 JUNE 1962.
SKLAREW S   THE MARQUARDT CORPORATION   VAN NUYS
CALIFORNIA
DDC
ASD-TDR-63-610/VOL 1/
AD-427180   16-43   1962

46912 THE SPECTRAL EMITTANCE OF NICKEL- AND OXIDE-COATED
NICKEL CATHODES.
MARTIN S L   WESTON G F
BRIT J APPL PHYS
1   318-24   1950

46914 MEASUREMENT OF THE REFLECTING POWER OF THICK METALLIC
FILMS /AU, PT, CR/ IN THE SCHUMANN REGION
DETERMINATION OF OPTICAL CONSTANTS.
ROBIN S
COMPT REND
236   674-6   1953

46921 MEASUREMENT OF THE THERMAL CONDUCTIVITY OF
MANUFACTURED CARBON PRODUCTS.
WILKENING S
ALUMINIUM /DUSSELDORF/
43   6   367-71   1967

46922 MEASUREMENT OF DIFFUSE REFLECTION FACTORS, AND A NEW
ABSOLUTE REFLECTOMETER.
TAYLOR A H
NATL BUR STANDARDS TECH NEWS BULL
16   421-36   1920

46924 THE INVESTIGATION OF THIN ABSORBING LAYERS WITH AID
OF THE ABSOLUTE PHASE.
SCHOPPER H
Z PHYSIK
130   565-84   1951

46935 EFFECTIVE AXIAL AND DIAMETRAL THERMAL EXPANSION OF
UO2 UNDER A RADIAL TEMPERATURE GRADIENT.
ASAMOTO R R   PERRY K J   ZEBROSKI E L   GENERAL
ELECTRIC CO   SAN JOSE CALIF
CFSTI AND AEC
GEAP-5284   N68-12733
                        1-49   1966   CA   67   69774

46949 EXCHANGE OF ENERGY BETWEEN TUNGSTEN AT HIGH
TEMPERATURES AND GASES AT ROOM TEMPERATURE.   PH.D.
THESIS.
LIM M-J   LAWRENCE RADIATION LAB CALIFORNIA UNIV
BERKELEY CALIFORNIA
CFSTI AND AEC
UCRL-17395 N67-38506
                        1-98   1967   PA   N67-5-23 4294
UNIV MICROFILMS PUBL
68-103   1-100   1967   CA   68   108203

46981 EMISSIVITY OF METALS HEATED IN AIR.
ZHOROV G A
TEPLOFIZ VYSOKIKH TEMPERATUR
5   3   450-7   1967   IAA 7-18 3064

46982 EMISSIVITY OF METALS HEATED IN AIR.   //ENGLISH
TRANSLATION OF TEPLOFIZ. VYSOKIKH TEMPERATUR 5 /3/
450-57, 1967.//
ZHOROV G A
HIGH TEMPERATURE
5   3   403-9   1967

46986 SPECTRAL EMISSIVITY OF HIGHLY DOPED SILICON.
LIEBERT C H   THOMAS R D   LEWIS RESERCH CENTER
CLEVELAND OHIO
NASA AND CFSTI
NASA-TN-D-4303
N68-21708   1-25   1968   RR   68-13 151

46990 DESIGN, TEST, AND PERFORMANCE OF THE MARINER V
TEMPERATURE CONTROL REFERENCE.
CARROLL W F   JET PROPULSION LABORATORY CALIFORNIA
INSTITUTE OF TECHNOLOGY   PASADENA CALIFORNIA
NASA
JPL-TR-32-1250
                        1-25   1968

46995 THE RESTRAHLEN OF ROCK SALT AND SYLVITE.
RUBENS H   ASCHKINASS E
ANN PHYSIK
65   6   241-56   1898

46996 THE TEMPERATURE RADIATION OF INCANDESCENT OXIDES AND
OXIDE MIXTURES IN THE ULTRARED SPECTRAL REGION.
RITZOW G
ANN PHYSIK
19   5   769-99   1934

46997 OPTICAL CONSTANTS, ELECTRICAL RESISTANCE AND
STRUCTURE OF THIN METALLAYERS.
KRAUTKRAMER J
ANN PHYSIK
32   5   537-76   1938

TPRC Number	Bibliographic Citation

47003 A CONTRIBUTION TO THE OPTICS OF PAINT COATINGS.
KUBELKA P   MUNK F
Z TEKHN PHYSIK
12   593-601   1931

47008 INVESTIGATION IN THE SPECTRAL REGION BETWEEN 20 AND
40 MICRONS.
STRONG J
PHYS REV
37   1565-72   1931

47009 ON THE STRUCTURE AND INTERPRETATION OF THE INFRARED
ABSORPTION SPECTRA OF CRYSTALS.
BARNES R B   BRATTAIN R R   SEITZ F
PHYS REV
48   582-602   1935

47024 DISCUSSION 15. FROM SUMMARY OF SECOND HIGH-
TEMPERATURE INORGANIC REFRACTORY COATINGS WORKING
GROUP MEETING. TECH. REPT.
WACHTELL R L   CHROMALLOY CORP WRIGHT-PATTERSON
BASE OHIO
DDC
WADC-TR-59-415
AD 232536   55-7   1959

47025 DISCUSSION 18. FROM SUMMARY OF SECOND HIGH-
TEMPERATURE INORGANIC REFRACTORY COATINGS WORKING
GROUP MEETING. TECH. REPT.
SKLAREW S   MARQUARDT AIRCRAFT CO WRIGHT-PATTERSON
AIR FORCE BASE OHIO
DDC
WADC-TR-59-415
AD-232536   62-6   1959

47034 THE SPECTRAL REFLECTANCE OF WATER AND CARBON DIOXIDE
CRYODEPOSITS FROM 0.36 TO 1.15 MICRONS. FINAL REPT.
MAY 1965-MAY 1967.
WOOD B E   SMITH A M   MC CULLOUGH B A   ARO INC
ARNOLD AIR FORCE STATION   TENNESSEE
CFSTI AND DDC
AEDC-TR-67-131 AD-857018
N67-38105   1-48   1967

47035 EFFECT OF THERMAL RESISTANCE OF JOINTS UPON THERMAL
STRESSES. TECHNICAL REPORT 56-6.
GATEWOOD B E   AIR FORCE INSTITUTE OF TECHNOLOGY
WRIGHT-PATTERSON AIR FORCE BASE OHIO
DDC
AD-106014   1-12   1956

47038 THERMAL RADIATION PROPERTIES OF SOME POLYMER BALLOON
FABRICS. TECH. REPT. VI.
DINGWELL I W   LITTLE /ARTHUR D/ INC   CAMBRIDGE
MASS
CFSTI AND DDC
AD-654744   1-53   1967

47049 RADIATIVE TRANSFER IN ANISOTROPICALLY SCATTERING
MEDIA. ALLOWANCE FOR FRESNEL REFLECTION AT THE
BOUNDARIES.
HOTTEL H C   SAROFIM A F   EVANS L B   VASALOS I A
TRANS ASME J HEAT TRANSFER
90   C   1   56-62   1968

47062 THE SPECTRAL REFLECTANCE OF ORDNANCE MATERIALS AT
WAVELENGTHS OF 1 TO 12 MICRONS. FINAL REPT. NO.
3196.
WILBURN D K   RENIUS O   DETROIT ARSENAL   CENTER
LINE MICHGAN
DDC
AD-87246   1-51   1955

47068 ADVANCED CERAMIC SYSTEMS FOR ROCKET NOZZLE
APPLICATIONS. THIRD QUARTERLY PROGRESS REPT.
MICCIOLI B R   SPECK D A   THE CARBORUNDUM COMP
NAVAL ORDANCE SYSTEMS COMMAND   NIAGARA FALLS NEW YORK
DDC
AD-822220   1-35   1966

47070 DIRECTORY OF GRAPHITE AVAILABILITY. SECOND EDITION.
GLASSER J   GLASSER W J   OHIO STATE UNIV
RESEARCH FOUNDATION COLUMBUS OHIO
DDC
AFML-TR-67-113
AD-821777   1-345   1967

47079 PARAMETRIC REPRESENTATIONS OF GAS-SURFACE INTERACTION
DATA AND THE PROBLEM OF SLIP-FLOW BOUNDARY CONDITIONS
WITH ARBITRARY ACCOMMODATION COEFFICIENTS.
SHEN S F
ENTROPIE
                     138-45   1967   IAA 8/5/ 865

47094 MATERIALS EVALUATION FOR SERVICEABILITY OF OPTICAL
GLASSES UNDER PROLONGED SPACE CONDITIONS. FINAL REPT
JULY, 1965-JUNE 1967.
HOLLAND W R   AVCO ELECTRONICS DIV AVCO CORP
TULSA OKLA
NASA
TR-67-G-109-F NASA-CR-65687
N67-34672   1-73   1967

47095 HEAT TRANSFER BETWEEN SURFACES IN CONTACT. THE
EFFECT OF LOW CONDUCTANCE INTERSTITIAL MATERIALS.
PART 1. EXPERIMENTAL VERIFICATION OF NASA TEST
APPARATUS.
SMUDA P A   FLETCHER L S   GYOROG D A   MECH ENG
DEPT ARIZONA STATE UNIVERSITY TEMPE ARIZONA
NASA AND CFSTI
NASA-CR-73122 ME-TR-033-1 /PT   1/
N67-37212   1-112   1967
N68-23134 /REVISED EDITION/
                     1-110   1968

47098 THERMOPHYSICAL PROPERTIES OF SOLIDS. FROM SUMMARY
OF THE ELEVENTH REFRACTORY COMPOSITES WORKING GROUP
MEETING.
PURCELL G V   THERMOPHYSICAL PROPERTIES UNIT AFML
WRIGHT-PATTERSON AIR FORCE BASE OHIO
DDC
AFML-TR-66-179
AD-804083   651-76   1966

47112 INFRARED ABSORPTION SPECTRUM OF CDTE.
BOTTGER G L   GEDDES A L
J CHEM PHYS
47   11   4858-9   1967

47142 CALORIMETRIC ABSORPTANCE MEASUREMENTS OF
THERMAL-CONTROL COATINGS BETWEEN ROOM AND CRYOGENIC
TEMPERATURES. FROM AMERICAN INSTITUTE OF
AERONAUTICS AND ASTRONAUTICS, INSTITUTE OF
ENVIRONMENTAL SCIENCES, AND AMERICAN SOCIETY FOR
TESTING AND MATERIALS, SPACE SIMULATION CONFERENCE,
HOUSTON, TEX., SEPTEMBER 7-9, 1966, TECHNICAL PAPERS,
P. 97-103.
SCHMIDT W F   OLENICK R W
JOURNAL OF SPACECRAFT AND ROCKETS
4   1495-9   1967   IAA 7-24 4526

47144 THE STABILIZATION OF OIL-IN-WATER EMULSIONS BY THE
NONIONIC SURFACTANT CETOMACROGOL 1000.
ELWORTHY P H   FLORENCE A T
J PHARM PHARMACOL SUPPL
19   140-54   1967   CA   68   33542

47145 EFFECTS OF ELECTRON BOMBARDMENT ON THE OPTICAL
PROPERTIES OF SPACECRAFT TEMPERATURE CONTROL
COATINGS.
BREUCH R A   DOUGLAS N J   VANCE D
AIAA JOURNAL
3   12   2318-27   1965

47170 CHECKERBOARD EFFECT AND THE EFFECT OF VARIOUS
FACTORS CAUSING IT ON THE SURFACE OF ANODIZED SHEETS.
RICHAUD M
REV ALUM
345   986-93   1966   CA   68   32556

47174 EFFECT OF CARBON DIOXIDE GAS PRESSURE ON THE THERMAL
CONDUCTIVITY OF SOME GRANULAR INSULATIONS.
SANDERSON P D
PROC INST MECH ENG /LONDON/
181   31   57-65   1966-7 NSA 22-4 790

47189 PREPARATION OF STANDARD COATINGS HAVING TIME-STABLE
RADIATION PROPERTIES.
NOVITSKII L A
TEPLOFIZ VYS TEMP
5   5   919-23   1967   CA   68   38138

47204 THE OPTICAL CONSTANTS OF GERMANIUM IN THE INFRA-RED
AND VISIBLE.
BRATTAIN W H   BRIGGS H B
PHYSICAL REVIEW
75   11   1705-10   1949

47205 MEASUREMENT OF THE EMISSIVITY IN NORMAL DIRECTION AND
IN DIRECTIONS CONTAINED WITHIN THE SOLID ANGLE 2 PI
FOR THE SURFACES OF SOME PLASTIC MATERIALS.
PAROLINI G
RICERCA SCIENTIFICA
24   5   1059-71   1954

47215 THE DEVELOPMENT AND EVALUATION OF GRAPHITE-MATRIX
FUEL COMPACTS FOR THE HIGH TEMPERATURE GRAPHITE
REACTOR.
CARPENTER F   ENGLE G   GOEDDEL W   GODSIN W
PONTELANDOLFO J   PYLE R   SHOFFNER J   JOHN JAY
HOPKINS LAB   SAN DIEGO CALIFORNIA
AEC
GA-2289   1-96   1961

47216 ENERGY TRANSFER IN PARTICLE-SURFACE INTERACTIONS IN
HYPERVELOCITY FLIGHT. COMMENTARY AND BIBLIOGRAPHY.
HURLBUT F C   RAND CORP   SANTA MONICA CALIF
CFSTI AND DDC
RM-4885-PR AD-655066
                     1-77   1967   RR   67-17 165

47231 THE USE OF DIRECTIONALLY DEPENDENT RADIATION
PROPERTIES FOR SPACECRAFT THERMAL CONTROL.
CLAUSEN O W   NEU J T
ASTRONAUTICA ACTA
11   5   339-30   1965

TPRC Number	Bibliographic Citation

47236 THERMAL CONDUCTIVITY OF BEDS OF COATED FUEL
PARTICLES.
STEVENS D W
TRANS AMER NUCL SOC
9   424-5   1966

47240 CERAMIC HONEYCOMB.
GOSS C L
PRODUCT ENGINEERING
33   26   52-5   1962

47244 THE OPTICAL AND ELECTRICAL PROPERTIES OF YTTERBIUM
THIN FILMS. FROM PROCEEDINGS OF THE 6TH RARE EARTH
RESEARCH CONFERENCE MAY 3-5, 1967.
OKORIE O A   SINGH S   HOWARD UNIV DEPT OF
PHYSICS   WASHINGTON D C
DDC AND CFSTI
AFOSR-67-1214 AD-653605
N67-36373   531-90   1967

47259 INTERFACIAL CONTACT RESISTANCES IN A FLOW REGIME.
M.S. THESIS.
MILLER M   COLUMBIA UNIVERSITY   NEW YORK CITY   NEW
YORK
COLUMBIA UNIVERSITY
         1-45   1965

47262 SPECTRAL EMISSIVITY OF HIGHLY DOPED SILICON. FROM
THERMOPHYSICS OF SPACECRAFT AND PLANETARY BODIES.
PROGRESS IN ASTRONAUTICS AND AERONAUTICS-VOL. 20.
LIEBERT C H
ACADEMIC PRESS   NEW YORK AND LONDON
        17-40   1967
AIAA THERMOPHYSICS SPECIALIST CONFERENCE
PAPER 67-302   1967
NASA AND CFSTI
NASA-TM-X-52254 N-67-12718

47263 INSPECTION TECHNIQUES FOR THE CHARACTERIZATION OF
SMOOTH, ROUGH, AND OXIDIZED SURFACES. FROM
THERMOPHYSICS OF SPACECRAFT AND PLANETARY BODIES.
PROGRESS IN ASTRONAUTICS AND AERONAUTICS-VOL. 20.
FUNAI A I   ROLLING R E
ACADEMIC PRESS   NEW YORK AND LONDON
        41-64   1967
AIAA THERMOPHYSICS SPECIALIST CONFERENCE
PAPER 67-318   1967

47267 A VACUUM INTEGRATING SPHERE FOR IN SITU REFLECTANCE
MEASUREMENTS AT 77 DEG. K FROM 0.5 TO 10 MU. FROM
THERMOPHYSICS OF SPACECRAFT AND PLANETARY BODIES.
PROGRESS IN ASTRONAUTICS AND AERONAUTICS-VOL. 20.
MC CULLOUGH B A   WOOD B E   SMITH A M
BIRKEBAK R C
ACADEMIC PRESS   NEW YORK AND LONDON
        137-50   1967
AIAA THERMOPHYSICS SPECIALIST CONFERENCE
PAPER 67-298   1967

47268 THE DEVELOPMENT AND TEST OF A LOW TO MODERATELY HIGH
TEMPERATURE EMISSOMETER. FROM THERMOPHYSICS OF
SPACECRAFT AND PLANETARY BODIES. PROGRESS IN
ASTRONAUTICS AND AERONAUTICS-VOL. 20.
ANDROULAKIS J G
ACADEMIC PRESS   NEW YORK AND LONDON
        151-76   1967
AIAA THERMOPHYSICS SPECIALIST CONFERENCE
PAPER 67-299   1967

47270 HEAT CAPACITY BY THE RADIANT ENERGY ABSORPTION
TECHNIQUE. FROM THERMOPHYSICS OF SPACECRAFT AND
PLANETARY BODIES. PROGRESS IN ASTRONAUTICS AND
AERONAUTICS-VOL. 20.
MAKAROUNIS O
ACADEMIC PRESS   NEW YORK AND LONDON
        203-18   1967
AIAA THERMOPHYSICS SPECIALIST CONFERENCE
PAPER 67-303   1967

47272 AN EXPERIMENTAL STUDY OF THE COMBINED SPACE
ENVIRONMENTAL EFFECTS ON A ZINC-OXIDE/POTASSIUM-
SILICATE COATING. FROM THERMOPHYSICS OF SPACECRAFT
AND PLANETARY BODIES. PROGRESS IN ASTRONAUTICS AND
AERONAUTICS-VOL. 20.
STREED E R
ACADEMIC PRESS   NEW YORK AND LONDON
        237-64   1967
AIAA THERMOPHYSICS SPECIALIST CONFERENCE
PAPER 67-339   1967

47273 SPECTRAL DEPENDENCE OF ULTRAVIOLET-INDUCED
DEGRADATION OF COATINGS FOR SPACECRAFT THERMAL
CONTROL. FROM THERMOPHYSICS OF SPACECRAFT AND
PLANETARY BODIES. PROGRESS IN ASTRONAUTICS AND
AERONAUTICS-VOL. 20.
ARVESEN J C
ACADEMIC PRESS   NEW YORK AND LONDON
        265-80   1967
AIAA THERMOPHYSICS SPECIALIST CONFERENCE
PAPER 67-340   1967

47274 IN SITU MEASUREMENTS OF SPECTRAL REFLECTANCE OF
THERMAL CONTROL COATINGS IRRADIATED IN VACUO. FROM
THERMOPHYSICS OF SPACECRAFT AND PLANETARY BODIES.
PROGRESS IN ASTRONAUTICS AND AERONAUTICS-VOL. 20.
OLLEVER P R   MILES J K   ROMANKO J
ACADEMIC PRESS   NEW YORK AND LONDON
        281-96   1967
AIAA THERMOPHYSICS SPECIALIST CONFERENCE
PAPER 67-342   1967   IAA 8/8/1400

47275 LOW SOLAR ABSORPTANCE SURFACES WITH CONTROLLED
EMITTANCE. A SECOND GENERATION OF THERMAL CONTROL
COATINGS. FROM THERMOPHYSICS OF SPACECRAFT AND
PLANETARY BODIES. PROGRESS IN ASTRONAUTICS AND
AERONAUTICS-VOL. 20.
GREENBERG S A   VANCE D A   STREED E R
ACADEMIC PRESS   NEW YORK AND LONDON
        297-314   1967
AIAA THERMOPHYSICS SPECIALIST CONFERENCE
PAPER 67-343   1967   IAA 8/8/1397

47276 DEVELOPMENT OF A BARRIER-LAYER ANODIC COATING
FOR REFLECTIVE ALUMINUM IN SPACE. FROM
THERMOPHYSICS OF SPACECRAFT AND PLANETARY BODIES.
PROGRESS IN ASTRONAUTICS AND AERONAUTICS-VOL. 20.
//AIAA/ASME 8TH STRUCTURAL DYNAMICS, AND MATERIALS
CONFERENCE.//
CLARKE D R   GILLETTE R B   BECK T R
ACADEMIC PRESS NEW YORK AND LONDON
        315-28   1967   IAA   8/8/1397

47278 SPACE-SIMULATION FACILITY FOR IN SITU REFLECTANCE
MEASUREMENTS. FROM THERMOPHYSICS OF SPACECRAFT AND
PLANETARY BODIES. PROGRESS IN ASTRONAUTICS AND
AERONAUTICS-VOL. 20.
ZERLAUT G A   COURTNEY W J
ACADEMIC PRESS   NEW YORK AND LONDON
        349-68   1967
AIAA THERMOPHYSICS SPECIALIST CONFERENCE
PAPER 67-312   1967

47279 ROLE OF FLIGHT EXPERIMENTS IN THE STUDY OF THERMAL-
CONTROL COATINGS FOR SPACECRAFT. FROM THERMOPHYSICS
OF SPACECRAFT AND PLANETARY BODIES. PROGRESS IN
ASTRONAUTICS AND AERONAUTICS-VOL. 20.
NEEL C B
ACADEMIC PRESS   NEW YORK AND LONDON
        411-38   1967
AIAA THERMOPHYSICS SPECIALIST CONFERENCE
PAPER 67-329   1967

47280 THERMAL CONTROL COATING DEGRADATION DATA FROM THE
PEGASUS EXPERIMENT PACKAGES. FROM THERMOPHYSICS
OF SPACECRAFT AND PLANETARY BODIES. PROGRESS IN
ASTRONAUTICS AND AERONAUTICS-VOL. 20.
SCHAFER C F   BANNISTER T C
ACADEMIC PRESS   NEW YORK AND LONDON
        457-73   1967
AIAA THERMOPHYSICS SPECIALIST CONFERENCE
PAPER 67-331   1967

47281 PREFLIGHT TESTING OF THE ATS-1 THERMAL COATINGS
EXPERIMENT. FROM THERMOPHYSICS OF SPACECRAFT AND
PLANETARY BODIES. PROGRESS IN ASTRONAUTICS AND
AERONAUTICS-VOL. 20.
REICHARD P J   TRIOLO J J
ACADEMIC PRESS   NEW YORK AND LONDON
        491-513   1967
AIAA THERMOPHYSICS SPECIALIST CONFERENCE
PAPER 67-333   1967

47285 THERMAL ENERGY TRANSPORT CHARACTERISTICS ALONG THE
LAMINATIONS OF MULTILAYER INSULATIONS. FROM
THERMOPHYSICS OF SPACECRAFT AND PLANETARY BODIES.
PROGRESS IN ASTRONAUTICS AND AERONAUTICS-VOL. 20.
COSTON R M   VLIET G C
ACADEMIC PRESS   NEW YORK AND LONDON
        909-23   1967
AIAA THERMOPHYSICS SPECIALIST CONFERENCE
PAPER 67-295   1967

47286 CRYOGENIC STORAGE ON THE MOON. FROM THERMOPHYSICS
OF SPACECRAFT AND PLANETARY BODIES. PROGRESS IN
ASTRONAUTICS AND AERONAUTICS-VOL. 20.
GLASER P E   STRONG P F   GABRON F
ACADEMIC PRESS   NEW YORK AND LONDON
        925-43   1967
AIAA THERMOPHYSICS SPECIALIST CONFERENCE
PAPER 67-296   1967

47289 HEAT TRANSFER IN A DISPERSE INSULATOR. //ENGLISH
TRANSLATION OF INZH. FIZ. ZH. AKAD. NAUK BELORUSSK.
S.S.R. 9 /3/ 377-83, 1965.//
KOSTYLEV V M   NABATOV V G
J ENGINEERING PHYSICS
9   3   259-63   1965

47298 A METHOD FOR THE MEASUREMENT OF INTERFACIAL TENSION
AT A LIQUID/LIQUID INTERFACE.
BROWN A H   HANSON C
CHEM IND /LONDON/
        46   1954-5   1967   CA   68   43466

TPRC Number	Bibliographic Citation						
47309	OPTICAL STUDY ON ALKALI HALIDE FILMS IN THE ULTRAVIOLET. WATANABE M KATO R JAPAN J APPL PHYS						
	7	1	21-6	1968	CA	68	44462
47314	SCATTERING AND ABSORPTION OF LIGHT BY CARBON BLACK. DONOIAN H C MEDALIA A I J PAINT TECHNOL						
	39 515		716-27	1967	CA	68	41037
47317	SUPERFICIAL AND INTERFACIAL TENSIONS OF DERIVED LAURYL SOLUTIONS. MOREL J KIKINDAI T COMPT REND						
	269	C	24	1440-3	1969		
47327	FIBERGLASS-REINFORCED PLASTICS BASED ON TETRAFURFURYLOXYSILANE OLIGOMER BINDERS. KAMENSKII I V SANIN I K AVTOKRATOVA N D SHOKINA R V BAIBAKOV K P PERTSOV L D BUTYAGA V A PLAST MASSY						
		12	43-5	1967	CA	68	40424
47332	INFRARED LATTICE VIBRATION OF VAPOR-GROWN ALUMINUM NITRIDE. AKASAKI I HASHIMOTO M SOLID STATE COMMUN						
	5	11	851-3	1967	CA	68	44376
47334	PRESTRESSED CONCRETE CAISSONS FOR NUCLEAR REACTORS. CAINE Y TRAVAUX						
	387		453-67	1965	NSA 22-10 2162		
47397	CHEMICAL VAPOR DEPOSITED MATERIALS FOR ELECTRON TUBES. QUARTERLY REPT. NO. 1 15 MAY - 14 AUG. 65. STEELE S R RAYTHEON CO WALTHAM MASS RESEARCH DIV DA DDC ECOM-01343-1 AD-472871						
		1-56	1965	TA	U65-23 A-48		
47410	SELF-EVACUATED MULTILAYER INSULATION OF LIGHTWEIGHT PREFABRICATED PANELS FOR CRYOGENIC STORAGE TANKS. PH.D. THESIS. PERKINS P J DENGLER R P NIENDORF L R NIES G E LEWIS RESEARCH CENTER CLEVELAND OHIO CFSTI AND NASA NASA-TN-D-4375 N68-18046						
		1-24	1968				
47412	EMISSION AND ABSORPTION STUDIES IN THE INFRARED /1-40 MICRONS/. NIELSEN H H RAO K N OHIO STATE UNIV RESEARCH FOUNDATION COLUMBUS DA DDC AROD-3155.12 AD-472915						
		1-140	1965	TA	65-23 A-145		
47418	PLASMA RESONANCE TRANSMISSION IN MAGNESIUM. TITTEL H O PHYS LETT						
	26	A	4	145-6	1968		
47421	REFLECTANCE SPECTROSCOPIC DETERMINATION OF COBALT, NICKEL, AND COPPER WITH PYRIDINE-2-ALDEHYDE-2-QUINOLY HYDRAZONE. FREI R W LIIVA R RYAN D E CAN J CHEM						
	46	2	167-73	1968			
47423	FINAL REPORT DETERMINATION OF THE EMISSIVITY OF MATERIALS. CLEARY R E EMANUELSON R LUOMA W AMMANN C PRATT AND WHITNEY AIRCRAFT EAST HARTFORD CONNECTICUT NASA PWA-3278						
		1-128	1968				
47427	THERMAL RESISTANCE OF METALLIC BOUNDARY SURFACES FOR LARGE TEMPERATURE JUMPS AT THOSE SURFACES /HELIUM TEMPERATURES/. BALCEREK K RAFALOWICZ J ACTA PHYS POL						
	32	6	935-48	1967	CA	69	81188
47434	HEMISPHERICAL REFLECTANCE OF METAL SURFACES AS A FUNCTION OF WAVE-LENGTH AND SURFACE ROUGHNESS. BIRKEBAK R C DAWSON J P MC CULLOUGH B A WOOD B E ARNOLD AIR FORCE STATION TENN USAF DDC AEDC-TR-65-170 AD-472767						
		1-24	1965	TA	65-23 A-144		
47435	PHYSICAL PROPERTIES OF A FOAM THERMAL INSULATION MATERIAL. FINAL REPT. CASE R ARMY TANK-AUTOMOTIVE CENTER WARREN MICH DDC AD-470249L						
		1-11	1962	TA	65-21 A-74		
47472	INTERFACIAL SURFACE TENSION OF GALLIUM AT A BOUNDARY WITH CERTAIN NONPOLAR ORGANIC LIQUIDS. ZADUMKIN S N KARASHAEV A A KUKHNO A I DOKL AKAD NAUK SSSR						
	175	6	1315-17	1967	CA	68	53595
47486	DETERMINATION OF THE SPECIFIC HEAT OF NON-CONDUCTING SOLIDS BY A TRANSIENT TECHNIQUE. M.S. THESIS. GARG S S M UNIVERSITY OF CINCINNATI CINCINNATI OHIO UNIVERSITY OF CINCINNATI						
		1-79	1965				
47488	THE EFFECTS OF CREEP ON THERMAL CONTACT CONDUCTANCE BETWEEN THIN PLATES IN VACUUM. M.S. THESIS. MAERSCHALK J C UNIVERSITY OF IOWA IOWA CITY IOWA UNIVERSITY OF IOWA						
		1-43	1965				
47499	THERMAL CONDUCTIVITY OF POROUS SYSTEMS. LUIKOV A V SHASHKOV A G VASILEV L L FRAIMAN YU E INTERN J HEAT MASS TRANSFER						
	11	2	117-40	1968	CA	68	72995
47502	REFLECTION OF MONODIRECTIONAL FLUX BY A COATING ON A SUBSTRATE. LOVE T J JR FRANCIS J E INTERN J HEAT MASS TRANSFER						
	11	2	369-74	1968			
47504	RAREFIED GAS DYNAMICS BETWEEN PARALLEL PLATES AND IN PIPES. PH. D. THESIS. MAREK C J ILLINOIS INSTITUTE OF TECHNOLOGY CHICAGO UNIV MICROFILMS PUBL 67-7025						
		1-141	1967				
47509	TEMPERATURE DEPENDENCE OF CONTACT ANGLE AND OF INTERFACIAL FREE ENERGIES IN THE NAPHTHALENE-WATER-AIR SYSTEM. JONES J B ADAMSON A W J PHYS CHEM						
	72	2	646-50	1968	CA	68	62979
	CORRECTION AND REPLY. LAVELLE J A ADAMSON ARTHUR W						
	72	6	2283-5	1968	CA	69	30374-5
47521	INTERFACIAL SURFACE TENSION OF GALLIUM AT A BOUNDARY WITH CERTAIN NONPOLAR ORGANIC LIQUIDS. //ENGLISH TRANSLATION OF DOKLADY AKAD. NAUK S.S.S.R. 175 /6/ 1315-17, 1967.// ZADUMKIN S N KARASHAEV A A KUKHNO A I DOKLADY PHYS CHEM						
	175	6	633-5	1967			
47526	MDM BANDPASS FILTERS FOR THE VACUUM ULTRAVIOLET. HARRISON D H APPL OPT						
	7	1	210	1968			
47527	MEASUREMENT OF DIRECTIONAL EMITTANCE OF ROUGHENED ALUMINUM SURFACES. M.S. THESIS. BAPAT S G GEORGIA INSTITUTE OF TECHNOLOGY ATLANTA GEORGIA GEORGIA INSTITUTE OF TECHNOLOGY						
		1-79	1966				
47531	A STUDY OF THE TEMPERATURE DROP ACROSS A SWEAT-COOLED WALL. PROGRESS REPORT. WHEELER H L JR JET PROPULSION LAB CALIFORNIA INSTITUTE OF TECHNOLOGY PASADENA CALIFORNIA AEC JPL-PR-4-71						
		1-12	1948				
47532	ADVANCED GRAPHITE FUEL SYSTEMS. QUARTERLY PROGRESS REPORT, PERIOD ENDING FEBRUARY 14, 1967. BOKROS J C ENGLE G B PRICE R J WHITE J L GENERAL DYNAMICS CORP SAN DIEGO CALIFORNIA CFSTI AND AEC GA-7827 N68-12745						
		1-50	1967				
47549	LABORATORY THERMAL EXPANSION MEASURING TECHNIQUES APPLIED TO BITUMINOUS CONCRETE. M. S. THESIS. HOOKS C C PURDUE UNIVERSITY LAFAYETTE INDIANA PURDUE UNIVERSITY						
		1-245	1965				
47554	EMITTANCE COATINGS FOR THE ENHANCEMENT OF RADIATION HEAT TRANSFER FROM SOLIDS. FROM SUMMARY OF THE FIFTH REFRACTORY COMPOSITES WORKING GROUP MEETING. AUGUST 1961. JOHNSON R L PLASMADYNE CORPORATION SANTA ANNA CALIFORNIA DDC ASD-TDR-63-96 AD-408720						
		1-8	1961				

TPRC Number	Bibliographic Citation

47565 THERMAL-SHOCK RESISTANCE FOR BUILT-UP MEMBRANES.
CULLEN W C   BOONE T H   BUILDING RESEARCH DIV
NATIONAL BUREAU OF STANDARDS   WASHINGTON D C
CFSTI AND DDC
NBS-BSS-9   1-19   1967   RR   67-21 101

47566 ANALYSIS OF SNAPTRAN REACTOR BEHAVIOR.
KESSLER W E   PRAEL R E   WEYDERT L N JR
ATOMIC ENERGY DIV PHILLIPS PETROLEUM CO   IDAHO FALLS
IDAHO
CFSTI
IDO-17204   N68-12202
                        1-19   1967   PA   N68-6-3 372

47570 INFLUENCE OF SURFACE ROGHNESS AND WAVINESS UPON
THERMAL CONTACT RESISTANCE.
YOVANOVICH M M   ROHSENOW W M   MASSACHUSETTS
INST OF TECH   CAMBRIDGE MASSACHUSETTS
CFSTI AND NASA
NASA-CR-91440
N68-13170   1-174   1967   PA   N68-6-4 594

47575 THE EFFECT OF TEMPERATURE AND PRESSURE ON THE
INTERFACIAL TENSION OF BENZENE-WATER AND NORMAL
DECANE-WATER.
JENNINGS H Y JR
J COLLOID INTERFACE SCI
24   3   323-9   1967   CA   68   53609

47578 STRUCTURAL STUDIES OF THE GE/100/-NA SYSTEM.   FROM
TWENTY-SEVENTH ANNUAL CONFERENCE ON PHYSICAL
ELECTRONICS. MARCH 1967.
CHEN J M   PHYSICAL ELECTRONICS LAB UNIV OF
MINNESOTA   MINNEAPOLIS MINNESOTA
AEC
CONF-870320
                        29-37   1967

47586 THE THERMAL ACCOMMODATION COEFFICIENT AND ADSORPTION
ON TUNGSTEN.   PH. D. THESIS.
WACHMAN Y H   UNIVERISTY OF MISSOURI   COLUMBIA AND
ROLLA MISSOURI
UNIV MICROFILMS PUBL
21108   1-174   1957

47599 FACTORS DETERMINING SURFACE TENSION.
ABRAMZON A A
KOLLOID ZH
29   4   467-73   1967   CA   68   53594

47608 SURFACE TENSION AND CONTACT ANGLE OF PARAFFINS.
GEVORKYAN B A
OBOGASHCH RUD
12   1   16-17   1967   CA   68   51698

47661 THERMAL CONDUCTIVITIES OF COMPOSITE MATERIALS.
BEHRENS E
JOURNAL OF COMPOSITE MATERIALS
2   2-17   1968   IAA 8-7 1205

47665 DETERMINATION OF THE THERMAL ACCOMMODATION
COEFFICIENTS OF GASES ON CLEAN SURFACES AT
TEMPERATURES ABOVE 300 K. BY THE TEMPERATURE JUMP
METHOD.
ROACH D V   THOMAS L B
DDC
AFOSR-67-1922 AD 657661
                        1-12   1967   RR   67-21 195
RAREFIED GAS DYNAMICS 5TH SYMPOSIUM
1   163-72   1967

47673 TRANSISTOR THERMAL-RESISTANCE MEASUREMENT USING THE
ULTRALINEAR-THERMOMETER PRINCIPLE.
AMBROZY A
ELECTRONICS LETTERS
5   5   100-1   1969

47687 INVESTIGATION OF INTERFACIAL TENSION AT THE BOUNDARY
BETWEEN LEAD-SODIUM ALLOYS AND SODIUM CHLORIDE MELTS.
//ENGLISH TRANSLATION OF ZHUR. PRIKLAD. KHIM. 40 /6/
1315-18, 1967.//
NARYSHKIN I I   MURACHEVSKII A G   PATROV B V
J APPL CHEM USSR
40   6   1264-6   1967

47688 CONDITIONS FOR OCCURRENCE OF SPONTANEOUS SURFACE
CONVECTION IN MASS TRANSFER.
OSTROVSKII M V   FRUMIN G T   KREMNEV L YA
ABRAMZON A A
ZHUR PRIKLAD KHIM
40   6   1319-27   1967

47689 CONDITIONS FOR OCCURRENCE OF SPONTANEOUS SURFACE
CONVECTION IN MASS TRANSFER.   //ENGLISH TRANSLATION
OF ZHUR. PRIKLAD. KHIM. 40 /6/ 1319-27, 1967.//
OSTROVSKII M V   FRUMIN G T   KREMNEV L YA
ABRAMZON A A
J APPL CHEM USSR
40   6   1267-74   1967

47690 VARIATION OF INTERFACIAL TENSION IN THE EXTRACTION
PROCESS.
FRUMIN G T   OSTROVSKII M V   ABRAMZON A A

47691 VARIATION OF INTERFACIAL TENSION IN THE EXTRACTION
PROCESS. //ENGLISH TRANSLATION OF ZHUR. PRIKLAD.
KHIM 40 /6/ 1328-35, 1967.//
FRUMIN G T   OSTROVSKII M V   ABRAMZON A A
J APPL CHEM USSR
40   6   1275-81   1967

47714 STRUCTURAL INSULATION THERMAL CONDUCTIVITY TESTS.
GRABER F M   TRIAS J   GENERAL DYNAMICS CONVAIR
SAN DIEGO CALIFORNIA
DDC
GDC-MP-58-457
AD-672199   1-7   1959   RR   68-18 100

47730 MEASUREMENT OF THERMAL CONDUCTIVITIES OF INSULATING
MATERIALS AT ELEVATED TEMPERATURES. M.S. THESIS.
FRANKE F R   UNIVERSITY OF FLORIDA   GAINESVILLE
FLORIDA
UNIVERSITY OF FLORIDA
                        1-115   1958

47737 RECENT ACTIVITIES IN HIGH TEMPERATURE COATING AND
MATERIAL PROGRAMS AT AMF.   FROM   SUMMARY OF THE SIXTH
REFRACTORY COMPOSITES WORKING GROUP MEETING.   VOLUME
II. 1962.
BROWNING M E   SCHATZ E A   LEAVENWORTH H W JR
AMERICAN MACHINE AND FOUNDRY CO   ALEXANDRIA VA
VIRGINIA
DDC
A3D-TDR-63-610 /VOL II/
AD-427181   460-76   1962

47738 NUMERICAL STUDY OF DENSITY-CURRENT SURGES.
DALY B J   PRACHT W E
PHYS FLUIDS
11   1   15-30   1968

47743 APPLICATION OF THE NOCILLA WALL REFLECTION MODEL TO
FREE-MOLECULE KINETIC THEORY.
HURLBUT F C   SHERMAN F S
DDC
AD 671420   1-11   1967
PHYS FLUIDS
11   3   486-96   1968   RR   68-17 149

47748 DROP FORMATION AT LOW VELOCITIES IN LIQUID-LIQUID
SYSTEMS.   PART 1. PREDICTION OF DROP VOLUME.
SCHEELE G F   MEISTER B J
A I CH E JOURNAL
14   1   9-15   1968

47750 DIRECT CONTACT HEAT TRANSFER BETWEEN IMMISCIBLE
LIQUIDS IN TURBULENT PIPE FLOW.
PORTER J W   GOREN S L   WILKE C R
A I CH E JOURNAL
14   1   151-8   1968

47754 USE OF THERMAL CONTACT RESISTANCE AS A PASSIVE
TEMPERATURE REGULATOR. M. S. THESIS.
HUANG P N-S   PURDUE UNIVERSITY   LAFAYETTE INDIANA
PURDUE UNIVERSITY
                        1-91   1965

47756 IN SITU ULTRAVIOLET RADIATION EFFECTS ON THERMAL
CONTROL COATINGS.   8TH QUARTERLY PROGRESS REPORT.
CANNADAY S S   FOGDALL L B   BOEING CO   SEATTLE
WASHINGTON
CFSTI AND NASA
NASA-CR-91351 N68-12858
D2-84118-8   1-111   1967

47759 DEVELOPMENT OF A COMBINED HIGH PERFORMANCE MULTILAYER
INSULATION AND MICROMETEOROID PROTECTION SYSTEM.
FROM   RESEARCH ACHIEVEMENTS REVIEW.   VOLUME 2 REPORT
NO. 7.
STUCKEY J M   MARSHALL SPACE FLIGHT CENTER
HUNTSVILLE ALABAMA
NASA
NASA-TM-X-53670 N68-14651
                        55-68   1967

47761 VIBRATIONAL EXCITATION OF CARBON MONOXIDE FOLLOWING
QUENCHING OF THE A-3-II STATE.
DONOVAN R J   HUSAIN D
TRANS FARADAY SOC
63   12   2879-87   1967

47767 CADMIUM TELLURIDE-MERCURY TELLURIDE HETEROSTRUCTURES.
ALMASI G S   SMITH A C
J APPL PHYS
39   1   233-45   1968

TPRC Number	Bibliographic Citation

47768 HEAT TRANSFER BETWEEN SURFACES IN CONTACT. THE EFFECT OF LOW CONDUCTANCE INTERSTITIAL MATERIALS. PT. 3. COMPARISON OF THE EFFECTIVE THERMAL INSULATION FOR INTERSTITIAL MATERIALS UNDER COMPRESSIVE LOADS. PTS. 4 AND 5. INVESTIGATION OF THERMAL ISOLATION MATERIALS AND THEIR APPLICATION IN FLANGE JOINTS.
SMUDA P A   GYOROG D A   MECHANICAL ENGINEERING DEPARTMENT ARIZONA STATE UNIV   TEMPE ARIZONA
NASA AND CFSTI
NASA-CR-73244 ME-TR-033-3

		N68-31671	1-82	1968	RR	68-21-206
	NASA-CR-73248 ME-TR-033-5 /PTS 4, 5/					
		N68-34080	1-68	1968	RR	68-23-161

47821 VERIFICATION OF THE ANOMALOUS-SKIN-EFFECT THEORY FOR SILVER IN THE INFRARED.
BENNETT H E   BENNETT J M   ASHLEY E J
MOTYKA R J
PHYS REV
165   3   755-64   1968

47827 OPTICAL PROPERTIES OF THIN FILMS OF INDIUM ARSENIDE.
HOWSON R P
BRIT J APPL PHYS
1   2   1   15-23   1968   CA   68   4446

47869 OPTICAL PROPERTIES OF TIN TELLURIDE IN THE VISIBLE AND INFRARED REGIONS.
SCHOOLAR R B   DIXON J R
J OPT SOC AMERICA
58   1   119-25   1968

47874 OPTICAL PROPERTIES OF AS2 SE3, AS/2X/ SB/2-2X/ SE3, AND SB/X/ S3.
EFSTATHIOU A   LEVIN E R
J OPT SOC AMERICA
58   3   373-7   1968   CA   68   91286

47905 KINETICS OF AN INTERFACIAL POLYCONDENSATION REACTION. PART. 1.-HYDROLYSIS OF TEREPHTHALOYL CHLORIDE.
CRAWFORD P J   BRADBURY J H
TRANS FARADAY SOC
64   1   185-91   1968

47907 THERMAL CONDUCTIVITY OF OIL-IMPREGNATED INSULATION.
ALLEN P H G
ELECTRICAL TIMES
22   243-4   1968

47908 MECHANISM OF CATALYTIC DECOMPOSITION OF FORMIC ACID ON A NICKEL SURFACE.
FUKUDA K   NAGASHIMA S   NOTO Y   ONISHI T
TAMARU K
TRANS FARADAY SOC
64   2   522-8   1968

47917 BIDIRECTIONAL REFLECTANCE MEASUREMENTS FOR SATELLITE CALIBRATION TARGET IN THE VISIBLE AND NEAR INFRARED.
SAIEDY F   JONES G D
APPL OPT
7   3   429-34   1968   CA   68   91553

47918 DETERMINATION OF OPTICAL CONSTANTS FROM INTENSITY MEASUREMENTS AT NORMAL INCIDENCE.
NILSSON P O
APPL OPT
7   3   435-42   1968

47919 THE EFFECT OF OPTICAL POLARIZATION ON REFLECTION SPECTRA.
REED A H   YEAGER E
APPL OPT
7   3   451-3   1968

47926 THE HEAT CONDUCTIVITY OF POWDERS.
NESSELMANN K   PFEIFFER H
BULL INTL INST REFRIG ANNEXE
83-7   1962

47928 THE CONDUCTIVE DISC METHOD OF MEASURING THE THERMAL CONDUCTIVITY OF INSULATIONS.
ROBINSON H E
BULL INTL INST REFRIG ANNEXE
43-50   1962

47930 MEASUREMENT OF THERMAL CONDUCTIVITY AT LOW TEMPERATURES.
VOS B H
BULL INTL INST REFRIG ANNEXE
179-90   1961

47935 SUPERINSULATION FOR THE LARGE SCALE STORAGE AND TRANSPORT OF LIQUEFIED GASES.
DUBS M A   DANA L I
BULL INTL INST REFRIG ANNEXE
71-83   1961

47938 RESEARCH IN PROTECTIVE COATINGS FOR REFRACTORY METALS. FROM SUMMARY OF THE SIXTH REFRACTORY COMPOSITES WORKING GROUP MEETING. VOL II. 1962.
CHAO P J   DORMER G J   PAYNE B S JR   PRIEST D K
ZUPAN J   THE PFAUDLER CO   ROCHESTER NEW YORK
DDC
ASD-TDR-63-610 VOL II
AD-427181   558-73   1962

47949 OPTICAL PROPERTIES OF VACUUM-EVAPORATED FILMS OF TELLURIUM AND AMORPHOUS SELENIUM. M. S. THESIS. /HAYES-VANDERBILT UNIV./
HAYES J D JR   ARAKAWA E T   WILLIAMS M W   OAK RIDGE NATIONAL LABORATORY   NASHVILLE TENN
CFSTI
ORNL-TM-2023 N68-18941
1-52   1968   NSA 22/5/955

47952 EMITTANCE MEASUREMENTS AT THE U. S. NAVAL RADIOLOGICAL DEFENSE LABORATORY. FROM SUMMARY OF THE SIXTH REFRACTORY COMPOSITES WORKING GROUP MEETING. VOLUME II. 1962.
ALVARES W J   U S NAVAL RADIOLOGICAL DEFENSE LABS
DDC
ASD-TDR-63-610 VOL II
AD-427181   682-3   1962

47953 METALLIC HEAT INSULATION.
NICHOLS J T
MECH ENG
57   521-4   1935
CORRECTION.
PARLETT R C
58   190-1   1936

47962 OPTICAL PROPERTIES OF ROKIDE COATINGS. FROM SUMMARY OF THE SIXTH REFRACTORY COMPOSITES WORKING GROUP MEETING. VOLUME II. 1962.
AULT N N   NORTON COMPANY   WORCESTER MASSACHUSETTS
DDC
ASD-TDR-63-610 VOL II
AD-427181   685-98   1962

47970 MEASURING HEAT TRANSMISSION IN BUILDING STRUCTURES AND A HEAT TRANSMISSION METER.
NICHOLLS P
J AM SOC HEAT-VENT ENGRG
35-70   1924

47972 PROPERTIES OF ALL-DIELECTRIC INTERFERENCE FILTERS. II. FILTERS IN PARALLEL BEAMS OF LIGHT INCIDENT OBLIQUELY AND IN CONVERGENT BEAMS.
LISSBERGER P H   WILCOCK W L
J OPT SOC AM
49   2   126-30   1959

47973 RESOLVING POWER OF MULTILAYER FILTERS IN NONPARALLEL LIGHT.
PIDGEON C R   SMITH S D
J OPT SOC AM
54   12   1459-65   1964

47979 DETERMINING THERMAL CONDUCTIVITY COEFFICIENTS OF VARIOUS INSULATORS.
MAXIMOV T F   BOITZOR S P
IZVEST TEPL INST
256-78   1932

47984 SOME RESULTS OF AN INVESTIGATION INTO CONTACT THERMAL RESISTANCE.
KHIZHNIAK P YE
IZVEST VYSSHIKH UCHEB ZAVEDENII ENERGET
9   2   69-76   1966

47991 THERMAL ANALYSIS OF DROP-CALORIMETER DETERMINATION OF SNAP FUEL SPECIFIC HEATS.
MAGEE P M   ATOMICS INTERNATIONAL   CANOGA PARK CALIFORNIA
AEC AND CFSTI
NAA-SR-MEMO-12324
1-38   1967   CA   68   45296

47994 MEASUREMENT OF THE THERMAL CONDUCTIVITY OF MATERIALS FOR NUCLEAR USE BETWEEN 100 AND 500 C. /ITALIAN/.
GIULIANI S   EUROPEAN ATOMIC ENERGY COMMUNITY   ISPRA ITALY
AEC
EUR-3644I   1-38   1967   CA   69   22734

47997 VACUUM ULTRAVIOLET ABSORPTION SPECTRA OF HOMO-POLY-NUCLEOTIDES.
ONARI S
J PHYS SOC JAPAN
26   1   214   1969

47998 INVESTIGATION OF THE EFFECT OF SURFACE CONDITIONS ON THE RADIANT PROPERTIES OF METALS. PART II, MEASUREMENTS ON ROUGHENED PLATINUM AND OXIDIZED STAINLESS STEEL.
ROLLING R E   FUNAI A I   LOCKHEED MISSILES AND SPACE CO   PALO ALTO CALIF
DDC AND CFSTI
AFML-TR-64-363/PT II/ I MSC-6-77-67-27

TPRC Number	Bibliographic Citation				
47999	STUDY OF FEREMOAGNETIC LIQUID. ROSENSWEIG R E KAISER R AVCO CORP SPACE SYSTEMS DIV WILMINGTON MASSACHUSETTS NASA AND CFSTI NASA-CR-91684 N68-14205				
	1-211	1967	PA	N68-6-5 698	
48001	DIELECTRIC DISPERSION AND THE STRUCTURES OF IONIC LATTICES. LOWNDES R P MARTIN D H PROC ROY SOC /LONDON/				
	308	A	473-96	1969	
48004	SUMMARY OF THE SIXTH MEETING OF THE REFRACTORY COMPOSITES WORKING GROUP. GIBEAUT W A MAYKUTH D J DEFENSE METALS INFORMATION CENTER BATTELLE MEMORIAL INSTITUTE COLUMBUS OHIO DDC AND CFSTI DMIC-175 AD-287029				
	1-62	1962			
48006	CRITICAL SURFACE TENSION OF URANIUM DIOXIDE. EBERHART J G J NUCL MATER				
	25	103-5	1968		
48008	THERMODYNAMIC PROPERTIES OF SUPERCONDUCTING CONTACTS. FULDE P MOORMANN W PHYS KONDENS MATER				
	6	5	403-15	1967	CA	69	30780
48032	LABORATORY AND FLIGHT TEST PROGRAM TO EVALUATE THE SPACE STABILITY OF HIGHLY SPECULAR REFLECTIVE SURFACES. MARSHALL K N DOUGLAS N J POLLARD H E ROLLING R E DDC AD 662968				
	1-8	1967			
	J SPACECRAFT ROCKETS				
	4	7	912-19	1967	TT	68-4 69
48034	THE TESTING OF THERMAL INSULATORS. DICKINSON H C VAN DUSEN M S REFRIG ENG				
	3	2	5-25	1916	
48051	INFLUENCE OF THE RATE OF FORMATION OF THIN SILVER LAYERS, OBTAINED BY THERMAL EVAPORATION, ON THEIR TRANSMISSIVITY AND REFLECTIVITY. PHILIP R J PHYS LE RADIUM				
	21	3	165-8	1960	
48084	THERMAL CONDUCTIVITY OF MONODISPERSED GRANULAR MATERIALS. FROM HEAT AND MASS TRANSFER VOL. 7. DULNEV G N SIGALOVA Z V ENERGIIA MOSCOW AND LENINGRAD				
	87-97	1966			
48092	THE TEMPERATURE FIELD OF A FOUR COMPONENT SYSTEM OCCURRING WHEN THERE IS RECIPROCAL THERMAL CONTACT BETWEEN THE COMPONENTS /COMMUNICATION I/. FROM HEAT AND MASS TRANSFER VOL. 7. VOLKENSHTEIN V S ENERGIIA MOSCOW AND LENINGRAD				
	185-98	1966			
48093	THE TEMPERATURE FIELD OF A FOUR COMPONENT SYSTEM OCCURRING WHEN THERE IS RECIPROCAL THERMAL CONTACT BETWEEN THE COMPONENTS /COMMUNICATION II/. FROM HEAT AND MASS TRANSFER VOL. 7. VOLKENSHTEIN V S ENERGIIA MOSCOW AND LENINGRAD				
	199-203	1966			
48097	DEVELOPMENT OF SILICON INFRARED OPTICAL COMPONENTS /TRANSMITTING WINDOWS/. COLE R L MITCHELL G HICKS J AERONAUTICAL SYSTEMS CENTER WRIGHT-PATTERSON AIR FORCE BASE OHIO DDC AMC-TR-60-7-719 AD-250513				
	1-211	1960			
48098	TRANSPARENT CONDUCTING COATINGS OF GOLD ON GLASS. PAWEL R E WRIGHT-PATTERSON AIR FORCE BASE OHIO DDC WADC-TN-58-302 AD-214386				
	1-13	1959			
48099	SPECTRALLY SELECTIVE COATINGS FOR TEMPERATURE CONTROL OF SPACE PROBES. TECH. REPT. VAN VLIET R M MATTICE J J CROSS R A WRIGHT-PATTERSON AIR FORCE BASE OHIO DDC WADD-TR-60-386 AD-249265L				
	1-23	1960			
48104	LABORATORY TECHNIQUES FOR STUDYING THERMALLY ABLATIVE PLASTICS. SCHWARTZ H S AERONAUTICAL SYSTEMS DIVISION WRIGHT-PATTERSON AIR FORCE BASE OHIO DDC				

TPRC Number	Bibliographic Citation				
48105	INVESTIGATION OF INFRARED TRANSMITTING MATERIALS. PART 3. KREIDL N J HAFNER H C HENSLER J R WEIDEL R A MATERIALS LABORATORY WRIGHT-PATTERSON AIR FORCE BASE OHIO DDC WADC-TR-55-500 /PT 3/ AD-202842				
	1-132	1958			
48106	DEVELOPMENT OF WHITE THERMALLY REFLECTIVE RAIN EROSION RESISTANT COATINGS. PART 1. VOGELSANG G K MATERIALS LABORATORY WRIGHT-PATTERSON AIR FORCE BASE OHIO DDC WADC-TR-57-158/PT 1/ AD-142033				
	1-156	1957			
48109	AN INVESTIGATION OF INFRARED TRANSMITTING MATERIALS. PART 1. KREIDL N J HAFNER H C HENSLER J R WEIDEL R A LETTER E C MATERIALS LABORATORY WRIGHT-PATTERSON AIR FORCE BASE OHIO DDC WADC-TR-55-500 PT I AD-130904				
	1-150	1957			
48116	THERMAL CONTACT CONDUCTANCE IN A VACUUM AND RELATED PARAMETER STUDY. ATKINS H L NATIONAL AERONAUTICS AND SPACE ADMINISTRATION MARSHALL SPACE FLIGHT CENTER HUNTSVILLE ALABAMA NASA AND CFSTI NASA-TM-X-60588 N68-10508				
	1-51	1964	RR	68-2 146	
48119	USE OF THE GIBBS EQUATION IN THE DETERMINATION OF THE LIMITED MOLECULAR OBSTRUCTIONS OF TENSION-ACTIVE STEROIDS AT THE OIL-WATER INTERFACE. BARET J-F ROUX R COMPT REND				
	266	C	4	243-5	1968
48122	INFRARED ABSORPTION SPECTRA OF THE CRYSTALLIZED MIXTURE BENZENE-DIOXANE. MARSAULT J-P COMPT REND				
	266	B	6	334-6	1968
48131	OPTICAL PROPERTIES OF CHROME AND LEAD IN THE FAR ULTRAVIOLET. GIRAULT P SEIGNAC A PRIOL M ROBIN S COMPT REND				
	266	B	11	688-90	1968
48135	FORMATION OF THIN POLYACRYLONITRILE FILMS AND THEIR ELECTRICAL PROPERTIES. HIRAI T NAKADA O JAPAN J APPL PHYS				
	7	2	112-21	1968	
48138	VAPOR DEPOSITION OF SILICON NITRIDE FILM ON SILICON AND PROPERTIES OF MANGANESE SULFIDE DIODES. SUGANO T HIRAI K KUROIWA K HOH K JAPAN J APPL PHYS				
	7	2	122-7	1968	
48139	EFFECT OF SURFACE ROUGHNESS ON THE 0.66 MICRON NORMAL SPECTRAL EMITTANCE OF VAPOR-DEPOSITED RHENIUM FROM 1500 TO 2100 K. CIPOLLONE P LEWIS RESEARCH CENTER CLEVELAND OHIO NASA AND CFSTI NASA-TM-X-1514 N68-19347				
	1-18	1968			
48153	DETERMINATION OF THE TOTAL EMITTANCE OF A NONGRAY SURFACE. WITTENBERG A M J APPL PHYS				
	39	4	1936-40	1968	
48163	SPECIFIC ADSORPTION AT THE MERCURY/SULPHOLANE INTERFACE. LAWRENCE J PARSONS R TRANS FARADAY SOC				
	64	3	751-70	1968	
48169	DEVELOPMENT OF WHITE HEAT RESISTANT AND WHITE ANTI-STATIC RAIN EROSION RESISTANT COATINGS. PART III. VOGELSANG G K MATERIALS LABORATORY WRIGHT-PATTERSON AIR FORCE BASE OHIO DDC WADC-TR-57-158 /PT III/ AD-142286				
	1-50	1958			
48170	INFLUENCE OF SURFACE ROUGHNESS AND WAVINESS UPON THERMAL CONTACT RESISTANCE. TECH. REPT. YOVANOVICH M M ROHSENOW W M ENGINEERING PROJECTS LABORATORY MASSACHUSETTS INSTITUTE OF TECHNOLOGY NASA NASA-TR-76361-48				
	1-137	1967			

TPRC Number	Bibliographic Citation

48171 DEVELOPMENT OF WHITE HEAT RESISTANT AND WHITE ANTI-STATIC RAIN EROSION RESISTANT COATINGS. PART II.
VOGELSANG G K   MATERIALS LABORATORY
WRIGHT-PATTERSON AIR FORCE BASE   OHIO
DDC
WADC-TR-158   /PT II/
AD-142185   1-164   1957

48179 OPTICAL TRANSMISSION AND ELECTRICAL RESISTANCE OF THIN FILMS OF STANNIC OXIDE PREPARED BY REACTIVE CATHODIC SPUTTERING.
BAILLOU J   BUGNET P   DEFORGES J   DURAND S
BATAILLER G
REVUE DE PHYSIQUE APPLIQUEE
3   1   78-82   1968

48184 THE EFFECT OF SURFACE CONDITION AND TEMPERATURE ON THE TOTAL HEMISPHERICAL EMITTANCE OF PLUTONIUM.
SEERY W N
J NUCL MATER
25   1   64-8   1968   CA   68   74429

48185 IRRADIATION DAMAGE AND IRRADIATION HARDENING IN SINGLE CRYSTAL MGO AFTER LOW NEUTRON DOSES LESS THAN 10 TO THE 18TH POWER NVT /GREATER THAN MEV/.
DAVIDGE R W
J NUCL MATER
25   1   75-86   1968

48225 CONTACT RESISTANCE BETWEEN METALS DURING TRANSIENT HEATING. M.S. THESIS.
SCHAUER D A   CALIFORNIA UNIV   BERKELEY CALIFORNIA
AEC
UCRL-14227-T
         1-59   1963

48227 AN EXPERIMENTAL AND THEORETICAL INVESTIGATION OF THE THERMAL CONTACT RESISTANCE. FINAL REPORT.
CLAUSING A M   ENGINEERING EXPERIMENT STATION
ILLINOIS UNIV   URBANA ILLINOIS
CFSTI AND NASA
ME-TN-242-3 NASA-CR-76867
N66-32804   1-81   1966   PA   N66-4-19 3888

48228 SOLAR-RADIATION-INDUCED DAMAGE TO OPTICAL PROPERTIES OF ZINC OXIDE-TYPE PIGMENTS. TECH. SUMMARY REPT. FOR PERIOD JULY 1966-FEB. 1968.
SKLENSKY A F   MAC MILLAN H F   GREENBERG S A
PALO ALTO RESEARCH LABORATORY
LOCKHEED RESEARCH LABORATORY   PALO ALTO CALIFORNIA
LMSD-4-17-68-1
         1-161   1968

48230 DEVELOPMENT OF SPACE-STABLE THERMAL-CONTROL COATINGS. TRIANNUAL REPORT.
ZERLAUT G A   FIRESTONE R F   RAZIUNAS V
SERWAY R   RUBIN G A   IIT RESEARCH INSTITUTE
CHICAGO ILLINOIS
NASA
IITRI-U6002-31
         1-42   1965

48231 DEVELOPMENT OF SPACE-STABLE THERMAL-CONTROL COATINGS. TRIANNUAL REPORT.
ZERLAUT G A   IIT RESEARCH INSTITUTE   CHICAGO ILLINOIS
NASA
IITRI-U6002-42
         1-29   1966

48232 DEVELOPMENT OF SPACE-STABLE THERMAL-CONTROL COATINGS. TRIANNUAL REPORT.
ZERLAUT G A   ROGERS F O   IIT RESEARCH
INSTITUTE   CHICAGO ILLINOIS
NASA
IITRI-U6002-47
         1-77   1966

48233 DEVELOPMENT OF SPACE-STABLE THERMAL-CONTROL COATINGS. MARCH 1-JULY 31, 1967.
ZERLAUT G A   NOBLE G   ROGERS F O   IIT
RESEARCH INSTITUTE   CHICAGO ILLINOIS
NASA
IIITRI-U6002-55
         1-71   1967

48234 DEVELOPMENT OF SPACE-STABLE THERMAL-CONTROL COATINGS. AUGUST 1-OCTOBER 31, 1967.
ZERLAUT G A   NOBLE G   ROGERS F O   IIT
RESEARCH INSTITUTE   CHICAGO ILLINOIS
NASA
IITRI-U6002-59
         1-75   1967

48236 AN IMPROVED APPARATUS FOR MEASURING THE THERMAL TRANSMISSION OF TEXTILES.
CLEVELAND R S
J NATIONAL BUR STANDARDS
19   675-84   1937

48258 THE REFLECTANCE, SOLAR ABSORPTION AND EMITTANCE OF ROKIDE A AT TEMPERATURES BETWEEN 83 AND 400 K.
BRANDENBERG W M   GENERAL DYNAMICS/ASTRONAUTICS
SAN DIEGO CALIFORNIA
DDC
GDA-AE61-0927
AD-677531   1-21   1961   RR   69-1 92

48283 THE CELITE TYPE HIGH-TEMPERATURE THERMAL CONDUCTIVITY APPARATUS.
WEINLAND C E
PROC AMERICAN SOC FOR TESTING MATERIAL
37 II   269-76   1937

48286 HEAT CONDITION OF CAPILLARY POROUS BODIES AT TEMPERATURES BELOW THE FREEZING POINT.
TANGANTSEVA T F   BUROV IU S
STROIT MAT
7   1   31-2   1961

48291 NEW INSULATION STUDIES. PART II.
QUEER E R   HECHLER F G
REFRIG ENG
35   247-52   1938

48292 VARIATION OF THERMAL CONDUCTIVITY WITH TEMPERATURE OF INSULATING POWDERS.
GLASER P E   KAYAN C F
REFRIG ENG
64   3   31-6   1956

48293 THERMAL EMISSIVITY AND CONDUCTIVITY OF ALUMINA HEATER COATINGS.
RUDNICK N   CARRONA J J
RCA REVUE
22   4   623-47   1961

48294 HEAT FLOW IN THE SOUTHERN KARROO.
GOUGH D I
PROC ROY SOC /LONDON/
272   A   207-30   1963

48295 THE WARMTH OF TEXTILE FIBRES. PART IV.
ILLINGWORTH J W
TEXTILE RECORDER
61   67-73   1944

48324 THE CONDUCTIVITY OF HEAT INSULATORS. PART A.
LAMB C G   WILSON W G
PROC ROYAL SOC
65   283-8   1899

48330 THERMAL CONDUCTIVITY DETERMINATION OF INSULATING MATERIALS BY TRANSIENT TEMPERATURE MEASUREMENTS. M.S. THESIS.
SHENHAV A   NORTHWESTERN UNIVERSITY   EVANSTON ILLINOIS
NORTHWESTERN UNIVERSITY
         1-32   1962

48344 OPTICAL PROPERTIES OF VERY THIN MEATAL LAYERS.
VOIGT W
ANN PHYSIK
25   95-114   1885

48345 THE REFLECTION AND REFRACTION OF LIGHT ON LAYERS OF ABSORBING ISOTROPIC MEDIA.
VOIGT W
ANN PHYSIK
35   76-100   1888

48349 A NEW APPARATUS FOR THE MEASUREMENT OF THE THERMAL CONDUCTIVITY OF WALLS.
LOCATELLI A
TERMOTECNICA /MILAN/
15   9   15-48   1961

48357 THERMAL CONDUCTIVITY OF AIR LAYERS
HARVALIK Z V
PROC MINN ACAD SCI
15   125-7   1947

48358 SOME THERMAL PROPERTIES OF CONCRETE.
NORTON C L
PROC NAT ASSO CEMENT USERS
7   78-101   1911

48359 FAR INFRARED REFLECTANCE OF SPACECRAFT COATINGS.
EDWARDS D K   HALL W M
PROGR ASTRONAUT AERONAUT
18   3-19   1966

48361 TOTAL NORMAL EMITTANCE MEASUREMENTS TO 2200 C IN AIR.
HEDGE J C
PROGR ASTRONAUT AERONAUT
18   33-46   1966   CA   69   14473

48362 SOLAR ABSORPTANCE AND THERMAL EMITTANCE OF ALUMINUM COATED WITH SURFACE FILMS OF EVAPORATED ALUMINUM OXIDE.
HASS G   RAMSEY J B   TRIOLO J J   ALBRIGHT H T
PROGR ASTRONAUT AERONAUT
18   47-60   1966   CA   69   14188

TPRC Number	Bibliographic Citation

48363 LOW-TEMPERATURE EMITTANCE DETERMINATIONS.
CAREN R P
PROGR ASTRONAUT AERONAUT
18　　　　61-73　　　1966

48365 DIRECTIONALLY REFLECTIVE COATING STUDY.
COX R L　　RAY J V
PROGR ASTRONAUT AERONAUT
18　　　　101-28　　　1966

48366 APPARATUS FOR SPECTRAL BIDIRECTIONAL REFLECTANCE MEASUREMENTS DURING ULTRAVIOLET IRRADIATION IN VACUUM.
MAC MILLAN H F　　SKLENSKY A F　　MC KELLAR L A
PROGR ASTRONAUT AERONAUT
18　　　　129-49　　　1966

48367 TOTAL HEMISPHERICAL THERMAL EMITTANCE OF NICKEL AS A FUNCTION OF OXIDE THICKNESS IN THE TEMPERATURE RANGE 400 - 900 C.
SHELTON J L　　AKERS W W
PROGR ASTRONAUT AERONAUT
18　　　　151-65　　　1966　　CA　　69　　13523

48370 FAR-INFRARED STUDIES OF SILICATE MINERALS.
ARONSON J R　　MC LINDEN H G
PROGR ASTRONAUT AERONAUT
18　　　　291-309　　　1966

48372 EXPLORATORY TRAPPED-PARTICLE AND TRAPPED-PARTICLE-PLUS-ULTRAVIOLET EFFECTS ON THE OPTICAL PROPERTIES OF SPACECRAFT THERMAL CONTROL COATINGS.
BREUCH R A
PROGR ASTRONAUT AERONAUT
18　　　　365-88　　　1966

48373 EFFECTS OF SIMULATED SOLAR-WIND BOMBARDMENT ON SPACECRAFT THERMAL CONTROL SURFACES.
JORGENSON G V
PROGR ASTRONAUT AERONAUT
18　　　　389-98　　　1966

48374 EFFECTS OF LOW ENERGY PROTONS ON THERMAL CONTROL COATINGS.
MILLER R A　　CAMPBELL F J
PROGR ASTRONAUT AERONAUT
18　　　　399-412　　　1966

48375 EFFECTS OF PROTONS AND ALPHA PARTICLES ON THERMAL PROPERTIES OF SPACECRAFT AND SOLAR CONCENTRATOR COATINGS.
GILLETTE R B　　BROWN R R　　SEILER R F
SHELDON W R
PROGR ASTRONAUT AERONAUT
18　　　　413-40　　　1966

48376 MARINER-MARS ABSORPTANCE EXPERIMENT.
LEWIS D W　　THOSTESEN T O
PROGR ASTRONAUT AERONAUT
18　　　　441-57　　　1966

48377 PRELIMINARY RESULTS FROM THE AMES EMISSIVITY EXPERIMENT ON OSO-II.
PEARSON B D JR
PROGR ASTRONAUT AERONAUT
18　　　　459-72　　　1966

48378 SNAP 10A THERMAL CONTROL COATINGS.
CROSBY J R　　PERLOW M A
PROGR ASTRONAUT AERONAUT
18　　　　473-91　　　1966

48379 EFFECTS OF UNCERTAINTIES IN THERMOPHYSICAL PROPERTIES ON ABLATION EFFICIENCY.
SHAW T E　　GARNER D C　　FLORENCE D E
PROGR ASTRONAUT AERONAUT
18　　　　513-48　　　1966

48381 THERMAL CONDUCTANCE OF METALLIC CONTACTS IN A VACUUM.
FRIED E　　KELLEY M J
PROGR ASTRONAUT AERONAUT
18　　　　697-718　　　1966

48382 MEASUREMENTS OF CONTACT COEFFICIENTS OF THERMAL CONDUCTANCE.
FRY E M
PROGR ASTRONAUT AERONAUT
18　　　　719-34　　　1966

48383 PERFORMANCE OF MULTILAYER INSULATION SYSTEMS FOR THE 300 TO 800 K TEMPERATURE RANGE.
STREED E R　　CUNNINGTON G R JR
ZIERMAN C A
PROGR ASTRONAUT AERONAUT
18　　　　735-71　　　1966

48387 AN ACCURATE METHOD FOR THE DETERMINATION OF THE THERMAL CONDUCTIVITY OF INSULATING SOLIDS.
MISCHKE C R　　FARBER E A
AMERICAN SOCIETY OF MECHANICAL ENGINEERS
ASME-53-A-185
1-6　　　1953

48399 MEASUREMENT OF SURFACE TENSION BETWEEN THE CALCIUM OXIDE-ALUMINUM OXIDE MOLTEN SLAG AND MOLTEN CHROMIUM-NICKEL-MOLYBDENUM STEEL.
ONECKI K　　SIKORA B　　ZIELINSKI M
HUTNIK
34　　10　　453-6　　1967　　CA　　68　　71441

48401 EFFECT OF SODIUM CONCENTRATION ON THE SURFACE TENSION OF MERCURY.
LEE Y
IND ENG CHEM PROD RES DEVELOP
7　　1　　66-8　　1968　　CA　　68　　72554

48405 ADSORPTION OF SHORT-CHAIN FATTY ACIDS AND THEIR IONS AT THE OIL-WATER INTERFACE.
CHATTERJEE A K　　CHATTORAJ D K
J COLLOID INTERFACE SCI
26　　1　　1-9　　1968　　CA　　68　　72575

48410 TRANSPORT OF IONS ACROSS LIPID MONOLAYERS. I. THE STRUCTURE OF DECYLAMMONIUM MONOLAYERS AT THE POLARIZED MERCURY-WATER INTERFACE.
BLANK M　　MILLER I R
J COLLOID INTERFACE SCI
26　　1　　26-33　　1968　　CA　　68　　72576

48415 THERMAL CONDUCTIVITY OF POROUS GRAPHITIZED MATERIALS AT 400-1400 DEGREES.
ZEIGARNIK V A　　PELETSKII V E　　TARABANOV A S
KHIM TVERD TOPL
3　　116-20　　1967　　CA　　68　　63359

48421 COMPATIBILITY OF POLYPROPYLENE WITH ULTRAVIOLET LIGHT ABSORBERS. SURFACE TENSION OF POLYPROPYLENE AND ULTRAVIOLET LIGHT ABSORBERS.
MARCINCIN A　　PIKLER A　　ONDREJMISKA K
PLAST HMOTY KAUC
4　　12　　360-1　　1967　　CA　　68　　60186

48433 ANALYTICAL DETERMINATION OF THE EFFECTIVE EMITTANCE OF AN INSULATED LOUVER SYSTEM.
OLLENDORF S　　NASA GODDARD SPACE FLIGHT CENTER GREENBELT MARYLAND
NASA AND CFSTI
NASA-TM-X-54804
N65-32120　　　1-18　　1965　　CA　　66　　117369

48434 DEVELOPMENT OF 400 TO 2200 F. FIBROUS-TYPE INSULATIONS FOR RADIOISOTOPE POWER SYSTEMS. SEVENTH QUARTERLY PROGR. REPT. JULY, 1968-SEPT. 1968.
COLLINS J O　　JUANARAJS K L　　REID D R
JOHNS-MANVILLE RESEARCH AND ENGINEERING CENTER MANVILLE NEW JERSEY
AEC
ALO-3633-10
1-56　　1968

48442 REFRACTORY CONCRETE WITH LOW THERMAL EXPANSION.
GUGEL E
SPRECHSAAL KERAM GLAS EMAIL SILIKATE
100　　21　　825-30　　1967　　CA　　68　　62411

48448 METHOD FOR SIMULTANEOUS AND INDEPENDENT MEASUREMENTS OF THERMAL AND ELECTRICAL PARAMETERS.
GIRAUDIER L
ENTROPIE
25　　　23-42　　1969

48463 USE OF THE IKS-14A DEVICE FOR MEASURING OF PLANE SURFACE REFLECTION SPECTRA.
RINAS V F　　AVDEEV S P　　MNUSKIN M G
ZH PRIKL SPEKTROSK
6　　5　　640-2　　1967　　CA　　68　　73709

48475 A SESSILE DROP STUDY OF LIQUID-SOLID ADHESION FOR THE SYSTEM LIQUID INDIUM-ALUMINUM USING ULTRAHIGH VACUUM TECHNIQUES.
ALDRICH R G　　KELLER D V JR
J PHYS CHEM
72　　4　　1092-9　　1968　　CA　　68　　99002

48481 THE CALCULATION OF COHESIVE AND ADHESIVE ENERGIES FROM INTERMOLECULAR FORCES AT A SURFACE.
PADDAY J F　　UFFINDELL N D
J PHYS CHEM
72　　5　　1407-14　　1968　　CA　　69　　5417

48487 EMITTANCE VALUES OF HAYNES-25 FOR A SELECTED REENTRY ENVIRONMENT.
DEVENEY J E　　SANDIA CORP　　ALBUQUERQUE N MEX
CFSTI AND AEC
SC-RR-66-576
N68-14377　　　1-41　　1967　　PA　　N68-6-5 674

48492 DEVELOPMENT OF SPACE-STABLE THERMAL-CONTROL COATINGS. TRIANNUAL REPORT.
ZERLAUT G A　　RUBIN G A　　IIT RESEARCH INSTITUTE CHICAGO ILLINOIS
CFSTI AND NASA
IITRI-U6002-36
1-46　　1966

TPRC Number	Bibliographic Citation

48493 THERMAL CONTACT CONDUCTANCE IN A VACUUM ENVIRONMENT AT LOW TEMPERATURES.
BLOOM M F   MISSILE AND SPACE SYSTEMS DIVISION
DOUGLAS AIRCRAFT COMP   SANTA MONICA CALIFORNIA
DOUGLAS MISSILE AND SPACE SYSTEMS DIVISION
SM-3200   1-35   1964

48494 PRILIMINARY RESULTS FROM A NEW THERMAL CONTACT CONDUCTANCE APPARATUS.
BLOOM M F   ENGINEER MISSILES AND SPACE DIVISION
DOUGLAS AIRCRAFT COMP   SANTA MONICA CALIFORNIA
DOUGLAS MISSILE AND SPACE SYSTEMS DIVISION
SM-1672   1-29   1963

48495 TEST RESULTS AND ANALYSIS OF TUNGSTEN-GRAPHITE THERMAL CONTACT RESISTANCE.
HENRY J J   ALLEGHANY BALLISTICS LAB   CUMBERLAND MARYLAND
DYNATECH CORPORATION CAMBRIDGE MASSACHUSETTS
DYNATEC   H NO 342
ABL-69021   1-15   1962

48496 STUDY OF INTERFACE THERMAL CONTACT CONDUCTANCE. SUMMARY REPORT MARCH 1965- JULY 1966.
FRIED E   GEORGE C MARSHALL SPACE FLIGHT CENTER HUNTSVILLE ALABAMA
NASA
GEN ELE   C DOC 66SD4471
                 1-73   1966

48497 HEAT TRANSFER ACROSS SURFACES IN CONTACT.   EFFECT OF AMBIENT PRESSURE CHANGES. PAPER 63-B-9.
AARON R L   BLUM H A
AM INST OF CHEM ENGRS
9   2   1-18   1963

48498 AN EXPERIMENTAL INVESTIGATION OF THERMAL CONTACT RESISTANCE IN A VACUUM. //PAPER NO. 63-WA-156//
PETRI F J
AM SOC MECH ENG
                 1-5   1963

48499 THERMAL INTERFACE CONDUCTANCE OF THERMOELECTRIC GENERATOR HARDWARE. //PAPER NO. 66-WA/NE-2//
HARGADON J M JR
AM SOC MECH ENG
                 1-9   1966

48500 CONTROLLING FACTORS OF THERMAL CONDUCTANCE ACROSS BOLTED JOINTS IN A VACUUM ENVIRONMENT. //PAPER NO. 63-WA-196//
ARON W   COLOMBO G
AM SOC MECH ENG
                 2-8   1963

48501 THERMAL CONDUCTANCE OF FILLED ALUMINUM AND MAGNESIUM JOINTS IN A VACUUM ENVIRONMENT. //PAPER NO. 64-WA/HT-40//
CUNNINGTON G R JR
AM SOC MECH ENG
                 1-9   1964

48502 THERMAL JOINT CONDUCTANCE IN A VACUUM. //PAPER NO. AHGT-18//
FRIED E
AM SOC MECH ENG
                 1-8   1962

48503 THERMAL CONDUCTION ACROSS ALUMINUM BOLTED JOINTS. //PAPER NO. 65-HT-53//
ELLIOTT D H
AM SOC MECH ENG
                 1-4   1965

48504 MEASUREMENTS OF THERMAL CONTACT CONDUCTANCE BETWEEN DISSIMILAR METALS IN A VACUUM. //PAPER NO. 64-HT-36//
KASPARECK W E   DAILEY R M
AM SOC MECH ENG
                 1-8   1964

48505 THE EFFECT OF INTERFACIAL METALLIC FOILS ON THERMAL CONTACT RESISTANCE. //PAPER NO. 65-HT-44//
KOH B   JOHN J E A
AM SOC MECH ENG
                 1-7   1965

48506 MEASUREMENTS OF THERMAL CONTACT CONDUCTANCE IN VACUUM //PAPER NO. 63-WA-150//
STUBSTAD W R
AM SOC MECH ENG
                 1-7   1963

48507 THERMAL CONDUCTANCE OF HIGH TEMPERATURE METALLIC HONEYCOMB COMPOSITES. FROM PROC. FOURTH THERMAL CONDUCTIVITY CONFERENCE, OCT. 1964,
MINGES M L   MATERIALS APPLICATIONS DIV
WRIGHT-PATTERSON AIR FORCE BASE   OHIO
DDC
                 1-35   1964

48508 THE CONDUCTIVITY OF JOINTS.   DIPLOMA THESIS.
ASHMEAD F A H   ROYAL NAVY COLLEGE OF AERONAUTICS
GREAT BRITAIN
GREAT BRITIAN-COLLEGE OF AERONAUTICS
                 1-47   1955

48509 THERMAL CONTACT RESISTANCE OF GRAPHITE.   MASTER THESIS.
HORN S S   MASSACHUSETTS INSTITUTE OF TECHNOLOGY CAMBRIDGE MASS
MASSACHUSETTS INSTITUTE OF TECHNOLOGY
                 1-62   1963

48510 SOME INFLUENCES OF MACROSCOPIC CONSTRICTIONS ON THE THERMAL CONTACT RESISTANCE.
CLAUSING A M   UNIVERSITY OF ILLINOIS   URBANA
NASA
ME-TN-242-2
                 1-42   1965

48513 FUNDAMENTAL ABSORPTION AND EXCITATION IN A SERIES OF IIA-VIB COMPOUNDS. //ENGLISH TRANSLATION OF IZV. AKAD. NAUK SSSR, SER. FIZ. 33 /6/ 974- , 1969.//
LEVSHIN V L   MIKHAILIN V V   SAULEVICH L K
BULL ACAD SCI USSR /PHYS SER/
33   6   900-2   1969

48523 EFFECT OF FOUNDRY COKE ASH CONTENT ON THE CARBON CONTENT OF CUPOLA CAST IRON.
TURBOVSKII M M   VYKHODETS A M   ENIKEEV E V
IZV AKAD NAUK UZB SSR SER TEKH NAUK
11   6   72-3   1967   CA   68   61885

48529 SURFACE TENSION IN THE NONANE-METHANE SYSTEM.
DEAM J R   BENNETT S E   MADDOX R N
PROC ANNU CONV NATUR GAS PROCESSORS ASS TECH PAP
             46   26-7   1967   CA   68   72559

48534 MOLECULAR FORCES AT INTERFACES.
FOWKES F M
SURFACES COATINGS RELATED PAPER WOOD SYMP SYRACUSE N Y
                 99-125   1967   CA   68   72577

48544 INTERFACIAL PHENOMENA AND WATER-REPELLENT FINISHING.
CLAUS L
TEXTILIS /GENT/
21   9   11-12 15-18 1965

48547 NON-DESTRUCTIVE MEASUREMENT OF SURFACE CONCENTRATIONS AND JUNCTION DEPTHS OF DIFFUSED SEMICONDUCTOR LAYERS.
ABE T   NISHI Y
JAPAN J APPL PHYS
7   4   397-403   1968

48556 THERMAL RESISTANCE OF IMPERFECT CONTACTS.   M.S. THESIS.
BROWN E C JR   NORTH CAROLINA STATE UNIVERSITY RALEIGH NORTH CAROLINA
NORTH CAROLINA STATE UNIVERSITY
                 1-47   1967

48566 ON THE DISJOINING PRESSURE OF THIN FILMS.
KUNI F M   RUSANOV A I
PHYS LETT
26   A   11   577-8   1968

48567 PLASMA RESONANCE IN THE REFLECTION SPECTRUM OF THIN ALUMINIUM FILMS.
FEUERBACHER B   GODWIN R P   SKIBOWSKI M
PHYS LETT
26   A   12   595-6   1968

48573 PROTON AND ULTRAVIOLET RADIATION EFFECTS ON SOLAR MIRROR REFLECTIVE SURFACES.
GILLETTE R B
J SPACECRAFT ROCKETS
5   4   454-60   1968   CA   68   109609
AIAA THERMOPHYSICS SPECIALIST CONFERENCE   1967
PAPER 67-341

48580 HEAT TRANSFER BELOW 0.2 K.
SUOMI M   ANDERSON A C   HOLMSTROM B
PHYSICA
38   1   67-80   1968
DDC
AFOSR-68-1461
AD 671359   1-16   1968   RR   68-17 174

48617 STUDY OF ZIRCONIUM MOLYBDATE BY SEVERAL PHYSICOCHEMICAL METHODS.
DENISOVA N E   BOICHINOVA E S
IZV AKAD NAUK SSSR NEORG MATER
3   6   1049-54   1967

48618 STUDY OF ZIRCONIUM MOLYBDATE BY SEVERAL PHYSICOCHEMICAL METHODS. //ENGLISH TRANSLATION OF IZV. AKAD. NAUK SSSR, NEORG. MATER. 3/6/ 1049-54, 1967.//
DENISOVA N E   BOICHINOVA E S
INORGANIC MATERIALS
3   6   933-7   1967

TPRC Number	Bibliographic Citation						
48654	ADSORPTION OF OCTANOATE IONS AT THE OIL-WATER INTERFACE. CHATTERJEE A K CHATTORAJ D K J COLLOID INTERFACE SCI						
	26	2	140-5	1968	CA	68	90167
48662	THERMAL CONDUCTIVITY OF HEAT-RESISTANT HIGH-ALUMINA CEMENT CONCRETE IN THE TEMPERATURE RANGE 200 - 1500 C. CHEKHOVSKOI V YA ROMANOV A I KAULENAS A A STAVROVSKII G I TEPLOFIZ VYSOKIKH TEMPERATUR						
	5	5	789-92	1967			
48663	THERMAL CONDUCTIVITY OF HEAT-RESISTANT HIGH-ALUMINA CEMENT CONCRETE IN THE TEMPERATURE RANGE 200 - 1500 C. //ENGLISH TRANSLATION OF TEPLOFIZ. VYSOKIKH TEMPERATUR 5 /5/ 789-92, 1967.// CHEKHOVSKOI V YA ROMANOV A I KAULENAS A A STAVROVSKII G I HIGH TEMPERATURE						
	5	5	704-7	1967			
48670	METALLIC SCREEN INSULATION IN AIR AND HELIUM. KORSHAKOV A I BOGDANOV F F TEPLOFIZ VYSOKIKH TEMPERATUR						
	5	5	821-6	1967			
48671	METALLIC SCREEN INSULATION IN AIR AND HELIUM. //ENGLISH TRANSLATION OF TEPLOFIZ VYSOKIKH TEMPERATUR 5 /5/ 821-6, 1967.// KORSHAKOV A I BOGDANOV F F HIGH TEMPERATURE						
	5	5	731-4	1967			
48674	PREPARATION OF STANDARD COATINGS WITH UNCHANGING RADIATION PROPERTIES. //ENGLISH TRANSLATION OF TEPLOFIZ. VYSOKIKH TEMPERATUR 5 /5/ 919-23, 1967.// NOVITSKII L A HIGH TEMPERATURE						
	5	5	817-20	1967			
48687	A STUDY ON THE HEAT CONDUCTION AT THE CONTACT OF METALS. I. THE CASE OF THE SAME CONDITIONS IN TWO CONTACTING SOLIDS. SANOKAWA K NIPPON KIKAI GAKKAI ROMBUNSHU						
	33		1097-107	1967		NSA 22/9/1741	
48688	A STUDY ON THE HEAT CONDUCTION AT THE CONTACT OF METALS. II. THE INFLUENCE OF THE THICKNESS OF A CONTACTING SOLID, AND THE DIFFERING CONDITIONS IN THE CONTACTING SOLIDS. SANOKAWA K NIPPON KIKAI GAKKAI ROMBUNSHU						
	33		1108-19	1967		NSA 22/9/1741	
48689	A STUDY ON THE HEAT CONDUCTION AT THE CONTACT OF METALS. III. THE CASE OF THE OXIDIZED CONTACTING SURFACES. SANOKAWA K NIPPON KIKAI GAKKAI ROMBUNSHU						
	33		1120-30	1967		NSA 22/9/174	
48690	A STUDY ON THE HEAT CONDUCTION AT THE CONTACT OF METALS. IV. THE INFLUENCE OF THE SURFACE CONFIGURATION, UNDULATIONS, AND THE APPROXIMATION METHODS FOR CALCULATION OF THE CONTACT HEAT-RESISTANCES. SANOKAWA K NIPPON KIKAI GAKKAI ROMBUNSHU						
	33		1131-7	1967		NSA 22/9/174	
48692	SURFACE ACTIVITY OF AQUEOUS AND HYDROCARBON SOLUTIONS OF SURFACTANTS AND THEIR ADSORPTION ON SANDS FROM A PRODUCTIVE BED OF THE APSHERON PENINSULA. KHALILOV E G MUSAEV R A KRAVCHENKO I I BABALYAN G A PRIMEN PROVERKH-AKTIV VESHCHESTV NEFT PROM TR VSES SOVESHCH 3RD /UFA/ USSR 1965						
			78-86	1966	CA	68	90183
48697	A THERMOCHEMICAL STUDY OF BORON TRIIODIDE. OWNBY P D GRETZ R D SURFACE SCI						
	9	1	37-56	1968	CA	68	90451
48712	SURFACE TENSION OF THE LIQUID ALKALI METALS MEASURED IN ULTRA-HIGH VACUUM. GERMER D MAYER H Z PHYS						
	210		391-402	1968	CA	68	81607
48720	OPTICAL CONSTANTS OF POLYSTYRENE IN THE VACUUM ULTRAVIOLET. SHAPIRO J T MADDEN R P J OPT SOC AM						
	58	6	771-5	1968			
48731	DIFFUSION IN SILICON NITRIDE FILMS ON ALUMINUM. KOVARSKII V YA ALEKSANYAN I T BONDARENKO O E ORLOV B M FIZ TVERD TELA						

TPRC Number	Bibliographic Citation						
48742	DEFORMATION OF FINE-GRAINED CONCRETE DURING HEATING. DORONIN L K MIKHAILOV N V IZV AKAD NAUK SSSR NEORG MATER						
	4	1	130-5	1968	CA	68	98381
48743	ELECTRON MICROSCOPIC STUDY OF THE SURFACE DIFFUSION OF BARIUM ON THE FACES OF A LARGE TUNSTEN SINGLE CRYSTAL. SMORODINOVA M I IZV AKAD NAUK UZB SSR SER FIZ-MAT NAUK						
	11	2	38-41	1967	CA	68	99416
48776	INFRARED FLUX AND SURFACE TEMPERATURE DETERMINATIONS FROM TIROS RADIOMETER MEASUREMENTS. WARK D Q YAMAMOTO G LIENESCH J METEOROLOGICAL SATELLITE LAB WASHINGTON D C NASA MSL-10 SUPPL.						
	N63-17446		1-10	1963			
48778	SIMULTANEOUS DETERMINATION OF SURFACE TENSION OF A SLAG AND OF INTERFACIAL TENSION ON THE METAL-SLAG BOUNDARY. BOBKOVA O PETUKHOV V SB TR TSENT NAUCH-ISSLED INST CHERN MET						
	56		51-2	1967	CA	68	97657
48786	ADSORPTION OF SULFUR CONTAINING SPECIES AT THE MERCURY/WATER INTERPHASE. PARSONS R SYMONS P C TRANS FARADAY SOC						
	64	4	1077-92	1968	CA	68	98985
48803	REFLECTION OF SEMICONDUCTING COPPER SULFIDE LAYERS IN THE INFRARED REGION OF THE SPECTRUM. KRYZHANOVSKII B P OPTIKA I SPEKTROSKOPIYA						
	24	2	263-6	1968	CA	68	109606
48805	THERMAL OXIDATION OF GALLIUM ARSENIDE. NAVRATIL K CZECHOSLOVAK JOURNAL OF PHYSICS SECTION B						
	18	2	266-74	1968	IAA 8/11/ 2089		
48808	THERMAL PROPERTIES OF ABLATIVE CHARS. FINAL REPORT. MAR. 1966-SEPT. 1967. CLAYTON W A KENNEDY P B EVANS R J COTTON J E FRANCISCO A C FABISH T J ELDRIDGE E A LAGEDROST J F DDC AND CFSTI AFML-TR-67-413 N69-11058						
	AD-875179		1-252	1968	CA	70	60412
48810	ROCKET VEHICLE SURFACE REFLECTANCE MEASUREMENT DURING RETRO-ROCKET FIRING. ADAMS R N IEEE TRANSACTIONS ON INSTRUMENTATION AND MEASUREMENT						
	IM-17		73-5	1968			
48811	EVALUATION OF THE THERMAL PROPERTIES OF MATERIALS. VOLUME 2. DATA HANDBOOK. FINAL REPORT, JUNE 1965-JUNE 1966. AVCO CORPORATION AVCO SPACE SYSTEMS DIVISION LOWELL MASSACHUSETS NASA AVSSD-0197-66-RR NASA-CR-65980						
	N68-18853		1-208	1966	RR	68-10 139	
48812	DETERMINATION OF THERMOPHYSICAL PROPERTIES OF ABLATIVE MATERIALS. PHASE 1. LABORATORY DETERMINATIONS. PART B. THERMAL CONDUCTIVITY AND THERMAL DIFFUSIVITY. NORDBERG R C LARSON R D CHARLAN G B TRW SYSTEMS REDONDO BEACH CALIF NASA AND CFSTI TRW-04812-6001-R001/PT B/ NASA-CR-92018						
	N68-19863		1-32	1967	CA	71	92334
48814	THERMAL ANALYSIS OF HERMETICALLY SEALED NICKEL-CADMIUM CELLS FOR SPACE APPLICATIONS. PREUSSE K E SHAIR R C GULTON INDUSTRIES INC METUCHEN N J NASA AIAA PAPER 64-751						
	N64-28849		1-21	1964	CA	62	15458
48824	THERMAL CHARACTERISTICS OF ADHESIVES. FROM PROCEEDINGS OF THE 4TH SYMPOSIUM ON THERMOPHYSICAL PROPERTIES, UNIV. MARYLAND COLLEGE PARK, MD. FADLER E C SAUER H J JR REMINGTON C R JR AMERICAN SOCIETY MECHANICAL ENGINEERS NEW YORK						
			155-60	1968	IAA 8-11 2043		
48826	EVALUATION OF TEMPERATURE-DEPENDENT EFFECTIVE THERMAL CONDUCTIVITIES FOR SPACECRAFT ROCKET ENGINE CHAMBER INSULATION MATERIALS. FROM PROCEEDINGS OF THE 4TH SYMPOSIUM ON THERMOPHYSICAL PROPERTIES, UNIV. MARYLAND COLLEGE PARK, MD. CARLSON L W HUEBNER A L CLAUSING A M AMERICAN SOCIETY MECHANICAL ENGINEERS NEW YORK						
			175-83	1968	IAA 8-11 2132		

TPRC Number	Bibliographic Citation

48831 THE THEORY OF ELECTROMAGNETIC TUNNELING OF THERMAL RADIATION BETWEEN HIGHLY ABSORBING MEDIA. FROM PROCEEDINGS OF THE 4TH SYMPOSIUM ON THERMOPHYSICAL PROPERTIES, UNIV. MARYLAND COLLEGE PARK, MD.
CAREN R P
AMERICAN SOCIETY MECHANICAL ENGINEERS NEW YORK
243-55   1968   IAA 8-11 2132

48832 APPLICATION OF MIE SCATTER THEORY TO THE REFLECTANCE OF PAINT-TYPE COATINGS. FROM PROCEEDINGS OF THE 4TH SYMPOSIUM ON THERMOPHYSICAL PROPERTIES, UNIV. MARYLAND COLLEGE PARK, MD.
SCHMIDT R N   TREUENFELS P M   MEEHAN E J
AMERICAN SOCIETY MECHANICAL ENGINEERS NEW YORK
256-69   1968   IAA 8-11 2133

48843 EVALUATION OF THE THERMAL PROPERTIES OF MATERIALS. VOLUME 1. TECHNICAL REPORT. FINAL REPORT, JUN. 1965-JUN. 1966.
AVCO CORPORATION AVCO SPACE SYSTEMS DIVISION LOWELL MASSACHUSETS
LOWELL MASSACHUSETTS
NASA AND CFSTI
AVSSD-0197-66-CR NASA-CR-65979
N68-18488   1-98   1966   RR   68-10 130

48844 STUDY ON HIGH-PERFORMANCE INSULATION THERMAL DESIGN CRITERIA. THIRD QUARTERLY PROG. REPT.
COSTON R M   BROGAN J J   GUILL J H   LOCKHEED MISSILES AND SPACE CO SUNNYVALE CALIFORNIA
NASA AND CFSTI
LMSC-A852904 NASA-CR-61456
N68-15757   1-61   1967

48855 THERMAL CONDUCTIVITY OF FIBROUS SILICA.
ROLINSKI E J   SWEENEY T L
J CHEM ENG DATA
13   2   203-6   1968

48858 RADIATION PROPERTIES OF A CDS SOLAR CELL AND VARIOUS METALS AT SPACE CONDITIONS.
JACK J R   SPISZ E W   LEWIS RESEARCH CENTER
CLEVELAND OHIO
NASA AND CFSTI
NASA-TM-X-52457
N68-33336   1-11   1968   RR   68   75

48862 SPACECRAFT ROCKET ENGINE CHAMBER INSULATION MATERIALS. INTERIM REPT. JULY 1967-JUNE 1968.
CARLSON L W   CARPENTER H   HUEBNER A L   MANSON L
TALMOR E   ROCKETDYNE CANOGA PARK CALIF
NASA AND CFSTI
R-7548   NASA-CR-96447
N68-33164   1-218   1968

48863 THE THERMAL ACCOMMODATION OF HELIUM, NEON, AND ARGON ON CLEAN TUNGSTEN FROM 77 TO 303 K. PH D THESIS.
SILVERNAIL W L   UNIVERSITY OF MISSOURI COLUMBIA
AND ROLLA MISSOURI
UNIV MICROFILMS PUBL
U7 10132   1-93   1954

48888 STUDY OF THE SURFACE EMISSIVITY OF TEXTILE FABRICS AND MATERIALS IN THE 1 - 15 MICRON RANGE. TECHNICAL REPT. MAR. 1965-MAR. 1966.
MASON M T   COLEMAN I   BLOCK ENGINEERING INC
CAMBRIDGE MASSACHUSETTS
DDC
USA-NLABS-TR-67-86-CM C/OM-TS-151
AD 665647   1-97   1967   RR   68-8 79

48897 REVIEW OF LITERATURE ON THERMAL CONTACT CONDUCTANCE IN A VACUUM. FINAL REPORT.
CARFAGNO S P   FRANKLIN INST RESEARCH LABS
PHILADELPHIA PA
NASA AND CFSTI
NASA-CR-61556 F-C1882
N68-16898   1-105   1967   RR   68-8 134

48912 OPTICAL PROPERTIES OF ALUMINUM OXIDE IN THE VACUUM ULTRAVIOLET.
ARAKAWA E T   WILLIAMS M W
J PHYS CHEM SOLIDS
29   5   735-44   1968   CA   69   14338

48919 OPTICAL ABSORPTION IN SILICON MONOXIDE.
RAWLINGS I R
BRIT J APPL PHYS
1   6   733-9   1968

48925 OPTICAL PROPERTIES OF ALPHA- AND BETA-TITANIUM TRICHLORIDE.
BALDINI G   POLLINI I   SPINOLO G
PHYS STAT SOL
27   1   95-100   1968

48926 THICKNESS DEPENDENCE OF THE QUANTUM YIELD OF CESIUM-ANTIMONY FILMS.
DEUTSCHER K   HIRSCHBERG K
PHYS STAT SOL
27   1   145-50   1968

48930 CHAMOTTE-VERMICULITE HEAT INSULATING CONCRETE.
PIROGOV A A   LEVE E N
OGNEUPORY
9   44-7   1967

48931 CHAMOTTE-VERMICULITE HEAT INSULATING CONCRETE. //ENGLISH TRANSLATION OF OGNEUPORY NO. 9, 44-7, 1967.//
PIROGOV A A   LEVE E N
REFRACTORIES
9   559-62   1967

48932 ON THE RESISTANCE OF CADMIUM SELENIDE FILMS.
ELLIS S G
J PHYS CHEM SOLIDS
29   7   1139-42   1968

48933 THE OPTICAL ABSORPTION SPECTRA OF METAL IODIDES WITH LAYER STRUCTURES.
TUBBS M R
J PHYS CHEM SOLIDS
29   7   1191-203   1968

48950 IDEAL BEHAVIOR OF SODIUM ALKYL SULFATES AT VARIOUS INTERFACES. THERMODYNAMICS OF ADSORPTION AT THE OIL-WATER INTERFACE.
GILLAP W R   WEINER N D   GIBALDI M
J PHYS CHEM
72   8   2222-7   1968

48974 EXTREME ULTRAVIOLET ABSORPTION OF ALKALI HALIDES.
SAITO H   SAITO S   ONAKA R   IKEO B
J PHYS SOC JAPAN
24   5   1095-8   1968

48978 OPTICAL ABSORPTION OF LONGITUDINAL AND TRANSVERSE EXCITONS IN KCL AND KBR.
EJIRI A
J PHYS SOC JAPAN
24   5   1181   1968

48979 COLORIMETRY.
NIMEROFF I   NATIONAL BUREAU OF STANDARDS
WASHINGTON D C
USGPO
NBS MONOGRAPH 104
1-47   1968

48990 CALCULATION OF THE THERMAL RESISTANCE OF THE ZONE OF SOLID CONTACT.
PRASOLOV R S
AT ENERG
24   1   86-7   1968   CA   68   83438

49006 AN INTEGRATING SPHERE REFLECTOMETER FOR THE DETERMINATION OF ABSOLUTE HEMISPHERICAL SPECTRAL REFLECTANCE.
ZERLAUT G A   KRUPNICK A C   NATIONAL AERONAUTICS AND SPACE ADMINISTRATION MARSHALL SPACE FLIGHT CENTER HUNTSVILLE ALA
NASA
AIAA   PAPER 64-255
N64-24981   1-11   1964

49007 ADVANCES IN HIGH-TEMPERATURE THERMAL PROTECTION SYSTEMS.
WECHSLER A E   ARTHUR D LITTLE INC CAMBRIDGE
MASS
NASA
N62-16604   1-19   1961   PA   62-2 1011

49010 ROLE OF INTERFACIAL TENSION IN THE PREPARATION AND STABILITY OF EMULSIONS STABILIZED BY HYDROPHILIC COLLOIDS.
MUKERJEE L N   SHUKLA S D
INDIAN J APPL CHEM
30 1/2   17-24   1967   CA   68   81616

49018 THE THERMAL CONDUCTIVITY OF GRANULAR QUARTZ AT LOW TEMPERATURES. FROM PROCEEDINGS OF THE 3RD CONFERENCE ON THERMAL CONDUCTIVITY. GATLINBURG, TENNESSEE. OCTOBER 16-18, 1963.
DE NEE P B   SAWYER R B   LEHIGH UNIVERSITY
BETHLEHEM PENNSYLVANIA
OAK RIDGE NATIONAL LAB OAK RIDGE TENNESSEE
529-56   1963

49023 A LINE HEAT SOURCE METHOD FOR THE DETERMINATION OF THERMAL CONDUCTIVITY OF POWDERS IN VACUUM. FROM PROCEEDINGS OF THE 3RD CONFERENCE ON THERMAL CONDUCTIVITY. GATLINBURG, TENNESSEE. OCTOBER 16-18, 1963.
WECHSLER A E   GLASER P E   LITTLE /ARTHUR D/ INC
CAMBRIDGE MASS
OAK RIDGE NATIONAL LAB OAK RIDGE TENNESSEE
613-27   1963

49027 A NOTE ON THE THERMAL CONDUCTIVITY OF POWDERS IN VACUUM. FROM PROCEEDINGS OF THE 3RD CONFERENCE ON THERMAL CONDUCTIVITY. GATLINBURG, TENNESSEE. OCTOBER 16-18, 1963.
LAUBITZ M J   NATIONAL RESEARCH COUNCIL OTTAWA
CANADA

TPRC Number	Bibliographic Citation

49030 DEVELOPMENT OF SPACE-STABLE THERMAL-CONTROL COATINGS /PAINTS WITH LOW SOLAR ABSORPTANCE/EMITTANCE RATIOS./ TRIANNUAL REPT. FEB. 1965.
ZERLAUT G A   KAYE B H   GEORGE C MARSHALL SPACE FLIGHT CENTER   HUNTSVILLE ALA
NASA
IITRI-C8014-21 NASA-CR-67559
N65-36558   1-34   1965

49031 DEVELOPMENT OF SPACE-STABLE THERMAL-CONTROL COATINGS /PAINTS WITH LOW SOLAR ABSORPTANCE/EMITTANCE RATIOS./ TRIANNUAL REPT. SEPT. 20, 1963-JAN. 20, 1964
ZERLAUT G A   TOMPKINS E H   HARADA Y   GEORGE C MARSHALL SPACE FLIGHT CENTER   HUNTSVILLE ALA
NASA
IITRI-C8014-8 NASA-CR-60355
N65-15712   1-55   1964

49033 THERMOPHYSICAL PROPERTIES OF REINFORCED COMPOSITIONS.
BIL V S   AVTOKRATOVA N D
PLAST MASSY
                    2   35-7   1968   CA   68   79466

49042 DOUBLE-BEAM TYPE REFLECTOMETER FOR DETERMINING OPTICAL CONSTANTS.
KUDO K   OGAWA T   YOSHIMOTO H
SCI LIGHT /TOKYO/
16   2   108-28   1967   CA   68   82663

49061 A DROP VOLUME METHOD FOR THE DETERMINATION OF THE EQUILIBRIUM INTERFACIAL TENSION BY MEANS OF THE HANGING MERCURY DROP ELECTRODE.
ROFFIA S   VIANELLO E
J ELECTROANAL CHEM INTERFACIAL ELECTROCHEM
17 1/2   13-20   1968   CA   68   83720

49110 STUDY OF THE OPTICAL PROPERTIES OF THIN FILMS OF YTTRIUM UNDER STATIC ULTRAVIOLET.
PETRAKIAN J-P
COMPT REND
269   B   10   434-7   1969

49118 OPTICAL PROPERTIES OF COPPER THIN FILMS IN THE 1.5-7-MICRON SPECTRAL RANGE.
VARENNE S
REV OPT
45   11   485-95   1965   CA   65   109629

49124 OPTICAL CHARACTERISTICS OF SILICON PHOTOCELL AND THE EFFICIENCY OF A THERMOPHOTOELECTRIC CONVERTER.
VASILEV A M   GOLOVNER T M   LANDSMAN A P
LIDORENKO N S
TEPLOFIZ VYS TEMP
5   6   1079-86   1967   CA   68   109615

49144 ZERO THERMAL EXPANSION COMPOSITES OF HIGH STRENGTH AND STIFFNESS.
DOW N F   ROSEN B W   GENERAL ELECTRIC COMP
PHILADELPHIA PA
NASA AND CFSTI
NASA-CR-1324 N69-24876
                    1-88   1969   RR   69-16 101

49152 DEVELOPMENT OF 400 TO 1800 F FIBROUS-TYPE INSULATION FOR RADIOISOTOPE POWER SYSTEMS.   PHASE 1, FINAL REPORT.   VOLUME 1.
COLLINS J O   JOHNS-MANVILLE RESEARCH AND ENGINEERING CENTER   MANVILLE NEW JERSEY
AEC AND CFSTI
ALO-3633-7 /VOL 1/
                    1-146   1967

49153 QUARTERLY PROGRESS REPORT IN THERMAL INSULATION STUDY, APRIL-JUNE 1967.
DE WITT W D   REID R L   UNION CARBIDE CORP
TONAWANDA N Y
AEC AND CFSTI
ALO-3632-17
                    1-115   1967

49160 URANIUM MONOCARBIDE SHAPING AND IRRADIATION. QUARTERLY REPORT NO. 23 JAN.-MARCH, 1967.
DEVILLARD J   MORLOT G   COMMISSARIAT A L ENERGIE ATOMIQUE   GRENOBLE FRANCE
AEC
EURAEC-1979
                    1-33   1967

49166 CONTACT HEAT TRANSFER IN GRANULAR MATERIAL IN A VACUUM.
KAGANER M G
INZH-FIZ ZH AKAD NAUK BELORUSSK SSR
11   1   30-6   1966   CA   65   14865

49167 CONTRIBUTION TO THE PROBLEM OF DETERMINING THE COMPONENTS OF THERMAL CONTACT RESISTANCE.
KRIZHNIAK P E
AVIATS TEKH
8   144-8   1965

49172 A GENERAL THEORY OF INTERFACIAL TENSION AND ITS APPLICATION TO ADSORPTION PHENOMENA.   PH.D. THESIS.
COTTON D J   HOWARD UNIV   WASHINGTON D C

49177 DIFFUSION IN EVAPORATED FILMS OF SILVER-ALUMINIUM.
WEAVER C   BROWN L C
PHIL MAG
17   881-97   1968   CA   68   117292

49181 OPTICAL AND ELECTRICAL PROPERTIES OF RARE EARTH THIN FILMS.   PH.D. THESIS.
OKORIE O A   HOWARD UNIVERSITY   WASHINGTON D C
UNIV MICROFILMS PUBL
67-13560   1-116   1967

49193 THE EFFECT OF POROSITY ON SHEARING RESISTANCE AND THERMAL CONDUCTIVITY FOR AMORPHOUS SOILS IN VACUUM. PH D THESIS.
WATERS R H   TEXAS A AND M UNIV   COLLEGE STATION TEXAS
UNIV MICROFILMS PUBL
67-9816   1-159   1967

49206 OPTICAL AND PHOTOELECTIRC PROPERTIES INCLUDING POLARIZATION EFFECTS, OF GOLD AND ALUMINUM IN THE EXTREME ULTRAVIOLET.   TECH. REPT. JAN. 1965-JAN. 1968.
MORSE A L   DEPT OF PHYSICS UNIVERSITY OF SOUTHERN CALIF   LOS ANGELES
NASA AND CFSTI
NASA-CR-93971 USC-VACUV-109
N68-21212   1-94   1968

49207 POSTIRRADIATION FISSION PRODUCT RELEASE FROM COATED FUEL PARTICLES.
ROSENBERG H S   MORRISON D L   TOWNELY C W
SUNDERMAN D N   BATTELLE MEMORIAL INSTITUTE COLUMBUS OHIO
AEC AND CFSTI
BMI-1734   1-49   1965

49228 EMITTANCE OF ABLATIVE MATERIALS.   FINAL REPT.
NEWMAN B E   TRW SYSTEMS   REDONDO BEACH CALIFORNIA
NASA AND CFSTI
NASA-CR-82063 N68-22415
                    1-57   1967   RR   68-14 142

49229 MEASUREMENTS REPORT.   THERMAL PROPERTY MEASUREMENTS OF MANNED SPACECRAFT CENTER SPACESUIT MATERIALS.
TURNBOW F J   TRW SYSTEMS   REDONODO BEACH CALIFORNIA
NASA AND CFSTI
TRW-68-3346.11JA-31 NASA-CR-92087
N68-22779   1-15   1968   RR   68-14 43

49233 INFRARED COATING STUDIES.   QUARTERLY REPORT 6, SEP. 15-DEC. 15, 1967.
MAIER R L   BAUSCH AND LOMB INC   ROCHESTER NEW YORK
DDC CFSTI AND NASA
AD 664786   N68-20132
                    1-16   1967

49250 PHASE 1.   QUARTERLY PROGRESS REPORT, SEPT.-NOV., 1967.
CERNI S   WESTINGHOUSE ASTRONUCLEAR LAB PITTSBURGH PA
NASA AND CFSTI
WANL-3800-9 WANL-PR-/SS/-010
N68-22714   1-119   1967

49251 DIRECTIONAL DISTRIBUTION IN THE REFLECTION OF HEAT RADIATION AND ITS EFFECT ON HEAT TRANSFER.   //ENGLISH TRANSLATION OF E. T. H. ZURICH /2434/ 89 PP. 1955.// PH. D. THESIS.
MUNCH B   SWISS TECHNICAL COLLEGE /ZURICH/
NASA
NASA-TT-F-497 N68-21536
                    1-90   1968

49254 INVESTIGATIONS OF ALUMINUM FILMS WITH SYNCHROTRON RADIATION OF WAVELENGTHS 500 TO 1000 A.   A.   POLARIZATION DEPENDENT TRANSMISSION AND REFLECTION.   B.   POLARIZATION DEPENDENT PHOTO-EFFECT.
SKIBOWSKI M   FEUERBACHER B   STEINMANN W
GODWIN R P   DEUTSCHES ELEKTRONEN-SYNCHROTRON HAMBURG GERMANY
NASA
DESY-67/33 N68-22636
                    1-51   1967

49264 KINETIC DESCRIPTION OF CYLINDRICAL HEAT CONDUCTION IN A POLYATOMIC GAS.
CIPOLLA J W JR   MORSE T F
PHYS FLUIDS
11   6   1292-300   1968

49265 EXPERIMENTAL STUDY OF RAREFIED ARGON CONTAINED BETWEEN CONCENTRIC CYLINDERS.
SHELDON D B   SPRINGER G S
PHYS FLUIDS
11   6   1312-20   1968

49266 THERMAL EXPANSION OF EPOXIES BETWEEN 2 AND 300 K.
HAMILTON W O   GREENE D B   DAVIDSON D E
REV SCI INSTR
39   5   645-8   1968

TPRC Number	Bibliographic Citation

49289 GLASS FILTERS FOR SIMULATING DAYLIGHT IN SENSITOMETRIC EXPOSURE.
VIETH G   HEILAND W
APPL OPT
7   6   1043-6   1968

49291 PROCEDURES FOR DETERMINING THE SPECTRAL RESPONSIVITY OF AN INFRARED RADIOMETER.
SCHNEIDER W E   GARVEY J A
APPL OPT
7   6   1141-8   1968

49310 INVESTIGATION OF A MOVING BELT RADIATOR.
SPEEDS J A   DULGEROFF C R   JOHNSON W K
JORTNER J   MADDOX J P   WRIGHT-PATTERSON AIR FORCE BASE   OHIO
DDC
AF-APL-TR-67-94
AD-819193   1-310   1967

49327 SIMULTANEOUS ELECTROCHEMICAL AND INTERNAL-REFLECTION SPECTROMETRIC MEASUREMENTS USING GOLD-FILM ELECTRODES.
PROSTAK A   MARK H B JR   HANSEN W N
J PHYS CHEM
72   7   2576-82   1968

49356 DIFFUSION CHROME PLATING OF MOLYBDENUM.
MULYAKAEV L M   DUBININ G N   RYUMIN V P
GOLUBEVA A S
IZV AKAD NAUK SSSR NEORG MATER
3   11   2114-17   1967

49357 DIFFUSION CHROME PLATING OF MOLYBDENUM.   //ENGLISH TRANSLATION OF IZV. AKAD. NAUK SSSR, NEORG. MATER. 3 /11/ 2114-7, 1967.//
MULYAKAEV L M   DUBININ G N   RYUMIN V P
GOLUBEVA A S
INORGANIC MATERIALS
3   11   1842-5   1967

49361 USE OF THE KRAMERS-KRONIG DISPERSION RELATIONS IN DETERMINING THE PHASE SHIFT OCCURRING UPON REFLECTION OF LIGHT FROM THIN DIELECTRIC LAYERS.
LUPASHKO E A   MILOSLAVSKII V K   SHKLYAREVSKII I N
OPTIKA I SPEKTROSKOPIYA
24   2   257-62   1968

49362 USE OF THE KRAMERS-KRONIG DISPERSION RELATIONS IN DETERMINING THE PHASE SHIFT OCCURRING UPON REFLECTION OF LIGHT FROM THIN DIELECTRIC LAYERS.   //ENGLISH TRANSLATION OF OPTIKA I SPEKTROSKOPIYA 24 /2/ 257-62, 1968.//
LUPASHKO E A   MILOSLAVSKII V K   SHKLYAREVSKII I N
OPT SPECTRY
24   2   132-4   1968

49363 REFLECTION OF SEMICONDUCTING COPPER SULFIDE LAYERS IN THE INFRARED REGION OF THE SPECTRUM.   //ENGLISH TRANSLATION OF OPTIKA I SPEKTROSKOPIYA 24 /2/ 263-6, 1968.//
KRYZHANOVSKII B P
OPT SPECTRY
24   2   135-7   1968

49367 HEAT-SHIELDING INTERFERENCE COATINGS.
ELSNER Z N
OPTIKA I SPEKTROSKOPIYA
24   3   437-41   1968

49368 HEAT-SHIELDING INTERFERENCE COATINGS.   //ENGLISH TRANSLATION OF OPTIKA I SPEKTROSKOPIYA 24 /3/ 437-41, 1968.//
ELSNER Z N
OPT SPECTRY
24   3   227-8   1968

49383 THERMAL CONDUCTIVITY OF POROUS MEDIA AS RELATED TO FREEZE-DRYING. M. S. THESIS.
TRIEBES T A   UNIVERSITY OF CALIFORNIA   BERKELEY CALIFORNIA
UNIVERSTIY OF CALIFORNIA
        1-80   1966

49394 EFFECT OF HIGH-TEMPERATURE SODIUM ON THE MECHANICAL PROPERTIES OF CANDIDATE ALLOYS FOR THE LMFBR PROGRAM. QUARTERLY PROG. REPT., OCT.-DEC. 1967.
MINES SAFETY APPLIANCES RESEARCH CORP   EVANS CITY PA
CFSTI
MSAR-68-14   1-19   1968   NSA 22-10 2025

49395 WINDOW AND WINDOW SCREENS AS MODIFIERS OF THERMAL RADIATION RELEASED IN NUCLEAR DETONATIONS. PROGRESS REPT.
BRACCIAVENTI J   NAVAL APPLIED SCIENCE LAB BROOKLYN N Y
DDC AND CFSTI
AD-643019   1-24   1966   NSA 21-13 2322

49397 PHOTOEMISSION AND OPTICAL STUDIES OF COBALT, PALLADIUM, PLATINUM, AND GADOLINIUM.   TECH. REPT.
YU A Y C   STANFORD ELECTRONICS LABS STANFORD UNIV CALIF
DDC AND CFSTI
SU-SEL-87-045 TR-5215-1
AD-667544   1-203   1967   CA   69   111772

49400 THERMOPHYSICAL PROPERTIES OF MATERIALS. FROM HEAT EXCHANGE IN EXTRUSION CASTING.
GALKIN M N   TARASUTIN T G   PUSHKIN I L
TR MOSK AVIATS TEKHNOL INST
58   81-99   1963

49401 THERMOPHYSICAL PROPERTIES OF MATERIALS. FROM HEAT EXCHANGE IN EXTRUSION CASTING.   //ENGLISH TRANSLATION OF TR. MOSK. AVIATS. TEKHNOL. INST. /58/ 81-99, 1963.//
GALKIN M N   TARASUTIN T G   PUSHKIN I L
FOREIGN TECH DIV WRIGHT PATTERSON AFB   OHIO
DDC
FTD-MT-64-257 N68-27362
AD-668233   73-90   1967   PA   N68-6-16 2668

49402 EFFECT OF INCOMPLETE ENERGY ACCOMMODATION ON FREE-MOLECULE RECOVERY TEMPERATURE.
EPSTEIN M   AEROSPACE CORP   EL SEGUNDO CALIF
DDC AND CFSTI
SAMSO-TR-67-18
AD-680533   1-36   1967   RR   68-1 153

49418 OPTICAL CHARACTERISTICS OF SILICON PHOTOCELLS AND THE EFFICIENCY OF A THERMOPHOTOELECTRIC CONVERTER.   //ENGLISH TRANSLATION OF TEPLOFIZ. VYSOKIKH TEMPERATUR 5 /6/ 1079-86, 1967.//
VASILEV A M   GOLOVNER T M   LANDSMAN A P
LIDORENKO N S
HIGH TEMPERATURE
5   6   967-73   1967

49482 DETERMINATION OF OPTICAL CONSTANTS FROM REFLECTANCE OR TRANSMITTANCE MEASUREMENTS ON BULK CRYSTALS OR THIN FILMS.
VERLEUR H W
J OPT SOC AM
58   10   1356-64   1968

49483 QUANTITATIVE PHENOMENOLOGICAL DESCRIPTION OF THE OPTICAL ABSORPTION OF SELENIUM IN A VERY DIVIDED STATE, STARTING FROM THE PERMITTIVITY OF THE BULK AMORPHOUS SOLID.
GONELLA J
COMPT REND
266   B   26   1611-13   1968

49510 REFLECTANCE OF BITUMINOUS COAL. III. EFFECT OF TEMPERATURE.
DE VRIES H A W   HABETS P J   BOKHOVEN C
BRENNST CHEM
49   4   105-10   1968   CA   69   4235

49515 APPARATUS FOR THE RAPID MEASURMENT OF THERMAL CONDUCTIVITY IN SOLIDS AT ROOM TEMPERATURE.
GIULIANI S
ENERG NUCL /MILAN/
14   7   430-5   1967   CA   69   5811

49526 STUDY OF NICKEL AS A DIFFUSION BARRIER BETWEEN URANIUM AND ALUMINUM.
BROSSA F   THEISEN R   HUET J I   TYTGAT D
CENTRE D ETUDE DE L ENERGIE NUCLEAIRE
AEC
EUR-17-F   1-51   1962

49527 RESEARCH AND DEVELOPMENT IN A THERMAL INSULATION STUDY. QUARTERLY PROG. REPT.. SEPT.-NOV. 1967.
DE WITT W D   REID R L   UNION CARBIDE CORP LINDE DIV   INDIANAPOLIS IND
NASA AND CFSTI
ALO-3632-23 N68-22549
        1-35   1968

49542 THERMAL CONTACT TO THE MIXING CHAMBER OF A DILUTION REFRIGERATOR.
MOTA A C   WHEATLEY J C
REV SCI INSTR
40   2   379-80   1969

49546 AN APPARATUS TO MEASURE THE THERMAL CONDUCTIVITY OF GRAPHITE NEAR ROOM TEMPERATURE.
VAN DE VELDE J   OCKFEN H   NOELS T   CENTRE D ETUDE DE L ENERGIE NUCLEAIRE MOL /BELGIUM/
AEC AND CFSTI
EUR-3909.E N68-24359
        1-20   1968

49551 ON THE QUESTION OF DETERMINING THE COMPONENTS OF CONTACT THERMAL RESISTANCE.
KHIZHNYAK P YE
IZV VYSSH UCHEB ZAVED AVIAT TEKH
        1   144-8   1965

TPRC Number	Bibliographic Citation

49552 ON THE QUESTION OF DETERMINING THE COMPONENTS OF CONTACT THERMAL RESISTANCE. FROM NEWS OF INSTITUTIONS OF HIGHER LEARNING, AERONAUTICAL ENGINEERING, NO. 1, 1965. //ENGLISH TRANSLATION OF IZV. VYSSH. UCHEB. ZAVED. AVIAT. TEKH. /1/ 144-48, 1965.//
KHIZHNYAK P YE   FOREIGN TECH DIV WRIGHT-PATTERSON AIR FORCE BASE   OHIO
DDC
FTD-MT-65-72 N68-18689
AD-663935   167-71   1967

49557 INTERFACIAL TENSION AND ADHESION AT THE BOUNDARY OF A HIGH MANGANESE SLAG-METAL SYSTEM WITH 3.05 PERCENT CARBON AND 1.02 PERCENT MANGANESE.
ERININ K   MALCHEVA N
KHIM IND /SOFIA/
1   52-5   1968   CA   69   4521

49568 INFRARED DIELECTRIC RESPONSE AND LATTICE VIBRATIONS OF CALCIUM AND STRONTIUM OXIDES.
JACOBSON J L   NIXON E R
J PHYS CHEM SOLIDS
29   6   967-76   1968

49571 ADSORPTION AT THE MERCURY/FORMIC ACID INTERPHASE.
LAWRENCE J   PARSONS R
TRANS FARADAY SOC
64   6   1656-78   1968

49574 STUDY OF SURFACE TENSION OF FUSED MIXTURES OF SODIUM FLUORIDE-ZIRCONIUM FLUORIDE-ZIRCONIUM OXIDE.
DECROLY C   FONTANA A   WINAND R
J NUCL MATER
27   1   36-47   1968   CA   69   69895

49580 MOMENTUM TRANSFER BETWEEN GAS MOLECULES AND METALLIC SURFACES IN FREE MOLECULE FLOW.
STICKNEY R E
PHYSICS FLUIDS
5   12   1617-24   1962

49589 REMOTE INFRARED TEMPERATURE-MAPPING OF HIGH-PERFORMANCE WEAPONS.
VAN DAMME G E   MC GARVEY J W   U S ARMY WEAPONS COMMAND   ROCK ISLAND ILL
DDC
AD-691132   1-27   1969

49590 DETERMINATION OF THE ABSORPTIVITY, TEMPERATURE AND THERMAL RESISTANCE OF THE IRRADIATED SURFACES IN COMBUSTION CHAMBERS.   //ENGLISH TRANSLATION OF TEPLOOBMEN, GIDRODINAMIKA I TEPLOFIZICHESKIE SVOISTVA VESHCHESTVA, MOSCOW, 1968, PP.   //
FILIMONOV S S   ADRIANOV V N   KRUSTALEV B A   KRYUKOVA M G
HEAT TRANSFER-SOVIET RESEARCH
1   4   143-53   1969

49591 EFFECTS OF THE PROPERTIES OF COMPONENTS AND THE GEOMETRIC CHARACTERISTICS OF THE STRUCTURE ON THE VALUES OF THERMAL CONDUCTIVITY COEFFICIENTS OF GLASS-REINFORCED PLASTICS.
VISHNEVSKII G E   SHLENSKII O F   MEKH POLIM
1   18-23   1968   CA   68   115308

49612 APPLICATION OF SHERESHEFSKYS EQUATION FOR THE DETERMINATION OF SOLID SURFACE DISPERSION ENERGY BY LIQUID PHASE ADSORPTION.
COTTON D J   NAVAL SHIP RESEARCH AND DEVELOPMENT CENTER
DDC
NSRDC-TR-2868
AD-690888   1-16   1969

49616 THERMAL CONDUCTIVITY OF ORDERED FIBROUS SYSTEMS. //ENGLISH TRANSLATION OF TEPLO MASSOPERENOS 7,   , 1968.//
DULNEV G N   ZARICHNYAK YU P   MURATOVA B L
DDC AND CFSTI
FTD-HT-23-820-68 PT. 1
AD-698517   85-97   1969

49627 PREPARATION OF MIRROR COATINGS FOR THE VACUUM ULTRA-VIOLET IN A 2-M EVAPORATOR.
BRADFORD A P   HASS G   OSANTOWSKI J F   TOFT A R
APPL OPT
8   6   1183-9   1969

49629 DEVICE FOR ABSOLUTE MEASUREMENT OF THE REFECTION COEFFICIENT OF MIRRORS.   IN FRENCH.
DUBAN M   CENTRE NATIONAL DE LA RECHERCHE SCIENTIFIQUE   MARSEILLES FRANCE
NASA AND CFSTI
N69-25814   1-28   1968   RR   69-16 172

49641 INTERFACIAL TENSION OF THE BOUNDARY OF ALUMINUM/ALKALI HYDROXIDE MELTS.
SHCHERBAKOV V K   KUZNETSOV S I
ZH PRIKL KHIM
41   3   505-9   1968   CA   69   5424

49647 INVESTIGATIONS OF ALUMINUM FILMS WITH SYNCHROTRON RADIATION OF WAVELENGTHS 500 TO 1000 A.   I. POLARIZATION-DEPENDENT TRANSMISSION AND REFLECTION.
SKIBOWSKI M   FEUERBACHER B   STEINMANN W   GODWIN R P
Z PHYS
211   4   329-41   1968   CA   69   6800

49660 EFFECT OF AN INTERFACE ON TRANSIENT TEMPERATURE DISTRIBUTION IN COMPOSITE AIRCRAFT JOINTS.
BARZELAY M E   HOLLOWAY G F   SYRACUSE UNIV SYRACUSE N Y
NASA
NACA-TN-3824 N62-55824
1-51   1957

49662 HEAT TRANSFER ACROSS SURFACES IN CONTACT.   EFFECT OF AMBIENT PRESSURE CHANGES.
AARON R L   BLUM H A
AEC
CONF-225-9   1-19   1963

49663 THERMAL INSULATION DEVELOPMENT.   FINAL REPT.
SPEIL S   JOHNS-MANVILLE RESEARCH CENTER   MANVILLE N J
AEC
TID-23683   1-21   1952

49665 IRRADIATION OF THERMAL CONTROL COATINGS.   FINAL REPORT, FEB. 1967-FEB. 1968.
SCANNAPIECO J F   GENERAL ELECTRIC CO MISSILE AND SPACE DIV   PHILADELPHIA PA
NASA AND CFSTI
GEC-R-68SD4224 NASA-CR-94684
N68-24465   1-220   1968

49666 LOW SOLAR ABSORPTANCE AND EMITTANCE SURFACES UTILIZING VACUUM DEPOSITED TECHNIQUES.   FINAL REPORT, JUNE 1967-MAR. 1968.
GREENBERG S A   VANCE D A   LOCKHEED MISSILES AND SPACE CO   PALO ALTO CALIF
NASA AND CFSTI
LMSC-4-06-68-1 NASA-CR-73228
N68-25328   1-83   1968

49687 THERMAL CONDUCTIVITY OF DRY POROUS SYSTEMS.
VASILEV L L
ISSLED TEPLOPROVODNOSTI INST TEPLO-MASSOOBMENA AKAD NAUK BELORUSS SSR
262-72   1987   CA   69   5807

49719 THERMAL-CONDUCTIVITY APPARATUS FOR LIQUIDS. HIGH-TEMPERATURE, VARIABLE-GAP TECHNIQUE.   PROGRESS REPORT.   FROM   PROCEEDINGS OF THE SIXTH CONFERENCE ON THERMAL CONDUCTIVITY.   DAYTON, OHIO.   OCT. 19-21, 1966.
MATOLICH J JR   DEEM H W   BATTELLE MEMORIAL , INSTITUTE COLUMBUS LABORATORIES   OHIO
AIR FORCE MATERIALS LABORATORY
39-51   1966

49725 ANALYTICAL INVESTIGATION OF OPTIMUM EXPERIMENTS FOR TRANSIENT DETERMINATION OF THERMAL CONTACT CONDUCTANCE.   FROM   PROCEEDINGS OF THE SIXTH CONFERENCE ON THERMAL CONDUCTIVITY.   DAYTON, OHIO. OCT. 19-21, 1966.
BECK J V   MECHANICAL ENGINEERING AND ENGINEERING RESEARCH MICHIGAN STATE UNIV   EAST LANSING MICHIGAN
AIR FORCE MATERIALS LABORATORY
301-29   1966

49738 HEAT TRANSFER BY RADIATION AND CONDUCTION IN EVACUATED POWDERS. FROM   PROCEEDINGS OF THE SIXTH CONFERENCE ON THERMAL CONDUCTIVITY.   DAYTON, OHIO. OCT. 19-21, 1966.
WECHSLER A E   LITTLE /ARTHUR D/ INC   CAMBRIDGE MASS
AIR FORCE MATERIALS LABORATORY
547-69   1966

49757 SEARCH FOR MAGNETITE IN LUNAR ROCKS AND FINES.
JEDWAB J   HERBOSCH A   WOLLAST R   NAESSENS G   VAN GREEN-PEERG N
SCIENCE
167   618-19   1970

49759 AN ANALYSIS OF THE EFFECTIVE THERMAL CONDUCTIVITY OF A DELAMINATED MATERIAL SUCH AS PYROLYTIC GRAPHITE. FROM   PROCEEDINGS OF THE SIXTH CONFERENCE ON THERMAL CONDUCTIVITY.   DAYTON, OHIO.   OCT. 19-21, 1966.
PEARS C D   SOUTHERN RESEARCH INSTITUTE   BIRMINGHAM ALABAMA
AIR FORCE MATERIALS LABORATORY
1051-84   1966

49761 THERMAL AND OPTICAL PROPERTY MEASUREMENTS WITHIN THE AIR FORCE MATERIALS LABORATORY.   PROGRESS REPORT. FROM   PROCEEDINGS OF THE SIXTH CONFERENCE ON THERMAL CONDUCTIVITY.   DAYTON, OHIO.   OCT. 19-21, 1966.
PURCELL G V   AIR FORCE MATERIALS LAB WRIGHT-PATTERSON AFB   OHIO
AIR FORCE MATERIALS LABORATORY
1127-41   1966

TPRC Number	Bibliographic Citation

49762 THERMAL PROPERTIES WORK AT ATOMICS INTERNATIONAL. FROM   PROCEEDINGS OF THE SIXTH CONFERENCE ON THERMAL CONDUCTIVITY.   DAYTON, OHIO.   OCT. 19-21, 1966.
SMITH C A   NORTH AMERICAN AVIATION INC   CANOGA PARK CALIFORNIA
AIR FORCE MATERIALS LABORATORY
1143-52   1966

49766 ORDERED STRUCTURE AND ION MIGRATION IN SILICON DIOXIDE FILMS.
SUGANO T   HOH K   KUDO K   HISHINUMA N
JAPAN J APPL PHYS
7   7   715-30   1968

49780 CHARACTERISTICS OF TEMPERATURE AND SHRINKAGE STRAINS OF CONCRETE AT TEMPERATURES BELOW FREEZING.
MOSKVIN V M   KAPKIN M M   ANTONOV L N
BETON ZHELEZOBETON
2   25-7   1968   CA   69   12707

49788 REFLECTION OF NICKEL IN THE ULTRAVIOLET ABOVE AND BELOW THE CURIE POINT.
YAROVAYA R G   TIMCHENKO L I
FIZ TVERD TELA
10   4   1237-9   1968   CA   69   14385

49791 ELECTRIC HEATING OF GAS-ASH CONCRETE WITH KERAMZIT AND PERLITE FILLERS.
DONCHEV ST   IORDANOV P   NENOV G
GOD VISSH KHIMIKOTEKHNOL INST 1965
12   2   417-22   1966   CA   69   12709

49800 ADHESION OF OXIDES TO INDUSTRIAL ALLOYS AND STEELS.
DERYABIN A A   POPEL S I   ZUPNIK A E
IZV AKAD NAUK SSSR METAL
2   78-84   1968   CA   69   12336

49810 CONTACT ANGLES AND ADSORPTION IN THE SYSTEM QUARTZ-WATER-DODECANE MODIFIED BY DODECYLAMMONIUM CHLORIDE.
GAUDIN A M   DECKER T G
J COLLOID INTERFACE SCI
24   2   151-8   1967   CA   69   13177

49813 A CONTROLLED ATMOSPHERE LANGMUIR TROUGH WITH SIMULTANEOUS AUTOMATIC RECORDING OF ELLIPSOMETRIC, CONTACT POTENTIAL, AND SURFACE TENSION MEASUREMENTS.
SMITH T
J COLLOID INTERFACE SCI
26   4   509-17   1968   CA   69   13136

49815 APPLICATION OF A THEORY OF BINARY SOLUTION SURFACE TENSION TO INTERFACIAL TENSION PHENOMENA.
COTTON D J   NAVAL SHIP RESEARCH AND DEVELOPMENT CENTER
DDC AND CFSTI
NSRDC-2867 N70-10990
AD-691244   1-23   1969   RR   69-18 61

49823 OPTICAL PROPERTIES OF MULTILAYER DIELECTRIC MIRRORS, RESULTING FROM CATHODE SPUTTERING.
MOTOVILOV O A
OPT-MEKH PROM
34   9   41-5   1967   CA   69   14449

49830 DEVELOPMENTS IN THE CALCULATION OF THE THERMAL CONDUCTIVITY OF POROUS CONCRETES.
BARANOV A T   BOBROV O D
STROIT MATER
14   3   18-19   1968   CA   69   12714

49842 MAXIMUM SUPERCOOLING OF AQUEOUS SOLUTIONS OF ORGANIC ACIDS.
KOMAROVA T A   KOROVKINA E K
ZH FIZ KHIM
42   1   123-7   1968   CA   69   13470

49874 THERMAL CONDUCTIVITY OF VERMICULITE-ASBESTOS MIXTURES AND OF THEIR CEMENT MORTARS.
TSUBAKI T   HATTORI M   TANAKA M
YOGYO KYOKAI SHI
74 849   178-80   1966   CA   69   89498

49882 EFFECT OF GAS ADSORPTION ON THE ABSORPTION SPECTRA OF CAROTENE FILMS.
MISRA T N   ROSENBERG B
J CHEM PHYS
48   12   5734-5   1968

49910 INTENSITY DISTRIBUTION OF REFLECTED MOLECULAR BEAMS IN RELATION TO THE ADSORPTION AND GRAIN STRUCTURE OF METALS.
GRAF U   NIKURADSE A
ENTROPIE
18   115-18   1967   CA   69   21985

49933 A THERMAL DIFFUSIVITY MEASUREMENT TECHNIQUE.   FROM PROCEEDINGS OF THE 3RD CONFERENCE ON THERMAL CONDUCTIVITY.   GATLINBURG, TENNESSEE.   OCTOBER 16-18, 1963.
PLUMMER W A   CORNING GLASS WORKS   CORNING NEW YORK
OAK RIDGE NATIONAL LAB OAK RIDGE TENNESSEE

49959 REFLECTANCE AND 1/EPSILON RESONANCE OF BERYLLIUM IN THE FAR ULTRAVIOLET.
TOOTS J   FOWLER H A   MARTON L
PHYS REV
172   3   670-6   1968   CA   69   72437

49962 OPTICAL AND ELECTRICAL PROPERTIES AND BAND STRUCTURE OF GERMANIUM TELLURIDE AND TIN TELLURIDE.
TSU R   HOWARD W E   ESAKI L
PHYS REV
172   3   779-88   1968   CA   69   62538

49963 OPTICAL PROPERTIES OF VO2 BETWEEN 0.25 AND 5 EV.
VERLEUR H W   BARKER A S JR   BERGLUND C N
PHYS REV
172   3   788-98   1968   CA   69   63220

49973 PREPARTATION AND OPTICAL PROPERTIES OF EPITAXIAL THIN FILMS OF SOME II-VI SEMICONDUCTING COMPOUNDS.   TECH. REPT.
LIDEKE R   HARVARD UNIVERSITY   CAMBRIDGE MASS
DDC
HP-22   N68-30427
AD-670084   1-248   1968   PA   N68-6-18 3154

49983 TABLES OF THE THERMAL RADIATION PROPERTIES OF MATERIALS.
BANAS C M   TAYLOR H D   UNITED AIRCRAFT CORPORATION RESEARCH LABORATORIES   EAST HARTFORD CONN
CFSTI
M-1501-1   1-89   1960

49984 SPACE MATERIALS HANDBOOK.
GOETZEL C G   SINGLETARY J B   EDITORS FOR LOCKHEED MISSILES AND SPACE COMPANY   SUNNYVALE CALIFORNIA
DDC
AD-284547   1-523   1962

49986 HEAT TRANSFER INSIDE NUCLEAR FUEL CONSISTING OF COMPACTED UO2 - PUO2 POWDERS.   SPECIAL REPORT NO. 8.   //ENGLISH TRANSLATION OF FRENCH REPORT.//
ANDRIESSEN H   CENTRE D ETUDE DE L ENERGIE NUCLEAIRE   BRUSSELS /BELGIUM/
AEC AND CFSTI
EURAEC-1998—EUR-3791
N68-28971   1-40   1967

50002 RADIATION PROPERTIES OF KH18N9T STAINLESS STEEL DURING HEATING IN AIR.
ZHOROV G A   SERGEEV V S
TEPLOFIZ VYS TEMP
6   2   340-2   1968   CA   69   21420

50016 METALLIC ABSORBING LAYERS OF BOLOMETERS FOR THE LONG-WAVE INFRARED REGION.
ZHUKOV A G
ZH PRIKL SPEKTROSK
8   2   337-9   1968   CA   69   23311

50019 THERMOPHYSICAL PROPERTIES OF PLASTIC MATERIALS AND COMPOSITES TO LIQUID HYDROGEN TEMPERATURE /-423 F/. QUARTERLY PROG. REPT. NO. 8, AUG.-NOV. 1964.
CAMPBELL M D   OBARR G L   HERTZ J   GENERAL DYNAMICS/ASTRONAUTICS   SAN DIEGO CALIFORNIA
NASA
BVJ-63-001-8
N65-20410   1-38   1964

50023 TOTAL HEMISPHERICAL EMISSIVITY MEASUREMENTS BY THE HEAT PIPE METHOD.
DEVERALL J E   LOS ALAMOS SCIENTIFIC LAB   N MEX
AEC AND CFSTI
LA-3834-MS   1-10   1967

50024 A SURVEY OF THE THERMAL CONDUCTANCE OF METALLIC CONTACTS.
WONG H Y   AERONAUTICAL RESEARCH COUNCIL   /GT BRITIAN/
NASA
ARC-26715   ARC-CP-973 AD 833463 N68-23173
N68-30153   1-38   1968

50030 DEVELOPMENT OF SPACE-STABLE THERMAL-CONTROL COATINGS /PAINTS WITH LOW SOLAR ABSORPTANCE/EMITTANCE RATIOS./ TRIANNUAL REPORT JAN. 1964-JUNE 1964.
ZERLAUT G A   HARADA Y   BERMAN L U   GEORGE C MARSHALL SPACE FLIGHT CENTER   HUNTSVILLE ALA
NASA
IITRI-C6014-13
1-49   1964

50031 DEVELOP 1800-400 F FIBROUS-TYPE INSULATION FOR RADIOISOTOPE POWER SYSTEMS.
COLLINS J O   JOHNS-MANVILLE RESEARCH AND ENGINEERING CENTER   MANVILLE N J
AEC AND CFSTI
ALO-3633-5 N68-22550
1-15   1967   PA   N68-6-12 1983

50036 DETERMINATION OF SURFACE TENSION OF FLUXES AND INTERFACIAL TENSION AT THE BOUNDARY WITH ARMCO IRON.
YAKOBASHVILI S B

TPRC Number	Bibliographic Citation

50040 THE EFFECT OF SELECTIVITY OF RADIATION ON HEAT TRANSFER IN FURNACES. //ENGLISH TRANSLATION OF TEPLOOBMEN. GIDRODINAMIKA I TEPLOFIZICHESKIE SVOISTVA VESHCHESTVA, MOSCOW, 1968, PP. //
ADRIANOV V N
HEAT TRANSFER-SOVIET RESEARCH
1   4   154-62   1969

50054 OPTICAL PROPERTIES OF THIN GERMANIUM FILMS IN THE INFRARED REGION OF THE SPECTRUM.
GISIN M A   IVANOV V A
OPTIKA I SPEKTROSKOPIYA
26   2   231-4   1969

50080 OPTICAL CONSTANTS OF CESIUM HALIDES AT LOW TEMPERATURES IN THE FAR INFRARED.
VERGNAT P   CLAUDEL J   HADNI A   STRIMER P
VERMILLARD F
J PHYS /PARIS/
30 8/9   723-35   1969   CA   72   26652

50073 EFFECT OF INTERPHASE PHENOMENA ON THE REMOVAL OF NONMETALLIC INCLUSIONS FROM A METAL.
EVSEEV P P   FILIPPOV A F
IZV AKAD NAUK SSSR METAL
3   57-64   1968   CA   69   38241

50083 COMPRESSIBILITY AND OTHER THERMODYNAMIC PROPERTIES OF POLYMERS.
SURLAND C C
J APPL POLYM SIC
12   6   1423-37   1968   CA   69   28423

50095 THERMAL CONDUCTIVITY OF COATINGS.
NOVICHENOK L N
LAKOKRASOCH MATER IKH PRIMEN
3   32-3   1968   CA   69   37178

50110 A PHASE TRANSITION IN THE QUINOL HYDROGEN CYANIDE CLATHRATE COMPOUND.
MATSUO T   SUGA H   SEKI S
J PHYS SOC JAPAN
25   2   641   1968

50111 FAR-INFRARED ABSORPTION DUE TO THE VIRTUAL MODES IN NACL.
KUBOTA T   HISANO K   MATUMURA O
J PHYS SOC JAPAN
25   2   642   1968

50114 IVI. THE INFLUENCE OF THE GLASSY BOND ON SOME PROPERTIES OF SILICA REFRACTORIES.
BADGER E H M   LEWCOCK W   WYLDE J H
TRANS BRIT CERAMIC SOC
45   269-89   1946

50117 APPARATUS FOR MEASURING THE SURFACE TENSION OF PETROLEUM AT A HIGH-PRESSURE GAS INTERFACE.
SHITIKOV L I
NEFTEPROM DELO NAUCH TEKH SB
1   26-30   1968   CA   69   29069

50141 SPECTRAL EMISSIVITY MEASUREMENTS OF ABLATING PHENOLIC GRAPHITE.
CHANG J H   SUTTON G W   ADVANCED RESEARCH LAB
EVERETT MASSACHUSETTS
DDC
AVCO-295   1-23   1968

50146 REFRACTORY EXPANDING CEMENT.
KUTATELADZE K S   GABADADZE T G   GARISHVILI B V
TSEMENT
34   2   16-18   1968   CA   69   29894

50153 IRRADIATION BEHAVIOR OF GRAPHITES. FROM RESEARCH ON GRAPHITE. QUARTERLY PROG. REPT. AUG.-NOV. 1967.
ENGLE G B   MORRIS W H   STETSON R F   GULF
GENERAL ATOMIC   SAN DIEGO CALIF
NASA AND CFSTI
GA-8376 N68-26534   18-21   1967

50168 THERMAL JOINT CONDUCTANCE. RESEARCH REPT. FEB. 1-MARCH 31, 1970.
HULTBERG J A   JET PROPULSION LAB CALIFORNIA
INSTITUTE OF TECHNOLOGY   PASADENA CALIFORNIA
JPL
JPL-SPS-37-62 VOL. III   198-202   1970

50174 COMMENTS ON METHODS, CONDITIONS AND INSTRUMENTATION. FROM THERMAL INSULATING MEASUREMENTS.
HICKMAN M J   GRIFFITH M V
INTERN INST REFRIG
48   1   32-59   1968

50176 THERMAL CONDUCTIVITY OF MOLYBDENUM THICK FILMS ON BERYLLIUM OXIDE.
COLE S S JR
AM CERAM SOC BULL
47   9   806-9   1968   JA   51   308

50190 ELECTROOPTICAL REMOTE SENSING METHODS AS NONDESTRUCTIVE TESTING AND MEASURING TECHNIQUES IN AGRICULTURE.
MYERS V I   ALLEN W A
APPL OPT
7   9   1819-38   1968

50218 THE INTERFACIAL TENSION OF THE BENZENE-WATER SYSTEM AND THE DECAHYDRONAPHTHALENE-WATER SYSTEM BY THE PENDANT DROP METHOD. M.S. THESIS.
WOSILAIT A A   STATE UNIVERSITY OF NEW YORK
BUFFALO NEW YORK
STATE UNIVERSITY OF NEW YORK
1-60   1967

50235 INFRARED RADIATION DETECTORS AND THEIR USE IN THE STUDY OF THERMAL RADIATION. M. S. THESIS.
BROWN T R   PURDUE UNIVERSITY   WEST LAFAYETTE INDIANA
PURDUE UNIVERSITY
1-135   1964

50239 SPECTRAL REFLECTANCE OF WATER AND CARBON DIOXIDE CRYODEPOSITS FROM 0.36 TO 1.15 MICRON.
WOOD B E   SMITH A M
AIAA J
6   7   1362-7   1968   CA   69   47929

50246 MEASUREMENTS OF OPTICAL PARAMETERS OF ANTIMONY SESQUISULFIDE FILMS.
KUCIREK J
CZECH J PHYS
18   6   795-800   1968   CA   69   47842

50252 MECHANISM OF INTERFACIAL POLYMERIZATION.
MAC RITCHIE F
TRANS FARADAY SOC
65   9   2503-7   1969

50271 KAPITZA RESISTANCE.
ABBE W J
NUOVO CIMENTO
58   B   1   187-9   1968

50272 ABSORPTION OF LIGHT IN THIN GRANULAR SILVER FILMS.
SHKLYAREVSKII I N   KORNEEVA T I
OPTIKA I SPEKTROSKOPIYA
24   5   744-50   1968   CA   69   47871

50282 OPTICAL PROPERTIES OF A FINELY DIVIDED ZINC OXIDE LAYER.
KOVALSKII L V   SAKHNOVSKII M YU
ZH PRIKL SPEKTROSK
8   4   629-33   1968   CA   69   47707

50288 SURFACE STUDIES IN THE VAPOR-SOLID SYSTEM BORON TRIIODIDE-TUNGSTEN. PH D THESIS.
OWNBY P D   OHIO STATE UNIVERSITY   COLUMBUS OHIO
UNIV MICROFILMS PUBL
67-16325   1-157   1967

50297 PHOTOELECTRIC AND OPTICAL PROPERTIES OF COMMERCIAL PLATINUM, GOLD AND PALLADIUM FOILS AND EVAPORATED ALUMINUM AND SILVER FILMS IN THE EXTREME ULTRAVIOLET. PH D THESIS.
STANFORD J L   UNIVERSITY OF TENNESSEE   KNOXVILLE TENNESSEE
UNIV MICROFILMS PUBL
67-1379   1-106   1966

50298 THE THERMAL RADIATION CHARACTERISTICS OF SOME HIGH-EMITTANCE COATINGS FOR SPACE APPLICATIONS.
LEWIS B W   WADE W R   SLEMP W S   PROGAR D J
NASA LANGLEY RESEARCH CTR   HAMPTON VA
NASA AND CFSTI
NASA-TM-X-59389
N68-27441   1-15   1966

50299 THE ACCOMMODATION COEFFICIENTS OF THE INERT GASES ON ALUMINUM, TUNGSTEN, PLATINUM, AND NICKEL AND THEIR DEPENDENCE ON SURFACE CONDITIONS. PH D THESIS.
FAUST J W JR   UNIVERSITY OF MISSOURI   COLUMBIA AND ROLLA MISSOURI
UNIV MICROFILMS PUBL
UM 10107   1-157   1954

50301 THE ACCOMMODATION COEFFICIENTS OF HELIUM, NEON, AND ARGON ON CLEAN POTASSIUM AND SODIUM SURFACES FROM 77 TO 298 K, AND A STUDY OF ADSORPTION OF POTASSIUM ON TUNGSTEN. PH D THESIS.
PETERSEN H L   UNIVERSITY OF MISSOURI   COLUMBIA AND ROLLA MISSOURI
UNIV MICROFILMS PUBL
58-7448   1-135   1958

50302 THE THERMAL ACCOMMODATION OF HELIUM AND NEON ON BERYLLIUM. PH D THESIS.
BROWN R E   UNIVERSITY OF MISSOURI   COLUMBIA AND ROLLA MISSOURI
UNIV MICROFILMS PUBL
UM 21099   1-77   1957

TPRC Number	Bibliographic Citation

50306 VACUUM THERMAL CONDUCTIVITY MEASUREMENTS OF NASA E4AL ELASTOMERIC HEAT SHIELD MATERIAL.
COMPARIN R A   WILSON W G   DEPT OF MECHANICAL ENGR VIRGINIA POLYTECHNIC INST   BLACKSBURG VIRGINIA
NASA AND CFSTI
NASA-CR-66629 N68-27226
1-57   1968

50307 DEVELOPMENT OF 400 TO 2200 F. FIBROUS-TYPE INSULATION FOR RADIOISOTOPE POWER SYSTEMS.   5TH QUARTERLY PROG. REPT. FOR JAN. 1-MARCH 29, 1968.
COLLINS J O   JAUNARAJS K L   REID D R
JOHNS-MANVILLE RESEARCH AND ENGR CTR   MANVILLE N J
AEC AND CFSTI
ALO-3633-8 N68-29115
1-48   1968   PA   N68-6-17 2904

50308 EVALUATION OF CANDIDATE GRAPHITES /AND CARBON/ FOR PSC CORE-SUPPORT APPLICATIONS.
MEYERS C   KOYAMA K   GULF GENERAL ATOMIC   SAN DIEGO CALIFORNIA
AEC AND CFSTI
GAMD-8371   N68-28973
1-135   1968   CA   69   102158

50317 AN EMISSIVITY STUDY OF VARIOUS MATERIALS.   M.S. THESIS.
KELLY D P   UNIVERSITY OF CINCINNATI   CINCINNATI OHIO
UNIVERSITY OF CINCINNATI
1-32   1960

50321 RADIANT ENERGY TRANSPORT IN CRYOGENIC CONDENSATES.
MC CONNELL D G   LEWIS RESEARCH CENTER   CLEVELAND OHIO
NASA
NASA-TM-X-52375
N68-12735   1-18   1967

50332 THE THERMAL CONDUCTANCE OF BOLTED JOINTS.   PH. D. THESIS.
FONTENOT J E JR   LOUISIANA STATE UNIV   BATON ROUGE LOUISIANA
CFSTI AND NASA
NASA-CR-94738 N68-24656
1-180   1968

50354 THE TOTAL NORMAL EMISSIVITY OF SEVERAL NEW ENGINEERING MATERIALS.   M.S. THESIS.
DAVIDSON D M   WASHINGTON UNIVERSITY   ST LOUIS MISSOURI
WASHINGTON UNIVERSITY
1-121   1962

50365 INFRARED REFLECTION SPECTRA OF GA/1-X/ IN/X/ AS.   A NEW TYPE OF MIXED-CRYSTAL BEHAVIOR.
BRODSKY M H   LUCOVSKY G
PHYS REV LETT
21   14   990-3   1968

50367 DETERMINATION OF THE MODULATION TRANSFER FUNCTION OF PHOTOGRAPHIC EMULSIONS FROM PHYSICAL MOVEMENTS.
WOLFE R N   MARCHAND E W   DEPALMA J J
J OPT SOC AM
58   9   1245-56   1968

50375 ELECTRICAL AND STRUCTURAL PROPERTIES OF MIXED CHROMIUM AND SILICON MONOXIDE FILMS.
MILGRAM A A   LU C-S
J APPL PHYS
39   9   4219-24   1968

50379 INFRARED LATTICE VIBRATIONS OF ZINC SELENIDE AND ZINC TELLURIDE.
RICCIUS H D
J APPL PHYS
39   9   4381-2   1968

50380 MEASUREMENT BY THE FLASH METHOD OF THERMAL DIFFUSIVITY, HEAT CAPACITY, AND THERMAL CONDUCTIVITY IN TWO-LAYER COMPOSITE SAMPLES.
LARSON K B   KOYAMA K
J APPL PHYS
39   9   4408-16   1968

50391 PREPARATION OF ELEMENTAL BORON AND MEASUREMENTS OF ITS EMITTANCE AT INCANDESCENT TEMPERATURES.   M.S. THESIS.
TALLEY C P   UNIVERSITY OF RICHMOND   RICHMOND VIRGINIA
UNIVERSITY OF RICHMOND
1-68   1960

50417 THE THERMOPHYSICAL PROPERTIES OF SOME MULTILAYER INSULATORS AT CRYOGENIC TEMPERATURES.   //ENGLISH TRANSLATION OF JINR, LAB. OF HIGH ENERGY, DUBNA, USSR.   PP. 3-20, 1967.//
GOLOVANOV L B   REDSTONE SCI INFORM CTR   REDSTONE ARSENAL ALABAMA
NASA AND CFSTI
NASA-TM-X-60999 RSIC-780 AD-671727
FTD-HT-23-12-68 AD-682236 N69-23100
N69-28903   1-10   1968

50418 THERMAL CONTACT CONDUCTANCE OF SURFACES IN CONTACT UNDER VACUUM CONDITONS.   M.S. THESIS.
BOLES M A   NORTH CAROLINA STATE UNIVERSITY   RALEIGH NORTH CAROLINA
NORTH CAROLINA STATE UNIVERSITY
1-80   1967

50420 SOME OPTICAL CONSTANTS OF ALKALI HALOGENIDE CRYSTALS.
KUBLITZKY A
ANN PHYSIK
20   5   793-808   1934

50422 SURFACE TENSION OF SOME MOLTEN POLYMERS.   PHYSICAL PROPERTIES OF MOLTEN POLYMERS FOR APPLICATION TO ADHESIVES.
HATA T
ASAHI GARASU KOGYO GIJUTSU SHOREI-KAI KENKYU HOKOKU
13   603-18   1967   CA   69   52583

50444 ON THE USE OF A LUMINESCENT SUBSTRATE FOR INVESTIGATING THE TRANSMISSION OF THIN FILMS IN THE VACUUM ULTRAVIOLET.
SOROKIN O M
OPTIKA I SPEKTROSKOPIYA
24   6   976-8   1968   CA   69   56010

50463 MEASUREMENT OF SURFACE TENSION AT THE PETROLEUM-SURFACTANT SOLUTION INTERFACE BY THE HANGING DROP METHOD.
SMIRNOV YU S   POZDNYSHEV G N   PETROV A A
TR GOS INST PROEKT ISSLED RAB NEFTEDOBYVAYUSHCHEI PROM
10   109-14   1967   CA   69   53368

50483 ABSORPTION OF LIGHT IN THIN GRANULAR SILVER FILMS.   //ENGLISH TRANSLATION OF OPTIKA I SPEKTROSKOPIYA 24 /5/ 744-50, 1968.//
SHKLYAREVSKII I N   KORNEEVA T I
OPT SPECTRY
24   5   398-401   1968

50484 THE RELATION OF INITIAL SPREADING PRESSURE OF POLAR COMPOUNDS ON WATER TO INTERFACIAL TENSION, WORK OF ADHESION, AND SOLUBILITY.
TIMMONS C O   ZISMAN W A   NAVAL RESEARCH LAB WASHINGTON D C
DDC AND CFSTI
NRL-6716 N69-12396
AD-672014   1-20   1968   RR   68-17 70

50490 ON THE THERMAL CONDUCTIVITY OF COATED PARTICLE CHARGES UNDER FILLER GAS AS FUNCTIONS OF GAS PRESSURE AND TEMPERATURE.
LEYERS H J   BINKELE L   BEUTH J
KERNFORSCHUNGSANLAGE   JUELICH /WEST GERMANY/
NASA AND CFSTI
JUL-5111-RX N68-28543
1-19   1968   RR   68-19 138

50496 MEASUREMENT OF THE FAR INFRARED OPTICAL PROPERTIES OF SOLIDS WITH A MICHELSON INTERFEROMETER USED IN THE ASYMMETRIC MODE.   PART II, THE VACUUM INTERFEROMETER.
RUSSELL E E   BELL E E
INFRARED PHYSICS
6   75-84   1965

50497 MEASUREMENT OF THE TOTAL NORMAL EMISSIVITY OF ALUMINUM SURFACES WITH DIFFERENT TREATMENTS.
TINCOLINA P
NATIONAL CONGRESS ATI SEPT 24-8
1-34   1968

50498 APPARATUS FOR THE DETERMINATION OF THE TOTAL NORMAL EMITTANCE OF SURFACES AT TEMPERATURES BETWEEN 50 AND 250 C.
ALFANO G   BETTA V
NATIONAL CONGRESS ATI SEPT 24-8
1-36   1968

50525 MEASUREMENTS REPORT.   THERMAL PROPERTY MEASUREMENTS OF MANNED SPACECRAFT CENTER SPACESUIT MATERIALS.
TURNBOW F J   TRW SYSTEMS   REDONDO BEACH CALIF
NASA AND CFSTI
NASA-CR-92198
N68-29984   1-4   1968

50531 THE THERMAL PROPERTIES AND BEHAVIOR OF NICKEL-CADMIUM AND SILVER-ZINC CELLS AND THEIR COMPONENTS.
BROOMAN E W   MC CALLUM J   BATTELLE MEMORIAL INST   COLUMBUS OHIO
DDC
BATT-7770-4 AFAPL-TR-68-41
AD-834300   1-114   1968

50539 EMISSIVITY AND REFLECTANCE OF SELECTED SURFACE COATINGS.
KLEMM R E   NORTH AMERICAN AVIATION INC   LOS ANGELES CALIFORNIA
NASA
NA-57-707-1 M64-83550
1-30   1958

TPRC Number	Bibliographic Citation

50554 ABSORPTION AND REFLECTION SPECTRA OF THE HALOGENIDES OF THALLIUM IN THE FAR INFRARED AT LOW TEMPERATURE.
CLAUDEL J   HADNI A   STRIMER P   VERGNAT P
J PHYS CHEM SOLIDS
29   9   1539-44   1968

50569 DEFORMATION AND BREAKUP OF LIQUID DROPLETS IN A SIMPLE SHEAR FIELD.
KARAM H J   BELLINGER J C
IND ENG CHEM FUNDAMENTALS
7   4   576-81   1968

50604 THERMAL EXPANSION OF COEFFICIENTS OF COMPOSITE MATERIALS. PH.D. THESIS.
ROSEN B W   PENNSYLVANIA UNIV   PHILADELPHIA PA
UNIV MICROFILMS PUBL
69-5661   1-138   1968

50636 ADHESION OF OXIDES TO INDUSTRIAL ALLOYS AND STEELS. //ENGLISH TRANSLATION OF IZV. AKAD. NAUK SSSR, METAL. /2/ 78-84, 1968.//
DERYABIN A A   POPEL S I   ZUPNIK A E
RUSSIAN METALLURGY /METALLY/
2   51-6   1968

50638 THERMAL CONDUCTIVITY TESTS OF CRYOGENIC INSULATION. FROM PROCEEDINGS OF THE 2ND CONFERENCE ON THERMAL CONDUCTIVITY. OTTAWA, ONTARIO. OCTOBER 10-12, 1962.
BLACK I A   GLASER P E   LITTLE /ARTHUR D/ INC
CAMBRIDGE MASS
DIVISION OF APPLIED PHYSICS NATIONAL RESEARCH COUNCIL
OTTAWA ONTARIO
111-31   1962

50642 MEASUREMENT OF THE EMISSIVITY OF SOME MATERIALS HEATED TO 100 DEGREES.
VOISHVILLO N A   MIRONOVA L R
ZH PRIKL SPEKTROSK
8   5   840-3   1968   CA   69   61698

50659 THERMAL INSULATING MATERIALS. HOW VARIOUS CHARACTERISTICS AFFECT THEIR EFFICIENCY.
DAWSON J W
PETROLEUM
24   286-9   1961

50660 METALLIC HEAT INSULATION.
NICHOLS J T
AM SOC MECH ENGR
58   190-1   1936

50663 RELATIONSHIP BETWEEN THE EFFECTIVE THERMAL CONDUCTIVITY AND ULTRASONIC TRANSMISSION FOR COPPER BRAZE BONDS.
DI NOVI R A
APPLIED MATERIALS RESEARCH
5   162-7   1966

50684 MEASUREMENT OF TRUE TEMPERATURE BY OPTICAL PYROMETRY.
EULER K J
CHEM ING TECH
38   2   154-9   1966

50688 HEAT TRANSFER THROUGH GLASSED STEEL.
ACKLEY E J
CHEM ENGNG
66   8   181-2   1959

50689 THE HEAT TRANSFER BETWEEN METALLIC SURFACES IN CONTACT.
VON KISS M
NEUE TECH
5   714-24   1963

50704 FLOW OF HEAT THROUGH CONTACT SURFACES.
HERING C
METALLURG AND CHEM ENGINEERING
10   1   40-4   1912

50724 FLAME-SPRAYED CERAMIC COATINGS IN SPACE TECHNOLOGY.
WHEILDON W M
SPACEFLIGHT
9   2   55-62   1967

50731 SPECTRAL EMISSION PROPERTIES OF NBS STANDARD PHOSPHOR SAMPLES UNDER PHOTO-EXCITATION.
SHELTON C F
USGPO
NBS-TN-417   1-29   1968

50741 INTERFACIAL TENSION MEASURMENT OF AMALGAMS BY MEANS OF A CAPILLARY ELECTROMETER.
POLYANOVSKAYA N S
ELEKTROKHIMIYA
4   5   549-53   1968   CA   69   69918

50764 DEPENDENCE OF THE APPARENT VISCOSITY OF PORTLAND CEMENT COMPOSITIONS ON THE COEFFICIENT OF SATURATION OF SILICON DIOXIDE BY LIME.
GOLDSHTEIN L YA   MANTSUROVA V N
IZV AKAD NAUK SSSR NEORG MATER
4   4   980-3   1968   CA   69   69479

50787 BINDING OF COUNTERIONS TO LECITHIN ON OIL-WATER INTERFACES.
WATANABE A   MATSUMOTO M   TAMAI H   GOTOH R
NIPPON KAGAKU ZASSHI
89   5   454-0   1968   CA   69   69911

50794 METHOD OF INTERPRETING THE SESSILE DROP FOR MEASURING INTERFACIAL TENSION AND CONTACT ANGLE.
LEFEBVRE DU PREY E
REV INST FR PETROLE ANN COMBUST LIQUIDES
23   3   365-73   1968   CA   69   69912

50802 ACCOMMODATION COEFFICIENT OF HYDROGEN, A SENSITIVE DETECTOR OF SURFACE FILMS.
BLODGETT K B   LANGMUIR I
PHYSICAL REV
40   78-104   1932

50803 THERMAL CONDUCTIVITY TESTS OF HONEYCOMB INSULATION PANELS.
TRIAS J A   GENERAL DYNAMICS CONVAIR   SAN DIEGO CALIFORNIA
DDC AND CFSTI
GDC-MP-59-146
AD-672200   1-12   1959   RR   68-18 100

50820 EFFECTS OF LITHIA SUBSTITUTION FOR SODA IN VITREOUS ENAMEL.
LEWIS M O
J AM CERAM SOC
26   3   77-83   1943

50856 ANTIREFLECTION PROPERTIES OF THERMALLY GROWN SILICON OXIDE ON SILICON OPTICAL ELEMENTS.
MILER M
INFRARED PHYS
7   2   117-9   1967   CA   69   6649

50857 STRUCTURAL SYNTHESIS OF COMPOSITE MATERIALS FOR ABLATIVE NOZZLE EXTENSIONS.
AU N N   SCHEYHING E R   SUMMERS G D   SPACE AND MISSILE SYSTEMS ORGANIZATION AIR FORCE SYSTEMS COMMAND   LOS ANGELES CALIF
DDC
SAMSO-TR-69-201
AD-691025   1-41   1969

50868 THE TEMPERATURE DEPENDENCE OF THE THERMAL CONTACT RESISTANCE ACROSS NON-METALLIC INTERFACES. PH.D. THESIS.
MINGES M L   OHIO STATE UNIVERSITY   COLUMBUS OHIO
UNIV MICROFILMS PUBL
68-15360   1-186   1968   PA   N69-7-13 12428
DDC
AFML-TR-69-1
AD-697988   1-148   1969

50878 PERLITOPHOSPHATE, A HIGH-TEMPERATURE HEAT-INSULATING MATERIAL.
SARTAKOV YU A
STROIT MATER
14   5   16-17   1968   CA   69   38433

50880 SOME FEATURES OF THE ELECTRICAL INSULATING PROPERTIES OF HEAT-RESISTANT CONCRETE. //ENGLISH TRANSLATION OF TEPLOFIZ. VYS. TEMP. 5 /1/ 155-60, 1967.//
SHAKHTAKHTINSKII T I   ROMANOV A I   SMIRNOVA L G
HIGH TEMPERATURE
5   1   134-8   1967

50884 PHYSICAL BASIS OF THERMAL CONDUCTIVITY IN FIBROUS INSULATORS.
PAYMAL J
VERRES REFRACTAIRES
19   6   403-15   1965   JA   51-6 173

50896 EXPERIMENTAL STUDIES ON SHADOW SHIELDS FOR THERMAL PROTECTION OF CRYOGENIC TANKS IN SPACE.
KNOLL R H   BARTOO E R   LEWIS RESEARCH CENTER CLEVELAND OHIO
NASA AND CFSTI
NASA-TN-D-4887   1-69   1968

50897 SLIDING NOBLE METAL CONTACTS. CONTACT RESISTANCE AND WEAR CHARACTERISTICS OF SOME NEW PALLADIUM ALLOY SLIDEWIRES.
DARLING A S   SELMAN G L
PLATINUM METALS REVIEW
12   4   122-8   1968

50898 PLATINUM METAL CONTACTS. PAPERS AT THE FOURTH INTERNATIONAL SYMPOSIUM ON ELECTRICAL CONTACT PHENOMENA.
PLATINUM METALS REVIEW
12   4   129-30   1968

50900 CHARACTERISTICS OF SILICON SILICON-DIOXIDE STRUCTURES FORMED BY DC REACTIVE SPUTTERING.
IWAUCHI S   TANAKA T
JAPAN J APPL PHYS
7   10   1193-201   1968

TPRC Number	Bibliographic Citation

50913 HEAT CONDUCTION IN A RAREFIED GAS BETWEEN CONCENTRIC CYLINDERS.
SU C-L   WILLIS D R
PHYS FLUIDS

11	10	2131-43	1968	

50921 RADIATION PROPERTIES OF TYPE KH18N9T STAINLESS STEEL WITH HEATING IN AIR.   /ENGLISH TRANSLATION OF TEPLOFIZ. VYS. TEMP. 6 /2/ 340-2, 1968./
ZHOROV G A   SERGEEV V S
HIGH TEMP

6	2	327-9	1968

50924 ON THE USE OF A LUMINESCENT SUBSTRATE FOR INVESTIGATING THE TRANSMISSION OF THIN FILMS IN THE VACUUM ULTRAVIOLET.   /ENGLISH TRANSLATION OF OPT. SPEKTROSK. 24 /6/ 976-8, 1968./
SOROKIN O M
OPT SPECTROSC

24	6	524-5	1968

50926 HEAT CONDUCTION FROM PLATINUM AND ALUMINUM SURFACES BY RARIFIED HELIUM, NEON AND ARGON.   M. S. THESIS.
FAUST J W JR   UNIVERSITY OF MISSOURI   ROLLA MISSOURI
UNIVERSITY OF MISSOURI

	1-92	1949

50943 EFFECT OF AN OXIDE FILM ON THE EMISSIVITY OF A METAL.
MUCHNIK G F   GUBKOV L A   SALNIKOV L A
TEPLOFIZ VYS TEMP

6	2	266-71	1968	CA	69	82031

50944 EFFECT OF AN OXIDE FILM ON THE EMISSIVITY OF A METAL. /ENGLISH TRANSLATION OF TEPLOFIZ. VYS. TEMP. 6 /2/ 266-71, 1968./
MUCHNIK G F   GUBKOV L A   SALNIKOV L A
HIGH TEMP

6	2	258-62	1968

50953 THE THERMAL CONDUCTIVITY OF TWO THERMAL INSULATING MATERIALS.   FINAL REPT.
TYE R P   DYNATECH CORP   CAMBRIDGE MASS
NASA AND CFSTI
NASA-CR-61915 DYNATECH NO. 798
N68-31508

	1-10	1968

50965 KINETICS OF ADSORPTION AT LIQUID/LIQUID INTERFACE. ROLE OF DIFFUSION, ENERGY BARRIER AND NUMBER OF FREE SITES WHEN THE DESORPTION IS SLIGHT.
BARET J F   ARMAND L   BERNARD M   DANOY G
TRANS FARADAY SOC

64	9	2539-48	1968

50966 EFFECT OF ULTRAVIOLET IRRADIATION ON THE PROPERTIES OF EVAPORATED SILICON OXIDE FILMS.
MICKELSEN R A
J APPL PHYS

39	10	4594-600	1968

50974 OPTICAL PROPERTIES OF AMORPHOUS SELENIUM IN THE VACUUM ULTRAVIOLET.
LEIGA A G
J OPT SOC AM

58	11	1441-5	1968

50975 CONDITIONS FOR ZERO REFLECTANCE OF THIN DIELECTRIC FILMS ON LASER MATERIALS.
SMILEY V N
J OPT SOC AM

58	11	1469-75	1968

51006 INTERFACIAL TENSION AT THE METAL-SALT MELT BOUNDARY.
MOZHAISKAYA G M   YUSFIN V S   GRIGOREV G A
IZV VYSSH UCHEB ZAVED CHERN MET

11	7	15-19	1968	CA	69	80520

51016 RELATION OF INITIAL SPREADING PRESSURE OF POLAR COMPOUNDS ON WATER TO INTERFACIAL TENSION, WORK OF ADHESION, AND SOLUBILITY.
TIMMONS C O   ZISMAN W A
J COLLOID INTERFACE SCI

28	1	106-17	1968	CA	69	80521

51025 THERMAL CONDUCTIVITY AND DIFFUSIVITY OF INSULATING MATERIALS.
DUDNIK D M   STEPANENKO A N
KHOLOD TEKH

45	1	27-9	1968	CA	69	81195

51049 CONNECTION BETWEEN THE CONSTITUTION AND CERTAIN PROPERTIES OF SURFACE-ACTIVE BENZENESULFONATES WITH HETERO ATOMS IN THE ALIPHATIC SIDE CHAIN.   I. SOLUBILITY, SURFACE ACTIVITY, CRITICAL MICELLE FORMATION CONCENTRATION, AND PAPER CHROMATOGRAPHY OF SULFONATES.
PUESCHEL F   TODOROV O
TENSIDE

5 7/8	193-8	1968	CA	69	78644

51088 A THEORY OF INTERFACIAL TENSION OF TWO-PHASE TERNARY LIQUID SYSTEMS.
COTTON D J

51101 A METHOD TO PREDICT THE THERMAL CONDUCTANCE OF A BOLTED JOINT.
FONTENOT J E JR   WHITEHURST C A   DIVISION OF ENGINEERING RESEARCH LOUISIANA STATE UNIVERSITY BATON ROUGE
NASA AND CFSTI
NASA-CR-96316 N68-33514

	1-37	1968	RR	68	147

51120 DEFECT EQUILIBRIA IN NONSTOICHIOMETRIC OXIDES.   FROM ARGONNE NATIONAL LABORATORY ANNUAL PROGRESS REPORT FOR 1966, METALLURGY DIVISION.
VOLPE M L   REDDY J F
AEC
ANL-7299

	272-4	1966

51127 A STUDY OF THE DIRECTIONAL RADIATION PROPERTIES OF SPECIALLY PREPARED V-GROOVE CAVITIES.
BLACK W Z   SCHOENHALS R J
J HEAT TRANSFER TRANS ASME

90	C	4	420-8	1968

51129 STUDY OF THE INFRARED VIBRATIONS OF TWO COPPER HYDRATE SALTS.
BREHAT F   WYNCKE B   HADNI A
COMPT REND

267	B	16	778-81	1968

51158 CONTACT ANGLES, SURFACE TENSIONS, AND ADHESION.
GRAY V R
ASPECTS ADHES 1964

2	42-8	1966	CA	69	89854

51207 EMITTANCE MEASUREMENTS.   FROM AEC FUELS AND MATERIALS DEVELOPMENT PROGRAM.   PROGRESS REPT. NO. 76.
JEUNKE E F   SJODAHL L H
AEC
GEMP-1008

	239-46	1968

51218 INTERFACE REACTIONS BETWEEN METALS AND CERAMICS.   IV. WETTING OF SAPPHIRE BY LIQUID-OXYGEN ALLOYS.
CHAKLADER A C D   ARMSTRONG A M   MISRA S K
J AM CERAM SOC

51	11	630-3	1968

51235 THERMAL CONDUCTIVITY OF SAND, MORTARS AND BRICK DEPENDING ON THE HUMIDITY OF MATERIALS.
CAMPAN T   SIMIONESCU A   ANGHELACHE D
BULETINUL INST POLITEH IASI

4 1/2	353-60	1958

51245 FACTORS INFLUENCING THE BEHAVIOR OF TITANIUM PIGMENTS IN INDUSTRIAL FINISHES.
KAMPFER W A
CHIM PEINTURES

22	4	93-102	1959

51247 THERMAL CONTACT CONDUCTANCE.
DEAN R A
TRANS AM NUCLEAR SOC

7	432	1964

51255 THE LIMITING DEGREE OF SUPERCOOLING OF ORGANIC ACID SOLUTIONS.   /ENGLISH TRANSLATION OF ZH. FIZ. KHIM. 42 /1/ 123-7, 1968./
KOMAROVA T A   KOROVKINA E K
RUSS J PHYS CHEM

42	1	61-4	1968

51271 THERMAL CONDUCTIVITY OF POROUS CATALYST PELLETS.
MASAMUNE S   SMITH J M
J CHEM ENGR DATA

8	1	54-8	1963

51273 BOILING POINTS AND SURFACE TENSIONS OF MIXTURES OF BENZYL ACETATE WITH DIOXANE, ANILINE, AND META-CRESOL.
KATTI P K   CHAUDHRI M M
J CHEM ENGR DATA

9	1	128-30	1964

51274 ACCOMMODATION COEFFICIENTS ON GAS COVERED PLATINUM.
AMDUR I   JONES M M   PEARLMAN H
J CHEM PHYS

12	5	159-66	1944

51275 THE ACCOMMODATION COEFFICIENTS OF GASES ON PLATINUM AS A FUNCTION OF PRESSURE.
THOMAS L B   BROWN R E
J CHEM PHYS

18	10	1367-72	1950

51276 THERMAL ACCOMMODATION COEFFIIENT OF HELIUM ON A BARE TUNGSTEN SURFACE.
THOMAS L B   SCHOFIELD E B
J CHEM PHYS

23	5	861-6	1955

51287 THE DYNAMICS OF SIMPLE CUBIC LATTICES.   I. APPLICATIONS TO THE THEORY OF THERMAL ACCOMMODATION COEFFICIENTS.
GOODMAN F O

TPRC Number	Bibliographic Citation

51290 CONSTITUTIVE RELATIONS IN THE WETTING OF LOW ENERGY SURFACES AND THE THEORY OF THE RETRACTION METHOD OF PREPARING MONOLAYERS.
SHAFRIN E G   ZISMAN W A
J PHYS CHEM
64   519-24   1960

51292 MEASUREMENT OF THE REFLECTING POWER OF PAINTS /FOR HEAT RADIATION/.
MERIGOUX M R
J PHYSIQUE RADIUM
15   12   67S-68S   1954

51300 MEASUREMENTS OF THE THERMAL CONTACT RESISTANCE FROM STAINLESS STEEL TO LIQUID SODIUM.   FROM DEVELOPMENTS IN HEAT TRANSFER.
SCHMIDT E H   JUNG E   MASSACHUSETTS INSTITUTE OF TECHNOLOGY
MIT PRESS
251-63   1963

51307 METHODS USED IN THE PERFORMANCE TESTING OF PAINT. IV.   GLOSS, HIDING POWER, REFLECTANCE.
MOORE W V
METAL FINISHING JOURNAL
3   34   389-95   1957

51331 VIBRATIONAL ENERGY EXCHANGE BETWEEN DIATOMIC MOLECULES AND A SURFACE.
HERMAN R   RUBIN R J
J CHEM PHYS
29   3   591-9   1958

51342 DETERMINATION OF THE EXISTENCE OF CERTAIN PROPERTIES PECULIAR TO MOLECULAR TRIATOMIC JETS OF HYDROGEN. //ENGLISH TRANSLATION OF COMPT. REND. 263B /25/ 1389-92, 1966.//
DEVIENNE F M   ROUSTAN J-C
NASA AND CFSTI
CTO-453 NP-TR-1638
N68-30266   1-4   1967

51373 ENERGY ACCOMMODATION COEFFICIENT FOR A GAS MIXTURE.
MIKAMI H
KAGAKU KOGAKU
32   7   715-16   1968   CA   69   97954

51381 ADSORPTION OF POLYHYDRIC ALCOHOLS AT SOLUTION-MERCURY AND SOLUTION-AIR INTERFACES.
KAGANOVICH R I   DAMASKIN B B   GANZHINA I M
ELEKTROKHIMIYA
4   7   867-71   1968   CA   69   99768

51385 EFFECT OF ADSORBED WATER ON THE CRITICAL SURFACE TENSION OF WETTING ON METAL SURFACES.
BERNETT M K   ZISMAN W A
J COLLOID INTERFACE SCI
28   2   243-9   1968   CA   69   99764

51403 OPTICAL SPHERE PAINT AND A WORKING STANDARD OF REFLECTANCE.
GRUM F   LUCKEY G W
APPL OPT
7   11   2289-94   1968

51438 INFLUENCE OF CAPILLARITY ON A DENSITY GRADIENT COLUMN.
MANSFIELD W W
TRANS FARADAY SOC
66   2   341-9   1970

51444 VIBRATION EFFECTS ON THERMAL CONTACT RESISTANCE. M. S. THESIS.
CHOONG P T-S   MASSACHUSETTS INSTITUTE OF TECHNOLOGY   CAMBRIDGE MASSACHUSETTS
MASSACHUSETTS INSTITUTE OF TECHNOLOGY
1-100   1967

51447 LIQUID-GAS INTERFACES STUDIED ON THE BASIS OF THE CLASSICAL SURFACE TENSION THEORY AND INTERMOLECULAR FORCE MODELS.   PH. D. THESIS.
FLACHSBART B B   STANFORD UNIVERSITY   STANFORD CALIFORNIA
UNIV MICROFILMS PUBL
66-6342   1-125   1965

51456 OPTICAL PROPERTIES OF AN ELECTRON GAS.   FURTHER STUDIES OF A NONLOCAL DESCRIPTION.
FUCHS R   KLIEWER K L
PHYS REV
185   3   905-13   1969

51470 REFLECTIONS IN A POOL OF MERCURY.   AN EXPERIMENTAL AND THEORETICAL STUDY OF THE INTERACTION BETWEEN ELECTROMAGNETIC RADIATION AND A LIQUID METAL.
BLOCH A N   RICE S A
PHYS REV
185   3   933-57   1969
DDC
AFOSR-70-0270-TR
AD-700838 /REPRINT/   1969

51490 RESEARCH AND DEVELOPMENT IN A THERMAL INSULATION STUDY.   QUARTERLY PROGRESS REPT., MARCH-MAY, 1968.
NOTARO F   REID R L   UNION CARBIDE CORP LINDE DIVISION   INDIANAPOLIS INDIANA
AEC AND CFSTI
ALO-3632-29 N68-37024
1-72   1968

51492 AN ULTRAHIGH VACUUM REFLECTOMETER FOR USE WITH EXTREME ULTRAVIOLET SYNCHROTRON RADIATION.
FEUERBACHER B   GODWIN R P   SKIBOWSKI M
DEUTSCHES ELEKTRONIN-SYNCHROTRON   HAMBURG /WEST GERMANY/
AEC
DESY-68-26 N68-37173
1-21   1968

51506 EMISSIVITY OF BEEF.
SEVCIK V J   SUNDERLAND J E
FOOD TECHNOLOGY
16   9   124-6   1962

51508 THE PRODUCTION OF A STANDARD COMPARATOR FOR THE SKIN COLOR OF MATURE CHERRIES.
BREARLEY N   BREEZE J E   CUTHBERT R M
FOOD TECHNOLOGY
18   9   232-3   1964

51516 THE EXCHANGE OF ENERGY BETWEEN A PLATINUM SURFACE AND GAS MOLECULES.
MANN W B
PROC ROY SOC
146   A   776-91   1934

51517 THE INTERACTION OF ATOMS AND MOLECULES WITH SOLID SURFACES.   VIII.   THE EXCHANGE OF ENERGY BETWEEN A GAS AND A SOLID.
DEVONSHIRE A F
PROC ROY SOC
158   A   269-79   1937

51518 THE EXCHANGES OF ENERGY BETWEEN A PLATINUM SURFACE AND HYDROGEN AND DEUTERIUM MOLECULES.
MANN W B   NEWELL W C
PROC ROY SOC
158   A   397-403   1937

51520 ON THE USE OF MICA CONDENSERS AS THERMAL SHUNTS IN CRYOGENIC APPARATUS.
ZIMMERMAN J E   ARROTT A   SKALYO J
REV SCI INSTRUMENTS
29   1148-9   1958

51522 THE THERMAL ACCOMMODATION COEFFICIENT OF GASES AND THEIR ADSORPTION ON IRON.
EGGLETON A E J   TOMPKINS F C
TRANS FARADAY SOC
48   8   738-49   1952

51525 ON THE EFFECT OF THE RETARDED EXCHANGE OF THE TRANSLATIONAL AND THE VIBRATIONAL ENERGY ON THE HEAT CONDUCTIVITY IN GASES.
SCHAFER K   RATING W   EUCKEN A
ANN PHYSIK
42   5   176-202   1942

51531 RADIANT COOLING OR SHIELDING.
RALL R M   YOSHIMOTO H
PRODUCT ENGINEERING
31   47   55-61   1960

51533 PROPERTIES AND USES OF PLASTICS.   DETERMINATION OF THE THERMAL CONDUCTIVITY OF GRPS FROM THE STRUCTURE AND PROPERTIES OF THE COMPONENTS.
SHLENSKII O F
SOVIET PLASTICS
35-9   1966

51560 METHOD FOR DETERMINING THE THERMAL CONDUCTIVITY AND TOTAL HEAT CAPACITY IN SOLIDS AND GRANULAR BODIES AS FUNCTIONS OF TEMPERATURES.
KOTSIUBINSKII O IU   KHINCHIN T A S
INZH FIZ ZHUR
1   11   125-9   1958

51562 THE INFLUENCE OF THE DEGREE OF RAREFACTION ON THE EFFECTIVE THERMAL CONDUCTIVITY OF REFRACTORY CERAMICS.
PUSTOVALOV V V
INZH FIZ ZHUR
3   10   57-9   1960

51572 THERMAL AND INSULATING QUALITY OF DOUBLE GLASS.
NEDOMLEL F   RAK M
SKLAR A KERAMIK
8   1   5-7   1958

51573 SELECTIVE RADIATION COATINGS.   PREPARATION AND HIGH TEMPERATURE STABILITY.
KOKOROPOULOS P   SALAM E   DANIELS F
SOLAR ENERGY
3   19-23   1959

TPRC Number	Bibliographic Citation						
51575	TEMPERATURE GRADIENTS AND THERMAL CONDUCTION OF STRAIGHT FINS MADE FROM MANY LAYERS OF METAL. HOHENHINNEBUSCH W TECH MITT KRUPP						
	23	3	85-96	1965			
51576	THERMAL CONDUCTIVITY OF SWAGED METAL FIBER-URANIUM DIOXIDE FUEL ELEMENT. NORIN M P TRANS AMER NUCL SOC						
	3	1	142	1960			
51581	INFLUENCE OF GLOSSY SURFACES ON REFLECTION MEASUREMENTS. BROCKES A FARBE						
	9 1/3		53-62	1960			
51592	PIGMENTATION STUDY OF ALKYD FLAT PAINTS. HALL R F BOZEMAN H OFFICIAL DIGEST						
	30		871-89	1958			
51593	EFFECT OF PARTICLE SIZE DISTRIBUTION OF PIGMENT AND PIGMENT VOLUME CONCENTRATION OF FLAT WALL PAINT PROPERTIES /SOLVENT TYPE/. HALL R F OFFICIAL DIGEST						
	31		788-99	1959			
51595	HIGH TEMPERATURE RESISTANT ORGANIC COATINGS. GLASER M A CAPLAN R E CLOPE R W OFFICIAL DIGEST						
	33		1197-214	1961			
51598	OPTICAL CHARACTERISTICS OF WATER AND ICE IN THE INFRARED AND RADIOWAVE REGIONS OF THE SPECTRUM. KISLOVSKII L D OPT I SPEKTROSKOPIYA						
	7	3	311-20	1959			
51599	REFLECTIVITY OF ALUMINUM PIGMENTS AND PAINT-II. RETHWISCH F B BABCOCK G M RIGGS E C PAINT OIL CHEM REV						
	116	24	22-4	1953			
51602	RESEARCHES ON HEAT CONDUCTION BY RAREFIED GASES. I. THE THERMAL ACCOMMODATION COEFFICIENT OF HELIUM, NEON, HYDROGEN AND NITROGEN ON GLASS AT 0 C. KEESOM W H SCHMIDT G PHYSICA						
	3	7	590-6	1936			
51603	RESEARCHES ON HEAT CONDUCTION BY RAREFIED GASES. III. THE THERMAL ACCOMMODATION COEFFICIENT OF HELIUM, NEON, AND HYDROGEN AT 12—20 K. KEESOM W H SCHMIDT G PHYSICA						
	4	10	828-34	1937			
51604	HEAT CONDUCTION BY POWDERS IN VARIOUS GASEOUS ATMOSPHERES AT LOW PRESSURE. PRINS J A SCHENK J SCHRAM A J G L PHYSICA						
	16	4	379-80	1950			
51605	OPTICAL PROPERTIES AND ELECTRON BAND STRUCTURE OF EUROPIUM AND BARIUM. MUELLER W E PHYS KONDENS MATER						
	6		243-68	1967	NSA 22/1/140		
51606	THE OPTICAL CONSTANTS OF CERTAIN LIQUIDS FOR SHORT ELECTRIC WAVES. TEAR J D PHYS REV						
	21		611-22	1923			
51615	THE HEAT-INSULATING PROPERTIES OF HEAT-PROOF CLOTH. MIYASAKA Y FIJIKURA Y SEN-I GAKKAISHI						
	22	1	7-13	1966			
51616	A NEW APPARATUS FOR MEASUREMENT OF THERMAL TRANSMITTANCE ON WALLS OF LARGE DIMENSIONS. CODEGONE C FERRO V TERMOTECNICA						
	21	9	463-7	1967			
51617	EFFECT OF WEAKLY CONCENTRATED AQUEOUS SOLUTIONS OF SULFONOL NP-1 ON RESIDUAL WATER SATURATION IN THE WELL-BORE ZONE OF ROCK BEDS. AKHMETSHIN M A SOLOMATIN G G TR TURKM FILIALA VSES NEFT NAUCH-ISSLED INST						
		8	96-107	1965	CA	67	83579
51618	MEASUREMENT OF THE SPECTRAL REFLECTANCE OF WHITE STANDARDS. I. MEASUREMENT OF THE INTERIOR PAINT OF AN ULBRICHT. BUDDE W FARBE						
	7 1/3		17-24	1958			

TPRC Number	Bibliographic Citation						
51629	THERMAL EXPANSION COEFFICIENTS OF COMPOSITE MATERIALS BASED ON ENERGY PRINCIPLES. SCHAPERY R A J COMPOSITE MATERIALS						
	2	3	380-404	1968			
	DDC AD-675519		1-28	1968	RR	68	79
51630	OPTICAL ABSORPTION OF POLYCRYSTALLINE FILMS OF ZINC TELLURIDE, CUBIC MODIFICATION. //ENGLISH TRANSLATION OF IZV. VUZ, FIZIKA 9 /4/ 12-7, 1966.// SHALIMOVA K V SPINULESCU-CARNARU I PIROGOVA N V SOVIET PHYS J						
	9	4	5-8	1966			
51633	HEAT EXCHANGE DURING CONDENSATION OF POTASSIUM AND SODIUM VAPOR. SUBBOTIN V I IVANOVSKII M N SOROKIN V P CHULKOV B A TEPLO-I MASOPERENOS						
			247-55	1966			
51634	HEAT EXCHANGE DURING CONDENSATION OF POTASSIUM AND SODIUM VAPOR //ENGLISH TRANSLATION OF TEPLO-I MASOPERENOS 247-55, 1966// SUBBOTIN V I IVANOVSKII M N SOROKIN V P CHULKOV B A DDC AND CFSTI FTD-MT-24-54-68 N69-10929						
	AD-675226		1-16	1968	RR	68	83
51637	THE MEASUREMENT OF SOLAR ABSORPTANCE AND THERMAL EMITTANCE. WESTCOTT M EUROPEAN SPACE TECHNOLOGY CENTRE NOORDWIJK THE NETHERLANDS NASA AND CFSTI ESRO-TN-23 N68-37036						
			1-43	1968			
51640	DIFFUSION IN SILICON NITRIDE-ALUMINUM FILMS. //ENGLISH TRANSLATION OF FIZ. TVERD. TELA 10 /1/ 272-4, 1968.// KOVARSKII V YA ALEKSANYAN I T BONDARENKO O E ORLOV B M SOVIET PHYS-SOLID STATE						
	10	1	209-10	1968			
51661	MEASUREMENT OF THE HEAT CONDUCTIVITY COEFFICIENT OF VACUUM-POWDER INSULATION AT HIGH TEMPERATURES. SEREBRYANYI G L ZARUDNYI L B SHORIN S N TEPLOFIZ VYS TEMP						
	6	3	547-8	1968			
51662	MEASUREMENT OF THE HEAT CONDUCTIVITY COEFFICIENT OF VACUUM-POWDER INSULATION AT HIGH TEMPERATURES. //ENGLISH TRANSLATION OF TEPLOFIZ. VYS. TEMP. 6 /3/ 547-8, 1968.// SEREBRYANYI G L ZARUDNYI L B SHORIN S N HIGH TEMP						
	6	3	523-4	1968			
51681	THE EXCHANGE OF ENERGY BETWEEN MONATOMIC GASES AND SOLID SURFACES. ZENER C PHYSICAL REV						
	40	3	335-9	1932			
51682	A QUANTUM MECHANICAL THEORY OF ENERGY EXCHANGES BETWEEN INERT GAS ATOMS AND A SOLID SURFACE. JACKSON J M PROC CAMBRIDGE PHIL SOC						
	28		136-64	1932			
51683	ENERGY EXCHANGE BETWEEN INERT GAS ATOMS AND A SOLID SURFACE. JACKSON J M MOTT N F PROC ROY SOC						
	137	A	703-17	1932			
51684	EXCHANGE OF ENERGY BETWEEN INERT GAS ATOMS AND A SOLID SURFACE. JACKSON J M HOWARTH A PROC ROY SOC						
	142	A	447-56	1933			
51685	THE EXCHANGE OF ENERGY BETWEEN GAS ATOMS AND SOLID SURFACES. III. THE ACCOMMODATION COEFFICIENT OF NEON. ROBERTS J K PROC ROY SOC						
	142	A	518-24	1933			
51686	THE ADSORPTION OF HYDROGEN ON TUNGSTEN. ROBERTS J K PROC ROY SOC						
	152	A	445-63	1935			
51693	EFFECT OF TEMPERATURE AND FLUID SATURATION ON ROCK THERMAL CONDUCTIVITY. M.S. THESIS. VILORIA G PENNSYLVANIA STATE UNIVERSITY UNIVERSITY PARK PENNSYLVANIA PENNSYLVANIA STATE UNIVERSITY						
			1-112	1968			

TPRC Number	Bibliographic Citation
51694	INVESTIGATION OF NEW METHODS FOR INCREASING THERMAL CONTACT CONDUCTANCE IN A VACUUM. FINAL REPORT. CARFAGNO S P FRANKLIN INSTITUTE RESEARCH LABORATORIES PHILADELPHIA PA NASA AND CFSTI NASA-CR-98014 F-C2076 N68-36869 1-58 1968 N 69-1 164
51698	CERAMICS REINFORCED BY FIBERS. HOFMANN U BERICHTE DEUT KERAM GES 43 5 337-45 1966
51699	CERAMICS REINFORCED BY FIBRES. //ENGLISH TRANSLATION OF BERICHTE DEUT. KERAM. GES. 43 337-45, 1966.// HOFMANN U AEC AND CFSTI NP-TR-1494 N67-25502 1-35 1966
51702	THERMAL CONDUCTANCE OF TWO DIMENSIONAL ECCENTRIC CONSTRICTIONS. VEZIROGLU T N HUERTA M A DEPT OF MECHANICAL ENGINEERING MIAMI UNIV CORAL GABLES FLORIDA NASA NASA-CR-96281 N68-32974 1-25 1968
51703	THERMAL CONDUCTIVITY AND THERMAL SHOCK QUALITIES OF ZIRCONIA COATINGS ON THIN GAGE HASTELLOY-X METAL. BUCKLEY J D NASA LANGLEY RESEARCH CENTER LANGLEY STATION HAMPTON VA NASA AND CFSTI NASA-TM-X-61277 N68-35768 1-19 1968
51706	ENERGY EXCHANGE BETWEEN COLD GAS MOLECULES AND A HOT TUNGSTEN SURFACE. M. S. THESIS. DEPOORTER G L UNIVERSITY OF CALIFORNIA BERKELEY CALIFORNIA UNIVERSITY OF CALIFORNIA UCRL-10504 1-30 1962
51708	THE EXCHANGE OF ENERGY BETWEEN GAS ATOMS AND SOLID SURFACES. ROBERTS J K PROC ROY SOC 129 A 146-61 1930
51719	AN ABSOLUTE EMISSIVITY CALORIMETER FOR TEMPERATURES TO 60 K. HAURY G L RUTNER E ADVAN CRYOG ENG 12 308-14 1967 CA 69 107822
51726	THERMAL ACCOMMODATION OF THE RARE GASES ON CLEAN METAL SURFACES. FROM FUNDAMENTALS OF GAS-SURFACE INTERACTIONS. PROC. OF THE SYMPOSIUM SAN DIEGO, CALIF., DEC. 14-16, 1966. MENZEL D KOUPTSIDIS J ACADEMIC PRESS NEW YORK AND LONDON 493-505 1967
51727	MEASUREMENTS OF MOMENTUM ACCOMMODATION OF GAS MOLECULES AT SURFACES. FROM FUNDAMENTALS OF GAS-SURFACE INTERACTIONS. PROC. OF THE SYMPOSIUM, SAN DIEGO, CALIF., DEC. 14-16, 1966. KOSTOFF R N ANDERSON J B FENN F B ACADEMIC PRESS NEW YORK AND LONDON 512-21 1967
51769	REFLECTION OF NICKEL IN THE ULTRAVIOLET REGION AT TEMPERATURES BELOW AND ABOVE THE CURIE POINT. //ENGLISH TRANSLATION OF FIZ. TVERD. TELA 10 /4/ 1237-39, 1968.// YAROVAYA R G TIMCHENKO L I SOVIET PHYS-SOLID STATE 10 4 983-4 1968
51798	CONTACT RESISTANCE IN SILICON CARBIDE CONTACTS. NAMBA M OYO BUTSURI 26 11 602-8 1957
51799	CONTACT RESISTANCE IN SILICON CARBIDE CONTACTS. //ENGLISH TRANSLATION OF OYO BUTSURI 26 /11/ 602-8, 1957.// NAMBA M NASA AND CFSTI NASA-TT-F-11736 N68-30214 1-13 1968
51800	THE MEASUREMENT OF THERMAL PROPERTIES OF REINFORCED PLASTICS AT TEMPERATURES UP TO 150 C. /INFRARED RADIATION METHOD/. OGAWA K NOGUCHI Y NATIONAL AEROSPACE LAB TOKYO JAPAN NASA AND CFSTI NAL-TR-150 N68-36636 1-23 1968 CA 71 71329

TPRC Number	Bibliographic Citation
51806	SURFACE TENSION OF TIN-GALLIUM ALLOYS. OFITSEROV A A PUGACHEVICH P P KUZNETSOV G M KUZMINA G N IZV VYSSH UCHEB ZAVED TSVET MET 11 2 130-2 1968 CA 69 90607
51818	METHOD FOR DETERMINING SURFACE-ACTIVE SUBSTANCES IN MEDICO-BIOLOGICAL INVESTIGATIONS. GANITKEVICH YA V LAB DELO 1 27-9 1967 CA 69 110111
51851	SOME ASPECTS CONCERNING THERMAL CONDUCTIVITY DATA OF LIQUIDS AND PROPOSALS FOR NEW STANDARD REFERENCE MATERIALS. FROM PROCEEDINGS OF THE SEVENTH CONFERENCE ON THERMAL CONDUCTIVITY. GAITHERSBURG, MARYLAND. NOV. 13-16, 1967. POLTZ H PHYSIKALISCH-TECHNISCHE BUNDESANSTALT BRAUNSCHWEIG GERMANY USGPO NBS-SP-302 47-56 1968 CA 71 25306
51855	GLASS BEADS—A STANDARD FOR THE LOW THERMAL CONDUCTIVITY RANGE. FROM PROCEEDINGS OF THE SEVENTH CONFERENCE ON THERMAL CONDUCTIVITY. GAITHERSBURG, MARYLAND. NOV. 13-16, 1967. WECHSLER A E ARTHUR D LITTLE INC CAMBRIDGE MASSACHUSETTS USGPO NBS-SP-302 89-96 1968 CA 71 7203
51864	HEAT LOSSES IN A CUT-BAR APPARATUS. EXPERIMENTAL ANALYTICAL COMPARISONS. FROM PROCEEDINGS OF THE SEVENTH CONFERENCE ON THERMAL CONDUCTIVITY. GAITHERSBURG, MARYLAND. NOV. 13-16, 1967. MINGES M L AIR FORCE MATERIALS LAB WRIGHT-PATTERSON AIR FORCE BASE DAYTON OHIO USGPO NBS-SP-302 197-206 1968
51867	METHOD FOR MEASURING TOTAL HEMISPHERIC EMISSIVITY OF PLANE SURFACES WITH CONVENTIONAL THERMAL CONDUCTIVITY APPARATUS. FROM PROCEEDINGS OF THE SEVENTH CONFERENCE ON THERMAL CONDUCTIVITY. GAITHERSBURG, MARYLAND. NOV. 13-16, 1967. HAGER N E JR ARMSTRONG CORK COMP LANCASTOR PENNSYLVANIA USGPO NBS-SP-302 241-6 1968 CA 70 107788
51868	DEVIATIONS FROM MATTHIESSENS RULE IN THE LOW TEMPERATURE THERMAL AND ELECTRICAL RESISTIVITIES OF VERY PURE COPPER. FROM PROCEEDINGS OF THE SEVENTH CONFERENCE ON THERMAL CONDUCTIVITY. GAITHERSBURG, MARYLAND. NOV. 13-16, 1967. SCHRIEMPF J T NAVAL RESEARCH LABORATORY WASHINGTON D C USGPO NBS-SP-302 249-52 1968 CA 70 11877
51884	AN INVESTIGATION OF THE MECHANISMS OF HEAT TRANSFER IN LOW-DENSITY PHENOLIC-NYLON CHARS. FROM PROCEEDINGS OF THE SEVENTH CONFERENCE ON THERMAL CONDUCTIVITY. GAITHERSBURG, MARYLAND. NOV. 13-16, 1967. SMYLY E D PYRON C M JR SOUTHERN RES INSTITUTE BIRMINGHAM ALABAMA USGPO NBS-SP-302 425-53 1968 CA 70 97513
51888	ON THE DEVELOPMENT OF METHODS FOR MEASURING HEAT LEAKAGE OF INSULATED WALLS WITH INTERNAL CONVECTION. FROM PROCEEDINGS OF THE SEVENTH CONFERENCE ON THERMAL CONDUCTIVITY. GAITHERSBURG, MARYLAND. NOV. 13-16, 1967. LORENTZEN G BRENDENG E FRIVIK P NORGES TEKNISKE HOGSKOLE TRONDHEIM NORWAY USGPO NBS-SP-302 495-506 1968
51890	A GUARDED HOT PLATE APPARATUS FOR MEASURING THERMAL CONDUCTIVITY FROM -80 TO +100 DEGREES. FROM PROCEEDINGS OF THE SEVENTH CONFERENCE ON THERMAL CONDUCTIVITY. GAITHERSBURG, MARYLAND. NOV. 13-16, 1967. ROUSSELLE J-C LABORATOIRE NATIONAL DESSAIS /L N E/ PARIS FRANCE USGPO NBS-SP-302 513-20 1968 CA 70 10774
51902	MEASUREMENTS OF THE THERMAL CONDUCTIVITY OF GRANULAR CARBON AND THE THERMAL CONTACT RESISTANCE AT THE CONTAINER WALLS. FROM PROCEEDINGS OF THE SEVENTH CONFERENCE ON THERMAL CONDUCTIVITY. GAITHERSBURG, MARYLAND. NOV. 13-16, 1967. FRITSCH C A PRETTYMAN P E BELL TELEPHONE LAB INC WHIPPANY NEW JERSEY USGPO NBS-SP-302 695-702 1968 CA 70 1188

TPRC Number	Bibliographic Citation

51903　THE THERMAL CONDUCTIVITY OF THORIA POWDER FROM 400 TO 1200 C. IN VARIOUS GASES AT ATMOSPHERIC PRESSURE. FROM　PROCEEDINGS OF THE SEVENTH CONFERENCE ON THERMAL CONDUCTIVITY.　GAITHERSBURG, MARYLAND.　NOV. 13-16, 1967.
FEITH A D　　NUCLEAR MATERIALS AND PROPULSION OPERATION GENERAL ELECTRIC COMP　CINCINNATI OHIO
USGPO
NBS-SP-302　　　　703-9　　1968　　CA　　70　　118832

51904　MEASUREMENTS OF THE THERMAL CONDUCTIVITY OF GLASS BEADS IN A VACUUM AT TEMPERATURES FROM 100 TO 500 K. FROM　PROCEEDINGS OF THE SEVENTH CONFERENCE ON THERMAL CONDUCTIVITY.　GAITHERSBURG, MARYLAND.　NOV. 13-16, 1967.
MERRILL R B　　SPACE SCIENCES LAB GEORGE C MARSHALL SPACE FLIGHT CENTER　HUNTSVILLE ALA
USGPO
NBS-SP-302　　　　713-20　　1968　　CA　　71　　7701

51905　THE MEASUREMENT OF THE EFFECTIVE THERMAL CONDUCTIVITIES OF WELL-MIXED POROUS BEDS OF DISSIMILAR SOLID PARTICLES BY USE OF THE THERMAL CONDUCTIVITY PROBE.　FROM　PROCEEDINGS OF THE SEVENTH CONFERENCE ON THERMAL CONDUCTIVITY.　GAITHERSBURG, MARYLAND.　NOV. 13-16, 1967.
BEROES C S　　HATTERS H D　　CHEM AND PETROLEUM ENGR DEPT UNIV OF PITTSBURGH　PITTSBURGH PENNSYLVANIA
USGPO
NBS-SP-302　　　　721-8　　1968　　CA　　70　　118813

51906　THERMAL CONDUCTIVITY MEASUREMENTS ON A FIBROUS INSULATION MATERIAL.　FROM　PROCEEDINGS OF THE SEVENTH CONFERENCE ON THERMAL CONDUCTIVITY. GAITHERSBURG, MARYLAND.　NOV. 13-16, 1967.
PETTYJOHN R R　　GENERAL DYNAMICS CORP CONVAIR DIVISION　SAN DIEGO CALIFORNIA
USGPO
NBS-SP-302　　　　729-36　　1968　　CA　　70　　11881

51909　A CORRELATION FOR THERMAL CONTACT CONDUCTANCE OF NOMINALLY-FLAT SURFACES IN A VACUUM.　FROM PROCEEDINGS OF THE SEVENTH CONFERENCE ON THERMAL CONDUCTIVITY.　GAITHERSBURG, MARYLAND.　NOV. 13-16, 1967.
TIEN C L　　UNIVERSITY OF CALIFORNIA　BERKELEY
USGPO
NBS-SP-302　　　　755-9　　1968

51910　THERMAL CONDUCTANCE OF IMPERFECT CONTACTS.　FROM PROCEEDINGS OF THE SEVENTH CONFERENCE ON THERMAL CONDUCTIVITY.　GAITHERSBURG, MARYLAND.　NOV. 13-16, 1967.
BROWN E C JR　　HOLT V E　　NORTH CAROLINA STATE UNIV　RALEIGH
USGPO
NBS-SP-302　　　　761-8　　1968　　CA　　70　　100400

51911　ULTRASONIC MEASUREMENT OF THE THERMAL CONDUCTANCE OF JOINTS IN VACUUM.　FROM　PROCEEDINGS OF THE SEVENTH CONFERENCE ON THERMAL CONDUCTIVITY.　GAITHERSBURG, MARYLAND.　NOV. 13-16, 1967.
WOLF L JR　　KOSTENKO C　　IIT RESEARCH INSTITUTE CHICAGO ILLINOIS
USGPO
NBS-SP-302　　　　769-76　　1968

51912　THERMAL RESISTANCE OF SAPPHIRE-SAPPHIRE CONTACT. FROM　PROCEEDINGS OF THE SEVENTH CONFERENCE ON THERMAL CONDUCTIVITY.　GAITHERSBURG, MARYLAND.　NOV. 13-16, 1967.
BAER Y　　LABORATORIUM FUR FESTKORPERPHYSIK ETH ZURICH SWITZERLAND
USGPO
NBS-SP-302　　　　777-86　　1968　　CA　　70　　10040

51913　HEAT TRANSFER ACROSS SURFACES IN CONTACT.　EFFECTS OF THERMAL TRANSIENTS ON ONE-DIMENSIONAL COMPOSITE SLABS.　FROM　PROCEEDINGS OF THE SEVENTH CONFERENCE ON THERMAL CONDUCTIVITY.　GAITHERSBURG, MARYLAND. NOV. 13-16, 1967.
MOORE C J JR　　BLUM H A
USGPO
NBS-SP-302　　　　787-98　　1968

51914　VARIATION OF WATER AND CARBON DIOXIDE CRYO-DEPOSIT REFLECTANCES WITH ANGLE OF INCIDENCE AND DEPOSIT THICKNESS.　FINAL REPT. MAY 1966-APR. 1968.
WOOD B E　　SMITH A M　　SEIBERT B A　　ARNOLD AIR FORCE STATION　TENNESSEE
DDC
AEDC-TR-68-144 N68-38174
AD-674742　　　　1-32　　1968　　CA　　70　　72566

51925　OPTICAL PROPERTIES OF VACUUM-EVAPORATED SELENIUM AND TELLURIUM.
HAYES J D　　ARAKAWA E T　　WILLIAMS M W
J APPL PHYS
39　　　　12　　5527-32　　1968

51926　PREPARATION, STRUCTURE, AND PROPERTIES OF SPUTTERED, HIGHLY NITRIDED TANTALUM FILMS.
COYNE H J JR　　TAUBER R N

51932　VARIATION OF INFRARED ABSORPTION OF SILICON OXIDE FILMS FROM OXIDATION OF SILANE WITH H2 TREATMENT.
IKEDA Y
JAP J APPL PHYS
7　　　　12　　1543-4　　1968

51937　REFELCTIVITY STUDIES OF DILUTE GOLD-IRON ALLOYS.
BEAGLEHOLE D　　HENDRICKSON T J
PHYS REV LETT
22　　　　4　　133-6　　1969

51944　INVESTIGATION OF THE OPTICAL PROPERTIES OF SIO2 FILMS OBTAINED BY THERMAL OXIDATION OF SILICON, IN THE 7.0-11.0 MICRON WAVELENGTH RANGE.
RAKOV A V　　POTAPOV E V　　MIZGIREVA L P
OPT SPEKTROSK
25　　　　1　　117-21　　1968

51945　INVESTIGATION OF THE OPTICAL PROPERTIES OF SIO2 FILMS OBTAINED BY THERMAL OXIDATION OF SILICON, IN THE 7.0-11.0 MICRON WAVELENGTH RANGE.　//ENGLISH TRANSLATION OF OPT. SPEKTROSK. 25 /1/ 117-21, 1968.//
RAKOV A V　　POTAPOV E V　　MIZGIREVA L P
OPT SPECTROSC
25　　　　1　　59-61　　1968

51949　TRANSPORT AND OPTICAL PROPERTIES OF THIN METALLIC FILMS.
ABELES F
J PHYSIQUE
29 SUPP　　　　L TO　　NO. 2/3
　　　　　　　　C2-37-49　　1968

51954　DEFORMATION OF FINE-GRAINED CONCRETE UPON HEATING. //ENGLISH TRANSLATION OF IZV. AKAD. NAUK SSSR, NEORG. MATER. 4 /1/ 130-5, 1968.//
DORONIN L K　　MIKHAILOV N V
INORGANIC MATERIALS
4　　　　1　　106-10　　1968

51985　THERMAL RADIATIVE AND ELECTRICAL PROPERTIES OF A CADMIUM SULFIDE SOLAR CELL AT LOW SOLAR INTENSITIES AND TEMPERATURES.
JACK J R　　SPISZ E W　　LEWIS RESEARCH CENTER CLEVELAND OHIO
NASA AND CFSTI
NASA-TN-D-4818
N68-35467　　　　1-16　　1968　　CA　　69　　100992

52030　INFRARED COATING STUDIES.　3RD QUARTERLY PROG. REPT.
MAIER R L　　TURNER A F　　RESEARCH AND DEVELOPMENT DIVISION BAUSCH AND LOMB INC　NEW YORK
DDC
AD-650222　　　　1-15　　1967

52031　OPTICAL CONSTANTS OF CHROMIUM.　M.S. THESIS.
CHOR L C-K　　MONTANA STATE UNIV　BOZEMAN MONTANA MONTANA STATE UNIVERSITY
　　　　　　　　1-32　　1968

52046　ENVIRONMENTAL MEASUREMENTS AND INSTRUMENTATION DATA IN SUPPORT OF NAVOCEANO AERIAL COLOR PHOTOGRAPHY TEST.　FINAL TECH. REPT.
YOST E　　WENDEROTH S　　ANDERSON R　　SCIENCE ENGINEERING RESEARCH GROUP　LONG ISLAND UNIVERSITY
DDC AND CFSTI
SERG-TR-02 AD-671847
　　　　　　　　1-90　　1968　　RR　　68-17 127

52057　FABRICATION OF THERMAL COATED EXTENDABLE BOOM. FINAL REPORT.
WESTINGHOUSE DEFENSE AND SPACE CENTER AEROSPACE DIVISION BALTIMORE MARYLAND
NASA AND CFSTI
NASA-CR-95805/REPT-7826A/
N68-29903　　　　1-42　　1968

52080　INFRARED REFLECTION FROM SEMICONDUCTING STANNIC OXIDE LAYERS.
KRYZHANOVSKII B P
OPTIKA I SPEKTROSKOPIYA
25　　　　3　　442-4　　1968

52081　INFRARED REFLECTION FROM SEMICONDUCTING STANNIC OXIDE LAYERS.　//ENGLISH TRANSLATION OF OPTIKA I SPEKTROSK. 25 /3/ 442-4, 1968//.
KRYZHANOVSKII B P
OPT SPECTRY
25　　　　3　　240-1　　1968

52090　THE THERMAL CONDUCTIVITY OF AN AIR-GLASS BEAD MIXTURE USING THE LINE HEAT SOURCE METHOD.　M.S. THESIS.
HRITZ G G　　UNIVERSITY OF PITTSBURGH　PITTSBURGH PENNSYLVANIA
UNIVERSITY OF PITTSBURGH
　　　　　　　　1-77　　1967

52091　STATIC THERMAL CONDUCTIVITIES OF PACKED BEDS OF COPPER PARTICLES.　M.S. THESIS.
AKSOY B R　　RENSSELAER POLYTECHNIC INSTITUTE　TROY NEW YORK
RENSSELAER POLYTECHNIC INSTITUTE

TPRC Number	Bibliographic Citation

52095 NORMAL SOLAR ABSORPTIVITY AND TOTAL HEMISPHERICAL
EMISSIVITY OF ENGINEERING SURFACES AND COATINGS.
CHRISTENSEN F H   CONVAIR ASTRONAUTICS DIVISION
GENERAL DYNAMICS CORPORATION
DDC
AD-843799   1-29   1961

52098 THERMAL PROPERTIES OF SOLIDS. VOLUME II.
WOLFE G W   LING-TEMCO-VOUGHT INC ASTRONAUTICS
DIVISION   DALLAS TEXAS
LING-TEMCO-VOUGHT INC DALAS TEXAS
LTV-E9R-12073
                             1-124   1959

52103 REFLECTANCE OF DIELECTRIC MIRRORS. M.S. THESIS.
REFERMAT S   UNIVERSITY OF ROCHESTER   ROCHESTER
NEW YORK
UNIVERSITY OF ROCHESTER
                             1-40   1966

52118 THERMAL INSULATION STUDY. FINAL REPORT. PHASE 1,
MARCH 1966-SEPT. 1967.
DE WITT W D   GIBBON N C   REID R L   WEBSTER D J
UNION CARBIDE CORP LINDE DIVISION   TONAWANDA N Y
CFSTI
ALO-3632-20 N69-11525
                             1-227   1967   PA   N69-7-2 389

52125 MATERIALS APPLICATIONS TO GLIDE RE-ENTRY STRUCTURE.
FROM   SYMPOSIUM ON PROCESSING MATERIALS FOR RE-ENTRY
STRUCTURES, 1960.
BRAUN M T   BOEING AIRPLANE COMPANY   SEATTLE
WASHINGTON
DDC
WADD-TR-60-58
AD-241597   29-46   1960

52128 CERAMIC AND COMPOSITE CERAMIC-METAL MATERIALS SYSTEMS
APPLICABLE TO RE-ENTRY STRUCTURES. FROM SYMPOSIUM
ON PROCESSING MATERIALS FOR RE-ENTRY STRUCTURES,
1960.
STERRY W M   BOEING AIRPLANE COMP AEROSPACE DIV
SEATTLE WASHINGTON
DDC
WADD-TR-60-58
AD-241597   333-74   1960

52151 DEVELOPMENT OF PHASE-CHANGE COATINGS FOR USE AS A
VARIABLE THERMAL CONTROL SURFACES. FINAL REPT.
GRIFFIN R N   LINDER B   GENERAL ELECTRIC CO
MISSILE AND SPACE DIV   PHILADELPHIA PENN
NASA
NASA-CR-66394
N67-36080   1-59   1967

52152 EMISSIVITY COATINGS FOR LOW-TEMPERATURE SPACE
RADIATORS. QUARTERLY PROG. REPT.
SMITH F J   GRAMMER J G   LOCKHEED MISSILES AND
SPACE CO   SUNNYVALE CALIFORNIA
NASA
NASA-CR-72059
N67-14886   1-30   1965

52153 THE INFLUENCE OF ULTRAVIOLET RADIATION ON THE
EMITTANCE AND SOLAR ABSORPTANCE OF A WHITE PAINT
COATING.
BRANDENBERG W M   GENERAL DYNAMICS/ASTRONAUTICS
SAN DIEGO CALIFORNIA
DDC
GDA-AE61-1223
AD-678151   1-13   1961   RR   69-2 87

52158 THERMODYNAMICS OF LIQUID SURFACES. ANNUAL REPT.
GOOD R J   OPDYCKE J D   CONVAIR SCIENTIFIC
RESEARCH LAB   SAN DIEGO CALIF
DDC
AD-677740   1-33   1961   RR   69-1 68

52160 PROPERTIES OF HIGH EMITTANCE MATERIALS.
CLEARY R E   EMANUELSON R   LUOMA W   AMMANN C
LEWIS RESEARCH CENTER NATIONAL AERONAUTICS AND SPACE
ADMINISTRATION
NASA AND CFSTI
NASA-CR-1278
                             1-120   1969

52182 THE OPTICAL PROPERTIES OF GOLD.
HODGSON J N
J PHYS CHEM SOLIDS
29   12   2175-81   1968

52199 THE INTEGRATING SPHERE REFLECTOMETER FOR EXPERIMENTAL
DETERMINATION OF THE REFLECTIVITY OF VARIOUS SURFACES
AT WAVELENGTHS IN THE .4 TO .9 MICRON RANGE. M.S.
THESIS.
ALLEN R C   UNIVERSITY OF DELAWARE   NEWARK
DELAWARE
UNIVERSITY OF DELAWARE
                             1-112   1968

52200 A STUDY OF THE EFFECTIVE THERMAL CONDUCTIVITIES OF
PACKED BEDS OF MIXTURES OF DISSIMILAR SOLID MATERIALS
BY MEANS OF THE THERMAL CONDUCTIVITY PROBE METHOD.
M.S. THESIS.
HATTERS H D   UNIVERSITY OF PITTSBURGH   PITTSBURGH
PENNSYLVANIA
UNIVERSITY OF PITTSBURGH
                             1-82   1967

52204 THE OPTICAL TRANSMITTANCE OF PARTIALLY OXIDIZED
COPPER FILMS. M.S. THESIS.
BOYKO F L   CLARKSON COLLEGE OF TECHNOLOGY
POTSDAM NEW YORK
CLARKSON COLLEGE OF TECHNOLOGY
                             1-51   1968

52206 VARIATIONS OF THE THERMAL CONDUCTIVITY COEFFICIENT
FOR FIBROUS INSULATION MATERIALS. M.S. THESIS.
MUMAW J R   OHIO STATE UNIVERSITY   COLUMBUS OHIO
OHIO STATE UNIVERSITY
                             1-129   1968

52210 A SILVER-GALLIUM ARSENIDE SCHOTTKY-BARRIER
ULTRAVIOLET DETECTOR.
BAERTSCH R D   RICHARDSON J R
J APPL PHYS
40   1   229-35   1969

52213 CORRELATION OF INTERFACIAL FREE ENERGY IN BINARY AND
TERNARY SYSTEMS.
PAUL G W   DE CHAZAL L E M
IND ENG CHEM FUNDAMENTALS
8   1   104-8   1969

52215 RELATIONSHIP BETWEEN APPARENT VISCOSITY OF PORTLAND
CEMENTS AND THE LIME SATURATION COEFFICIENT OF
SILICA. //ENGLISH TRANSLATION OF IZV. AKAD. NAUK
SSSR, NEORG. MATER. 4 /6/ 980-3, 1968.//
GOLDSHTEIN L YA   MANTSUROVA V N
INORGANIC MATERIALS
4   6   861-4   1968

52232 THERMAL BEHAVIOR OF FIBROUS INSULATING MATERIALS.
GARCIA A   INSTITUTO EDUARDO TORROJA DE LA
CONSTRUCCION Y   DEL CEMENTO MADRID SPAIN
NASA
N68-28756   1-56   1967   RR   68-19 95

52260 DETERMINATION OF THE THERMAL PROPERTIES OF MATERIALS
AT ELEVATED TEMPERATURES. SUMMARY REPORT JUNE
1957-JUNE 1960.
MOELLER C E   WILSON D R   MIDWEST RESEARCH
INSTITUTE   KANSAS CITY MISSOURI
MIDWEST RESEARCH INSTITUTE
MRI-2059-E   1-95   1960

52267 THE EFFECT OF ARC PLASMA DEPOSITION ON THE STABILITY
OF NON METALLIC MATERIALS. FINAL REPORT APRIL
1960-MAY 1961.
KRAMER B E   LEVINSTEIN M A   GRENIER J W
FLIGHT PROPULSION LAB GENERAL ELECTRIC COMPANY
CINCINNATI OHIO
DDC
AD-260536   1-57   1961

52271 STUDIES IN THE UO2-ZRO2 SYSTEM.
WRIGHT T R   KIZER D E   KELLER D L   BATTELLE
MEMORIAL INSTITUTE   COLUMBUS OHIO
AEC AND CFSTI
BMI-1689   1-34   1964

52274 INVESTIGATION OF HIGH TEMPERATURE RESISTANT
MATERIALS. QUARTERLY REPT. NO. 17.
MASON C R   WALTON J D   BOWEN M D   TEAGUE W T
MURPHY C A   ENGINEERING EXPERIMENT STATION
GEORGIA INSTITUTE OF TECHNOLOGY
DDC
AD-238214   1-45   1960

52276 RADIATION-RESISTANT MAGNET WIRE FOR USE IN AIR AND
VACUUM AT 850 C.
ANACONDA WIRE AND CABLE COMPANY
MAGNET WIRE RESEARCH LABORATORIES
MUSKEGON MICHIGAN
DDC
ASD-TDR-63-164
AD-414198   1-104   1963

52282 INVESTIGATIONS OF GLASS FIBER-METAL COMPOSITE
MATERIALS. FINAL REPT. MARCH 1955- OCT. 1960.
LOCKWOOD P A   OWENS-CORNING FIBERGLAS
CORPORATION   GRANVILLE OHIO
DDC
AD-274530   1-164   1960

52287 METAL REINFORCED CERAMIC RADOME.
ATLAS L M   ARMOUR RESEARCH FOUNDATION OF ILLINOIS
INSTITUTE OF TECHNOLOGY
DDC
WADC-TR-58-329
AD-155872   1-41   1958

TPRC Number	Bibliographic Citation

52292 A THEORETICAL STUDY OF THE THERMAL CONDUCTANCE OF JOINTS WITH VARYING AMBIENT PRESSURE. M.S. THESIS.
AARON R L   SOUTHERN METHODIST UNIVERSITY   DALLAS TEXAS
SOUTHERN METHODIST UNIVERSITY
       1-65   1963

52293 DEVELOPMENT OF SPACE-STABLE THERMAL-CONTROL COATINGS. TRIANNUAL REPORT.
ZERLAUT G A   ASHFORD N   IIT RESEARCH INSTITUTE TECHNOLOGY CENTER   CHICAGO ILLINOIS
NASA
IITRI-U6002-73
       1-68   1969

52298 THE CHANGE OF REFLECTIVE PROPERTIES OF THREE SELECTED THERMAL CONTROL COATINGS DUE TO EXTREME ULTRAVIOLET RADIATION AND VACUUM. M.S. THESIS.
STROUD D E   AIR FORCE INSTITUTE OF TECHNOLOGY WRIGHT-PATTERSON AFB   OHIO
AIR FORCE INSTITUTE OF TECHNOLOGY
GSF-MECH-68-11
       1-99   1968

52300 THERMAL CONDUCTION THROUGH AN EVACUATED IDEALIZED POWDER OVER THE TEMPERATURE RANGE OF 100 TO 500 K.
MERRILL R B   GEORGE C MARSHALL SPACE FLIGHT CENTER   HUNTSVILLE ALABAMA
NASA AND CFSTI
NASA-TN-D-5063
       1-76   1969

52319 FUSED QUARTZ FIBER OPTICS FOR ULTRAVIOLET TRANSMISSION.
LI P C   PONTARELLI D A   OLSON O H   SCHWARTZ M A
BULL AM CERAM SOC
48   2   214-20   1969

52324 DIRECTIONAL EFFECT IN THERMAL CONDUCTANCE OF METALLIC CONTACTS. M.S. THESIS.
PATEL J D   UNIVERSITY OF MIAMI   MIAMI FLORIDA
UNIVERSITY OF MIAMI
       1-133   1968

52328 TEMPERATURES DURING VACUUM DEPOSITION OF METAL FILMS.
YODA E
JAPAN J APPL PHYS
8   2   191-201   1969

52329 INFRARED ABSORPTION OF BORON SILICATE FILMS PREPARED BY SIH4-B2H6-O2 AND -NO2 SYSTEMS.
HANETA Y
JAPAN J APPL PHYS
8   2   274   1969

52342 ON THE CONTOUR OF THE ABSORPTION LINES IN CU2O.
UENO T
J PHYS SOC JAPAN
26   2   438-46   1969

52343 VACUUM ULTRAVIOLET ABSORPTION SPECTRA OF SYNTHESIZED POLYMER FILMS.
ONARI S
J PHYS SOC JAPAN
26   2   500-4   1969

52387 ABSORPTION BANDS OF GADOLINIUM IN THE FERROMAGNETIC AND PARAMAGNETIC STATES.
HODGSON J N   CLEYET B
J PHYS C /SOLID STATE PHYS/
2   1   97-101   1969

52417 FOAM CERAMICS DEVELOPMENT.
SIMPSON F H   BOEING AIRPLANE COMPANY   SEATTLE WASHINGTON
DDC
D2-10339 AD-411657
       1-54   1962

52421 MECHANICAL AND PHYSICAL PROPERTIES OF TANTALUM AND CARBURIZED TANTALUM ABOVE APPROXIMATELY 3500 F.
SALDINGER I L   GLASIER L F JR   MATERIALS AND PROCESSES DEPT AEROJET-GENERAL CORP   AZUSA CALIFORNIA
AEROJET-GENERAL CORP
M-1795
       1-66   1959

52529 EFFECT OF HEAT-TREATMENT ON THE CONSTITUTION AND MECHANICAL PROPERTIES OF SOME HYDRATED ALUMINOUS CEMENTS.
SCHNEIDER S J
J AM CERAMIC SOC
42   4   184-93   1959

52540 TITANIUM ENAMELS WITH A LOW TIO2 CONTENT.
VARGIN V V   SMIRNOVA G P
STEKLO I KERAMIKA
19   8   35-7   1962

52541 TITANIUM ENAMELS WITH A LOW TIO2 CONTENT. //ENGLISH TRANSLATION OF STEKLO I KERAMIKA 19 /8/ 35-7, 1962.//
VARGIN V V   SMIRNOVA G P
GLASS AND CERAMICS
19   8   439-42   1962

52546 A METHOD OF DETERMINING THE THERMAL CONDUCTIVITY OF POWDERED AND FIBROUS INSULATION AS A FUNCTION OF FILLER GAS PRESSURE. //ENGLISH TRANSLATION OF INZH.-FIZ. ZH. 9 /6/ 751-6, 1965.//
DULNEV G N   PLATUNOV E S   MURATOVA B L   SIGALOVA Z V
J ENG PHYS
9   6   460-3   1965

52560 COMPILATION OF MATERIALS RESEARCH DATA. FOURTH QUARTERLY PROGRESS REPT. PHASE 1. DEC. 1961-FEB. 1962.
BERGSTEDT P   GENERAL DYNAMICS ASTRONAUTICS   SAN DIEGO CALIFORNIA
DDC AND CFSTI
MRG-160 AD-273065
       1-211   1962

52562 COMPILATION OF MATERIALS RESEARCH DATA. FOURTH QUARTERLY PROG. REPT. PHASE 1. MARCH 1963-MAY 1963.
HOOPER A F   CONVAIR ASTRONAUTICS DIVISION GENERAL DYNAMICS CORPORATION
DDC
MRG-327 AD-405184
       1-4   1963

52569 THE PROTECTION OF CARBON AND GRAPHITE AGAINST OXIDATION AT TEMPERATURES OF 1200 DEGREES.
SAZONOVA M V   SITNIKOVA A YA   APPEN A A
ZHUR PRIKLAD KHIM
34   3   505-12   1961

52570 THE PROTECTION OF CARBON AND GRAPHITE AGAINST OXIDATION AT TEMPERATURES OF 1200 DEGREES. //ENGLISH TRANSLATION OF ZHUR. PRIKLAD. KHIM. 34 /3/ 505-12, 1961.//
SAZONOVA M V   SITNIKOVA A YA   APPEN A A
DDC AND CFSTI
MCL-1285/1 AND 2/
AD-269653
       1-13   1961

52588 ORIGIN OF WATER-INDUCED TOUGHENING IN MGO CRYSTALS.
SHOCKEY D A   GROVES G W
J AM CERAM SOC
52   2   82-5   1969

52592 THIN FILM EMISSION OF KBR, RBCL, RBBR AND NACL IN THE FAR INFRA-RED.
MOOIJ J E   VAN DE BUNT W B   SCHRIJVERS J E
PHYS LETT
28   A   8   573-4   1969

52594 NEAR-INFRA-RED DIFFUSE REFLECTIVITIES OF NATURAL AND MAN-MADE MATERIALS.
NICOLLE R L   IRVINE J   BOWDEN F G
BRIT J APPL PHYS /J PHYS D/
2   2   201-4   1969

52604 DEPENDENCE OF THE ABSORPTION EDGE OF FILMS OF INDIUM ANTIMONIDE ON THICKNESS.
FILATOV O N   KARPOVICH I A
FIZ TVERD TELA
10   9   2886-7   1968

52605 DEPENDENCE OF THE ABSORPTION EDGE OF FILMS OF INDIUM ANTIMONIDE ON THICKNESS. //ENGLISH TRANSLATION OF FIZ. TVERD. TELA 10 /9/ 2886-7, 1968.//
FILATOV O N   KARPOVICH I A
SOVIET PHYSICS-SOLID STATE
10   9   2284-5   1969

52629 INTERFACIAL SURFACE ENERGIES AND CONTACT ANGLES OF WETTING OF SOLIDS BY LIQUID IN EQUILIBRIUM AND NON-EQUILIBRIUM SYSTEMS.
NAIDICH YU V
ZH FIZ KHIM
42   8   1946   1968

52630 INTERFACIAL SURFACE ENERGIES AND CONTACT ANGLES OF WETTING OF SOLIDS BY LIQUID IN EQUILIBRIUM AND NON-EQUILIBRIUM SYSTEMS. //ENGLISH TRANSLATION OF ZH. FIZ. KHIM. 42 /8/ 1946-, 1968.//
NAIDICH YU V
RUSS J PHYS CHEM
42   8   1023-6   1968

52643 SOME OPTICAL AND ELECTRICAL PROPERTIES OF THIN FILMS OF CU3 P SE, CU3 P SE S/X/ AND CU3 P SE I/X/.
KRYZHANOVSKII B P
OPTIKA I SPEKTROSKOPIYA
25   4   613-15   1968   CA   70   24340

52644 SOME OPTICAL AND ELECTRICAL PROPERTIES OF THIN FILMS OF CU3 P SE, CU3 P SE S/X/ AND CU3 P SE I/X/. //ENGLISH TRANSLATION OF OPTIKA I SPEKTROSKOPIYA 25 /4/ 613-5, 1968.//
KRYZHANOVSKII B P
OPT SPECTROSC
25   4   343-4   1968

52645 PROTON EXCITATION OF THE ARGON ATOM.
HURST G S   BORTNER T E   STRICKLER T D
PHYS REV

TPRC Number	Bibliographic Citation
52654	RESEARCH OF TECHNIQUES OF METHODS ON WHICH TO BASE DEVELOPMENT OF HIGH TEMPERATURE GLAZING ATTACHMENTS. PARTAIN G K PISCOPO F A WILSON R E WRIGHT AIR DEVELOPMENT DIVISION WRIGHT-PATTERSON AIR FORCE BASE OHIO DDC WADD-TR-60-119 AD-240357 1-63 1960
52655	THE ACCOMMODATION COEFFICIENTS OF HE, NE, A, H2, D2, O2, CO2, AND HG ON PLATINUM AS A FUNCTION OF TEMPERATURE. THOMAS L B OLMER F J AM CHEM SOC 65 1036-43 1943
52660	INVESTIGATION OF GRAPHITE BODIES. BORTZ S A LUND H H NAKAMURA H H ARMOUR RESEARCH FOUNDATION ILLINOIS INSTITUTE OF TECHNOLOGY ARMOUR RESEARCH FOUNDATION ILLINOIS INSTITUTE OF TECHNOLOGY DDC WADC-TR-59-706 AD-283026 1-107 1960
52661	RESEARCH ON ELEVATED TEMPERATURE RESISTANT CERAMIC STRUCTURAL ADHESIVES. PART VI. FORLANO R J KRUMWIEDE D M BENZEL J F THORNTON H R LEFORT H G DDC WADC-TR-55-491/PT IV/ AD-282302 1-72 1962
52663	ABSORPTION OF INFRARED RADIATION BY ICE CRYODEPOSITS. PEPPER S V LEWIS RESEARCH CENTER CLEVELAND OHIO NASA AND CFSTI NASA-TN-D-5181 1-20 1969 CA 71 8043
52668	STRESSES IN POLYCOMPONENT SOLID CERAMICS OR CERMETS. QUARTERLY REPT. AUG. 1959-OCT. 1959. GILES T M SHEVLIN T S EVERHART J O RESEARCH FOUNDATION OHIO STATE UNIVERSITY DDC RF-806-7 AD-231031 1-8 1959
52669	STRESSES IN POLYCOMPONENT SOLID CERAMICS OR CERMETS. FINAL SUMMARY REPORT. GILES T M SHEVLIN T S EVERHART J O RESEARCH FOUNDATION OHIO STATE UNIVERSITY DDC RF-806 AD-236828 1-14 1959
52740	THERMAL ACCOMMODATION COEFFICIENTS ON GAS-COVERED TUNGSTEN, NICKEL AND PLATINUM. AMDUR I GUILDNER L A J AM CHEM SOC 79 311-15 1957
52756	APPARATUS FOR DETERMINING LINEAR THERMAL EXPANSIONS OF MATERIALS IN VACUUM OF CONTROLLED ATMOSPHERE. FULKERSON S D OAK RIDGE NATIONAL LABORATORY OAK RIDGE TENNESSEE AEC ORNL-2856 1-39 1960
52757	INFRARED-SPECTROSCOPIC INVESTIGATION OF THE VITRIFICATION AND CRYSTALLIZATION OF LEAD GLASS WITH A COMPOSITION CORRESPONDING TO THE COMPOUND PBO.SIO2. //ENGLISH TRANSLATION OF IZV. AKAD. NAUK SSSR, NEORG. MATER. 4 /7/ 1124-8, 1968.// SMIRNOVA E V INORGANIC MATERIALS 4 7 985-8 1968
52771	STATUS REPORT ON NON-METALLIC FIBROUS REINFORCED METAL COMPOSITES. MACHLIN E S MATERIALS RESEARCH CORPORATION YONKERS NEW YORK DDC AD-265943 1-243 1961
52782	ELECTRODEPOSITION OF EROSION AND OXIDATION RESISTANT COATINGS FOR GRAPHITE. SECOND QUARTERLY PROGRESS REPORT. WRIGHT C H BARGERO G F HUMINIK J JR VALUE ENGINEERING COMPANY ALEXANDRIA VIRGINIA DDC AD-267110 1-10 1961
52783	INVESTIGATION OF GLASS-METAL COMPOSITE MATERIALS. ELEVENTH QUARTERLY PROG. REPT. WHITEHURST H B WILEY R B OWENS-CORNING FIBERGLAS CORP NEWARK OHIO DDC AD-223057 1-28 1958
52784	TOTAL NORMAL SPECTRAL CHARACTERISTICS OF CENTAUR PAINT COATINGS. SHINKLE F J GENERAL DYNAMICS/ASTRONAUTICS SAN DIEGO CALIFORNIA
52787	ELECTRODEPOSITION OF EROSION AND OXIDATION RESISTANT COATINGS FOR GRAPHITE. FINAL SUMMARY REPT. GOODMAN E BARGERO G F HUMINIK J VALUE ENGINEERING COMPANY ALEXANDRIA VIRGINIA DDC AD-273853 1-59 1962
52788	COMPOSITE CERAMIC RADOME MANUFACTURE BY MOSAIC TECHNIQUES. INTERIM ENGINEERING PROGRESS REPORT MAY 1963-JULY 1963. FILIPPI F J NARMCO RESEARCH AND DEVELOPMENT SAN DIEGO CALIFORNIA DDC IR-7-984-/III/ ASD-7-984 AD-415269 1-107 1963
52830	NONLOCAL THEORY OF THE OPTICAL PROPERTIES OF THIN METALLIC FILMS. JONES W E KLIEWER K L FUCHS R PHYS REV 178 3 1201-3 1969
52841	WETTING OF BINARY ALUMINUM ALLOYS IN CONTACT WITH BERYLLIUM, CARBON TETRABORIDE, AND GRAPHITE. MANNING C R JR GURGANUS T B J AM CERAM SOC 52 3 115-18 1969
52850	INFRARED COATING STUDIES. 2ND QUARTERLY REPT. SEPT. 1966-DEC. 1966. MAIER R L RESEARCH AND DEVELOPMENT DIVISION. BAUSCH AND LOMB INC ROCHESTER N Y DDC AD-648425 1-16 1966
52855	TARGET SIGNATURE MEASUREMENTS LABORATORY STUDY. 3RD INTERIM TECH. REPT. ANDRYCHUK D BENN M TEXAS INSTRUMENTS INC DALLAS TEXAS DDC AD-822204 1-121 1967
52864	PRODUCTION OF HIGH-PURITY THERMAL CONTROL COATINGS. INTERIM ENGR. PROG. REPT. JUNE 1967-AUG. 1967. BAILIN L J LOCKHEED MISSILES AND SPACE COMP PALO ALTO CALIFORNIA DDC AD-820643 1-32 1967
52865	INFRARED EMISSION FROM THE SURFACE OF THE MOON. WINTER D F BOEING SCIENTIFIC RESEARCH LABS SEATTLE WASHINGTON DDC AD-672074 1-78 1968
52866	INTERNALLY LUBRICATED RTPS FOR GEARS AND BEARINGS. THEBERGE J E MOD PLASTICS 47 3 104-21 1970
52870	INFRARED COATING STUDIES. 7TH QUARTERLY REPORT. DEC. 1967-MARCH 1968. RESEARCH AND DEVELOPMENT DIVISION BAUSCH AND LOMB INC ROCHESTER NEW YORK DDC AD-666787 1-13 1968
52871	THE OPTICAL PROPERTIES OF METAL BLACKS AND CARBON BLACKS. HARRIS L MASSACHUSETTS INSTITUTE OF TECHNOLOGY CAMBRIDGE MIT PRESS CAMBRIDGE MASS MONO-SERIES NO. 1 DDC AND NASA AD-831788 N68-19784 1-116 1967
52872	STUDY OF DEPOSITED INSULATING LAYERS ON SILICON. NUTTALL R ROWBOTHAM C EASTWOOD E FERRANTI LIMITED WYTHENSHAWE MANCHESTER ENGLAND DDC AD-827049 1-23 1967
52873	ADVANCED, FIBER-REINFORCED TUNGSTEN NOZZLE. 3RD QUARTERLY REPT. OCT. 1966-DEC. 1966. OWEN L J CLEVITE CORPORATION MECHANICAL RESEARCH DIVISION CLEVELAND OHIO DDC AD-379508 1-40 1967
52883	INVESTIGATION OF COLD MIRROR COATINGS. WALLS J J JR FIRE CONTROL DEVELOPMENT AND ENG LABS FRANKFORD ARSENAL PHILADELPHIA PA DDC FA-M-68-13-1 AD-826180 1-73 1967
52885	RADIATIVE HEAT TRANSFER IN ABSORBING, EMITTING AND SCATTERING MEDIA. LOVE T J STOCKHAM L W LEE F C MUNTER W A TSAI Y W SCHOOL OF AERO AND MECHANICAL ENGINEERING UNIV OF OKLAHOMA NORMAN OKLAHOMA

TPRC Number	Bibliographic Citation

52890 VARIABLE THERMAL CONDUCTIVITY HEAT TRANSFER RESEARCH FOR PERSONAL PROTECTIVE ASSEMBLIES. FINAL REPT. JUNE 1966-AUG. 1967.
OLSON R B   FELDER J W   LOMBARD C F   NORTHROP CORPORATE LABS   HAWTHORNE CALIFORNIA
DDC
AMRL-TR-67-190
AD-668765   1-43   1967

52891 HIGH-TEMPERATURE THERMOELECTRIC RESEARCH.
COOLEY R A   JANOWIECKI R J   SONNENSCHEIN G   STROP H R   WILLSON M C   MONSANTO RESEARCH CORP   DAYTON OHIO
DDC
AFAPL-TR-66-51
AD-484771   1-278   1966

52892 BIOLOGICAL METHODS FOR CAMOUFLAGE CONCEALMENT, AND TONEDOWN. FINAL REPT. JUNE 1966-DEC. 1967.
SCAFF W L JR   MACIASR F M   BELL R T   NORTHROP CORPORATE LABS   HAWTHORNE CALIFORNIA
DDC
AFAPL-TR-67-153 NCL-67-162
AD-830415   1-69   1968

52893 EVALUATION OF CHARACTERISTICS AFFECTING ATTAINMENT OF OPTIMUM PROPERTIES OF ABLATIVE PLASTICS. VOLUME II.
DAVIS H O   LAMPMAN J A   WARGA J J   AEROJET-GENERAL CORP   SACRAMENTO CALIFORNIA
DDC
AFRPL-TR-68-29 /VOL II/
AD-830834   1-560   1968

52895 INFRA-RED REFLECTION SPECTRA OF ADSORBED CARBON MONOXIDE ON COPPER.
PRITCHARD J   SIMS M L
TRANS FARADAY SOC
66   2   427-33   1970

52896 ULTRASONIC AND THERMAL STUDIES OF SELECTED PLASTICS LAMINATED MATERIALS, AND METALS.
ASAY J R   URZENDOWSKI S R   GUENTHER A H   AIR FORCE WEAPONS LAB KIRTLAND AIR FORCE BASE   NEW MEXICO
DDC
AFWL-TR-67-91
AD-827596   1-266   1968

52897 NARROW-BAND ULTRAVIOLET FILTER PHOTOGRAPHY.
MANGOLD V L   AIR FORCE FLIGHT DYNAMICS LAB WRIGHT-PATTERSON AIR FORCE BASE   OHIO
DDC
AFFDL-TR-66-140
AD-808792   1-31   1966

52899 CONSIDERATIONS FOR THE APPLICATION OF 254 NANOMETER ULTRAVIOLET PHOTOGRAPHY TO NIGHT AERIAL RECONNAISSANCE. TECH. DOCUMENTARY REPT. OCT. 1966-JULY 1967.
HUBBARD R N   SYSTEMS RESEARCH LABS INC   DAYTON OHIO
DDC
AFAL-TR-67-252
AD-822402   1-45   1967

52900 HIGH PERFORMANCE AIRCRAFT WINDSHIELD DEVELOPMENT PROGRAM.
SHOEMAKER A F   CORNING GLASS WORKS NEW YORK
DDC
AFML-TR-66-69
AD-812203   1-175   1966

52901 HIGH MODULUS, HIGH STRENGTH FILAMENTS AND COMPOSITES.
WITUCKI R M   ASTRO RESEARCH CORPORATION   SANTA BARBARA CALIFORNIA
DDC
AFML-TR-66-187
AD-658872   1-132   1967

52903 INFRARED DIFFUSE REFLECTOR COATING. PART I. THEORETICAL. FINAL TECH. REPT. JULY 1966-SEPT. 1967.
SCHMIDT R N   TREUENFELS P M   MEEHAN E J   HONEYWELL INC SYSTEMS AND RESEARCH DIVISION   ST PAUL MINN
DDC
AFML-TR-67-337
AD-828583   1-112   1967

52904 INVESTIGATION TO PRODUCE METAL MATRIX COMPOSITES WITH HIGH-MODULUS, LOW DENSITY CONTINUOUS FILAMENT REINFORCEMENTS. SUMMARY TECH. REPT. JULY 1965-APRIL 1967.
ALEXANDER J A   CUNNINGHAM A L   CHUANG K C   GENERAL TECHNOLOGIES CORP   RESTON VA
DDC
AFML-TR-67-391
AD-829129   1-148   1968

52905 EFFECTS OF VACUUM-ULTRAVIOLET ENVIRONMENT ON OPTICAL PROPERTIES OF BRIGHT ANODIZED ALUMINUM TEMPERATURE CONTROL COATINGS.
WEAVER J H   AIR FORCE MATERIALS LAB WRIGHT-PATTERSON AIR FORCE BASE   OHIO
DDC

52906 AIR DRYING HIGH TEMPERATURE RESISTANT, SILICONE PROTECTIVE COATINGS.
STOUT R L   AIR FORCE MATERIALS LAB WRIGHT-PATTERSON AIR FORCE BASE   OHIO
DDC
AFML-TR-67-433
AD-839537   1-108   1968

52909 SURFACE CHARACTERISTICS EFFECT ON THERMAL REGIME. PHASE 1.
WECHSLER A E   GLASER P E   LITTLE /ARTHUR D/ INC   CAMBRIDGE MASS
DDC
CRREL-SR-88 AD-485168
                             1-32   1966

52912 CADMIUM OXIDE. /DATA SHEETS/
NEUBERGER M   ELECTRONIC PROPERTIES INFORMATION CENTER   CULVER CITY CALIFORNIA
DDC
EPIC-DS-149 AD-486595
                             1-83   1966

52913 LEAD SULFIDE. /DATA SHEETS/
NEUBERGER M   HUGHES AIRCRAFT COMP   CULVER CITY CALIFORNIA   ELECTRONIC PROPERTIES INFORMATION CENTER
DDC
EPIC-DS-150 AD-803886
                             1-114   1966

52917 PSEUDO-BREWSTER ANGLE TECHNIQUE FOR DETERMINING OPTICAL CONSTANTS.
POTTER R F   NAVAL ORDNANCE LAB   CORONA CALIFORNIA
DDC
NOLC-699 AD-647058
                             1-42   1967

52920 REFLECTANCE, SOLAR ABSORPTIVITY, AND THERMAL EMISSIVITY OF SIO2-COATED ALUMINUM.
HASS G   RAMSEY J B   HEANEY J B   TRIOLO J J
APPL OPT
8   2   275-81   1969

52925 RADIANT POWER FLOW AND ABSORPTANCE IN THIN FILMS.
BAUMEISTER P W
APPL OPT
8   2   423-36   1969

52926 AN INTRACAVITY INTERFERENCE FILTER LASER WAVELENGTH SELECTOR.
HANES G R   DOBROWOLSKI J A
APPL OPT
8   2   482-3   1969

52929 OPTICAL PROPERTIES AND PLASMA OSCILLATIONS OF A SELF-SUPPORTING METALLIC FILM COVERED BY OXIDE LAYERS.
SHIEH S Y
PHYS LETT
29   A   1   46-7   1969

52940 EFFECT OF HEAT TREATMENT ON ELECTRON-GUN-DEPOSITED SIO2 FILMS.
RAMSEY T H   RICE S B
J AM CERAM SOC
52   4   225-6   1969

52998 FORMATION AND STABILITY OF TRIBUTYL PHOSPHATE-LITHIUM NITRATE ASSOCIATIONS.
CHIFU E   GABRIELLI G
GAZZ CHIM ITAL
98   10   1213-23   1968   CA   70   41373

52999 EFFECTS OF ATMOSPHERE AND MINOR CONSTITUENTS ON THE SURFACE TENSION OF GLASS MELTS.
AKHTAR S   CABLE M
GLASS TECHNOL
9   5   145-51   1968   JA   52   95

53006 COMPARATIVE THERMAL CONDUCTIVITY METHOD FOR SELF-GUARDING DISK SPECIMENS. FROM PROCEEDINGS OF THE 1ST CONFERENCE ON THERMAL CONDUCTIVITY. BATTELLE MEMORIAL INSTITUTE. COLUMBUS, OHIO. OCTOBER 26-28, 1961.
LUCKS C F   DEEM H W   BATTELLE MEMORIAL INSTITUTE   COLUMBUS OHIO
BATTELLE MEMORIAL INSTITUTE   COLUMBUS OHIO
                             1-9   1961

53012 THERMAL CONDUCTIVITY OF THIN FILMS OF WATER.
METSIK M S   TIMOSHCHENKO G T
ISSLED OBL POVERKH SIL SB DOKL KONF 3RD 1966
                       41-2   1967   CA   70   51391

53017 NONEQUILIBRIUM VALUES OF INTERFACIAL TENSION AND ADHESION IN A METAL-OXIDE MELT SYSTEM.
DERYABIN A A   POPEL S I   SABUROV L N
IZV AKAD NAUK SSSR METAL
                5   51-9   1968   CA   70   6777

TPRC Number	Bibliographic Citation

53036 ADSORPTION OF NORMAL ALKANES ON MERCURY. EXPERIMENTAL CHECK OF THE GIBBS ADSORPTION EQUATION AND THEORETICAL CONTACT POTENTIAL.
SMITH T
J COLLOID INTERFACE SCI
28 3/4   531-42   1968   CA   70   23121

53052 PARTICLE DEPOSITION FROM TURBULENT STREAMS BY MEANS OF THERMAL FORCE.
BYERS R L   CALVERT S
IND ENG CHEM FUNDAMENTALS
8   4   646-55   1969

53068 REFLECTANCE OF TITANIA OPACIFIED PORCELAIN ENAMELS.
EPPLER R A
AM CERAM SOC BULL
48   5   549-54   1969

53071 RAY TRACES THROUGH HOLLOW METAL LIGHT-PIPE ELEMENTS.
LOEWENSTEIN E V   NEWELL D C
J OPT SOC AM
59   4   407-14   1969

53075 COMPARISON OF CONTINUUM AND DISCRETE LATTICE RESULTS FOR GAS-SURFACE INTERACTIONS.
KARAMCHETI K   SCOTT L B JR
J CHEM PHYS
50   6   2364-71   1969

53085 APPLICATIONS TECHNOLOGY SATELLITE THERMAL DESIGN. FROM   PROC. OF THE JOINT AIR FORCE NASA THERMAL CONTROL WORKING GROUP.   16, 17 AUG. 1967.
WENSLEY R J   KANE G A
DDC
AFML-TR-68-198
AD-841387   34-81   1968

53086 ADVANCED PIONEER THERMAL CONTROL DEVELOPMENTS. FROM   PROC. OF THE JOINT AIR FORCE NASA THERMAL CONTROL WORKING GROUP.   16, 17 AUG. 1967.
STREED E R   KIRKPATRICK J P
DDC
AFML-TR-68-198
AD-841387   285-330   1968

53087 SURVEYOR SPACECRAFT THERMAL CONTROL.   FROM   PROC. OF THE JOINT AIR FORCE NASA THERMAL CONTROL WORKING GROUP. 16, 17, AUG. 1967.
KNUDSON H E   TUCHSCHER J S
DDC
AFML-TR-68-198
AD-841387   369-457   1968

53088 THERMAL DESIGN OF THE ORBITING VEHICLE TYPE ONE /OV1/ SATELLITE SERIES.   FROM   PROC. OF THE JOINT AIR FORCE NASA THERMAL CONTROL WORKING GROUP. 16, 17 AUG. 1967.
HOWELL G A
DDC
AFML-TR-68-198
AD-841387   458-84   1968

53089 THE BEHAVIOR OF SEVERAL WHITE PIGMENTS AS DETERMINED BY IN SITU REFLECTANCE MEASURMENTS OF IRRADIATED SPECIMENS.   FROM   PROC. OF THE JOINT AIR FORCE NASA THERMAL CONTROL WORKING GROUP.   16, 17 AUG. 1967.
DDC
AFML-TR-68-198
AD-841387   529-82   1968

53091 ENGINEERING DATA ON NEWLY DEVELOPED STRUCTURAL MATERIALS.   FINAL REPT. APRIL 1967-APRIL 1968.
DEEL O L   HYLER W S   BATTELLE MEMORIAL INSTITUTE   COLUMBUS OHIO
DDC
AFML-TR-68-211
AD-840065   1-139   1968

53094 ABLATIVE COMPOSITES FOR LIFTING REENTRY THERMAL PROTECTION.   SUMMARY TECH. REPT. JUNE 1967-MAY 1968.
THOMAS H K   RECESSO J V   AVCO MISSILE SPACE AND ELECTRONICS GROUP LOWELL INDUSTRIAL PARK   LOWELL MASSACHUSETTS
DDC
AFML-TR-67-270 PT. III
AD-835223   1-216   1968

53098 INTEGRATED APPROACH TO RESEARCH ON AND DESIGN METHODS FOR CARBON COMPOSITE MATERIALS.   PROGRESS REPT. NO. 5.   MAY 1966-OCT. 1966.
SPENCE G B   SCHMIT L A   ANTHONY F M   UNION CARBIDES CORP CARBON PRODUCTS DIVISION   LAWRENCEBURG TENN
DDC
AD-831493   1-205   1966

53103 THE APPLICATION OF HEAT-INSULATION HONEYCOMB PLASTIC /SOTOPLAST/.
LIFANOV B V
KHOLODILNAYA TEKHNIKA
          3   45-7   1964

53104 THE APPLICATION OF HEAT-INSULATION HONEYCOMB PLASTIC /SOTOPLAST/.   //ENGLISH TRANSLATION OF KHOLODILNAYA TEKHNIKA /3/ 45-47, 1964.//
LIFANOV B V
DDC
FTD-HT-66-248 AD-639341
          1-8   1966

53111 LINEAR THERMAL EXPANSION MEASUREMENTS FOR SELECTED GRAPHITES AND CARBONS.   PHASE REPT. JAN. 1965-SEPT. 1965.
HECHT N L   RESEARCH INSTITUTE UNIV OF DAYTON DAYTON OHIO
DDC
AD-476314   1-19   1965

53113 FEASIBILITY STUDY OF CONTROL MECHANISM AND COOLANT PROPERTIES OF A TERRESTRIAL UNATTENDED REACTOR POWER SYSTEMS.
MAGLADRY R   ZINDLER G F   MARTIN MARIETTA CORP NUCLEAR DIVISION   MIDDLE RIVER MARYLAND
DDC
AFWL-TR-67-114 AD-836951
          1-174   1968

53122 COMPILATION OF SPECTRAL EMITTANCES OF BACKGROUND AND TARGET CONSTITUENTS IN THE 8 TO 14 MICRON RANGE.
CZARNIK J W   LEE T H P   NAVAL WEAPONS CENTER CHINA LAKE CALIFORNIA
DDC
NWC-TP-4624 AD-841158
          1-188   1968

53123 APPLICATIONS OF BIOPHYSICAL HEAT TRANSFER STUDIES IN PROTECTION SYSTEMS.
STOLL A M   CHIANTA M A   U S NAVAL AIR DEVELOPMENT CENTER AEROSPACE MEDICAL RESEARCH CENTER JOHNSVILLE WARMINSTER PA
DDC
NADC-MR-6722 AD-832339
          1-15   1967

53136 MOLECULAR COMPOSITION OF BLACK HYDROCARBON FILMS IN AQUEOUS SOLUTIONS.
COOK G M W   REDWOOD W R   TAYLOR A R   HAYDON D A
KOLLOID-Z Z POLYM
227 1/2   28-37   1968   CA   70   40967

53144 DETERMINATION OF THE THERMAL CONDUCTIVITY OF DAMP MATERIALS BY THE REGULAR-CONDITIONS METHOD.
DIMITRIU-VILCEA E
MATERIALPRUEFUNG
10   9   293-8   1968   CA   70   12201

53166 NONMETALLIC ABSORBING FILMS AND THE ANTIREFLECTION COATING.   I.
KATSUBE S   KATSUBE Y   MITOME K   FURUTA S
OSAKA KOGYO GIJUTSU SHIKENJO KIHO
19   3   167-73   1968   CA   70   50089

53168 THERMAL LENGTH CHANGES OF SOME REFRACTORY CASTABLES.
NBS-RP-2768.
SCHNEIDER S J   MONG L E
J RESEARCH NAT BUR STANDARDS
59   1   1-8   1957

53177 THE VARIATION OF THERMAL CONTACT RESISTANCE WITH INTERFACE PRESSURE AND CRYOGENIC TEMPERATURE.   M.S. THESIS.
DRANEY W G   UTAH STATE UNIVERSITY   LOGAN UTAH UTAH STATE UNIVERSITY
          1-72   1967

53178 A FIELD EVALUATION OF CONVENTIONAL PREDICTION EQUATIONS FOR THERMAL CONDUCTIVITY IN HIGHWAY PAVEMENT MATERIALS.   M.S. THESIS.
HELEY W   WEST VIRGINIA UNIVERSITY   MORGANTOWN WEST VIRGINIA
WEST VIRGINIA UNIVERSITY
          1-179   1968

53100 ABSOLUTE INTENSITY OF THIN-FILM C8H6 /SOLID/ FROM REFLECTANCE AND TRANSMITTANCE NEAR 880 CM/-1/.
GLOVER D E   HOLLENBERG J L
J PHYS CHEM
73   4   889-94   1969

53225 SURFACE TENSION AND DENSITY OF PALLADIUM-IRON, PALLADIUM-CHROMIUM, AND PALLADIUM-SILICON MOLTEN ALLOYS.
UKHOV V F   DUBININ E L   ESIN O A   VATOLIN N A
ZH FIZ KHIM
42   10   2631-4   1968   CA   70   49937

53226 SURFACE TENSIONS AND DENSITIES OF LIQUID PALLADIUM-IRON, PALLADIUM-CHROMIUM, AND PALLADIUM-SILICON ALLOYS.   //ENGLISH TRANSLATION OF ZH. FIZ. KHIM. 42 /10/ 2631-4, 1968.//
UKHOV V F   DUBININ E L   ESIN O A   VATOLIN N A
RUSS J PHYS CHEM
42   10   1391-3   1968

TPRC Number	Bibliographic Citation

53231 THE HIGH TEMPERATURE, HIGH VACUUM VAPORIZATION AND
THERMODYNAMIC PROPERTIES OF URANIUM DIOXIDE.
ACKERMANN R J   ARGONNE NATIONAL LAB   LEMONT
ILLINOIS
AEC
ANL-5482   1-114   1955

53250 INTERFACIAL PHENOMENA AND EXCHANGE KINETICS ON
SOLVENT EXTRACTION WITH A QUATERNARY AMMONIUM SALT.
SCIBONA G   DANESI P R   ORLANDINI F   SCUPPA B
MAGINI M
SOLVENT EXTR CHEM PROC INST CONF GOTEBORG 1966
                              547-51   1967   CA   70   6776

53277 EFFECT OF OXIDATION ON THE RADIATION POWER OF
MOLYBDENUM DISILICIDE COATINGS.
ZHOROV G A   SIVAKOVA E V
TEPLOFIZ VYS TEMP
6   6   1040-3   1968   CA   70   49932

53318 OPTICAL PROPERTIES OF THIN EVAPORATED FILMS OF
SILVER AND GOLD.
PARIKH D O   NAIK Y G
INDIAN J PURE APPL PHYS
7   1   22-6   1969

53332 SURFACE TENSION AT THE /HE-3/-/HE-4/ INTERFACE.
BROUWER W   PATHRIA R K
PHYS REV
179   1   209-11   1969

53354 THERMAL ACCOMMODATION OF THE HELIUM ISOTOPES ON CLEAN
TUNGSTEN SURFACES.
KOUPTSIDIS J   MENZEL D
Z NATURFORSCH
24   A   3   479-80   1969

53355 THERMAL CONDUCTIVITY OF SILICON IN THE BOUNDARY
SCATTERING REGIME.   PH.D. THESIS
HURST W S   PENNSYLVANIA STATE UNIV   UNIVERSITY
PARK PENNSYLVANIA
UNIV MICROFILMS PUBL
69-14529   1-163   1968

53357 EXPERIMENTAL DETERMINATION OF THERMAL ACCOMMODATION
COEFFICIENTS USING A TWO DIRECTIONAL GUARDED
CALORIMETER.   M.S. THESIS.
KLETT D E   UNIVERSITY OF FLORIDA   GAINESVILLE
FLORIDA
UNIVERSITY OF FLORIDA
                              1-72   1968

53374 PROJECT THROMBUS.   SURFACE PROPERTIES OF MATERIALS
FOR PROSTHETIC IMPLANTS.
BAIER R E   CORNELL AERONAUTIC LAB INC   BUFFALO
NEW YORK
DDC
CAL-173 AD-695307
                              1-96   1969   RR   69-23 37

53413 PROPERTIES OF MATERIALS.   I.   APPLICATION OF
ADHESIVES.
ASHWORTH T   RECHOWICZ M
CRYOGENICS
8   6   361-3   1968   CA   70   58675

53436 CHANGE IN THE INTERFACIAL TENSION IN A METAL-SLAG
NON-EQUILIBRIUM SYSTEM.
DERYABIN A A   POPEL S I   SABUROV L N
IZV AKAD NAUK SSSR METAL
                       1   129-35   1969   CA   70   70380

53459 OPTICAL BEHAVIOR OF IRON OXIDE PIGMENTS.
BUTTIGNOL V
J PAINT TECHNOL
40 526   479-93   1968   CA   70   30105

53485 PREPARATION AND CHARACTERISTICS OF PHOTOCONDUCTIVE
SILVER SULPHIDE LAYERS.
MANGALAM M J   RAO K N   RANGARAJAN N
SURYANARAYANA C V
INDIAN J PURE APPL PHYS
7   9   628-30   1969

53488 THERMOPHYSICAL CHARACTERISTICS OF HEAT SENSITIVE
PAINTS.
ABRAMOVICH B G   NOVICHENOK L N
TEPLOFIZ VYS TEMP
6   5   839-43   1968   CA   70   30077

53491 OPTICAL SOLAR REFLECTOR.   A HIGHLY STABLE, LOW
AS/E SPACECRAFT THERMAL CONTROL SURFACE.
MARSHALL K N   BREUCH R A
J SPACECRAFT ROCKETS
5   9   1051-6   1968
DDC
AD-678799   1968

53492 A USEFUL INFRARED SOURCE.
CARLON H R   EDGEWOOD ARSENAL   MARYLAND
DDC
AD-662698   1-15   1967

53496 THE RADIATIVE PROPERTIES OF TITANIUM AND STAINLESS
STEEL WITH DIFFERENT SURFACE CONDITIONS.
BRANDENBERG W M   ASTRONAUTICS DIVISION GENERAL
DYNAMICS CORPORATION   SAN DIEGO CALIF
DDC
GDA-AE62-0289
AD-681747   1-33   1962   RR   69-7 105

53497 EMISSIVITY OF THE ATLAS STAINLESS STEEL SKIN.
SHINKLE F J   ASTRONAUTICS DIVISION GENERAL
DYNAMICS CORPORATION   SAN DIEGO CALIF
DDC
GDA-ERR-AN-037
AD-681760   1-30   1961   RR   69-7 106

53498 THE TOTAL NORMAL ABSORPTIVITIES AND TOTAL NORMAL
EMISSIVITIES OF SHERWIN WILLIAMS FLAT BLACK ACRYLIC
PAINT AT -320, O AND 100 F.
SHINKLE F J   ASTRONAUTICS DIVISION GENERAL
DYNAMICS CORPORATION   SAN DIEGO CALIF
DDC
GDA-ERR-AN-056
AD-681775   1-14   1961   RR   69-7 103

53509 MULTILAYER FILTERS FOR THE REGION 0.8 TO 100 MICRONS.
FINAL REPT. NOV. 1964-MAY 1968.
SMITH S D   SEELEY J S   AIR FORCE CAMBRIDGE
RESEARCH LABORATORIES   BEDFORD MASS
DDC
AFCRL-68-0496 /VOLS 1,2/
AD-879565   1-198   1968

53513 AN INVESTIGATION OF THE REFLECTION OF HIGHLY ALLOYED
GALLIUM ARSENIDE IN A WIDE SPECTRAL RANGE.   //ENGLISH
TRANSLATION OF ZH. PRIKL. SPEKTROSK. 5 /6/ 770-3,
1966.//
KAGAN M B   KOLTUN M M   LANDSMAN A P   FOREIGN
TECHNOLOGY DIVISION WRIGHT-PATTERSON AIR FORCE BASE
OHIO
DDC
FTD-HT-23-1612-67
AD-679792   1-6   1968

53519 APPROXIMATE MEASUREMENTS OF SOME PHYSICAL PROPERTIES
OF CARBON FIBRE REINFORCED POLYMERS, RELEVANT TO
THEIR APPLICATIONS IN TRIBOLOGY.
LANCASTER J K   ROYAL AIRCRAFT ESTABLISHMENT
FARNBOROUGH /ENGLAND/
DDC AND CFSTI
RAE-TR-68-123/UL N69-34403
AD-847814   1-34   1968   CA   72   22214

53543 THERMAL CONDUCTIVITY OF THIN METAL FILMS AND WIRES AT
CRYOGENIC TEMPERATURES.   FROM   PROCEEDINGS OF THE
EIGHTH CONFERENCE ON THERMAL CONDUCTIVITY.
THERMOPHYSICAL PROPERTIES RESEARCH CENTER.   PURDUE
UNIVERSITY.   WEST LAFAYETTE, INDIANA.   OCT. 7-10,
1968.
TIEN C L   ARMALY B F   JAGANNATHAN P S   DEPT
OF MECHANICAL ENGINEERING UNIV OF CALIFORNIA
BERKELEY
PLENUM PRESS NEW YORK
                              13-19   1969

53568 A STATISTICAL THEORY OF INTERFACIAL THERMAL
CONDUCTIVITY.   FROM   PROCEEDINGS OF THE EIGHTH
CONFERENCE ON THERMAL CONDUCTIVITY.   THERMOPHYSICAL
PROPERTIES RESEARCH CENTER.   PURDUE UNIVERSITY.   WEST
LAFAYETTE, INDIANA.   OCT. 7-10, 1968.
SEXL H   SEXL R U   BURKHARD D G   SCHOCKEN K
UNIV OF GEORGIA DEPT OF PHYSICS AND ASTRONOMY   ATHENS
PLENUM PRESS NEW YORK
                              467-76   1969

53569 CORRELATION AND PREDICTION OF THERMAL CONTACT
CONDUCTANCE FOR NOMINALLY FLAT SURFACES.   FROM
PROCEEDINGS OF THE EIGHTH CONFERENCE ON THERMAL
CONDUCTIVITY.   THERMOPHYSICAL PROPERTIES RESEARCH
CENTER.   PURDUE UNIVERSITY.   WEST LAFAYETTE, INDIANA.
OCT. 7-10, 1968.
HSIEH C K   TOULOUKIAN Y S   THERMOPHYSICAL
PROPERTIES RES CTR PURDUE UNIV   WEST LAFAYETTE IND
PLENUM PRESS NEW YORK
                              477-94   1969

53570 FUNDAMENTAL STUDIES OF THE THERMAL CONDUCTANCE OF
METALLIC CONTACTS.   FROM   PROCEEDINGS OF THE EIGHTH
CONFERENCE ON THERMAL CONDUCTIVITY.   THERMOPHYSICAL
PROPERTIES RESEARCH CENTER.   PURDUE UNIVERSITY.
WEST LAFAYETTE, INDIANA.   OCT. 7-10, 1968.
WONG H Y   DEPT OF AERONAUTICS AND FLUIDS
MECHANICS UNIV OF GLASGOW   GLASGOW SCOTLAND
PLENUM PRESS NEW YORK
                              495-511   1969

53571 APPARATUS FOR MEASURING THERMAL AND ELECTRICAL
CONTACT RESISTANCE OF MICROSCOPICALLY SIZED CONTACTS.
FROM   PROCEEDINGS OF THE EIGHTH CONFERENCE ON THERMAL
CONDUCTIVITY.   THERMOPHYSICAL PROPERTIES RESEARCH
CENTER.   PURDUE UNIVERSITY.   WEST LAFAYETTE, INDIANA.
OCT. 7-10, 1968.
GALE E H JR   SYRACUSE UNIVERSITY   SYRACUSE
NEW YORK

TPRC Number	Bibliographic Citation

53572 THE EFFECT OF PHYSICALLY ADSORBED GASES ON THERMAL CONTACT RESISTANCE. FROM PROCEEDINGS OF THE EIGHTH CONFERENCE ON THERMAL CONDUCTIVITY. THERMOPHYSICAL PROPERTIES RESEARCH CENTER. PURDUE UNIVERSITY. WEST LAFAYETTE, INDIANA. OCT. 7-10, 1968.
HEIMBURG R W   TONG K N   DEPT OF MECHANICAL AND AEROSPACE ENGR SYRACUSE UNIV   SYRACUSE NEW YORK
PLENUM PRESS NEW YORK
527-40   1969

53583 THERMAL CONDUCTIVITY OF THIN SOLID FILMS. FROM PROCEEDINGS OF THE EIGHTH CONFERENCE ON THERMAL CONDUCTIVITY. THERMOPHYSICAL PROPERTIES RESEARCH CENTER. PURDUE UNIVERSITY. WEST LAFAYETTE, INDIANA. OCT. 7-10, 1968.
MATHAD G S   ING P W   IBM COMPONENTS DEVELOPMENT LAB   EAST FISHKILL NEW YORK
PLENUM PRESS NEW YORK
713-23   1969

53585 AN AUTOMATIC PLATE APPARATUS FOR MEASUREMENTS OF THERMAL CONDUCTIVITY OF INSULATING MATERIALS AT HIGH TEMPERATURES. FROM PROCEEDINGS OF THE EIGHTH CONFERENCE ON THERMAL CONDUCTIVITY. THERMOPHYSICAL PROPERTIES RESEARCH CENTER. PURDUE UNIVERSITY. WEST LAFAYETTE, INDIANA. OCT. 7-10, 1968.
FERRO V   SACCHI A   ISTITUTO FISICA TECNICA POLITECNICO   TORINO ITALY
PLENUM PRESS NEW YORK
737-60   1969

53594 TRANSIENT SYSTEM FOR MEASUREMENT OF THERMAL PROPERTIES OF NUCLEAR FUEL POWDERS OF VARYING DENSITIES. FROM PROCEEDINGS OF THE EIGHTH CONFERENCE ON THERMAL CONDUCTIVITY. THERMOPHYSICAL PROPERTIES RESEARCH CENTER. PURDUE UNIVERSITY. WEST LAFAYETTE, INDIANA. OCT. 7-10, 1968.
SCHILMOELLER N H   WHITE D   METALLURGY DIVISION ARGONNE NATL LAB   ARGONNE ILLINOIS
PLENUM PRESS NEW YORK
857-70   1969

53597 EXPERIMENTS ON THE SEPARATION OF HEAT TRANSFER MECHANISMS IN LOW-DENSITY FIBROUS INSULATION. FROM PROCEEDINGS OF THE EIGHTH CONFERENCE ON THERMAL CONDUCTIVITY. THERMOPHYSICAL PROPERTIES RESEARCH CENTER. PURDUE UNIVERSITY. WEST LAFAYETTE, INDIANA.
PELANNE C M   JOHNS-MANVILLE RESEARCH AND ENGINEERING CENTER   MANVILLE NEW JERSEY
PLENUM PRESS NEW YORK
897-911   1969

53605 THERMAL DIFFUSIVITY OF POLY/TETRAFLUOROETHYLENE/ BETWEEN -140 AND 125 C. FROM PROCEEDINGS OF THE EIGHTH CONFERENCE ON THERMAL CONDUCTIVITY. THERMOPHYSICAL PROPERTIES RESEARCH CENTER. PURDUE UNIVERSITY. WEST LAFAYETTE, INDIANA. OCT. 7-10, 1968.
SHELLEY D L   HUBER S F   ARMSTRONG CORK COMPANY RESEARCH AND DEVELOPMENT CENTER   LANCASTER PENNSYLVANIA
PLENUM PRESS NEW YORK
1067-77   1969

53606 THERMAL DIFFUSIVITY MEASUREMENTS FROM PERIODIC HEAT FLOW DUE TO A SQUARE PULSE INPUT FLUX INTO A MULTI-LAYER INFINITE SLAB. FROM PROCEEDINGS OF THE EIGHTH CONFERENCE ON THERMAL CONDUCTIVITY. THERMOPHYSICAL PROPERTIES RESEARCH CENTER. PURDUE UNIVERSITY. WEST LAFAYETTE, INDIANA. OCT. 7-10, 1968.
HALTEMAN E K   GERRISH R W JR   PITTSBURGH CORNING CORPORATION   PITTSBURGH PENNSYLVANIA
PLENUM PRESS NEW YORK
1087-97   1969

53607 A TRANSIENT METHOD FOR MEASURING THERMAL CONDUCTIVITY OF MULTI-LAYER INSULATIONS. FROM PROCEEDINGS OF THE EIGHTH CONFERENCE ON THERMAL CONDUCTIVITY. THERMOPHYSICAL PROPERTIES RESEARCH CENTER. PURDUE UNIVERSITY. WEST LAFAYETTE, INDIANA. OCT. 7-10, 1968.
GRUNERT W E   NOTARO F   SUCKOW D H   UNION CARBIDE CORP LINDE DIVISION   TONAWANDA N Y
PLENUM PRESS NEW YORK
1101-14   1969

53608 EXPERIMENTAL APPARATUS FOR THERMAL TRANSMITTANCE MEASUREMENT THROUGH LARGE SIZE WALLS. FROM PROCEEDINGS OF THE EIGHTH CONFERENCE ON THERMAL CONDUCTIVITY. THERMOPHYSICAL PROPERTIES RESEARCH CENTER. PURDUE UNIVERSITY. WEST LAFAYETTE, INDIANA. OCT. 7-10, 1968.
CODEGONE C   FERRO V   ISTITUTO FISICA TECNICA POLITECNICO   TORINO ITALY
PLENUM PRESS NEW YORK
1115-23   1969

53609 EXPERIMENTS OF THERMAL OSCILLATIONS ON LARGE WALLS. FROM PROCEEDINGS OF THE EIGHTH CONFERENCE ON THERMAL CONDUCTIVITY. THERMOPHYSICAL PROPERTIES RESEARCH CENTER. PURDUE UNIVERSITY. WEST LAFAYETTE, INDIANA. OCT. 7-10, 1968.
BONDI P   SACCHI A   ISTITUTO FISICA TECNICA POLITECNICO   TORINO ITALY
PLENUM PRESS NEW YORK
1125-39   1969

53611 STUDY OF THERMAL CONDUCTIVITY REQUIREMENTS. VOLUME 1. INTERIM REPORT. HIGH PERFORMANCE INSULATION THERMAL CONDUCTIVITY TEST PROGRAM.
HALE D V   LOCKHEED MISSILES AND SPACE COMP HUNTSVILLE ALABAMA
NASA
LMSC/HREC-D   148611-1 NASA-CR-61279 NASA-CR-102602 N69-25282 N70-24178
1-93   1969   PA   N69-7-13 2428

53612 A STUDY OF THERMAL CONDUCTIVITY REQUIREMENTS. FINAL REPORT.
HALE D V   SIMS W H   LANE J H   LOCKHEED MISSILES AND SPACE COMP   HUNTSVILLE ALABAMA
NASA
LMSC/HREC-A-784841 NASA-CR-61442
1-80   1967

53616 A COMPARATIVE STUDY OF THE TOTAL DIRECTIONAL EMITTANCE OF A DIELECTRIC AND A METAL AS A FUNCTION OF SURFACE ROUGHNESS. M.S. THESIS.
SIEGFRIED R G   UNIVERSITY OF KENTUCKY   LEXINGTON KENTUCKY
UNIVERSITY OF KENTUCKY
1-104   1968

53680 THERMAL CONDUCTIVITY OF THIN-LAYER REFRACTORY COATINGS.
GERARD-HIRNE J   LAZENNEC Y
TRANS INT CERAM CONGR 10TH 1966
413-30   1967   CA   70   81703

53681 EFFECT OF SOME PHYSICAL-CHEMICAL PROPERTIES OF MELTS ON THE OUTPUT OF CONVERTERS DURING THE PROCESSING OF NICKEL-CONTAINING COPPER MATTES.
MECHEV V V   CHEDZHEMOV M M   SHUSTITSKII V D VASILEV M G   ROMANOV V D   GORDEEV A P BELOUSOV V A
TSVET METAL
42   1   24-6   1969   CA   70   80137

53701 SOME RELATIONS IN ETHYLENE OXIDE CONDENSATE SYSTEMS.
HOLLIS G L
CHEM PHYS APPL SURFACE ACTIVE SUBST PROC INT CONGR 4TH 1964
1   183-98   1967   CA   70   79400

53707 PREPARATION AND OPTICAL PROPERTIES OF CU2S FILMS.
RAMOIN M   SORBIER J-P   BRETZNER J-F MARTINUZZI S
COMPT REND
268   B   16   1097-100   1969

53717 THERMAL CONTACT RESISTANCE BETWEEN CONCENTRIC ANNULI. M.S. THESIS.
SHAH P R   BRIGHAM YOUNG UNIVERSITY   PROVO UTAH BRIGHAM YOUNG UNIVERSITY
1-88   1968

53722 BIFACIAL TENSION OF BLACK LIPID MEMBRANES.
TIEN H TI
ADVAN CHEM SER
84   104-14   1968   CA   70   23122

53733 STABILIZATION OF OIL-IN-WATER EMULSIONS BY NONIONIC DETERGENTS. PROPERTIES OF SYNTHETIC DETERGENTS AT ANISOLE- AND CHLOROBENZENE-WATER SURFACES.
ELWORTHY P H   FLORENCE A T
J PHARM PHARMACOL
21   2   72-8   1969   CA   70   60779

53753 EFFECTS OF TEMPERATURE AND COMPOSITION ON THE DENSITY AND SURFACE ENERGY OF IRON AND ALUMINUM MELTS.
AYUSHINA G D   LEVIN E S   GELD P V
ZH FIZ KHIM
42   11   2799-804   1968   CA   70   40212

53763 OPTICAL AND PHOTOEMISSIVE PROPERTIES OF NICKEL IN THE VACUUM-ULTRAVIOLET SPECTRAL REGION.
VEHSE R C   ARAKAWA E T
PHYS REV
180   3   695-700   1969

53828 INFRARED RADIOMETER FOR THE 1969 MARINER MISSION TO MARS.
CHASE S C JR
APPL OPT
8   3   639-43   1969

53829 THE DESIGN AND PREPARATION OF INDUCED TRANSMISSSION FILTERS.
HOLLOWAY R J   LISSBERGER P H
APPL OPT

TPRC Number	Bibliographic Citation						
53852	CONTINUOUS ELLIPSOMETRIC DETERMINATION OF THE OPTICAL CONSTANTS AND THICKNESS OF A SILVER FILM DURING DEPOSITION. YAMAGUCHI T YOSHIDA S KINBARA A JAPAN J APPL PHYS						
	8	5	559-67	1969			
53853	THE PREDICTION OF THE THERMAL CONDUCTIVITY OF TWO AND THREE PHASE SOLID HETEROGENEOUS MIXTURES. CHENG S C VACHON R I INTERN J HEAT MASS TRANSFER						
	12	3	249-64	1969	CA	70	109705
53854	THERMAL CONTACT CONDUCTANCE. COOPER M G MIKIC B B YOVANOVICH M M INTERN J HEAT MASS TRANSFER						
	12	3	279-300	1969			
53871	EFFECT OF TEMPERATURE AND COMPOSITION ON THE DENSITIES AND SURFACE TENSIONS OF IRON-ALUMINIUM ALLOYS. //ENGLISH TRANSLATION OF ZH. FIZ. KHIM. 42 /11/ 2799-804, 1968.// AYUSHINA G D LEVIN E S GELD P V RUSS J PHYS CHEM						
	42	11	1489-92	1968			
53875	CALCULATION OF HEAT CONDUCTION IN POLYMER MATERIALS DURING THERMAL DEGRADATION. SHLENSKII O F INZH-FIZ ZH						
	10	1	101-5	1966			
53876	CALCULATION OF HEAT CONDUCTION IN POLYMER MATERIALS DURING THERMAL DEGRADATION. //ENGLISH TRANSLATION OF INZH.-FIZ.ZH. 10 /1/ 101-5, 1966.// SHLENSKII O F J ENG PHYS						
	10	1	68-71	1966			
53887	DETERMINATION OF THE EFFECTIVE ABSORPTIVITY OF BOILER FURNACE RADIANT HEAT-ABSORBING SURFACES. FILIMONOV S S ADRIANOV V N KHRUSTALEV B A KRYUKOVA M G TEPLOENERGETIKA						
	15	8	39-42	1968			
53888	DETERMINATION OF THE EFFECTIVE ABSORPTIVITY OF BOILER FURNACE RADIANT HEAT-ABSORBING SURFACES. //ENGLISH TRANSLATION OF TEPLOENERGTIKA 15 /8/ 39-42, 1968.// FILIMONOV S S ADRIANOV V N KHRUSTALEV B A KRYUKOVA M G THERMAL ENGINEERING						
	15	8	58-61	1968			
53903	THERMOPHYSICAL CHARACTERISTICS OF HEAT SENSITIVE PAINTS. //ENGLISH TRANSLATION OF TEPLOFIZ. VYS. TEMP. 6 /5/ 839-43, 1968.// ABRAMOVICH B G NOVICHENOK L N HIGH TEMPERATURE						
	6	5	801-4	1968			
53923	EFFECT OF OXIDATION ON THE EMISSIVITY OF MOLYBDENUM DISILICIDE COATINGS. //ENGLISH TRANSLATION OF TEPLOFIZ. VYS. TEMP. 6 /6/ 1040-3, 1968.// ZHOROV G A SIVAKOVA E V HIGH TEMPERATURE						
	6	5	995-8	1968			
53974	OPTICAL AND PHOTOELECTRIC PROPERTIES OF GOLD AND ALUMINUM IN THE EXTREME ULTRAVIOLET. PH. D. THESIS. MORSE A L UNIVERSITY OF SOUTHERN CALIFORNIA LOS ANGELES CALIFORNIA UNIV MICROFILMS PUBL						
	68-12050		1-94	1968			
53982	THERMAL CONTACT RESISTANCE AT SEMICONDUCTOR METAL INTERFACES. VUTZ N CARNEGIE-MELLON UNIV PITTSBURG PA UNIV MICROFILMS PUBL						
	68-17619		1-69	1968	CA	70	62348
53985	PHOTON-PHONON INTERACTION IN THIN FILMS. BALKANSKI M LE TOULLEC R LABORATOIRE DE PHYSIQUE DES PARIS UNIV /FRANCE/ DDC						
	AD-684159		1-21	1969			
53988	REMOTE MEASUREMENT OF SURFACE TEMPERATURE AND ITS APPLICATION TO ENERGY BALANCE AND EVAPORATION STUDIES OF BARE SOIL SURFACES. CONAWAY J VAN BAVEL C H M AGRICULTURAL RESEARCH SERVICE TEMPE ARIZONA DDC						
	AD-650286		1-130	1967			
54014	INFRARED MEASUREMENTS ON CADMIUM SULFIDE THIN FILMS DEPOSITED ON ALUMINUM. PROIX F BALKANSKI M PHYS STATUS SOLIDI						
	32	1	119-26	1969	CA	70	91932
54038	ABSORPTANCE OF FILMS AT THE SURFACE OPPOSITE THE SURFACE OF INCIDENCE. JONES J E J OPT SOC AM						
	59	7	877-9	1969			
54041	ELECTRICAL AND OPTICAL PROPERTIES OF SPUTTERED CDO FILMS. TANAKA K KUNIOKA A SAKAI Y JAPAN J APPL PHYS						
	8	6	681-91	1969			
54044	NA /+1/L2, 3 ABSORPTION SPECTRA OF SODIUM SPECTRA OF SODIUM HALIDES. NAKAI S SAGAWA T J PHYS SOC JAPAN						
	26	6	1427-34	1969			
54049	CLASSICAL THEORY OF SMALL ENERGY ACCOMMODATION COEFFICIENTS. APPLICATION TO HELIUM-3 AND HELIUM-4 ON TUNGSTEN. GOODMAN F O J CHEM PHYS						
	50	9	3855-63	1969	CA	71	16051
54052	OPTICAL ABSORPTION SPECTRA OF CRYSTAL-FIELD TRANSITIONS IN MNO. HUFFMAN D R WILD R L SHINMEI M J CHEM PHYS						
	50	9	4092-4	1969	CA	71	17297
54056	ABSORPTIVITY OF ICE I IN THE RANGE 4000-30 /CM-1/. BERTIE J E LABBE H J WHALLEY E J CHEM PHYS						
	50	10	4501-20	1969			
54060	POSSIBLE BRAZE COMPOSITIONS FOR PYROLYTIC GRAPHITE. TAKAMORI T AKANUMA M AM CERAM SOC BULL						
	48	7	734-6	1969			
54062	THERMAL CONDUCTIVITY OF DISPERSED MATERIAL IN VARIOUS GASES AT ELEVATED TEMPERATURES. //ENGLISH TRANSLATION OF AN ARTICLE ORIGINALLY PUBLISHED IN AKAD. NAUK BYEL. SSR NO. 2, 1968.// NIKITIN V S ZABRODSKIY S S ANTONISHIN N V HEAT TRANSFER-SOVIET RESEARCH						
	1	3	39-42	1969			
54071	DETERMINATION OF OPTIMUM, TRANSIENT EXPERIMENTS FOR THERMAL CONTACT CONDUCTANCE. BECK J V INTERN J HEAT MASS TRANSFER						
	12	5	621-33	1969			
54072	THERMAL DIFFUSIVITY MEASUREMENT BY THE MODULATED ELECTRON BEAM METHOD. THERMAL DIFFUSIVITY OF IRON BETWEEN 280 AND 1100 C. WHEELER M J HIGH TEMPERATURES-HIGH PRESSURES						
	1	1	13-20	1969			
54085	ABSORPTION-SPECTROPHOTOMETRY BY TESTS OF SLIGHT TRANSPARENCY. BRAUNBECK J ANGEW CHEM						
	72	1	31-3	1960			
54099	SPECTROSCOPY AT EXTREME INFRA-RED WAVELENGTHS. I. TECHNIQUE. BLOOR D DEAN T J JONES G O MARTIN D H MAWER P A PERRY C H PROC ROY SOC						
	260	A 1	510-22	1961			
54104	SURFACE PROPERTIES OF FUSED SALTS AND GLASSES. I. SESSILE-DROP METHOD FOR DETERMINING SURFACE TENSION AND DENSITY OF VISCOUS LIQUIDS AT HIGH TEMPERATURES. ELLEFSON B S TAYLOR N W J AM CERAM SOC						
	21	6	193-213	1938			
54118	EFFECT OF FINAL DEOXIDATION ON THE SURFACE PROPERTIES OF STEELS. SHITIKOV V S GEDEREVICH N A IZMAILOV O P LITEINOE PROIZVOD						
		5	25-6	1968			
54120	RADIATIVE PROPERTIES OF TANTALUM, MOLYBDENUM, NIOBIUM, GRAPHITE AND NIOBIUM CARBIDE AT HIGH TEMPERATURES. FROM TEPLOOBMEN, GIDRODIN I TEPLOFIZICH. SVOISTV VESHCHESTVA. /I.T. ALADEV, EDITOR./ KHRUSTALEV B A RAKOV A M NAUKA PRESS MOSCOW						
			198-219	1968	CA	71	107242
54128	ENERGY EXCHANGE BETWEEN HELIUM, NEON AND ARGON ATOMS AND A METALLIC SURFACE. ZACHARJIN G SPIVAK G PHYSIK Z SOWJETUNION						
	10		495-509	1936			

TPRC Number	Bibliographic Citation
54137	RESEARCHES ON HEAT CONDUCTION BY RAREFIED GASES. II. THE THERMAL ACCOMMODATION COEFFICIENT OF HELIUM, NEON, HYDROGEN, AND NITROGEN ON GLASS AT 70-90 K. COMMUN. KAMERLINGH ONNES LAB. UNIV. LEIDEN NO. 245B KEESOM W H SCHMIDT G PROC ACAD SCI AMSTERDAM 39 1048-9 1936
54143	THERMAL RESISTANCE OF VACUUM CONTACTS BETWEEN METALLIC SURFACES OF DISSIMULAR SURFACE ROUGHNESS. KAGANER M G ZHUKOVA R I TRUDY VSES NAUCHN ISSLED INST KISLORODN MASHINOSTR 11 100-15 1967
54149	THE PARTIAL THERMAL ACCOMMODATION OF DIFFERENT GASES BY THE BAND-DRAHT METHOD ON PLATINUM BETWEEN O AND 100 C. SCHAFER K KLINGENBERG M Z ELEKTROCHEMIE 58 828-36 1954
54153	THE VARIATION OF TOTAL NORMAL EMITTANCE AND TOTAL DIRECTIONAL EMITTANCE AS A FUCTION OF DIELECTRIC FILM THICKNESS ON A METAL SUBSTRATE. M.S. THESIS. BRANNON R R JR UNIVERSITY OF MINNESOTA MINNEAPOLIS MINNESOTA UNIVERSITY OF MINNESOTA 1-124 1968
54175	SPACE MATERIALS HANDBOOK. SECOND EDITION-SUPPLEMENT 2. RITTENHOUSE J B SINGLETARY J B AIR FORCE MATERIALS LAB WRIGHT-PATTERSON AIR FORCE BASE OHIO DDC AFML-TR-64-40 SUPPL. 2 1-190 1967
54179	RESEARCH AND DEVELOPMENT IN A THERMAL INSULATION STUDY. QUARTERLY PROGRESS REPT. JUNE-AUG. 1968. NOTARO F GRUNERT W E UNION CARBIDE CORP LINDE DIVISION TONAWANDA N Y AEC AND CFSTI ALO-3632-33 N69-19628 1-100 1968
54189	ANALYSIS OF A STUDY OF SUBMICRON POWDERED MEDIA RELATIVE TO THEIR USE AS THERMAL INSULATORS OF PRESSURE TUBES IN A NUCLEAR REACTOR, SIO2 POWDERS AND POWDER-FIBER MIXTURES. DORRA L EIDGENOESSICHES INSTITUT FUR REAKTORFORSCHUNG WURENGEN /SWITZERLAND/ AEC AND CFSTI EIR-138 N69-19157 1-125 1968 CA 70 102113
54194	VAPOR DEPOSITION IN THE FABRICATION OF FISSION THERMOCOUPLES. FAIRCHILD C I BERTINO J P MC CREARY W J SALGADO P G LOS ALAMOS SCI LAB LOS ALAMOS N MEXICO AEC AND CFSTI LA-DC-9004 CONF-680202-5 1-12 1966 CA 70 25018
54195	THERMAL DIFFUSION OF SX-5 GRAPHITE FROM 800 TO 2800 C. MORRISON B H LOS ALAMOS SCIENTIFIC LAB NEW MEXICO AEC AND CFSTI LA-DC-9969 CONF-681009-9 N69-17143 1-19 1968 CA 70 100393
54201	ELECTRONIC PROPERTIES OF LIQUID WATER IN THE VACUUM ULTRAVIOLET REGION. M.S. THESIS. /SEE ALSO TPRC NO. 54887./ PAINTER L R BIRKHOFF R D ARAKAWA E T OAK RIDGE NATIONAL LAB OAK RIDGE TENNESSEE AEC AND CFSTI ORNL-TM-2261 1-107 1968 CA 70 72447
54211	EFFECTIVE OPTICAL CONSTANTS OF THIN GRANULAR SILVER FILMS. SHKLYAREVSKII I N KORNEEVA T I ZOZULYA K N OPTIKA I SPEKTROSKOPIYA 27 2 332-8 1969
54212	EFFECTIVE OPTICAL CONSTANTS OF THIN GRANULAR SILVER FILMS. //ENGLISH TRANSLATION OF OPTIKA I SPEKTROSK. 27 /2/ 332-8, 1969.// SHKLYAREVSKII I N KORNEEVA T I ZOZULYA K N OPT SPECTROSC 27 2 174-6 1969
54227	REGIONAL HEAT LOSS IN MAN IN A STATE OF THERMAL COMFORT. CLIFFORD J M MINISTRY OF DEFENSE LONDON DDC AD-847692 1-22 1966
54228	ELECTRICAL AND THERMAL TRANSPORT MODELS FOR ANALYSIS OF REINFORCED COMPOSITES. DUGA J J BATTELLE MEMORIAL INSTITUTE COLUMBUS OHIO DDC AD-486667 1-41 1966
54229	ACTION OF THE SURFACE FORCES ON THIN FILMS OF ORGANIC POLAR SUBSTANCES. INFRARED ABSORPTION SPECTRA, IN POLARIZED LIGHT, OF PYRROLE AT 77 K. MELLIER A COMPT REND 270 B 5 331-4 1970
54231	RADIATIVE PROPERTIES OF TANTALUM, MOLYBDENUM, NIOBIUM, GRAPHITE AND NIOBIUM CARBIDE AT HIGH TEMPERATURES. //ENGLISH TRANSLATION OF TEPLOOBMEN, GIDRODINAMIKA I TEPLOFIZICHESKIE SVOISTVA VESHCHESTVA, MOSCOW, 198-219, 1968.// KHRUSTALEV B A RAKOV A M HEAT TRANSFER-SOVIET RESEARCH 1 4 187-206 1969 NASA AND CFSTI N69-23250 1-23 1969 PA N69-7-11 1909
54237	THE SURFACE TENSION AND DENSITY PROFILE IN THE VAN KAMPEN MODEL OF A VAN DER WAALS FLUID IN A TWO-PHASE STATE. MC GUIRE D W HARRY DIAMOND LABORATORY WASHINGTON D C DDC AND CFSTI HDL-TR-1425 N69-34121 AD-687284 1-30 1969 RR 69-14 59
54242	FAILURE MECHANISMS IN SEALED BATTERIES. MC CALLUM J FAUST C L BATTELLE MEMORIAL INSTITUTE COLUMBUS OHIO DDC AFAPL-TR-67-48/PT. V/ BATT-7770-V AD-851058 1-67 1969
54248	VISCOELASTIC BEHAVIOR OF FIBER-REINFORCED COMPOSITE MATERIALS. FINAL REPT. FEB.-DEC. 1967. LOU Y C SCHAPERY R A PURDUE UNIVERSITY LAFAYETTE INDIANA DDC AND CFSTI AFML-TR-68-90 N69-20492 AD-681136 1-50 1968 PA 69-7-9 1538
54256	RESEARCH AND DEVELOPMENT OF VACUUM FOIL-TYPE INSULATION FOR RADIOISOTOPE POWER SYSTEMS. QUARTERLY PROG. REPT. JUNE-AUG. 1968. CARVALHO J DUNLAY J B FRONDUTO J PAQUIN M L POIRIER V L NASA ALO-3634-10 N69-15424 1-64 1968
54260	INFRARED MEASUREMENTS ON CDS THIN FILMS DEPOSITED ON ALUMINUM. PROIX F BALKANSKI M LABORATOIRE DE PHYSIQUE DES SOLIDES UNIVERSITE DE PARIS /FRANCE/ DDC AND CFSTI ARL-69-0026 N69-28787 AD-684160 1-12 1968 PA N69-7-15 2833
54277	SPECIFIC FEATURES IN MEASURING HEAT CONDUCTIVITY COEFFICIENTS OF MULTILAYER INSULATIONS. GOLOVANOV L B AIR FORCE SYSTEMS COMMAND WRIGHT-PATTERSON AIR FORCE BASE OHIO DDC AND CFSTI FTD-HT-23-13-68 N69-17682 AD-679687 1-18 1968 PA N69-7-7 1240
54282	INTERIM SUMMARY OF WORK APPLICABLE TO ECC ON PHYSICO-CHEMICAL STUDIES OF CLAD UO2 UNDER REACTOR ACCIDENT CONDITIONS. WHITE J F GENERAL ELECTRIC MISSILE AND SPACE DIVISION CINCINNATI OHIO CFSTI GEMP-619 N69-15295 1 41 1968
54286	PROPERTIES OF ORGANIC MATERIALS AT LOW TEMPERATURE INCLUDING COMPATIBILITY WITH LIQUID OXYGEN. WATSON J F GENERAL DYNAMICS/ASTRONAUTICS SAN DIEGO CALIFORNIA DDC MRG-80 AD-836129 1-77 1959
54291	MEASUREMENT OF OPTICAL PROPERTIES OF FABRICS. STOLL A M CHIANTA M A NAVAL AIR DEVELOPMENT CENTER JOHNSVILLE WARMINSTER PA DDC NADC-MR-6902 AD-851333 1-15 1969
54292	ULTRAVIOLET-PROTON RADIATION EFFECTS ON SOLAR CONCENTRATOR RELECTIVE SURFACES. FINAL REPT. GILLETTE R B BOEING COMP SEATTLE WASH NASA AND CFSTI NASA-CR-1024 N68-23560

TPRC Number	Bibliographic Citation

54293 LOSS FACTOR AND RESONANT FREQUENCY OF VISCOELASTIC
SHEAR-DAMPED STRUCTURAL COMPOSITES.
DERBY T F   RUZICKA J E   BARRY WRIGHT
CORPORATION   WATERTOWN MASS
NASA
NASA-CR-1269 N69-17766
                                    1-220   1969

54294 DIRECTIONALLY REFLECTIVE COATING STUDY. TECH.
SUMMARY REPT. JUNE 1964-MAY 1965.
COX R L   LING-TEMCO-VOUGHT ASTRONAUTICS DIVISION
DALLAS TEXAS
NASA
NASA-CR-63772 X85-17370
                                    1-151   1965

54296 THE THERMAL INSULATION OF CARIBOU PELTS.
MOOTE I
TEXTILE RESEARCH JOURNAL
25   832-7   1955

54297 VACUUM ULTRAVIOLET ABSORPTION SPECTRA OF POLYAMIDES.
ONARI S
JAPAN J APPL PHYS
9   2   227   1970

54298 AN INVESTIGATION OF SOME THERMAL AND MECHANICAL
PROPERTIES OF A LOW-DENSITY PHENOLIC-NYLON ABLATION
MATERIAL.   FINAL REPORT.
SANDER H G   SMYLY E D   PEARS C D   SOUTHERN
RESEARCH INST   BIRMINGHAM ALABAMA
NASA AND CFSTI
NASA-CR-66731 N69-18664
                          1-459   1969   PA   N69-7-8 1356

54303 IN SITU ELECTRON, PROTON, AND ULTRAVIOLET RADIATION
EFFECTS ON THERMAL CONTROL COATINGS. FINAL REPORT,
SEPT. 1965-JUL. 1968.
FOGDALL L B   CANNADAY S S   BROWN R R
AEROSPACE GROUP BOEING COMP   SEATTLE WASHINGTON
NASA AND CFSTI
NASA-CR-100146
N69-23865   1-149   1968   PA   N69-7-12 2132

54306 QUANTUM THEORY OF A PARTICLE OSCILLATOR COLLISION
APPLIED TO GAS-SURFACE INTERACTIONS AT CRYOGENIC
TEMPERATURES.
SMITH J D   LEWIS RESEARCH CENTER   CLEVELAND OHIO
NASA
NASA-TN-D-4982
N69-14288   1-40   1968   CA   71   64177

54307 CALCULATION OF TEMPERATURE HISTORIES IN ABLATING
MATERIALS.
HARTMANN B   U S NAVAL ORDNANCE LABORATORY   WHITE
OAK   SILVER SPRING MARYLAND
DDC
NOLTR-68-107
AD-687708   1-22   1968

54313 THE RELATIVE REFLECTANCE OF A MONOLAYER-COVERED
WATER SURFACE. PH.D. THESIS.
BEARD J T   OKLAHOMA STATE UNIVERSITY   STILLWATER
OKLAHOMA
CFSTI
PB-180338   1-167   1965   RR   69-3 124

54315 QUARTERLY PROGRESS REPORT OF RESEARCH AND DEVELOPMENT
IN A THERMAL INSULATION STUDY, SEPT.-NOV. 1968.
GRUNERT W E   LINDE DIVISION UNION CARBIDE CORP
INDIANAPOLIS INDIANA
CFSTI
ALO-3632-36 PB-182644
N69-29047   1-71   1968   PA   N69-7-16 2974

54318 ALLEVIATION OF LEADING-EDGE HEATING BY CONDUCTION AND
RADIATION.
CAPEY E C   ROYAL AIRCRAFT ESTABLISHMENT
FARNBOROUGH /ENGLAND/
DDC
RAE-TR-66311
AD-809637   1-37   1966

54327 OPTICAL PROPERTIES OF THIN GERMANIUM FILMS IN THE
INFRARED REGION OF THE SPECTRUM. //ENGLISH
TRANSLATION OF OPTIKA I SPEKTROSKOPIYA 26 /2/ 231-4,
1969.//
GISIN M A   IVANOV V A
OPT SPECTROSC
26   2   124-5   1969

54346 TEMPERATURE DEPENDENCE OF THE THERMAL CONDUCTIVITY OF
LEAD AND IRON PELLETS.
VASILEV A N   POLEKHINA T K
AT ENERG
26   3   296-7   1969   CA   70   118830

54349 POLAR EFFECTS AT SOLID/LIQUID INTERFACES.
LOWE A C   RIDDIFORD A C
CAN J CHEM
48   5   865-6   1970

54365 MEASUREMENT OF THE EMISSIVITY OF INDUSTRIAL SURFACES
BY A SIMPLE TECHNIQUE.
DALLMEYER H
CHEM-ING-TECH
41 5/6   323-8   1969   CA   70   98308

54375 THERMAL CONDUCTIVITY OF PERLITE CONCRETE.
ADAMS L
CRYOG TECHNOL
5   1   10-11   1969   CA   70   90497

54400 CONTACT HEAT TRANSFER IN GRANULAR MATERIAL UNDER
VACUUM. //ENGLISH TRANSLATION OF INZH. FIZ. ZHUR. 11
/1/ 30-36, 1966.//
KAGANER M G
J ENGINEERING PHYS
11   1   19-22   1966

54423 APPARATUS FOR THE STUDY OF HEAT- AND MASS-TRANSFER
PROPERTIES OF INSULATING MATERIALS AT LOW MEAN
TEMPERATURES.
AGRAWAL K N   VERMA V V
INDIAN J TECHNOL
7   1   4-8   1969   CA   70   98212

54469 VERIFICATION OF THE SZYSZKOWSKI LAW FOR ADSORPTION OF
STEROIDS AT THE OIL-WATER INTERFACE.
ROUX R   BARET J F
J CHIM PHYS PHYSICOCHIM BIOL
65   11   2163-5   1968   CA   70   99911

54476 ON THE RESISTIVITY AND SURFACE TENSION OF THE
EUTECTIC ALLOY OF GALLIUM AND INDIUM.
ZRNIC D   SWATIK D S
J LESS-COMMON METALS
18   1   67-8   1969   CA   70   119236

54481 IMPURITY EFFECTS ON HETEROGENEOUS NUCLEATION FROM THE
VAPOR. III. MERCURY ON PYREX GLASS.
KINAWI A A   HUDSON J B
J VAC SCI TECHNOL
6   1   68-73   1969   CA   70   118351

54483 EFFECTIVE THERMAL CONDUCTIVITY OF PRESSURIZED PACKED
BED.
SUGIYAMA S   HASATANI M   YATA A   TSUCHIYA S
KAGAKU KOGAKU
33   2   163-8   1969   CA   70   116535

54502 THERMAL EXPANSION, INTERNAL STRESS, AND STRENGTH OF
GLASS-CRYSTAL COMPOSITES.
TUMMALA R R
UNIV MICROFILMS PUBL
UM69-15410   1-109   1969   CA   72   114566

54503 THERMAL CONDUCTIVITY MEASUREMENTS OF GAPLESS BEHAVIOR
INDUCED BY THE PROXIMITY EFFECT. PH.D. THESIS.
MC CONNELL R D   RUTGERS STATE UNIV   NEW BRUNSWICK
N J
UNIV MICROFILMS PUBL
69-14742   1-125   1969   CA   72   105099

54519 OPTICAL CHARACTERISTICS OF TITANIUM OXIDE FILMS
PREPARED BY THE REACTIVE-EVAPORATION OF TI2O3.
KATSUBE S   KATSUBE Y
OSAKA KOGYO GIJUTSU SHIKENJO KIHO
19   4   203-8   1968   CA   70   101342

54542 ZISMANS CRITICAL SURFACE TENSION. TOOL IN THE
PRINTING INK, PAINT, AND ADHESIVE INDUSTRY.
STROEBECH C
SKAND TIDSKR FAERG LACK
15   1   5-16   1969   CA   70   107530

54556 IMPURITY EFFECTS ON HETEROGENEOUS NUCLEATION FROM
THE VAPOR. II. CADMIUM ON NITROGEN-CONTAMINATED
TUNGSTEN.
HUDSON J B   SANDEJAS J S
SURFACE SCI
15   1   27-36   1969   CA   70   100421

54566 THERMAL CONDUCTIVITY OF METAL SCREEN INSULATION MADE
OF CRUMPLED STEEL AND ALUMINUM FOILS IN AIR.
KORSHAKOV A I   BOGDANOV F F
TEPLOFIZ VYS TEMP
7   1   81-4   1969   CA   70   109711

54568 INTEGRAL HEMISPHERICAL EMISSIVITY OF PYROLYTIC
SILICON CARBIDE.
PETROV V A   CHEKHOVSKOI V YA   DYMOV B K
EMYASHEV A V
TEPLOFIZ VYS TEMP
7   1   179-80   1969   CA   70   100856

54578 THERMOPHYSICAL PROPERTIES OF SOME ELECTRODE CARBON
MATERIALS.
AGROSKIN A A   GONCHAROV E I   LOVETSKII L V
MAKEEV L A   ATMANSKII A I
TSVET METAL
41   12   59-82   1968   CA   70   91403

TPRC Number	Bibliographic Citation

54627 DETERMINATION OF THE OPTICAL CONSTANTS OF THIN FILMS
BY CONSTANT TRANSMISSION CURVES.
KRETSCHMANN E
Z PHYS
221   4   346-56   1969   CA   70   101355

54642 THE REFLECTIVE POWER OF LAMP- AND PLATINUM-BLACK.
ROYDS T
PHIL MAGAZINE
21   167-72   1911

54664 THERMAL CONTACT RESISTANCE.
MIKIC B B   ROHSENOW W M   DEPT OF MECHANICAL
ENGINEERING MASSACHUSETTS INST TECHNOLOGY
NASA
MIT-TR-4542-41 DSR-74542-41 NASA-CR-78319
N66-37518   1-129   1966

54683 KINETICS OF GRAIN BOUNDARY GROOVING IN CHROMIUM,
MOLYBDENUM, AND TUNGSTEN.
ALLEN B C
TRANS MET SOC AIME
245   7   1621-32   1969

54701 A RADIOMETRIC INVESTIGATION OF EMISSIVITIES AND
EMITTANCES OF SELECTED MATERIALS AT CRYOGENIC
TEMPERATURES. PH.D. THESIS.
HAWKS K H   PURDUE UNIVERSITY   WEST LAFAYETTE
INDIANA
PURDUE UNIVERSITY
UM-8901   1-235   1969   CA   74   25752

54703 SPECTRAL AND DIRECTIONAL THERMAL RADIATION
CHARACTERISTICS OF SURFACES FOR HEAT REJECTION BY
RADIATION.
EDWARDS D K   RODDICK R D
PROGRESS ASTRONAUTICS AERONAUTICS
11   427-46   1963

54711 EFFECT OF ILLUMINATING AND VIEWING GEOMETRY ON THE
COLOR COORDINATES OF SAMPLES WITH VARIOUS SURFACE
TEXTURES.
BILLMEYER F W JR   MARCUS R T
APPL OPT
8   4   763-8   1969

54712 SOME ACCESSORIES FOR MAKING COLOR MEASUREMENTS WITH
THE CARY 14 SPECTROPHOTOMETER.
MITCHELL W N   WOLFE R W
APPL OPT
8   4   785-91   1969

54714 INFRARED REFLECTANCE OF PAINTS.
KREWINGHAUS A B
APPL OPT
8   4   807-12   1969

54747 INTEGRAL HEMISPHERICAL EMISSIVITY OF PYROLYTIC
ZIRCONIUM CARBIDE.
PETROV V A   CHEKHOVSKOI V YA   DYMOV B K
KILIN V S
TEPLOFIZ VYS TEMP
7   2   260-4   1969   CA   71   16049

54811 REFRACTORY COMPOSITES EVALUATION PROGRAMS AT
UNIVERSITY OF DAYTON RESEARCH INSTITUTE. FROM
SUMMARY OF THE THIRTEENTH REFRACTORY COMPOSITES
WORKING GROUP MEETING. SEATTLE, WASHINGTON. 1967.
WURST J C   UNIV OF DAYTON RESEARCH INST   DAYTON
OHIO
DDC
AFML-TR-68-84 AD-838781
362-79   1968

54815 SUBMICROSCOPIC SILICON CARBIDE WHISKER TECHNOLOGY.
FROM SUMMARY OF THE THIRTEENTH REFRACTORY COMPOSITES
WORKING GROUP MEETING. SEATTLE, WASHINGTON. 1967.
HARRINGTON R V   CORNING GLASS   CORNING NEW YORK
DDC
AFML-TR-68-84 AD-838781
552-75   1968

54816 THE FABRICATION OF IRIDIUM AND IRIDIUM-ALLOY COATINGS
ON GRAPHITE BY PLASMA-ARC DEPOSITION AND GAS-PRESSURE
BONDING.
WRIGHT T R   BRAECKEL R T   KIZER D E   BATTELLE
MEMORIAL INST   COLUMBUS OHIO
DDC
AFML-TR-68-6 AD-831448
1-144   1968

54817 THE EVALUATION OF AEROSPACE MATERIALS.
GERDEMAN D A   WURST J C   CHERRY J A   BERNER W E
UNIV OF DAYTON RESEARCH INST   DAYTON OHIO
DDC
AFML-TR-68-53 AD-835768
UDRI-TR-68-09
1-94   1968

54818 SNAP-23A, PHASE 1. QUARTERLY PROGRESS REPT.,
SEPT.-NOV. 1968.
WILGUS W S   WESTINGHOUSE ELECTRIC CORP
PITTSBURGH PA
CFSTI
WANL-3800-25 N69-25682
WANL-PR-DD-22
1-150   1968   PAN 69-7-13 2367

54820 PROTON BOMBARDMENT-INDUCED TARGET TEMPERATURES AND
THERMAL ACCOMMODATION COEFFICIENTS OF ROCK POWDERS.
MEASUREMENTS BY INFRARED PYROMETRY.
NASH D B   JET PROPULSION LAB CALIFORNIA INST OF
TECHNOLOGY
NASA AND CFSTI
NASA-CR-100922 N69-25406
JPL-TM-33-413
1-22   1969   CA   71   84613

54848 INFRARED SPECTRA OF THE DICHLORO- AND
DIBROMOPHOSPHINYL RADICALS IN SOLID ARGON.
ANDREWS L   FREDERICK D L
J PHYS CHEM
73   8   2774-8   1969

54883 SPACE MATERIALS HANDBOOK THIRD EDITION.
RITTENHOUSE J B   SINGLETARY J B   LOCKHEED
MISSILES AND SPACE COMPANY
DDC
AFML-TR-68-205
NASA-SP-305 I AD-692333
1-734   1968   RR   69-20 102

54887 OPTICAL MEASUREMENTS OF LIQUID WATER IN THE VACUUM
ULTRAVIOLET. /SEE ALSO TPRC NO. 54201./
PAINTER L R   BIRKHOFF R D   ARAKAWA E T
J CHEM PHYS
51   1   243-51   1969

54893 SPECTRA OF BENZIL.
BERA S C   MUKHERJEE R   CHOWDHURY M
J CHEM PHYS
51   2   754-61   1969

54898 METALLIC MESH BANDPASS FILTERS AND FABRY-PEROT
INTERFEROMETER FOR THE FAR INFRARED.
SAKAI K   FUKUI T   TSUNAWAKI Y   YOSHINAGA H
JAPAN J APPL PHYS
8   8   1046-55   1969

54959 SOLAR ABSORPTANCE AND TOTAL HEMISPHERICAL EMITTANCE
OF SURFACES FOR SOLAR ENERGY COLLECTION.
BUTLER C P   JENKINS R J   PARKER W J
DDC
ASD-TR-61-558
AD-284862   1-80   1962

54974 INTERFACIAL TENSION IN THE NEIGHBORHOOD OF THE
ELECTROCAPILLARY MAXIMUM.
VERDIER E T
COMPT REND
268   C   25   2146-8   1969

54978 MARINER MARS ABSORPTIVITY STANDARD.
THOSTESEN T O   LEWIS D W   JET PROPULSION LAB
PASADENA CALIFORNIA
CFSTI
NASA-CR-93489 N68-18700
JPL-TR-32-734
1-46   1967

54983 TESTING OF HIGH-EMITTANCE COATINGS.
CLEARY R E   AMMANN C   LEWIS RESEARCH CENTER
CLEVELAND OHIO
NASA AND CFSTI
NASA-CR-1413
1-104   1969

54985 INTERACTION OF MOLECULES WITH THE SURFACE OF A SOLID.
EROFEEV A I
ZH PRIKL MEKH I TEKH FIZ
7   2   42-9   1966

54986 INTERACTION OF MOLECULES WITH THE SURFACE OF A SOLID.
//ENGLISH TRANSLATION OF ZH. PRIKL. MEKH. I TEKH.
FIZ. 7 /2/ 42-9, 1966.//
EROFEEV A I
J APPL MECH TECH PHYS
7   2   26-30   1966

54996 SURFACE VISCOSITY AND ELASTICITY. SIGNIFICANT
PARAMETERS IN INDUSTRIAL PROCESSES.
KANNER B   GLASS J E
IND ENG CHEM
61   5   31-41   1969   CA   71   31713

55009 TRANSIENT-FLOW METHOD OF MEASURING THERMAL
CONDUCTIVITY AND DIFFUSIVITY.
BALL E F
PROC INST REFRIG /LONDON/
63   42-8   1967   CA   71   31724

TPRC Number	Bibliographic Citation

55028 DEVELOPMENT OF 400 TO 2200 F FIBROUS-TYPE INSULATIONS FOR RADIOISOTOPE POWER SYSTEMS. QUARTERLY PROG. REPT. OCT.-DEC. 1968.
COLLINS J O   JAUNARAJS K L   REID D R
JOHNS-MANVILLE RESEARCH AND ENGINEERING CENTER
MANVILLE NEW JERSEY
AEC AND CFSTI
ALO-2661-11 N69-24622
         1-31   1969

55037 INFLUENCE OF TRIVALENT RARE-EARTH OXIDE LAYER ON KERR EFFECTS IN EUROPIUM OXIDE FILMS.
AHN K Y
J APPL PHYS
40   8   3193-5   1969

55041 EXPERIMENTS CONCERNING INFRARED DIFFUSE REFLECTANCE STANDARDS IN THE RANGE 0.8 TO 20.0 MICRONS.
AGNEW J T   MC QUISTAN R B
J OPT SOC AMER
43   11   999-1007   1953

55042 OPTICAL PROPERTIES OF EVAPORATED INDIUM ANTIMONIDE FILMS.
POTTER R F   KRETSCHMAR G G
J OPT SOC AMER
51   693-6   1961

55043 THE ABSORPTION SPECTRUM OF SOLID ANTIMONY TRISULPHIDE.
DOYLE W P
J PHYS CHEM SOLIDS
3 1/2   156-7   1957

55048 QUARTERLY PROGRESS REPORT IN A THERMAL INSULATION STUDY. REPT. FOR DEC. 1968-FEB. 1969.
GRUNERT W E   LINDE DIVISION UNION CARBIDE CORP
INDIANAPOLIS INDIANA
CFSTI
PB-183977   ALO-3632-39
         1-94   1969   RR   69-14 91

55057 AN INVESTIGATION INTO THE THERMAL CONDUCTIVITY OF SOFT STEEL AND ALUMINUM FOIL INSULATION IN AIR. //ENGLISH TRANSLATION OF TEPLOFIZ. VYS. TEMP. 7 /1/ 81-4, 1969.//
KORSHAKOV A I   BOGDANOV F F
HIGH TEMPERATURE
7   1   74-7   1969

55069 INTEGRAL HEMISPHERICAL EMISSIVITY OF PYROLYTIC SILICON CARBIDE. //ENGLISH TRANSLATION OF TEPLOFIZ. VYS. TEMP. 7 /1/ 179-80, 1969.//
PETROV V A   CHEKHOVSKOI V YA   DYMOV B K
EMYASHEV A V
HIGH TEMPERATURE
7   1   167-8   1969

55070 EMISSIVITY COATINGS FOR LOW-TEMPERATURE SPACE RADIATORS.
CUNNINGTON G R JR   GRAMMER J R   SMITH F J
LOCKHEED PALO ALTO RESEARCH LAB LOCKHEED MISSILES AND SPACE COMPANY   SUNNYVALE CALIF
NASA AND CFSTI
NASA-CR-1420
         1-156   1969

55080 MEASUREMENTS REPORT. THERMAL PROPERTY MEASUREMENTS OF MANNED SPACECRAFT CENTER SPACESUIT MATERIALS.
TURNBOW F J   LUEDKE E E
NASA AND CFSTI
NASA-CR-99535 N69-19405
         1-9   1969

55081 SURFACE PROPERTIES OF THE INTERFACE OF IRON-CARBON MELTS WITH ALUMINUM MANGANESE SILICATES.
MIKIASHVILI SH M   SAMARIN A M
KONFERENTSIIA PO FIZIKO KHIMICHESKIN OSNOVAM PROZVODSTVA STALI 7TH
         29-32   1962

55084 SOLAR-RADIATION-INDUCED DAMAGE TO OPTICAL PROPERTIES OF ZINC OXIDE-TYPE PIGMENTS. TECH. SUMMARY REPT. JUNE 1964-JUNE 1965.
MC KELLAR L A   BLAKEMORE J S   CUFF K F
GREENBERG S A   KUGLIN C D   MAC MILLAN H F
SKLENSKY A F   SPICER W E   WASHWELL E R
WILLIAMS L R   LOCKHEED MISSILES AND SPACE CO
SUNNYVALE CALIF
NASA AND CFSTI
N66-16083   NASA-CR-69713
         1-153   1965

55085 DEVELOPMENT OF S-13G-TYPE COATINGS AS ENGINEERING MATERIALS. FINAL REPORT SEPT. 1966-AUG. 1968.
ROGERS R O   ZERLAUT G A   IIT RESEARCH INST TECHNOLOGY CENTER   CHICAGO ILL
NASA AND CFSTI
NASA-CR-100771 N69-24390
         1-78   1969   PAN 69-7-12 2132

55086 CASCADED THERMOELECTRIC TEST GENERATOR, PHASE 2. QUARTERLY PROG. REPT. DEC. 1968-FEB. 1969.
ASTRONUCLEAR LAB WESTINGHOUSE
ELECTRIC CORP   PITTSBURGH PA
NASA AND CFSTI
NASA-CR-100775 N69-24505
WANL-PD-005   1-18   1969   PAN 69-7-12 2029

55090 OPTICAL CHARACTERISTICS OF WATER AND ICE IN THE INFRARED AND RAAIO-WAVE REGIONS OF THE SPECTRUM. //ENGLISH TRANSLATION OF OPT. I SPEKTROSKOPIYA 7 /3/ 311-20, 1959.//
KISLOVSKII L D
OPTICS SPECTROSCOPY
7   3   201-5   1959

55091 THE CONSTANT OF THE RADIATION LAW OF STEFAN.
WESTPHAL W H
VERHANDL DEUT PHYSIK GES
       21   987-1012   1912

55094 DEGREE OF BLACKNESS OF A TUBULAR RADIATOR IN THE PRESENCE OF NONISOTHERMY. //ENGLISH TRANSLATION OF TEPLOOBMEN, GIDRODIN. I TEPLOFIZICH. SVOISTVA VESHCHESTVA, MOSCOW 219-30, 1968.//
KHRUSTALEV B A
NASA
N69-19396   1-12   1968
HEAT TRANSFER-SOVIET RESEARCH
1   4   207-17   1969

55120 THERMODYNAMIC PROPERTIES OF UREA PLUS HYDROCARBON ADDUCTS FROM 12 TO 300 K. HEAT CAPACITIES AND ENTROPIES OF THE ADDUCTS OF THE N-PARAFFIN C11H20, C16H32, AND C20H40.
COPE A F G   PARSONAGE N G
J CHEM THERMODYN
1   1   99-110   1969   CA   71   42996

55130 BULK MODULUS SHRINKAGE AND THERMAL EXPANSION OF A TWO-PHASE MATERIAL.
HOBBS D W
NATURE
222   849-51   1969   CA   71   41910

55149 ADSORPTION ACTIVITY OF SOME DIAMINES AT INTERFACIAL SURFACES OF WATER-AIR AND WATER-CHLOROFORM.
NIKONOV V Z   SOKOLOV L B
ZH FIZ KHIM
43   4   1039-41   1969   CA   71   42619

55175 THERMAL CONTROL SURFACE RESEARCH AT THE ROYAL AIRCRAFT ESTABLISHMENT. TECH. REPT.
SMITH A E   ROYAL AIRCRAFT ESTABLISHMENT
FARNBOROUGH ENGLAND
DDC AND CFSTI
RAE-RR-68276 AD-688908
         1-37   1968   RR   69-15 205

55189 LIQUID METALS. PART XII. THE SURFACE TENSION OF LIQUID BISMUTH IN PURE AND IMPURE HYDROGEN, IN OTHER GASES, AND IN VACUO.
ADDISON C C   RAYNOR J B
J CHEM SOC
       8   965-7   1966

55195 OPTICAL CONSTANTS OF GERMANIUM IN THE REGION O-27 EV.
RUSTGI OM P   NODVIK J S   WEISSLER G L
PHYS REV
122   4   1131-4   1961

55198 TEMPERATURE CONTROL TECHNIQUES FOR SOLAR ENERGY CONVERTERS.
BAKER J K   GENERAL ELECTRIC COMP MISSILE AND SPACE VEHICLE REPT
DDC
ASD-TR-61-689
AD-274922   1-189   1962

55203 SUMMARY OF THERMAL CONDUCTIVITY TEST RESULTS FOR BEDS OF COATED FUEL PARTICLES.
STEVENS D W   GULF GENERAL ATOMIC INC   SAN DIEGO CALIF
AEC AND CFSTI
GAMD-8324   1-18   1967   NSA 23-9 1690

55206 LOW TEMPERATURE THERMAL EXPANSION OF PLASTICS.
LAQUER H L   HEAD E L   LOS ALAMOS SCIENTIFIC LABORATORY   LOS ALAMOS N M
AEC
AECU-2161   1-24   1952

55220 THERMAL RESISTANCE DETERMINATION ON DISC-CELL TYPE SILICON RECTIFIERS AND SILICON THYRISTORS. PH.D. THESIS.
GLOECKEL R   TECHNISCHE HOCHSCHULE   HANNOVER
WEST GERMANY
NASA AND CFSTI
N69-29501   1-177   1968   PAN 69-7-16 2926

TPRC Number	Bibliographic Citation

55221 STUDY OF SPACE ENVIRONMENT EFFECTS ON THERMAL CONTROL COATINGS. DEPENDENCE OF THERMAL CONTROL COATING DEGRADATION UPON ELECTRON ENERGY. FINAL REPORT.
FOGDALL L B   CANNADAY S S   AEROSPACE GROUP
BOEING COMP   SEATTLE WASH
NASA AND CFSTI
NASA-CR-103205 D126114-1
N69-30549   1-125   1969

55223 A SIMPLE METHOD OF MEASURING LIQUID INTERFACIAL TENSIONS, ESPECIALLY AT HIGH TEMPERATURES, WITH MEASUREMENTS OF THE SURFACE TENSION OF TELLURIUM.
SMITH C S   SPITZER D P
J PHYS CHEM
66   5   946-7   1962

55224 AN INFRARED REFLECTOMETER.
OLDHAM M S
J OPTICAL SOC AMERICA
41   10   673-5   1951

55228 THERMODYNAMIC DESIGN FUNDAMENTALS OF HIGH-PERFORMANCE INSULATION.
FOLKMAN N R   LEE T G
J SPACECRAFT ROCKETS
5   8   954-9   1968

55245 DETERMINATION OF SURFACE TENSION OF FLUXES AND INTERFACIAL TENSION AT THE BOUNDARY WITH ARMCO IRON. //ENGLISH TRANSLATION OF AVT. SVARKA 21 /3/ 17-19, 1968.//
YABOBASHVILI S B
AUTOMATIC WELDING
21   18-21   1968

55248 ON THERMAL CONDUCTIVITY OF COMPOSITES.
DANDREA G   WATERVLIET ARSENAL   WATERVLIET NEW YORK
DDC
WVT-691S AD-691330
     1-175   1969   CA   72   15112

55263 EFFECT OF RADIATION ON THE PROPERTIES OF CONCRETE.
DUBROVSKII V B   KORENEVSKII V V
PERGAMENSHCHIK B K
BETON ZHELEZOBETON
15   5   33-4   1969   CA   71   53194

55288 THERMAL CONTACT RESISTANCE OF THICK OXIDE LAYERS ON STEEL.
KHAN E U   DIX R C   KALPAKJIAN S
J IRON STEEL INST LONDON
207   4   457-60   1969   CA   71   52630

55303 RADIATION CHEMICAL PROCESSING. FROM ANNUAL REPORT, BROOKHAVEN NATIONAL LAB., NUCLEAR ENGINEERING DEPARTMENT.
STEINBERG M   KUKACKA L   COLOMBO P   FONTANA J
VARELA J   KANICE K
AEC AND CFSTI
BNL-50149   45-50   1968

55318 SELECTIVE FILMS ON GLASS, AND THEIR CHARACTERISTICS.
REKANT N B   SHEKLEIN A V
PREOBRAZOVATELI SOLN ENERG POLUPROV
     190-9   1968   CA   71   55230

55320 OPTICAL STUDY OF THIN LAYERS.
LOSTIS P
PUBL SCI TECH MIN AIR FR NOTES TECH
162   1-57   1968   CA   71   54987

55330 DEGREE OF BLACKNESS OF A TUBULAR RADIATOR IN THE PRESENCE OF NONISOTHERMY.
KHRUSTALEV B A
TEPLOOBMEN GIDRODIN I TEPLOFIZICH SVOISTVA VESTCHESTVA MOSCOW
     219-30   1968   CA   71   51654

55331 METHOD FOR THE SIMULTANEOUS MEASUREMENT OF THERMAL EXPANSION, SETTING TEMPERATURE, AND YOUNGS MODULUS OF A VITREOUS ENAMEL IN THE FORM OF A THIN LAYER.
STIRLING J F   HARDY L E
TRANS BRIT CERAM SOC
68   3   119-24   1969   CA   71   53100

55348 DETERMINATION OF MICRO REFLECTION AND OPTICAL CONSTANTS OF METALS AND REFRACTORY COMPOUNDS.
KNOSP H
Z METALLK
60   6   526-31   1969   CA   71   55219

55356 THERMAL CONDUCTIVITY OF HETEROGENEOUS MATERIALS.
GORRING R L   CHURCHILL S W
CHEM ENGR PROGRESS
57   7   53-9   1961

55357 FORMATION AND STUDY OF THE PROPERTIES OF HIGH ENERGY MOLECULAR JETS OBTAINED FROM TRIATOMIC IONS OF HYDROGEN, DEUTERIUM AND NITROGEN.
DEVIENNE F M
ENTROPIE
24   35-50   1968

55464 HIGH-TEMPERATURE REDUCTION OF SILICON DIOXIDE TO THE MONOXIDE.
SAKHAROV B A   MARIN K G   LYUBIMOV V K
FEDOROVA D L   KULAGIN I D   SOROKIN L M
IZV AKAD NAUK SSSR NEORG MATER
4   11   2035-6   1968

55465 HIGH-TEMPERATURE REDUCTION OF SILICON DIOXIDE TO THE MONOXIDE. //ENGLISH TRANSLATION OF IZV. AKAD. NAUK SSSR, NEORG. MATER. 4 /11/ 2035-6, 1968.//
SAKHAROV B A   MARIN K G   LYUBIMOV V K
FEDOROVA D L   KULAGIN I D   SOROKIN L M
INORGANIC MATERIALS
4   11   1769-70   1968

55471 PHOTOGRAPHIC METHOD FOR STUDYING THE SPECTRAL EMISSIVITY AND THE SPATIAL DISTRIBUTION OF THE INTENSITIES OF THE SURFACE RADIATION OF BODIES AT HIGH TEMPERATURES.
KHRUSTALEV B A   RAKOV A M
TEPLOOBMEN GIDRODIN TEPLOFIZ SVOISTVA VESHCHESTV
     191-8   1968

55489 WETTING OF ALUMINUM OXIDE BY MOLTEN ALUMINUM AND OTHER METALS.
CHAMPION JOHN A   KEENE B J   SILLWOOD J M
J MATER SCI
4   1   39-49   1969   JA   52-10 363

55497 THERMAL CONDUCTIVITY OF ADHESIVES AT LOW TEMPERATURES.
DENNER H
CRYOGENICS
9   4   282-3   1969   CA   71   71325

55500 REFLECTANCE OF BALL COALS.
CHANDRA D   CHATTERJEE K K   GUPTA M L
FUEL
48   3   271-6   1969   CA   71   72725

55503 DETERMINING THERMAL CONDUCTIVITY OF POROUS THERMAL INSULATION IMPREGNATED WITH LIQUID HARDENING MEDIA.
VENERAKI I E   TOPOLNITSKII G G   MELNICHENKO V S
ROMANKO K S
IZV VYSSH UCHEB ZAVED ENERG
12   5   78-83   1969   CA   71   74963

55512 SURFACE TENSION OF BERYLLIUM.
MILOV I V   SKOROV D M
MET METALLOVED CHIST METAL
     7   174-7   1968   CA   71   74340

55534 MEASUREMENT OF OPTICAL PROPERTIES OF FABRICS.
STOLL A M   CHIANTA M A
TEXT RES J
39   7   657-62   1969   CA   71   71783

55549 GLARE MEASUREMENTS FROM BRIGHT AND SATIN CHROMIUM CURVED SURFACES.
MELDRUM J F
APPL OPT
8   9   1791-8   1969

55550 A BONDING MATERIAL USEFUL IN THE 2-14 MILLIMICRON SPECTRAL RANGE.
PACKARD R D
APPL OPT
8   9   1901-3   1969

55551 COMPUTER DESIGN AND FABRICATION TECHNIQUES FOR A WIDEBAND DIELECTRIC MIRROR.
BERTHOLD J
APPL OPT
8   9   1919-24   1969

55558 THE OPTICAL ABSORPTION OF LITHIUM IODIDE.
BACHRACH R Z
PHYS LETT
30   A   5   318-19   1969

55561 STUDIES OF THERMAL CONDUCTANCE AT INTERFACE OF FUEL AND CLADDING FROM FUELS AND MATERIALS DEVELOPMENT PROGRAM. QUARTERLY PROGRESS REPT. FOR PERIOD ENDING DECEMBER 31, 1968.
WILLIAMS R K   MC ELROY D L
AEC
ORNL-4390   55-7   1969

55566 SURFACE-CHEMICAL ASPECTS OF THE RUSTING OF STEEL SURFACES IN OIL CONTAINING AN ENTRAINED AQUEOUS PHASE.
HUGHES R I
CORROS SCI
9   7   535-56   1969   CA   71   83812

55575 THE INTEGRATED HEMISPHERICAL EMISSIVITY OF ZIRCONIUM CARBIDE. //ENGLISH TRANSLATION OF TEPLOFIZ. VYS. TEMP. 7 /2/ 260-4, 1969.//
PETROV V A   CHEKHOVSKOI V YA   DYMOV B K
KILIN V S
HIGH TEMPERATURE
7   2   239-42   1969

TPRC Number	Bibliographic Citation

55599 ELECTROCAPILLARY STUDIES OF THE ADSORPTION OF SOME ORGANIC MOLECULES AT THE MERCURY-ELECTROLYTIC SOLUTION INTERFACE.
BROADHEAD D E   HANSEN R S   POTTER G W JR
J COLLOID INTERFACE SCI
31   1   61-72   1969   CA   71   84826

55600 MARS. INTERPRETATION OF SPECTRAL REFLECTIVITY OF LIGHT AND DARK REGIONS.
ADAMS J B   MC CORD T B
J GEOPHYS RES
74   20   4851-6   1969   CA   71   86214

55608 GAS DIFFUSION IN POLYURETHANE FOAM AND ITS EFFECT ON THERMAL CONDUCTIVITY.
MITTASCH H
PLASTE KAUT
16   8   589-92   1969   CA   71   81986

55626 THERMOPHYSICAL CHARACTERISTICS OF THIN-LAYERED POLYMERIC COATINGS.
NOVICHENOK L N
TEPLO-MASSOOBMENA NENYUTONOVSK ZHIDK
                   217-30   1968   CA   71   82694

55627 MEASURING THE THERMOPHYSICAL CHARACTERISTICS OF LIQUID COATINGS.
SMOLSKII B M   SHULMAN Z P   GORISLAVETS V M
GORODKIN R G
TEPLO-MASSOOBMENA NENYUTONOVSK ZHIDK
                   239-46   1968   CA   71   82686

55634 PREPARATION AND PROPERTIES OF BERYLLIUM OXIDE THIN FILMS.
NAGAI H
JAPAN J APPL PHYS
8   10   1221-8   1969

55636 MODIFICATION OF THE DEVONSHIRE FORMULA FOR THE THERMAL ACCOMMODATION COEFFICIENT OF HELIUM ON TUNGSTEN.
ROACH D V   HARRIS R E
J CHEM PHYS
51   8   3404-6   1969

55640 THERMAL CONTACT RESISTANCE OF SELECTED LOW-CONDUCTANCE INTERSTITIAL MATERIALS.
FLETCHER L S   SMUDA P A   GYOROG D A
AIAA J
7   7   1302-9   1969   CA   71   93018

55659 X-RAY DIFFRACTION STUDIES OF A GRAPHITIZED CARBON.
WALKER P L JR   MC KINSTRY H A   WRIGHT C C
INDUSTRIAL ENGINEERING CHEM
45   8   1711-15   1953

55660 SURFACE AND INTERFACIAL TENSIONS OF POLYMER MELTS. I. POLYETHYLENE, POLYISOBUTYLENE, AND POLY/VINYL ACETATE/.
WU S
J COLLOID INTERFACE SCI
31   2   153-61   1969   CA   71   92012

55663 INTERFACIAL TENSION BETWEEN POLYMER LIQUIDS.
ROE R-J
J COLLOID INTERFACE SCI
31   2   228-35   1969   CA   71   95128

55679 INTERFACIAL TENSION IN HYDROCARBON SYSTEMS.
DEAM J R   MADDOX R M
PROC ANNU CONV NATUR GAS PROCESS ASS TECH PAP
            48   41-4   1969   CA   71   95127

55698 EFFECT OF THE GAS PHASE ON SURFACE TENSION /OF WATER/.
MISNIAKIEWICZ W   POKRZYK S
ZESZ NAUK POLITECH SLASK CHEM
            47   103-10   1969   CA   71   95117

55722 TEMPERATURE AND ANGLE DEPENDENCE OF THE TOTAL EMISSIVITY OF POOR HEAT CONDUCTORS IN THE TEMPERATURE RANGE OF -60 TO 250 C. PH.D. THESIS.
LOHRENGEL J   RHEINISCH-WESTFALISIHEN TECHNISCHEN HOCHSCHULE AACHEN
RHEINISCH-WESTFALISIHEN TECHNISCHEN HOCHSCHULE AACHEN
                   1-92   1969

55723 KAPITZA RESISTANCE.
POLLACK G L
REV MOD PHYS
41   1   48-81   1969   CA   71   61154

55735 GROWTH AND PROPERTIES OF HG/1-X/CD/X/TE EPITAXIAL LAYERS.
TUFTE O N   STELZER E L
J APPL PHYS
40   11   4559-68   1969

55741 A COMPARISON OF INFRARED-EMITTANCE MEASUREMENTS AND MEASUREMENT TECHNIQUES.
MILLARD J P   STREED E R
APPL OPT
8   7   1485-92   1969

55747 THE OPTIMUM CONDITIONS FOR THE EVAPORATION OF CADMIUM SULPHIDE FILMS FOR THE GENERATION OF MICROWAVE ULTRASONICS.
KING P J
BRIT J APPL PHYS /J PHYS D/
2   9   1349-52   1969

55801 EFFECT OF COEFFICIENT OF EXPANSION OF GROUND- AND COVER-COAT ENAMELS ON THERMAL-SHOCK AND IMPACT RESISTANCE.
PETERSEN F A   ANDREWS A I
J AM CERAM SOC
29   10   288-95   1946

55805 THE THERMAL CONDUCTIVITIES OF MOLYBDENUM AND TUNGSTEN KNITMESH PELLETS USED IN NUCLEAR FUEL PINS.
HOWL D A
J NUCL MATER
33   2   138-48   1969

55806 SULFUR AS A STANDARD OF REFLECTANCE IN THE INFRARED.
KRONSTEIN M   KRAUSHAAR R J   DEACLE R E
J OPT SOC AM
53   4   458-65   1963

55810 OPTICAL AND ELECTRICAL PROPERTIES OF THIN FILMS OF ALPHA-LEAD AZIDE.
FAIR H D JR   FORSYTH A C
J PHYS CHEM SOLIDS
30   11   2559-70   1969

55812 EXPERIMENTAL INVESTIGATION OF RADIATION AND REFLECTION CHARACTERISTICS OF SOME SOLID BODIES.
MITOR V V   KONOPELKO I N
TEPLOENERGETIKA
16   5   29-32   1969

55813 EXPERIMENTAL INVESTIGATION OF RADIATION AND REFLECTION CHARACTERISTICS OF SOME SOLID BODIES. //ENGLISH TRANSLATION OF TEPLOENERGETIKA 16 /5/ 29-32, 1969.//
MITOR V V   KONOPELKO I N
THERMAL ENGINEERING
16   5   47-52   1969

55814 OPTICAL PROPERTIES OF SINGLE-CRYSTAL FILMS OF CADMIUM DICHROMIUM TETRASULFIDE.
BERGER S B   EKSTROM L
PHYS REV LETT
23   26   1499-503   1969

55821 FUNDAMENTAL ABSORPTION BAND OF AMORPHOUS GERMANIUM.
TAUC J   ABRAHAM A
CZECH J PHYS
19   10   1246-54   1969   CA   71   118087

55822 OPTICAL PROPERTIES OF GEO2 IN THE ULTRAVIOLET REGION.
PAJASOVA L
CZECH J PHYS
19   10   1265-70   1969   CA   71   118048

55826 ELECTRICAL AND THERMAL CONDUCTIVITY OF METAL-CONTAINING POLYIMIDE FILMS.
LIDORENKO N S   GINDIN L G   EGOROV B N
KONDRATENKOV V I   RAVICH I YA   TOROPTSEVA T N
DOKL AKAD NAUK SSSR
187   3   581-4   1969   CA   71   113542

55829 THERMAL CHARACTERIZATION OF LOW-DENSITY CHARRING ABLATORS FOR PLANETARY LANDER VEHICLES.
MAAG C R   LEONE J E   BRAZEL J P
ENERG MATER-ENERGY BENEFIT AEROSP MANKIND SOC AEROSP MATER PROCESS ENG NAT SYMP EXHIB 13TH
13   263-87   1968   CA   71   113788

55847 QUANTUM THEORY OF THERMAL ACCOMMODATION.
DRAUGLIS E
MOL PROCESSES SOLID SURFACES BATTELLE INST MATER SCI COLLOQ 3RD 1968
                   367-85   1969   CA   71   116661

55853 ANOMALIES IN THE THERMAL PROPERTIES OF WATER.
CINI R   LOGLIO G   FICALBI A
NATURE /LONDON/
223   1148-9   1969   CA   71   116824

55859 PHYSICAL PROPERTIES OF MOLTEN REFRACTORY METALS AND OXIDES.
ELYUTIN V P   KOSTIKOV V I   MAURAKH M A
MITIN B S   PENKOV I A
POVERKH YAVLENIYA RASPLAVAKH
                   155-9   1988   CA   71   116775

55869 PROPERTIES OF SOLAR ENERGY REFLECTING FILMS OF TITANIUM DIOXIDE AND NOBLE METALS ON GLASS.
FURUUCHI S
SYMP SURFACE VERRE SES TRAIT MOD C R
                   297-309   1967   CA   71   118269

55879 NONSTATIONARY METHOD OF MEASURING CONTACT THERMAL RESISTANCES AND THERMAL CONDUCTIVITY COEFFICIENTS.
MEEROVICH I G   KERTSELLI I YU
TEPLO MASSOPERENOS

TPRC Number	Bibliographic Citation

55899 THE SPECIFIC HEAT OF GADOLINIUM MOLYBDATE.
FOUSKOVA A
J PHYS SOC JAPAN
27   6   1699   1969

55924 THE SELECTIVE REFLECTANCE OF MAGNESIUM OXIDE.
PRIEST I G   RILEY J O
J OPT SOC AMERICA
20   156-7   1930

55945 A STUDY OF THE THERMAL CONDUCTIVITY OF FROST FORMED ON A PLATE AT SUBFREEZING TEMPERATURES. M.S. THESIS.
TODARO F R   LOUISIANA POLYTECHNIC INST   RUSTON LOUISIANA
LOUISIANA POLYTECHNIC INSTITUTE
1-45   1969

55981 DEVELOPMENT OF THE THERMAL CONDUCTIVITY PROBE.
HOOPER F C   CHANG S C
TRANS AM SOC HEATING VENTILATING ENGR
59   463-72   1953

56006 DETERMINATION OF EXTRATERRESTRIAL SOLAR SPECTRAL IRRADIANCE FROM A RESEARCH AIRCRAFT.
ARVESEN J C   GRIFFIN R N JR   PEARSON B D JR
APPL OPT
8   11   2215-32   1969

56007 SIMULTANEOUS RECORDING OF NEAR-FIELD AND FAR-FIELD PATTERNS OF LASERS.
BIRKY M M
APPL OPT
8   11   2249-53   1969

56008 REFLECTANCE OF SEMITRANSPARENT PLATINUM FILMS ON VARIOUS SUBSTRATES IN THE VACUUM ULTRAVIOLET.
HASS G   RAMSEY J B   HUNTER W R
APPL OPT
8   11   2255-9   1969   CA   72   7630

56009 AN INFRARED COMB FILTER DESIGN.
TOWNSEND C L
APPL OPT
8   11   2341   1969

56024 EFFECTIVE THERMAL CONDUCTIVITY OF STEEL FOIL INSULATION.
NARINSKII D A   SHEININ B I
TEPLOFIZ VYS TEMP
7   3   433-7   1969

56025 EFFECTIVE THERMAL CONDUCTIVITY OF STEEL FOIL INSULATION. //ENGLISH TRANSLATION OF TEPLOFIZ. VYS. TEMP. 7 /3/ 433-7, 1969.//
NARINSKII D A   SHEININ B I
HIGH TEMPERATURE
7   3   394-7   1969

56027 PERFORMANCE OF A MECHANICAL HEAT SWITCH AT LOW TEMPERATURES.
COLWELL J H
REV SCI INSTRUM
40   9   1182-6   1969   CA   71   85280

56033 AMORPHOUS VERSUS CRYSTALLINE GERMANIUM TELLURIDE FILMS. II. OPTICAL PROPERTIES.
BAHL S K   CHOPRA K L
J APPL PHYS
40   12   4940-7   1969

56041 OPTICAL ABSORPTION AND PHOTOCONDUCTIVITY IN THIN FILMS OF UNSTABLE AZIDES. PART 1. CUPROUS AZIDE.
DEB S K
TRANS FARADAY SOC
65   11   3074-80   1969

56046 INFRARED SPECTRA OF CRYSTALLINE CYCLOPENTANE AND CYCLOPENTANE-D SUB 10.
SCHETTINO V   MARZOCCHI M P   CALIFANO S
J CHEM PHYS
51   12   5264-76   1969   CA   72   60871

56051 METAMERISM IN COLORED GLAZES.
SEABRIGHT C A
AM CERAM SOC BULL
49   3   269-71   1970

56053 PREDICTED DEPTH OF FREEZE OR THAW IN SOILS BY CLIMATOLOGICAL ANALYSIS OF CUMULATIVE HEAT FLOW.
SCOTT R F   COLD REGIONS RESEARCH AND ENGINEERING LAB   HANOVER N H
DDC
CRREL-TR-195 AD-696414
1-54   1969

56054 DETERMINATION OF HEMISPHERICAL EMITTANCE BY MEASUREMENTS OF INFRARED BI-HEMISPHERICAL REFLECTANCE. M.S. THESIS.
SHERRELL F G   ARNOLD ENGINEERING DEVELOPMENT CENTER ARNOLD AIR FORCE STATION   TENNESSEE
DDC AND CFSTI
AEDC-TR-69-213 N70-17035
AD-696621   1-82   1969   RR   70-1 225

56056 ON THE INFRA-RED RESPONSE OF SILICON SOLAR CELLS AS A FUNCTION OF THICKNESS.
JENKINS R M   ROYAL AIRCRAFT ESTABLISHMENT FARNBOROUGH /ENGLAND/
DDC
RAE-TR-69126 AD-696850
1-33   1969

56057 CONTRAST-ENHANCEMENT IN IMAGING DEVICES BY SELECTION OF INPUT PHOTOSURFACE SPECTRAL RESPONSE. FROM 4TH SYMPOSIUM PHOTOELECTRIC IMAGING DEVICES. IMPERIAL COLLEGE, LONDON. SEPT. 16-20, 1968.
RICHARDS E A   SIGNALS RESEARCH AND DEVELOPMENT ESTABLISHMENT
DDC
SRDE-69009 AD-696930
1-20   1969

56060 MEASUREMENT OF THE INTERFACIAL TENSION AND THE DIFFERENTIAL CAPACITY OF THE DOUBLE FILM OF MERCURY-SOLUTION OF SODIUM PERCHLORATE ANHYDRIDE IN ACETONITRILE.
CHAMPION P
COMPT REND
269   C   20   1159-62   1969

56078 ELECTRICAL SWITCHING IN AMORPHOUS LAYERS OF PYROACTIVATED QUARTZ CONTAINING IMPURITIES.
KORZO V F
FIZ TVERD TELA
11   7   1758-62   1969

56079 ELECTRICAL SWITCHING IN AMORPHOUS LAYERS OF PYROACTIVATED QUARTZ CONTAINING IMPURITIES. //ENGLISH TRANSLATION OF FIZ. TVERD. TELA 11 /7/ 1758-62, 1969.//
KORZO V F
SOVIET PHYSICS-SOLID STATE
11   7   1425-8   1970

56100 QUANTITATIVE OPTICAL AND ELECTRON-PROBE STUDIES OF THE OPAQUE PHASES.
SIMPSON P R   BOWIE S H U
SCIENCE
167   619-21   1970

56105 THERMAL RADIATION PROPERTIES AND THERMAL CONDUCTIVITY OF LUNAR MATERIAL.
BIRKEBAK R C   CREMERS C J   DAWSON J P
SCIENCE
167   724-6   1970

56122 ENHANCEMENT OF SURFACE PLASMA RESONANCE ABSORPTION IN MIRRORS BY OVERCOATING WITH DIELECTRICS.
STANFORD J L   BENNETT H E
APPL OPT
8   12   2556-7   1969

56125 PHOTOINJECTION INTO SIO2. USE OF OPTICAL INTERFERENCE TO DETERMINE ELECTRON AND HOLE CONTRIBUTIONS.
POWELL R J
J APPL PHYS
40   13   5093-101   1969

56145 METHODS OF DECREASING THERMAL CONDUCTIVITY OF HOLLOW CERAMIC BUILDING UNITS.
BALINT P   BAKOS J
EPITOANYAG
19   12   466-70   1967   JA   53-1 8

56163 LATTICE VIBRATIONS OF KN3, RBN3 AND CSN3.
MALHOTRA M L   MOELLER K D   IQBAL Z
PHYS LETT
31   A   2   73-4   1970

56167 ACCUMULATION OF IONS AT WATER-NITROBENZENE INTERFACES DURING TRANSFERENCE.
BLANK M
CHEM PHYS APPL SURFACE ACTIVE SUBST PROC INT CONGR 4TH 1964
2   233-43   1967   CA   71   129050

56168 THERMAL CONDUCTIVITY OF GENERAL ELECTRIC NO. 7031 VARNISH.
MC TAGGART J H   SLACK G A
CRYOGENICS
9   5   384-5   1969   CA   71   126060

56174 RESULTS OF COMPLEX HYDROGEOLOGICAL STUDIES OF THE PETROLEUM POTENTIALS OF THE KAZAN REGION IN RELATION TO ESTIMATION.
GERASIMOV V G   BULYCHEV M M   YULMETOV SH F
DORONKIN K N   KHALIKOVA G F
GEOL NEFTI GAZA
13   7   46-9   1969   CA   71   127219

56180 SPECIFIC HEAT OF WOOL CONTAINING AN ADDITIVE AND THE HEAT OF FUSION OF THE ABSORBED WATER.
HALY A R
J TEXT INST
60   10   403-10   1969   CA   71   125813

TPRC Number	Bibliographic Citation						

56184 ADSORPTION OF SODIUM LAURYL SULFATE AT THE BENZENE-WATER INTERFACE.
CHATTERJEE A K   CHATTORAJ D K
KOLLOID-Z Z POLYM
233 1/2   966-71   1969   CA   71   129046

56185 KINETICS STUDY OF ADSORPTION AT THE OIL-WATER INTERFACE USING A NEW SUSPENDED DROP TENSIOMETER WHEN DIFFUSION PLAYS A DETERMINING ROLE.
BARET J F
KOLLOID-Z Z POLYM
233 1/2   971-9   1969   CA   71   129042

56189 ELECTROCAPILLARY PHENOMENA AT OIL-WATER INTERFACES. IV. INTERACTION BETWEEN SURFACTANTS AND DYES AT OIL-WATER INTERFACES.
WATANABE A   TAMAI H   MATSUMOTO M   GOTOH R
NIPPON KAGAKU ZASSHI
90   8   738-42   1969   CA   71   129048

56190 OPTICAL EXCITATION OF NONRADIATING SURFACE PLASMONS ON ROUGH SILVER FILMS.
SCHROEDER E
OPT COMMUN
1   1   13-6   1969   CA   71   130446

56194 SPONTANEOUS DISPERSION OF METALS IN MOLTEN ALKALIES, AND INTERFACIAL TENSION AT THE METAL-ALKALI INTERFACE IN THE MELTS.
SHCHERBAKOV V K   KUZNETSOV S I
POVERKH YAVLENIYA RASPLAVAKH
441-5   1968   CA   71   129045

56209 EFFECTS OF FILLERS ON THERMAL CONDUCTIVITY, ELECTRICAL CONDUCTIVITY, AND DIELECTRIC PROPERTIES OF THIN POLYMER COATINGS.
SMEKHOV F M   NOVICHENOK L N   NITSBERG L V
NEPOMNYASHCHII A I
TEPLO MASSOPERENOS
7   560-6   1968   CA   71   126075

56230 CONTACTS FORMED FROM ALLOYS IN A LIQUID-SOLID STATE.
SILIN L L   TEREKOV A YA   TIPIKIN V V
SHUBIN YU B
FIZ KHIM OBRAB MATER
5   122-7   1969   CA   72   5676

56233 SURFACE ACTIVITY OF CRUDE OILS OF THE PETROLEUM INDUSTRY AGENCY SHIRVANNEFT UNDER THE INFLUENCE OF SURFACTANTS.
ASHIMOV M A   MAMEDOVA M A   MURSALOVA M A
RAGIMOV F M   NURIEVA Z D
ISSLED NEFTEI NEFTEPROD RAZRAB PROTSESSOV IKH PERERAB
89-98   1968   CA   72   4862

56239 SPECTRAL REFLECTING POWER OF SOME PAINT AND VARNISH COATINGS IN THE 0.25-15.0 MICRON REGION.
AFANASEVA G D   VINOGRADOVA L M   ILLYASOV S G
FRIDZON M B   TYURIN B F
LAKOKRASOCH MATER IKH PRIMEN
4   38-41   1969   CA   72   4371

56291 METHODS OF NUMERICAL DIFFERENTIATION IN THE ANALYSIS OF THERMODYNAMIC DATA FOR THE MERCURY-SOLUTION INTERFACE.
FAWCETT W R   KENT J E
CAN J CHEM
48   1   47-53   1970

56300 SPECIFIC HEAT MEASUREMENT AT VERY LOW TEMPERATURES.
WAKI S
DENKI SHIKENJO IHO
32   12   1151-6   1968   CA   72   16248

56318 HEAT-INSULATING PROPERTIES OF MODERN GAS-EXPANDED PLASTICS.
DUDNIK D M   STEPANENKO A N
KHOLOD TEKH TEKHNOL
8   101-4   1969   CA   72   13327

56323 LIGHT-COLORED TITANIUM ENAMELS FOR STEEL KITCHENWARE.
VARGIN V V   POPOVA L B
NEORG STEKLOVIDYNYE POKRYTIYA MATER
303-7   1969   CA   72   15282

56337 SPECIFIC HEAT OF POLY/ETHYLENE TEREPHTHALATE/ /PET/ AND THE FUSION OF ITS ABSORBED WATER.
HALY A R   SNAITH J W
TEXT RES J
39   10   906-11   1969   CA   72   13675

56347 RADIATIVE PROPERTIES OF THERMISTOR BEADS. M.S. THESIS.
TOSCANO W M   MASSACHUSETTS INSTITUTE OF TECHNOLOGY   CAMBRIDGE MASSACHUSETTS
MASSACHUSETTS INSTITUTE OF TECHNOLOGY
1-78   1969

56350 THERMAL CONDUCTIVITY OF A NON-EQUILIBRIUM CHEMICALLY REACTING GAS MIXTURE. PH.D. THESIS.
CHANG T-C   MASSACHUSETTS INSTITUTE OF TECHNOLOGY
CAMBRIDGE MASSACHUSETTS
MASSACHUSETTS INSTITUTE OF TECHNOLOGY

56351 HYDROGEN TANKAGE APPLICATION TO MANNED AEROSPACE SYSTEMS. PHASES II AND III. VOLUME I. DESIGN AND ANALYTICAL INVESTIGATIONS.
HEATHMAN J H   GENERAL DYNAMICS/CONVAIR   SAN DIEGO CALIFORNIA
DDC
GDC-DCB68-008-VOL 1 AFFDL-TR-68-75-VOL 1
AD-833232   1-171   1968

56352 THERMAL CONDUCTIVITY OF HELIUM-PERMEATED STYROFOAM.
GUTZMER H A   GENERAL DYNAMICS/ASTRONAUTICS   SAN DIEGO CALIFORNIA
DDC
GDA-7D2339 AD-832274
1-14   1959

56353 RADIATION SHIELD INSULATION WITH APPLICATION TO CRYOGENIC TANKS IN SPACE.
CHRISTENSEN E H   GENERAL DYNAMICS/ASTRONAUTICS SAN DIEGO CALIFORNIA
DDC
GDA-AZJ-55-003-TN
AD-832455   1-18   1959

56354 AN EVALUATION OF THE MECHANICAL PROPERTIES OF ADHESIVES AT CRYOGENIC TEMPERATURES AND THEIR CORRELATION WITH MOLECULAR STRUCTURE.
HERTZ J   GENERAL DYNAMICS/ASTRONAUTICS   SAN DIEGO CALIF
DDC
GDA-ERR-AN-196
AD-831537   1-54   1962

56355 SNOW ALBEDO MODIFICATION. A REVIEW OF LITERATURE.
SLAUGHTER C W   COLD REGIONS RESEARCH AND ENGINEERING LAB   HANOVER N H
DDC
CRREL-TR-217 AD-698023
1-31   1969

56356 THERMAL CONDUCTIVITY OF INSULATING MATERIALS BY THE GUARDED HOT PLATE METHOD.
DI MATTEO G A   KREISLER R I   WOODS J A
GENERAL DYNAMICS/ASTRONAUTICS   SAN DIEGO CALIFORNIA
DDC
GDA-7-D-476 AD-830210
1-30   1958

56357 ANGULAR DISTRIBUTION OF RADIATION REFLECTED FROM CARBON DIOXIDE CRYODEPOSITS FORMED ON 77 K. SURFACES.
THERMAL RADIATION SECTION ARNOLD ENGINEERING DEVELOPMENT CENTER ARNOLD AIR FORCE STATION
DDC
AEDC-TR-68-46
AD-668432   1-63   1968

56360 A THEORY FOR THE ESTIMATION OF SURFACE AND INTERFACIAL ENERGIES. I. DERIVATION AND APPLICATION TO INTERFACIAL TENSION.
GIRIFALCO L A   GOOD R J
J PHYS CHEM
61   904-9   1957

56364 LARGE CIRCULATING DROPS IN VERTICAL TUBES.
O BRIEN V   APPLIED PHYSICS LAB JOHNS HOPKINS UNIV SILVER SPRING MD
DDC
APL-TG-1084 AD-697907
1-85   1969

56366 HANDBOOK OF THERMAL DESIGN DATA FOR MULTILAYER INSULATION SYSTEMS. VOLUME II.
COSTON R M   LOCKHEED MISSILES AND SPACE CO SUNNYVALE CALIF
NASA
LMSC-A847882 VOL 2 NASA-CR-87485
N67-34910   1-180   1967

56374 ANALYTICAL INVESTIGATION OF THERMAL DEGRADATION OF HIGH-PERFORMANCE MULTILAYER INSULATION IN THE VICINITY OF A PENETRATION.
JOHNSON W R   SPRAGUE E L   LEWIS RESEARCH CENTER   CLEVELAND OHIO
NASA
NASA-TN-D-4778
N68-33167   1-60   1968

56375 THERMAL PROTECTION SYSTEM X-15A-2. DESIGN REPORT.
PRICE A B   MARTIN COMPANY   DENVER COLORADO
NASA
ER-14535 NASA-CR-82003
N68-25717   1-287   1968

56378 A DAY-NIGHT HIGH RESOLUTION INFRARED RADIOMETER EMPLOYING TWO-STAGE RADIANT COOLING, VOL. II.
ITT INDUSTRIAL LABORATORIES FORT WAYNE IND
NASA
NASA-CR-77541
N66-35229   1-58   1966

TPRC Number	Bibliographic Citation

56379 DEVELOPMENT OF MANUFACTURING TECHNIQUES FOR
APPLICATION OF HIGH PERFORMANCE CRYOGENIC INSULATION.
FINAL REPT. JUNE 21-OCT. 20, 1967.
LOFGREN C L   GIESEKING D E   MISSILE AND
INFORMATION SYSTEMS DIV BOEING CO   SEATTLE WASH
NASA
NASA-CR-61557
N68-17099   1-58   1967

56385 INSULATION.   FROM STUDY OF INTEGRATED CRYOGENIC
FUELED POWER GENERATING AND ENVIRONMENTAL CONTROL
SYSTEMS.   VOLUME II.   CRYOGENIC TANKAGE
INVESTIGATION.
BEECH AIRCRAFT CORPORATION BOULDER
DIVISION   BOULDER COLORADO
DDC
ASD-TR-61-327 VOL. II
AD-270474   4.1-4.32   1961

56416 STUDY ON ELECTROCONDUCTIVE GLASS AND ITS USES.
PART 2. OPTICAL PROPERTIES.
AKEYOSHI K   KANAI E
ASAHI GARASU KENKYU HOKOKU
8   1-12   1958

56418 THERMAL RADIATIVE PROPERTY MEASUREMENTS USING CYCLIC
RADIATION.
JACK J R   SPISZ E W   LEWIS RESEARCH CENTER
CLEVELAND OHIO
NASA
NASA-TN-D-5651
                        1-34   1970

56441 COMPARATIVE STUDIES ON INTERFACIAL TENSION OF
DIFFERENT OINTMENT BASES WITH WATER-OIL EMULSIFIERS.
NEUWALD F   FETTING K E
CHEM PHYS APPL SURFACE ACTIVE SUBST PROC INT CONGR
4TH 1964
3   351-60   1967   CA   72   24582

56447 COATING NORMALIZES IR EMISSIVITY.
CARTER G W
ELECTRO-TECHNOL /NEW YORK/
79   4   89-90   1967   CA   72   22628

56458 TRANSMISSION OF FAR-INFRARED RADIATION THROUGH THIN
FILMS OF SUPERCONDUCTING AMORPHOUS BISMUTH AND
GALLIUM AND BETA-PHASE GALLIUM.
HARRIS R E   GINSBERG D M
PHYS REV
188   2   737-44   1969

56491 CORRELATION OF MECHANICAL AND THERMAL PROPERTIES OF
THE LUNAR SURFACE.
HALAJIAN J D   REICHMAN J
ICARUS
10   179-96   1969

56499 AUTOMATIC RECORDING REFLECTOMETER FOR MEASURING
DIFFUSE REFLECTANCE IN THE VISIBLE AND INFRARED
REGIONS.
DERKSEN W L   MONAHAN T I   LAWES A J
J OPT SOC AMERICA
47   11   995-9   1957

56502 A DEVICE FOR THE RAPID MEASUREMENT OF TOTAL
EMITTANCE.
NELSON K E   LUEDKE E E   BEVANS J T
J SPACECRAFT ROCKETS
3   5   758-60   1966

56504 SOLAR ABSORPTANCE AND THERMAL EMITTANCE OF
ALUMINUM COATED WITH SURFACE FILMS OF EVAPORATED
ALUMINUM OXIDE.   FROM PROC. CONF. AIAA THERMOPHYSICS
SPECIALIST, MONTEREY, CALIFORNIA, SEPT. 13-15, 1965.
HASS G   RAMSEY J B   TRIOLO J J   ALBRIGHT H T
ACADEMIC PRESS   NEW YORK AND LONDON
AIAA   PAPER NO. 65-656
                        1-12   1965

56505 AN ANALYTICAL AND EXPERIMENTAL STUDY OF THE
RADIATION PROPERTIES OF OXIDE FILMS ON METAL
SUBSTRATES.   FROM PROC. CONF. AIAA THERMOPHYSICS
SPECIALIST, NEW ORLEANS, LOUISIANA, APRIL 17-20,
1967.
HASSAN S A   DHANAK A M   BUELOW F H
AMERICAN INSTITUTE AERONAUTICS AND ASTRONAUTICS
NEW YORK
AIAA   PAPER NO. 67-286
                        1-10   1967

56508 GENERALIZATION OF THEORY FOR ESTIMATION OF
INTERFACIAL ENERGIES.
GOOD R J
IND AND ENG CHEM
62   3   54-78   1970

56515 THE INFLUENCE OF THE DYNAMIC SURFACE EFFECT AND
SURFACE PROPERTIES ON THE INTERPHASE TRANSFER OF
MACROPARTICLES.
SHVINDLERMAN L N   GRIGORYAN V A   ROGOV A I
ZH FIZ KHIM
43   6   1460-   1969

56516 THE INFLUENCE OF THE DYNAMIC SURFACE EFFECT AND
SURFACE PROPERTIES ON THE INTERPHASE TRANSFER OF
MACROPARTICLES.   //ENGLISH TRANSLATION OF ZH. FIZ.
KHIM. 43 /6/ 1460-, 1960.//
SHVINDLERMAN L N   GRIGORYAN V A   ROGOV A I
RUSS J PHYS CHEM
43   6   812-16   1969

56525 OPTICAL AND PHOTOEMISSIVE PROPERTIES OF EVAPORATED
FILMS OF PALADIUM, NICKEL, AND COPPER IN THE VACUUM
ULTRAVIOLET SPECTRAL REGION.   PH.D. THESIS.
VEHSE R C   UNIVERSITY OF TENNESSEE   KNOXVILLE
TENNESSEE
UNIV MICROFILMS PUBL
69-1269   1-236   1968

56530 SIMULTANEOUS HEAT AND MASS TRANSFER IN POROUS MEDIA.
PH.D. THESIS.
RIBACK W J   RENSSELAER POLYTECHNIC INSTITUTE
TROY NEW YORK
UNIV MICROFILMS PUBL
69-2477   1-94   1968   PA   N69-7-18 3493

56535 CORRELATION AND PREDICTION OF THERMAL CONTACT
CONDUCTANCE FOR NOMINALLY FLAT SURFACES.
PH.D. THESIS.
HSIEH C K   PURDUE UNIVERSITY   LAFAYETTE INDIANA
UNIV MICROFILMS PUBL
69-2931   1-179   1968   PA   N69-7-24 4678

56536 A STUDY OF RADIATIVE CHARACTERISTICS OF CONDENSED GAS
DEPOSITS ON COLD SURFACES.   PH.D. THESIS.
MERRIAM R L   PURDUE UNIVERSITY   LAFAYETTE INDIANA
UNIV MICROFILMS PUBL
69-2957   1-171   1968   PA   N69-7-24 4625

56537 THERMAL CONDUCTION THROUGH AN EVACUATED IDEALIZED
POWDER OVER THE TEMPERATURE RANGE OF 100 TO 500 K.
PH.D. THESIS.
MERILL R B   BRIGHAM YOUNG UNIVERSITY   PROVO UTAH
UNIV MICROFILMS PUBL
69-3006   1-99   1968

56538 OPTICAL REFLECTANCE STUDIES OF SILVER FOILS IN THE
WAVELENGTH RANGE 3000 TO 6000 ANGSTROM.   PH.D.
THESIS.
JASPERSON S N   PRINCETON UNIVERSITY   PRINCETON
NEW JERSEY
UNIV MICROFILMS PUBL
69-3297   1-111   1968

56550 ELLIPSOMETRIC STUDY OF AMORPHOUS SELENIUM ON
VACUUM-EVAPORATED GOLD.   PH.D. THESIS.
WEITZENKAMP L A   UNIVERSITY OF NEBRASKA   LINCOLN
NEBRASKA
UNIV MICROFILMS PUBL
68-18039   1-225   1968   PA   N69-7-14 2541

56569 DETERMINATION OF THERMAL CONDUCTIVITY OF EXTREMELY
THIN LAYERS OF VARIOUS MATERIALS.   //ENGLISH
TRANSLATION OF TEPLO MASSOPERENOS 7,
VOLKENSHTEIN V S   MEDVEDEV N N
DDC AND CFSTI
FTD-HT-23-820-68 PT. 2
AD-698518   275-9   1969

56570 THERMOMETRIC DETERMINATION OF THERMOPHYSICAL
CHARACTERISTICS.   //ENGLISH TRANSLATION OF TEPLO
MASSOPERENOS 7 1968.//
GERASHCHENKO O A   GRISHCHENKO T G   PILIPENKO A M
FEDOROV V G
DDC AND CFSTI
FTD-HT-23-820-68-PT2
AD-698518   280-94   1969

56577 NONSTEADY-STATE METHOD OF MEASURING CONTACT THERMAL
RESISTANCES AND THERMAL CONDUCTIVITY COEFFICIENTS.
//ENGLISH TRANSLATION OF TEPLO MASSOPERENOS 7,
354-60, 1968.//
MEEROVICH I G   KERTSELLI I YU
DDC AND CFSTI
FTD HT 23 820 68 PT. 2
AD-698518   376-83   1969

56578 METHODS OF MEASURING THE COEFFICIENT OF THERMAL
CONDUCTIVITY OF MATERIALS WITH LOW THERMAL
CONDUCTIVITY.   //ENGLISH TRANSLATION OF TEPLO
MASSOPERENOS 7,
PETROV I N
DDC AND CFSTI
FTD-HT-23-820-68 PT. 2
AD-698518   384-91   1969

56579 METHODS OF QUASI-STEADY-STATE REGULAR AND
STEADY-STATE REGIMES IN DETERMINING HEAT TRANSFER
COEFFICIENTS AT ELEVATED TEMPERATURES.   //ENGLISH
TRANSLATION OF TEPLO MASSOPERENOS 7,
PETROV-DENISOV V G   ZASEDATELEV I B
MASLENNIKOV L A
DDC AND CFSTI
FTD-HT-23-820-68 PT. 2
AD-698518   392-9   1969

TPRC Number	Bibliographic Citation

56585 THERMOPHYSICAL CHARACTERISTICS OF HEAT-SENSITIVE
PAINTS. //ENGLISH TRANSLATION OF TEPLO MASSOPERENOS
7,
ABRAMOVICH B G   NOVICHENOK L N
DDC AND CFSTI
FTD-HT-23-820-68 PT. 3
AD-698519   463-9   1969

56600 EFFECT OF THE COMPOSITION OF CLOTH AND THE FORM OF
ABSORBED MOISTURE ON ITS THERMOPHYSICAL
CHARACTERISTICS. //ENGLISH TRANSLATION OF TEPLO
MASSOPERENOS 7,   , 1968.//
SALIVON N I   KAZANSKII M F
DDC AND CFSTI
FTD-HT-23-820-68 PT. 3
AD-698519   598-602   1969

56601 EFFECTS OF FILLING ON THERMAL CONDUCTIVITY,
ELECTROCONDUCTIVITY AND DIELECTRIC CONSTANT OF
THIN-LAYER POLYMER COATINGS. //ENGLISH TRANSLATION
OF TEPLO MASSOPERENOS 7, 560-6, 1968.//
SMEKHOV F M   NOVICHENOK L N   NITSBERG L V
NEPOMNYASHCHII A I
DDC AND CFSTI
FTD-HT-23-820-68 PT. 3
AD-698519   603-10   1969

56611 EFFECT OF AGING ON TWO-DIMENSIONAL PRESSURE AND
VISCOSITY OF A MONOLAYER OF OCTADECYL ALCOHOL AT A
WATER/AIR INTERFACE.
OGAREV V A   IVANOVA T N
ZH FIZ KHIM
43   7   1815-21   1969   CA   71   84827

56612 EFFECT OF AGING ON THE TWO-DIMENSIONAL PRESSURE AND
VISCOSITY OF AN OCTADECYL ALCOHOL MONOLAYER AT A
WATER-AIR INTERFACE. //ENGLISH TRANSLATION OF ZH.
FIZ. KHIM. 43 /7/ 1815-21, 1969.//
OGAREV V A   IVANOVA T N
RUSS J PHYS CHEM
43   7   1015-19   1969

56621 REFLECTION OF MATT PAPERS.
KLINGELHOEFFER H
ALLG PAP-RUNDSCH
40   1393-4   1969   CA   72   33394

56626 NATURE OF INTERFACIAL TENSION OF THE INTERFACE
BETWEEN TWO LIQUID PHASES.
NEUMANN H J
CHEM PHYS APPL SURFACE ACTIVE SUBST PROC INT CONGR
4TH 1964
2   65-73   1967   CA   72   36076

56636 HEAT CAPACITY OF BASIC MATERIALS OF PAPER POWER
CONDENSERS.
CHEKHOVSKII I R
IZV VYSSH UCHEB ZAVED ENERG
12   5   112-16   1969   CA   72   33454

56642 THERMAL AND COMPOSITION EXPANSION OF CLATHRATES IN
THE ETHYLENE OXIDE-WATER SYSTEM. FROM MOLECULAR
DYNAMICS AND STRUCTURE OF SOLIDS. 2ND MATERIALS
RESEARCH SYMPOSIUM. GAITHERSBURG, MD. OCT. 16-19,
1967.
MC INTYRE J A   PETERSEN D R
USGPO
NBS-SP-301   407-10   1969   CA   72   10483

56670 EFFECT OF SALTS ON THE SURFACE/INTERFACIAL TENSION
AND CRITICAL MICELLE CONCENTRATION OF SURFACTANTS.
WAN L S C   POON P K C
J PHARM SCI
58   12   1562-7   1969   CA   72   36075

56695 THE KAPITZA CONDUCTANCE OF LEAD. FROM PROCEEDINGS
OF THE ELEVENTH INTERNATIONAL CONFERENCE ON LOW
TEMPERATURE PHYSICS. VOLUME 1. ST. ANDREWS,
SCOTLAND. AUG. 21-28, 1968.
CHEEKE J D N
ST ANDREWS UNIVERSITY PRESS SCOTLAND
              567-70   1968   CA   73   92256

56696 THE MODIFICATION OF THE KAPITZA RESISTANCE DUE TO A
SURFACE FILM. FROM PROCEEDINGS OF THE ELEVENTH
INTERNATIONAL CONFERENCE ON LOW TEMPERATURE PHYSICS.
VOLUME 1. ST. ANDREWS, SCOTLAND. AUG. 21-28, 1968.
WHELAN M F   OSBORNE D V
ST ANDREWS UNIVERSITY PRESS SCOTLAND
              575-8   1968   CA   73   92270

56697 THE KAPITZA EFFECT IN THE LIQUID AND VAPOR PHASES OF
HE-3 AND HE-4. FROM PROCEEDINGS OF THE ELEVENTH
INTERNATIONAL CONFERENCE ON LOW TEMPERATURE PHYSICS.
VOLUME 1. ST. ANDREWS, SCOTLAND. AUG. 21-28, 1968.
MATE C F   SAWYER S P
ST ANDREWS UNIVERSITY PRESS SCOTLAND
              579-84   1968   CA   73   92269

56708 THERMAL CONDUCTIVITY MEASUREMENTS OF GAPLESS BEHAVIOR
INDUCED BY THE PROXIMITY EFFECT. FROM PROCEEDINGS
OF THE ELEVENTH INTERNATIONAL CONFERENCE ON LOW
TEMPERATURE PHYSICS. VOLUME 2. ST ANDREWS,
SCOTLAND. AUG. 21-28, 1968.
DEUTSCHER G   LINDENFELD P   MC CONNELL R D
ST ANDREWS UNIVERSITY PRESS SCOTLAND
2   993-6   1968   CA   72   137578

56733 ATTEMPTS AT PREPARATION AND PROPERTIES OF INDIUM
PHOSPHIDE FILMS PREPARED BY CATHODIC PULVERIZATION.
SORBIER J-P   GAAL S   MARTINUZZI S
COMPT REND
270   4   285-7   1969

56746 CONDITIONS FOR THE FORMATION OF STABLE EMULSIONS IN
THE PARAFFIN-HYDROXYQUINOLINE SYSTEMS NEAR THE
CRITICAL REGION OF MIXING.
SHCHUKIN E D   FEDOSEEVA N P   KOCHANOVA L A
REBINDER P A
DOKL AKAD NAUK SSSR
189   1   123-6   1969   CA   72   47761

56770 SPECTRAL REFLECTANCE OF CARBON DIOXIDE-WATER FROSTS.
KIEFFER H
J GEOPHYS RES
75   3   501-9   1970   CA   72   49335

56793 EFFECTIVE THERMAL CONDUCTIVITY OF INSULATING PAPER.
TERADA T   ITO N   GOTO Y
KAGAKU KOGAKU
33   8   807-9   1969   CA   72   25863

56794 DEFORMATIONS DURING HARDENING AND THE COEFFICIENT OF
LINEAR EXPANSION OF PHOSPHATE BINDERS.
SYCHEV M M
KHIM OSN TEKHNOL PRIMEN FOSFATNYKH SVYAZOK POKRYTII
              106-15   1968   CA   72   35357

56833 REFLECTANCE OF EVAPORATED ALUMINUM FILMS IN THE
1050-1600 A. REGION, AND THE INFLUENCE OF THE
SURFACE PLASMON.
FEUERBACHER B   STEINMANN W
OPT COMMUN
1   2   81-5   1969   CA   72   49342

56847 THERMAL CONDUCTIVITY OF BORAZONE POWDERS.
LEZHENIN F F   OSITINSKAYA T D   VISHNEVSKII A S
POROSH MET
9   11   57-60   1969   CA   72   36591

56848 INFLUENCE OF LITHIUM FLUORIDE AND CERIUM DIOXIDE
FILMS ON THE OPTICAL PROPERTIES OF INDIUM AND
ALUMINUM LAYERS.
DOBIERZEWSKA-MOZRZYMASOWA E
PROC COLLOQ THIN FILMS 2ND 1967
              233-40   1968   CA   72   49306

56849 OPTICAL PROPERTIES OF SILVER-CHROMIUM AND
CHROMIUM-SILVER DOUBLE LAYERS IN 0.4-1-MU WAVELENGTH
RANGE.
IDCZAK E
PROC COLLOQ THIN FILMS 2ND 1967
              241-8   1968   CA   72   49287

56852 TEMPERATURE DEPENDENCE OF THE BASIC PARAMETERS OF THE
INTERFACE ENERGY.
NEUMANN A W
CHEM PHYS APPL SURFACE ACTIVE SUBST PROC INT CONGR
4TH 1964
2   335-41   1967   CA   72   47758

56858 EFFECT OF OIL COMPOSITION, ACTIVE COMPONENTS AND
PRESSURE ON THE WETTING OF A QUARTZ.
PASHAEV N G   TAIROV N D
RAZRAB NEFT GAZOV MESTOROZHD AZERB
              118-29   1969   CA   72   45699

56864 ELECTRICAL RELIABILITY OF PARYLENE FILMS FOR DEVICE
PASSIVATION.
LEE S M   LICARI J J   LITANT I
MET TRANS
1   3   701-11   1970

56865 INFRARED ABSORPTION CHARACTERISTICS OF SOLUBLE AND
INSOLUBLE PHOSPHOROUS-BEARING OXIDE LAYERS.
CORL E A   REESE W E
MET TRANS
1   3   747-8   1970

56874 INCANDESCENCE FROM THIN SHEETS OF SILICON.
GOLDSMID H J   MONK R W   MOYS B A
HIGH TEMPERATURES-HIGH PRESSURES
1   4   429-36   1969

56965 OPTICAL ABSORPTION AND PHOTOCONDUCTIVITY IN THIN
FILMS OF UNSTABLE AZIDES. PART 3. CADMIUM AZIDE AND
MERCURIC AZIDE.
DEB S K
TRANS FARADAY SOC
66 571   1802-8   1970

TPRC Number	Bibliographic Citation					
56969	INFLUENCE OF MAGNESIUM IONS ON THE ADSORPTION OF SODIUM LAURYL SULFATE, SUCROSE MONOLAURATE, AND LAURYL HEXAETHYLENE OXIDE MONOETHER AT THE OIL/WATER INTERFACE. MILLER G CHEM PHYS APPL SURFACE ACTIVE SUBST PROC INT CONGR 4TH 1964					
	2		827-39	1967	CA	72	59518
56971	LIQUID-LIQUID EXTRACTION. V. STUDY OF INTERFACIAL CONCENTRATION BY MEASURING INTERFACIAL TENSION. KUGA M KUMAKAWA Y HOKKAIDO DAIGAKU KOGAKUBU KENKYU HOKOKU					
	47		75-120	1968	CA	72	59458
56988	SOLUBLE SURFACE FILMS OF SHORT-CHAIN MONOCARBOXYLIC ACIDS ON ORGANIC AND AQUEOUS SUBSTRATES. WRIGHT E H M AKHTAR B A J CHEM SOC					
	B	1	151-7	1970	CA	72	59451
56990	INFLUENCE OF ADSORBED LAYERS ON THE DETERMINATION OF INTERFACIAL TENSION, USING DIFFERENT METHODS. SONNTAG H STRENGE K J COLLOID INTERFACE SCI					
	32	1	159-82	1970	CA	72	59459
57005	EFFECT OF A TITANIUM SUBSTRATE ON THE PROPERTIES OF ALUMINUM LAYERS INTENDED FOR THE PREPARATION OF DIFFRACTION GRATINGS. EGOROV V M KOZLOV V N SKORODUMOV V N FEDOTOV A I OPT-MEKH PROM					
	36	9	9-13	1969	CA	72	58318
57009	THERMAL CONDUCTIVITY AND TEMPERATURE DISTRIBUTION IN THE MICROSTRUCTURE OF COMPOSITE MATERIALS. VAN P P OGILKO T F PRIKL MEKH					
	5	10	8-14	1969	CA	72	58597
57015	THERMOREFLECTANCE OF SILVER AT THE PLASMA FREQUENCY. BALDINI G NOBILE M SOLID STATE COMMUN					
	8	1	7-11	1970	CA	72	61025
57018	EFFECT OF VIBRATION SWELLING ON SOME PROPERTIES OF AERATED SILICATES. SPEKTOR B V LOZHKINA T V STROIT MATER DETALI IZDELIYA					
		12	175-9	1969	CA	72	58748
57019	CALCULATED VALUES OF THE THERMAL CONDUCTIVITY OF LIGHTWEIGHT CONCRETES AND METHODS FOR THEIR DETERMINATION. SPEKTOR B V STROIT MATER DETALI IZDELIYA					
	12		115-18	1969	CA	72	58721
57031	INVESTIGATION OF DIELECTRIC OPTICAL COATINGS FOR LASERS. MOYS B A WOOD R M WARR P D GENERAL ELECTRIC COMP LIMITED CENTRAL RESEARCH LABS WEMBLEY /ENGLAND/ DDC C.V.D.-RP-4-67 AD-867413					
			1-12	1969		
57052	INTERFACIAL TENSIONS AGAINST WATER OF SOME C10-C15 HYDROCARBONS WITH AROMATIC OR CYCLOALIPHATIC RINGS. JASPER J J NAKONECZNYJ M SWINGLEY C S LIVINGSTON H K J PHYS CHEM					
	74	7	1535-9	1970		
57056	FREEZE DURABILITY AND REQUIREMENTS IN THE CASE OF BOTH SIDES OF A FROZEN FISH FILET. WATZINGER A KALTETECHNIK					
	1	8	189-94	1949		
57070	PHONONS IN MIXED CD/X/ ZN/1-X/ S SEMICONDUCTORS. LISITSA M P VALAKH M YA KONOVETS N K PHYS STAT SOLIDI					
	34	1	269-78	1969		
57094	HIGHLY REFRACTORY ALUMINOPHOSPHATE BONDED CONCRETE. SALMANOV G D ALEKSANDROVA G N IZV AKAD NAUK SSSR NEORG MATER					
	5	1	148-51	1969		
57095	HIGHLY REFRACTORY ALUMINOPHOSPHATE BONDED CONCRETE. //ENGLISH TRANSLATION OF IZV. AKAD. NAUK SSSR, NEORG. MATER. 5 /1/ 148-51, 1969.// SALMANOV G D ALEKSANDROVA G N INORGANIC MATERIALS					
	5	1	120-2	1969		
57102	THE MUTUAL DIFFUSION OF THE MATERIALS OF THE FILM AND SUBSTRATE IN THE EPITAXIAL GROWTH OF HETEROJUNCTIONS. LISENKER B S MARONCHUK I E MARONCHUK YU E SHUMSKII V N SHIPILOVA S I					

TPRC Number	Bibliographic Citation					
57103	THE MUTUAL DIFFUSION OF THE MATERIALS OF THE FILM AND SUBSTRATE IN THE EPITAXIAL GROWTH OF HETEROJUNCTIONS. //ENGLISH TRANSLATION OF IZV. AKAD. NAUK SSSR, NEORG. MATER. 5 /2/ 295-300, 1969.// LISENKER B S MARONCHUK I E MARONCHUK YU E SHUMSKII V N SHIPILOVA S I INORGANIC MATERIALS					
	5	2	245-9	1969		
57110	INVESTIGATION OF THE STRUCTURAL, ELECTRICAL, AND OPTICAL PROPERTIES OF TIN DIOXIDE AND INDIUM SESQUIOXIDE FILMS. BURBULYAVICHUS L I VAINSHTEIN V M GERASIMOVA L G DANCHEVSKAYA M N ZARIFYANTS YU A PANASYUK G P FIGUROVSKAYA E N KRUSTALEVA S B IZV AKAD NAUK SSSR NEORG MATER					
	5	3	551-5	1969		
57111	INVESTIGATION OF THE STRUCTURAL, ELECTRICAL, AND OPTICAL PROPERTIES OF TIN DIOXIDE AND INDIUM SESQUIOXIDE FILMS. //ENGLISH TRANSLATION OF IZV. AKAD. NAUK SSSR, NEORG. MATER. 5 /3/ 551-5, 1969.// BURBULYAVICHUS L I VAINSHTEIN V M GERASIMOVA L G DANCHEVSKAYA M N ZARIFYANTS YU A PANASYUK G P FIGUROVSKAYA E N KRUSTALEVA S B INORGANIC MATERIALS					
	5	3	462-5	1969		
57171	EFFECT OF GAS-CARRIER ON THE FORMATION AND PROPERTIES OF PYROLYTIC FILMS OF SIO2. KALNYNYA R P FELTYN I A FREIBERGA L A EGLITIS I E IZV AKAD NAUK SSSR NEORG MATER					
	5	9	1540-5	1969		
57172	EFFECT OF GAS-CARRIER ON THE FORMATION AND PROPERTIES OF PYROLYTIC FILMS OF SIO2. //ENGLISH TRANSLATION OF IZV. AKAD. NAUK SSSR. NEORG. MATER. 5 /9/ 1540-5, 1969.// KALNYNYA R P FELTYN I A FREIBERGA L A EGLITIS I E INORGANIC MATERIALS					
	5	9	1306-10	1969		
57187	COMPARISON OF LASER REFLECTIVITY FROM TYPICAL WALL MATERIALS AND PAINTS. EBBERS R W RODRIGUEZ T L SPROUFFSKE J F AEROMEDICAL RESEARCH LAB HOLLOMAN AFB NEW MEXICO DDC ARL-TR-69-13 AD-699570					
			1-17	1969	RR	70-6 179
57191	DEVELOPMENT OF 400-2200 F FIBROUS-TYPE INSULATIONS FOR RADIOISOTOPE POWER SYSTEMS. SEVENTH QUARTERLY PROGRESS REPORT. JULY 1, 1968-SEPT. 30, 1968. COLLINS J O JAUNARAJS K L REID D R JOHNS-MANVILLE RESEARCH AND ENGR CENTER MANVILLE NEW JERSEY NASA AND CFSTI ALO-3633-10 N69-41277					
			1-65	1968	PA	N69-7-24 4596
57221	TRANSMISSION EFFECTS ON PLASTIC FILMS IRRADIATED WITH ULTRAVIOLET LIGHT, ELECTRONS, AND PROTONS. ANAGNOSTOU E SPAKOWSKI A E LEWIS RESEARCH CENTER NASA AND CFSTI NASA-TM-X-1905 N69-40339					
			1-12	1969		
57222	THERMAL CONDUCTIVITY. FROM THERMOPHYSICAL AND CHEMICAL CHARACTERIZATION OF CHARRING ABLATIVE MATERIALS. FINAL REPORT. LAGEDROST J F FABISH T J NASA AND CFSTI NASA-CR-73399 N70-14131					
			13-8	1968	RR	70-7 71
57223	INFLUENCE OF PORE STRUCTURE ON THERMAL CONDUCTIVITY OF POLYBENZIMIDAZOLE CHAR MATERIAL. FROM THERMOPHYSICAL AND CHEMICAL CHARACTERIZATION OF CHARRING ABLATIVE MATERIALS. FINAL REPORT. FABISH T J NASA AND CFSTI NASA-CR-73399 N70-14131					
			18-20	1968	PAN	70-8-4 772
57224	ENTHALPY, SPECIFIC HEAT. FROM THERMOPHYSICAL AND CHEMICAL CHARACTERIZATION OF CHARRING ABLATIVE MATERIALS. FINAL REPORT. ELDRIDGE E A DEEM H W NASA AND CFSTI NASA-CR-73399 N70-14131					
			21-30	1968		
57225	LINEAR THERMAL EXPANSION. FROM THERMOPHYSICAL AND CHEMICAL CHARACTERIZATION OF CHARRING ABLATIVE MATERIALS. FINAL REPORT. ELDRIDGE E A DEEM H W NASA AND CFSTI NASA-CR-73399 N70-14131					

TPRC Number	Bibliographic Citation

57227 PROPULSION BEAM DIVERGENCE EFFECTS. THIRD QUARTERLY TECHNICAL REPORT.
HALL D F   TRW SYSTEMS GROUP   REDONDO BEACH CALIFORNIA
NASA AND CFSTI
NASA-CR-107431
N70-14485   1-29   1969   PA   N70-8-4 4707

57233 OIL-VAPOR CONTAMINATION OF SATELLITE OPTICAL SURFACES.
WILLIAMS T N   ROYAL AIRCRAFT ESTABLISHMENT
AEC
RAE-TR-69055
AD-699674   1-20   1969

57237 THERMAL CONTACT RESISTANCE OF RIVETED JOINTS.
AMBROSIO A   LINDH K G   DEPARTMENT OF ENGINEERING UNIVERSITY OF CALIFORNIA
UCLA
UCLA-DEPT ENGR REPT C55-4
         1-11   1955

57246 SOME PROPERTIES OF SILICON CARBIDE THIN FILMS PREPARED BY ELECTRON BEAM EVAPORATION.
BUNTON G V
J PHYS D /APPL PHYS/
3   2   232-5   1970

57251 ON A PHENOMENON OF ABNORMAL ABSORPTION PRODUCED IN THIN FILMS.
PETRAKIAN J-P
COMPT REND
270   B   9   624-7   1970

57267 NEW THERMAL INSULATION MATERIAL OF HIGH REFRACTORINESS—LIGHT INSULATING BRICK MADE OF CARBON.
HOYNANT G
SILICATES IND
34   6   185-6   1969 JA   53-4 92

57276 HIGHLY REFLECTING MULTILAYER COATINGS WITH LARGE DISPERSION OF PHASE DISCONTINUITY.
SHKLYAREVSKII I N   UMEROV R I   USOSKIN A I
OPTIKA I SPEKTROSK
27   3   497-500   1969

57277 HIGHLY REFLECTING MULTILAYER COATINGS WITH LARGE DISPERSION OF PHASE DISCONTINUITY. //ENGLISH TRANSLATION OF OPTIKA I SPEKTROSKOPIYA 27 /3/ 497-500, 1969.//
SHKLYAREVSKII I N   UMEROV R I   USOSKIN A I
OPT SPECTROSC
27   3   266-7   1969

57316 EXPERIMENTAL EVIDENCE FOR EXCITATION OF LONGITUDINAL PLASMONS BY PHOTONS.
LINDAU I   NILSSON P O
PHYS LETT
31   A   7   352-3   1970

57323 AN ELECTRON CONTRIBUTION TO THE THERMAL CONDUCTION ACROSS A METAL-SOLID DIELECTRIC INTERFACE.
WOLFMEYER M W   FOX G T   DILLINGER J R
PHYS LETT
31   A   7   401-2   1970

57342 ULTRAVIOLET OPTICAL CONSTANTS OF THIN FILMS DETERMINED BY REFLECTANCE MEASUREMENTS.
BALDINI G   RIGALDI L
J OPT SOC AM
60   4   495-8   1970

57363 APPROXIMATE EXPRESSIONS FOR COEFFICIENTS OF REFLECTION AND TRANSMISSION OF THIN FILMS.
SHKLYAREVSKII I N   SHKLYAREVSKII O I
OPTIKA I SPEKTROSKOPIYA
27   4   654-60   1969

57364 APPROXIMATE EXPRESSIONS FOR COEFFICIENTS OF REFLECTION AND TRANSMISSION OF THIN FILMS. //ENGLISH TRANSLATION OF OPTIKA I SPEKTROSKOPIYA 27 /4/ 654-60, 1969.//
SHKLYAREVSKII I N   SHKLYAREVSKII O I
OPT SPECTRY
27   4   353-6   1969

57365 ABSORPTION OF LIGHT BY SMALL PARTICLES OF AG, CU, AL, AND SE.
PETROV YU I
OPTIKA I SPEKTROSKOPIYA
27   4   665-73   1969

57366 ABSORPTION OF LIGHT BY SMALL PARTICLES OF AG, CU, AL, AND SE. //ENGLISH TRANSLATION OF OPTIKA I SPEKTROSK. 27 /4/ 665-73, 1969.//
PETROV YU I
OPT SPECTRY
27   4   359-64   1969

57378 INTERFACIAL TENSION IN HYDROCARBON SYSTEMS.
DEAM J R   MADDOX R N
J CHEM ENG DATA

57389 NEW TYPES OF CONCRETE RESISTANT TO HIGH TEMPERATURES FOR THE PROTECTION OF NUCLEAR REACTORS. FROM SYMPOSIUM ON ADVANCED AND HIGH TEMPERATURE GAS-COOLED REACTORS, JUELICH, GERMANY.
DUBOIS F   MAUNY P   RAPPENEAU J
CFSTI
CEA-CONF-1091 CONF-881008-2
        1-40   1968   NSA 23-6 1057

57392 MEASUREMENT OF THE HEAT CURRENT FUNCTION OF SEVERAL CARBON AND GRAPHITE FELTS.
REISS F E   KERNFORSCHUNGSZENTRUM KARLSRUHE WEST GERMANY
CFSTI AND AEC
KFK-1062   1-20   1969 CA   73   111328

57408 INFRARED REFLECTANCE MEASUREMENTS. FINAL REPT. JAN. 1967-JUNE 1969.
RICHMOND J C   GEIST J C
NASA AND CFSTI
NBS-10071   NASA-CR-107844
N70-17622   1-95   1969 PA N70-8-6 1055

57413 HIGH EFFICIENCY REFLECTING SURFACES FOR THE VACUUM ULTRAVIOLET.
FEUERBACHER B   FITTON B   STEINMANN W
ELDO/ESRO TECH REV
1   3   385-403   1969 PA N70-8-6 1102

57437 EMITTANCE MEASUREMENT STUDY.
HEINISCH R P   ANDERSSON J K   SCHMIDT R N
CFSTI
NASA-CR-1583 N70-25687
      1-65   1970 PA N70-8-12 2264

57439 TWO FACTORS INFLUENCING TEMPERATURE DISTRIBUTIONS AND THERMAL STRESSES IN STRUCTURES.
BROOKS W A JR   GRIFFITH G E   STRASS H K
DDC
NACA-TN-4052 AD-133067
      1-13   1957

57450 HEAT TRANSFER ACROSS NON-METALLIC INTERFACES AT HIGH TEMPERATURES. FROM EUROPEAN CONFERENCE ON THERMO-PHYSICAL PROPERTIES OF SOLIDS AT HIGH TEMPERATURES IN BADEN-BADEN. NOV. 11-13, 1968.
MINGES M L   AIR FORCE MATERIALS LAB WRIGHT-PATTERSON AIR FORCE BASE   OHIO
ZENTRALSTELLE FUR ATOMKERNENERGIE-DOKUMENTATION /ZAED/ KARLSRUHE
      248-62   1968

57455 MEASUREMENT OF THE THERMAL CONDUCTIVITY OF ULTRA THIN SINGLE OR DOUBLE LAYER SAMPLES. FROM EUROPEAN CONFERENCE ON THERMO-PHYSICAL PROPERTIES OF SOLIDS AT HIGH TEMPERATURES IN BADEN-BADEN. NOV. 11-13, 1968.
GILCHRIST K E   REACTOR MATERIALS LABORATORY CULCHETH NR WARRINGTON   LANCASHIRE
ZENTRALSTELLE FUR ATOMKERNENERGIE-DOKUMENTATION /ZAED/ KARLSRUHE
      368-92   1968

57510 MEASUREMENT OF THICKNESS AND OPTICAL CONSTANTS OF THIN FILMS.
VASICEK A   KUCIREK J
BASIC PROBL THIN FILM PHYS PROC INT SYMP 1965
      258-61   1966 CA 72 84522

57511 LEAD TELLURIDE-TIN TELLURIDE.
NEUBERGER M   HUGHES AIRCRAFT CO ELECTRONIC PROPERTIES INFORMATION CENTER   CULVER CITY CALIFORNIA
DDC AND CFSTI
EPIC-DS-164 AD-701075
      1-211   1970 RR 70-7 197

57513 THERMAL CONDUCTIVITY OF SETTING MORTARS AND CONCRETES.
ZASEDATELEV I B   MISHIN G V
BETON ZHELEZOBETON
15   10   32-5   1969 CA 72 82610

57514 COMBINED PERFORMANCE OF A CEMENT MORTAR AND A STEEL CASING DURING A DROP IN TEMPERATURE.
KLYACHIN A Z   KATAEV G N
BETON ZHELEZOBETON
15   12   21-2   1969 CA 72 93071

57533 THERMOPHYSICAL CHARACTERISTICS OF THIN-LAYERED POLYMERIC COATINGS. //ENGLISH TRANSLATION OF TEPLO-MASSOOBMENA NENYUTONOVSK. ZHIDK. 217-30, 1968.//
NOVICHENOK L N
DDC
FTD-MT-24-306-68
AD-697615   258-76   1969

57537 MEASURING THE THERMOPHYSICAL CHARACTERISTICS OF LIQUID COATINGS. //ENGLISH TRANSLATION OF TEPLO-MASSOOBMENA NENYUTONOVSK. ZHIDK. 239-46, 1968.//
SMOLSKII B M   SHULMAN Z P   GORISLAVETS V M
GORODKIN R G
DDC

TPRC Number	Bibliographic Citation						
57565	HIGH-TEMPERATURE PROPERTIES OF SELF-HARDENING SAND MOLD. HAMADA S OHASHI A NAKABAYASHI T IGUCHI T OKABAYASHI K IMONO						
	41	5	393-401	1969	CA	71	73331
57569	SPECIAL TECHNIQUE FOR MANUFACTURING INSULATING COLLECTOR MULTILAYER TUBES. PEEHS M SCHOERNER H STEHLE H INT CONF THERMION ELEC POWER GENERATION 2ND						
			647-54	1968	CA	72	71957
57578	SPECTRAL EMITTANCE OF SOOT. HIBBARD R R LIEBERT C H LEWIS RESEARCH CENTER CLEVELAND OHIO NASA AND CFSTI NASA-TN-D-5647						
	N70-19178		1-12	1970	PA	N70-8-7 1201	
57581	STUDY OF THERMAL CONDUCTIVITY REQUIREMENTS. HIGH PERFORMANCE INSULATION. FINAL REPORT. HALE D V ONEILL M J LOCKHEED MISSILES AND SPACE CO RESEARCH AND ENGINEERING CENTER HUNTSVILLE ALABAMA NASA AND CFSTI NASA-CR-102600						
	N70-24181		1-79	1970	PA	N70-8-11 2128	
57595	EFFECT OF AGING ON THERMAL CONDUCTIVITY OF CELLULAR MATERIALS. DIXON R R EDELMAN L E MC LAIN D K J CELL PLAST						
	6	1	44-7	1970	CA	72	79760
57597	PLANETARY ENVIRONMENT SIMULATION. EROSION AND DUST COATING EFFECTS. ADLON G L RUSERT E L ALLEN T H MC DONNELL-DOUGLAS ASTRONAUTICS CO ST LOUIS MO NASA AND CFSTI MDC-E0038-VOL.-1 NASA-CR-66878						
	N70-19860		1-39	1969	PA	N70-8-8 1464	
57598	DEVELOPMENT OF ADVANCED MATERIALS FOR INTEGRATED TANK INSULATION SYSTEM FOR THE LONG TERM STORAGE OF CRYOGENS IN SPACE. FINAL REPORT. GILLE J P MARTIN MARIETTA CORP DENVER COLORADO NASA AND CFSTI MCR-69-405 NASA-CR-102570						
	N70-23348		1-171	1969			
57632	THERMAL CONDUCTIVITY OF MOIST CAPILLARY-POROUS SYSTEMS. FROM PROCEEDINGS OF THE NINTH CONFERENCE ON THERMAL CONDUCTIVITY. IOWA STATE UNIVERSITY, AMES, IOWA. OCTOBER 6-8, 1969. SHASHKOV A G VASILIEV L L TANAEVA S A HEAT AND MASS TRANSFER INSTITUTE BSSR ACADEMY OF SCIENCES MINSK USSR CFSTI						
	CONF-691002		279-87	1970			
57639	THE THERMAL CONDUCTIVITY OF MIN-K-2000 THERMAL INSULATION IN DIFFERENT ENVIRONMENTS TO HIGH TEMPERATURES. FROM PROCEEDINGS OF THE NINTH CONFERENCE ON THERMAL CONDUCTIVITY. IOWA STATE UNIVERSITY, AMES, IOWA. OCTOBER 6-8, 1969. TYE R P DYNATECH R/D COMPANY CAMBRIDGE MASSACHUSETTS CFSTI						
	CONF-69-1002	9	341-51	1970	CA	75	11262
57642	THE THERMAL CONDUCTIVITY OF PYRO-CARB 406. A CARBON-CARBON COMPOSITE. FROM PROCEEDINGS OF THE NINTH CONFERENCE ON THERMAL CONDUCTIVITY. IOWA STATE UNIVERSITY, AMES, IOWA. OCTOBER 6-8, 1969. BRAZEL J P DUBIN P KENNEDY B S GENERAL ELECTRIC COMPANY PHILADELPHIA PENN CFSTI						
	CONF-691002		393-419	1970			
57643	THE EFFECTIVE THERMAL CONDUCTIVITY OF PYROLYTIC GRAPHITE CYLINDERS IN VARIOUS GAS ENVIRONMENTS. FROM PROCEEDINGS OF THE NINTH CONFERENCE ON THERMAL CONDUCTIVITY. IOWA STATE UNIVERSITY, AMES, IOWA. OCTOBER 6-8, 1969. PYRON C M JR SMYLY E D SOUTHERN RESEARCH INSTITUTE BIRMINGHAM ALABAMA CFSTI						
	CONF-691002		421-48	1970			
57644	INVESTIGATION OF THE THERMAL CONDUCTIVITY OF PHENOLIC-NYLON CHAR. FROM PROCEEDINGS OF THE NINTH CONFERENCE ON THERMAL CONDUCTIVITY. IOWA STATE UNIVERSITY, AMES, IOWA. OCTOBER 6-8, 1969. SMYLY E D PYRON C M JR SOUTHERN RESEARCH INSTITUTE BIRMINGHAM ALABAMA CFSTI						
	CONF-69-1002	9	448-93	1970			

TPRC Number	Bibliographic Citation						
57647	A RADIAL HEAT-FLOW METHOD FOR POOR CONDUCTORS. FROM PROCEEDINGS OF THE NINTH CONFERENCE ON THERMAL CONDUCTIVITY. IOWA STATE UNIVERSITY, AMES, IOWA. OCTOBER 6-8, 1969. ASHWORTH T SMITH M G PHYSICS DEPT SOUTH DAKOTA SCHOOL OF MINES AND TECHNOLOGY CFSTI						
	CONF-69-1002	9	524-9	1970	CA	75	37122
57652	AN ANALYTICAL STUDY OF THERMAL CONTACT CONDUCTANCE FOR TWO ROUGH AND WAVY SURFACES UNDER PRESSURE CONTACT. FROM PROCEEDINGS OF THE NINTH CONFERENCE ON THERMAL CONDUCTIVITY. IOWA STATE UNIVERSITY, AMES, IOWA. OCTOBER 6-8, 1969. HSIEH C K YEDDANAPUDI K M TOULOUKIAN Y S THERMOPHYSICAL PROPERTIES RESEARCH CENTER PURDUE UNIVERSITY WEST LAFAYETTE IND CFSTI						
	CONF-69-1002	9	554-70				
57658	GUARDED FLAT PLATE THERMAL CONDUCTIVITY APPARATUS FOR TESTING MULTI-FOIL INSULATIONS IN THE 20-1000 C. RANGE. FROM PROCEEDINGS OF THE NINTH CONFERENCE ON THERMAL CONDUCTIVITY. IOWA STATE UNIVERSITY, AMES, IOWA. OCTOBER 6-8, 1969. GRUNERT W E MORIHARA H REID R L MASSING P N UNION CARBIDE CORPORATION LINDE DIVISION TONAWANDA NEW YORK CFSTI						
	CONF-69-1002	9	658-72	1970	CA	75	26174
57660	EFFECT OF THERMAL RADIATION ON THERMAL CONDUCTIVITY OF INSULATION WITH RESPECT TO ITS THICKNESS. FROM PROCEEDINGS OF THE NINTH CONFERENCE ON THERMAL CONDUCTIVITY. IOWA STATE UNIVERSITY, AMES, IOWA. OCTOBER 6-8, 1969. WATANABE T NIPPON ASBESTOS CO LTD TOKYO JAPAN CFSTI						
		9	704	1970			
57661	A STUDY OF SOME OF THE FACTORS AFFECTING THERMAL PROPERTIES IN PACKED BED SYSTEMS. FROM PROCEEDINGS OF THE NINTH CONFERENCE ON THERMAL CONDUCTIVITY. IOWA STATE UNIVERSITY, AMES, IOWA. OCTOBER 6-8, 1969. SCHILMOELLER N H ARGONNE NATIONAL LAB CFSTI						
	CONF-69-1002	9	705-11	1970			
57663	THE THERMAL CONDUCTIVITY OF SOME ELECTRICAL ENGINEERING MATERIALS. FROM PROCEEDINGS OF THE NINTH CONFERENCE ON THERMAL CONDUCTIVITY. IOWA STATE UNIVERSITY, AMES, IOWA. OCTOBER 6-8, 1969. ROBERTS T J ALLEN P H G CFSTI						
	CONF-69-1002	9	719-34	1970			
57666	INVESTIGATION OF SPACECRAFT COATINGS. MAYER R A ZARING M L KEMP H T BATTELLE MEMORIAL INSTITUTE NASA AND CFSTI NASA-CR-61267						
	N69-21673		1-252	1969	CA	71	71951
57764	MAGNETIC AND OPTICAL STUDIES OF CHROMIUM OXIDES. PART 1. CALCINATION OF CHROMIUM TRIOXIDE SUPPORTED ON ALUMINA. ELLISON A OUBRIDGE J O V SING K S W TRANS FARADAY SOC						
	66	4	1004-14	1970			
57767	STATISTICAL MODELS FOR SURFACE RENEWAL IN HEAT AND MASS TRANSFER. PART IV. WALL TO FLUIDIZED BED HEAT TRANSFER COEFFICIENTS. KOPPEL L B PATEL R D HOLMES J T A I CH E JOURNAL						
	16	3	464-71	1970			
57793	DESIGN OF FABRY-PEROT REFLECTORS FOR THE VACUUM ULTRAVIOLET. THETFORD A BATES B MAC DONALD J APPL OPT						
	9	1	35-9	1970	CA	72	60960
57794	ABSOLUTE SPECULAR REFLECTANCE MEASUREMENTS OF HIGHLY REFLECTING OPTICAL COATINGS AT 10.6 MICRON. KELSALL D APPL OPT						
	9	1	85-90	1970	CA	72	84821
57796	MEASURED TIME VARIATIONS OF REFLECTANCE FOR CARBON DIOXIDE LASER BEAM SPLITTERS. MONSON D J APPL OPT						
	9	1	224-5	1970	CA	72	84809

TPRC Number	Bibliographic Citation

57797 SOLAR ABSORPTIVITY AND THERMAL EMISSIVITY OF ALUMINUM COATED WITH SILICON OXIDE FILMS PREPARED BY EVAPORATION OF SILICON MONOXIDE.
BRADFORD A P   HASS G   HEANEY J B   TRIOLO J J
TRIOLO J J
APPL OPT
9   2   339-44   1970   CA   72   94946

57800 FLUORESCENT INTEGRATING SPHERE FOR THE VACUUM ULTRAVIOLET.
BRANDENBERG W M
APPL OPT
9   2   451-7   1970

57801 ON THE FABRICATION AND EVALUATION OF AN INTEGRATING HEMIELLIPSOID.
HEINISCH R P   BRADEC F J   PERLICK D B
APPL OPT
9   2   483-7   1970

57802 PREPARATION OF OPTICAL SURFACES ON BERYLLIUM.
BLOXSOM J T   SCHROEDER J B
APPL OPT
9   3   539-43   1970

57838 THERMAL-CONDUCTIVITY MEASUREMENTS OF GAPLESS BEHAVIOR PRODUCED BY THE PROXIMITY EFFECT.
DEUTSCHER G   LINDENFELD P   MC CONNELL R D
PHYS REV
1   B   5   2169-76   1970

57851 STUDY OF EXTERNAL INSULATIONS ON CRYOGENIC TANKAGE IN A THERMAL VACUUM CHAMBER. FINAL REPORT JAN. 1968-JULY 1969.
BENDIX CORP INSTRUMENTS AND LIFE
DIVISION   DAVENPORT IOWA
NASA AND CFSTI
NASA-CR-101868
N69-35571   1-248   1969   RR   69-23 127

57853 RESEARCH AND DEVELOPMENT OF VACUUM FOIL-TYPE INSULATION FOR RADIOISOTOPE POWER SYSTEMS. QUARTERLY PROG. REPT. MARCH-MAY, 1969.
CARVALHO J   DUNLAY J B   PAGUIN M L   POIRIER V L
THERMO ELECTRON CORP   WALTHAM MASS
CFSTI
ALO-3634-13 N70-15855
     1-27   1969

57854 METAL FABRIC AND SILICON MATERIALS. THERMAL PROPERTY MEASUREMENTS REPORT.
TRW SYSTEMS GROUP   REDONDO BEACH
CALIFORNIA
NASA AND CFSTI
TRW-70-8526.161-03 NASA-CR-102169
N70-18705   1-19   1970   RR   70-9 94

57859 DETERMINATION OF PHYSICAL AND TECHNOLOGICAL PROPERTIES OF SUPERINSULATION.
MUELLER E   BOELKOW G M B H   OTTOBRUNN BEI
MUENCHEN/WEST GERMANY/
NASA AND CFSTI
BMWF-FB-W-69-24
N70-11236   1-40   1969   RR   70-4 179

57860 ULTRAVIOLET ABSORPTION AND STIMULATED LUMINESCENCE INVESTIGATIONS. PART 1. ULTRAVIOLET VIDEO IMAGING SYSTEM. PART 2. SPECTRAL DISTRIBUTION OF ULTRAVIOLET STIMULATED LUMINESCENCE. PART 3. MEASUREMENT OF ULTRAVIOLET REFLECTANCE INTERIM REPORT.
DANIELS D L   FISCHER W A   GAWARECKI S J
GERHARZ R   HEMPHILL W R
NASA AND CFSTI
NASA-TM-X-61721
N69-29390   1-120   1964   RR   69-19 77

57887 EMITTANCE OF OXIDE LAYERS ON A METAL SUBSTRATE.
BRANNON R R JR   GOLDSTEIN R J
TRANS ASME /J HEAT TRANSFER/
92   C   2   257-63   1970

57940 THERMAL CONDUCTIVITY OF INSULATING FOAMS BY A RADIAL FLOW METHOD. M.S. THESIS.
SMITH M G   SOUTH DAKOTA SCHOOL OF MINES AND
TECHNOLOGY   RAPID CITY SOUTH DAKOTA
SOUTH DAKOTA SCHOOL OF MINES AND TECHNOLOGY
     1-33   1969

57942 MEASUREMENTS OF THE THERMAL ACCOMMODATION OF HELIUM ON TUNGSTEN AT ROOM TEMPERATURE. M.S. THESIS.
BOUCHEZ J-P F A   UNIVERSITY OF MINNESOTA
MINNEAPOLIS MINNESOTA
UNIVERSITY OF MINNESOTA
     1-69   1969

57943 THE DETERMINATION OF SURFACE TEMPERATURES. FROM ESRO ENVIRONMENTS AND THEIR ROLE IN SPACECRAFT TECHNOLOGY. VOL. 2. THERMAL ASPECTS.
JANES M
NASA AND CFSTI
ESRO-SP-41 N69-37526
     187-218   1969

57946 THERMAL CONDUCTIVITY OF PHENOLIC-CARBON CHARS.
CLAYTON W A   FABISH T J   LAGEDROST J F
AEROSPACE SYSTEMS DIV BOEING COMP   SEATTLE WASHINGTON
COMP   SEATTLE WASHINGTON
DDC AND CFSTI
AFML-TR-69-313
AD-702112   1-192   1969   RR   70-9 101

57948 FAIRING COMPOSITIONS FOR AIRCRAFT SURFACES.
TURNER P S   DORAN J   REINHART F W
CFSTI
NACA-TN-958
     1-41   1944

57953 OPTICAL CONSTANTS OF SILVER SULFIDE TARNISH FILMS.
BENNETT J M   STANFORD J L   ASHLEY E J
J OPT SOC AMER
60   2   224-32   1970   CA   72   72862

57957 INTERFACIAL TENSION OF 3-METHYLPENTANE-NITROETHANE NEAR THE CRITICAL POINT.
WIMS A M   SENGERS J V   MC INTYRE D
SHERESHEFSKY J
J CHEM PHYS
52   6   3042-9   1970   CA   72   104134

57980 THERMAL ACCOMMODATION COEFFICIENTS DETERMINED FROM OBSERVATIONS OF ARGON BEAMS SCATTERED FROM SILVER.
FISHER S S   BISHARA M N
ENTROPIE
30   113-19   1969   CA   72   104824

58015 THE DETERMINATION OF THE RADIAL EXPANSION OF DUMET WIRES WITH THE AID OF THE DATA OF CHEMICAL ANALYSIS.
VAN LIEMPT J A M
REC TRAV CHIM PAYS-BAS
52   399-402   1933

58105 AUTOGRAPHIC DILATOMETER FOR USE WITH PYROPHORIC AND ALPHA-ACTIVE MATERIALS.
YAGGEE F L   STYLES J W   ARGONNE NATL LAB
ILLINOIS
CFSTI AND AEC
ANL-7643 N70-30066
     1-19   1969   NSA 24-10 1906

58147 THERMOPHYSICAL PROPERTIES OF FLUID MOLDING MIXTURES.
COSNEANU C   COHN E   CATANA V
METALURGIA /BUCHAREST/
21   11   572-6   1969   CA   72   103080

58165 PRECISION MEASUREMENTS IN THIN-FILM OPTICS.
BENNETT H E   BENNETT J M
PHYS THIN FILMS
4   1-96   1967   CA   72   94682

58202 THERMAL AND ELECTRICAL PROPERTIES OF POLYCRYSTALLINE SILICON IN THE DIELECTRIC ISOLATION PROCESS.
BEAN K E   HENTZSCHEL H P   COLMAN D
SEMICOND SILICON INT SYMP PAP 1ST
747-57   1969   CA   71   106813

58221 THERMAL EXPANSIVITY AND ELASTICITY OF PAPER AND BOARD.
KUBAT J   MARTIN-LOF S   SOREMARK C
SV PAPPERSTIDN
72   23   763-7   1969   CA   72   89552
DDC
AD-712149   REPRINT   1969

58255 INTERACTION OF FINITE SODIUM CHLORIDE CRYSTALS WITH INFRARED RADIATION.
MARTIN T P
PHYS REV
1   B   8   3480-8   1970

58261 PHYSICOMECHANICAL PROPERTIES OF GLASS FIBER REINFORCED PLASTICS ON A POLYESTER BINDER WHICH ARE HARDENED IN A HIGH-FREQUENCY ELECTRIC FIELD.
ENGALYCHEV S A   FEDOROVA I G   VODOPYANOV M YA
TR VSES NAUCH-ISSLED INST TOKOV VYS CHASTOTY
10   180-9   1969   CA   72   101294

58264 OPTICAL ABSORPTION AND VACUUM-ULTRAVIOLET REFLECTANCE OF GALLIUM NITRIDE THIN FILMS.
KOSICKI B B   POWELL R J   BURGIEL J C
PHYS REV LETT
24   25   1421-3   1970

58286 CERAMMED COATINGS WITH A HIGH COEFFICIENT OF THERMAL EXPANSION.
PEVZNER B Z   APPEN A A   ANTONOVA E A
ZHAROSTOIKIE TEPLOSTOIKIE POKRYTIYA TR VSES SOVESHCH
4TH 1968
205-10   1969   CA   72   82397

58297 MAGNETO-OPTIC PROPERTIES OF MANGANESE ARSENIDE FILMS.
STOFFEL A M   SCHNEIDER J
J APPL PHYS
41   3   1405-7   1970

TPRC Number	Bibliographic Citation
58301	INFRARED REFLECTANCE SPECTRA OF CARBON MONOXIDE ADSORBED ON NICKEL AND RHODIUM /110/ SURFACES IN HIGH AND ULTRAHIGH VACUA. ECKSTROM H C POSSLEY G G HANNUM S E SMITH W H J CHEM PHYS 52 10 5435-41 1970
58314	MEASUREMENTS OF THE KAPITZA CONDUCTANCE BETWEEN METALS AND LIQUID HELIUM II BY THE TRANSMISSION OF SECOND SOUND. CHALLIS L J SHERLOCK R A J PHYS C /SOLID ST PHYS/ 3 5 1193-206 1970
58317	STRESS IN SILICON FILMS DEPOSITED HETEROEPITAXIALLY ON INSULATING SUBSTRATES WITH PARTICULAR REFERENCE TO CORUNDUM. JEFKINS D M J PHYS D /APPL PHYS/ 3 5 770-7 1970
58320	THERMAL INSULATION STUDY. RESEARCH AND DEVELOPMENT PROGRESS REPORT. JUNE-NOV. 1969. GRUNERT W E UNION CARBIDE CORP INDIANAPOLIS IND LINDE DIVISION CFSTI ALO-2832-42 N70-25559 1-77 1969 PA N70-8-12 2236
58321	RESEARCH AND DEVELOPMENT IN A THERMAL INSULATION STUDY. QUARTERLY PROGRESS REPT., MARCH-MAY 1969. GRUNERT W E MORIHARA H UNION CARBIDE CORP INDIANAPOLIS IND LINDE DIVISION CFSTI ALO-3632-41 N70-12575 1-51 1969
58322	ELECTRICAL AND OPTICAL PROPERTIES OF LEAD-TIN TELLURIDE SEMICONDUCTING ALLOYS. BIS R F DIXON J R NAVAL ORDNANCE LAB WHITE OAK MD CFSTI NOLTR-69-146 N70-22394 AD-697647 1-110 1969 PA 70-8-9 1730
58324	CRYOGENIC SOLID OXYGEN STORAGE AND SUBLIMATION INVESTIGATION. AHERN J E LAWSON T W JR AEROJET GENERAL CORP ELECTRONICS DIVISION AZUSA CALIF DDC AMRL-TR-68-105 AGC-3545 AD-687852 1-185 1968
58326	DEVELOPMENT OF AN SPS/DPS HYDROGEN SHROUDED CRYOGENIC HELIUM STORAGE SYSTEM. FINAL REPT. JULY 1967-SEPT. 1968. BALD W B BENDIX CORP DAVENPORT IOWA NASA NASA-CR-92441 N69-14774 1-135 1968
58327	PLASTIC AND ELASTOMERIC FOAM MATERIALS. ARDEN B NORTHROP CORP PALOS VERDES ESTATES CALIFORNIA NASA AND CFSTI NASA-CR-100463 N69-21173 1-186 1966
58328	INTER-RELATIONS BETWEEN ADVANCED PROCESSING TECHNIQUES, INTEGRATED CIRCUITS, MATERIALS DEVELOPMENT AND ANALYSIS. QUARTERLY REPORT, MARCH-MAY, 1969. COCHRUN B L CARLSON W FINE S NOVAK W B NASA AND CFSTI NASA-CR-105631 N69-36276 1-15 1969
58330	RADIANT COOLER DESIGN AND EMISSIVITY STUDY. PART 2. EMISSIVITY STUDY. FINAL REPORT, APRIL-DEC. 1969. IIT AEROSPACE/OPTICAL DIVISION FT WAYNE IND NASA AND CFSTI NASA-CR-108772 N70-20114 1-37 1969 RR 70-11 123

58335	THE MEASUREMENT OF THERMAL SURFACE PROPERTIES.
	DOWNEY M J SCHAMLE G EUROPEAN SPACE RESEARCH AND TECHNOLOGY CENTER NOORDWIJK
	NASA AND CFSTI
	ESRO-SP-41 N69-37526
	          155-86 1969
	ESRO-TN-74 N69-37632
	          1-36 1969

58340	SOME THERMAL TRANSPORT PROPERTIES OF A LIMESTONE CONCRETE. MOORE J P GRAVES R S STRADLEY J G HANNAH J H MC ELROY D L OAK RIDGE NATIONAL LAB TENN AEC ORNL-TM-2644 1-25 1969 RR 70-2 97

TPRC Number	Bibliographic Citation
58359	THE ACCOMMODATION COEFFICIENTS OF SIMPLE GASES ON CARBON AT HIGH TEMPERATURE. MIDOL-MONNET C DUVAL X COMPT REND 270 C 18 1493 5 1970
58397	THE PREDICTION OF THE THERMAL CONDUCTIVITY OF HETEROGENEOUS MIXTURES. PH.D. THESIS. CHENG S-C AUBURN UNIVERSITY AUBURN ALABAMA UNIV MICROFILMS PUBL 69-5506 1-77 1968
58400	A MODEL FOR THE DRYING-HARDENING PROCESS IN CEMENT PASTES. PH.D. THESIS. JERIC M Z UNIVERSITY OF CALIFORNIA LOS ANGELES CALIFORNIA UNIV MICROFILMS PUBL 69-5317 1-196 1968
58406	THE STRUCTURE AND OPTICAL PROPERTIES OF CO-EVAPORATED NOBLE METAL-RARE EARTH ALLOY FILMS. PH.D. THESIS. TSUBOI Y UNIVERSITY OF PENNSYLVANIA PHILADELPHIA PENNSYLVANIA UNIV MICROFILMS PUBL UM69-15135 1-176 1968
58408	THERMAL EXPANSION AND MELTING ANOMALIES OF SMALL ALUMINUM CRYSTALS. PETROV YU I FIZ TVERD TELA 5 0 2462-76 1963
58409	THERMAL EXPANSION AND MELTING ANOMALIES OF SMALL ALUMINUM CRYSTALS. //ENGLISH TRANSLATION OF FIZ. TVERD. TELA 5 /9/ 2462-76, 1963.// PETROV YU I SOVIET PHYSICS-SOLID STATE 5 9 1793-805 1964
58436	THE EFFECTIVE THERMAL PROPERTIES OF TWO PHASE SOLIDS. BEN-AMOZ M INTERNATIONAL JOURNAL OF ENGINEERING SCIENCE 8 39-47 1970 CA 72 115402
58438	EFFECT OF MOISTURE CONTENT AND ULTRAVIOLET RADIATION ON THE PROPERTIES OF VARNISH FILMS. CHESNOKOV V F GORENKOV M P IZV VYSSH UCHEB ZAVED LES ZH 12 6 94-5 1969 CA 72 112853
58439	USE OF A QUASI-STEADY STATE HEAT FLOW METHOD FOR EVALUATING THE THERMOPHYSICAL CHARACTERISTICS OF MASTIC FLOOR COVERINGS. BERKHOER I D NOVOZHILOV A F IZV VYSSH UCHEB ZAVED STROIT ARKHITEKT 12 12 87-90 1969 CA 72 112386
58452	EASY, QUANTITATIVE HIDING POWER MEASUREMENTS. MITTON P B J PAINT TECHNOL 42 542 159-83 1970 CA 72 112848
58453	DETERMINATION OF THE THERMAL CONDUCTIVITY OF PROTECTIVE COATINGS ON LONG CYLINDERS OF FINITE DIAMETERS. I. MAYHAN K G ALLEN S C MONTLE J F J PAINT TECHNOL 42 542 184-8 1970 CA 72 112881
58454	DETERMINATION OF THE THERMAL CONDUCTIVITY OF PROTECTIVE COATINGS ON LONG CYLINDERS OF FINITE DIAMETERS. II. /ALSO PUBLISHED P. 887-94 IN PROCEEDINGS OF THE EIGHTH CONFERENCE ON THERMAL CONDUCTIVITY PUBLISHED BY PLENUM PRESS, NEW YORK./ MITCHELL J W BECKMAN W A J PAINT TECHNOL 42 542 189-92 1970 CA 72 112880
58455	CRYOSORPTION PUMPING OF HELIUM ON POROUS SILVER AT 4.2 K. HOBSON J P WILLIAMS B R J VAC SCI TECHNOL 6 6 965-7 1969 CA 72 115105
58474	VIBRATIONAL RELAXATION OF NITROUS OXIDE OPTICALLY EXCITED TO THE/00 DEGREE 1/ LEVEL. ARDITI I MARGOTTIN-MACLOU M GUEGUEN H DOYENNETTE L COMPT REND 270 7 477-80 1970 CA 72 126881
58510	SPECTRAL AND TOTAL ABSORPTION OF SOLAR RADIATION BY SOME CONSTRUCTION MATERIALS. BABAEV CH IZV AKAD NAUK TURKM SSR SER FIZ-TEKH KHIM GEOL NAUK 6 110-13 1969 CA 72 122249
58524	RELATED OPTICAL AND ELECTRICAL PROPERTIES OF THIN FILMS OF INDIUM ARSENIDE. HOWSON R P MALINA V J PHYSICS-APPLIED PHYSICS /PROC PHYS SOC/ 3 D 6 854-62 1970

TPRC Number	Bibliographic Citation							

58530 EFFECT OF THE SURFACE STATE OF A FILLER ON THE PROPERTIES OF COMPOSITIONS BASED ON POLYETHYLENE.
AINBINDER S B   ANDREEVA N G   VORONKOV M G
RASTRIGINA E F
MEKH POLIM
5   6   1038-45   1969   CA   72   122295

58535 RAPID METHOD FOR DETERMINING THE COLOR CHARACTERISTICS OF WHITE PAPERBOARD.
ANELIUNAS A E
PULP PAP MAG CAN
71   2   T45-9   1970   CA   72   123170

58560 THERMAL CONTACT RESISTANCE OF METALLIC INTERFACES. AN ANALYTICAL AND EXPERIMENTAL STUDY. PH.D. THESIS.
FLETCHER L S   ARIZONA STATE UNIVERSITY   TEMPE ARIZONA
UNIV MICROFILMS PUBL
69-5724   1-190   1969

58562 HIGH QUALITY SPUTTERED MULTILAYER COATINGS FOR INFRARED LASER APPLICATIONS.
GAVER R L   SEGUIN H J
REV SCI INSTRUM
41   3   427-9   1970

58563 ANOTHER COMPARISON OF THERMAL BONDING AGENTS.
ANDERSON A C   RAUCH R B   KREITMAN M M
REV SCI INSTRUM
41   3   469-70   1970
DDC
ARL-70-0315-W AD-716898
REPRINT   1970

58567 OPTICAL ABSORPTION PROPERTIES OF VANADATE GLASSES.
ANDERSON G W   COMPTON W D
J CHEM PHYS
52   12   6166-74   1970   CA   73   50454

58573 LIGHT ABSORPTION IN SEMICONDUCTOR AMORPHOUS FILMS OF ALKALI-METAL THIO- AND SELENOANTIMONITES.
ZORINA E L   GNIDASH N I   FINKELSHTEIN YA G
BERUL S I   LUZHNAYA N P
IZV AKAD NAUK SSSR NEORG MATER
5   12   2099-104   1969   CA   72   72264

58574 LIGHT ABSORPTION IN AMORPHOUS SEMICONDUCTIVE FILMS OF ALKALI-METAL THIOANTIMONITES AND SELENOANTIMONITES. //ENGLISH TRANSLATION OF IZV. AKAD. NAUK SSSR, NEORG. MATER. 5 /12/ 2099-2104, 1969.//
ZORINA E L   GNIDASH N I   FINKELSHTEIN YA G
BERUL S I   LUZHNAYA N P
INORGANIC MATERIALS
5   12   1788-92   1969

58585 THERMAL CONDUCTANCE AT THE INTERFACE OF A SOLID AND HELIUM II /KAPITZA CONDUCTANCE/.
SNYDER N S
USGPO
NBS-TN-385 PB-190548
1-98   1969

58590 CALIBRATION CHANGES IN EXTREME ULTRAVIOLET SOLAR SATELLITE INSTRUMENTS.
REEVES E M   PARKINSON W H
APPLIED OPTICS
9   5   1201-8   1970

58591 A MODE CONTROLLED Q-SWITCHED TUNEABLE RUBY LASER.
SCHOTLAND R M
APPLIED OPTICS
9   5   1211-13   1970

58622 TEMPERATURE DEPENDENCE OF THE VISCOSITY OF VERY THIN WATER FILMS BETWEEN QUARTZ GLASS SURFACES.
PESCHEL G   ADLFINGER K H
NATURWISSENSCHAFTEN
56   11   558   1969   CA   72   136651

58623 DETERMINATION OF THE OPTICAL CONSTANTS OF THIN FILMS FROM REFLECTANCE AND TRANSMITTANCE MEASUREMENTS BY A CURVE-FITTING PROCEDURE.
RIVORY J
OPT COMMUN
1   7   334-8   1970   CA   72   138065

58659 MEASUREMENTS OF THE STABILITY OF BLEACHED PHOTOGRAPHIC PHASE HOLOGRAMS.
MC MAHON D H   MALONEY W T
APPLIED OPTICS
9   6   1363-8   1970

58660 OPTICAL INTERFERENCE FILTERS FOR THE ADJUSTMENT OF SPECTRAL RESPONSE AND SPECTRAL POWER DISTRIBUTION.
DOBROWOLSKI J A
APPLIED OPTICS
9   6   1396-402   1970

58661 A ONE-SOLAR-CONSTANT IRRADIANCE STANDARD.
SCHNEIDER W E
APPLIED OPTICS
9   6   1410-18   1970

58662 AN ABSORBING SURFACE FOR MEASURING INTENSE THERMAL RADIATION.
DAVIES J M   PETER P H   GOFF R J
APPLIED OPTICS
9   6   1473-4   1970

58663 THREE-LAYER BROADBAND ANTIREFLECTION COATINGS FOR LITHIUM NIOBATE.
BERTHOLD J W III
APPLIED OPTICS
9   6   1490-1   1970

58664 A NEW OPTICAL FILTER DESIGNED FOR 01 1300-A TRIPLET DETECTION.
TOMIKI T   MIYATA T
APPLIED OPTICS
9   6   1492-3   1970

58670 PREPARATION OF THIN BARIUM TITANATE FILMS BY DC DIODE SPUTTERING.
SHINTANI Y   TADA O
J APPL PHYS
41   6   2376-80   1970

58672 MNBI FILMS. HIGH-TEMPERATURE PHASE PROPERTIES AND CURIE-POINT WRITING CHARACTERISTICS.
CHEN D   AAGARD R L
J APPL PHYS
41   6   2530-4   1970

58678 SPECTROPHOTOMETRIC THICKNESS MEASUREMENT FOR VERY THIN SILICON DIOXIDE FILMS ON SILICON.
RAND M J
J APPL PHYS
41   2   787-90   1970   CA   72   84363

58681 RADIATION OF ENERGY BY OXIDE FILMS ON METALS.
ZHOROV G A   PAVLOVSKAYA T G
TEPLOFIZ VYS TEMP
7   6   1107-11   1969   CA   72   114946

58682 EMISSION OF ENERGY OF OXIDE FILMS ON METALS. //ENGLISH TRANSLATION OF TEPLOFIZ. VYS. TEMP. 7 /6/ 1107-11, 1969.//
ZHOROV G A   PAVLOVSKAYA T G
HIGH TEMPERATURE
7   6   1031-4   1969

58708 THE CHANGE IN THICKNESS AND OTHER OPTICAL PROPERTIES OF ULTRAVIOLET IRRADIATED SILICON OXIDE FILMS.
HODGKINSON I J
APPLIED OPTICS
9   7   1577-86   1970

58712 MODE SELECTION IN A PULSE TRANSMISSION MODE RUBY SYSTEM FOR HOLOGRAPHY.
GREGOR E   GUSCOTT B R   MYERS J J
APPLIED OPTICS
9   7   1723-4   1970

58721 THERMAL CONTACT RESISTANCE. THE DIRECTIONAL EFFECT AND OTHER PROBLEMS.
THOMAS T R   PROBERT S D
INT J HEAT MASS TRANSFER
13   5   789-807   1970

58726 DETERMINATION OF THE APPLICABILITY RANGE OF THE NEW METHOD FOR SIMULTANEOUS EVALUATION OF EFFECTIVE THERMAL CONDUCTIVITY AND WALL HEAT TRANSFER COEFFICIENT IN PACKED BEDS ON GAS FLOW.
CYBULSKI A
BULL ACAD POLON SCI SER SCI CHIM
18   2   109-13   1970

58744 THERMAL CONDUCTIVITY OF RIGID URETHANE FOAMS.
BALL G W   HURD R   WALKER M G
J CELL PLAST
6   2   66-78   1970   CA   73   4419

58753 TEMPERATURE DEPENDENCE OF THE HEAT CAPACITIES OF REGENERATOR PACKINGS.
KOMISSAROV V M   KENDYS P N
TEPLOFIZ SVOISTVA TVERD TEL VYS TEMP TR VSES KONF 1966
1   170-7   1969   CA   73   5310

58767 SPECTROPHOTOMETRIC METHOD FOR DETERMINING SMALL AMOUNTS OF ADSORBED SUBSTANCES.
MAJOR G   ALESKOVSKII V B
IZV VYSSH UCHEB ZAVED KHIM KHIM TEKHNOL
13   1   34-7   1970   CA   73   21125

58803 INFRARED SPECTRA OF MATRIX-ISOLATED HYDROGEN SULFIDE IN SOLID NITROGEN.
TURSI A J   NIXON E R
J CHEM PHYS
53   2   518-21   1970

58814 APPLICATION OF DISPERSION RELATIONS TO THE DETERMINATION OF OPTICAL PARAMETERS OF SELENIUM.
FROISSART C
COMPT REND
270   8   24   1544-7   1970

TPRC Number	Bibliographic Citation

58821 THERMAL INTERFACE CONDUCTANCE IN A VACUUM. FROM PROC. 1ST AIAA ANNUAL MEETING, WASHINGTON, D. C., JUNE 29-JULY 2, 1964.
ATKINS H L   FRIED E
AMERICAN INSTITUTE AERONAUTICS AND ASTRONAUTICS   NEW YORK
AIAA   PAPER 64-253
                        1-23   1964

58822 MEASUREMENTS OF CONTACT COEFFICIENTS OF THERMAL CONDUCTANCE. FROM   PROC. CONF. AIAA THERMOPHYSICS SPECIALIST CONFERENCE, MONTEREY, CALIFORNIA. SEPT. 13-15, 1965.
FRY E M
AMERICAN INSTITUTE AERONAUTICS AND ASTRONAUTICS   NEW YORK
AIAA   PAPER 65-662
                        1-16   1965

58823 VACUUM INTERGRATING SPHERES FOR MEASURING CRYODEPOSIT REFLECTANCES FROM 0.35 TO 15 MICRONS. FROM PROC. CONF. AIAA THERMOPHYSICS SPECIALIST, MONTEREY, CALIFORNIA, SEPT. 13-15, 1965.
WOOD B E   MC CULLOUGH B A   DAWSON J P
BIRKEBAK R C
AMERICAN INSTITUTE AERONAUTICS AND ASTRONAUTICS   NEW YORK
AIAA   PAPER 65-674
                        1-14   1965

58842 HEAT PROCESSING OF TAR-BONDED REFRACTORIES.
BORISOV V G   KRAVETS L V
OGNEUPORY
                9   37-43   1969

58843 HEAT PROCESSING OF TAR-BONDED REFRACTORIES. //ENGLISH TRANSLATION OF OGNEUPORY /9/ 37-43, 1969.//
BORISOV V G   KRAVETS L V
REFRACTORIES
                9   562-8   1969

58844 CREEP IN REFRACTORY CONCRETES WITH ALUMINOSILICATE AGGREGATES AND VARIOUS BONDS.
MAMYKIN P S   KOKSHAROV V D   PURGIN A K
OGNEUPORY
                9   46-50   1969

58845 CREEP IN REFRACTORY CONCRETES WITH ALUMINOSILICATE AGGREGATES AND VARIOUS BONDS. //ENGLISH TRANSLATION OF OGNEUPORY /9/ 46-50, 1969.//
MAMYKIN P S   KOKSHAROV V D   PURGIN A K
REFRACTORIES
                9   572-5   1969

58874 THERMAL CONTACT OF SOLIDS.
THOMAS T R   PROBERT S D
CHEM PROCESS ENGINEERING
47   11   51-60   1966

58876 TRANSIENT HEAT FLOW BETWEEN CONTACTING SOLIDS.
HEASLEY J H
INT J HEAT MASS TRANSFER
8   147-54   1965

58878 THE COMPUTATION OF MECHANICAL AND THERMAL PROPERTIES OF COMPOSITE LAMINATES. M.S. THESIS.
FRYE D E JR   UNIVERSITY OF TENNESSEE   KNOXVILLE TENNESSEE
UNIVERSITY OF TENNESSEE
                        1-65   1969

58882 THERMAL CONTACT CONDUCTANCE OF NOMINALLY-FLAT, ROUGH SURFACES IN A VACUUM ENVIRONMENT. FROM PROC. CONF. AIAA 3RD AEROSPACE SCIENCES MEETING, NEW YORK, NEW YORK, JAN. 24-26, 1966.
YOVANOVICH M M   FENECH H
AMERICAN INSTITUTE AERONAUTICS AND ASTRONAUTICS NEW YORK
AIAA   PAPER NO. 66-42
                        1-30   1966

58891 DETERMINATION OF DIFFUSION AND ACCOMMODATION COEFFICIENTS AT THE WALL OF N2O MOLECULES, EXCITED TO THE VIBRATIONAL LEVEL /00 DEGREE 1/.
GUEGUEN H   DOYENNETTE L   ARDITI I
MARGOTTIN-MACLOU M
COMPT REND
270   B   26   1668-71   1970

58897 SURFACE ADSORPTION OF LIGHT GAS ATOMS.
HOLLENBACH D   SALPETER E E
J CHEM PHYS
53   1   79-86   1970

58903 INFRARED INVESTIGATION OF STRUCTURAL AND ORDERING CHANGES IN AMMONIUM CHLORIDE AND BROMIDE.
SCHUMAKER N E   GARLAND C W
J CHEM PHYS
53   392-407   1970   CA   73   50334

58931 THERMAL BOUNDARY RESISTANCE BETWEEN SOLIDS AND HELIUM BELOW 1 K.   /TID REPORT 20473./
ANDERSON A C   CONNOLLY J I   WHEATLEY J C

58962 GLIDE RE-ENTRY NOSE CAP GRAPHITE DEVELOPMENT TEST PLAN RESULTS.
ROGERS D C
SYMPOSIUM MATERIALS SPACE VEHICLE USE
1   7   1-39   1963

58969 DIELECTRIC PROPERTIES OF Y2O3 THIN FILMS PREPARED BY VACUUM EVAPORATION.
TSUTSUMI T
JAPAN J APPL PHYS
9   7   735-9   1970

58986 HEAT TRANSPORT THROUGH HELIUM II.   KAPITZA CONDUCTANCE.
SNYDER N S
CRYOGENICS
10   2   89-95   1970

58988 A GONIOREFLECTOMETER FACILITY USING COHERENT AND INCOHERENT SOURCES.
BAIR M E   CARMER D C   STEWART S R   WILLOW RUN
LAB INSTITUTE OF SCI AND TECHNOLOGY UNIVERSITY OF MICHIGAN
DDC
AFAL-TR-70-161
                        1-30   1970

59010 MEASUREMENTS OF OPTICAL PARAMETERS OF ANTIMONY TRISULPHIDE FILMS.
KUCIREK J
CZECH J PHYS
18   B   6   795-800   1968

59046 THERMAL DIFFUSIVITY OF ALUMINOSILICATE REFRACTORIES IN THE RANGE 200-1600 C.
LITOVSKII E YA   LANDA YA A   MILSHENKO R S
OGNEUPORY
35   5   17-19   1970   CA   73   48231

59083 ADSORPTION OF MIXTURES OF GASES ON ISOELECTRONIC ANALOGUES OF GERMANIUM.
KIROVSKAYA I A
ZH FIZ KHIM
44   1   159-   1970

59084 ADSORPTION OF MIXTURES OF GASES ON ISOELECTRONIC ANALOGUES OF GERMANIUM. //ENGLISH TRANSLATION OF ZH. FIZ. KHIM. 44 /1/ 159-, 1970.//
KIROVSKAYA I A
RUSSIAN J PHYS CHEM
44   1   87-90   1970

59085 CATALYTIC ACTIVITY OF PLATINUM METALS AND REACTIVITY OF SURFACE CARBONYLS.
KAVTARADZE N N   SOKOLOVA N P
ZH FIZ KHIM
44   1   171-   1970

59086 CATALYTIC ACTIVITY OF PLATINUM METALS AND REACTIVITY OF SURFACE CARBONYLS. //ENGLISH TRANSLATION OF ZH. FIZ. KHIM. 44 /1/ 171-, 1970.//
KAVTARADZE N N   SOKOLOVA N P
RUSSIAN J PHYS CHEM
44   1   93-6   1970

59107 EFFECTIVE THERMAL EXPANSION COEFFICIENTS AND SPECIFIC HEATS OF COMPOSITE MATERIALS.
ROSEN B W   HASHIN Z
INTERNATIONAL JOURNAL OF ENGINEERING SCIENCE
8   157-73   1970

59115 SPECTRAL PROPERTIES OF FLUOREL L-3203-6 AND 1059 AND THEIR DERIVATIVES WHEN MONOCHROMATICALLY IRRADIATED WITH ENERGIES IN THE SOLAR SPECTRUM.
FRENCH B O
SAMPE QUARTERLY
1   31-7   1970

59117 HYPERSONIC NEAR-FREE MOLECULE FLOW OVER A SHARP-LEADING-EDGED FLAT PLATE WITH A SMALL THERMAL ACCOMMODATION COEFFICIENT.
ORTLOFF C R
ASTRONAUTICA ACTA
15   215-29   1970

59133 THERMAL DESIGN OF HIGH-ALTITUDE BALLOONS AND INSTRUMENT PACKAGES.
KREITH F
J HEAT TRANSFER TRANS ASME
92   C   3   307-32   1970

59138 THE EFFECT OF THERMAL CONDUCTIVITY OF PLATING MATERIAL ON THERMAL CONTACT RESISTANCE.
MIKIC B B   CARNASCIALI G
J HEAT TRANSFER TRANS ASME
92   C   3   475-82   1970

59139 STEADY STATE HEAT TRANSFER IN PARTIALLY LIQUID FILLED POROUS MEDIA.
HANSEN D   BREYER W H   RIBACK W J
J HEAT TRANSFER TRANS ASME
92   C   3   520-7   1970

TPRC Number	Bibliographic Citation

59140 FROST DEPOSITION ON COLD SURFACES.
BRIAN P L T   REID R C   SHAH Y T
IND ENG CHEM FUNDAMENTALS
9   3   375-80   1970

59141 LABORATORY EXPERIMENTS TO STUDY SURFACE CONTAMINATION
AND DEGRADATION OF OPTICAL COATINGS AND MATERIALS IN
SIMULATED SPACE ENVIRONMENTS.
HASS G   HUNTER W R
APPLIED OPTICS
9   9   2101-10   1970   CA   73   93110

59143 OPTICAL PROPERTIES OF THIN LAYERS OF YTTRIUM UNDER
STATIC ULTRAVACUUM.
PETRAKIAN J P   PALMARI J-P   RASIGNI G
APPLIED OPTICS
9   9   2115-18   1970

59149 INVESTIGATION OF THE THERMAL CONDUCTIVITY OF FINELY
FIBROUS MATERIALS BASED ON KAOLIN AND BASALT FIBERS.
EGOROV B N   KONDRATENKOV V N
TEPLOFIZ VYSOKIKH TEMPERATUR
8   1   209-11   1970

59150 INVESTIGATION OF THE THERMAL CONDUCTIVITY OF FINELY
FIBROUS MATERIALS BASED ON KAOLIN AND BASALT FIBERS.
//ENGLISH TRANSLATION OF TEPLOFIZ. VYSOKIKH
TEMPERATUR 8 /1/ 209-11, 1970.//
EGOROV B N   KONDRATENKOV V N
HIGH TEMPERATURE
8   1   198-200   1970

59160 ELECTRONIC PROPERTIES OF LIQUID WATER IN THE VACUUM
ULTRAVIOLET REGION. PH.D. THESIS.
PAINTER L L R   UNIVERSITY OF TENNESSEE   KNOXVILLE
TENNESSEE
UNIV MICROFILMS PUBL
UM69-7173   N70-20326
                      1-109   1968   PAN 70-8-8 1394

59163 THERMAL CONDUCTANCES AND PEAK HEAT CURRENTS OF
IMMERSED GRAPHITE HEATERS IN LIQUID HELIUM II.
PH.D. THESIS.
CHAPMAN R C   UNIVERSITY OF CALIFORNIA   LOS
ANGELES CALIFORNIA
UNIV MICROFILMS PUBL
69-8093   1-184   1968

59164 THERMAL RADIATION PROPERTIES OF COMPACTED ALUMINA
POWDER. PH.D. THESIS.
DILLENIUS M F E   UNIVERSITY OF CALIFORNIA
BERKELEY
UNIV MICROFILMS PUBL
69-10278   1-152   1968

59169 A MASS SPECTROMETRIC STUDY OF VAPOR-SOLID
INTERACTIONS. PH.D. THESIS.
KINAWA A A   RENSSELAER POLYTECHNIC INST   TROY N Y
UNIV MICROFILMS PUBL
UM69-13685 N70-22745   1969   PAN 70-8-10 1785

59176 THERMAL PROPERTIES OF SELECTED EVACUATED POWDER
INSULATIONS AT CRYOGENIC TEMPERATURES.
KNIGHT B L
UNIV MICROFILMS PUBL
UM69-19554   1-322   1969   CA   72   113253

59184 INFRARED SPECTRUM OF WATER ADSORBED ON HECTORITE.
PROST R   CHAUSSIDON J
CLAY MINER
8   2   143-9   1969   CA   73   60813

59191 THERMAL CONDUCTIVITY AND FIGURE OF MERIT OF CUPROUS
SULFIDE.
ROUTIE R   MAHENC J
ELECTROCHIM ACTA
15   7   1201-7   1970   CA   73   60355

59192 RELATION BETWEEN EXTINCTION OR TRANSMISSION AND
REFLECTANCE OF NONSCATTERING COATINGS ON A WHITE
SUBSTRATE.
HOFFMANN K
FARBE LACK
76   7   665-73   1970   CA   73   57244

59196 STUDY OF ENVIRONMENTAL EFFECTS UPON PARTICULATE
RADIATION INDUCED ABSORPTION BANDS IN SPACECRAFT
THERMAL CONTROL COATING PIGMENTS.
MC CARGO M   GREENBERG S A   BREUCH R A
LOCKHEED MISSILES AND SPACE CO   PALO ALTO CALIFORNIA
NTIS
NASA-CR-73289 N69-16868
                      1-104   1969

59204 DEMONSTRATION OF MANUFACTURING TECHNIQUES FOR
APPLICATION OF HIGH PERFORMANCE CRYOGENIC INSULATION.
LOFGREN C L   GIESEKING D E   BOEING CO   SEATTLE
WASHINGTON   NTIS
NTIS
N69-26898   NASA-CR-98459
D2-139267-1
                      1-47   1968

TPRC Number	Bibliographic Citation

59206 STUDY OF THERMAL CONDUCTIVITY REQUIREMENTS. MSFC
20-INCH AND 105-INCH CRYOGENIC TANK ANALYSES.
HALE D V   LOCKHEED MISSILES AND SPACE CO
HUNTSVILLE ALABAMA   NTIS
NTIS
NASA-CR-61288 N69-35811
                      1-28   1969

59219 PROPERTIES OF THIN FILMS OF LEAD TELLURIDE AND TIN
TELLURIDE DEPOSITED AT TEMPERATURES BETWEEN 4.2 AND
300 K.
BROWN R W   MILLNER A R   ALLGAIER R S
THIN SOLID FILMS
5   3   157-68   1970   CA   73   60411

59244 COMPARISON OF BIDIRECTIONAL REFLECTANCE MEASUREMENTS
AND MODEL FOR ROUGH METALLIC SURFACES.
SMITH T F   HERING R G
PROC SYMP THERMOPHYS PROP 5TH
                      429-35   1970

59259 THERMAL CONTACT RESISTANCE OF MACHINED METAL SURFACES
IN A VACUUM ENVIRONMENT.
MALKOV V A
HEAT TRANSFER-SOVIET RESEARCH
2   4   24-33   1970

59274 THERMAL RESISTANCE BETWEEN POWDERED CERIUM MAGNESIUM
NITRATE AND LIQUID HELIUM AT VERY LOW TEMPERATURES.
BLACK W C   MOTA A C   WHEATLEY J C   BISHOP J H
BREWSTER P M   CALIFORNIA UNIV   SAN DIEGO
NTIS
UCSD-34-P-143-35
N71-25463   1-13   1970

59276 DERIVATION OF ONE-DIMENSIONAL HEAT TRANSFER RELATION
IN MULTILAYER COMPOSITES. FROM HIGH-PERFORMANCE
THERMAL PROTECTION SYSTEMS, VOLUME 2, FINAL REPORT.
LOCKHEED MISSILES AND SPACE COMPANY
SUNNYVALE CALIFORNIA
NASA AND CFSTI
N70-30604   LMSC-A964947
NASA-CR-102624
                      1-20   1969

59278 CONTINUOUS WAVE CHEMICAL LASER CAVITY STUDIES.
SPENCER D J   BIXLER H A   DURRAN D A   THE
AEROSPACE CORP   EL SEGUNDO CALIFORNIA
DDC
SAMSO-TR-70-255
AD-709202   1-8   1970

59290 INFRARED REFLECTANCE OF CO2 CRYODEPOSITS. FINAL
REPORT, JULY 1968-SEPT. 1969.
WOOD B E   SMITH A M   SEIBER B A   ARO INC
ARNOLD ENGINEERING DEVELOPMENT CTR
DDC
AEDC-TR-70-108
AD-708509   1-38   1970

59299 THERMAL CONDUCTIVITY MEASUREMENTS OF A CANDIDATE
VIKING HEAT-SHIELD MATERIAL AFTER STERILIZATION, AND
DURING EXPOSURE TO VACUUM, AND TO A SIMULATED MARTIAN
ATMOSPHERE. FROM PROC. OF THE SPACE SIMULATION
CONFERENCE. NBS, GAITHERSBURG, MARYLAND. SEPTEMBER
14-16, 1970.
GREENWOOD L R   FLEMING R M
NTIS
NASA-TM-X-66938
COM71-50046
N71-20214   197-208   1970

59319 PROPERTIES OF MULTILAYER FILTERS. INTERIM REPT. NO.
1, SEPT. 1969-FEB. 1970.
BAUMEISTER P W   INST OF OPTICS ROCHESTER UNIV
N Y
NASA AND CFSTI
NASA-CR-110405
N70-30121   1-26   1970

59331 INTERFERENCE IN PHOTO-GRAPHIC LAYERS.
HOESCHEN D   VIETH G
Z NATURFORSCH
25   A   7   1158-9   1970

59347 THE ROLE OF TRANSITION ELEMENTS IN COLORATION OF THE
GLASS PHASE IN CEMENT CLINKER.
PONOMAREV I F   GRACHYAN A N   ROTYCH N V
ZH PRIKL KHIM
41   5   965-8   1968

59348 THE ROLE OF TRANSITION ELEMENTS IN COLORATION OF THE
GLASS PHASE IN CEMENT CLINKER. //ENGLISH TRANSLATION
OF ZH. PRIKL. KHIM. 41 /5/ 965-68, 1968.//
PONOMAREV I F   GRACHYAN A N   ROTYCH N V
J APPLIED CHEM USSR
41   5   916-19   1968

59361 INTERFACIAL TENSION BETWEEN ALUMINUM AND CAUSTIC
ALKALI METLTS. //ENGLISH TRANSLATION OF ZH. PRIKL.
KHIM. 41 /3/ 505-9, 1968.//
SHCHERBAKOV V K   KUZNETSOV S I
J APPLIED CHEM USSR

TPRC Number	Bibliographic Citation
59381	ELECTRICAL AND OPTICAL PROPERTIES OF TITANIUM DIOXIDE FILMS. TRAVINA T S MUKHIN YU A IZV VYS UCH ZAV FIZ 9 6 74 80 1066
59382	ELECTRICAL AND OPTICAL PROPERTIES OF TITANIUM DIOXIDE FILMS. //ENGLISH TRANSLATION OF IZV. VYS. UCH. ZAV. FIZ. 9 /6/ 74-80, 1966.// TRAVINA T S MUKHIN YU A SOVIET PHYSICS JOURNAL 9 6 40-3 1966
59387	TEMPERATURE DEPENDENCE OF THE OPTICAL PROPERTIES OF DIELECTRIC NARROW-BAND FILTERS. FURMAN SH A LEVINA M D OPT SPEKTROSK 28 4 766-74 1970 CA 73 20011
59388	INVESTIGATIONS OF TEMPERATURE DEPENDENCE OF OPTICAL PROPERTIES OF DIELECTRIC NARROW-BAND FILTERS. //ENGLISH TRANSLATION OF OPT. SPEKTROSK. 28 /4/ 766-74, 1970.// FURMAN SH A LEVINA M D OPTICS SPECTROSCOPY 28 4 412-16 1970
59389	WIDE-BAND REFLECTORS WITH MULTILAYER DIELECTRIC COATINGS. KOROLEV F A KLEMENTEVA A YU MESHCHERYAKOVA T F RAMAZINA I A OPT SPEKTROSK 28 4 775-80 1970
59390	WIDE-BAND REFLECTORS WITH MULTILAYER DIELECTRIC COATINGS. //ENGLISH TRANSLATION OF OPT. SPEKTROSK. 28 /4/ 775-80, 1970.// KOROLEV F A KLEMENTEVA A YU MESHCHERYAKOVA T F RAMAZINA I A OPTICS SPECTROSCOPY 28 4 416-19 1970
59427	GUARDED ELECTRICAL CYLINDRICAL CALORIMETER FOR MEASURING THERMAL CONDUCTIVITY OF MULTILAYER INSULATION. HALE D V RENY G D HYDE E H ADVAN CRYOG ENG 1969 15 324-31 1970 CA 73 67923
59491	DEPENDENCE OF THE CONTACT HEAT CONDUCTIVITY OF GRANULAR SYSTEMS ON THE EXTERNAL LOAD. //ENGLISH TRANSLATION OF INZH. FIZ. ZH. 11 /2/ 202-6, 1966.// DULNEV G N ZARICHNYAK YU P MURATOVA B L SIGALOVA Z V J ENGINEERING PHYS 11 2 110-12 1966
59492	DETERMINATION OF THE THERMAL RESISTANCE IN VACUUM OF CONTACTS BETWEEN METALLIC SURFACES WITH VARIOUS DEGREES OF ROUGHNESS. //ENGLISH TRANSLATION OF INZH. FIZ. ZH. 11 /3/ 329-37, 1966.// KAGANER M G ZHUKOVA R I J ENGINEERING PHYS 11 3 182-8 1966
59494	AUTOMATIC DETERMINATION OF THE THERMAL DIFFUSIVITY OF HEAT INSULATORS. VLASOV V V DOROGOV N N INZH FIZ ZH 11 3 354-8 1966
59495	AUTOMATIC DETERMINATION OF THE THERMAL DIFFUSIVITY OF HEAT INSULATORS. //ENGLISH TRANSLATION OF INZH. FIZ. ZH. 11 /3/ 354-58, 1966.// VLASOV V V DOROGOV N N J ENGINEERING PHYS 11 3 199-201 1966
59500	VACUUM-TUBE AND TRANSISTOR INSTRUMENTS FOR INVESTIGATING THERMAL CONDUCTIVITY BY MEANS OF A LINEAR HEAT SOURCE. YANKELEV L F BLUVSHTEIN I M INZH FIZ ZH 11 6 756-60 1966
59501	VACUUM-TUBE AND TRANSISTOR INSTRUMENTS FOR INVESTIGATING THERMAL CONDUCTIVITY BY MEANS OF A LINEAR HEAT SOURCE. //ENGLISH TRANSLATION OF INZH. FIZ. ZH. 11 /6/ 756-60, 1966.// YANKELEV L F BLUVSHTEIN I M J ENGINEERING PHYS 11 6 406-9 1966
59508	AN EXPERIMENTAL INVESTIGATION OF THE THERMAL CONDUCTIVITY AND ELECTRICAL RESISTIVITY OF THREE POROUS 304L STAINLESS STEEL RIGIMESH MATERIAL TO 1300 K. TYE R P DYNATECH CORPORATION NASA NASA-CR-72710 1-23 1970

TPRC Number	Bibliographic Citation
59520	A STUDY OF AN ARRAY OF SQUARE OPENINGS. BELL R J ROMERO H V APPLIED OPTICS 9 10 2341-9 1970
59521	THEORY AND EXPERIMENTS FOR MULTIELEMENT GRID FILTERS IN A DIELECTRIC. BELL R J ROMERO H V BLEA J M APPLIED OPTICS 9 10 2350-8 1970
59527	THEORY OF METALLIC THIN FILM OPTICS. M.S. THESIS. NESTELL J E JR DARTMOUTH COLLEGE HANOVER NEW HAMPSHIRE DARTMOUTH COLLEGE 1-59 1970
59536	CONTRIBUTION TO THE STUDY OF THE OPTICAL PROPERTIES OF COPPER HALOGENIDES AT LOW TEMPERATURES. REISS R CAHIERS PHYS 13 104 129-72 1959
59537	TABLES FOR THE THERMAL EXPANSIONS OF DIVERS SIMPLE METALLIC OR NON-METALLIC BODIES, AND OF SOME COMPOUNDS OF HYDROCARBON. FIZEAU M H COMPT REND 68 20 1125-31 1869
59542	INFRARED ABSORPTION AT LONGITUDINAL OPTICAL FREQUENCY IN CUBIC CRYSTAL FILMS. BERREMAN D W PHYS REV 130 6 2193-8 1963
59552	THE INFRARED EIGEN FREQUENCIES OF THE ALKALI HALOGENIDE CRYSTALS. BARNES R B Z PHYSIK 75 723-34 1932
59561	THERMAL CONDUCTIVITY OF PARTICULATE BASALT AS A FUNCTION OF DENSITY IN SIMULATED LUNAR AND MARTIAN ENVIRONMENTS. FOUNTAIN J A WEST E A JOURNAL OF GEOPHYSICAL RESEARCH 75 4063-9 1970
59568	THERMAL CONDUCTIVITY OF REINFORCED PLASTICS. SUNDSTROM D W CHEN S Y J COMPOSITE MATERIALS 4 113-17 1970 IAA 10-7 1320
59587	MEASUREMENTS IN THE LONG WAVE-LENGTH INFRARED FROM 20 EPSILON TO 135 EPSILON. BARNES R B PHYS REV 39 4 562-75 1932
59599	THERMAL AND THERMOELASTIC PROPERTIES OF ISOTROPIC COMPOSITES. BUDIANSKY B JOURNAL OF COMPOSITE MATERIALS 4 286-95 1970
59607	EFFECTIVE CONDUCTANCE ALONG PARALLEL RADIATION SHIELDS. POGSON J T MAC GREGOR R K AMERICAN INSTITUTE OF AERONAUTICS AND ASTRONAUTICS THERMOPHYSICS CONFERENCE 5TH LOS ANGELES CALIF JUNE 29-JULY 1 1970 1-8 PAPER-70-847 1970
59608	THERMAL CHARACTERISTICS OF MULTILAYER INSULATION. CAU H MOY H C AMERICAN INSTITUTE OF AERONAUTICS AND ASTRONAUTICS THERMOPHYSICS CONFERENCE 5TH LOS ANGELES CALIF JUNE 29-JULY 1 1970 5 1-12 PAPER-70-850 1970
59609	EXTRAVEHICULAR SPACE SUIT THERMAL INSULATIONS. RICHARDSON D L RUCCIA F E AMERICAN INSTITUTE OF AERONAUTICS AND ASTRONAUTICS THERMOPHYSICS CONFERENCE 5TH LOS ANGELES CALIF JUNE 29-JULY 1 1970 1-13 PAPER-70-851 1970
59610	PREDICTION OF THERMAL CONTACT CONDUCTANCE BETWEEN SIMILAR METAL SURFACES. FLETCHER L S AMERICAN INSTITUTE OF AERONAUTICS AND ASTRONAUTICS THERMOPHYSICS CONFERENCE 5TH LOS ANGELES CALIF JUNE 29-JULY 1 1970 5 1-10 PAPER-70-852 1970

TPRC Number	Bibliographic Citation

59611 EXPERIMENTAL CONFIRMATION OF CYCLIC THERMAL JOINT
CONDUCTANCE.
MC KINZIE D J JR
AMERICAN INSTITUTE OF AERONAUTICS AND ASTRONAUTICS
THERMOPHYSICS CONFERENCE 5TH LOS ANGELES CALIF JUNE
29-JULY 1 1970
     5   1-13
PAPER-70-853   1970

59613 AN INVESTIGATION OF THE VALIDITY OF KIRCHOFFS LAW FOR
FREELY RADIATING METALLIC SURFACES.
GRIMM T C
AMERICAN INSTITUTE OF AERONAUTICS AND ASTRONAUTICS
THERMOPHYSICS CONFERENCE 5TH LOS ANGELES CALIF JUNE
29-JULY 1 1970
     5   1-16
PAPER-70-858   1970

59614 MEASUREMENT OF BIDIRECTIONAL REFLECTANCE USING A
PHOTOGRAPHIC TECHNIQUE.
LOEHRLEIN J E   WINTER E R F   VISKANTA R
AMERICAN INSTITUTE OF AERONAUTICS AND ASTRONAUTICS
THERMOPHYSICS CONFERENCE 5TH LOS ANGELES CALIF JUNE
29-JULY 1 1970
     5   1-9
PAPER-70-859   1970

59615 EMITTANCE MEASUREMENT WITH A DIFFERENTIAL SCANNING
CALORIMETER. M.S. THESIS.
KHAN A M   VILLANOVA UNIVERSITY   VILLANOVA
PENNSYLVANIA
VILLANOVA UNIVERSITY
      1-70   1969

59627 OPTICAL EXCITATION OF SOLID NEON IN THE VACUUM
ULTRAVIOLET.
HAENSEL R   KEITEL G   KOCH E E   KOSUCH N
SKIBOWSKI M
PHYS REV LETTERS
25   18   1281-3   1970

59632 ABSORPTION SPECTRA OF TELLURIUM IN THE SPECTRAL
RANGE FROM 18 TO 460 CM/-1/.
GROSSE P   RICHTER W
PHYS STATUS SOLIDI
41   1   239-46   1970

59637 DEVELOPMENT OF SPACE-STABLE THERMAL-CONTROL COATINGS.
TRIANNUAL REPORT.
ZERLAUT G A   GILLIGAN J E   IIT RESEARCH
INSTITUTE TECHNOLOGY CENTER CHICAGO ILLINOIS
NASA
IITRI-U6002-90
      1-53   1970

59639 PHOTOCONDUCTIVITY OF CDSE FILMS PREPARED BY A VAPOR
EVAPORATING-REACTIVE SPUTTERING METHOD.
TANAKA K
JAPANESE J APPLIED PHYS
9   9   1070-7   1970

59640 A CURRENT INSTABILITY IN TIO2 THIN FILM.
HADA T   HAYAKAWA S   WASA K
JAPANESE J APPLIED PHYS
9   9   1078-84   1970

59641 ELECTRICAL AND OPTICAL PROPERTIES OF SILICON-TIN
DIOXIDE HETEROJUNCTIONS.
NISHINO T   HAMAKAWA Y
JAPANESE J APPLIED PHYS
9   9   1085-90   1970

59646 INVESTIGATION OF A MODEL FOR BIDIRECTIONAL
REFLECTANCE OF ROUGH SURFACES.
TREAT C H   WILDIN M W
PROGRESS ASTRONAUT AERON
23   77-92   1970
NTIS
AIAA-69-84   1970

59647 RADIATION-INDUCED ABSORPTION BANDS IN SPACECRAFT
THERMAL CONTROL COATING PIGMENTS.
MC CARGO M   GREENBERG S A   DOUGLAS N J
PROGRESS ASTRONAUT AERON
23   189-218   1970
NTIS
AIAA-69-642   1970

59648 ELECTRON ENERGY DEPENDENCE FOR IN-VACUUM DEGRADATION
AND RECOVERY IN THERMAL CONTROL SURFACES.
FOGDALL L B   CANNADAY S S   BROWN R R
PROGRESS ASTRONAUT AERON
23   219-48   1970
NTIS
AIAA-69-643   1970

59649 RESULTS FROM THE ATS-3 REFLECTOMETER EXPERIMENT.
HEANEY J B
PROGRESS ASTRONAUT AERON
23   249-74   1970
NTIS
AIAA-69-644   1970

59654 SOLAR PANEL TEST SET.
RAY W E   THE JOHNS HOPKINS UNIVERSITY APPLIED
PHYSICS LAB   SILVER SPRING MD
DDC
AD-707345   1-27   1970

59668 INVESTIGATION OF THE OPTICAL AND THERMAL BEHAVIOR OF
SURFACE COATINGS FOR THE HELIOS SPACE PROBES.
HOERSTER H   PHILIPS ZENTRALLABORATORIUM AACHEN
NASA AND CFSTI
N70-23877   1-22   1969

59672 HIGH-SPEED INFRARED TEMPERATURE-MAPPING OF
HIGH-PERFORMANCE WEAPON COMPONENTS.
VAN DAMME G E   AMORUSO M J   IVERSEN R J
MC GARVEY J W   U S ARMY WEAPONS COMMAND ROCK
ISLAND ILLINOIS
DDC
RE-70-158   AD-710229
      1-28   1970

59676 THERMAL CONDUCTANCE OF ALUMINA-NICKEL INTERFACES AT
ELEVATED TEMPERATURES.
HAYS L G
INTERN J HEAT MASS TRANSFER
13   8   1293-7   1970 CA   73   81443

59686 DEVELOPMENT OF SPACE STABLE, LOW SOLAR ABSORPTANCE,
PIGMENTED THERMAL CONTROL COATINGS. FINAL REPORT
OCT. 1968-OCT. 1969.
GILLIGAN J E   BRZUSKIEWICZ J   ZERLAUT G A
YAMATE G   IIT RESEARCH INST CHICAGO ILL
NASA AND CFSTI
NASA-CR-66917 N70-30822
IITRI-C6166-12
      1-192   1969

59711 SPECTRAL FILTERING POSSIBILITIES OF SURFACE PLASMA
OSCILLATIONS IN THIN METAL FILMS.
SALWEN A   STENSLAND L
OPT COMMUN
2   1   9-13   1970 CA   73   82448

59712 OBSERVATION OF PLASMA RESONANCE AT 790 ANGSTROM IN
REFLECTION FROM THIN ZIRCONIUM FILMS.
RUDISILL J E   MATSUI A   WEISSLER G L
OPT COMMUN
2   1   39-40   1970 CA   73   82284

59738 AN INFRARED SPECTROSCOPY STUDY OF THE ADSORPTION OF
CERTAIN MOLECULES ON ZIRCONIUM DIOXIDE.
TRETYAKOV N E   POZDNYAKOV D V   ORANSKAYA O N
FILIMONOV V N
ZH FIZ KHIM
44   4   1077-   1970

59739 AN INFRARED SPECTROSCOPIC STUDY OF THE ADSORPTION OF
CERTAIN MOLECULES ON ZIRCONIUM DIOXIDE. //ENGLISH
TRANSLATION OF ZH. FIZ. KHIM. 44 /4/ 1077- , 1970./
TRETYAKOV N E   POZDNYAKOV D V   ORANSKAYA O N
FILIMONOV V N
RUSSIAN J PHYS CHEM
44   4   596-600   1970

59740 EFFECT OF THE CARRIER ON THE INFRARED SPECTRA OF
CARBON MONOXIDE ADSORBED ON RHODIUM AND COPPER.
KAVTARADZE N N   SOKOLOVA N P
ZH FIZ KHIM
44   4   1088-   1970

59741 EFFECT OF THE CARRIER ON THE INFRARED SPECTRA OF
CARBON MONOXIDE ADSORBED ON RHODIUM AND COPPER.
//ENGLISH TRANSLATION OF ZH. FIZ. KHIM. 44 /4/
1088- , 1970.//
KAVTARADZE N N   SOKOLOVA N P
RUSSIAN J PHYS CHEM
44   4   603-5   1970

59832 STUDIES OF THE OPTICAL PROPERTIES OF SOLIDS BY
ELLIPSOMETRY. PH.D. THESIS.
KNAUSENBERGER W H   PENNSYLVANIA STATE UNIVERSITY
PHILADELPHIA PENNSYLVANIA
PENNSYLVANIA STATE UNIVERSITY
      1-65   1969

59833 DESIGN AND USE OF INTERFERENCE PASSBAND FILTERS WITH
WIDE-ANGLE LENSES FOR MULTISPECTRAL PHOTOGRAPHY.
MC KENNEY D B   SLATER P N
APPLIED OPTICS
9   11   2435-40   1970

59835 VARIABLE METAL MESH COUPLER FOR FAR INFRARED LASERS.
ULRICH R   BRIDGES T J   POLLACK M A
APPLIED OPTICS
9   11   2511-16   1970

59874 THE COEFFICIENT OF EXPANSION OF CONCRETE.
PENCE WM D
J WEST SOC ENGRS
6   549-75   1901

TPRC Number	Bibliographic Citation
59892	THE RATIO OF THE THERMAL ACCOMMODATION COEFICIENTS OF HE4 TO HE3 AS OBSERVED AND AS PREDICTED ON THE BASIS OF VARIOUS EXISTING THEORIES. FROM ADVANCES IN APPLIED MECHANICS. /RAREFIED GAS DYNAMICS, PROC. OF THE 6TH INTERNAT. SYMP., M.I.T. CAMBRIDGE, MASS. JULY 22-6, 1968. VOL. 2./ THOMAS L B KRUEGER C L HARRIS R E MISSOURI UNIVERSITY COLUMBIA MO ACADEMIC PRESS NEW YORK SUPPL. 5 1015-23 1969
59893	EXPERIMENTAL INVESTIGATION OF MOMENTUM EXCHANGE BETWEEN ARGON AND KRYPTON BEAMS INCIDENT ON CLEAN GOLD SURFACES. FROM ADVANCES IN APPLIED MECHANICS. /RAREFIED GAS DYNAMICS, PROC. OF THE 6TH INTERNAT. SYMP., M.I.T. CAMBRIDGE, MASS. JULY 22-6, 1968. VOL. 2./ MC GINN J H GE SPACE SCIENCES LABORATORY KING OF PRUSSIA PA ACADEMIC PRESS NEW YORK SUPPL. 5 1025-34 1969
59894	HIGH ENERGY SCATTERING OF INERT GASES FROM WELL-CHARACTERIZED SURFACES. II-THEORETICAL. FROM ADVANCES IN APPLIED MECHANICS. /RAREFIED GAS DYNAMICS, PROC. OF THE 6TH INTERNAT. SYMP., M.I.T. CAMBRIDGE, MASS. JULY 22-6, 1968. VOL. 2./ JACKSON D P FRENCH J B TORONTO UNIVERSITY TORONTO CANADA ACADEMIC PRESS NEW YORK SUPPL. 5 1119-34 1969
59896	STUDIES OF THE VELOCITY DISTRIBUTIONS OF INCIDENT AND REFLECTED A BEAMS. FROM ADVANCES IN APPLIED MECHANICS. /RAREFIED GAS DYNAMICS, PROC. OF THE 6TH INTERNAT. SYMP., M.I.T. CAMBRIDGE, MASS. JULY 22-6, 1968. VOL. 2./ BROWN R F POWELL H M TRAYER D M ARO INC ARNOLD ENGINEERING DEVELOPMENT CENTER ARNOLD AIR FORCE STATION TENN ACADEMIC PRESS NEW YORK SUPPL. 5 1187-90 1969
59935	THE COEFFICIENTS OF THERMAL EXPANSION OF WOOD AND WOOD PRODUCTS. WEATHERWAX R C STAMM A J TRANS ASME 69 421-32 1947
59944	EXPANSION OF BORON TRIOXIDE. BONNELL D G R HARPER F C J INST CIVIL ENGINEERS 33 4 320-30 1950
59952	MECHANICAL PROPERTIES OF EPOXY RESINS AND GLASS/EPOXY COMPOSITES AT CRYOGENIC TEMPERATURES. SOFFER L M MOLHO R PROC NASA-CASE CONF PROPERTIES OF POLYMERS AT CRYOGENIC TEMPERATURES CLEVELAND OHIO APRIL 25-27 1967 87-117 1968
59973	A. MULTILAYER INSULATION TESTING. CROSBY J R NASA JPL-37-64 /VOL III/ 79-80 1970
59975	INVESTIGATION OF TRANSIENT DEGRADATION/CONTAMINATION OF THERMAL COATINGS. QUARTERLY PROGRESS REPT. NO. 2. AUG. 1970-OCT. 1970. MC CARGO M GREENBERG S A ROLLING R E MC DONALD S L PALO ALTO RESEARCH LAB LOCKHEED MISSILES AND SPACE COMP PALO ALTO CALIF LMSC LMSC-TP-3139 1-40 1970
59985	EFFECT OF ERROR IN SPECTRAL MEASUREMENTS OF SOLAR SIMULATORS ON SURFACE RESPONSE. CURTIS H B LEWIS RESEARCH CENTER CLEVELAND OHIO NASA AND CFSTI NASA-TN-D-5904 E-4913 N70-33050 1-20 1970
59986	MECHANISMS OF RECOVERY OF RADIATION DAMAGED SILICATE-TREATED ZINC OXIDE THERMAL CONTROL COATINGS. COLONY J A BASS J A GODDARD SPACE FLIGHT CENTER GREENBELT MD NASA AND CFSTI NASA-TM-X-63965 X-713-70-251 N70-33138 1-16 1970
59993	INITIAL THERMAL EXPANSION CHARACTERISTICS OF INSULATION REFRACTORY CONCRETES. CROWLEY M S AM CERAMIC SOC BULL 35 12 465-8 1956
59998	SOUND VELOCITY AND ABSORPTION IN LOW-PRESSURE GASES CONFINED TO TUBES OF CIRCULAR CROSS SECTION. SHIELDS F D FAUGHN J

TPRC Number	Bibliographic Citation
60025	FLAT PLATE CALORIMETER APPARATUS. FROM HIGH-PERFORMANCE THERMAL PROTECTION SYSTEMS, VOLUME 2, FINAL REPORT. LOCKHEED MISSILES AND SPACE COMPANY SUNNYVALE CALIFORNIA NASA AND CFSTI N70-30605 NASA-CR-102624 LMSC-A964947 1-11 1969
60026	CALORIMETRIC EMITTANCE AND REFLECTANCE MEASURING EQUIPMENT. FROM HIGH-PERFORMANCE THERMAL PROTECTION SYSTEMS, VOLUME 2, FINAL REPORT. LOCKHEED MISSILES AND SPACE COMPANY SUNNYVALE CALIFORNIA NASA AND CFSTI N70-30606 NASA-CR-102624 LMSC-A964947 1-14 1969
60027	PROPOSED SPECIFICIATION FOR REFLECTIVE-SHIELD EMITTANCE. FROM HIGH-PERFORMANCE THERMAL PROTECTION SYSTEMS, VOLUME 2, FINAL REPORT. LOCKHEED MISSILES AND SPACE COMPANY SUNNYVALE CALIFORNIA NASA AND CFSTI N70-30607 NASA-CR-102624 LMSC-A96447 1-4 1969
60028	A METHOD FOR ESTIMATING AS-INSTALLED THERMAL CONDUCTIVITY VALUES. FROM HIGH-PERFORMANCE THERMAL PROTECTION SYSTEMS, VOLUME 2, FINAL REPORT. LOCKHEED MISSILES AND SPACE COMPANY SUNNYVALE CALIFORNIA NASA AND CFSTI N70-30611 NASA-CR-102624 LMSC-A964947 1-5 1969
60029	LOCALIZED AND SELF-TRAPPED EXCITONS IN CESIUM IODIDE. LAMATSCH H ROSSEL J SAURER E PHYSICA STATUS SOLIDI 41 2 605-14 1970
60031	INTRINSIC OPTICAL PROPERTIES OF CHROMIUM TRICHLORIDE. POLLINI I SPINOLO G PHYSICA STATUS SOLIDI 41 2 691-701 1970
60035	DETERMINATION OF THE DIFFUSION AND ACCOMMODATION ON THE WALL OF CO2 MOLECULES EXITED ON THE VIBRATIONAL LEVEL /00 DEGREES 1/. MARGOTTIN-MACLOU M ARDITI I GUEGUEN H DOYENNETTE L COMPT REND 271 B 12 563-6 1970
60037	DETERMINATION OF THE DIFFUSION COEFFICIENT AND THE ACCOMMODATION ON THE WALLS OF HYDROCHLORIDE GAS MOLECULES EXCITED IN THE FIRST VIBRATIONAL STATE. HENRY A DOYENNETTE L MARGOTTIN-MACLOU M COMPT REND 271 B 13 634-7 1970
60045	DETERMINATION OF THE THERMAL CONTACT RESISTANCE OF PLANE-ROUGH SURFACES WITH THE ROUGHNESSES DEFORMING IN DIFFERENT MANNERS. POPOV V M HEAT TRANSFER-SOVIET RESEARCH 2 5 26-31 1970
60049	THE EFFECTIVE THERMAL CONDUCTIVITY OF BEDS OF UNCONSOLIDATED PARTICLES. TAMARIN A I HEAT TRANSFER-SOVIET RESEARCH 2 5 190-4 1970
60085	PROPERTIES OF THIN FILMS OF ZINC OXIDE PREPARED BY A CHEMICAL SPRAY METHOD. NOBBS J MC K GILLESPIE F C J PHYS CHEM SOLIDS 31 10 2353-9 1970
60092	OPTICAL PROPERTIES OF SINGLE-CRYSTAL FILMS OF PBS, PBSE, AND PBTE. //ENGLISH TRANSLATION OF KRISTALLOGRAFIYA 10 /4/ 515-19, 1965.// SEMILETOV S A VORONINA I P KORTUKOVA E I SOVIET PHYSICS-CRYSTALLOGRAPHY 10 4 429-32 1966
60099	CARBON FIBER REINFORCED PLASTICS-AN INITIAL EVALUATION. PHILLIPS L N ROYAL AIRCRAFT ESTABLISHMENT FARNBOROUGH ENGLAND DDC AND CFSTI RAE-TR-67088 AD-655890 1-32 1967
60103	INSTRUMENTATION FOR COLOR AERIAL PHOTOGRAPHY. ROBERTSON K D ARMY ENGINEER TOPOGRAPHIC LABS FORT BELVOIR VA DDC DDC

TPRC Number	Bibliographic Citation

60107 THERMAL AND ELCTRICAL PROPERTIES OF COATED CONDUCTIVE
SUBSTRATES FOR INTEGRATED-CIRCUIT CHIP MOUNTING.
MATTES H G
BELL SYST TECH J
49   6   1151-82   1970   CA   73   103451

60111 PROPERTIES OF PALARITE THERMAL INSULATION.
WEBER T J   SANDIA LABS   LIVERMORE CALIF   NTIS
NTIS
SCL-DR-69-150
          1-26   1970   CA   74   45056

60152 INFLUENCE OF DEFECTS ON THE PROPERTIES OF SILICON
NITRIDE FILM.
SYNOROV V F   ALEINIKOV N M   BITYUTSKAYA L A
UGAI YA A
IZV AKAD NAUK SSSR NEORG MATER
6   5   1002-3   1970

60153 INFLUENCE OF DEFECTS ON THE PROPERTIES OF SILICON
NITRIDE FILM. //ENGLISH TRANSLATION OF IZV. AKAD.
NAUK SSSR, NEORG. MATER. 6 /5/ 1002-3, 1970.//
SYNOROV V F   ALEINIKOV N M   BITYUTSKAYA L A
UGAI YA A
INORGANIC MATERIALS
6   5   878-9   1970

60164 AMBIENT AND HIGH TEMPERATURE EXPERIMENTS ON
BORON-DOPED POLYCRYSTALLINE GRAPHITES.
WAGNER P   DICKINSON J M
CARBON
8   3   313-20   1970   CA   73   113816

60173 PRACTICAL APPLICATION OF INFRARED REFLECTION
SPECTROSCOPY TO THE STUDY OF THE GAS-METAL INTERFACE.
DRMAJ D T   HAYES K E
J CATAL
19   2   154-61   1970   CA   73   113268

60177 THERMOPHYSICS RESEARCH. FROM PROC. OF THE THIRD
SOUTHEASTERN SEMINAR ON THERMAL SCIENCES, MAY 8-9,
1967. HUNTSVILLE, ALABAMA.
HELLER G B   GEORGE C MARSHALL SPACE FLIGHT CENTER
HUNTSVILLE ALABAMA   NTIS
NTIS
NASA-TM-X-62639
N70-18690   143-91   1968

60204 SPECTRAL STUDIES OF ORGANIC DYES. I. REFLECTANCE
SPECTRA OF ACRIDINE ORANGE IN BASE FORM ADSORBED ON
SOLID SURFACES.
YAMAOKA K   DEMOTO J   MIURA M
J SCI HIROSHIMA UNIV
33   A2   311-23   1969

60240 OPTICAL PROPERTIES OF THIN LAYERS OF RHODIUM AND
VANADIUM IN THE FAR ULTRAVIOLET.
SEIGNAC A   ROBIN S
COMPT REND
271   B 17   919-22   1970

60242 UTILIZATION OF A NIOBIUM ALLOY IN THE REALIZATION OF
A HYPERSONIC VEHICLE.
PEREZ   SYRE   BILLON   PICHOIR   GUYOT
REV PHYSIQUE APPLIQUEE
5   3   455-65   1970

60244 THERMAL ACCOMMODATION OF CSCL MOLECULES DURING THEIR
SURFACE IONIZATION ON CARBON-COVERED IRIDIUM.
ZANDBERG E YA   TONTEGODE A YA   YUSIFOV F K
FIZ TVERD TELA
12   6   1740-4   1970

60245 THERMAL ACCOMMODATION OF CSCL MOLECULES DURING THEIR
SURFACE IONIZATION ON CARBON-COVERED IRIDIUM.
//ENGLISH TRANSLATION OF FIZ. TVERD. TELA 12 /6/
1740-44, 1970.//
ZANDBERG E YA   TONTEGODE A YA   YUSIFOV F K
SOVIET PHYSICS-SOLID STATE
12   6   1376-9   1970

60262 THERMAL EXPANSION OF BORON FIBER-ALUMINUM
COMPOSITES.
KREIDER K G   PATARINI V M
TRANS MET SOC AIME
1   12   3431-5   1970

60268 A STUDY OF ADSORPTION ON SINGLE CRYSTALS BY INTERNAL
REFLECTANCE SPECTROSCOPY.
HALLER G L   RICE R W
J PHYS CHEM
74   25   4386-93   1970

60282 THERMAL EXPANSION OF COMPOSITE MATERIALS.
TUMMALA R R   FRIEDBERG A L
J APPL PHYS
41   13   5104-7   1970

60284 THERMAL EXPANSION OF GRAPHITE-EPOXY COMPOSITES.
FAHMY A A   RAGAI A N
J APPL PHYS
41   13   5112-15   1970

60299 OPTICAL PROPERTIES OF MOLYBDENUM AND RUTHENIUM.
KRESS K A   LAPEYRE G J
J OPT SOC AMER
60   12   1681-4   1970

60313 A METHOD OF MEASURING THERMAL DIFFUSIVITY OF HIGH
EXPLOSIVE MATERIALS.
MURRAY R C   COOPER T E   LAWRENCE RADIATION LAB
CALIFORNIA UNIV   LIVERMORE
NTIS
UCRL-50827 N70-35817
          1-27   1970

60314 RESULTS OF ERROR ANALYSIS FOR IN PILE THERMAL
CONDUCTIVITY MEASUREMENT USING THE BALANCED
OSCILLATOR TECHNIQUES.
DAENNER W   KERNFORSCHUNGSZENTRUM   KARLSRUHE
WEST GERMANY
NTIS
KFK-1140 EURFNR-755
N71-27631   1-45   1970

60370 IX. NOTE ON THE EXPANSION CHARACTERISTICS OF GLASSES
FOR SODIUM VAPOUR LAMP SEALS.
STANWORTH J E
J SOC GLASS TECHNOLOGY
25   159-63   1941

60443 THERMAL EXPANSION OF SILICA MORTARS AFTER FIRING AT
950, 1200, AND 1500 C.
COLE S S   LYNN D C
J AM CERAM SOC
14   12   906-12   1931

60495 EXCITON ABSORPTION AND PHOTOCONDUCTIVITY IN MERCUROUS
HALIDES.
DEB S K
PHYS REV
2   B 12   5003-7   1970

60498 OPTICAL CONSTANTS OF RUBIDIUM AND CESIUM FROM 0.5 TO
4.0 EV.
SMITH N V
PHYS REV
2   B 8   2840-8   1970

60518 THE INFRA-RED SPECTRUM OF MAGNESIUM OXIDE BETWEEN 1
AND 21 MICRONS.
MOMIN A U
PROC INDIAN ACAD SCI
37   A   254-9   1953

60523 OXIDE AND SILICON CARBIDE COATINGS FOR HIGH
TEMPERATURE RESISTANCE.
NORTON CO   WOOSTER MASS
INDUSTRIAL HEATING
22   11   2352-60   1955

60526 DILATOMETER METHOD FOR DETERMINATION OF THERMAL
COEFFICIENT OF EXPANSION OF FINE AND COARSE
AGGREGATE.
VERBECK G J   HASS W E
HIGHWAY RES BOARD PROC
30   187-93   1951

60580 CORRECTIONS TO PLANCKS FORMULA OF BLACK-BODY
RADIATION FOR SMALL CAVITIES AND ITS IMPLICATIONS ON
FAR INFRARED STANDARD SOURCES.
BALTES H P   MURI R   KNEUBUHL F K
REV INT HAUTES TEMP REFRACT
7   3   192-6   1970

60590 THE DIRECT DETERMINATION OF THERMAL CONDUCTIVITY BY
THE FLASH TECHNIQUE.
PEGGS I D   MILLS R W
REV INT HAUTES TEMP REFRACT
7   3   264-7   1970   CA   74   80525

60598 OPTICAL TRANSMISSION AND EXCITATION OF PLASMA
OSCILLATIONS IN FREE FILMS OF AL, IN, SN, MG, GE, SE,
AND TE IN THE VACUUM ULTRAVIOLET REGION OF THE
SPECTRUM.
SOROKIN O M   BLANK V A
OPTIKA I SPEKTROSKOPIYA
28   6   1178-85   1970

60599 OPTICAL TRANSMISSION AND EXCITATION OF PLASMA
OSCILLATIONS IN FREE FILMS OF AL, IN, SN, MG, GE, SE,
AND TE IN THE VACUUM ULTRAVIOLET REGION OF THE
SPECTRUM. //ENGLISH TRANSLATION OF OPTIKA I
SPEKTROSKOPIYA 28 /6/ 1178-85, 1970.//
SOROKIN O M   BLANK V A
OPTICS SPECTROSCOPY
28   6   834-7   1970

60608 HEAT TRANSFER OF CONDENSABLE VAPOUR DIFFUSING THROUGH
POROUS MEDIA.
KITO M   SUGIYAMA S
INT J HEAT MASS TRANSFER
13   11   1705-13   1970

TPRC Number	Bibliographic Citation

60616 DESIGN AND PERFORMANCE EVALUATION OF A HIGH ACCURACY REFLECTOMETER FOR USE IN THE NEAR ULTRAVIOLET TO FAR INFRARED REGION OF THE SPECTRUM. M.S. THESIS.
HERNICZ R S   PURDUE UNIVERSITY   LAFAYETTE IND
PURDUE UNIVERSITY
1-120   1970

60680 PREPARATION OF SNO2 FILMS BY OXYDIZING EVAPORATED SN FILMS.
WATANABE HIDEO
JAPANESE J APPLIED PHYS
9   12   1551-2   1970

60692 EFFECTS OF MICROMETEOROIDS ON OPTICAL SURFACES.
MANDEVILLE J C   OFFICE NATIONAL D ETUDES ET DE RECHERCHES AEROSPATIALES   TOULOUSE FRANCE
NTIS
ONERA-DERTS-NT-01-12
N71-17218   1-37   1970

60695 ENERGY EXCHANGE BETWEEN A COLD GAS AND A CARBON SURFACE AT HIGH TEMPERATURES.
MIDOL M C   DUVAL X
J CHIM PHYS PHYSICOCHIM BIOL
67   7   1351-9   1970

60699 NOTE ON THE CONSTANT-REFLECTIVITY METHOD OF DETERMINING SUBSTRATE OPTICAL CONSTANTS.
SHEWCHUN J   ROWE E C
OPTICS COMMUNICATIONS
2   1   22-4   1970

60706 THERMAL CONDUCTIVITY OF SILVER SULFIDE.
ROUTIE R   MAHENC J
REV INT HAUTES TEMP REFRACT
7   2   165-9   1970   CA   73   134557

60713 EFFECTS OF SURFACE FILMS ON THERMAL CONTACT CONDUCTANCE. I. MICROSCOPIC EXPERIMENTS. FROM SPACE SYSTEMS AND THERMAL TECHNOLOGY FOR THE SEVENTIES. AMERICAN SOCIETY OF MECHANICAL ENGINEERS, SPACE TECHNOLOGY AND HEAT TRANSFER CONFERENCE, LOS ANGELES, CALIF, JUNE 21-24, 1970 PROCEEDINGS PART 2.
GALE E H JR
ASME   NEW YORK
1-9/22 REFS/1970

60715 EFFECTIVE THERMAL CONDUCTIVITY OF DRY AND LIQUID-SATURATED SINTERED FIBER METAL WICKS. FROM SPACE SYSTEMS AND THERMAL TECHNOLOGY FOR THE SEVENTIES. AMERICAN SOCIETY OF MECHANICAL ENGINEERS, SPACE TECHNOLOGY AND HEAT TRANSFER CONFERENCE, LOS ANGELES, CALIF, JUNE 21-24, 1970 PROCEEDINGS PART 2.
SOLIMAN M M   GRAUMANN D W   BERENSON P J
ASME   NEW YORK
1-10/9 REFS/1970

60718 REFLECTANCE OF GALLIUM FILMS IN THE WAVELENGTH RANGE 0.25 TO 25 MICRONS.
STOLECKI B   WESOLOWSKA C
ACTA PHYS POL
37   A   5   759-65   1970

60756 INFRARED REFLECTANCE SPECTRA OF BARRIER-TYPE ANODIC OXIDE FILMS FORMED ON PURE ALUMINUM.
TAKAMURA T   KIHARA-MORISHITA H   MORIYAMA U
THIN SOLID FILMS
6   1   R17-19   1970   CA   74   8021

60790 INFRARED REFLECTION OF HIGHLY DOPED INDIUM OXIDE FILMS.
VAINSHTEIN V M   FISTUL V I
FIZ TEKH POLUPROV
4   8   1495-9   1970   CA   73   125263

60817 DESIGN AND EXPERIMENTAL INVESTIGATION OF A HIGH-VOLTAGE PHOTOELECTRIC ENERGY CONVERTER.
LANDSMAN A P   STREBKOV D S
FIZ TEKH POLUPROV
4   10   1922-8   1970

60818 DESIGN AND EXPERIMENTAL INVESTIGATION OF A HIGH-VOLTAGE PHOTOELECTRIC ENERGY CONVERTER. //ENGLISH TRANSLATION OF FIZ. TEKH. POLUPROV. 4 /10/ 1922-28, 1970.//
LANDSMAN A P   STREBKOV D S
SOVIET PHYSICS-SEMICONDUCTORS
4   10   1647-52   1971

60823 THE DETERMINATION OF OPTICAL CHARACTERISTICS OF DOUBLE LAYERS DEPOSITED ON SUBSTRATES.
TEKUCHEVA I A
OPT SPEKTROSK
29   1   198-202   1970

60824 THE DETERMINATION OF OPTICAL CHARACTERISTICS OF DOUBLE LAYERS DEPOSITED ON SUBSTRATES. //ENGLISH TRANSLATION OF OPT. SPEKTROSK. 29 /1/ 198-202, 1970.//
TEKUCHEVA I A
OPTICS SPECTROSCOPY
29   1   103-5   1970

60836 THERMAL DIFFUSIVITY OF ALUMINOSILICATE REFRACTORIES IN THE RANGE 200-1600 C. //ENGLISH TRANSLATION OF OGNEUPORY /5/ 17-19, 1970.//
LITOVSKII E YA   LANDA YA A   MILSHENKO R S
REFRACTORIES
5   284-6   1970

61035 THE ELECTRIC REFLECTANCE PHOTOMETER IN THE PULP AND PAPER INDUSTRY.
STENIUS A S   SVENSKA TRAFORSKININGSINSTITET STOCKHOLM
DDC
MEDDELANDE-42
AD-711990   1-6   1969

61037 FROST INVESTIGATION 1944-1945. APPENDIX 13. REPORT ON LABORATORY TESTS ON FROST PENETRATION AND THERMAL CONDUCTIVITY OF COHESIONLESS SOILS.
ARCTIC CONSTRUCTION AND FROST EFFECTS LAB   BOSTON MASS
DDC
ACFEL-TR-6-APPENDIX-13
AD-712471   1-45   1945

61042 DESIGN AND ANALYSIS OF A SYSTEM TO DETERMINE SPECTRAL EMISSIVITIES OF CESIUM COATED METAL SURFACES. M.S. THESIS.
BORYS R S   CLARKSON COLLEGE OF TECHNOLOGY POTSDAM NEW YORK
CLARKSON COLLEGE OF TECHNOLOGY
1-54   1970

61044 THE EFFECT OF TEMPERATURE AND GASEOUS SPECIES ON THERMAL ACCOMMODATION COEFFICIENTS. M.S. THESIS.
COCHRAN M E   UNIVERSITY OF FLORIDA   GAINESVILLE FLORIDA
UNIVERSITY OF FLORIDA
1-67   1970

61051 HEAT TRANSFER IN FLOW PROCESSES FROM DEEP BOREHOLES. FROM THERMOPHYSICS AND THERMOENGINEERING. MINING THERMOPHYSICS.
DOBRIANSKII IU P   BARATOV Z I   CHERNIAK V P
NTIS
N70-16593   59-65   1968

61071 COMPARISON OF THE SPECTRAL REFLECTIVITY OF MARS WITH OXIDIZED METEORITIC MATERIAL.
GIBSON E K JR
ICARUS
13   1   96-9   1970   CA   74   15043

61078 THERMAL CONTACT RESISTANCE IN CYLINDRICAL ELEMENTS IN THERMIONIC ENERGY CONVERTERS.
DANILOV YU I   KOSHKIN V K   MIKHAILOVA T V
MIKHEEV YU S   ORLIN S A
INT CONF THERMION ELEC POWER GENERATION 2ND
683-98   1968   CA   74   17049

61079 THERMAL RESISTANCE OF MULTILAYER CYLINDRICAL ELEMENTS IN THERMIONIC CONVERTERS.
ARKIN E S A   KUKUSHKIN S F   MURINSON E A
OGLOBLIN B G   CHEREPANOV P Z   CHEKHOVICH V S
INT CONF THERMION ELEC POWER GENERATION 2ND
2   721-30   1968   CA   74   17069

61087 ANOMALOUS CERIUM MAGNESIUM NITRATE-HELIUM-3 THERMAL BOUNDARY RESISTANCE.
LEGGETT A J   VUORIO M
J LOW TEMP PHYS
3   4   359-76   1970   CA   74   16303

61104 DETERMINING THE OPTICAL CONSTANTS OF THIN FILMS BY INTERNAL REFLECTION. M.S. THESIS.
ANDERSON W J   UTAH STATE UNIVERSITY   LOGAN UTAH
UTAH STATE UNIVERSITY
1-56   1970

61109 THERMAL CONDUCTIVITY OF FILLED PLASTICS. M.S. THESIS.
CHEN S Y   UNIVERSITY OF CONNECTICUT   STORRS CONNECTICUT
UNIVERSITY OF CONNECTICUT
1-41   1969

61174 STUDY OF CORUNDUM PHOSPHATE-BONDED CONCRETE.
KHOROSHAVIN L B   DOLGIKH G V   USTYANTSEV V M
OGNEUPORY
8   43-4   1970

61175 STUDY OF CORUNDUM PHOSPHATE-BONDED CONCRETE. //ENGLISH TRANSLATION OF OGNEUPORY /8/ 43-44, 1970.//
KHOROSHAVIN L B   DOLGIKH G V   USTYANTSEV V M
REFRACTORIES
8   509-10   1970

61181 THERMAL ISOLATION WITH LOW-CONDUCTANCE INTERSTITIAL MATERIALS UNDER COMPRESSIVE LOADS.
SMUDA P A   GYOROG D
PROGRESS ASTRONAUT AERON
23   3-22   1970

TPRC Number	Bibliographic Citation

61182 THERMAL CONTACT RESISTANCE MEASUREMENTS AT AMBIENT PRESSURES OF ONE ATMOSPHERE TO 3 TIMES 10/-12/ MM HG AND COMPARISON WITH THEORETICAL PREDICITONS.
CASSIDY J F   MARK H
PROGRESS ASTRONAUT AERON
23   23-44   1970

61184 PORTABLE REFLECTOMETER.
FINKEL M W
PROGRESS ASTRONAUT AERON
23   65-76   1970

61186 OPACIFIED FIBROUS INSULATIONS.
GRUNERT W E   NOTARO F   REID R L
PROGRESS ASTRONAUT AERON
23   127-51   1970

61187 TECHNIQUES FOR IMPROVING THE THERMAL PERFORMANCE OF LOW-DENSITY FIBROUS INSULATION.
LOPEZ E L
PROGRESS ASTRONAUT AERON
23   153-72   1970

61190 EFFECT OF THIN SURFACE FILMS ON THE RADIATIVE PROPERTIES OF METAL SURFACES.
CRAVALHO E G   DRAZEN E L C
PROGRESS ASTRONAUT AERON
23   363-83   1970

61191 SCALE MODELING OF A MULTILAYER INSULATED SPACECRAFT FOR USE IN A PRELIMINARY DESIGN STUDY.
MARSHALL K N   ROLLING R E
PROGRESS ASTRONAUT AERON
23   437-60   1970

61192 THERMAL CONSIDERATIONS OF A LANDED VEHICLE ON THE SURFACE OF MARS.
ROSENBERG M J
PROGRESS ASTRONAUT AERON
23   491-514   1970

61193 MARS LANDER THERMAL CONTROL SYSTEM PARAMETRIC STUDIES.
TRACEY T R   MOREY T F
PROGRESS ASTRONAUT AERON
23   515-46   1970

61199 ABSORPTION SPECTRA OF THIN FERRITE LAYERS.
ZAKHAROV V P   KIREI G G   MATVEEVA I A
SLOBODYANYUK A I
SPEKTROSK AT MOL
           413-6   1969   CA   74   17385

61245 SELECTION OF A THERMAL BONDING AGENT FOR TEMPERATURES BELOW 1 K.
ANDERSON A C   PETERSON R E
CRYOGENICS
10   5   430-3   1970   CA   74   23271

61246 RELIABLE LOW-THERMAL-RESISTANCE BOND BETWEEN DIELECTRICS AND METALS FOR USE AT LOW TEMPERATURES.
BROWN M A
CRYOGENICS
10   5   439-40   1970   CA   74   25746

61323 MULTICOMPONENT DISPERSIVE INTERFERENCE-POLARIZATION FILTERS.
VINOGRADOVA T A
OPTIKA SPEKTROSKOPIYA
29   2   395-400   1970

61324 MULTICOMPONENT DISPERSIVE INTERFERENCE-POLARIZATION FILTERS. //ENGLISH TRANSLATION OF OPTIKA SPEKTROSKOPIYA 29 /2/ 395-400, 1970.//
VINOGRADOVA T A
OPTICS SPECTROSCOPY
29   2   209-11   1970

61438 EFFECT OF INTERFERENCE PHASE ON LIGHT ABSORPTION IN A SEMITRANSPARENT METAL LAYER.
KULAGIN E S   MERKULOV A V
OPTIKA I SPEKTROSKOPIYA
29   3   587-93   1970

61439 EFFECT OF INTERFERENCE PHASE ON LIGHT ABSORPTION IN A SEMITRANSPARENT METAL LAYER. //ENGLISH TRANSLATION OF OPTIKA I SPEKTROSKOPIYA 29 /3/ 587-93, 1970.//
KULAGIN E S   MERKULOV A V
OPTICS SPECTROSCOPY
29   3   313-6   1970

61485 ABSORPTION EFFECTS IN HIGH REFLECTANCE FILMS.
M.S. THESIS.
GEHRKE W L   UNIVERSITY OF UTAH   SALT LAKE CITY UTAH
UNIVERSITY OF UTAH
           1-121   1970

61547 TOTAL EMITTANCE MEASUREMENTS OF THIN METALLIC FILMS AT CRYOGENIC TEMPERATURES.
CUNNINGTON G R   BELL G A   ARMALY B F   TIEN C L
JOURNAL OF SPACECRAFT AND ROCKETS
7   1496-9   1970

61567 PHYSICOMECHANICAL PROPERTIES OF FERROMAGNETIC DIELECTRIC PLASTICS.
GERASHCHENKO O A   GRISHCHENKO T G   ROLIK A N
LAKOVLEV A I
PROBLEMY TEKHNICHESKOI ELEKTRODINAMIKI
23   98-100   1970

61693 PREPARATION AND PROPERTIES OF THIN LAYERS OF CARBIDIZED CHROMIUM.
KRYZHANOVSKII B P   KRUGLOV B M
IZV AKAD NAUK SSSR NEORG MATER
6   12   2119-22   1970

61694 PREPARATION AND PROPERTIES OF THIN LAYERS OF CARBIDIZED CHROMIUM. //ENGLISH TRANSLATION OF IZV. AKAD. NAUK SSSR, NEORG. MATER. 6 /12/ 2119-22, 1970.//
KRYZHANOVSKII B P   KRUGLOV B M
INORGANIC MATERIALS
6   12   1859-62   1970

61704 EXPERIMENTAL STUDY OF THE HEAT CONDUCTIVITY OF LAMINAR-VACUUM INSULATION. //ENGLISH TRANSLATION OF INZH. FIZ. ZHUR. 12 /4/ 426-32, 1967.//
MIKHALCHENKO R S   GERZHIN A G   PERSHIN N P
KLIPACH L V
J ENGINEERING PHYS
12   4   220-4   1967
NTIS
FTD-HT-23-342-68
AD-715233   1-12   1970

61705 HEAT TRANSMISSION OF INSULATED, OIL IMPREGNATED TRANSFORMER COILS IN LAMINATED OIL FLOW.
BRECHNA H   TSCHUDI H
SCHWIZERISCHE ELEKTROTECHNISCHE VEREIN BULLETIN
53   1069-81   1962

61706 LABORATORY RESEARCH FOR THE DETERMINATION OF THE THERMAL PROPERTIES OF SOILS.
KERSTEN M S   ENGINEERING EXPERIMENT STATION UNIV OF MINNESOTA
DDC
CRREL-TR-23 AD-712516
           1-227   1949

61715 INFRARED EMISSION SPECTRA. ENHANCEMENT OF DIAGNOSTIC FEATURES BY THE LUNAR ENVIRONMENT.
LOGAN L M   HUNT G R
SCIENCE
169   865-6   1970
NTIS
AFCRL70-0622 AD-715145   1970

61716 OPTICAL PROPERTIES OF REAL METALLIC FILMS AND MICROTHEORY.
ROZENBERG G V
OPTIKA TONKOSLOYNIKH POKRYTIY
           268-354   1958

61717 OPTICAL PROPERTIES OF REAL METALLIC FILMS AND MICROTHEORY. //ENGLISH TRANSLATION OF OPTIKA TONKOSLOYNIKH POKRYTIY 268-354, 1958.//
ROZENBERG G V   ARMY FOREIGN SCIENCE AND TECHNOLOGY CENTER   WASHINGTON D C
DDC AND NTIS
FSTC-HT-23-1187-70 AD-715296
           1-72   1970

61718 CHARACTERISTIC ENERGY ABSORPTION SPECTRA OF SOLIDS. FROM FAR INFRARED PROPERTIES OF SOLIDS.
PLENDL J N
PLENUM PRESS
AFCRL-M70-0673 AD-715586
           387-449   1970

61719 REFLECTANCE STUDIES OF THE GOLD-ELECTROLYTE INTERFACE.
CAHAN B D   HORKANS J   YEAGER E   CASE WESTERN RESERVE UNIV   CLEVELAND OHIO
NTIS
AD-715666   1-20   1970

61720 SPECULAR REFLECTIVITY STUDIES OF THE ADSORPTION OF HALIDE ANIONS ON GOLD ELECTRODES IN 0.2 M HCLO4.
TAKAMURA T   TAKAMURA K   YEAGER E   CASE WESTERN RESERVE UNIV   CLEVELAND OHIO
NTIS
AD-715667   1-35   1970

61721 INFRARED REFLECTANCE OF WATER FROSTS CONDENSED ON LIQUID NITROGEN-COOLED SURFACES IN VACUUM.
WOOD B E   SMITH A M   SEIBER B A   ROUX J A
ARNOLD ENGINEERING DEVELOPMENT CENTER ARNOLD AIR STATION   TENNESSEE
DDC AND NTIS
AEDC-TR-70-215 N71-18682
AD-715915   1-41   1970

61753 RATIO OF THE THERMAL ACCOMMODATION COEFFICIENTS OF HELIUM-4 AND HELIUM-3 ISOTOPES ON BARE TUNGSTEN FROM 77 TO 513 K.
KRUEGER C J   UNIV OF MISSOURI   COLUMBIA MISSOURI

TPRC Number	Bibliographic Citation

61779 THE BEHAVIOR OF SEVERAL WHITE PIGMENTS AS DETERMINED BY IN SITU REFLECTANCE MEASUREMENTS OF IRRADIATED SPECIMENS.
ZERLAUT G A   ROGERS F O   IIT RESEARCH INST
CHICAGO ILL
NTIS
NASA-CR-102930
N71-12815   1-47   1970

61781 A COMPARISON OF TWO TRANSIENT METHODS OF MEASURING THERMAL CONDUCTIVITY OF PARTICULATE SAMPLES.
SCOTT R W   FOUNTAIN J A
NTIS
NASA-TM-X-64559
N71-12280   1-29   1970

61782 THE VACUUM UV OPTICAL CONSTANTS OF ALUMINIUM AND OF MAGNESIUM FLUORIDE, AND THE PERFORMANCE OF MAGNESIUM FLUORIDE COATED MIRRORS.
FREEMAN G H C   REJMAN M A Z   NATIONAL PHYSICAL LAB TEDDINGTON /ENGLAND/
NTIS
NPL-QU-12   N71-13319
                                   1-26   1970

61785 DETERMINATION OF THE OPERATING THERMAL CONDUCTIVITY OF CERAMIC FUEL AND THE COEFFICIENT OF THERMAL CONDUCTIVITY OF THE FISSION GAS BALANCED OSCILLATOR METHOD.
DABNNER W   KERNFORSCHUNGSZENTRUM KARLSRUHE /WEST GERMANY/
AEC
KFK-1125 N71-13752
                                   1-171   1970

61793 THERMAL CONTACT CONDUCTANCE DUE TO PLANE LAYERS OF ADSORBED GASES AT THE INTERFACE OF SMOOTH METALLIC CONTACTS.
WADHWA S K   SYRACUSE UNIV   SYRACUSE N Y
UNIV MICROFILMS PUBL
UM-70-10401   1-82   1969   CA   74   25754

61852 LOW-VACUUM SILVER CONDENSATES.
BONDAR V V   LYSENKO V S   MALNEV A F
RYZHKOV YU T   SIVAK B V
FIZ METAL METALLOVED
28   5   799-803   1969

61853 LOW-VACUUM SILVER CONDENSATES. //ENGLISH TRANSLATION OF FIZ. METAL. METALLOVED. 28 /5/ 799-803, 1969.//
BONDAR V V   LYSENKO V S   MALNEV A F
RYZHKOV YU T   SIVAK B V
PHYSICS METALS METALLOGRAPHY
28   5   33-7   1969

61878 EFFECTS OF EDGE LOSSES ON THE THERMAL CONDUCTIVITY OF THERMAL INSULATIONS AT HIGH TEMPERATURE.
TYE R P
REV INT HAUTES TEMP REFRACT
7   4   308-12   1970

61942 REFLECTANCE AND MICROHARDNESS OF BORNITE, GALENA, PYRITE, AND MAGNETITE.
HAUSMANN K   VON GEHLEN K
NEUES JAHRB MINERAL MONATSH
             11   498-506   1970   CA   74   33412

61948 AN INVESTIGATION OF THE EMISSIVITY OF BUILDING MATERIALS.
VOZNESENSKII A A   FERT A R
INZH FIZ ZHUR
12   5   610-14   1967

61949 AN INVESTIGATION OF THE EMISSIVITY OF BUILDING MATERIALS. //ENGLISH TRANSLATION OF INZH. FIZ. ZHUR. 12 /5/ 610-14, 1967.//
VOZNESENSKII A A   FERT A R
J ENGINEERING PHYSICS
12   5   327-30   1967

61959 INFRARED REFLECTION OF HEAVILY DOPED FILMS OF IN2 O3. //ENGLISH TRANSLATION OF FIZ. TEKH. POLUPROV. 4 /8/ 1495-9, 1970.//
VAINSHTEIN V M   FISTUL V I
SOVIET PHYSICS-SEMICONDUCTORS
4   8   1278-81   1971

61986 INFLUENCE OF AN ELECTRIC FIELD ON THE REFRACTIVE INDEX OF AMORPHOUS SELENIUM.
PEROV P I   AVDEEVA L A   ELINSON M I
STEPANOV G V
FIZ TEKH POLUPROV
3   2   183-7   1969

61987 INFLUENCE OF AN ELECTRIC FIELD ON THE REFRACTIVE INDEX OF AMORPHOUS SELENIUM. //ENGLISH TRANSLATION OF FIZ. TEKH. POLUPROV. 3 /2/ 183-87, 1969.//
PEROV P I   AVDEEVA L A   ELINSON M I
STEPANOV G V
SOVIET PHYSICS-SEMICONDUCTORS
3   2   153-6   1969

62042 EFFECT OF THE ROUGHNESS FACTOR ON RADIATION PROPERTIES OF SOLIDS /EXPERIMENTAL CHECK/.
AGABABOV S G
TEPLOFIZ VYS TEMP
8   4   770-3   1970

62043 EFFECT OF THE ROUGHNESS FACTOR ON RADIATION PROPERTIES OF SOLIDS /EXPERIMENTAL CHECK/. //ENGLISH TRANSLATION OF TEPLOFIZ. VYS. TEMP. 8 /4/ 770-73, 1970.//
AGABABOV S G
HIGH TEMPERATURE
8   4   728-31   1970

62054 EXPERIMENTAL METHODS FOR DETERMINING THE THERMAL CONDUCTIVITY COEFFICIENT AND EMISSIVE POWER OF COATINGS.
SMIRNOV E V
TEPLOFIZ VYS TEMP
8   4   875-8   1970   CA   73   124268

62055 EXPERIMENTAL METHOD OF DETERMINING THERMAL CONDUCTIVITY AND EMITTANCE OF COATINGS. //ENGLISH TRANSLATION OF TEPLOFIZ. VYS. TEMP. 8 /4/ 875-8, 1970.//
SMIRNOV E V
HIGH TEMPERATURE
8   4   819-21   1970

02179 HEAT TRANSFER BY NATURAL CONVECTION IN POROUS INSULANTS.
MARTIN G   HASELDEN G G
PROC INTERN CONGRESS OF REFRIGERATION 11TH MUNICH 1963 /PROG REFRIG SCI TECHNOL/
1   241-5   1965

62180 INVESTIGATION OF THE INFLUENCE OF FREE THERMAL CONVECTION ON HEAT TRANSFER THROUGH GRANULAR MATERIAL.
SCHNEIDER K-J
PROC INTERN CONGRESS OF REFRIGERATION 11TH MUNICH 1963 /PROG REFIRG SCI TECHNOL/
1   247-54   1965

62228 THERMAL CONDUCTIVITY OF CARBIDE THERMAL INSULATION /MATERIAL/.
GORIN A I   MOISEEV A M
TEPLOFIZ VYS TEMP
8   5   998-1001   1970   CA   74   16367

62229 THERMAL CONDUCTIVITY OF CARBIDE THERMAL INSULATION MATERIALS. //ENGLISH TRANSLATION OF TEPLOFIZ. VYS. TEMP. 8 /5/ 998-1001, 1970.//
GORIN A I   MOISEEV A M
HIGH TEMPERATURE
8   5   935-7   1970

62230 APPROXIMATE METHOD OF SOLVING A KINETIC EQUATION CLOSE TO THE BOUNDARY II. TEMPERATURE DISCONTINUITY.
ABRAMOV YU YU
TEPLOFIZ VYS TEMP
8   5   1013-17   1970

62231 APPROXIMATE METHOD OF SOLVING A KINETIC EQUATION CLOSE TO THE BOUNDARY II. TEMPERATURE DISCONTINUITY. //ENGLISH TRANSLATION OF TEPLOFIZ. VYS. TEMP. 8 /5/ 1013-17, 1970.//
ABRAMOV YU YU
HIGH TEMPERATURE
8   5   947-50   1970

62244 EXCHANGE OF ENERGY AND MOMENTUM BETWEEN GAS PARTICLES AND THE SURFACE OF A SOLID.
EROFEEV A I
ZH PRIKL MEKH TEKH FIZ
8   2   135-40   1967

62245 EXCHANGE OF ENERGY AND MOMENTUM BETWEEN GAS PARTICLES AND THE SURFACE OF A SOLID. //ENGLISH TRANSLATION OF ZH. PRIKL. MEKH. TEKH. FIZ. 8 /2/ 135-40, 1967.//
EROFEEV A I
J APPLIED MECHAN TECH PHYS
8   2   88-91   1967

62304 RESULTS WITH THE DTA AND DTA-DS-APPARATUS.
THORMANN P
TONIND-ZEITUNG KERAM RUNDSCHAU
85   17   408-10   1961

62305 THE INVESTIGATION OF RAW MATERIALS WITH THE AID OF COMBINED DILATATOMETRIC AND DTA.
LEHMANN H   THORMANN P
TONIND-ZEITUNG KERAM RUNDSCHAU
86   24   606-12   1962

62417 INFRARED REFLECTION STUDIES OF METAL SURFACES.
POLING G W
J COLLOID INTERFACE SCI
34   3   365-74   1970   CA   74   5 8904

62437 DETERMINATION OF THE COEFFICIENT OF THERMAL CONDUCTIVITY IN AN ASYMMETRICAL HEATING MODE.
KAGANER M G

TPRC Number	Bibliographic Citation

62501 INFRARED REFLECTANCE SPECTRA /IRRS/ OF ANODIC OXIDE
FILMS FORMED ON PURE TANTALUM.
KIHARA-MORISHITA H   TAKAMURA T   TAKEDA T
THIN SOLID FILMS
6   3   R29-R31   1970   CA   74   69709

62533 THERMAL CONDUCTIVITY OF EXPANDED CORKBOARD.
ANDRADE A
PROGR REFRIG SCI TECHNOL PROC INT CONGR REFRIG 12TH
1967
2   233-6   1969   CA   74   77611

62587 A THEORETICAL AND EXPERIMENTAL STUDY OF LIGHT
SCATTERING IN THERMAL CONTROL MATERIALS. FROM HEAT
TRANSFER AND SPACECRAFT THERMAL CONTROL. PROGRESS IN
ASTRONAUTICS AND AERONAUTICS. VOL. 24.
GILLIGAN J E   BRZUSKIEWICZ J   IIT RESEARCH
INSTITUTE   CHICAGO ILL
MIT PRESS CAMBRIDGE MASSACHUSETTS
24   69-92   1970

62590 HEMISPHERICAL THERMAL EMITTANCE OF COPPER AS A
FUNCTION OF OXIDATION CONDITIONS. FROM HEAT
TRANSFER AND SPACECRAFT THERMAL CONTROL. PROGRESS IN
AEROSPACE SCIENCE MEETING.
REID R L   COON C W
MIT PRESS CAMBRIDGE MASSACHUSETTS
24   8
PAPER-70-66   184-204   1970

62676 THE STUDY OF THE REFLECTIVITY OF INORGANIC MATERIALS
IMPORTANT FOR REMOTE SENSING APPLICATIONS. FINAL
REPT. SEPT. 1968-SEPT. 1970.
PERRY C H   LOWNDES R P   PHYSICS DEPT
NORTHEASTERN UNIV   BOSTON MASS
NTIS
AFCRL-70-0512
AD-720874   1-118   1970

62683 CALCULATION OF ELASTIC PROPERTIES OF COMPOSITES BY
MEANS OF THE VAN DER POEL THEORY.
SCHWARZL F R   VAN DEN EIKHOFF J   CENTRAL
LABORATORY   TNO-DELFT NETHERLANDS
NTIS
CL-71-55 ONR-TR-9
AD-721376   1-31   1971

62781 DEVELOPMENT OF TECHNIQUES AND ASSOCIATED
INSTRUMENTATION FOR HIGH TEMPERATURE EMISSIVITY
MEASUREMENTS.
CUNNINGTON G R   FUNAI A I   LOCKHEED MISSILES
AND SPACE CO   SUNNYVALE CALIFORNIA
NTIS
NASA-CR-103071 N71-21025
N-JF-71-1   1-77   1970

62782 INVESTIGATION OF INTEGRATING SPHERE MEASUREMENT
PARAMETERS.
NEWMAN B E   BROWN G L   TRW SYSTEMS GROUP
REDONDO BEACH CALIFORNIA
NTIS
NASA-CR-117402 N71-20732
                 1-139   1970

62784 HYPERVELOCITY OXIDATION TESTS OF THORIA DISPERSED
NICKEL CHROMIUM ALLOYS.
CENTOLANZI F J   AMES RESEARCH CENTER   MOFFETT
FIELD CALIFORNIA
NTIS
NASA-TM-X-62015
N71-20390   1-52   1971

62794 INVESTIGATION OF THE INFLUENCE EXERTED ON A SOLID
BY LUMINOUS RADIATION FROM A SOURCE OF THE EXPLOSIVE
TYPE.
ZHARIKOV I F   NEMCHINOV I V   TSIKULIN M A
ZHUR PRIKLAD MEKH TEKH FIZ
8   1   31-44   1967

62795 INVESTIGATION OF THE INFLUENCE EXERTED ON A SOLID BY
LUMINOUS RADIATION FROM A SOURCE OF THE EXPLOSIVE
TYPE. //ENGLISH TRANSLATION OF ZHUR. PRIKLAD. MEKH.
TEKH. FIZ. 8 /1/ 31-44, 1967.//
ZHARIKOV I F   NEMCHINOV I V   TSIKULIN M A
J APPL MECHAN TECH PHYS
8   1   20-8   1967

62857 THERMAL CONDUCTIVITY OF SEVERAL ORIENTED PLASTICS
WHEN COMPRESSED AND UNDER TEMPERATURE 180 TO 200 K.
FROM   TEPLOFIZICHESKIE SOVOISTVA VESHCHESTV I
MATERIALOV.   /THERMOPHYSICAL PROPERTIES OF SUBSTANCES
AND MATERIALS, VOLUME 2/.
ZHUKOVA R I   KAGANER M G   MARKELOVA N V
COMMITTEE OF STANDARDS MEASURES AND MEASURING DEVICES
PRESS MOSCOW
2   281-2   1970

62859 OBTAINING THIN FILMS BY EVAPORATING RARE-EARTH
METALS, AND STUDY OF SOME OF THEIR PHYSICO-CHEMICAL
PROPERTIES. FROM RARE-EARTH METALS AND THEIR
COMPOUNDS.
SAMSONOV G V   SERVETSKAIA M G   ISAEVA L P
SEREBRIAKOVA T E

62938 METHOD FOR DETERMINING THE EMISSIVITY OF A ROD WITH
INTERNAL HEAT SOURCES.
SMIRNOV E V   IONIN V E
ZAVODSKAYA LAB
36   11   1358-60   1970

62939 METHOD FOR DETERMINING THE EMISSIVITY OF A ROD WITH
INTERNAL HEAT SOURCES. //ENGLISH TRANSLATION OF
ZAVODSKAYA LAB. 36 /11/ 1358-60, 1970.//
SMIRNOV E V   IONIN V E
IND LAB USSR
36   11   1731-3   1970

62951 TEMPERATURE DEPENDENCES OF THE THERMAL CONDUCTIVITIES
OF LEAD AND IRON SHOT. //ENGLISH TRANSLATION OF AT.
ENERG. 26 /3/ 296-7, 1969.//
VASILEV A N   POLEKHINA T K
SOV AT ENERGY
26   3   332-3   1969

62953 EXPERIMENTAL DETERMINATION OF THE HEAT-CONDUCTIVITY
COEFFICIENT FOR NONMETALLIC TUBES /EXCHANGE OF
EXPERIENCE/.
VOLKOV D I   DMITRIEV G I   ROMANOV V A
TURLAKOV A S
ZAVOD LAB
35   9   1096-7   1969

62954 EXPERIMENTAL DETERMINATION OF THE HEAT-CONDUCTIVITY
COEFFICIENT FOR NOMETALLIC TUBES /EXCHANGE OF
EXPERIENCE/. //ENGLISH TRANSLATION OF ZAVOD. LAB. 35
/9/ 1096-97, 1969.//
VOLKOV D I   DMITRIEV G I   ROMANOV V A
TURLAKOV A S
IND LAB USSR
35   9   1320-1   1969

62974 THE STRUCTURE OF CONTINUOUSLY EVAPORATED CERMET
RESISTORS.
ALLAM D S   WATKINS J   PITT K E G
THIN SOLID FILMS
3   1-3   1969

62975 OPTICAL PROPERTIES OF THIN FILMS OF NI, CU, AND
NI + 60 PCT. CU IN THE WAVELENGTH RANGE
3000-6000 A.
MURR L E
THIN SOLID FILMS
3   321-32   1969

62976 OPTICAL ABSORPTION AND DISPERSION IN EVAPORATED FILMS
OF COBALT DIFLUORIDE.
YOUNG P A
THIN SOLID FILMS
4   25-33   1969

62977 THE STRUCTURE SENSITIVITY OF THE ULTRAVIOLET
ABSORBANCE OF ELECTROPLATED NICKEL LAYERS.
HEGEDUS R   ROTHENSTEIN B F   TUDOR V
THIN SOLID FILMS
4   46-8   1969

62978 PROGRESS IN THE PRODUCTION OF OPTICAL MULTI-LAYER
FILMS.
CLAPHAM P B
THIN SOLID FILMS
4   291-305   1969

62979 REALISATION OF A CHAMBER FOR THE STUDY OF THIN
METALLIC FILMS UNDER STATIC ULTRA HIGH VACUUM.
PETRAKIAN J P   PALMARI J P
THIN SOLID FILMS
4   423-7   1969

63038 EFFECT OF INNER SURFACE AIR VELOCITY AND TEMPERATURE
UPON HEAT GAIN AND LOSS THROUGH GLASS FENESTRATIONS.
PAPER NO. 2156.
PENNINGTON C W   MC DUFFIE D E JR
ASHRAE TRANS
76   2   190-200   1970

63039 EFFECT OF AIR DENSITY ON THE HEAT TRANSMISSION
COEFFICIENTS OF AIR FILMS AND BUILDING MATERIALS.
PAPER NO. 2158.
SWARD G R   HARRIS W S
ASHRAE TRANS
76   2   227-39   1970

63107 PREDICTION OF THE THERMAL CONDUCTIVITY OF FILLED AND
REINFORCED PLASTICS.
THORNBURGH J D   PEARS C D
ASME
ASME-65-WA/HT-4
                 1-11   1965

63131 INVESTIGATION OF THE EMITTANCE OF COATED REFRACTORY
METALS. FROM HEAT TRANSFER AND SPACECRAFT THERMAL
CONTROL. PROGRESS IN ASTRONAUTICS AND AERONAUTICS.
VOL. 24. 8TH AEROSPACE SCIENCE MEETING.
BARTSCH K O   HUDGINS W P   GEYER N M
MIT PRESS CAMBRIDGE MASSACHUSETTS
24   8
70-88   205-30   1970

TPRC Number	Bibliographic Citation

63132 INVESTIGATION OF THERMAL ISOLATION MATERIALS FOR CONTACTING SURFACES. FROM HEAT TRANSFER AND SPACECRAFT THERMAL CONTROL. PROGRESS IN ASTRONAUTICS AND AERONAUTICS. VOL. 24.
GYONOG D A   UNIVERSITY OF MISSOURI   ROLLA MO
MIT PRESS CAMBRIDGE MASSACHUSETTS
24   8   310-36
PAPER-70-13   1970

63133 NUMERICAL EVALUATION OF MULTILAYER INSULATION SYSTEM PERFORMANCE. FROM HEAT TRANSFER AND SPACECRAFT THERMAL CONTROL. PROGRESS IN ASTRONAUTICS AND AERONAUTICS. VOL. 24.
MAC GREGOR R K   POGSON J T   RUSSELL D J   THE BOEING COMPANY   SEATTLE WASH
MIT PRESS CAMBRIDGE MASSACHUSETTS
24   502-18   1970

63134 THERMAL TESTING OF INFLATABLE SOLAR SHIELDS FOR CRYOGENIC SPACE VEHICLES. FROM HEAT TRANSFER AND SPACECRAFT THERMAL CONTROL. PROGRESS IN ASTRONAUTICS AND AERONAUTICS. VOL. 24.
DOUGHTY R O   JONES L R   GENERAL DYNAMICS CORPORATION   FORT WORTH TEXAS
MIT PRESS CAMBRIDGE MASSACHUSETTS
24   580-600   1971

63135 MARTIAN SOFT LANDER INSULATION STUDY. FROM HEAT TRANSFER AND SPACECRAFT THERMAL CONTROL. PROGRESS IN ASTRONAUTICS AND AERONAUTICS. VOL. 24.
WILBER3 O J   SCHELDEN B J   CONTI J C
MIT PRESS CAMBRIDGE MASSACHUSETTS
24   630-58   1970

63152 RADIATION DAMAGE IN ORDINARY CONCRETE.
DUBROVSKII V G   IBRAGIMOV SH SH   KULAKOVSKII M YA   LADYGIN A YA   PERGAMENSHCHIK B K
ATOMNAYA ENERGIYA USSR
23   4   310-6   1967

63153 RADIATION DAMAGE IN ORDINARY CONCRETE. //ENGLISH TRANSLATION OF ATOMNAYA ENERGIYA, USSR 23 /4/ 310-16, 1967.//
DUBROVSKII V B   IBRAGIMOV SH SH   KULAKOVSKII M YA   LADYGIN A YA   PERGAMENSHCHIK B K
SOVIET ATOMIC ENERGY
23   4   1053-8   1967

63154 CALCULATING THE THERMAL RESISTANCE AT THE ZONE OF CONTACT BETWEEN SOLID BODIES. //ENGLISH TRANSLATION OF ATOMNAYA ENERGIYA, USSR 24 /1/ 86-87, 1968.//
PRASOLOV R S
SOVIET ATOMIC ENERGY
24   1   100-2   1968

63164 EFFECT OF NEUTRON IRRADIATION ON SOME PROPERTIES OF HEAT-RESISTANT CONCRETES. //ENGLISH TRANSLATION OF ATOMNAYA ENERGIYA, USSR 21 /2/ 108-12, 1966.//
DUBROVSKII V B   IBRAGIMOV SH SH   LADYGIN A YA   PERGAMENSHCHIK B K
SOVIET ATOMIC ENERGY
21   2   740-4   1966

63236 ABSORPTION, EXCITATION AND INFRARED STIMULATED FLASH IN CALCIUM AND STRONTIUM SULFIDE PHOSPHORS.
LEVSHIN V L   MIKHAILIN V V   NIZOVTSEV V V
IZV AKAD NAUK SSSR SER FIZ
30   9   1552   1966

63237 ABSORPTION, EXCITATION AND INFRARED-STIMULATED FLASH IN CALCIUM AND STRONTIUM SULFIDE PHOSPHORS. //ENGLISH TRANSLATION OF IZV. AKAD. NAUK SSSR, SER. FIZ. 30 /9/ 1552- , 1966.//
LEVSHIN V L   MIKHAILIN V V   NIZOVTSEV V V
BULL ACAD SCI USSR /PHYS SER/
30   9   1618-20   1966

63247 THE OPTICS OF REFLECTING MIRRORS.
KOZLENKOV A I
IZV AKAD NAUK SSSR SER FIZ
31   6   931   1967

63248 THE OPTICS OF REFLECTING MIRRORS. //ENGLISH TRANSLATION OF IZV. AKAD. NAUK SSSR, SER. FIZ. 31 /6/ 931- , 1967.//
KOZLENKOV A I
BULL ACAD SCI USSR /PHYS SER/
31   6   945-50   1967

63268 MANUFACTURING AND PROPERTIES OF FILMS OF Y2O3 AND RARE EARTH OXIDES ON GLASS.
FRANK B   GROTH R
THIN SOLID FILMS
3   41-50   1969

63348 THERMAL CONDUCTIVITY OF BORAZON POWDERS. //ENGLISH TRANSLATION OF POROSH. MET., USSR 9 /11/ 57-60, 1969.//
LEZHENIN F F   OSITINSKAYA T D   VISHNEVSKII A S
SOVIET POWDER MET METAL CERAMICS
11   911-3   1969

63384 SPECTRAL ABSOLUTE-REFLECTANCE MEASUREMENTS OF CARBON DIOXIDE FROSTS IN THE 0.5 TO 12.0 MICRON REGION. FROM FIFTH SPACE SIMULATION CONFERENCE. NATIONAL BUREAU OF STANDARDS$ GAITHERSBURG, MD. SEPT. 14-16, 1970.
WOOD B E   SMITH A·M   SEIBER B A
NTIS
NBS-SP-336 COM-71-50046
165-83   1970   CA   74   132580

63386 USE OF ALBEDOMETER IN DETERMINING OPTICAL COEFFICIENTS OF TECHNICAL MATERIALS.
GRIGOREV B A   FOMICHEV S N
INZH FIZ ZHUR
1   1   34-40   1958

63387 USE OF ALBEDOMETER IN DETERMINING OPTICAL COEFFICIENTS OF TECHNICAL MATERIALS. //ENGLISH TRANSLATION OF INZH. FIZ. ZHUR. 1 /1/ 34-40, 1958.//
GRIGOREV B A   FOMICHEV S N
NTIS
JPRS-51385   1-10   1958

63393 REFLECTIVITY OF ALUMINUM, COPPER, ZINC, AND NICKEL AND ITS COMPARISON WITH THEORY. PH.D. THESIS.
HSIA J J G   PURDUE UNIV   LAFAYETTE IND
NTIS
AD-711624   1-131   1968   CA   74   148729

63400 OPTICAL MATERIALS STUDY PROGRAM.
GOGGIN W R   PAQUIN R A   PERKIN-ELMER CORP OPTICAL OPERATIONS DIVISION   NORWALK CONN
NTIS
AD-865842L   1-79   1970

63421 A LOW CONDUCTANCE OPTICAL SLIT FOR WINDOWLESS VACUUM ULTRAVIOLET LIGHT SOURCES.
GEORGE R A   ROBERTS I   WILSON E G
J PHYSICS SCI INSTR
4   5   384-7   1971

63445 INFLUENCE OF INTERFACIAL PHENOMENA ON THE ELIMINATION OF NON-METALLIC INCLUSIONS FROM METALS. //ENGLISH TRANSLATION OF IZV. AKAD. NAUK SSSR, METAL. /3/ 57-64, 1968.//
EVSEYEV P P   FILIPPOV A F
RUSSIAN METALLURGY /METALLY/
3   41-6   1968

63448 NON-EQUILIBRIUM VALUES OF INTERFACIAL TENSION AND ADHESION IN A METAL-OXIDE SYSTEM. //ENGLISH TRANSLATION OF IZV. AKAD. NAUK SSSR, METAL. /5/ 51-9, 1968.//
DERYABIN A A   POPEL S I   SABUROV L N
RUSSIAN METALLURGY /METALLY/
5   37-42   1968

63449 VARIATIONS IN INTERFACIAL TENSION IN A NON-EQUILIBRIUM METAL-SLAG SYSTEM. //ENGLISH TRANSLATION OF IZV. AKAD. NAUK SSSR, METAL. /1/ 129-35, 1969.//
DERYABIN A A   POPEL S I   SABUROV L N
RUSSIAN METALLURGY /METALLY/
1   50-5   1969

Part D
AUTHOR INDEX

USE OF AUTHOR INDEX

The *Author Index* is given in two parts, namely, personal authors and coauthors followed by corporate authors. The numbers shown along with the name of each author refer to the TPRC serial numbers of the references cited in the *Bibliography*.

For simplification in the automated production of the *Author Index* it was decided to use only author's last name and his initials. Naturally, in the case of some of the more popular last names having the same initials, this practice is likely to lead to the unfortunate result of improperly crediting papers to similarly named different authors. For this the Editors express their sincere apology, and efforts will be made to correct this unfortunate situation in future publications. In such instances, the users of this *Author Index* should inspect the titles of the documents cited in the *Bibliography* to properly identify those technical papers which may have been improperly credited.

The term "corporate author" applies to industrial organizations, government agencies, colleges, universities, etc. In alphabetizing corporate names, the main listing is made under the key word of the parent company or organization's name, and names of divisions, laboratories, departments, etc., are cross-referenced under the key words of their names.

In the case of universities, the word "university" always appears second in abbreviated form. In identification of corporate names, information is listed in the following sequence: name of organization, city, state division, laboratory, etc.

Because of past limitations at TPRC it was not feasible to print lower-case letters or symbols such as the apostrophe. It has therefore been necessary to adopt nonconventional printout formats in the case of certain names. Examples of such instances are: McLennan is written as MC LENNAN, de Haas is written as DE HAAS, O'Brien is written as O BRIEN, and Frontas'ev is written as FRONTAS EV.

PERSONAL AUTHOR INDEX

AAGARD R L	58672		
AARON R L	48497	49662	52292
ABAUF N	45253		
ABBA F	39286	39717	
ABBE W J	50271		
ABE T	48547		
ABELES F	39840	41989	46856
51949			
ABRAHAM A	43043	55821	
ABRAHAM L H	36469		
ABRAMOV YU YU	62230	62231	
ABRAMOVICH B G	53488	53903	56585
ABRAMS L	40530		
ABRAMZON A A	46360	47599	47688
47689 47690	47691		
ACCOMAZZO M A	40709		
ACHTZIGER J	37646		
ACKERMANN R J	53231		
ACKLEY E J	50688		
ADACHI A	44482		
ADAMS J B	55600		
ADAMS J L	36470		
ADAMS L	54375		
ADAMS R N	48810		
ADAMSON A W	47509		
ADDISON C C	55189		
ADELBERG M	45042		
ADLFINGER K H	58622		
ADLON G L	57597		
ADRIANOV V N	49590	50040	53887
53888			
AFANASEVA G D	56239		
AFIFY M Y M	40671		
AGABABOV S G	62042	62043	
AGALARZADE P S	38771	38772	
AGNEW J T	55041		
AGRAWAL K N	54423		
AGROSKIN A A	54578		
AHERN J E	58324		
AHN K Y	55037		
AIDANOVA O S	38189		
AILES H B	43566		
AINBINDER S B	58530		
AKANUMA M	54060		
AKASAKI I	47332		
AKERS W W	48367		
AKEYOSHI K	56416		
AKHMETSHIN M A	51617		
AKHTAR B A	56988		
AKHTAR S	52999		
AKHUNDOV S K	41506		
AKOROV V M	36893	42262	

ALBRIGHT H T	48362	56504	
ALDRICH R G	48475		
ALEINIKOV N M	60152	60153	
ALEKSANDROV A N	43989	43990	
ALEKSANDROVA G N		57094	57095
ALEKSANYAN I T	48731	51640	
ALEKSEENKO M P	46368		
ALEKSEEV P G	45800		
ALESKOVSKII V B	40565	40566	41075
41076 58767			
ALEXANDER J A	52904		
ALFANO G	50498		
ALLAM D S	41612	62974	
ALLCUT E A	44094		
ALLEN B C	54683		
ALLEN J M	36212		
ALLEN P H G	47907	57663	
ALLEN R C	52199		
ALLEN R D	34740	42628	
ALLEN R T	34354		
ALLEN S C	58453		
ALLEN T H	45552	57597	
ALLEN W A	42185	50190	
ALLEN W C	43489		
ALLGAIER R S	59219		
ALMASI G S	47767		
ALOFS D J	38104		
ALVARES N J	37477		
ALVARES W J	47952		
ALVAREZ G H	33974	35840	35907
AMBROSIO A	57237		
AMBROZY A	47673		
AMDUR I	51274	52740	
AMMANN C	47423	52160	54983
AMMAR A S	38030		
AMORUSO M J	59672		
ANAGNOSTOU E	35021	37595	57221
ANASHKIN O P	41336		
ANDERS J	36681	43024	
ANDERSEN H H	37409		
ANDERSON A	34480		
ANDERSON A C	37014	48580	58563
58931 61245			
ANDERSON G W	58567		
ANDERSON J B	51727		
ANDERSON R	52046		
ANDERSON R B	39754		
ANDERSON R V	44554		
ANDERSON W J	61104		
ANDERSSON J K	57437		
ANDRADE A	62533		

ANDREEVA E A	40308		
ANDREEVA N G	58530		
ANDREEVA N V	45226		
ANDREW I D C	37381		
ANDREWS A I	55801		
ANDREWS L	40362	54848	
ANDRIESSEN H	49986		
ANDROULAKIS J G	47268		
ANDRYCHUK D	42740	52855	
ANELIUNAS A E	58535		
ANGHELACHE D	51235		
ANTHONY A J	45475		
ANTHONY F M	53098		
ANTONISHIN N V	54002		
ANTONOV I N	42392	44182	
ANTONOV L N	49780		
ANTONOVA E A	58286		
AOKI S	45579		
APPEN A A	52569	52570	58286
ARAGONES M	40133		
ARAKAWA E T	33799	34456	34548
37970 39086	42451	42894	43185
43684 44376	46004	47949	48912
51925 53763	54201	54887	
ARDEN B	58327		
ARDITI I	58474	58891	60035
ARKADEV B A	38609	38610	
ARKIN E S A	61079		
ARMALY B F	53543	61547	
ARMAND G	43464		
ARMAND L	50965		
ARMSTRONG A M	51218		
ARMSTRONG C S	36144		
ARNDT J	40620		
ARON W	48500		
ARONSON J R	48370		
ARRIGHI J	46882		
ARROTT A	51520		
ARVESEN J C	36528	40417	47273
56006			
ASAMOTO R R	46935		
ASAY J R	52896		
ASCHKINASS E	46995		
ASHFORD N	52293		
ASHIMOV M A	56233		
ASHLEY E J	36036	36686	37677
39314 47821	57953		
ASHMEAD F A H	48508		
ASHWORTH T	53413	57647	
ATKINS H L	43788	48116	58821
ATKINSON W H	37275		

BILENKY B F 35099
BILLMEYER F W JR   36348 54711
BILLON 60242
BINKELE L 50490
BIRKEBAK R C 34041 34345 36229
  44754 47267 47434 56105 58823
BIRKHOFF R D 54201 54887
BIRKY M M 56007
BIS R F 34488 58322
BISHARA M N 57980
BISHOP J H 59274
BISHOP P H H 41450
BITYUTSKAYA L A 60152 60153
BIXLER H A 59278
BLACK I A 33988 35537 41217
  50638
BLACK W C 59274
BLACK W E 34752 43744
BLACK W Z 51127
BLACKMAN L C F 38827
BLAIR P M JR 43527
BLAKEMORE J S 55084
BLAKNEY T L 40343
BLANC R 36963 41348 42312
  43129
BLANK M 48410 56167
BLANK V A 60598 60599
BLASINGAME J M 39084
BLAU H H JR 36499
BLEA J M 59521
BLEVIN W R 35723
BLIFFORD I H JR 34627
BLIZNYUKOV S A 41870
BLOCH A N 51470
BLOCK M J 34137
BLODGETT A J JR 44308
BLODGETT K B 50802
BLOOM M F 48493 48494
BLOOR D 54099
BLOXSOM J T 57802
BLUENVEN J 41927
BLUM H A 39352 40474 41427
  42596 46396 48497 49662 51913
BLUMENTHAL J L 40399
BLUVSHTEIN I M 59500 59501
BOBKOVA O 34146 48778
BOBROV O D 49830
BOCK R O 36754
BOEBEL C 40358
BOEBEL C P 35545
BOGANOV A G 36903 38627
BOGATYRENKO B B 41872
BOGDANOV F F 48670 48671 54566
  55057
BOGUSLAVSKII L I   44263

BOILEAU A R 33880
BOITZOR S P 47979
BOKHOVEN C 49510
BOKROS J C 43100 43937 43938
  47532
BOLES M A 50418
BOLLER T J 35110
BOLOTIN N K 45275
BONDAR V V 61852 61853
BONDARENKO O E 48731 51640
BONDI P 44995 53609
BONNELL D G R 59944
BONNIAUD R 37370
BOONE T H 47565
BOOTY M J 34623
BOR J 42092
BORDELON F M 38727
BORISOV V G 58842 58843
BORISOV V I 33797 33798
BORISOVA I I 41412
BORODULIN E K 39946
BORODULYA V A 35922 35923
BOROSIC J 36109
BORTNER T E 52645
BORTZ S A 52660
BORYS R S 61042
BOTTGER G L 43447 47112
BOTTOMS W T 39262
BOUCHEZ J-P F A 57942
BOUGNOT J 38833
BOURG A 42151 43021
BOURG M 42151 43021
BOURG M 40114
BOURGEOIS P 43296
BOURICIUS G M B 34262
BOUSQUET P 37146 37633
BOVENKERK H P 42824
BOVKUN O P 45105 46077
BOWDEN F G 52594
BOWEN M D 52274
BOWIE S H U 56100
BOYKO F L 52204
BOZEMAN H 51592
BRACCIAVENTI J 49395
BRADBURY J H 47905
BRADEC F J 57801
BRADFORD A P 36688 49627 57797
BRADLEY D 39622
BRADLEY D J 39013 42890
BRADSHAW W G 40343
BRAECKEL R T 54816
BRAGER N N 41839
BRAHMS S 39391

BRANDENBERG W M 34454 34455 36516
  48258 52153 53496 57800
BRANNON R R JR 54153 57887
BRATTAIN R R 47009
BRATTAIN W H 47204
BRAU M 44340
BRAUN M T 52125
BRAUNBECK J 54085
BRAZEL J P 55829 57642
BREARLEY N 51508
BRECHNA H 40148 61705
BREEZE J E 51508
BREHAT F 51129
BREHM R K 39185
BRENDENG E 51888
BRESLICH F N 36144
BRETSZNAJDER S 34076
BRETT D A 38512
BRETZNER J-F 53707
BREUCH R 46337
BREUCH R A 36523 39363 47145
  48372 53491 59196
BREWER W D 40476
BREWSTER P M 59274
BREYER W H 59139
BRIAN P L T 59140
BRIDGES T J 59835
BRIGGS H B 47204
BROADHEAD D E 55599
BROCKES A 51581
BROCKETT R I 34851
BRODIE D E 33811 38834
BRODSKY M H 50365
BROGAN J J 48844
BROOKS W A JR 57439
BROOM R F 44703
BROOMAN E W 50531
BROSSA F 49526
BROUWER W 53332
BROWN A H 47298
BROWN C R 45517
BROWN E C JR 48556 51910
BROWN G L 43493 62782
BROWN L C 48177
BROWN M A 61246
BROWN R E 50302 51275
BROWN R F 59896
BROWN R G 42204
BROWN R R 36046 41589 44096
  44224 48375 54303 59648
BROWN R W 59219
BROWN T R 50235
BROWN W J 35723
BROWNING M E 37399 39338 46366
  47737

CHURAKOVA R S	46606	46607	
CHURCH P V	33879	39008	
CHURCHILL S W	55356		
CINI R	55853		
CIPOLLA J W JR	49264		
CIPOLLONE P	48139		
CLAPHAM P B	41161	62978	
CLARK H E	39212		
CLARK S P JR	42284		
CLARKE D R	47276		
CLAUDEL J	34542	39806	42321
50060	50554		
CLAUS L	48544		
CLAUSEN O W	34454	36516	47231
CLAUSEN W	34333		
CLAUSING A M	36943	40017	43491
48227	48510	48826	
CLAYTON W A	48806	57946	
CLEARY R E	47423	52160	54983
CLEMENS D L JR	37443		
CLEVELAND R S	48236		
CLEYET B	52387		
CLIFFORD J M	54227		
CLINTON W C	45003	46357	
CLOPE R W	51595		
CLOUGH D J	36290		
COCHRAN M E	61044		
COCHRUN B L	58328		
CODACCIONI J P	39286	39717	
CODEGONE C	51616	53608	
CODY G D	41664		
COEN E	40738		
COFFMAN A W	39700		
COHEN A D	45669		
COHEN I	37438	45477	
COHEN L	38028	42359	
COHN E	58147		
COLARUSSO P E	37595		
COLE R L	40083	48097	
COLE S S	38880	60443	
COLE S S JR	50176		
COLEMAN C F	35526		
COLEMAN I	48888		
COLES W D	43791		
COLLINS J O	37570	48434	49152
50031	50307	55028	57191
COLLINS J R	36754		
COLMAN D	58202		
COLOMBO G	48500		
COLOMBO P	55303		
COLONY J A	59986		
COLWELL J H	56027		
COLYER B	42814		
COMBS L L	45972		
COMPARIN R A	50306		
COMPTON W D	58567		
CONAWAY J	53986		
CONDAS G A	35311	39032	
CONNOLLY J I	34706	58931	
CONRAD A G	46167		
CONTI J C	63135		
COOK G M W	53136		
COOK J L	34059		
COOKE B A	46261		
COOLEY R A	52891		
COON C W	62590		
COOPER J L	41664		
COOPER M G	53854		
COOPER T E	60313		
COPE A F G	55120		
COPELAND R L	36084	37509	
COPLIN D H	37631	40368	
CORDIER H	33905	39296	
CORL E A	56865		
CORRIGAN F R	37971		
CORRUCCINI R J	45699		
COSNEANU C	46438	58147	
COSTICH V R	35113	36325	
COSTON R M	41169	47285	48844
56366			
COTTON D J	49172	49612	49815
51088			
COTTON J E	48806		
COTTS R M	39801		
COULSON K L	34262		
COULTER J K	39894		
COUNTS C R III	33974	35840	35907
37398	40351		
COURTNEY W J	47278		
COWIE J M	39036		
COWIN M	39362		
COX D W JR	35909		
COX J D	43458		
COX J T	36466	42270	46843
COX R L	36491	36506	39347
48365	54294		
COYNE H J JR	51926		
CRAVALHO E G	61190		
CRAWFORD M F	39837		
CRAWFORD P J	47905		
CREMERS C J	56105		
CRONIN J F	34138		
CROSBY J R	40136	48378	59973
CROSS R A	48099		
CROSS R I	34752	43744	
CROUCHER T R	37493		
CROWLEY M S	59993		
CUFF K F	55084		
CULLEN W C	47565		
CUNNINGHAM A L	52904		
CUNNINGTON G R	61547	62781	
CUNNINGTON G R JR		40343	44986
40000	40501	55070	
CURTIS H B	38109	43530	46810
59985			
CUSUMANO J A	35888		
CUTHBERT R M	51508		
CUTRIGHT R C	39496		
CYBULSKI A	58726		
CZANDERNA A W	34487	35963	
CZARNECKI K R	36212		
CZARNIK J W	53122		
D ANDREA G	35338		
DAENNER W	60314	61785	
DAHL J P	39391		
DAHM T J	36538		
DAILEY R M	43789	45498	48504
DALLMEYER H	54365		
DALY B J	47738		
DAMASKIN B B	51381		
DANA L I	47935		
DANCHEVSKAYA M N		57110	57111
DANDH K V	45991		
DANDREA G	55248		
DANESI P R	53250		
DANIEL J H JR	36418	37489	
DANIEL R C	37438	45477	
DANIELS D L	57860		
DANIELS F	51573		
DANILOV I B	41336		
DANILOV YU I	61078		
DANOY G	50965		
DANTER E	40908		
DARLING A S	50897		
DARTER M I	44400		
DAS M B	34401		
DASSU G	40194		
DATZ S	42599		
DAUDE A	38996	39944	42325
44041			
DAUNOIS A	39181		
DAUNT J G	38460	42071	
DAVEY J E	33966	35988	
DAVID J P	40957		
DAVIDGE R W	48185		
DAVIDO K W	38663		
DAVIDSON D E	49266		
DAVIDSON D M	50354		
DAVIDSON H R	39479		
DAVIDSON K W.	43479		
DAVIES G A	44349		
DAVIES J M	36320	58662	
DAVIS H O	52893		

DAVIS R M	44560		
DAWSON J P	34040	34041	34345
47434	56105	58823	
DAWSON J W	50659		
DAYET J	36823		
DE CARLO V A	35739	39264	
DE CHAZAL L E M	52213		
DE LA PERRELLE E T		45667	
DE NEE P B	49018		
DE REUS M E	38901		
DE VENUTO G	46814		
DE VORE J R	37272		
DE VRIES H A W	49510		
DE WAARD R	36371		
DE WITT D P	34683		
DE WITT W D	45482	46397	49153
49527	52118		
DEACLE R E	55806		
DEAM J R	48529	55679	57378
DEAN R A	40689	51247	
DEAN T J	54099		
DEARTH L R	34550	44551	
DEB S K	35809	40159	56041
56965	60495		
DECIUS J C	38697		
DECKER T G	49810		
DECROLY C	49574		
DEDRICK R L	44767		
DEEL O L	53091		
DEEM H W	49719	53006	57224
57225			
DEFORGES J	48179		
DEHAAN J R	41218		
DEISS J L	39181		
DEITCH M E	34251		
DEL PIANO V N JR		38673	
DELEUIL R	37146		
DELMORE J E	42529		
DELPONT J P	42921		
DEMOTO J	60204		
DENGLER R P	47410		
DENISOVA N E	48617	48618	
DENMAN G L	43742		
DENNER H	55497		
DENNISON D M	44326		
DEPALMA J J	50367		
DEPOORTER G L	51706		
DERBENEVA S S	43797	43798	
DERBY T F	54293		
DERICK B N	42204		
DERKSEN W L	34050	56499	
DERYABIN A A	37711	46667	46748
46749	49800	50636	53017	53436
63448	63449		
DESANTIS V J	35821		

DEUTSCHER K	48926		
DEVENEY J E	48487		
DEVERALL J E	35810	50023	
DEVIENNE F M	36706	40173	42840
51342	55357		
DEVILLARD J	49160		
DEVONSHIRE A F	51517		
DEWAR J	43063		
DHANAK A M	56505		
DI MATTEO G A	56356		
DI NOVI R A	36480	50663	
DI PHILIPPO P	43636		
DIANA A L	38962		
DICKINSON H C	48034		
DICKINSON J M	60164		
DIEDRICH J H	43530		
DILLENIUS M F E	59164		
DILLINGER J R	34309	37580	43616
57323			
DIMITRIU-VILCEA E		53144	
DIMITROFF J M	40788		
DINGWELL I W	47038		
DINKLAGE J B	44688		
DISTLER G I	39136	39137	
DIX R C	55288		
DIXON J R	35221	39169	47869
58322			
DIXON R R	57595		
DMITRIEV G I	62953	62954	
DOBIERZEWSKA-MOZRZYMASOWA E		56848	
DOBRIANSKII IU P		61051	
DOBROWOLSKI J A	52926	58660	
DOERFFLER W W	34394		
DOHERTY P	35537		
DOKUKINA E S	40996	40997	
DOKUKINA L F	46360		
DOLGIKH G V	61174	61175	
DOLGOPOLOV D G	34195	34196	
DONCHEV ST	49791		
DONOIAN H C	47314		
DONOVAN R J	47761		
DONOVAN T M	36036	37677	
DORAN J	57948		
DOREMUS R H	39111		
DORMER G J	47938		
DOROGOV N N	59494	59495	
DORONIN L K	48742	51954	
DORONKIN K N	56174		
DORRA L	54189		
DORSEY H G	40328		
DOUGHTY D W	45873		
DOUGHTY R O	63134		
DOUGLAS N J	36514	47145	48032
59647			

DOWNEY M J	58335		
DOYENNETTE L	58474	58891	60035
60037			
DOYLE W P	55043		
DOZIER J B JR	46042		
DRAGER J	37044		
DRANEY W G	53177		
DRAUGLIS E	55847		
DRAZEN E L C	61190		
DREISBACH G W	43735		
DRICHKO N M	46683	46684	
DRISKO R W	34682		
DRMAJ D T	60173		
DRUMHELLER C E	35029		
DRUMMETER L F JR		36386	
DU PRE F K	39631		
DUBAN M	49629		
DUBIN P	57642		
DUBININ E L	53225	53226	
DUBININ G N	33797	33798	49356
49357			
DUBOIS F	57389		
DUBROVSKII V B	41240	55263	63153
63164			
DUBROVSKII V G	63152		
DUBS C W	40583		
DUBS M A	47935		
DUDLEY L V	45669		
DUDNIK D M	51025	56318	
DUEVEL C O JR	40619		
DUGA J J	54228		
DUGEON E H	40018		
DULGEROFF C R	49310		
DULNEV G N	41480	45325	48084
49616	52546	59491	
DUMEZ P	45257		
DUNCAN J C	43078		
DUNDERLEY F J	33974		
DUNKERLEY F J	37399	46366	
DUNKLE R V	34038		
DUNLAY J B	34726	37375	54256
57853			
DUNN S T	34683	35323	39212
DUPEYRAT M	42876		
DURAND J L	34724		
DURAND S	48179		
DURKA K	41280		
DURRAN D A	59278		
DUSHCHENKO V P	45790		
DUTTON D B	38288		
DUVAL X	58359	60695	
DWYER H A	43692		
DYBAN E F	40020	43548	43997
DYBAN E P	43552		
DYBBS A	35851	38716	

EASTWOOD E — 52872
EATHER R H — 39178
EBBERS R W — 57187
EBERHART J G — 48006
EBERLY P E JR — 43179
EBY J E — 38288
ECKERT E R G — 36229
ECKSTROM H C — 58301
ECONOMY J — 44554
EDAGAWA H — 44419
EDELMAN L E — 57595
EDNERAL F P — 35050 46668
EDWARDS D K — 34928 36425 38391 43112 48359 54703
EDWARDS D O — 38460
EFSTATHIOU A — 47874
EGGLETON A E J — 51522
EGLITIS I E — 57171 57172
EGOROV B N — 55826 59149 59150
EGOROV V M — 57005
EGOROVA E I — 46360
EHRHARDT H — 34936
EICHINGER R — 36681 43024
EINHAUS R — 34936
EISENMAN W L — 44415
EJIRI A — 48978
EKSTROM L — 55814
EL-SAHRIGI A F — 37447
ELDRIDGE E A — 48806 57224 57225
ELIASON L K — 36285
ELINSON M I — 61986 61987
ELLEFSON B S — 54104
ELLIOTT D H — 40019 48503
ELLIOTT J J — 33925
ELLIS S G — 42885 48932
ELLISON A — 57764
ELSNER N B — 40006
ELSNER Z N — 49367 49368
ELWORTHY P H — 47144 53733
ELYUTIN V P — 55859
EMANUELSON R — 47423 52160
EMANUELSON R C — 35889 38618 40400 42780 46336
EMERIC A — 34981 36819 36960 39279 39299
EMERIC N — 34981 36819 36960 39299
EMYASHEV A V — 54568 55069
ENEVER R P — 43670
ENGALYCHEV S A — 58261
ENGEL N N — 36797
ENGELER W E — 44305
ENGELKE W — 34936
ENGELKE W T — 36419 44987
ENGLE G D — 36213

ENGLE G B — 43937 47532 50153
ENGLE G P — 43938
ENIKEEV E V — 48523
ENTWISTLE A G — 39622
EPIK A P — 39080
EPPLER R A — 53068
EPSTEIN M — 40323 49402
EREMENKO V N — 33892 33893 34610 40985 41114 41872 43310 43842
EREMENKO V V — 41117
ERININ K — 49557
ERMOLENKO I N — 41097 41098
EROFEEV A I — 54985 54986 62244 62245
EROLES A J — 36544
ERSHOV O A — 39993 39994
ESAKI L — 49962
ESIN O A — 46713 46714 46748 46749 53225 53226
EUCKEN A — 51525
EULER K J — 50684
EVANS J L — 39242
EVANS L B — 47049
EVANS M V — 37597
EVANS R J — 48806
EVEREST A — 35535
EVERHART J O — 52668 52669
EVSEEV P P — 50073
EVSEYEV P P — 63445
FABISH T J — 48806 57222 57223 57946
FABRE D — 37148 43128
FADLER E C — 48824
FAHMY A A — 60284
FAHRENWALD F A — 43414
FAIR H D JR — 55810
FAIRCHILD C I — 54194
FARBER E A — 38247 48387
FARON M J — 38309
FAUGERE J F — 35934 42921
FAUGHN J — 59998
FAUST C L — 54242
FAUST J W JR — 50299 50926
FAWCETT W R — 56291
FEDER R — 40770
FEDOROV V G — 56570
FEDOROVA D L — 55464 55465
FEDOROVA I G — 58261
FEDOROVICH E D — 42690 42691
FEDOSEEVA N P — 56746
FEDOTOV A I — 57005
FEINLEIB J — 39120
FEITH A D — 51903
FELDER J W — 52890
FELTYN I A — 57171 57172

FENN F B — 51727
FEREBAUER R — 39734
FERRO V — 41292 44990 44995 51616 53585 53608
FERT A R — 61948 61949
FESENKO V V — 33893 40985 41114
FETISOVA L I — 42017
FETTING K E — 56441
FEUER P — 34354
FEUERBACHER B — 48567 49254 49647 51492 56833 57413
FICALBI A — 55853
FIGUROVSKAYA E N — 57110 57111
FIJIKURA Y — 51615
FILATOV O N — 52604 52605
FILIMONOV S S — 49590 53887 53888
FILIMONOV V N — 59738 59739
FILIPPI F J — 52700
FILIPPOV A F — 50073 63445
FILIPPOV S I — 37734
FINCH H L — 38392
FINCK C — 34972
FINE S — 58328
FINKE D D — 34834
FINKEL M W — 61184
FINKELSHTEIN YA G — 58573 58574
FINKENRATH H — 37015
FIRESTONE R F — 42290 48230
FISCHER W A — 57860
FISCHER W H — 46542
FISHER E I — 39239 39977 40471 45458
FISHER S A — 39362
FISHER S S — 57980
FISTUL V I — 60790 61959
FITTON B — 57413
FIZEAU M H — 59537
FLACHSBART B B — 51447
FLECHSIG W — 43384
FLEISHER H — 39604
FLEMING R M — 59299
FLETCHER L S — 47095 55640 58560 59610
FLINT O — 43408
FLORENCE A T — 47144 53733
FLORENCE D E — 48379
FLORENTINE R A — 38028 42359
FLOWERS W — 39951
FOCSA V — 37815
FOGDALL L B — 36046 41589 47756 54303 55221 59648
FOGEL V O — 45800
FOLKMAN N R — 55228
FOLMAN M — 42109
FOMICHEV S N — 63386 63387

FONTANA A	49574		
FONTANA J	55303		
FONTENOT J E JR	40318	50332	51101
FORCHT B A	40298	40301	
FORD C G	39619		
FOREST J D	34631		
FORLANO R J	52661		
FORSYTH A C	55810		
FOSTER C F	41266		
FOUCHE F	35200	39296	41389
FOUNTAIN J A	40413	41217	59561
61781			
FOURNY J	37153		
FOUSKOVA A	55899		
FOWKES F M	48534		
FOWLER H A	49959		
FOX G T	57323		
FRAENZ I	43767	43768	
FRAIMAN YU E	42535	43098	43468
47499			
FRANCIS H A	36499		
FRANCIS J E	47502		
FRANCISCO A C	48806		
FRANK B	63268		
FRANKE F R	47730		
FREDERICK D L	54848		
FREEMAN G H C	36700	38296	61782
FREI R W	35573	47421	
FREIBERGA L A	57171	57172	
FRENCH B O	59115		
FRENCH J B	59894		
FRIDZON M B	56239		
FRIED E	44681	45768	46639
48381	48496	48502	58821
FRIEDBERG A L	36544	60282	
FRIICHTENICHT J F		35806	
FRITSCH C A	51902		
FRIVIK P	51888		
FRODYMA M M	35573		
FROISSART C	38143	58814	
FROMHOLD A T JR	33844		
FRONDUTO J	34726	37375	54256
FRUMIN G T	47688	47689	47690
47691			
FRY E M	48382	58822	
FRYE D E JR	58878		
FUCHS R	51456	52830	
FUCHS R A	43123		
FUERSTENAU D W	36813		
FUJIOKA T	37043		
FUJITA I	39239	39977	40471
FUKUDA K	47908		
FUKUI T	54898		
FUKUTANI H	44314		
FULDE P	48009		

FUNAI A I	44986	47263	47998
62781			
FURMAN SH A	39653	39654	40099
40100	40600	40601	59387	59388
FURUTA S	53166		
FURUUCHI S	55869		
GAAL S	56733		
GABADADZE T G	50146		
GABOR J D	36367		
GABRIELLI G	52998		
GABRON F	47286		
GADZUK J W	41113		
GAGE P R	38717	43741	
GAGER H M	39620		
GALE E H JR	53571	60713	
GALKIN M N	49400	49401	
GANIN E A	37283	43634	
GANITKEVICH YA V		51818	
GANNON R E	42351		
GANSIEVSKAYA YA I		45382	
GANZHINA I M	51381		
GARA M	44490		
GARCIA A	52232		
GARDNER K A	36088		
GARFINKEL M	44305		
GARG S S M	47486		
GARISHVILI B V	50146		
GARLAND C W	42237	58903	
GARNER D C	48379		
GARVEY J A	49291		
GARZA J J	42975		
GASPARINI J-P	39288		
GASQUET R	36058		
GASTAUD A	37146		
GATEWOOD B E	47035		
GAUDIN A M	49810		
GAULIER C	40612	40613	
GAVER R L	58562		
GAW W	38578		
GAWARECKI S J	57860		
GEACINTOV N	39670		
GEDDES A L	43447	47112	
GEDEREVICH N A	54118		
GEHRKE W L	61485		
GEIST J C	39212	57408	
GELD P V	53753	53871	
GEORGE R A	63421		
GERARD-HIRNE J	53680		
GERASHCHENKO O A		56570	61567
GERASIMOV V G	56174		
GERASIMOVA L G	57110	57111	
GERDEMAN D A	42738	54817	
GERHARZ R	57860		

GERRISH R W JR	53606		
GERSTACKER H	43349	43357	
GERY A	46480		
GERZHIN A G	38190	41482	45570
61704			
GEVIRZMAN R	42109		
GEVORKYAN B A	47608		
GEYER N M	63131		
GIBALDI M	44385	48950	
GIBBON N C	41210	52118	
GIBEAUT W A	39081	48004	
GIBSON E K JR	61071		
GIER J T	34038	43112	
GIESEKING D E	56379	59204	
GILBERT J	46119		
GILCHRIST K E	57455		
GILCREST A S	36535		
GILES T M	52668	52669	
GILETTE R B	44224		
GILLAP W R	44385	48950	
GILLE J P	57598		
GILLESPIE F C	60085		
GILLESPIE J S	39620		
GILLETTE R B	38317	44096	47276
48375	48573	54292	
GILLHAM E J	43603		
GILLIGAN J E	34819	36522	43525
43840	45545	59637	59686	62587
GILPIN T M	45987		
GINDIN L G	55826		
GINSBERG D M	39016	56458	
GIRAUDIER L	36824	48448	
GIRAULT P	45286	48131	
GIRIFALCO L A	56360		
GISIN M A	38765	38766	50054
54327			
GITTLEMAN J I	41664		
GITZEN W H	42294		
GIULIANI S	39353	47994	49515
GLADYSHEV S A	37701		
GLASER M A	51595		
GLASER P E	33988	35535	35537
38646	41217	42374	45639	45753
47286	48292	49023	50638	52909
GLASIER L F JR	52421		
GLASS J E	54996		
GLASSER J	47070		
GLASSER W J	47070		
GLEBOVA L I	42504		
GLEIM P S	45177		
GLOECKEL R	55220		
GLOSSMAN N	45271		
GLOVER D E	41836	53180	
GNIDASH N I	58573	58574	
GODSIN W	47215		

Name	Refs
KANICE K	55303
KANNER B	54996
KAO HS	42695
KAPANY NS	36758 39750
KAPKIN MM	49780
KAPLUNOV YA N	45800
KAPPEL G	42978
KARAFIATH LL	44944
KARAM HJ	50569
KARAMCHETI K	53075
KARASHAEV AA	46641 46642 47472 47521
KARIBAEV KK	41095 41096
KARIOTIS AH	35110
KARLSSON RH	34258 42635
KARP GS	41168 44681 46639
KARPOVICH IA	52604 52605
KASHIMA T	41159
KASPAR J	46860
KASPARECK WE	43789 45498 48504
KATAEV GN	57514
KATAOKA Y	37596
KATAYAMA K	38272
KATO R	34733 41599 47309
KATSUBE S	53166 54519
KATSUBE Y	53166 54519
KATTI PK	51273
KATZ S	34327 42783
KAULENAS AA	45494 46646 48662 48663
KAVTARADZE NN	59085 59086 59740 59741
KAYAN CF	48292
KAYE BH	34327 43811 49030
KAZAKOV VP	43797 43798
KAZANSKII MF	56600
KAZANSKII VB	40863 40864
KAZNACHEEV YU I	36356 36357
KEAGY BJ	40699
KEATING GM	37355
KECK PH	40934
KECK RR	37493
KEDESDY H	40934
KEEGAN HJ	33962 36500
KEENE BJ	55489
KEENE PA	36327
KEESOM WH	51602 51603 54137
KEITEL G	59627
KEKELIDZE MA	42042
KELLER DL	52271
KELLER DV JR	48475
KELLEY DL	40764
KELLEY MJ	48381
KELLIHER WC	40333 46335
KELLY FJ	36493 40764
KELLY FM	39837
KELSALL D	57794
KEMME JE	35810
KEMP HT	57666
KENDYS PN	58753
KENNEDY BS	57642
KENNEDY PB	48806
KENT JE	56291
KERLEE C	37194
KERN CD	34138 35414 35856 42764
KERR JR	43458
KERSTEN MS	61706
KERTSELLI I YU	55879 56577
KESSLER WE	47566
KETCHMAN JJ	36550
KEVORKIAN V	33925
KHABAROV YU M	46368
KHABAROVA EA	46367
KHALIKOVA GF	56174
KHALILOV EG	48692
KHAN AM	59615
KHAN EU	55288
KHAN IH	44155
KHANNA RK	38697
KHARLAMOV AG	38586 42531
KHILYA GP	41872
KHIM JH	43413
KHINCHIN TAS	51560
KHIRIN VN	39609 39610
KHIZHNIAK PYE	47984
KHIZHNYAK PYE	49551 49552
KHOLEVA MN	45829 45830
KHOROSHAVIN LB	61174 61175
KHRUSTALEV BA	53887 53888 54120 54231 55094 55330 55471
KIEFFER H	56770
KIHARA-MORISHITA H	60756 62501
KIKINDAI T	47317
KILIN VS	54747 55575
KIMURA T	38272
KINAWA AA	59169
KINAWI AA	54481
KINBARA A	53852
KING CJ	43690
KING P	40783
KING PJ	55747
KING PP	34740
KINGERY WD	34611 39447 39695 40992
KINZER ET	43207
KIREI GG	61199
KIRILLOV GA	46695 46696
KIRK DD	45698
KIRKPATRICK JP	53086
KIROVSKAYA IA	59083 59084
KIRSH YU E	33976 33977
KISELEV AV	36080 41386
KISHII T	40706 40707
KISLOVSKII LD	40598 40599 51598 55090
KITO M	60608
KIUCHI Y	44708
KIZER DE	52271 54816
KJELBY AS	36109
KLAIBER RJ	33854
KLEMENTEVA A YU	59389 59390
KLEMM R	41149
KLEMM RE	50539
KLETT DE	53357
KLIEWER KL	51456 52830
KLINGELHOEFFER H	56621
KLINGENBERG M	54149
KLIPACH LV	45570 61704
KLYACHIN AZ	57514
KNAUSENBERGER WH	59832
KNEISSL GJ	40764
KNEUBUHL FK	60580
KNIGHT BL	59176
KNOLL RH	50896
KNOSP H	55348
KNOWLES D	34753
KNUDSON HE	53087
KOBAYASHI T	37043
KOBAYASHI Y	45579
KOCH DFA	42612
KOCH EE	59627
KOCH GE	36539
KOCHANOVA LA	56746
KOENIG NR	36876
KOENITZER LH	38383
KOH B	40683 48505
KOKIN GA	37792 37793
KOKOROPOULOS P	37597 51573
KOKSHAROV VD	58844 58845
KOLB RP	38615
KOLESNIKOVA NA	36356 36357
KOLOSKOVA LA	34195 34196
KOLTUN MM	38116 38117 39661 39662 40954 40955 44625 45829 45830 53513
KOMAROVA TA	49842 51255
KOMISSAROV VM	58753
KONDAK NM	43552 43997
KONDO S	36085
KONDO Y	41159
KONDRASHOV YU A	44552 45327

LYONS M F	37631	40368	
LYSENKO V S	61852	61853	
LYUBIMOV V K	55464	55465	
MAAG C R	55829		
MAC DONALD J	57793		
MAC DONALD R E	35225		
MAC GREGOR R K	59607	63133	
MAC KAY H A	46528		
MAC MILLAN H F	48228	48366	55084
MAC RAE R A	33799	37970	43684
MAC RITCHIE F	50252		
MACHALICKY J	44873	44874	
MACHLIN E S	52771		
MACIASR F M	52892		
MACK E B	40006		
MACK F E	41182		
MACQUERON J L	46480		
MADDEN R P	44278	44317	44319
48720			
MADDOX J P	49310		
MADDOX R M	55679		
MADDOX R N	48529	57378	
MADSEN J	34740		
MAERSCHALK J C	47488		
MAGEE P M	47991		
MAGINI M	53250		
MAGLADRY R	53113		
MAHENC J	59191	60706	
MAIDICH YU V	34610		
MAIER R L	44931	49233	52030
52850			
MAIR W N	41932		
MAISEL L	37886		
MAJOR G	58767		
MAKAROUNIS O	47270		
MAKAROV L P	36903	38627	
MAKEEV L A	54578		
MAKHKAMOV K	41097	41098	
MAKI K	39919		
MAKUUCHI M	45975		
MALCHEVA N	49557		
MALE D	44281		
MALHOTRA M L	56163		
MALINA V	58524		
MALKOV V A	59259		
MALLOY E S	45308		
MALNEV A F	61852	61853	
MALONEY W T	58659		
MAMEDOVA M A	56233		
MAMULA P	45080		
MAMYKIN P S	58844	58845	
MANABE A	42590	44016	
MANDAL R P	37491		
MANDEVILLE L C	60692		
MANGOLD V L	41573	52897	
MANN W B	51516	51518	
MANNING C R JR	52841		
MANSFIELD W W	51418		
MANSON L	48862		
MANSUR L C	34542		
MANTSUROVA V N	50764	52215	
MANZHELII V G	34195	34196	
MARCHAND E W	50367		
MARCINCIN A	48421		
MARCUS R T	54711		
MAREK C J	47504		
MARGOTTIN-MACLOU M		58474	58891
60035 60037			
MARIN K G	55464	55465	
MARK H	36530	61182	
MARK H B JR	49327		
MARKELOVA N V	62857		
MARKINA Z N	45105	46077	
MARKOV M N	36352	36353	
MARKS B S	43794		
MARONCHUK I E	57102	57103	
MARONCHUK YU E	57102	57103	
MARSAULT J-P	48122		
MARSDEN D G H	45253		
MARSHALL K N	48032	53491	61191
MARTENS H E	39516		
MARTENSSON J O	34621		
MARTIN D H	46830	48001	54099
MARTIN G	62179		
MARTIN S L	46912		
MARTIN T P	33970	36467	36684
58255			
MARTIN-LOF S	58221		
MARTINUZZI S	36961	37153	40957
53707 56733			
MARTON L	44080	49959	
MARY D J	36100		
MARZOCCHI M P	56046		
MASAMUNE S	51271		
MASLENNIKOV L A	56579		
MASON C R	52274		
MASON M T	48888		
MASON R B	39866		
MASSEY W M JR	46059		
MASSING P N	57658		
MATE C F	33811	38834	56697
MATHAD G S	53583		
MATHUR K C	34450		
MATOLICH J JR	49719		
MATSCH L C	41210		
MATSUI A	59712		
MATSUMOTO M	50787	56189	
MATSUO C	37596		
MATSUZAWA A	37043		
MATTES H G	60107		
MATTHEWS P W	45517		
MATTICE J J	40099		
MATUMURA O	38036	46135	50111
MATUYAMA Y	44091		
MATVEEVA I A	61199		
MAUNY P	57389		
MAURAKH M A	55859		
MAURI R E	35892		
MAWER P A	46830	54099	
MAX E	39604		
MAXIMOV T F	47979		
MAYER H	48712		
MAYER J E	46551		
MAYER R	40164		
MAYER R A	57000		
MAYHAN K G	58453		
MAYKUTH D J	48004		
MAZE R C	42893		
MC ALISTER A J	39497		
MC CALLUM J	50531	54242	
MC CAMONT J W	38717		
MC CANDLESS C	39338		
MC CARGO M	59196	59647	59975
MC CONNELL D G	50321		
MC CONNELL R D	54503	56708	57838
MC CONVILLE G T	41664		
MC CORD T B	55600		
MC CREARY W J	54194		
MC CUBBIN T K	40784		
MC CULLOUGH B A	34040	34041	34345
43801 44754 47034 47267 47434			
58823			
MC DANIEL R H	36884	40294	
MC DERMOTT M N	33965		
MC DONALD S L	59975		
MC DOWELL W J	35526		
MC DUFFIE D E JR	63038		
MC ELROY D L	36797	44453	55561
58340			
MC FARLAND M	36688		
MC GARVEY J W	49589	59672	
MC GINN J H	59893		
MC GOWAN R J	42976		
MC GUIRE D W	54237		
MC HENRY E R	41910		
MC INTYRE D	57957		
MC INTYRE G W	36517		
MC INTYRE J A	56642		
MC KEE T B	36213		
MC KEEHAN W	40788		
MC KELLAR L A	36526	44772	48366
55084			
MC KENNEY D B	59833		

MC KINNEY A R	40298	40301	
MC KINSTRY H A	55659		
MC KINZIE D J JR		37173	59611
MC LAIN D K	57595		
MC LINDEN H G	48370		
MC MAHON D H	58659		
MC MASTER D G	38320		
MC NARY R O	45796		
MC NAUGHTAN I I	36327		
MC NICHOLAS H J	40591		
MC QUISTAN R B	55041		
MC QUISTON F C	40298		
MC SWAIN B D	39896		
MC TAGGART J H	56168		
MECHAM W J	36367		
MECHEV V V	53681		
MEDALIA A I	47314		
MEDFORD J A	40298		
MEDVEDEV N N	56569		
MEE C D	35225		
MEEHAN E J	48832	52903	
MEEROVICH I G	45155	55879	56577
MEISTER B J	47748		
MELDRUM J F	55549		
MELLIER A	54229		
MELLNER M	35537		
MELNICHENKO V S	55503		
MELNICK A M	38028	42359	
MENARD W A	33989		
MENZEL D	45745	51726	53354
MERDY H	37154 38139 41228 45678		
MERIGOUX M R	51292		
MERIGOUX R	45834		
MERILL R B	56537		
MERING J	39942		
MERKULOV A V	61438	61439	
MERRIAM R L	56536		
MERRILL R B	51904	52300	
MESHCHERYAKOVA T F		59389	59390
MESSMER J H	44339	45653	
METCALF A G	40006		
METHERELL A F	45237		
METSIK M S	38189	53012	
MEYER B	39181		
MEYER K H	34333		
MEYERS C	50308		
MICCIOLI B R	47068		
MICCIOLI G R	35862		
MICHAUD-BONNET J	44051	39286	39717
MICHEL J	42876		
MICKELSEN R A	39447	50966	

MIDOL M C	60695		
MIDOL-MONNET C	58359		
MIKAMI H	51373		
MIKHAILIN V V	46485 63237	48513	63236
MIKHAILOV N V	48742	51954	
MIKHAILOVA T V	61078		
MIKHALCHENKO R S	45570 61704	38190	41482
MIKHALIK E	46721	46722	
MIKHEEV YU S	61078		
MIKIASHVILI SH M		42042	55081
MIKIC B B	38270 54664 59138	42688	53854
MILEK J T	43519		
MILER M	50856		
MILES J K	41590	47274	
MILEWSKI C	44560		
MILGRAM A A	50375		
MILIANSHUK M V	35099		
MILLARD J P	41919	43687	55741
MILLER E	36532	41123	41149
MILLER E R	34449		
MILLER G	56969		
MILLER I R	48410		
MILLER J	40709		
MILLER M	47259		
MILLER R A	48374		
MILLER R E	45044		
MILLER W D	43493		
MILLNER A R	59219		
MILLS R W	60590		
MILOSLAVSKII V K		49361	49362
MILOV I V	55512		
MILSHENKO R S	59046	60836	
MIMURA T	37595		
MINAKOV V A	37016		
MINGES M L	33858 48507 50868 51864 57450	36413	39356
MINKO N I	37016		
MINLOV I M	40952	40953	
MIRONOVA L R	50642		
MIRTICH M J	36530		
MISCHKE C R	48387		
MISHIN G V	57513		
MISNIAKIEWICZ W	55698		
MISRA S K	51218		
MISRA T N	49882		
MISSENARD F A	42373		
MITCHEL B J	37053		
MITCHELL G	48097		
MITCHELL J W	58454		
MITCHELL L J	43480		
MITCHELL W N	54712		

MITOR V V	39236 55813	41003	55812
MITSUISHI A	42590	44016	
MITTASCH H	55608		
MITTON P B	58452		
MIURA M	60204		
MIYASAKA Y	51615		
MIYATA T	58664		
MIZGIREVA L P	51944	51945	
MNUSKIN M G	48463		
MOCH P	43296		
MOCHEL J	35591		
MOCHEL J M	40056		
MOELLER C E	52260		
MOELLER K D	37884	38042	56163
MOISEEV A M	62228	62229	
MOISEEV I	46438		
MOLGAARD J	41397		
MOLHO R	59952		
MOMIN A U	60518		
MONAHAN T I	34050	56499	
MONFORE G E	41593		
MONG L E	53168		
MONK R W	56874		
MONMA K	43057		
MONSON D J	57796		
MONTET G L	39677		
MONTLE J F	58453		
MONTOVILOV O A	45018	45019	
MOOIJ J E	52592		
MOOK C P	44949		
MOORE C J JR	40474 51913	41427	42596
MOORE D G	36493	39212	
MOORE G E	42599		
MOORE H R	43736		
MOORE J P	58340		
MOORE W V	51307		
MOORMANN W	48008		
MOOTE I	54296		
MORACHEVSKII A G		45820	47687
MORAN J P	45307		
MORDCHELLES G	45087		
MOREL J	47317		
MOREY T F	61193		
MORGAN H	45669		
MORGAN J G	39264		
MORGAN R M	42889		
MORIHARA H	57658	58321	
MORITA N	35440		
MORITA Y	39779	39780	44419
MORIYAMA K	45954		
MORIYAMA U	60756		

PETERSON A W	40934		
PETERSON D W	39638		
PETERSON E W	35116		
PETERSON L E	40421		
PETERSON R E	61245		
PETHERBRIDGE P	40903	40904	40905
PETRAKIAN J P	39717	59143	62979
PETRAKIAN J-P	39286	49110	57251
PETRI F J	48498		
PETROV A A	50463		
PETROV I N	56578		
PETROV V A	54568	54747	55069
55575			
PETROV YU I	57365	57366	58408
58409			
PETROV-DENISOV V G	56579		
PETTYJOHN R R	51906		
PETUKHOV V	34140	48778	
PEVZNER B Z	58286		
PEZDIRTZ G F	36527	37193	40333
40414	40416	42295	46335
PFAHL R C JR	37053		
PFEIFFER H	47926		
PHILIP R	34981	39288	48051
PHILLIPS L N	60099		
PHILLIPS R P	40269		
PIAZZESI G	40194		
PICHOIR	60242		
PIDGEON C R	47973		
PIEPER S C	35113		
PIEROWAY C S	37122		
PIERSON A H	40324		
PIKE E W	37790		
PIKLER A	48421		
PIKUS I F	43733		
PILIPENKO A M	56570		
PILLER H	42966		
PILLING B	39383		
PILPEL N	43670		
PILSWORTH M N JR		34640	
PINION E C	40529		
PIPER E L	41910		
PIROGOV A A	48930	48931	
PIROGOV YU A	36903	38627	
PIROGOVA N V	36054	51630	
PISCOPO F A	52654		
PITT K E G	62974		
PITTMAN C M	40476		
PLATO G	43837		
PLATUNOV E S	45325	46172	52546
PLENDL J N	34542	61718	
PLISKIN I	46625		
PLISKIN W A	40269		
PLUMMER W A	49933		

POE T L	36285		
POGSON J T	59607	63133	
POIRIER V L	34726	37375	54256
57853			
POKRZYK S	55698		
POLEKHINA T K	54346	62951	
POLING G W	62417		
POLLACK G L	55723		
POLLACK M A	59835		
POLLARD H E	36523	40343	48032
POLLINI I	48925	60031	
POLLOCK D H	34820		
POLTZ H	51851		
POLUKHIN V N	46367		
POLYANOVSKAYA N S		50741	
POMERANTZ P	45003	46357	
PONAMAREVA G I	34367	34368	
PONOMAREV I F	59347	59348	
PONTARELLI D A	52319		
PONTELANDOLFO J	47215		
PONTER A B	44349		
POON P K C	56670		
POPE M	39670		
POPEL S I	37711	46667	46713
46714	46748	46749	49800	50636
53017	53436	63448	63449
POPOV V M	60045		
POPOVA L B	56323		
PORRECA F	42151		
PORRECA F	40114		
PORTER J	34045		
PORTER J W	47750		
POSSLEY G G	58301		
POTAPOV E V	51944	51945	
POTTER G W JR	55599		
POTTER J H	40678		
POTTER R F	34622	52917	55042
POTTIER M	45087		
POULOS N E	42974		
POVOLOTSKII L V	38609	38610	
POWELL H M	59896		
POWELL R J	56125	58264	
POWELL R L	38525		
POWELL R W	40268		
POWERS C T	34059		
POZDNYAKOV D V	59738	59739	
POZDNYSHEV G N	50463		
PRACHT W E	47738		
PRADOS J W	35739		
PRAEL R E	47566		
PRASINOS T	40724		
PRASOLOV R S	40744	40745	48990
63154			
PRATT A W	33864		
PRESKI R J	41334		

PRETTYMAN P E	51902		
PREUSSE K E	48814		
PRICE A B	56375		
PRICE R J	42100	42937	42938
47532			
PRIEST D K	47938		
PRIEST I G	55924		
PRIMAK W	39040		
PRINS J A	51604		
PRIOL M	35321	36710	37831
39944	42325	44041	48131
PRIOR B W	40018		
PRITCHARD J	52895		
PROBERT S D	33800	43583	58721
58874			
PROCHASKA F O JR		45911	
PROGAR D J	45665	50298	
PROIX F	54014	54260	
PROST R	59184		
PROSTAK A	49327		
PU S L	37689		
PUESCHEL F	51049		
PUGACHEVICH P P	51806		
PURCELL G V	39358	46648	47098
49761			
PURGIN A K	58844	58845	
PURSLOW B W	40677		
PUSHKIN I L	49400	49401	
PUSTOVALOV V V	51562		
PUTNAM B R	37795		
PUTNER T	35580		
PUTTNAM N A	36810		
PY B	46277		
PYLE R	47215		
PYRON C M JR	44987	45925	51884
57643	57644		
QASHOU M S	38021		
QUEER E R	48291		
QUESADA A F	38673		
RADU A	37815		
RADUL N M	46073	46074	
RAFALOWICZ J	37035	42558	47427
RAGAI A N	60284		
RAGIMOV F M	56233		
RAJAGOPALAN R	41466		
RAK M	51572		
RAKOV A M	54120	54231	55471
RAKOV A V	51944	51945	
RALL R M	51531		
RAMAN C V	38456		
RAMAZINA I A	59389	59390	
RAMOIN M	53707		
RAMSEY J B	36466	45808	48362
52920	56008	56504	
RAMSEY J W	34297		
RAMSEY T H	52940		

SKRIPAK V N	46589	46590	
SLACK G A	56168		
SLATER P N	59833		
SLAUGHTER C W	56255		
SLAUGHTER J I	40418		
SLAWSKY Z I	33944		
SLEMP W S	41768	41934	43388
50298			
SLOBODYANYUK A I	61199		
SMEKHOV F M	44457	56209	56601
SMELTZER W W	41397		
SMILEY V N	50975		
SMIRNOV A I	45773		
SMIRNOV E V	44552	45327	62054
62055	62938	62939	
SMIRNOV YU S	50463		
SMIRNOVA E V	34183	52757	
SMIRNOVA G P	52540	52541	
SMIRNOVA L G	45279	50880	
SMITH A C	47767		
SMITH A E	45666	55175	
SMITH A M	43801	47034	47267
50239	51914	59290	61721	63384
SMITH C A	49762		
SMITH C S	55223		
SMITH E F	34838		
SMITH E T	45044		
SMITH F J	39238	43481	44270
52152	55070		
SMITH F M	46055		
SMITH G T	46217		
SMITH J D	54306		
SMITH J M	39858	51271	
SMITH J W	40398		
SMITH M G	57647	57940	
SMITH N V	60498		
SMITH R A	39504		
SMITH R N	39504		
SMITH S D	34626	47973	53509
SMITH T	38575	42354	49813
53036			
SMITH T F	59244		
SMITH W H	58301		
SMITH W K	35374		
SMOLAK G R	46217		
SMOLSKII B M	55627	57537	
SMOLYARENKO V D	35050	46668	
SMORODINOVA M I	48743		
SMUDA P A	47095	47768	55640
61181			
SMYLY E D	45925	51884	54298
57643	57644		
SNAITH J W	56337		
SNITKO O V	36066	45752	
SNODDY W	36532		
SNYDER N S	58585	58986	
SOFFER L M	59952		
SOGIN H H	38727		
SOKOLOV L A	43124	43140	
SOKOLOV L B	38400	55140	
SOKOLOVA N P	59085	59086	59740
59741			
SOLIMAN M M	60715		
SOLLER W	45700		
SOLOMATIN G G	51617		
SOMA Y	42114		
SOMMERS R D	43791		
SONNENSCHEIN G	52891		
SONNTAG H	56990		
SORBIER J-P	53707	56733	
SOREMARK C	58221		
SOROKIN L M	55464	55465	
SOROKIN O M	50444	50924	60598
00600			
SOROKIN V P	51633	51634	
SOUTHERLAN R E	43820		
SOWELL E F	40918		
SPADE G L	36126		
SPAKOWSKI A E	57221		
SPARROW E M	33896		
SPECK D A	47068		
SPEEDS J A	49310		
SPEIL S	49663		
SPEKTOR B V	57018	57019	
SPENCE G B	53098		
SPENCER D J	59278		
SPENCER W T	40283		
SPICER W E	37677	44308	55084
SPINOLO G	48925	60031	
SPINULESCU-CARNARU I		36054	51630
SPISZ E W	48858	51985	56418
SPITZER D P	55223		
SPIVAK G	54128		
SPIVAK G V	40135		
SPIVAK N YA	36809		
SPORTON T M	44702		
SPRAGUE E L	56374		
SPRINGER G S	35851	38104	38716
42858	45491	49265	
SPROUFFSKE J F	57187		
SRIVASTAVA S K	42169		
STACY E F	42908		
STACY J T	36144		
STAHL D	35812		
STAMM A J	59935		
STANFORD J L	42894	50297	56122
57953			
STANGELAND B	36367		
STANWORTH J E	60370		
STARK R L	44933		
STARKEY R E	44772		
STAUFFER C E	46632		
STAVROVSKII G I	48662	48663	
STEELE S R	47397		
STEERE R D	39442		
STEERE R C	33971		
STEHLE H	57569		
STEHMAN W J	36418	37489	
STEINBERG M	55303		
STEINBERG S	35832		
STEINMANN W	49254	49647	56833
57413			
STELZER E L	55735		
STENIUS A S	61035		
STENSLAND L	59711		
STEPANENKO A N	51025	56318	
STEPANOV G V	61986	61987	
STEPHAN D	36947		
STEPHAN G	44045	44048	
STERN E A	39497		
STERRY W M	52128		
STETSON A R	37378		
STETSON R F	50153		
STEUDEL TH	46328		
STEVENS D W	45825	47236	55203
STEVENSON J R	46004		
STEVISON D F	41849		
STEWART J E	40798		
STEWART J V	36526		
STEWART S R	58988		
STICKNEY R E	49580		
STIEGLITZ C B V	46106		
STIERWALT D L	39952	45698	
STIMLER F J	42914		
STIRLING J F	55331		
STOCKHAM L W	52885		
STOECKER W F	35539		
STOFAN A J	46217		
STOFFEL A M	58297		
STOLECKI B	60718		
STOLL A M	38249	53123	54291
55534			
STOTZ D S	46814		
STOUT R L	52906		
STRADLEY J G	58340		
STRANSKI I N	43152		
STRASS H K	57439		
STRASSER A	35812		
STRATTON W K	35556	36144	
STREBKOV D S	60817	60818	
STREED E R	36536	40415	47272
47275	48383	53086	55741
STREIB H	41631		
STRENGE K	56990		
STRICKLER T D	52645		

STRINDEHAG O M	34814		
STROBL W W	42759		
STROEBECH C	54542		
STRONG H M	42824		
STRONG J	40580	47008	
STRONG P F	47286		
STROP H R	52891		
STROUD C W	35894		
STROUD D E	52298		
STUBSTAD W R	48506		
STUCKEY J M	47759		
STURDY G	46542		
STYHR K H	36123	37338	37393
STYLES J W	58105		
SU C-L	50913		
SUBBOTIN V I	51633	51634	
SUCKOW D H	53607		
SUEOKA O	44314		
SUETAKA W	34547		
SUFFREDINI J R	35527		
SUGA H	41776	50110	
SUGA K	39779	39780	
SUGANO T	48136	49766	
SUGAWARA A	34829		
SUGIYAMA S	54483	60608	
SUIKOVSKAYA N V	45016	45017	
SUKHAREVA L A	44594		
SULLIVAN E J	38512		
SULZBACH F	40528		
SUMMERS G D	50857		
SUNDERLAND J E	35120	51506	
SUNDERMAN D N	49207		
SUNDSTROM D W	59568		
SUOMI M	48580		
SURLAND C C	50083		
SURYANARAYANA C V		53485	
SUTHERLAND J C	42451	43185	46004
SUTO H	43057		
SUTTON G W	50141		
SUZUKI M	45975		
SVIRIDOVA A A	45016	45017	
SWANSON P	36113		
SWARD G R	63039		
SWATIK D S	54476		
SWEENEY T L	48855		
SWIFT D L	39869		
SWINGLEY C S	57052		
SWOFFORD D D	41573		
SYCHEV M M	56794		
SYMONS P C	48786		
SYNOROV V F	60152	60153	
SYRE	60242		

TABOR H	35417	36534	
TADA O	58670		
TAIROV N D	56858		
TAITS N YU	45766		
TAKAGI S	45441		
TAKAMORI T	54060		
TAKAMURA K	61720		
TAKAMURA T	60756	61720	62501
TAKAZAWA K	40796		
TAKEDA T	62501		
TALBERT S G	43297		
TALLEY C P	50391		
TALMOR E	48862		
TAMAI H	50787	56189	
TAMAI Y	45975		
TAMARIN A I	35922	35923	60049
TAMARU K	42114	47908	
TANAEVA S A	57632		
TANAKA A	35620		
TANAKA K	54041	59639	
TANAKA M	49874		
TANAKA T	50900		
TANGANTSEVA T F	48286		
TANGER G E	35797	46334	
TANZILLI R A	38028	42359	
TARABANOV A S	48415		
TARASEVICH YU I	46073	46074	
TARASUTIN T G	49400	49401	
TARBELL D W	44938		
TAUBER R N	51926		
TAUC J	43043	55821	
TAYLOR A H	46922		
TAYLOR A R	53136		
TAYLOR B N	41182		
TAYLOR E H	42599		
TAYLOR H D	49983		
TAYLOR L B	42916		
TAYLOR N W	54104		
TAYLOR T S	43637		
TEAGAN W P	42858		
TEAGARDEN K	42867		
TEAGUE W T	52274		
TEAR J D	51606		
TEEGARDEN K J	38288		
TEICHNER S J	38309		
TEKUCHEVA I A	37504 44373 60824	38248 60823	38924
TEMPLIN P R	43309		
TEPPER F	37391		
TERADA T	56793		
TEREKOV A YA	56230		
THEBERGE J E	52866		
THEISEN R	49526		

THEYE M L	41989		
THOMAS G M	33989		
THOMAS H K	53094		
THOMAS L B	46400 51275 52655	46879 51276 59892	47665
THOMAS R D	46986		
THOMAS T R	33800 58874	43583	58721
THORMANN P	62304	62305	
THORN J	38417		
THORNBURGH J D	36419	63107	
THORNLEY P G	44349		
THORNTON H R	45217	52661	
THOSAS R	38505		
THOSTESEN T O	40746	48376	54978
TIEMANN J J	44305		
TIEN C L	51909	53543	61547
TIEN H TI	38962	53722	
TIMCHENKO L I	49788	51769	
TIMMONS C O	50484	51016	
TIMOSHCHENKO G T		53012	
TINCOLINA P	50497		
TINGWALDT C P	43597		
TIPIKIN V V	56230		
TITTEL H O	47416		
TODARO F R	55945		
TODOROV O	51049		
TOFT A R	49627		
TOKUDA H	34669		
TOLKSDORF S	43153		
TOMIKI T	58664		
TOMITIKA T	40796		
TOMPKINS E H	36385 49031	36525	42292
TOMPKINS F C	51522		
TOMPKINS L M	36385		
TOMSIC M	37066		
TONG K N	53572		
TONTEGODE A YA	60244	60245	
TOOTS J	44080	49959	
TOPOLNITSKII G G		45275	55503
TOPORETS A S	40630		
TORBORG R	36477		
TORBORG R H	37390		
TOROPTSEVA T N	55826		
TORRANCE K E	33896		
TOSCANO W M	46420	56347	
TOTH L W	35110	35899	
TOULOUKIAN Y S	33865	53569	57652
TOUSEY R	40608		
TOWNELY C W	49207		
TOWNSEND C L	56009		
TOWNSHEND B	39844		
TRACEY T R	61193		

CORPORATE AUTHOR INDEX

CONDENSED MATERIALS GROUP INDEX TO
SIX-VOLUME RETRIEVAL GUIDE SUPPLEMENT I*

*Material classes such as woods, pharmaceuticals, foods, natural products, etc., or references concerning major compendia, are not reported in this six-volume **Supplement I.** Special bibliographic searches on these may be requested from TPRC directly.

CONDENSED INSTRUCTIONS ON USE OF VOLUME 6*

INQUIRY EXAMPLE 1: Our technical staff would like to find **all** references which give information/data on the **thermal conductivity** of **Foamed Concrete.**

SEARCH STRATEGY AND RESULTS: On page A1 of the Materials Directory, looking under **Concrete, Foamed** you will find the notation "see Concrete, Expanded......525–." Now go to **Concrete, Expanded** and note property codes **a, d, e** with substance number 526-0034. Continue the search in Part B, Chapter 1— Thermal Conductivity, page B1. Here, under substance number 526-0034, you will find fifteen entries, which may be examined to determine which TPRC numbers are to be selected. Finally, find the bibliographic citations in Part C. Using this same substance number 526-0034, a total of twenty-five more entries can be found in the **Basic Edition** (Plenum, 1967). Books 1, 2, and 3 of the **Basic Edition** correspond directly to Parts A, B, and C of this six-volume **Supplement I.**

INQUIRY EXAMPLE 2: My engineering staff would like **thermal diffusivity** data on **Plaster of Paris.**

SEARCH STRATEGY AND RESULTS: On page A2 of the Materials Directory, you will find **Plaster of Paris,** but, property **d** is not listed. However, note other cross references, such as: "see also Gypsum Board .....661–." Now in this same volume, on page A47, you will find **Gypsum Board** property **d** and substance number 661-0470. Continue the search in Part B, Chapter 4—Thermal Diffusivity, page B17. Here you will find substance 661-0470 and there is one reference, TPRC number 37058. Continue to Part C to find the bibliographic citation for 37058.

Here are the relationships of classes to volumes in this publication series that must be borne in mind when you notice a "see also" substance class cross reference: 100–127, Vol. 1; 200–227, 606, and 631, Vol. 2; 300–482, Vol. 3; 501, 504, 507, and 521, Vol. 4; 511 and 516, Vol. 5; 526–551, 621, 651, and 661, Vol. 6.

INQUIRY EXAMPLE 3: The materials engineer at our company would like to locate references on **thermal conductivity** for crumpled **Aluminum Foil.**

SEARCH STRATEGY AND RESULTS: On page A2 of the Materials Directory, note the substance name **Aluminum Foil, Corrugated** or **Crumpled,** property **a** and substance number 528-0164. Continue the search on page B1, and you will find four entries opposite substance number 528-0164 with TPRC numbers 39866, 40228, 54566, and 55057. The complete bibliographic citation can then be found in Part C in this volume.

INQUIRY EXAMPLE 4: References are needed for **reflectance** for **PV-100 White Paint.**

SEARCH STRATEGY AND RESULTS: On page A40 you will find **Paint, White PV-100** with properties **g, h, j** and substance number 551-0949. Go to page B31 for substance 551-0949 and you will find reference TPRC number 43527. Continue the search in Part C to locate the bibliographic citation for 43527.

INQUIRY EXAMPLE 5: Our aerospace engineering staff would like to find a certain technical paper. The only information known about this paper is that it contains data on **charring materials** and the author's last name is **Wilson.**

SEARCH STRATEGY AND RESULTS: On page D26 of the Author Index, you will find the following: **WILSON D R** 52260; **WILSON E G** 63421; **WILSON R G** 35530 36511 37899; **WILSON W G** 40082 48324 50306. Now, go to Part C to look up the bibliographic citations for these nine different TPRC numbers. Through a process of elimination you will find that only the paper with TPRC **35530** fits the description given and that the author's name is **R. G. Wilson.**

INQUIRY EXAMPLE 6: Our design group would like references on the properties of carbon-reinforced **Poly(adipic acid-1,6-hexanediamine) Resin.**

SEARCH STRATEGY AND RESULTS: You will find this substance's name, properly systematized as **Resin, Poly(adipic acid-1,6-hexanediamine),** starts at the bottom of page A15 and continues at the top of page A16 with property codes **a** and **n** and substance number 531-1507. Conclude the search in the manner described in the examples above.

---

*For additional details on the use of this volume, see the introductory remarks for Parts A, B, C, and D, scan the **Contents,** and also note the **Condensed Materials Group Index** on the opposite page.